Electrical Circuits in Biomedical Engineering

Ali Ümit Keskin

Electrical Circuits in Biomedical Engineering

Problems with Solutions

Springer

Ali Ümit Keskin
Department of Biomedical Engineering
Yeditepe University
Istanbul
Turkey

Additional material to this book can be downloaded from http://extras.springer.com.

ISBN 978-3-319-55100-5 ISBN 978-3-319-55101-2 (eBook)
DOI 10.1007/978-3-319-55101-2

Library of Congress Control Number: 2017935392

Printed on acid-free paper

This Springer imprint is published by Springer Nature
The registered company is Springer International Publishing AG
The registered company address is: Gewerbestrasse 11, 6330 Cham, Switzerland

Preface

The book has basically emerged from my lectures on "Electrical Circuits in Biomedical Engineering" at Yeditepe University since 2004 and it is aimed to assist the students in solving general and application-specific problems of Electrical Circuits in Biomedical Engineering at undergraduate and graduate levels.

Majority of the problems given in this book are self-contained and have varying levels of difficulty to encourage the student. Problems that deal with SPICE or MATLAB simulations are particularly intended to guide the student to understand the nature and specific aspects of these problems.

Due to lack of space and their secondary significance in biomedical engineering field, some topics such as three-phase power circuits, time-varying and nonlinear networks, state-space analysis, graph theory, and a relatively new topic of memristors are not included in this book. Neither nonstandard electrical network concept, nor hyper-real currents or voltages are within the scope of this book. Nevertheless, typical infinite electrical circuits are illustrated with sample problems, since they appeal some interest among the students in different engineering and science fields.

Attempt is made to show all steps of calculations in the solutions of problems. Relevant references are included at the end of each chapter. A separate bibliography of useful textbooks related to specific topics of interest is provided at the end of the book, as well.

Here you will also find software that supplements this book. These include EXCEL (xlsx files), MATLAB (.m files), and SPICE (.cir) files. The names of the files used for the solution of a problem are indicated at the end of each corresponding problem statement.

In order to run these software applications, one needs to have Microsoft's EXCEL (2013) and MathWorks' MATLAB (version R2009a or higher, along with symbolic and curve fitting toolboxes). On the other hand, student versions of SPICE main operating software source codes (such as PSPICE, HSPICE, AIMSPICE, ISPICE, 5SPICE, XSPICE, ADICE, LTSPICE, NGSPICE) can be freely downloaded via the net.

It is hoped that the book will be helpful for students, as well as for engineers in analyzing and/or designing practical biomedical instrumentation circuits.

Acknowledgements: The author thanks particularly the research assistants, Kübra Öztürk, Anil Ozdemirli, Betül Yardibi, Ibrahim Kapici, Ahmet Yetkin, Surhan Bozkurt, Ercument Cenap Turan, and Sinan Yagcioglu of Biomedical and Electrical Engineering Departments at Yeditepe, who supported this work with critical comments and discussions at various stages of its preparation.

Finally, I wish to express my deep appreciation to my wife, Naciye, for her encouragement, continued support, and understanding.

Istanbul, Turkey Ali Ümit Keskin

Contents

About the Author

Ali Ümit Keskin received his BSEE degree from Bogazici University, MSEE from Yildiz Technical University and Ph.D. degree from the Institute of Science and Technology, Istanbul Technical University, and then he joined Siemens AG, Turkey. He was actively engaged in the fields of medical instrumentation, diagnostic imaging, and radiotherapy throughout his professional career.

Since 2002, he is a staff member of Yeditepe University, Faculty of Engineering. Professor Keskin is one of the founders and recently the head of Biomedical Engineering Department at Yeditepe University, where he teaches Biomedical Engineering, Circuit Theory and Electrical Circuits in Biomedical Engineering courses. His main research interests include analog circuits and signal processing, sensors, transducers, and their computer simulation, medical instrumentation theory and its applications.

The following are trademarks or registered trademarks of respective holders:

MATLAB	The MathWorks, Inc.
EXCEL	Microsoft Inc.
Windows	Microsoft Inc.
PSPICE	OrCAD-MicroSim Corporation
ADAPTIV Biphasic	Physio Control
LIFEPAK	Physio Control
SMART Biphasic	Philips Medical
HeartStart	Philips Medical
Multipulse Biowave	Schiller AG
Rectilinear Biphasic	Zoll Medical Corporation
Actibiphasic	Nihon Kohden
Cardiolife	Nihon Kohden
Force FX	Covidien-Medtronic Inc.
Instant Response	Covidien-Medtronic Inc.
Force Argon II	Valleylab (Covidien-Medtronic Inc.)
Vio	Erbe Elektromedizin GmbH
APC	Erbe Elektromedizin GmbH
Beamer	Martin
Maxium	Martin
MarVAC	Martin
S-PILOT	Karl Storz

Chapter 1
Basic Concepts

1.1 Electric Charge, Current, Voltage, Power, Energy, and Sources

Problem 1.1.1 How much charge is made out of 1 Terra electrons?

Solution

$$\begin{aligned} q &= -1.602 \times 10^{-19} \times 10^{12} = -1.602 \times 10^{-7} \\ &= -1.602 \times 10^{-1}\,\mu\text{C} = -0.1602\,\mu\text{C} \end{aligned}$$

Problem 1.1.2 How much charge passes through a cross section of a conductor in 60 s if a dc current value is measured as 0.1 mA?

Solution

$$q = it = 10^{-7} \times 60 = 6 \times 10^{-6}\,\text{C} = 6\,\mu\text{C}$$

Problem 1.1.3 How much charge passes through a cross section of a conductor in 60 s if a current value is given as $i(t) = 5t$ mA, t in s?

Solution

$$q = \int_0^{60} i(t)\text{d}t = \int_0^{60} (5t) \times 10^{-3} \times \text{d}t = 5 \times 10^{-3} \times \left.\frac{t^2}{2}\right|_0^{60} = 5 \times 10^{-3} \times \frac{3600}{2} = 9\,\text{C}$$

A.Ü. Keskin, *Electrical Circuits in Biomedical Engineering*,
DOI 10.1007/978-3-319-55101-2_1

Problem 1.1.4 Determine the total charge entering a terminal between $t = 0$ s and $t = 10$ s if the current passing the terminal is

$$i(t) = \frac{1}{\sqrt{5t + 2}} \text{ A}$$

Solution

$$q = \int_0^{10} i(t)\mathrm{d}t = \int_0^{10} \frac{1}{\sqrt{5t+2}}\mathrm{d}t,$$

Let $u = 5t + 2$, then

$$\frac{\mathrm{d}u}{\mathrm{d}t} = 5 \quad \rightarrow \quad \mathrm{d}t = \frac{1}{5}\mathrm{d}u$$

$$q = \frac{1}{5}\int_0^{10} u^{-\frac{1}{2}}\,\mathrm{d}u = \frac{2}{5}\left(u^{\frac{1}{2}}\right)\Big|_0^{10} = \frac{2}{5}(5t+2)^{\frac{1}{2}}\Big|_0^{10} = 2.319\,\mathrm{C}$$

Problem 1.1.5 A 12 V battery requires a total charge of 60 Ah during charging. How many joules are absorbed by the battery?

Solution

$$E = pt = (vi)t = v(it) = 12\,\mathrm{V} \times (60\,\mathrm{Ah} \times 3600\,\mathrm{s}) = 2{,}592{,}000\,\mathrm{J}$$

Problem 1.1.6 A 12 V battery is rated at 60 Ah. What is the maximum current it can supply for 30 h? How many hours will it last if it is connected to 1 kΩ load resistor?

Solution

The maximum current this battery can supply for 30 h,

$$i = \frac{60}{30} = 2\,\mathrm{A}$$

$$t = \frac{60\,\mathrm{Ah}}{\frac{12\,\mathrm{V}}{1000\,\Omega}} = \frac{60}{12 \times 10^{-3}} = 5000\,\mathrm{h} = \frac{5000}{24} = 208.3\,\mathrm{days}$$

Problem 1.1.7 A 6 V battery is rated at 2.4 Ah has a lifetime of 10 h.

(a) What is the maximum current it can supply?
(b) How much power can it deliver?
(c) How much energy is stored in this battery?

Solution

(a) The maximum current this battery can supply for 10 h is

$$i = \frac{2.4}{10} = 240\,\text{mA}$$

(b) Amount of power it can deliver is $p = i\,v = 0.24 \times 6 = 1.44\,\text{W}$
(c) The energy stored in this battery is $E = p\,t = 1.44 \times 10 = 14.4\,\text{Wh}$

Problem 1.1.8 Total energy supplied by a pacemaker to a load in 5 years is 6417.6 J.

Determine the capacity of its battery in Ah, (ampere-hours), if the battery is a 5.6 V lithium type and 40% of its energy is spent for the pacing in 5 years period.

Solution

Battery energy:

$$E_B = 6417.6\,\text{J} = \text{VA s} = 0.4 \times 5.6 \times (\text{A s}), \quad \text{A s} = \text{capacity, in seconds}$$
$$6417.6 = 2.24 \times \text{Capacity}$$
$$\text{Capacity} = \frac{6417.6}{2.24}\,\text{A s} = 2865\,\text{A s}$$
$$\text{Capacity in A h} = \frac{2865(\text{A s})}{3600(\text{s/h})} = 0.796\,\text{A h}.$$

Problem 1.1.9 To move a charge q from point a to point b, requires (−10) J. Find the voltage drop if,

(a) $q = 2$ C, (b) $q = -5$ C.

Solution

(a) $V_{ab} = \frac{W}{q} = \frac{-10}{2} = -5\,\text{V}$

(b) $V_{ab} = \frac{W}{q} = \frac{-10}{-5} = 2\,\text{V}$

Problem 1.1.10 Prove the electrical power relationship, $P = v \cdot i$, where v and i are symbols for instantaneous voltage and current on a load.

Solution

If q represents electric charge, the current is defined by simple differential equation,

$$i = \frac{\partial q}{\partial t}$$

If W represents energy acting on the electric charge, the equation of voltage is

$$v = \frac{\partial W}{\partial q}$$

If these two formulas are multiplied,

$$i \cdot v = \frac{\partial q}{\partial t} \cdot \frac{\partial W}{\partial q} = \frac{\partial W}{\partial t}$$

which is the definition of power. Therefore, instantaneous power value is obtained by multiplying instantaneous current and voltage values on an element, $P = v \cdot i$.

Problem 1.1.11 How much energy in kWh does a 1 HP motor deliver in 1 h? How much does it cost monthly if this motor operates 10 h each day at 13 cents/kWh electricity tariff?

Solution

$$p = 1\,\text{HP} = 0.746\,\text{kW}$$
$$E = (p)t = 0.746\,\text{kW} \times 1\,\text{h} = 0.746\,\text{kWh}$$
$$\cos t = 0.746 \times 10 \times 30 \times 13 = \$\,29.09$$

Problem 1.1.12 A lightning bolt with 10 kA strikes an arrester for 14 μs. How much charge is transferred to the arrester?

Solution

$$q = it = 10{,}000 \times 14 \times 10^{-6} = 0.14\,\text{C}$$

Problem 1.1.13 A 75 W electrical appliance operates at 220 V. How much electrical charge flows through the appliance in 10 h? How many electrons does this charge flow correspond?

Solution

$$p = vi \quad \rightarrow \quad i = \frac{p}{v} = \frac{75}{220} = 0.341\,\text{A}$$
$$q = it = 0.341 \times 10 \times 3600 = 12{,}272\,\text{C}$$
$$N = 12{,}272 \times 6.24 \times 10^{18} = 76{,}581 \times 10^{18} = 7.6581 \times 10^{22}\ \text{electrons}$$

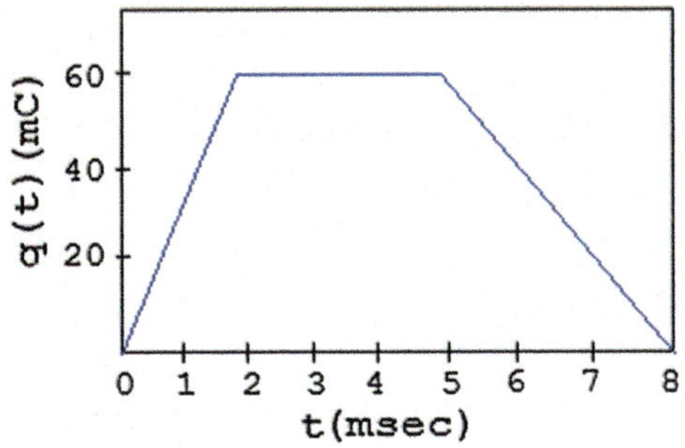

Fig. 1.1 Electrical charge variation through a cross section of a conductor

Problem 1.1.14 Electrical charge variation through a cross section of a conductor is shown in Fig. 1.1. Find the current at,

(a) $t = 0$ s, (b) $t = 1$ s, (c) $t = 4$ s, (d) $t = 6$ s, (e) $t = 7$ s.

Solution

(a) At $t = 0$ s, $i = 0$ A,

(b) At $t = 1$ s, $i = \frac{dq}{dt} = \frac{\Delta q}{\Delta t} = \frac{60}{2} = 30\,\text{A}$

(c) At $t = 4$ s, $i = \frac{\Delta q}{\Delta t} = \frac{0}{3} = 0\,\text{A}$

(d) At $t = 6$ s, $i = \frac{\Delta q}{\Delta t} = -\frac{60}{3} = -20\,\text{A}$

(e) At $t = 7$ s, $i = \frac{\Delta q}{\Delta t} = -\frac{60}{3} = -20\,\text{A}$

Problem 1.1.15 Electric power consumption of an operating room in one working day is shown in Fig. 1.2.

(a) What is the total energy consumed in kWh?
(b) Find the average power consumption in one day.

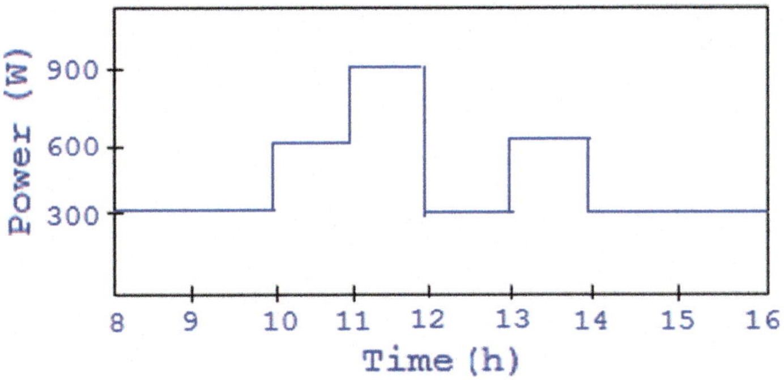

Fig. 1.2 Electric power consumption of an operating room

Solution

(a) $E = \sum_{i=8}^{16} Pt = 300 \times 2 + 600 \times 1 + 900 \times 1 + 300 \times 1 + 600 \times 1 + 300 \times 2$

$E = 3600\,\text{Wh} = 3.6\,\text{kWh}$

(b) $P_{\text{av}} = \dfrac{3.6}{24} = 0.15\,\text{kW}$

Problem 1.1.16 The resistance of a wire having linear temperature dependance is measured at 20 °C as 98.0 Ω. Determine the resistance of the wire at, (a) 30 °C, (b) 0 °C, (c) 500 °C $\left(\rho_0 = 100 \times 10^{-8}\,\Omega\,\text{m},\ \alpha = 0.0004(°\text{C})^{-1}\right)$.

Solution

(a) $R = R_0[1 + \alpha(T - T_0)] = 98[1 + 0.0004 \times (37 - 20)] = 98.6664\,\Omega$
(b) $R = 98[1 + 0.0004 \times (0 - 20)] = (1 - 0.008) + 98 = 97.2160\,\Omega$
(c) $R = 98[1 + 0.0004 \times (500 - 20)] = (1 + 0.192) \times 98 = 116.8160\,\Omega$

Problem 1.1.17 Input resistance of the heater section of a laboratory furnace is 20 Ω when it is cold. It is connected to 230 V AC plug and draws 10 A. Determine the temperature of the heater wire during its operation, if temperature coefficient of its resistivity is $0.0004(°\text{C})^{-1}$.

Solution

The resistance of the heater in operation, $(\cos\varphi = 1)$

$$R = \frac{V}{I} = \frac{230\,\text{V(rms)}}{10\,\text{A(rms)}} = 23\,\Omega$$
$$R = R_0[1 + \alpha(T - T_0)]$$

"Cold" means environment is at 20 °C;

$$23 = 20[1 + 0.0004(T - 20\,°\text{C})] = 20 + 0.0008(T - 20) = 0.008T + 19.99872$$
$$T = \frac{23 - 19.99872}{0.008} \cong 375\,°\text{C}$$

Problem 1.1.18 During stimulation of an excitable cell, current and voltage variation as functions of time are given as

$$i(t) = 3t \quad [\text{A}] \quad t \geq 0$$
$$v(t) = 10\,e^{6t}, \quad [\text{V}] \quad t \geq 0$$

Determine the energy consumed (in micro joules) between 0 and 2 ms.

Solution

$$W = \int_0^{2\times10^{-3}} i(t)v(t)\mathrm{d}t = \int_0^{2\times10^{-3}} 3t.10\mathrm{e}^{6t}\mathrm{d}t = 30\int_0^{2\times10^{-3}} t\mathrm{e}^{6t}\mathrm{d}t$$

Using "integration by parts" rule,

$$\int_a^b u\,\mathrm{d}v = u\,v|_a^b - \int_a^b v\,\mathrm{d}u$$

with

$$u = t, \quad du = \mathrm{d}t, \quad dv = \mathrm{e}^{6t}\,\mathrm{d}t, \quad v = \frac{1}{6}\mathrm{e}^{6t}$$

$$W = 30\left[t\cdot\frac{1}{6}\mathrm{e}^{6t}\Big|_0^{2\times10^{-3}} - \frac{1}{6}\int_0^{2\times10^{-3}} \mathrm{e}^{6t}\,\mathrm{d}t\right]$$

$$= 30\left[t\cdot\frac{1}{6}\mathrm{e}^{6t}\Big|_0^{2\times10^{-3}} - \frac{1}{36}\mathrm{e}^{6t}\Big|_0^{2\times10^{-3}}\right]$$

$$= 30\left[\frac{2\times10^{-3}}{6}\mathrm{e}^{12\times10^{-3}} - \frac{1}{36}\left(\mathrm{e}^{12\times10^{-3}} - 1\right)\right]$$

$$= 30\left[3.374\times10^{-4} - 3.353\times10^{-4}\right]$$

$$= 30\times0.021\times10^{-4} = 630\times10^{-7}\,\mathrm{J} = 63\,\mu\mathrm{J}$$

Problem 1.1.19 Power relationship in a circuit is defined as

$$p_n = \frac{1}{3^n}|\sin t|, \quad \text{for } n\pi \le t \le (n+1)\pi, \quad n = 0, 1, 2, \ldots$$

Determine the energy consumed in this circuit.

Solution

$$W = \int_0^{\infty} p_n(t)\mathrm{d}t = \int_0^{\infty} \frac{1}{3^n}|\sin t|\mathrm{d}t$$

$$p_n(t) = \frac{1}{3^0}|\sin t| + \frac{1}{3^1}|\sin t| + \frac{1}{3^2}|\sin t| + \cdots = |\sin t|\cdot\sum_{n=0}^{\infty}\frac{1}{3^n} = |\sin t|\cdot\frac{1}{1-\frac{1}{3}} = \frac{3}{2}$$

Note that, inside of the summation above equals to the sum of a geometric series. Finally,

$$W = \frac{3}{2}\int_{0}^{\pi} |\sin t| \mathrm{d}t = \frac{3}{2}|\cos t|\Big|_{0}^{\pi} = \frac{3}{2}(-1-1) = \frac{3}{2}\times 2 = 3\,\mathrm{J}$$

Problem 1.1.20 Current passing through an electrical element is defined as $i(t) = A\,\sin\omega t$. Determine energy on this element if the voltage across this element is $v(t) = B\cos\omega t, \quad t \geq 0$.

Plot current, voltage, and energy as functions of time (plot_t_function2.m).

Solution

Power relationship, $p(t) = i(t)\cdot v(t) = A\sin\omega t \cdot B\cos\omega t = AB\sin\omega t\cdot\cos\omega t$

Energy,

$$w = \int p(t)\mathrm{d}t = \int_{t0}^{t} AB\sin\omega\tau\cdot\cos\omega\tau\,\mathrm{d}\tau$$

$$\sin(2u) = 2\sin(u)\cdot\cos(u), \quad \rightarrow \quad \sin(u)\cdot\cos(u) = \frac{\sin(2u)}{2} \tag{1.1}$$

$$\text{Let } \omega\tau = u, \quad \rightarrow \quad \omega d\tau = \mathrm{d}u \quad \rightarrow \quad \mathrm{d}\tau = \frac{1}{\omega}\mathrm{d}u$$

Substituting these in (1.1):

$$\begin{aligned} w &= \int AB\frac{\sin(2u)}{2}\cdot\frac{1}{\omega}\mathrm{d}u = \frac{AB}{2\omega}\int \sin(2u)\mathrm{d}u = \frac{AB}{2\omega}\cdot\frac{-\cos(2u)}{2}\Big|_{t0=0}^{t} \\ &= \frac{AB}{4\omega}[1-\cos(2\omega t)] \end{aligned}$$

The waveforms are shown in Fig. 1.3.

Following is the MATLAB script (plot_t_function2.m);

```
t = 0:0.1:2*pi;
w=1; A=1; B=1; K=(A*B)/(4*w);
i=A*sin(w*t); v=B*cos(w*t); w=K*(1-cos(2*w*t));
plot(t,w,'r',t,i,'b',t,v,'k','linewidth',2.5);
grid on; axis([0  2*pi  -A  A]); xlabel('time [s]');
title('voltage,current and energy waveforms');legend('w','i','v');
```

Problem 1.1.21 During an in vitro (petri dish) experiment, the peak electric power that a group of stem cells can tolerate without some serious functional

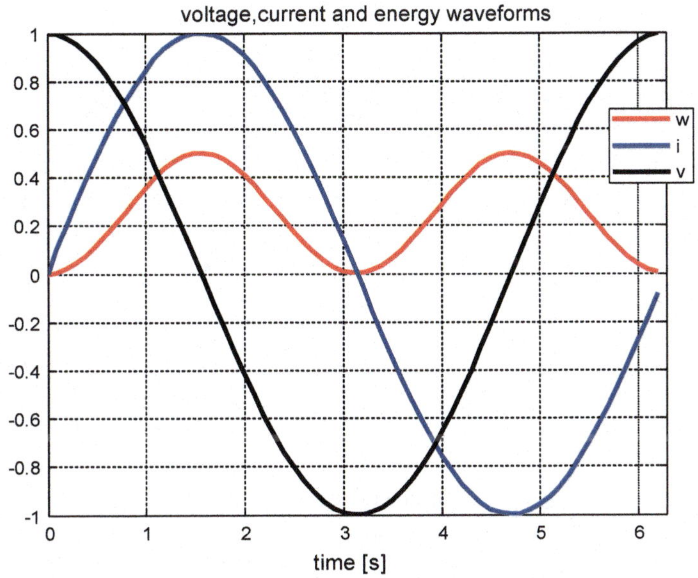

Fig. 1.3 Voltage, current, and energy functions

consequences is known to have a threshold of 1 mW. If the power delivered to this group of cells is $p(t) = 2e^{-t}\sin(5t)[\text{mW}]$, will the cells be harmed? (petridish.xlsx).

Solution

$$p(t) = 2e^{-t}\sin(5t)[\text{mW}]$$

$$\frac{d}{dt}[p(t)] = 2e^{-t}[-\sin(5t) + 5\cos(5t)] = 0 \quad \rightarrow \quad \sin(5t) = 5\cos(5t)$$

$$\tan(5t) = 5 \quad \rightarrow \quad t = \frac{\text{atan}(5)}{5} = \frac{1.3734}{5} = 0.275\,\text{s}$$

$$p(0.275) = 2e^{-0.275}\sin(5 \times 0.275) = 1.49\,[\text{mW}] > 1\,[\text{mW}]$$

The cells will be harmed.

Alternative solution is to plot power versus time curve (see, Fig. 1.4) and check for the threshold value. It is noted here that both solutions verify the conclusion.

Problem 1.1.22 A DC electrical pump brings out m kg water out of h meters deep well, and ejects it with a speed of v m/s. If this job is to be finished in T s, determine the minimum current (I) of the pump if it is operating at U volts (Assume that efficiency = 1).

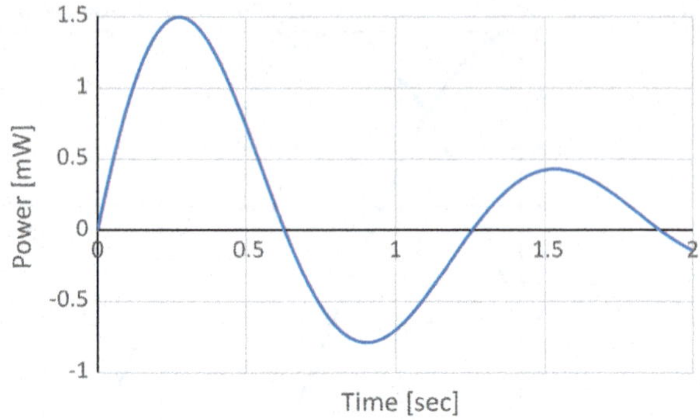

Fig. 1.4 Power versus time graph

Solution

$$P = \text{Power} = \frac{\text{Total work done}}{\text{Time}} = \frac{W}{T}$$

$$\text{Total work done} = W = \frac{1}{2}\,\text{mv}^2 + \text{mgh}$$

$$P = \frac{\frac{1}{2}\,\text{mv}^2 + \text{mgh}}{T} = \frac{\text{m}\left(\frac{\text{v}^2}{2} + \text{gh}\right)}{T} = \frac{\text{m}(\text{v}^2 + 2\,\text{gh})}{2T}$$

On the other hand,

$$P = \text{IU}$$

$$I = \frac{P}{U} = \frac{\text{m}(\text{v}^2 + 2\,\text{gh})}{2\text{UT}} \quad [\text{A}]$$

Problem 1.1.23 A hospital has a dynamic uninterrupted power system (D-UPS) consisting of a diesel generator, synchronous machine, and a kinetic energy unit. If the energy to hospital fails, kinetic energy unit continues to feed the 600 kW electrical load of the hospital for 20 s, then diesel engine and synchronous machine take the load over and feed the load.

(a) Determine total energy capacity of the kinetic energy unit.
(b) If the load is 240 kW at a specific power failure instant, how long the unit can feed load?

Solution

(a) Assuming no energy losses,

$$E = Pt = 600\,\text{kW}\,20\,\text{s} = 12{,}000\,\text{kJ} = 12\,\text{MJ}$$

(b) Since actual load is less than the available power, it takes longer to feed the load,

$$12{,}000\,\text{kJ} = 240\,\text{kW}\,t\,\text{s} \quad \rightarrow \quad t = 50\,\text{s}$$

Problem 1.1.24 The amount of a substance produced at one of the electrodes in an electrolytic cell is directly proportional to the amount of electrical charge that passes through the cell. This is known as Faraday's law.

Faraday's constant describes the number of coulombs of charge carried by a mole of electrons, F = 96,500 C/mol.

(a) Calculate the mass of sodium metal that will form at the cathode of an electrolytic cell when a 5-amp current is passed through molten sodium chloride for a period of 2 h. (Balanced equation at the cathode of this cell gives one mole of sodium for every mole of electrons. 1 mol of sodium mass is 23 g).
(b) What mass of copper can be deposited at the cathode in an electrolytic cell using aqueous copper sulfate solution if a current of 5 A flows for 2 h? (Balanced equation at the cathode of this cell gives one mole of copper for 2 mol of electrons. 1 mol of copper mass is 63.5 g).
(c) How much energy in (kWh) is required to produce 1 kg of chlorine gas from aqueous NaCl solution if the cell operates at 2.5 V for 10 h? The electrolysis of aqueous NaCl produces hydrogen gas, H_2 at the cathode and chlorine gas, Cl_2 at the anode. Balanced equation at the anode of this cell gives one mole of chlorine gas for 2 mol of electrons. 1 mol of atomic chlorine mass is 35.5 g).

Solution

(a) $q = It = 5\,\text{A} \times 2\,\text{h} = 5\,\text{A} \times 2 \times 3600\,\text{s} = 36{,}000\,\text{C}$
The number of moles of electrons transferred when 36,000 coulombs of electric charge flow through the electrolytic cell

$$n = \frac{36{,}000}{96{,}500} = 0.373\,\text{mol electrons}$$

Balanced equation at the cathode of this cell gives one mole of sodium for every mole of electrons,

$$\text{Na}^+ + \text{e}^- \rightarrow \text{Na}$$

Therefore, 0.373 mol of sodium is collected at the cathode in 2 h. The mass of this metal is

$$m = 0.373 \times 23 = 8579\,\text{g}$$

(b) $q = It = 5\,\text{A} \times 2\,\text{h} = 5\,\text{A} \times 2 \times 3600\,\text{s} = 36{,}000\,\text{C}$
The number of moles of electrons transferred when 36,000 coulombs of electric charge flow through the electrolytic cell

$$n = \frac{36{,}000}{96{,}500} = 0.373\,\text{mol electrons}$$

Balanced equation at the cathode of this cell gives 1 mol of copper for 2 mol of electrons,

$$\text{Cu}^{+2} + 2\text{e}^- \rightarrow \text{Cu}$$

Therefore, 0.373 mol of copper is collected at the cathode in 2 h. The mass of this metal is

$$m = 0.373 \times 63.5 \times 0.5 = 11.843\,\text{g}$$

(c) Balanced equation at the anode of this cell gives one mole of chlorine gas for 2 mol of electrons,

$$2\text{Cl}^- - 2\text{e}^- \rightarrow \text{Cl}_2$$

Moles of 1 kg chlorine gas,

$$n = \frac{1000\ \text{g}}{35.5\left(\frac{\text{g}}{\text{mol}}\right) \times 2} = 14.085\,\text{mol}$$

$$\text{Faradays of charge} = 2 \times 14.085 = 28.17\ \text{F}$$

Electrical charge,

$$q = 96,500 \times 28.17 = 2,718,405\,\text{C}$$

Current is calculated as

$$I = \frac{2{,}718{,}405\,\text{C}}{3600 \times 10\,\text{s}} = \frac{2{,}718{,}405}{36{,}000} = 75.511\,\text{A}$$

Power,

$$P = IV = 75.511 \times 2.5 = 188.8\,\text{W}$$

Energy,

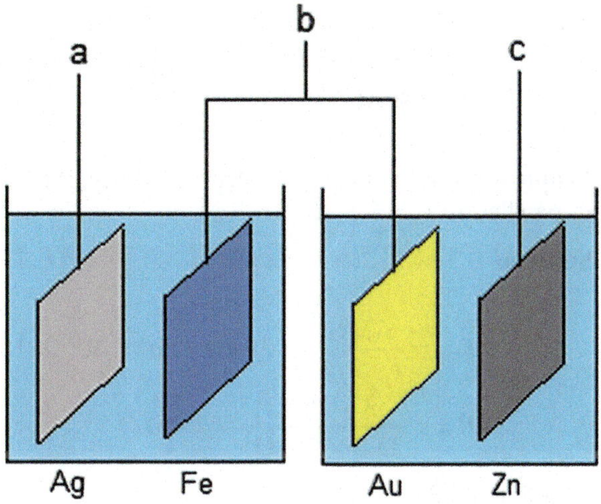

Fig. 1.5 Electrochemical cells for Problem 1.1.25

$$E = Pt = 188.8 \times 10 = 1888\,\text{Wh} = 1.888\,\text{kWh}$$

Problem 1.1.25 Determine the potential difference between a and c terminals of Fig. 1.5, considering the half-cell potentials; $\text{Ag} = 0.799\,\text{V}$, $\text{Au} = 1.680\,\text{V}$, $\text{Zn} = -0.763\,\text{V}$, $\text{Fe} = -0.509\,\text{V}$.

Solution

$$\begin{aligned} V_{\text{ac}} &= V_{\text{ab}} + V_{\text{bc}} = 0.799 - (-0.509) + 1.680 - (-0.763) \\ &= 1.308 + 2.443 = 3.751\,\text{V} \end{aligned}$$

Problem 1.1.26 Defining equation of a pH electrode is $V_{\text{m}} = -60\log[H^+] + k\,(\text{mV})$.

When a pH electrode is immersed in a solution of pH = 4, the attached voltmeter measures a voltage of $V_{\text{m1}} = 0.210\,\text{V}$. The pH electrode is then immersed in an unknown solution. The voltmeter now reads $V_{\text{m2}} = 400\,\text{mV}$. What is the pH of the unknown solution?

Solution

$$\begin{aligned} &V_{\text{m}} = -60\log[H^+] + k\,(\text{mV}), \quad \text{pH}_1 = -\log[H^+] = 4 \\ &210 = -60(-4) + k = 240 + k \quad \rightarrow \quad k = -30\,\text{mV} \\ &V_{\text{m2}} = 400 = -60\log[H^+] - 30 \quad \rightarrow \quad \log[H^+] = -430/60 = -7.167 \\ &\text{pH}_2 = -\log[H^+] = -(-7.167) = 7.167 \end{aligned}$$

1.2 Ohm's Law, Kirchhoff's Laws, Parallel–Series Connection of Resistors and Sources, Delta–Wye Conversion

Problem 1.2.1 Find the equivalent resistance, if a resistor R_1 = 10 kΩ is connected in parallel to R_2 = 3.3 kΩ.

Solution

$$R_{eq} = \frac{R_1.R_2}{R_1 + R_2}, \quad 10\,\text{k}\Omega = 10 \times 10^3\,\Omega, 3.3\,\text{k}\Omega = 3.3 \times 10^3\,\Omega$$

$$R_{eq} = \frac{33 \times 10^6\,\Omega^2}{13.3 \times 10^3\,\Omega^2} = 2.48\,\text{k}\Omega$$

Problem 1.2.2

(a) What is the equivalent resistance of three parallel-connected resistors with equal values?
(b) If $R = R_1 = R_2 = R_3 = 10\,\text{k}\Omega$, R_{eq} = ?

Solution

(a) $$\frac{1}{R_{eq}} = \frac{1}{R} + \frac{1}{R} + \frac{1}{R} = \frac{3}{R}, \quad \rightarrow \quad 3R_{eq} = R \quad \rightarrow \quad R_{eq} = \frac{R}{3}$$

(b) $$R_{eq} = \frac{10\,\text{k}\Omega}{3} = 3.333\,\text{k}\Omega$$

Problem 1.2.3 Four resistors are connected in parallel.
$R_1 = 10\,\text{k}\Omega$, $R_2 = 1\,\text{k}\Omega, R_3 = 5\,\text{k}\Omega, R_4 = 3\,\text{k}\Omega$. Calculate their equivalent resistance.

Solution

$$\frac{1}{R_{eq}} = \frac{1}{R_1} + \frac{1}{R_2} + \frac{1}{R_3} + \frac{1}{R_4} = 10^{-4} + 10^{-3} + 0.2 \times 10^{-3} + 0.3333 \times 10^{-3}$$
$$= (0.1 + 1 + 0.2 + 0.3333) \times 10^{-3} = 1.6333 \times 10^{-3}$$

$$R_{eq} = \frac{1}{1.6333 \times 10^{-3}} = \frac{10^3}{1.6333} = 612.26\,\Omega$$

Problem 1.2.4 Find the values of equivalent resistors in the circuit of Fig. 1.6 ($Y \rightarrow \Delta$ conversion).

$$R_1 = 10\,\Omega, \quad R_2 = 10\,\Omega, \quad R_3 = 5\,\Omega.$$

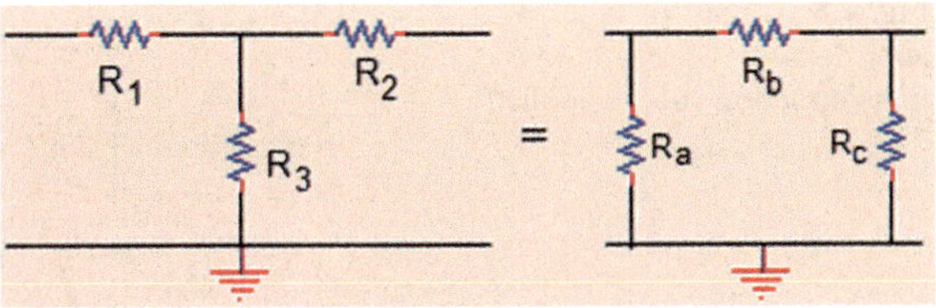

Fig. 1.6 *Y* to Δ conversion

Solution

$$Y \rightarrow \Delta \, \text{conversion,}$$

$$R_a = \frac{R_1R_2 + R_2R_3 + R_1R_3}{R_1} = \frac{10 \times 10 + 10 \times 5 + 10 \times 5}{10} = 20\,\Omega$$
$$R_b = \frac{R_1R_2 + R_2R_3 + R_1R_3}{R_2} = \frac{10 \times 10 + 10 \times 5 + 10 \times 5}{10} = 20\,\Omega$$
$$R_c = \frac{R_1R_2 + R_2R_3 + R_1R_3}{R_3} = \frac{10 \times 10 + 10 \times 5 + 10 \times 5}{5} = 40\,\Omega$$

Problem 1.2.5

(a) In the circuit shown in Fig. 1.7, use delta–wye transformation rule, and determine the power dissipated by resistor $R_1.(R_1 = 10\,\Omega, R_2 = R_3 = R_4 = 20\,\Omega, R_5 = R_6 = 40\,\Omega, U = 40\,\text{V})$.
(b) Find the voltage at node a.
(c) Check the results using SPICE and print a netlist (delta_wye1.cir).

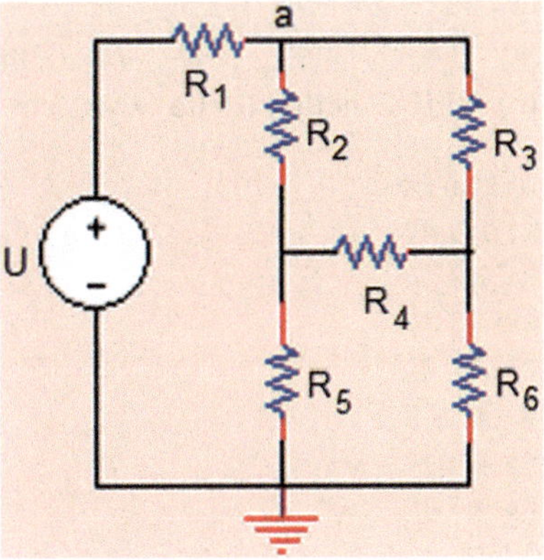

Fig. 1.7 Circuit for the application of delta–wye transformation rule

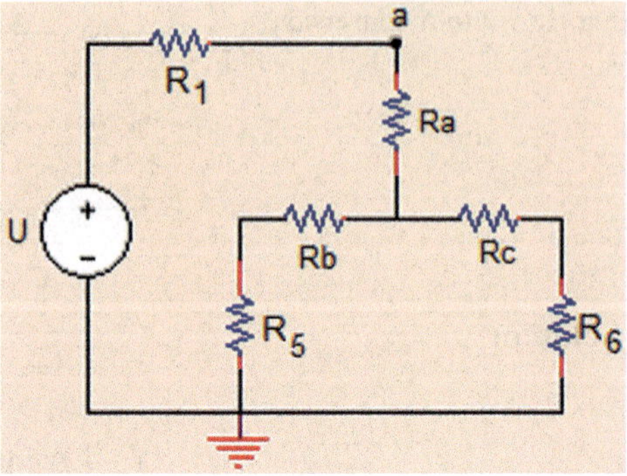

Fig. 1.8 Circuit of Fig. 1.7 after delta–wye transformation rule is applied

Solution

(a) Delta–wye conversion can be applied to $R_2 - R_3 - R_4$ mesh, which is a delta form (Fig. 1.8).

$$R_a = R_b = R_c = \frac{R_2R_3}{R_2 + R_3 + R_4} = \frac{20 \times 20}{20 + 20 + 20} = \frac{400}{60} = \frac{20}{3}\,\Omega$$

$$R_b + R_5 = R_c + R_6 = \frac{20}{3} + 40 = \frac{20 + 120}{3} = \frac{140}{3}\,\Omega$$

$$R_p = (R_b + R_s) \parallel (R_c + R_6) = \frac{140}{6}\,\Omega$$

$$R = R_1 + R_a + R_p = 10 + \frac{20}{3} + \frac{140}{6} = \frac{60 + 40 + 140}{6} = 40\,\Omega$$

$$I = \frac{U}{R} = \frac{40}{40} = 1\,\text{A}$$

$$P_{R1} = I^2R = 1 \times 10 = 10\,\text{W}$$

(b) $V_a = U - V_{R1} = U - IR_1 = 40 - 1 \times 10 = 30\,\text{V}$

(c) SPICE netlist (delta_wye1.cir) is given below.

```
Delta-Wye1
V1 1 0 40
R1 1 2 10
R2 2 3 20
R3 2 4 20
R4 3 4 20
R5 3 0 40
R6 4 0 40
*.op
```

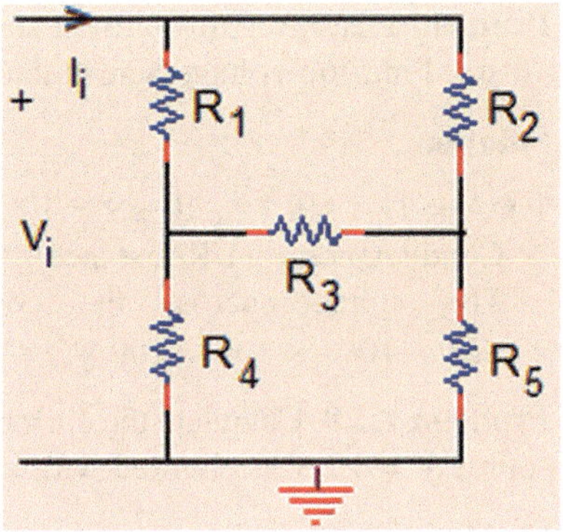

Fig. 1.9 Unbalanced bridge circuit

Problem 1.2.6

(a) Determine the equivalent resistance of an unbalanced bridge circuit shown in Fig. 1.9. (Z_symbolic2.m).
(b) If $R_1 = R_2 = R_3 = R_4 = R_5 = 1\ \Omega$, $R_{eq} = ?$

Solution

(a) Apply $Y - \Delta$ transformation to R_2, R_3, R_5:

$$R_a = \frac{R_2R_3 + R_3R_5 + R_2R_5}{R_5}, \quad R_b = \frac{R_2R_3 + R_3R_5 + R_2R_5}{R_2}, \quad R_c = \frac{R_2R_3 + R_3R_5 + R_2R_5}{R_3}$$

$$R_d = R_1 \parallel R_a = \frac{R_1 \cdot R_a}{R_1 + R_a}, \quad R_e = R_4 \parallel R_b = \frac{R_4 \cdot R_b}{R_4 + R_b}, \quad R_f = R_d + R_e$$

$$R_{eq} = R_f \parallel R_c = \frac{R_f \cdot R_c}{R_f + R_c}$$

$$R_{eq} = \frac{R_1R_2R_3 + R_1R_2R_4 + R_1R_2R_5 + R_1R_3R_5 + R_2R_3R_4 + R_1R_4R_5 + R_2R_4R_5 + R_3R_4R_5}{R_1R_3 + R_1R_4 + R_1R_5 + R_2R_3 + R_2R_4 + R_2R_5 + R_3R_4 + R_3R_5}$$

(b) $R_1 = R_2 = R_3 = R_4 = R_5 = 1\ \Omega$ is a balanced case; then

$$R_{eq} = (R_1 + R_4)//(R_2 + R_5) = 2//2 = 1\,\Omega.$$

Following is the MATLAB script (Z_symbolic2.m).

```
.m
%Equivalent resistance of an unbalanced bridge
syms s Z1 Z2 Z3 Z4 Z5
A=Z3*Z5+Z2*Z3+Z2*Z5; Za=A/Z5; Zb=A/Z2; Zc=A/Z3;
Zd=Z1*Za/(Z1+Za); Ze=Z4*Zb/(Z4+Zb); Zf=Zd+Ze;
Z=Zf*Zc/(Zc+Zf);factor(Z)
```

Problem 1.2.7 A 1 mA current source is connected to a conductor with a value of 10 μS. Find the voltage across the conductor.

Solution

1 mA = 1×10^{-3} A, 10 μS = 10×10^{-6} S,

Conductance = (1/Resistance), then, $R = 1/(10 \times 10^{-6}) = 10^5\ \Omega$

The voltage across the conductor is calculated using Ohm's Law: $v = iR = 10^{-3} \times 10^5 = 100\,\text{V}$

Problem 1.2.8 Calculate the current in mA, conductance and the power if a voltage source V = 10 V is shunted with a resistor of R_1 = 5 kΩ.

Solution

$$i = \frac{V}{R} = 2\,\text{mA}, \quad G = \frac{1}{R} = 2 \times 10^{-4}\,\text{S} = 0.2\,\text{mS}$$
$$P = i \cdot V = 2\,\text{mA} \cdot 10\,\text{V} = 20\,\text{mW}$$

Problem 1.2.9 Find the power consumed over a resistor, if the voltage across the resistor is $v(t)$ = 10 sin πt V, and R = 5 kΩ.

Solution

$$i = \frac{V}{R} = 2 \sin \pi t\,(\text{mA}), \quad \rightarrow \quad P = i \cdot V = 20 \sin^2 \pi t\,(\text{mW})$$

Problem 1.2.10 V_1 = 2 V, V_4 = 4 V, V_5 = 2 V and $R_1 = R_2$ = 1 kΩ. Find the current flowing in the circuit of Fig. 1.10.

KVL (Selecting the current direction counterclockwise),

$$V_1 + V_4 + V_5 = V_3 + V_2$$
$$2 + 4 + 2 = 8 = i(R_1 + R_2) = 2000i$$

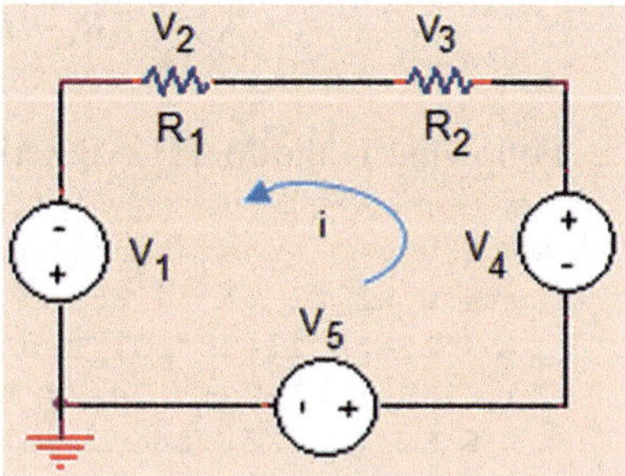

Fig. 1.10 The circuit for Problem 1.2.10

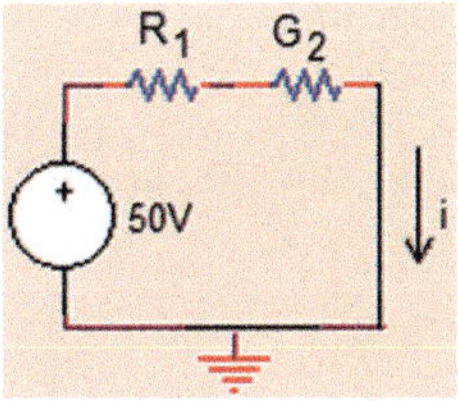

Fig. 1.11 The circuit for Problem 1.2.11

Therefore,

$$i = \frac{8\,\text{V}}{2000\,\Omega} = 4\,\text{mA}$$

Problem 1.2.11 Find the current in the circuit of Fig. 1.11 ($R_1 = 0.1\ \Omega$, $G_2 = 0.05$ S).

Solution

$$i = \frac{50}{R_1 + R_2} = \frac{50}{0.1 + \frac{1}{0.05}} = \frac{50}{0.1 + 20} = \frac{50}{20.1} = 20.48\,\text{A}$$

Problem 1.2.12 For the circuit in Fig. 1.12, what is

(a) the voltage between the point a and b (V_{ab})?
(b) the power (P) consumed on resistor, in mW?
(c) the conductance (G)?

Solution

$$i = 3\,\text{mA} = 3 \times 10^{-3}\,\text{A}, \quad R = 1\,\text{k}\Omega = 1 \times 10^{3}\,\Omega$$

(a) By Ohm's Law, $V_{ab} = 3 \times 10^{-3}\,\text{A} \times 10^{3}\,\Omega = 3\,\text{V}$
(b) $P = V \cdot i = 3\,\text{V} \times 3 \times 10^{-3} = 9 \times 10^{-3}\,\text{W} = 9\,\text{mW}$

Problem 1.2.13 For the circuit of Fig. 1.13, what is the voltage drop across the resistor?

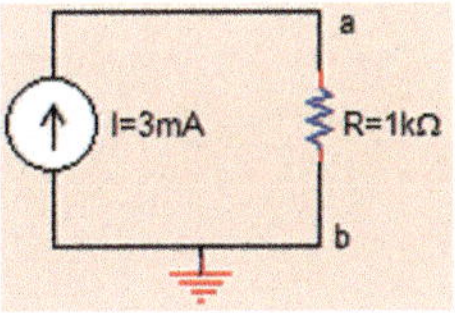

Fig. 1.12 The circuit for Problem 1.2.12

Fig. 1.13 The circuit for Problem 1.2.13

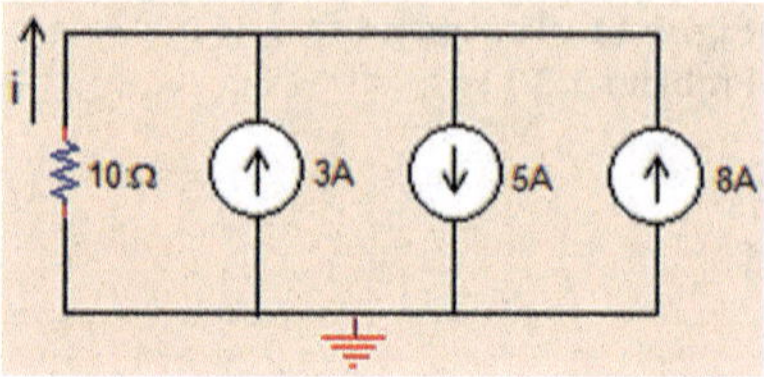

Solution

Use Kirchhoff's current law, and let the current flowing through the resistance be i,

$$\sum_{n=1}^{N} i_n = 3\,\text{A} - 5\,\text{A} + 8\,\text{A} + i = 0 \quad \rightarrow \quad i + 6\,\text{A} = 0 \quad \rightarrow \quad i = -6\,\text{A}$$

Voltage drop across the resistor is $v = i \times R = 6\,\text{A} \times 10\,\Omega = 60\,\text{V}$

Problem 1.2.14 Find the total current for the circuit of Fig. 1.14.

Solution

By KCL,

$$1 + 2 - 5 + 4 + i = 0 \quad \rightarrow \quad i = -2\,\text{A}$$

Problem 1.2.15 In the circuit shown in Fig. 1.15, if $i = 1$ A, $R_1 = 2\ \Omega$, $R_2 = 4\ \Omega$, $R_3 = 6\ \Omega$, $V_4 = ?$

Fig. 1.14 The circuit for Problem 1.2.14

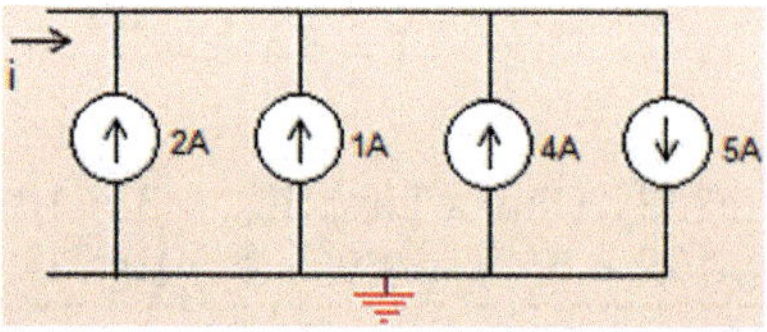

Fig. 1.15 The circuit for Problem 1.2.15

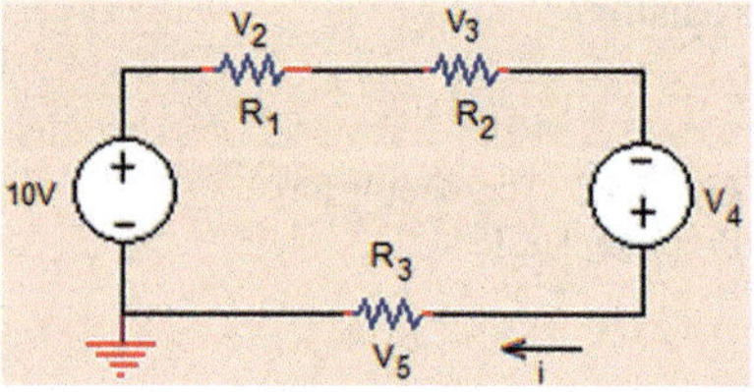

Solution

Apply the Kirchhoff's voltage law to the circuit,

$$-V_1 + V_2 + V_3 - V_4 + V_5 = -V_1 + R_1 \cdot i + R_2 \cdot i - V_4 + R_3 \cdot i = 0$$
$$-10 + 2\,\Omega \cdot 1\,\mathrm{A} + 4\,\Omega \cdot 1\,\mathrm{A} - V_4 + 6\,\Omega \cdot 1\,\mathrm{A} = 0$$

$$V_4 = 2\,\mathrm{V}$$

Problem 1.2.16 Design a voltage divider circuit (Fig. 1.16) to measure voltage values up to 20 kV (DC) with the following constraints:
$R_1/R_2 = 1000$, maximum current flow through the circuit = 200 microamperes.

Solution

$$V_0 = V_1 \frac{R_1}{R_1 + R_2} \approx V_1 \frac{R_2}{R_1} = 20 \times 10^3 \times 10^{-3} = 20\,\mathrm{V} \quad (R_1 \gg R_2)$$
$$R_2 = \frac{20}{200 \times 10^{-6}} = 100\,\mathrm{k}\Omega \quad \rightarrow \quad R_1 = 1000 \times R_2 = 10^3 \times 100 \times 10^3 = 100\,\mathrm{M}\Omega$$

Problem 1.2.17 Find the voltage V_1 in the circuit of Fig. 1.17. Comment on the possible values of k.

Solution

By KVL, $-18 + kV_1 + (1 + 2 + 3)RI = 0$

By Ohm's law, $V_1 = RI$,

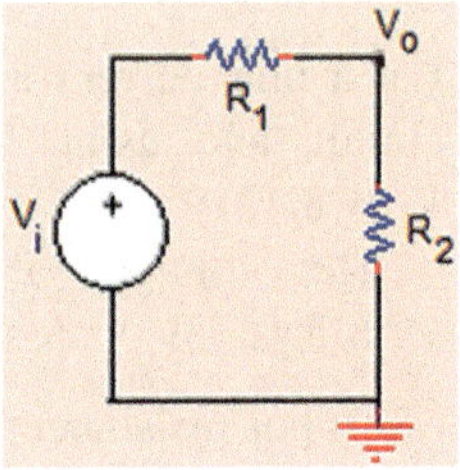

Fig. 1.16 The circuit for Problem 1.2.16

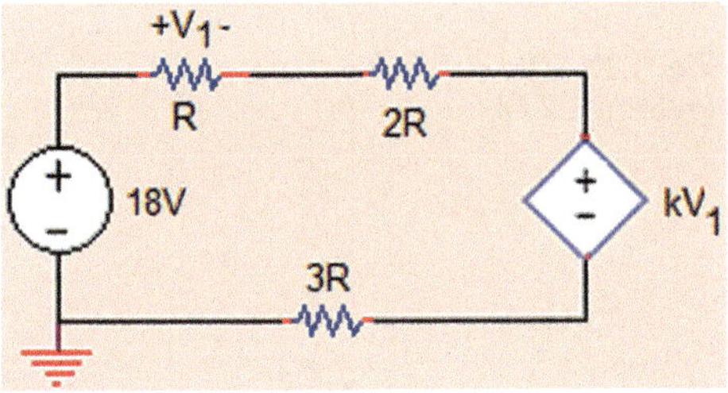

Fig. 1.17 The circuit for Problem 1.2.17

$$-18 + kRI + (1+2+3)RI = 0 \quad \rightarrow \quad -18 + (k+6)RI$$
$$= 0 \quad \rightarrow \quad I = \frac{18}{(k+6)R}$$
$$V_1 = R \times \frac{18}{(k+6)R} = \frac{18}{k+6}, \quad k \neq -6$$

Problem 1.2.18

(a) Find the input resistance seen by the voltage source in Wheatstone bridge of Fig. 1.18.
(b) Determine the power dissipated in the circuit in terms of (U, R).

Solution

(a) $R_{eq} = \frac{(2R)(2R)}{(4R)} = \frac{4R^2}{4R} = R\,(\Omega)$

(b) $P = \frac{U^2}{R_{eq}} = \frac{U^2}{R}\,(\mathrm{W})$

Problem 1.2.19

(a) Determine input resistance of the circuit of Fig. 1.19, using SPICE analysis

$$R_1 = R_2 = R_4 = 10\,\Omega, R_3 = R_5 = 5\,\Omega \quad (\text{WB ridge1.cir}).$$

(b) Find the value of current (in mA) and its direction of flow through resistor R_3.

Solution

Let a unit current source be used as the excitation at the input terminals of the circuit, and using SPICE analysis, it is found that the voltage at node 1 is V_1 = 8.42105 V.

The voltages at nodes 2 and 3 are computed as V_2 = 3.68421 V and V_3 = 3.15789 V, respectively.

(a) Input resistance of the circuit is $R_{in} = V_1 / I_{in}$ = 8.42105 Ω
(b) The current through resistor R_3

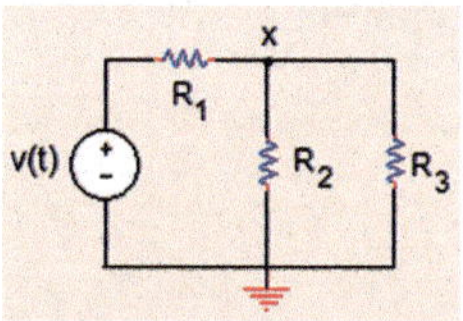

Fig. 1.18 The circuit for Problem 1.2.18

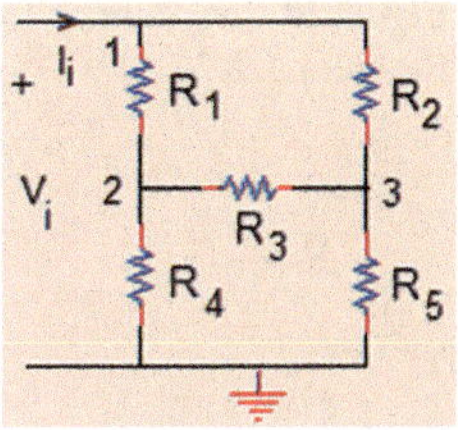

Fig. 1.19 The circuit for Problem 1.2.19

$$I_{R_3} = \frac{V_2 - V_3}{R_3} = \frac{3.68421 - 3.15789}{5} = 0.105264\,\text{A} = 105.264\,\text{mA}$$

The current flow direction is from node 2 to node 3. The SPICE list is given below.

```
*WBridge1.cir
*Wheatstone resistive circuit1
*OP analysis
Iin 0 1 1
R1 1 2 10
R2 1 3 10
R3 2 3 5
R4 2 0 10
R5 3 0 5
```

Problem 1.2.20

(a) A dc voltage source of $V_i = U$ volts feeds a load consisting of two series-connected resistors as in Fig. 1.16. Determine the current in the circuit in terms of G_1, G_2 and U.
(b) Determine the voltage at the junction between the two resistors.
(c) Calculate numerical values of unknowns if $G_1 = 200\,\text{mS}$, $G_2 = 100\,\text{mS}$, $U = 5\,\text{V}$.

Solution

(a) $I = UG = U \times \dfrac{G_1 G_2}{G_1 + G_2}$

(b) $U_A = U \dfrac{\frac{1}{G_2}}{\frac{1}{G_1} + \frac{1}{G_2}} = U \times \dfrac{\frac{1}{G_2}}{\frac{G_1 + G_2}{G_1 G_2}} = \dfrac{U G_1 G_2}{G_2(G_1 + G_2)} = \dfrac{G_1}{G_1 + G_2} \times U$

(c) $I = 5 \dfrac{0.2 \times 0.1}{0.2 + 0.1} = 5 \dfrac{0.02}{0.3} = 0.333\,A \quad \rightarrow \quad U_A = \dfrac{0.1}{0.2 + 0.1} \times 5 = \dfrac{0.5}{0.3} = 1.667\,V$

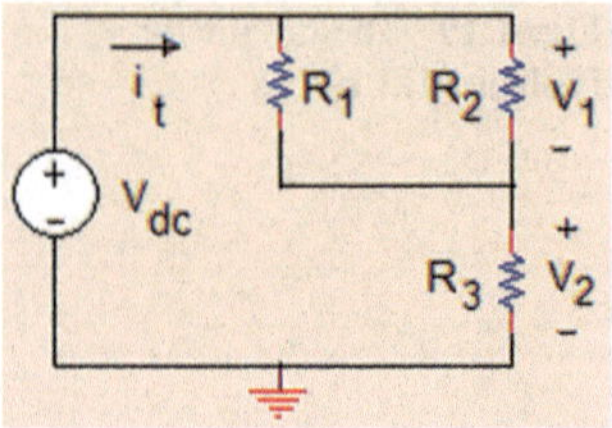

Fig. 1.20 The circuit for Problem 1.2.21

Problem 1.2.21 Find the current flowing through R_1 in the circuit shown in Fig. 1.20.

V_{dc} = 12 V, $R_1 = R_2$ = 1 kΩ, R_3 = 10 kΩ.

Solution

Equivalent resistance of parallel branch is

$$R_p = \frac{1\,\mathrm{k\Omega} \times 1\,\mathrm{k\Omega}}{1\,\mathrm{k\Omega} + 1\,\mathrm{k\Omega}} = 0.5\,\mathrm{k\Omega}$$

Total resistance is $R_t = 0.5\,\mathrm{k\Omega} + 10\,\mathrm{k\Omega} = 10.5\,\mathrm{k\Omega}$
Main current in the circuit is

$$i_t = \frac{V_{dc}}{R_t} = \frac{12\,\mathrm{V}}{10.5\,\mathrm{k\Omega}} = 1.143\,\mathrm{mA}$$

Current in a parallel branch is half of this value, $i = 0.571\,\mathrm{mA}$.

Problem 1.2.22 Explain why the "Preferred Value" system is used for resistors. Use data for E12 Resistance series, and prepare a numerical presentation (in MATLAB or EXCEL program) to illustrate your reasoning.

Solution

It would be impractical and uneconomical to manufacture resistors with every possible value.

Resistors are manufactured in a limited range of values. Each value is a specific number of resistance value within some tolerance (percentage). Manufacturing of resistors with values between these limits is not necessary.

For example, in E12 series, a 100 Ω resistor has a tolerance of ±10%, in other words, it might have any value between 90 and 110 Ω. The E12 series consists of 12 preferred resistor values in a decade. Each of these values has a tolerance of ±10%. The sequence in each series can be scaled up or down. For example, for E12 series, a preferred value of 1.5 in the decade range may represent {0.15 Ω 1.5 Ω 15 Ω 150 Ω 1.5 kΩ 15 kΩ 150 kΩ 1.5 MΩ}. Table 1.1 demonstrates E12 series performance.

Table 1.1 E12 resistance series

	−10%	R	+10%
1	0.9	1.0	1.1
2	1.08	1.2	1.32
3	1.35	1.5	1.65
4	1.62	1.8	1.98
5	1.98	2.2	2.42
6	2.43	2.7	2.97
7	2.97	3.3	3.63
8	3.51	3.9	4.29
9	4.23	4.7	5.17
10	5.04	5.6	6.16
11	6.12	6.8	7.48
12	7.38	8.2	9.02

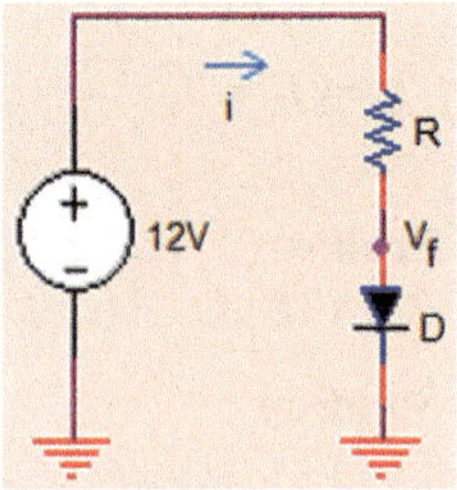

Fig. 1.21 The circuit for Problem 1.2.23

Problem 1.2.23 A red light-emitting diode (LED) has a rated forward current of 20 mA at its operating voltage of $V_F = 20$ V. Determine the value of resistor in E12 series that is required to operate the LED, when the power supply voltage is 12 V, see Fig. 1.21.

Solution

By KVL: $12 = i \cdot R + V_F$

$$12 - 2 = 20 \times 10^{-3} \times R$$

$$R = \frac{10}{20 \times 10^{-3}} = 0.5\ \text{k}\Omega \rightarrow \text{next closest E12 series value is } 560\,\Omega\,(\text{see appendix}).$$

Note that 470 Ω is not selected, since it does not provide a safe operating region for the LED. (Considering 10% tolerance, minimum possible resistance would be 470 − 47 = 423, which is quite below the calculated resistance value.)

Problem 1.2.24 Three resistors each having a value of 2.2 kΩ are connected in parallel. Find the current through each resistor if the system is supplied with a current source of 10 mA.

Solution

Since equal valued resistors are connected in parallel,

$$I_x = I_z = I_t = \frac{10}{3} = 3.33\,\text{mA}$$

Alternative solution;

$$i = 10\,\text{mA} = 0.01\,\text{A} = vG = v\left(3.\frac{1}{R}\right) = v\left(3.\frac{1}{2.2\times 10^3\,\Omega}\right) = 1.363\times 10^{-3}\,\text{V}$$

$$v = \frac{0.01}{1.363\times 10^{-3}} = 7.34\,\text{V}$$

$$I_x = I_z = I_t = \frac{v}{R} = \frac{7.34\,\text{V}}{2.2\times 10^3\,\Omega} = 3.33\,\text{mA}$$

Problem 1.2.25 In the circuit of Fig. 1.22, $R_1 = 10\,\text{k}\Omega$, $R_2 = 30\,\text{k}\Omega$, $R_3 = 10\,\text{k}\Omega$, $v(t) = 10\,\sin(314t)$, $v_x(t) = ?$

Solution

$$R_2 \parallel R_3 = \frac{30\times 10}{40}\,\text{k}\Omega = \frac{300}{40}\,\text{k}\Omega = 7.5\,\text{k}\Omega$$
$$R_1 + (R_2 \parallel R_3) = (10+7.5)\text{k}\Omega = 17.5\,\text{k}\Omega$$

$$v_x(t) = v_1(t)\cdot\frac{R_2 \parallel R_3}{R_1 + R_2 \parallel R_3} = 10\sin(314t)\times\frac{7.5}{17.5} = 4.286\sin(314t)\text{V}$$

Problem 1.2.26 Electrical conductivity of a fiber composite material can be predicted using the so-called "the rule of mixtures-ROM-" and with the following assumptions: (1) Fibers in the composite material are continuous, unidirectional, and uniformly distributed throughout the base matrix. (2) Perfect bonding between fibers and matrix exists. (3) Matrix is void free. (4) Applied currents are either

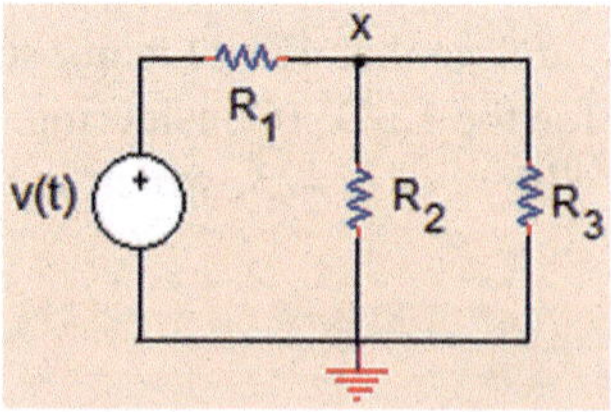

Fig. 1.22 The circuit for Problem 1.2.25

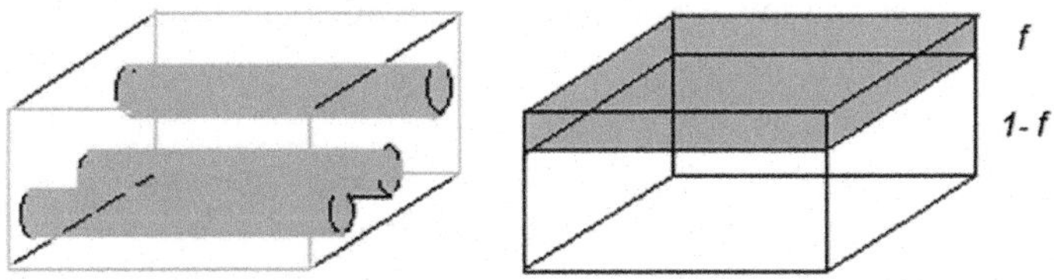

Fig. 1.23 The circuit for Problem 1.2.26

parallel or perpendicular to the fiber direction. (5) Linearity applies to both fiber and the matrix. See, Fig. 1.23.

$$\sigma_f = \text{Conductivity of fibers},$$

$$\sigma_m = \text{Conductivity of matrix},$$

$$\sigma_c = \text{Conductivity of composite},$$

$$f = \text{The volume fraction of fibers} = \frac{V_f}{V_f + V_m}$$

$$V_s = \text{Volume of inclusions}, \quad V_m = \text{Volume of matrix}$$

In the case of $\sigma_f > \sigma_m$, upper bound of conductivity for the composite material is

$$\sigma_c = f\sigma_f + (1 - f)\sigma_m \tag{1.2}$$

This corresponds to current flow parallel to the fibers (Voigt model).

The lower bound of electrical conductivity for the composite material, which corresponds to current flow perpendicular to the direction of fibers (Reuss model or inverse ROM model) is

$$\sigma_c = \frac{1}{\frac{f}{\sigma_f} + \frac{1-f}{\sigma_m}} = \frac{\sigma_f\sigma_m}{f\sigma_m + (1 - f)\sigma_f} \tag{1.3}$$

Therefore,

$$\frac{1}{\frac{f}{\sigma_f} + \frac{1-f}{\sigma_m}} \le \sigma_c \le f\sigma_f + (1 - f)\sigma_m \quad \sigma_f > \sigma_m \tag{1.4}$$

If $\sigma_f < \sigma_m$, upper and lower bounds of electrical conductivity for the composite material become

$$\frac{1}{\frac{f}{\sigma_f}+\frac{1-f}{\sigma_m}} \geq \sigma_c \geq f\sigma_f + (1-f)\sigma_m \quad \sigma_f < \sigma_m \tag{1.5}$$

(a) Derive Eq. (1.2).
(b) Derive Eq. (1.3).
(c) Calculate the conductivity parallel to alignment of the fibers for an aluminum matrix with 40% volume fraction aluminum oxide fibers $\left(\sigma_m = \sigma_{Al} = 35.36 \times 10^6\ \mathrm{Sm^{-1}},\ \sigma_f = 10^{-11}\ \mathrm{Sm^{-1}}\right)$.
(d) Calculate the conductivity perpendicular to the alignment of the fibers for an aluminum matrix with 40% volume fraction aluminum oxide fibers $\left(\sigma_m = \sigma_{Al} = 35.36 \times 10^6\ \mathrm{Sm^{-1}}, \sigma_f = 10^{-11}\ \mathrm{Sm^{-1}}\right)$.
(e) Plot σ_c versus f, for both of the Voigt (parallel resistors) and Reuss (series resistors) models (Rmix.m), (fiber mixing.xlsx).

Solution

(a) Assume that resistance of the composite material is made out of two resistors connected in parallel. In this case, direction of current flow is parallel to the alignment of fibers, see, Fig. 1.24. (Voigt model of fiber composites).

$$R = R_f \parallel R_m,$$

where

R_f fiber equivalent resistance with resistivity ρ_f, conductivity σ_f
R_m matrix equivalent resistance with resistivity ρ_m, conductivity σ_m
R composite bulk resistance with conductivity σ
A composite cross-sectional area
A_f fiber cross-sectional area
A_m matrix cross-sectional area
T, x dimensions of rectangular cross section of composite material

Fig. 1.24 Voigt model of fiber composites

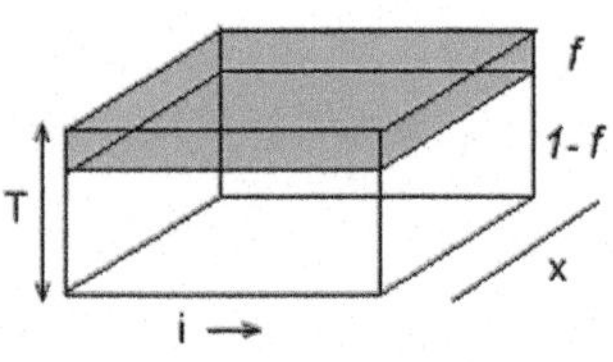

$$f = \frac{V_\mathrm{f}}{V_\mathrm{f} + V_\mathrm{m}} = \frac{A_\mathrm{f}}{A}, \quad A = xT = A_\mathrm{f} + A_\mathrm{m}, \quad A_\mathrm{f} = fxT, \quad A_\mathrm{m} = (1-f)xT$$

$$R = R_\mathrm{f} \parallel R_\mathrm{m}, \quad G = \frac{1}{R} = G_\mathrm{f} + G_\mathrm{m}$$

$$\frac{\sigma A}{L} = \frac{\sigma_\mathrm{f} A_\mathrm{f}}{L} + \frac{\sigma_\mathrm{m} A_\mathrm{m}}{L}$$

$$\sigma A = \sigma_\mathrm{f} A_\mathrm{f} + \sigma_\mathrm{m} A_\mathrm{m}, \quad \rightarrow \quad \sigma xT = \sigma_\mathrm{f} fxT + \sigma_\mathrm{m}(1-f)xT,$$
$$\sigma = \sigma_\mathrm{f} f + \sigma_\mathrm{m}(1-f),$$

which is Eq. (1.2).

(b) Assume that resistance of the composite material is made out of two resistors connected in series. In this case, direction of current flow is perpendicular to the alignment of fibers (Reuss model of fiber composites),

$$R = R_\mathrm{f} + R_\mathrm{m}$$

where,

- R_f fiber equivalent resistance with resistivity ρ_f, conductivity σ_f
- R_m matrix equivalent resistance with resistivity ρ_m, conductivity σ_m
- R composite bulk resistance with resistivity ρ,
- A area (perpendicular to current flow direction)
- L_f fiber thickness
- L_m matrix thickness

$$R = \frac{\rho_\mathrm{f} L_\mathrm{f}}{A} + \frac{\rho_\mathrm{m} L_\mathrm{m}}{A} = \frac{L_\mathrm{f}}{\sigma_\mathrm{f} A} + \frac{L_\mathrm{m}}{\sigma_\mathrm{m} A} = \frac{L_\mathrm{f}}{\sigma_\mathrm{f} A} + \frac{L - L_\mathrm{f}}{\sigma_\mathrm{m} A} = \frac{L}{A}\left(\frac{L_\mathrm{f}}{\sigma_\mathrm{f} L} + \frac{L - L_\mathrm{f}}{\sigma_\mathrm{m} L}\right)$$
$$R = \frac{L}{A}\left(\frac{L_\mathrm{f}}{\sigma_\mathrm{f} L} + \frac{L - L_\mathrm{f}}{\sigma_\mathrm{m} L}\right) = \frac{L}{A}\left[\frac{L_\mathrm{f}}{\sigma_\mathrm{f} L} + \frac{1}{\sigma_\mathrm{m}} \cdot \left(1 - \frac{L_\mathrm{f}}{L}\right)\right]$$

Considering volume fractions of fiber and matrix over the total volume of the composite, and realizing that the current flows through the same area A,

$$f = \frac{V_\mathrm{f}}{V_\mathrm{f} + V_\mathrm{m}} = \frac{L_\mathrm{f}}{L}$$
$$R = \frac{L}{A}\left[\frac{f}{\sigma_\mathrm{f}} + \frac{1}{\sigma_\mathrm{m}} \cdot (1-f)\right] = \frac{L}{A}\left[\frac{f}{\sigma_\mathrm{f}} + \frac{1-f}{\sigma_\mathrm{m}}\right] \tag{1.6}$$

Using the relationship between the resistance of the composite material and its bulk conductivity,

$$R = \frac{L}{\sigma A} \quad \rightarrow \quad \sigma = \frac{L}{AR}$$

Substituting (1.6) into this equation,

$$\sigma = \frac{L}{A \cdot \frac{L}{A}\left[\frac{f}{\sigma_\mathrm{f}} + \frac{1-f}{\sigma_\mathrm{m}}\right]} = \frac{1}{\left[\frac{f}{\sigma_\mathrm{f}} + \frac{1-f}{\sigma_\mathrm{m}}\right]}$$

which is Eq. (1.3).

(c) The conductivity *parallel* to the fibers for an aluminum matrix with 40% volume fraction

aluminum oxide fibers $\left(\sigma_\mathrm{m} = \sigma_\mathrm{Al} = 35.36 \times 10^6\,\mathrm{Sm}^{-1},\ \sigma_\mathrm{f} = 10^{-11}\,\mathrm{Sm}^{-1}\right)$,

$$\begin{aligned}\sigma_\mathrm{c} &= f\sigma_\mathrm{f} + (1-f)\sigma_\mathrm{m} = 0.4 \times 10^{-11} + (1-0.4) \times 35.36 \times 10^6 \\ &= 21.22 \times 10^6\,\mathrm{Sm}^{-1}\end{aligned}$$

(d) The conductivity *perpendicular* to alignment of the fibers for an aluminum matrix with 40% volume fraction aluminum oxide fibers ($\sigma_\mathrm{m} = \sigma_\mathrm{Al} = 35.36 \times 10^6\,\mathrm{Sm}^{-1},\ \sigma_\mathrm{f} = 10^{-11}\,\mathrm{Sm}^{-1}$),

$$\sigma_\mathrm{c} = \frac{1}{\frac{f}{\sigma_\mathrm{f}} + \frac{1-f}{\sigma_\mathrm{m}}} = 2.5 \times 10^{-11}\,\mathrm{Sm}^{-1}$$

Note that,

$$2.5 \times 10^{-11} \le \sigma_\mathrm{c} \le 21.22 \times 10^6, \quad \sigma_\mathrm{f} = 10^{-11} < \sigma_\mathrm{m} = 35.36 \times 10^6\,(\mathrm{Sm}^{-1})$$

(e) Bulk conductivities of composite material as a function of fiber volume fraction for two cases of current flow-fiber alignment direction are given in Figs. 1.25 and 1.26.

Problem 1.2.27 A reference voltage is often used by electronic circuit designers. Here is a different type of reference voltage implementation, as shown in Fig. 1.27.

(a) Determine the expression for voltage Vx.
(b) Use SPICE DC Sweep analysis and prove the result found above.

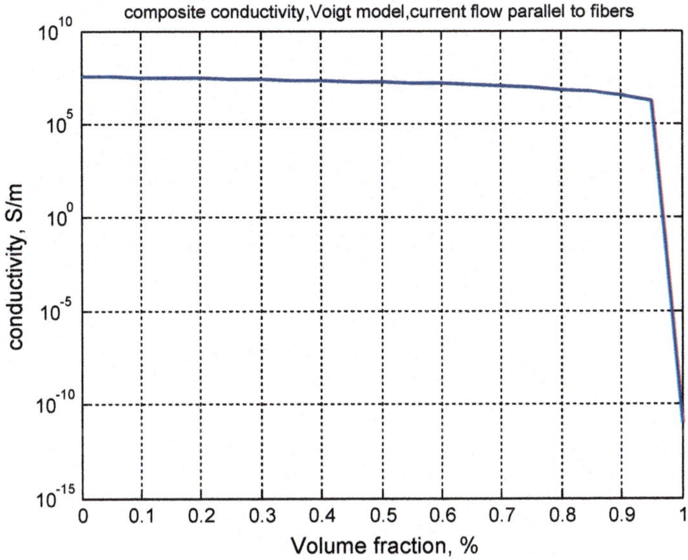

Fig. 1.25 Electrical conductivity of fiber composite. Voigt model, when current flow is parallel to fibers

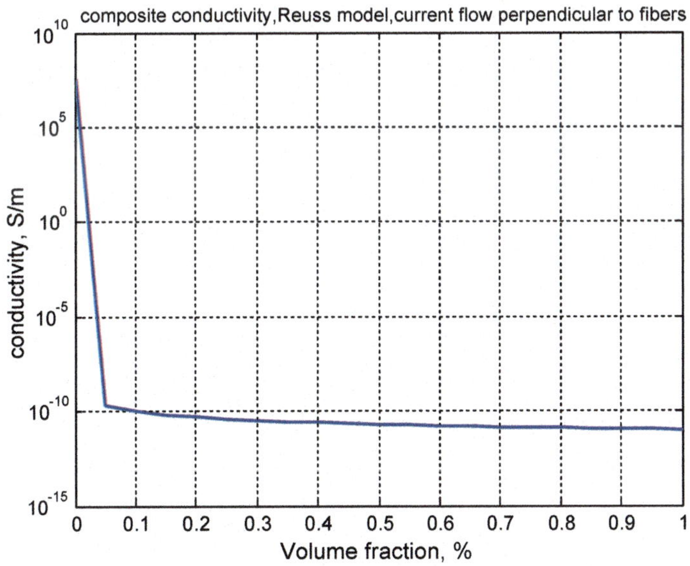

Fig. 1.26 Electrical conductivity of fiber composite. Reuss model, when current flow is perpendicular to fibers

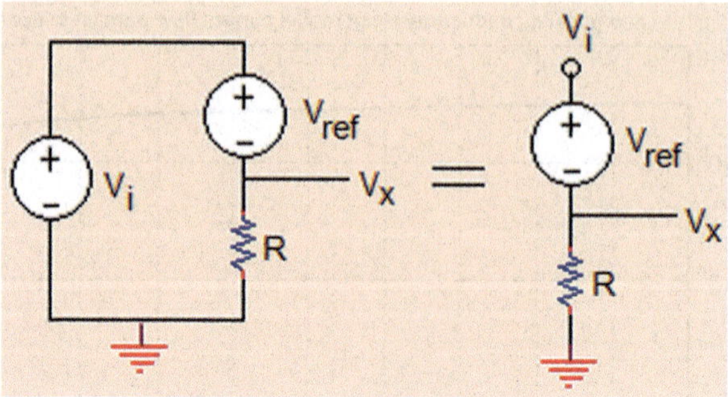

Fig. 1.27 A reference voltage implementation, Problem 1.2.27

Solution

(a) Applying KVL around the loop,

$$V_i - V_{\text{ref}} - V_x = 0 \rightarrow V_x = V_i - V_{\text{ref}}$$

This is a voltage subtraction circuit with the resulting voltage

$$V_x \geq 0 \quad \text{if} \quad V_i - V_{\text{ref}} \geq 0 \quad \rightarrow \quad V_x \geq 0 \quad \textit{if} \quad V_i \geq V_{\text{ref}}$$

In other words, output voltage is positive if input voltage is greater than the reference voltage.

(b) Using SPICE DC Sweep analysis, the result found above is demonstrated in Fig. 1.28.

Input voltage is swept between 1 and 1 V. Reference voltage is set to 0.3 V.

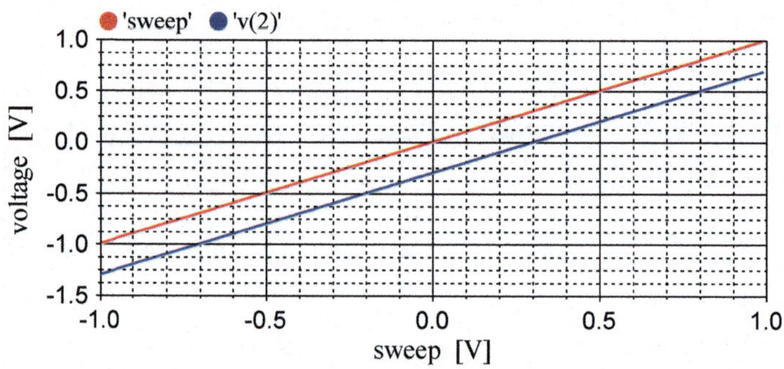

Fig. 1.28 SPICE Sweep analysis for the circuit of Problem 1.2.27

Once the input voltage level exceeds the reference voltage value, the voltage at node 2 becomes positive, increasing with the same slope of the input voltage.

```
SPICE net-list Analysis: DC Transfer Curves
Reference voltage is subtracted from Input voltage
Vi 1 0 1
R1 2 0 1000
Vref 1 2  0.3
*.dc Vi -1 1 0.01
```

Problem 1.2.28 In a stirred electrolytic solution voltammetry experiment using three electrodes as shown in Fig. 1.29, external dc voltage (= V_{dc}) causes a potential difference ($E = V_{rw}$) between the reference and working electrodes. CE is the counter electrode [1, 2].

Voltammogram (the graph of current flow i vs $E = V_{rw}$) obtained in this redox experiment is also given.

(a) What is the standard-state reduction potential?
(b) If $k = 0.5$ (μA/mol/l), determine the concentration of analyte.
(c) What is the impedance between reference and the working electrode at standard-state potential?
(d) If power supply value is 1.5 V, determine the impedance between reference and counter electrodes at standard-state potential.

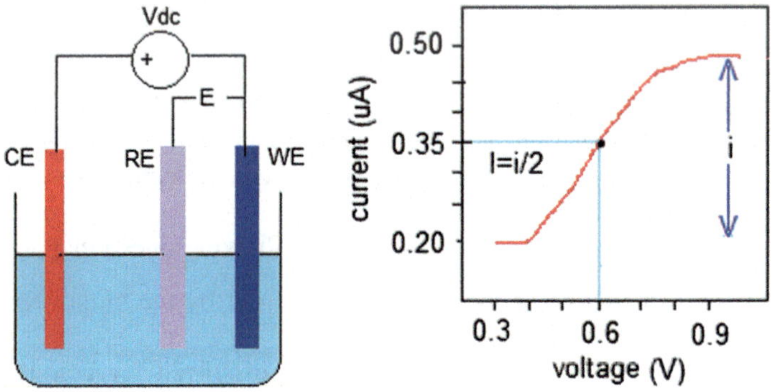

Fig. 1.29 Voltammetry experiment using three electrodes and its current-voltage diagram

Solution

(a) The standard-state potential is obtained from the curve,

$$I = \frac{i}{2} = 0.2 + \frac{0.5 - 0.2}{2} = 0.35\,\mu\text{A}$$

corresponding voltage is $E_{1/2} = 0.6$ V

(b) If $k = 0.5 \times 10^{-6}$ (A/mol/l), the concentration (c) of analyte is

$$I = kc \quad \rightarrow \quad c = \frac{I}{k} = \frac{0.35}{0.5} = 0.7\,\text{mol/l}$$

(c) The impedance between reference and the working electrode at standard-state potential is

$$Z_{RW} = \frac{0.6}{0.35 \times 10^{-6}} = 1.714\,\text{M}\Omega$$

(d) When power supply value is 1.5 V, the impedance between reference and counter electrodes at standard-state potential is

$$Z_{CR} = \frac{1.5 - 0.6}{0.35 \times 10^{-6}} = 2.571\,\text{M}\Omega$$

(e) When power supply value is 1.5 V, the impedance between counter and the working electrode at standard-state potential is

$$Z = Z_{RW} + Z_{CR} = 1.714 + 2.571 = 4.285\,\text{M}\Omega$$

Problem 1.2.29 Measurement of arterial blood oxygen level is performed on patients in the operating room and the intensive care unit to provide information about the metabolic and respiratory state in the body reflecting the sufficiency of blood oxygenation.

Indicating the reference sources, answer the following questions:

(a) What is the percentage of the oxygen carried out by the blood transported in dissolved form in the plasma?
(b) What type of sensor is used to determine dissolved oxygen level (pO_2)? Describe its operating principles.
(c) For a pO_2 measuring sensor, the parameters of the sensor are given below. This sensor is used in a water (calibration) solution saturated with dissolved oxygen at normal temperature and pressure. Determine the test current and time to reach its plate.

D_m membrane diffusion constant = 1.05×10^{-12} mol atm^{-1} s^{-1}
F Faraday constant = 9.64×10^4 C/mol
b membrane thickness = 12.7 μm
A cathode area = 0.030 cm^2

Partial oxygen pressure for air saturated water = 0.2095 atm = 159 mm Hg
O_2 concentration = 2.4×10^{-4} mol/dm^3 = 8 ppm = 8 mg/dm^3

Solution

(a) Approximately 2% of the oxygen carried out by the blood is transported in dissolved form in the plasma, and remaining 98% is transported to tissues in erythrocytes [3].
(b) Clark electrode is used to measure the partial oxygen pressure (pO_2) in a blood sample. It is an amperometric type sensor which requires an external voltage source. Its principle is illustrated in Fig. 1.30.

Here, dissolved oxygen molecules passing from blood through an oxygen permeable membrane (PTFE, PP-polypropylene-, PE-polyethylene-, PVC or rubber) into the aqueous electrolyte (KCl) react with water in the presence of electrons to produce hydroxyl ions so that the redox reaction generates a current in the presence of external polarizing voltage supply around 0.6 V.

Oxygen is reduced at the surface of platinum cathode,

$$O_2 + 2H_2O + 4e^- \leftrightarrow 4OH^-$$

The resulting hydroxyl ions react with Ag/AgCl anode material (oxidation)

$$Ag \leftrightarrow Ag^+ + e^-$$

$$Ag^+ + Cl^- \leftrightarrow AgCl \downarrow$$

This means that silver from the anode material is oxidized to silver ions while liberating electrons to the anode. Silver ions then combine with chloride ions and

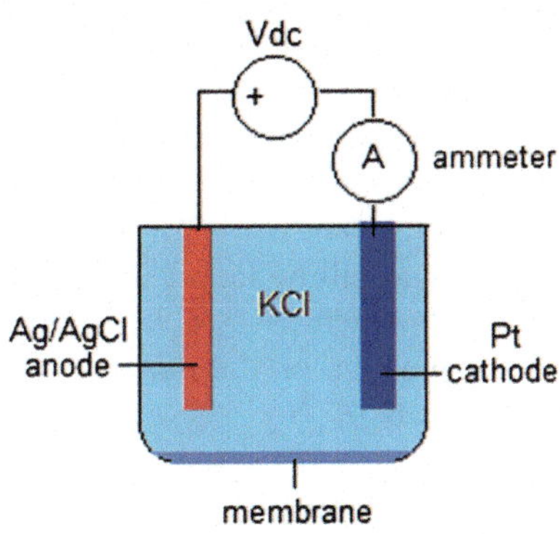

Fig. 1.30 Principle of Clark electrode for measuring partial oxygen pressure (pO2) in blood

precipitate as silver chloride on the surface of anode. Magnitude of external current flow is directly proportional to the number of oxygen molecules that are reduced on the cathode.

The amount of dissolved oxygen in blood can be determined by measuring this current [4, 5].

Lowest limit of detection for a Clark sensor is 0.5 ppm (mg O_2/L). The time necessary for current to reach the plate is

$$t = \frac{b^2}{D_m},$$

where b is the thickness, and D_m is the diffusion constant (permeability) of the sensor membrane [3].

(c) Given sensor parameters:

D_m membrane diffusion constant = 1.05×10^{-12} mol atm^{-1} s^{-1}
F Faraday constant = 9.64×10^4 C/mol
b membrane thickness = 12.7 μm
A cathode area = 0.030 cm^2
pO_2 partial oxygen pressure for air-saturated water = 0.2095 atm

The time necessary for current to reach the plate is

$$t = \frac{b^2}{D_m} = \frac{(12.7 \times 10^{-6})^2}{1.05 \times 10^{-12}} = \frac{161.29}{1.05} = 153.61\ \text{s}$$

The current is $i = k \cdot pO_2$,

$$k = \frac{4FD_mA}{b} = \frac{4 \times 9.64 \times 10^4 \times 1.05 \times 10^{-12} \times 0.030 \times 10^{-2}}{12.7 \times 10^{-6}} = 0.0956 \times 10^{-4}\ \text{A/atm}$$

$$i = 0.0956 \times 10^{-4} \times 0.2095 = 2\ \mu\text{A}$$

Problem 1.2.30

(a) How is electrical conductivity of a salt solution measured? What is the practical unit of these types of measurements?
(b) Electrical conductivity of a salt solution changes 1.9% per celsius degree increase in temperature.
(i) Determine the conductivity at 25 °C if measured conductivity at 18 °C is 100 μS/cm.
(ii) What will be the measured conductivity at 35 °C, if conductivity at 25 °C is 100 μS/cm?
(e) Comment on the practical application of these results.

Solution

(a) Electrical conductivity (EC) is measured by measuring the current flow between the two stainless steel or platinum electrodes in the liquid sample. The more dissolved salt in the water, the higher the current flow and the EC. Measurements of EC can be used to give an estimate of TDS (total dissolved solids) in much quicker and simpler way. EC measures the conductance of liquid in a specific measuring cell in microSiemens per centimeter (μS/cm) at 25 °C.

(b, i) $G_{25} = G_{18}(1 + 0.019\,\Delta T) = 100[1 + 0.019(25 - 18)] = 113.3\,\mu\text{S/cm}$

(b, ii) $100 = G_{35}(1 + 0.019\,\Delta T) = G_{35}[1 + 0.019.(25 - 35)] = G_{35} \times 0.81$

$$G_{35} = \frac{100}{0.81} = 123.5\,\mu\text{S/cm}$$

(c) Most conductivity meters measure conductivity and temperature simultaneously by using the same probe. They also make correction to standard temperature. The standard temperature is 25 °C in US and European Pharmacopoeia, ASTM D 1125-95 and ISO 7888-1985 standards. (Nevertheless, the corrected reading may not be as accurate as taking a reading of the sample's conductivity at 25 °C).

Problem 1.2.31

(a) Search three methods of determining the salt content (salinity) of electrolyte solutions.
(b) Determine the salinity of electrolyte solution S at 18 °C, if the measured conductivity of electrolyte is, (i) G_x = 50,000 μS/cm, (ii) G_x = 45893.8 μS/cm.
(c) Determine the amount of total dissolved salts in part b, if equipment is programmed for electrical conductivity changes of 2.5% per °C increase in temperature.

Solution

(a) The three methods of determining the salinity of water are the as follows [6]:

1. Total dissolved salts (Solids) (TDS) is measured by evaporating a known volume of water to dryness, and weighing the remaining residue. TDS is recorded in milligrams of dissolved solid in one liter of water (mg/L). Parts per million (ppm) is equivalent to mg/L.
2. Electrical Conductivity (EC) is measured in the liquid sample. The more dissolved salt in the water, the higher the current flow and the EC. Measurements of EC can be used to give an estimate of TDS in much quicker and simpler way.

EC can then be converted to TDS: TDS (mg/L) $\approx$ EC (μS/cm at 25 °C) $\times$ k

(Here, $k = 0.635$ when EC > 720 μS/cm. For more precise calculations, EC versus TDS of KCl solution at 25 °C is plotted and proper curve fitting equations must be derived).

Electrical conductivity of a solution at specific temperature T (in °C) is measured ($= G_x$), and this value is substituted in the conductivity ratio R:

$$R = \frac{G_x}{G_{\mathrm{KCl}}}$$

$$S = a_0 + a_1 R^{1/2} + a_2 R + a_3 R^{3/2} + a_4 R^2 + a_5 R^{5/2} + \Delta S$$

$$\Delta S = K.(b_0 + b_1 R^{1/2} + b_2 R + b_3 R^{3/2} + b_4 R^2 + b_5 R^{5/2})$$

$$K = \frac{T - 15}{1 + 0.0162(T - 15)}$$

$$\begin{aligned}
a_0 &= 0.0080, & a_1 &= -0.1692, & a_2 &= 25.3851,\\
a_3 &= 14.0941, & a_4 &= -7.0261, & a_5 &= 2.7081\\
b_0 &= 0.0005, & b_1 &= -0.0056, & b_2 &= -0.0066,\\
b_3 &= -0.0375, & b_4 &= 0.0636, & b_5 &= -0.0144
\end{aligned}$$

G_{KCl} is a measure of the reference conductivity of KCl solution containing a mass of 32.4356 g KCl in a mass of 1 kg of solution. G_{KCl} can be calculated using the following equation (in uS/cm):

$$\begin{aligned}
G_{\mathrm{KCl}} = {} & -0.0267243T^3 + 4.6636947T^2 + 861.302764T\\
& + 29035.1640851
\end{aligned}$$

For example, at $T = 18$ °C, the reference (KCl) conductivity is found by substituting $T = 18$ in this equation: $G_{\mathrm{KCl}} = 45893.8$ μS/cm.

This is valid for $2 \le S \le 42$. It should also be noted here that if the EC meter is automatically correcting the reading taken at water temperature to a reading at 25 °C, then the value of corrected conductivity for $T = 25$ °C must be substituted in place of G_x.

3. The third method of determining the salinity of water is a method where the density of solution is measured by some hydrometer or refractometer and corresponding salinity is looked up from tables.

(b) (i) Substituting $T = 18$ °C, $G_x = 50{,}000$ μS/cm in the above given equations yields $S = 38.538$.

(ii) Substituting $T = 18$ °C, $G_x = 45893.8$ μS/cm in the above given equations yields $S = 35.000$.

(c) The amount of approximate total dissolved salts (TDS) at 25 °C are,

(i) $G_{25} = G_{18}(1 + 0.025\,\Delta T) = 50{,}000[1 + 0.025(25 - 18)]$
$= 58{,}750\ \mu\text{S/cm}$

$\text{TDS} = 58{,}750 \times 0.635 = 37306.25\ \text{mg/L}$

$\text{TDS} = 37.3063\ \text{g/L}$. This is approximately 3.73% by mass of solution.

(ii) $G_{25} = 45893.8[1 + 0.025(25 - 18)]$
$= 53925.215\ \mu\text{S/cm}$

$\text{TDS} = 53925.215 \times 0.635 = 34242.51\ \text{mg/L}$

$\text{TDS} = 34.243\ \text{g/L}$. This is approximately 3.42% by mass of solution.

1.3 Resistivity and Resistance, Voltage and Current Division, Resistive Sensors and Mixture Models, Blood Cell Counting, Wheatstone Bridge, Infinite Ladders

Problem 1.3.1

(a) A conducting square prismatic object has length, width, and resistivity of L, a, and ρ, respectively (Fig. 1.31). Determine the ratio of resistance values between (parallel) square faces and rectangular faces.
(b) If L = 100a, determine the ratio, numerically.

Solution

(a) Resistance between square faces is

$$R_s = \rho \frac{L}{a^2}$$

Resistance between parallel rectangular faces is

$$R_p = \rho \frac{a}{aL} = \frac{\rho}{L}$$

Their ratio:

$$\frac{R_p}{R_s} = \frac{\frac{\rho}{L}}{\frac{\rho L}{a^2}} = \frac{a^2}{L^2} = \left(\frac{a}{L}\right)^2$$

Fig. 1.31 A conducting square prismatic object for Problem 1.3.1

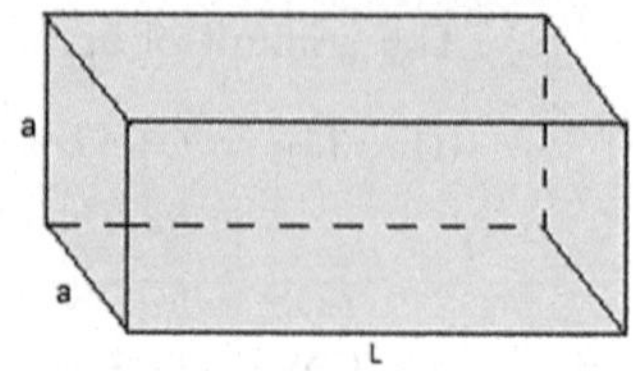

(b) If $L = 100a$,

$$\frac{R_p}{R_s} = \left(\frac{a}{100a}\right)^2 = \frac{1}{10^4}$$

in other terms, $R_s = 10^4 \cdot R_p$

Problem 1.3.2 Determine the resistance of a homogenous and isotropic tissue with a cross-sectional area which can be described by the functions $y = 8 - x^2$, $x = -2\,\text{cm}$ and $x = 2\,\text{cm}$ on x–y plane, its length $L = 10\,\text{cm}$, parallel to z axis, and its resistivity $\rho = 80\,\Omega\text{m}$.

Solution

$$R = \rho L \frac{1}{A}$$

$$A = \int_{-2}^{2} (8 - x^2)\text{d}x = 8x - \frac{x^3}{3}\bigg|_{-2}^{2} = 32 - \frac{16}{3} = \frac{80}{3}\,[\text{cm}^2]$$

Substituting given values in resistance equation yields

$$R = 80 \times 0.1 \times \frac{1}{\frac{80}{3} \times 10^{-4}} = 80 \times 0.1 \times \frac{3}{80} \times 10^4 = 3\,\text{k}\Omega$$

Problem 1.3.3 A potential difference U is applied between the inner and outer surfaces of a hollow cylinder conductor with resistivity ρ, length L, inner and outer radii of a and b, respectively. Determine the resistance (see, Fig. 1.32).

Solution

Consider a differential thin cylinder of inner radius r and outer radius $(r + \text{d}r)$ and length L. Its differential resistance,

$$\text{d}R = \rho \cdot \frac{\text{d}L}{A} = \rho \cdot \frac{\text{d}r}{2\pi r L} \quad \rightarrow \quad R = \int_a^b \text{d}R = \frac{\rho}{2\pi L}\int_a^b \frac{\text{d}r}{r} = \frac{\rho}{2\pi L}\text{lnr}\Big|_a^b = \frac{\rho}{2\pi L}\ln\left(\frac{b}{a}\right)$$

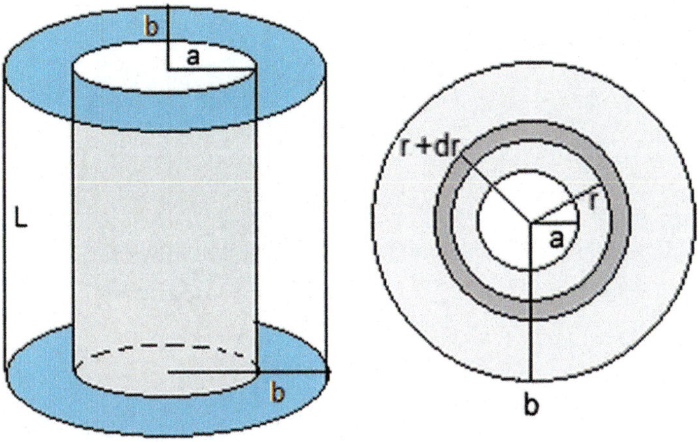

Fig. 1.32 Hollow cylinder conductor for Problem 1.3.3

Problem 1.3.4 A truncated cone shaped element of resistivity ρ, height h and upper and lower truncated circular areas with radii a, b caries uniformly distributed current through its cross section. Determine the resistance between the upper and lower ends of this element (Fig. 1.33).

Solution

Let a thin disk with radius r be located at a distance x from the base,

$$\frac{b-a}{h} = \frac{b-r}{x} \rightarrow r = \frac{(a-b)x}{h} + b$$

$$\text{since } R = \rho L/A, \quad \rightarrow \quad \mathrm{d}R = \rho \cdot \frac{\mathrm{d}x}{\pi r^2} = \rho \cdot \frac{\mathrm{d}x}{\pi \left[\frac{a-b}{h} \cdot x + b\right]^2}$$

$$R = \int_0^h \mathrm{d}R = \frac{\rho}{\pi} \int_0^h \frac{\mathrm{d}x}{\left(\frac{a-b}{h} \cdot x + b\right)^2},$$

$$\text{Let} \quad A = \frac{a-b}{h},$$

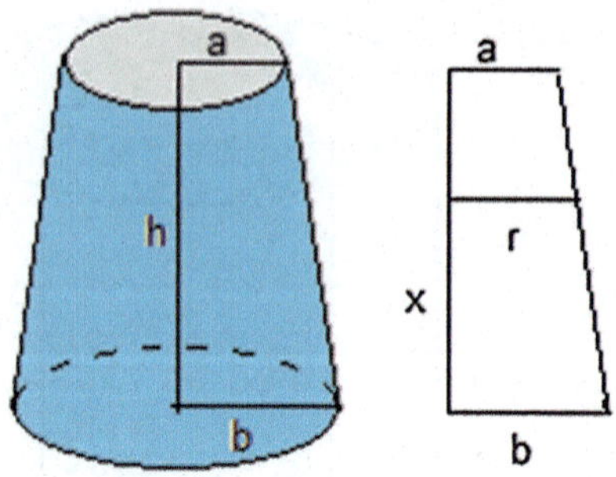

Fig. 1.33 A truncated cone shaped conducting element, Problem 1.3.4

$$R = \frac{\rho}{\pi}\int_0^h \frac{\mathrm{d}x}{(Ax+b)^2} = \frac{\rho}{\pi}\int_0^h (Ax+b)^{-2}\cdot \mathrm{d}x = \frac{\rho}{\pi}\frac{(Ax+b)^{-1}}{-A}\Bigg|_0^h = \frac{-\rho}{\pi}\left[\frac{1}{A(Ah+b)} - \frac{1}{Ab}\right]$$

$$= \frac{-\rho}{\pi}\left[\frac{b-(Ah+b)}{Ab(Ah+b)}\right] = \frac{\rho}{\pi}\cdot\frac{h}{b(Ah+b)} = \frac{\rho h}{\pi b\left[\frac{a-b}{h}\cdot h + b\right]} = \frac{\rho h}{\pi b[(a-b)+b]}$$

$$R = \frac{\rho h}{\pi b a}$$

Problem 1.3.5 A homogenous solid sphere of Radius = a, and resistivity ρ is symmetrically truncated with two flat parallel planes at Radius = b. Determine the resistance of the object between the two truncating planes (Fig. 1.34).

Solution

The equation of sphere in cylindrical coordinates

$$r^2 + z^2 = \theta^2$$

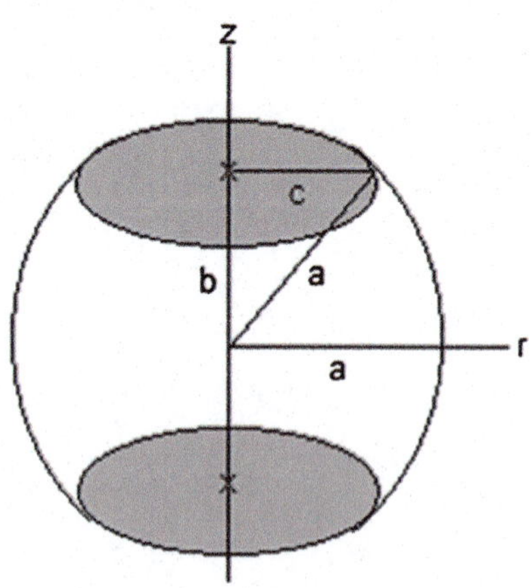

Fig. 1.34 A homogenous solid conducting sphere, Problem 1.3.5

r radial distance from z-axis

$$R = \rho \frac{l}{A} \quad \rightarrow \quad \mathrm{d}R = \rho \frac{\mathrm{d}l}{A}$$

At one truncating plane (top),

$$a^2 - z^2 = c^2$$

$$A = \pi c^2 = \pi\left(\theta^2 - z^2\right), \quad \mathrm{d}l = \mathrm{d}z$$

$$\mathrm{d}R = \rho \frac{\mathrm{d}z}{\pi\left(\theta^2 - z^2\right)}$$

$$R = \int_{-b}^{b} \frac{\rho}{\pi} \frac{\mathrm{d}z}{\left(\theta^2 - z^2\right)} = 2\frac{\rho}{\pi} \int_{0}^{b} \frac{\mathrm{d}z}{\theta^2 - z^2} = 2\frac{\rho}{\pi} \int_{0}^{b} \frac{\mathrm{d}z}{(a-z)(a+z)}$$

Applying partial fraction expansion inside of the integral results in the following,

$$R = 2\frac{\rho}{\pi} \int_{0}^{b} \left(\frac{\frac{1}{2a}}{a-z} + \frac{\frac{1}{2a}}{a+z}\right)\mathrm{d}z = \frac{\rho}{\pi a} \int_{0}^{b} \left(\frac{1}{a-z} + \frac{1}{a+z}\right)\mathrm{d}z$$

$$= \frac{\rho}{\pi a}\left[-\ln(a-z) + \ln(a+z)\right]\Big|_{a}^{b} = \frac{\rho}{\pi a} ln\left(\frac{a+z}{a-z}\right)\Big|_{a}^{b} = \frac{\rho}{\pi a} ln\left(\frac{a+b}{a-b}\right)$$

Problem 1.3.6 An aluminum wire with the cross-sectional area of 0.5 mm^2 has the resistance of 20 Ω. A 1.5 V battery is connected across this wire ends.

(a) What is the length of the wire?
(b) Determine the power dissipated in the wire. (mW) ($\rho_{\mathrm{Al}} = 2.65 \times 10^{-8}\,\Omega\mathrm{m}$)

Solution

(a) $R = \rho_{\mathrm{Al}} \frac{l}{A} \quad \rightarrow \quad l = \frac{RA}{\rho_{\mathrm{Al}}} = \frac{20 \times 0.5 \times 10^{-6}}{2.65 \times 10^{-8}} = 377.36\,\mathrm{m}$

(b) $P = \frac{V^2}{R} = \frac{1.5^2}{20} = \frac{2.25}{20} = 0.1125\,\mathrm{W} = 112.5\,\mathrm{mW}$

Problem 1.3.7 A silver wire and an aluminum wire have equal lengths (= 1 km) and cross-sectional areas (= 1 mm^2). If they are connected in series, what is their total resistance? $\left(\rho_{\mathrm{Ag}} = 1.64 \times 10^{-8}\,\Omega\mathrm{m},\ \rho_{\mathrm{Al}} = 2.65 \times 10^{-8}\,\Omega\mathrm{m}\right)$.

Solution

$$R_{\mathrm{T}} = \rho_{\mathrm{Ag}} \times \frac{l_1}{A_1} + \rho_{\mathrm{Al}} \times \frac{l_2}{A_2} = (\rho_{\mathrm{Ag}} + \rho_{\mathrm{Al}})\frac{l}{A} = (1.64 + 2.65) \times 10^{-8} \times \frac{10^3}{10^{-6}}$$

$$R_{\mathrm{T}} = (1.64 + 2.65) \times 10 = 4.29 \times 10 = 42.9\,\Omega$$

Problem 1.3.8 A copper wire (1 mm^2 area, 1 km length) is connected in series with an aluminum wire of the same size with the copper wire. Find the voltage at their junction point, neglecting Seebeck effect, if 12 V power supply feeds the series connection of these resistors.

$$\rho_{\mathrm{Al}} = 2.65 \times 10^{-8}\,\Omega\mathrm{m}$$
$$\rho_{\mathrm{Cu}} = 1.78 \times 10^{-8}\,\Omega\mathrm{m}$$

Solution

$$R_{\mathrm{Cu}} = 1.78 \times 10^{-8} \times \frac{10^3\,(\mathrm{m})}{10^{-6}\,(\mathrm{m}^2)} = 17.8\,\Omega,$$
$$R_{\mathrm{Al}} = 2.65 \times 10^{-8} \times \frac{10^3\,(\mathrm{m})}{10^{-6}\,(\mathrm{m}^2)} = 26.5\,\Omega$$

Voltage division:

$$V_x = V_{\mathrm{in}} \cdot \frac{R_{\mathrm{AL}}}{R_{\mathrm{Al}} + R_{\mathrm{Cu}}} = 12 \cdot \frac{26.5}{26.5 + 17.8} = 12 \cdot 0.5982 = 7.178\,\mathrm{V}$$

Problem 1.3.9 A copper wire (1 mm^2 area, 1 km length) and an aluminum wire of the same size and length of copper wire are connected in parallel, and a current of 1A feeds the parallel circuit. Find the current through the copper wire. $(\rho_{\mathrm{Al}} = 2.65 \times 10^{-8}\,\Omega\mathrm{m},\ \rho_{\mathrm{Cu}} = 1.78 \times 10^{-8}\,\Omega\mathrm{m})$

Solution

$$R_{\mathrm{cu}} = 1.78 \times 10^{-8} \times \frac{10^3\,\mathrm{m}}{10^{-6}\,\mathrm{m}^2} = 17.8\,\Omega,$$
$$R_{\mathrm{Al}} = 2.65 \times 10^{-8} \times \frac{10^3\,\mathrm{m}}{10^{-6}\,\mathrm{m}^2} = 26.5\,\Omega$$

Using current division rule,

$$I_{\mathrm{cu}} = 1 \times \frac{26.5}{26.5 + 17.8} = \frac{26.5}{44.3} = 0.598\,\mathrm{A}$$

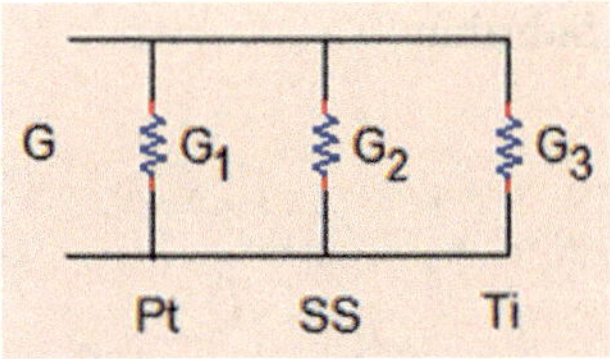

Fig. 1.35 Circuit for Problem 1.3.10

Problem 1.3.10 A 10 cm long implant wire consists of concentric cylindrical shells. The core is titanium, coated with stainless steel and outermost is platinum. The radii of shells are r_1 = 3 mm, r_2 = 2 mm and r_3 = 1 mm, respectively. Find the equivalent conductance of the implant wire.

$\rho_1 = 1.06 \times 10^{-7}$ Ωm (*Pt*), $\rho_2 = 6.90 \times 10^{-7}$ Ωm (S. Steel), $\rho_3 = 4.20 \times 10^{-7}$ Ωm (*Ti*) (Fig. 1.35).

Solution

$$R_1 = \rho_1 \cdot \frac{L}{A_1} = \rho_1 \cdot \frac{L}{\pi\left(r_1^2 - r_2^2\right)}, \quad R_2 = \rho_2 \cdot \frac{L}{\pi\left(r_2^2 - r_3^2\right)}$$

Innermost titanium section,

$$R_3 = \rho_3 \cdot \frac{L}{\pi r_3^2}$$

$$G = G_1 + G_2 + G_3 = \frac{1}{R} = \frac{1}{R_1} + \frac{1}{R_2} + \frac{1}{R_3}$$

$$G = \frac{\pi\left(r_1^2 - r_2^2\right)}{\rho_1 L} + \frac{\pi\left(r_2^2 - r_3^2\right)}{\rho_2 L} + \frac{\pi r_3^2}{\rho_3 L} = \frac{\pi}{L}\left[\frac{r_1^2 - r_2^2}{\rho_1} + \frac{r_2^2 - r_3^2}{\rho_2} + \frac{r_3^2}{\rho_3}\right]$$

$$= \frac{\pi}{10^{-1} \times 10^{-7}}\left[\frac{10^{-6}}{1.06} + \frac{10^{-6}}{6.9} + \frac{10^{-6}}{4.2}\right] = 100\pi \times 1.3264 = 416.7\,\text{S}$$

Problem 1.3.11 Conductance of a sodium channel of a cell membrane at a specific time is (G/A =) 10 mS/cm^2. If channel length is 100 nm, find the value of its conductivity in (μS/m).

Solution

$$R = \rho\frac{L}{A}, \quad \sigma = \frac{1}{\rho} \quad \rightarrow \quad G = \frac{1}{R} = \frac{1}{\rho}\cdot\frac{A}{L} = \sigma\frac{A}{L} \quad \rightarrow \quad \frac{G}{A} = \frac{\sigma}{L} \quad \rightarrow \quad \sigma = \left(\frac{G}{A}\right)\cdot L$$

$$\sigma = 10\frac{\text{mS}}{\text{cm}^2} \times 100 \times 10^{-\rho}\text{m} = 10\frac{\text{mS}}{10^{-4}\text{m}^2}\cdot 10^{-7}\text{m} = 10^{-2}\frac{\text{mS}}{\text{m}} = 10^{-2} \times 10^{-3}\frac{\text{S}}{\text{m}}$$

$$\sigma = 10^{-5} \times 10^{6}\frac{\mu\text{S}}{\text{m}}$$

$$\sigma = 10\frac{\mu\text{S}}{\text{m}}$$

Problem 1.3.12 A resistive-type hydrogen sulfide (H_2S) gas sensor is to be designed. The sensor characteristics are as follows [7]:

Concentration	1 ppm	5 ppm	10 ppm	50 ppm	90 ppm
Resistance (kΩ)	300	90	50	90	6

A voltage divider circuit is to be realized with a grounded load resistor and sensor floating. One pin of the sensor is connected to +5 V DC voltage source and the other pin is the output terminal. Determine the load resistance if H_2S concentration above 5 ppm is the threshold level, with output voltage set at: (a) 2.5 V, (b) 3.5 V, (c)1.5 V.

Solution

(a) By voltage division,

$$V_{\text{O}} = \frac{R_{\text{L}}}{R_{\text{L}} + R_{\text{S}}} V_{\text{DC}}$$

at 5 ppm H_2S, $R_S = 90\,\text{k}\Omega$

$$\frac{V_{\text{O}}}{V_{\text{DC}}} = \frac{R_L}{R_{\text{L}} + 90\,(\text{k}\Omega)} \quad \rightarrow \quad \frac{2.5}{5} = 0.5 = \frac{R_{\text{L}}}{R_{\text{L}} + 90\,(\text{k}\Omega)}$$
$$R_{\text{L}} = 90\,\text{k}\Omega$$

(b) Applying voltage division rule at given gas concentration,

$$\frac{3.5}{5} = 0.7 = \frac{R_{\text{L}}}{R_{\text{L}} + 90\,(\text{k}\Omega)} \rightarrow 0.7R_{\text{L}} + 63 = R_{\text{L}}$$
$$0.3R_{\text{L}} = 63 \quad \rightarrow \quad R_{\text{L}} = \frac{63}{0.3} = 210\,\text{k}\Omega$$

(c) Repeating previous procedure for the new alarm threshold voltage at given gas concentration,

$$\frac{1.5}{5} = 0.3 = \frac{R_L}{R_L + 90\,(\mathrm{k\Omega})} \quad \rightarrow \quad 0.3R_L + 27 = R_L$$
$$0.7R_L = 27 \quad \rightarrow \quad R_L = \frac{27}{0.7} = 38.5\,\mathrm{k\Omega}$$

Problem 1.3.13 Show that for an isotropic material with equally spaced four electrodes aligned along a straight line, the resistivity is calculated as

$$\rho = 2\pi d \frac{V}{I}, \tag{1.7}$$

where d is the spacing between the two electrodes.

Solution

The current injected by first electrode spreads spherically. The current density at a distance r_1 is;

$$J = \frac{I}{2\pi r_1} \tag{1.8}$$

The electric field at r_1 is

$$E = -\frac{\mathrm{d}V}{\mathrm{d}r} \tag{1.9}$$

$$E = \rho J = \rho \frac{I}{2\pi r_1^2} \tag{1.10}$$

Then,

$$-\frac{\mathrm{d}V}{\mathrm{d}r} = \rho \frac{I}{2\pi r_1^2} \tag{1.11}$$

The electric potential at point x becomes

$$\int_0^V \mathrm{d}V = -\frac{\rho I}{2\pi} \int_0^{r_1} \frac{\mathrm{d}r}{r^2} \tag{1.12}$$

$$V = \frac{\rho I}{2\pi r_1} \tag{1.13}$$

The voltage drop between the two inner probes is

$$V = V_2 - V_3 = \frac{\rho I}{2\pi}\left[\frac{1}{d_1} - \frac{1}{d_2} - \left(\frac{1}{d_3} - \frac{1}{d_4}\right)\right] \tag{1.14}$$

Since

$$d_2 = d_3 = 2d_1 = 2d_4 = 2d \tag{1.15}$$

$$V = \frac{\rho I}{2\pi}\left[\frac{1}{d} - \frac{1}{2d} - \left(\frac{1}{2d} - \frac{1}{d}\right)\right] = \frac{\rho I}{2\pi}\left[-\frac{2}{2d} + \frac{1}{d} + \frac{1}{d}\right] = \frac{\rho I}{2\pi d} \tag{1.16}$$

Pulling out resistivity term from last equation yields

$$\rho = 2\pi d\frac{V}{I},$$

which is Eq. (1.2).

Problem 1.3.14 Show that when the thickness of a sample (= t) in Fig. 1.36 is smaller than the electrode spacings (i.e., $t \ll d$), the sheet resistivity is calculated as

$$\rho = \frac{\pi t}{\ln 2} \times \frac{V}{I} \tag{1.17}$$

Comment on this formula.

Solution

In this case, material is an infinite two-dimensional sheet. The current from first electrode spreads out cylindrically and the current density is

$$J = \frac{I}{2\pi rt} \tag{1.18}$$

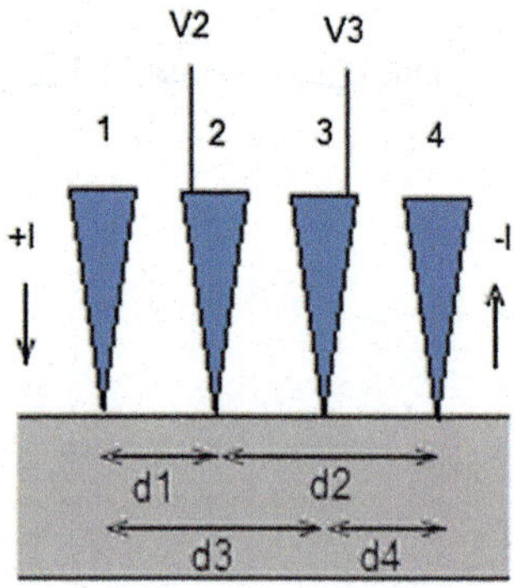

Fig. 1.36 Four electrodes resistivity measurement

The electric field and potential relationship,

$$E = -\frac{\mathrm{d}V}{\mathrm{d}r} = \frac{\rho I}{2\pi rt} \tag{1.19}$$

Integrating both sides of (3)

$$\int_0^V \mathrm{d}V = -\frac{\rho I}{2\pi t}\int_0^{r_1} \frac{\mathrm{d}r}{r} \tag{1.20}$$

$$V = -\frac{\rho I}{2\pi t}\ln|r| \tag{1.21}$$

The voltage drop between two inner electrodes,

$$= V_2 - V_3 = \frac{I\rho}{2\pi t}(\ln d_2 - \ln d_1) - \frac{I\rho}{2\pi t}(\ln d_4 - \ln d_3) \tag{1.22}$$

$$= \frac{I\rho}{2\pi t}(\ln d_2 - \ln d_1 - \ln d_4 - \ln d_3) \tag{1.23}$$

$$= \frac{I\rho}{2\pi t}\ln\left(\frac{d_2 d_3}{d_1 d_4}\right) \tag{1.24}$$

With $d_2 = d_3 = 2d_1 = 2d_4 = 2d$

$$V = \frac{I\rho}{2\pi t}\ln\left(\frac{2d \times 2d}{d \times d}\right) \tag{1.25}$$

$$2\pi tV = I\rho \cdot \ln(4) \tag{1.26}$$

$$\rho = \frac{2\pi tV}{I \times \ln(4)}$$

Since $\ln(a^b) = b \cdot \ln(a)$

$$\rho = \frac{\pi t}{\ln 2} \cdot \frac{V}{I},$$

which is Eq. (1.17), or

$$\rho = \frac{3.415926 \times t}{0.69315} \times \frac{V}{I} = 4.5323 \times t \times \frac{V}{I}$$

It can be concluded that sheet resistivity is independent of electrode spacings. The sheet resistance can be given as

$$R_{\text{sheet}} = \frac{\rho}{t} = 4.5323 \times \frac{V}{I} \tag{1.27}$$

Problem 1.3.15 Porosity is defined as a fraction of the volume of voids over the total volume of a solid material. In other words, porosity is the percentage of solid that is void of material and thus can be filled with gas or fluid. The larger the pore space, the higher the porosity. Therefore, its numerical value is between 0 and 1.

A sintered porous titanium material (with average pore size of 100 μm) is evaluated as a part of its biomedical applicability tests by performing conductivity analysis. Bulk conductivity of titanium is 2.34 MS/m, and the following empirical equations can be tested.

What is the void volume and resistance of a bar made out of this material with L = 50 cm, cross-sectional area A = 0.5 cm^2 and porosity factor p = 0.40? (porosity_conductance.m)

$$\sigma_1 = \sigma_0 \cdot (1-p)^{3.26} \tag{1.28}$$

$$\sigma_2 = \sigma_0 \cdot \left(1 - \frac{p}{2.31}\right)^{9.2} \tag{1.29}$$

$$\sigma_3 = \sigma_0 \cdot \frac{1-p}{1+7.15p} \tag{1.30}$$

$$\sigma_4 = \sigma_0 \cdot \frac{1 - \dfrac{p}{2.31}}{1+7.15p} \tag{1.31}$$

These equations are due to the following sources:
(1.28): Sevostianov [8], (1.29): Zhou [9], (1.30): Mori [10], (1.31): Zhu [11].

Solution

By definition of porosity,

$$p = \frac{V_v}{V} \quad \rightarrow \quad V_v = pV = p(AL) = 0.4 \times \left(0.5 \times 10^{-4}\,\text{m}^2 \times 0.5\,\text{m}\right) = 10^{-5}\,\text{m}^3$$

$$\sigma_0 = 2.34 \times 10^6\ \text{S/m},$$

$$\rho_0 = \frac{1}{\sigma_0} = 4.2735 \times 10^{-7}\,\Omega\text{m}$$

Conductivity ratios as functions of porosity factor for all four methods are plotted on the same graph in Fig. 1.37.

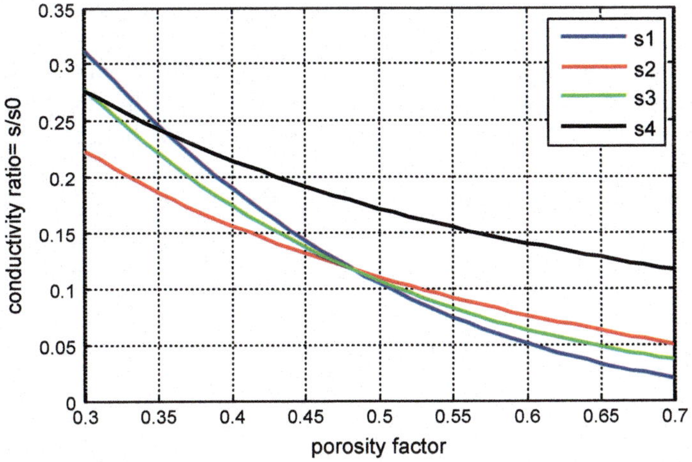

Fig. 1.37 Conductivity ratio versus porosity due to (s1) Sevostianov I. et al.; (s2) D. Zhou, et al.; (s3) Mori, T., Tanaka, K.; (s4) Zhu, et al.

Computations at $p = 0.4$ yield the following results;

$$\sigma_1 = 4.4258 \times 10^5 \text{ S/m}$$
$$\sigma_2 = 3.6373 \times 10^5 \text{ S/m}$$
$$\sigma_3 = 4.0690 \times 10^5 \text{ S/m}$$
$$\sigma_4 = 5.0124 \times 10^5 \text{ S/m}$$

$$\frac{\sigma_1}{\sigma_0} = 0.1891, \quad \frac{\sigma_2}{\sigma_0} = 0.1554, \quad \frac{\sigma_3}{\sigma_0} = 0.1739, \quad \frac{\sigma_4}{\sigma_0} = 0.2142$$

$$\rho_1 = \frac{1}{\sigma_1} = 2.26 \times 10^{-6}\,\Omega\text{m},$$
$$\rho_2 = \frac{1}{\sigma_2} = 2.75 \times 10^{-6}\,\Omega\text{m},$$
$$\rho_3 = \frac{1}{\sigma_3} = 2.46 \times 10^{-6}\,\Omega\text{m}$$
$$\rho_4 = \frac{1}{\sigma_4} = 2.00 \times 10^{-6}\,\Omega\text{m}$$

$$\frac{\rho_1}{\rho_0} = 5.287, \quad \frac{\rho_2}{\rho_0} = 6.433, \quad \frac{\rho_3}{\rho_0} = 5.751, \quad \frac{\rho_4}{\rho_0} = 4.669$$

As porosity increases, conductivity decreases and resistivity increases.

All models predict different conductivity values for a given porosity factor (p = 0.4).

These are approximately 19, 15.5, 17.4, and 21.4% of bulk conductivity of titanium, respectively. On the other hand, computed resistivities for these models at p = 0.4 level are 5.287, 6.433, 5.751, and 4.669 times the bulk resistivity of titanium, respectively.

Resistance of the bulk titanium rod having specified dimensions will be

$$R_0 = \rho_0 \frac{L}{A} = 4.2735 \times 10^{-7} \times \frac{0.5}{0.5 \times 10^{-4}} = 4.3\,\mathrm{m}\Omega$$

Based upon these models, resistance value of the given porous Ti specimen can be predicted as

$$R_1 = \rho_1 \frac{L}{A} = 2.26 \times 10^{-6} \times \frac{0.5}{0.5 \times 10^{-4}} = 22.6\,\mathrm{m}\Omega$$
$$R_2 = \rho_2 \frac{L}{A} = 2.75 \times 10^{-6} \times \frac{0.5}{0.5 \times 10^{-4}} = 27.5\,\mathrm{m}\Omega$$
$$R_3 = \rho_3 \frac{L}{A} = 2.46 \times 10^{-6} \times \frac{0.5}{0.5 \times 10^{-4}} = 24.6\,\mathrm{m}\Omega$$
$$R_4 = \rho_4 \frac{L}{A} = 2.00 \times 10^{-6} \times \frac{0.5}{0.5 \times 10^{-4}} = 20.0\,\mathrm{m}\Omega$$

Problem 1.3.16 Human body can be considered as a mixture of cells and extracellular fluid (Fig. 1.38). Under DC conditions (for zero frequency), electrical current cannot enter the cells and it flows in the extracellular fluid (ECF) with resistivity of ρ_{ECF}.

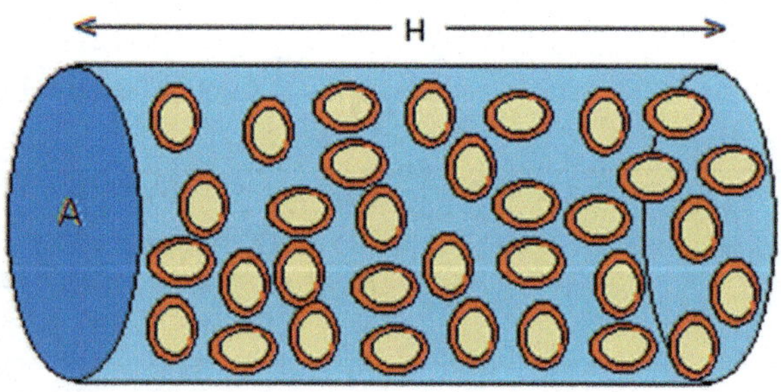

Fig. 1.38 Body model as a mixture of cells and extracellular fluid

If the cells take up the part c (percent volume) of total body, then the overall body resistivity is [12]:

$$\rho = \frac{\rho_{\mathrm{ECF}}}{(1-c)^{3/2}} \tag{1.32}$$

Wrist to ankle body resistance is

$$R = \frac{K\rho H^2}{V} \tag{1.33}$$

The extracellular fluid volume is

$$V_{\mathrm{ECF}} = k_{\mathrm{ECF}} \cdot \left(\frac{H^2\sqrt{W}}{R_{\mathrm{E}}}\right)^{2/3} \tag{1.34}$$

where,

$$k_{\mathrm{ECF}} = \left(\frac{K^2\rho_{\mathrm{ECF}}^2}{d}\right)^{1/3} \tag{1.35}$$

where K is the body geometry factor (De Lorenzo factor [13]), H is the body height, W is Body mass in kg, d is body density, V is the body volume, in m^3.

(a) What does the term $(1 - c)$ indicate in terms of V_{ECF} and V?
(b) Check for the unit dimension of k_{ECF} and then prove that V_{ECF} is indeed cm^3, if CGS units are used in calculations.
(c) Derive Eq. (1.3).
(d) Let $c = 0.7$, $c = 0.5$ and $c = 0.3$. Calculate tissue resistivity for each case, if $\rho_{\mathrm{ECF}} = 100\,\Omega\mathrm{m}$. What can be concluded about the tissue resistivity for a mixture of fluid and cells?
(e) Let $\rho_{\mathrm{ECF}} = 40.5(\Omega\,\mathrm{cm}), K = 4.3, d = 1.05\,\mathrm{g\,cm}^{-3}, R_{\mathrm{E}} = 800\,\Omega, H = 1.7\,\mathrm{m}$, $W = 60\,\mathrm{kg}$. Determine the volume of extracellular fluid for this person using CGS units.
(f) Determine the percentage of extracellular fluid and total body resistivity in part (e).
(g) Determine percent change of V_{ECF} if extracellular fluid resistivity value deviates 3%.

Solution

(a) If c is the percentage of cells, then $(1 - c)$ indicates the percentage of extracellular fluid,

$$1 - c = \frac{V_{\text{ECF}}}{V} \quad (1.36)$$

(b) $$k_{\text{ECF}} = \left[\frac{1 \times (\Omega \times \text{cm})^2}{\left(\frac{g}{\text{cm}^3}\right)}\right]^{1/3} = \frac{\Omega^{2/3} \times \text{cm}^{5/3}}{g^{1/3}}$$

$$V_{\text{ECF}} = \frac{\Omega^{2/3} \times \text{cm}^{5/3}}{g^{1/3}} \times \frac{\text{cm}^{4/3} \times g^{1/3}}{\Omega^{2/3}} = \text{cm}^3$$

(c) Substituting (1.36) into (1.32),

$$\rho = \frac{\rho_{\text{ECF}}}{(1-c)^{3/2}} = \frac{\rho_{\text{ECF}}}{\left(\frac{V_{\text{ECF}}}{V}\right)^{3/2}} = \rho_{\text{ECF}}\left(\frac{V}{V_{\text{ECF}}}\right)^{3/2} \quad (1.37)$$

Inserting (1.37) in (1.33),

$$R = R_{\text{E}} = \frac{K\rho H^2}{V} = \frac{K \cdot \rho_{\text{ECF}} \cdot \left(\frac{V}{V_{\text{ECF}}}\right)^{3/2} \cdot H^2}{V} = K \cdot \rho_{\text{ECF}} \cdot \frac{\sqrt{V}}{V_{\text{ECF}}^{3/2}} \cdot H^2 \quad (1.38)$$

or,

$$V_{\text{ECF}}^{3/2} = K \cdot \rho_{\text{ECF}} \cdot \frac{\sqrt{V}}{R_{\text{E}}} \cdot H^2 \quad (1.39)$$

Assuming that the body density is independent from person to person $(d = W/V)$, this equation can be rewritten as,

$$V_{\text{ECF}}^{3/2} = K \cdot \rho_{\text{ECF}} \cdot \frac{\sqrt{W/d}}{R_{\text{E}}} \cdot H^2 = K \cdot \rho_{\text{ECF}} \cdot \frac{\sqrt{W}}{\sqrt{d}\,R_{\text{E}}} \cdot H^2 \quad (1.40)$$

Removing power term in the left side yields,

$$V_{\text{ECF}} = \left(\frac{K^2 \rho_{\text{ECF}}^2}{d}\right)^{\frac{1}{3}} \cdot \left(\frac{H^2\sqrt{W}}{R_{\text{E}}}\right)^{\frac{2}{3}} \quad (1.41)$$

This is the same Eq. (1.3), with the coefficient k_{ECF} given by (1.35).

Therefore, if height, weight, and DC resistance R_E of a person is measured, his or her extracellular fluid volume is calculated using Eqs. (1.3) and (1.3).

(d) $\rho_1 = \dfrac{\rho_{ECF}}{(1-0.7)^{3/2}} = 6.1\,\rho_{ECF} = 610\,\Omega\text{m}$

$\rho_2 = \dfrac{\rho_{ECF}}{(1-0.5)^{3/2}} = 2.8\,\rho_{ECF} = 280\,\text{m}$

$\rho_3 = \dfrac{\rho_{ECF}}{(1-0.3)^{3/2}} = 1.7\,\rho_{ECF} = 170\,\Omega\text{m}$

The tissue resistivity under DC operating conditions ($f = 0$ Hz) is higher for a mixture of fluid and cells than for pure (unmixed) extracellular fluid.

(e) Using CGS units,

$$k_{ECF} = \left[\frac{18.49 \times 1640.25}{1.05}\right]^{1/3} = 30.682\left[\frac{\Omega^{2/3} \times \text{cm}^{5/3}}{9^{1/3}}\right]$$

$$V_{ECF} = k_{ECF} \cdot \left(\frac{H^2\sqrt{W}}{R_E}\right)^{\frac{2}{3}} = 30.682 \times \left(\frac{170^2\sqrt{60,000}}{800}\right)^{\frac{2}{3}}$$

$$= 30.682 \times \left(\frac{28,900 \times 244.95}{800}\right)^{\frac{2}{3}}$$

$$V_{ECF} = 30.682 \times (8848.78)^{\frac{2}{3}} = (30.682) \times (429.11)$$

$$= 13166.03\,\text{cm}^3 = 13.1661$$

(f) The percentage of extracellular fluid is

$$1 - c = \frac{V_{ECF}}{V} \tag{1.42}$$

$$V_{ECF} = 13.1661$$

$$V = \frac{W}{d} = \frac{60{,}000}{1.05} = 57142.86\,\text{cm}^3$$

$$1 - c = \frac{13.166}{57.143} = 0.2304$$

The percentage of extracellular fluid is 23%.

If the cells take up 23% (volume) of total body, then the overall body resistivity is given by (1),

$$\rho = \frac{\rho_{\text{ECF}}}{(1-c)^{3/2}} = \frac{40.5}{(0.23)^{3/2}} = \frac{40.5}{0.1103} = 367.167\,\Omega\,\text{cm}$$

(g) Since V_{ECF} is proportional to $\rho_{\text{ECF}}^{2/3}$, $(1.03)^{2/3} = 1.02$; ΔECV, $(0.97)^{2/3} = 0.98$ In other words, 3% increase/decrease of extracellular resistivity causes 2% increase/decrease in the value of extracellular volume.

Problem 1.3.17 Find,

(a) The value of voltage, V_x,
(b) Power consumed in the circuit of Fig. 1.39a.

Solution

(a) The current supplied by the voltage source is $R_{\text{eq}} = 1 + 2 \parallel (1+1) = 1 + 1 = 2\,\Omega$

$$I = \frac{2\,\text{V}}{2\,\Omega} = 1\,\text{A}$$

By current division, 0.5 A flows through series (1Ω, 1Ω) resistors branch. Then,

$$V_x = \frac{1}{2}\,\text{A} \times 1\,\Omega = \frac{1}{2}\,\text{V}$$

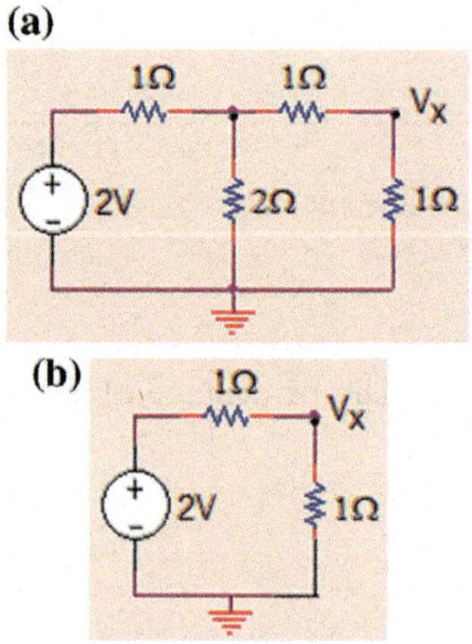

Fig. 1.39 Circuit of Problem 1.3.17 (a) , its reduced form (b)

(b) Power consumed = power delivered.

$$I = 1\,\text{A}$$

$$P = I \cdot V = 1 \cdot 2 = 2\,\text{W}$$

Problem 1.3.18 Basic ECG derivation of "Einthoven Triangle" gives electrical potential differences between the right arm, left arm, and left leg.

$V_{\text{I}} = V_{\text{LA}} - V_{\text{RA}}$, $V_{\text{II}} = V_{\text{RA}} - V_{\text{LL}}$, $V_{\text{III}} = V_{\text{LL}} - V_{\text{LA}}$

$V_{\text{II}} = V_{\text{I}} + V_{\text{III}}$

Below is an ECG record showing derivation I (top) and derivation II (below). What is the peak amplitude of derivation III? (Each small square indicates 100 mV) (Fig. 1.40).

Solution

Since $V_{\text{II}} = V_{\text{I}} + V_{\text{III}}$,

$V_{\text{II}} - V_{\text{I}} = V_{\text{III}}$

14 – 4 = 10 small divisions = 10 × 100 mV = 1 mV

Thus, the peak amplitude of derivation III is 1 mV in this case.

Problem 1.3.19 ECG signal measured under the skin is 1 mV at a certain time. Skin resistance is 100 kΩ. If a monitor instrument with input resistance of (a) 500 kΩ, (b) 1 MΩ, (c) 2 MΩ is used, determine the voltages measured at the input of monitor equipment. Comment on these results (Fig. 1.41).

Solution

The voltage measured is V_{m} while $V_i = 1\,\text{mV}, R_s = 100\,\text{k}\Omega$. By voltage division rule,

(a) $V_{\text{m1}} = \dfrac{500}{500} V_i = \dfrac{5}{6} V_i = 0.83\,\text{mV}$

(b) $V_{\text{m2}} = \dfrac{1000}{1000 + 100} V_i = \dfrac{1}{1.1} V_i = 0.91\,\text{mV}$

Fig. 1.40 ECG signal for Problem 1.3.18

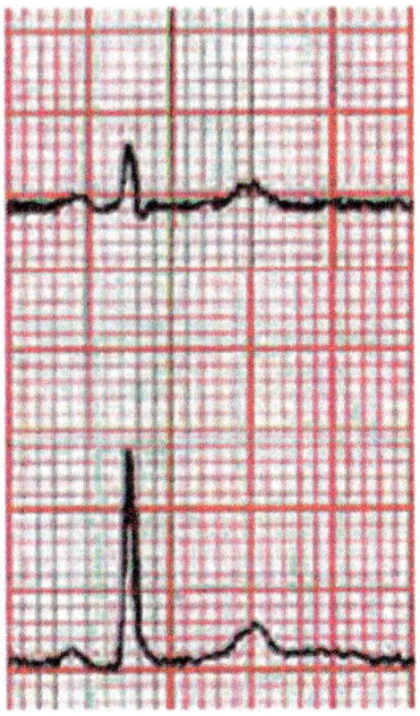

Fig. 1.41 Circuit for Problem 1.3.19

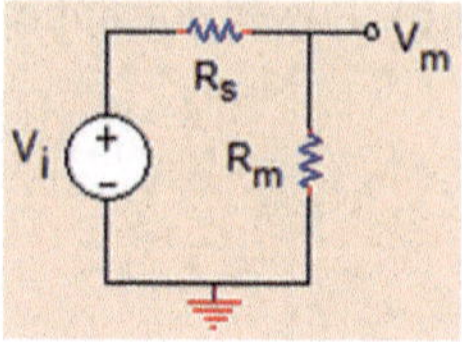

(c) $V_{m3} = \dfrac{2000}{2000+100} V_i = \dfrac{2}{2.1} V_i = 0.95\,\mathrm{mV}$

It is noted that higher the input impedance, higher the voltage available at the input of monitoring equipment.

Problem 1.3.20 When $V_x = 3\,\mathrm{V}$, what is the ratio of R_1 to R_2, in the circuit of Fig. 1.42?

Solution

By voltage division,

$$V_x = V_1 \cdot \frac{R_2}{R_1+R_2} \rightarrow \frac{V_x}{V_1} = \frac{R_2}{R_1+R_2} \rightarrow \frac{R_1}{R_2} = \frac{V_1}{V_x} - 1 = \frac{12}{3} - 1 = 3$$

Problem 1.3.21 Find the values of V_b, I_1, I_2, I_3 in the circuit shown in Fig. 1.43. R_1 = 68 kΩ, R_2 = 22 kΩ, R_3 = 33 kΩ, R_4 = 15 kΩ, V_{dc} = 12 V.

Fig. 1.42 Circuit for Problem 1.3.20

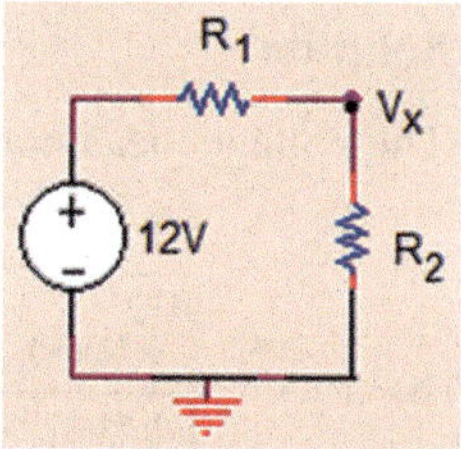

Fig. 1.43 Circuit for Problem 1.3.21

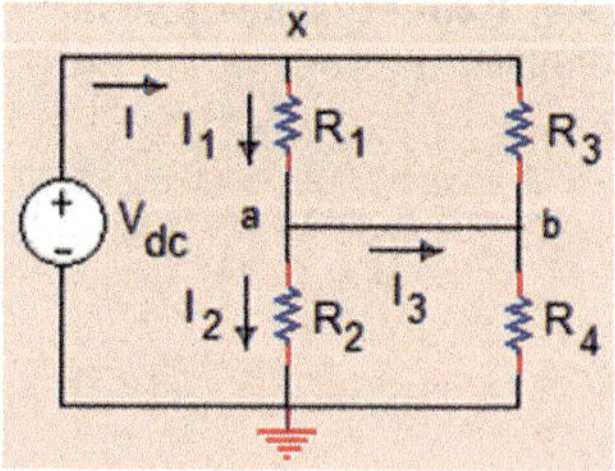

Solution

$$R_{p1} = \frac{R_1.R_3}{R_1 + R_3} = \frac{68\,\text{k}\Omega \cdot 33\,\text{k}\Omega}{68\,\text{k}\Omega + 33\,\text{k}\Omega} = 22.22\,\text{k}\Omega,$$
$$R_{p2} = \frac{R_2.R_4}{R_2 + R_4} = \frac{22\,\text{k}\Omega \cdot 15\,\text{k}\Omega}{22\,\text{k}\Omega + 15\,\text{k}\Omega} = 8.92\,\text{k}\Omega,$$
$$I = \frac{V_{dc}}{R_{p1} + R_{p2}} = \frac{12}{22.22\,\text{k}\Omega + 8.92\,\text{k}\Omega} = 0.385\,\text{mA},$$
$$V_{xa} = R_{p1}.I = (22.22\,\text{k}\Omega)(0.385\,\text{mA}) = 8.563\,\text{V}$$
$$V_a = V_b = R_{p2}.I = (8.92\,\text{k}\Omega)(0.385\,\text{mA}) = 3.434\,\text{V}$$

(Note that this could also be calculated as

$$V_a = V_b = V_{dc} - V_{xa} = 12 - 8.563 = 3.437\,\text{V}$$

while 3 mV difference is due to computational rounding off.)

$$I_1 = \frac{V_{xa}}{R_1} = \frac{8.563\,\text{V}}{68\,\text{k}\Omega} = 0.126\,\text{mA}$$
$$I_2 = \frac{V_a}{R_2} = \frac{3.434\,\text{V}}{22\,\text{k}\Omega} = 0.156\,\text{mA}$$
$$I_3 = I_1 - I_2 = 0.126 - 0.156 = -0.03\,\text{mA}$$

As a proof of this,

$$I_{R_3} = \frac{V_{xa}}{R_3} = \frac{8.563\,\text{V}}{33\,\text{k}\Omega} = I - I_1 = 0.385\,\text{mA} - 0.126\,\text{mA} = 0.259\,\text{mA}$$
$$I_{R_4} = \frac{V_b}{R_4} = \frac{3.434\,\text{V}}{15\,\text{k}\Omega} = 0.229\,\text{mA}$$
$$I_3 = I_{R_3} - I_{R_4} = 0.259 - 0.229 = 0.03\,\text{mA}$$

This is also a proof of the sign of the current I_3, entering into the node *a*, leaving the node *b*.

Problem 1.3.22 Other than common mode rejection, describe the advantage of using a full bridge circuit as compared to a voltage divider circuit for the measurement of nonelectrical quantities (strain, pressure, temperature…).

Fig. 1.44 Circuit for Problem 1.3.22

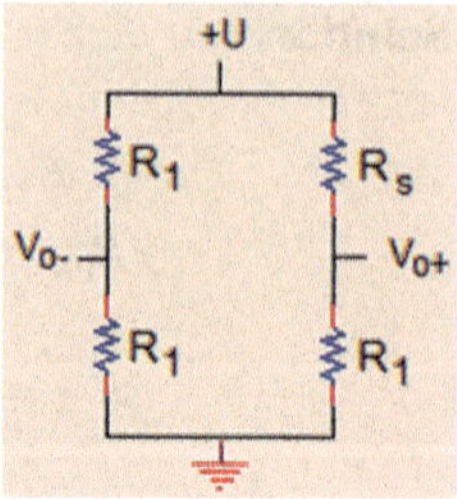

Solution

Consider the circuit shown in Fig. 1.44. The right side of the bridge is a voltage divider with

$$V_{0+} = U\left(\frac{R_1}{R_1 + R_s}\right)$$

Rearranging this equation for R_s (sensor resistance),

$$R_s = R_1\left(\frac{U}{V_{0+}} - 1\right)$$

This is a nonlinear relationship between R_s and V_{0+}. The voltage divider circuit produces small changes in voltage, embedded in a large offset. If the offset value is unknown, it is difficult to measure the actual measurand.

Now, include the left side of the bridge into the analysis,

$$V_{0-} = \frac{R_1}{R_1 + R_1}U = \frac{U}{2}$$

$$V_0 = V_{0+} = V_{0-} = U\left(\frac{R_1}{R_1 + R_s} - \frac{1}{2}\right)$$

Assuming that $R_s = R_1$ at rest, the last equation shows that the offset is reduced (eliminated) by the addition of a voltage divider and measuring the output differentially.

Problem 1.3.23

(a) Balanced output of a Wheatstone bridge becomes unbalanced by a slight variation in the resistance value of one of its arms. Determine the voltage variation in terms of resistance variation, another bridge arm resistance and power supply voltage. Assume that arm resistance change is much smaller than the resistance of the arm and there is no load resistance at the output terminals.

(b) Let V_i = 12 V, $R_1 = R_2 = R_3 = R_4 = 100\ \Omega$ at 0 °C. A Pt100 temperature sensor element with $\alpha = 0.0039$/K, and $R_T = Ro(1 + \alpha T)$ first-order characteristic equation is used in place of R_2. Find the voltage change at the output of

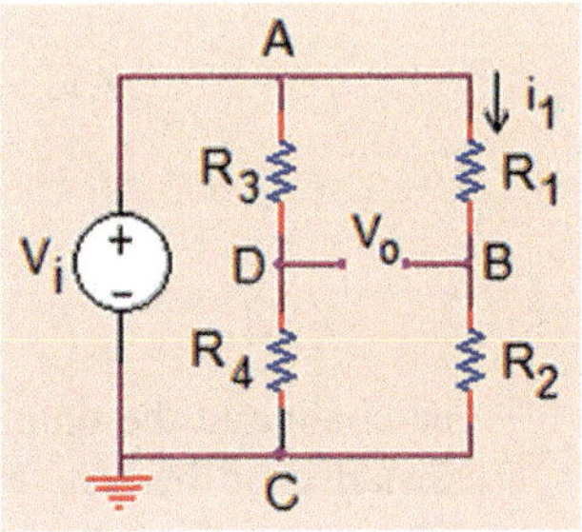

Fig. 1.45 Wheatstone bridge

this bridge corresponding to one degree centigrade change at temperature. What is the value of output voltage at 20 °C? (Fig. 1.45).

Solution

(a) Balanced bridge:

$$V_B = V_D, \quad V_o = 0$$

$$I_1 R_1 = I_2 R_3 \tag{1.43}$$

$$I_1 R_2 = I_2 R_4 \tag{1.44}$$

Dividing (1.43) by (1.44):

$$\frac{R_1}{R_2} = \frac{R_3}{R_4}$$

$$V_o = V_{BC} - V_{DC} = V_i\left(\frac{R_2}{R_1 + R_2} - \frac{R_4}{R_3 + R_4}\right)$$

Unbalanced bridge:

$$R_2 \to R_2 + \Delta R_2, \quad V_o \to V_o + \Delta V_o$$

$$V_o + \Delta V_o = V_i\left(\frac{R_2 + \Delta R_2}{R_1 + R_2 + \Delta R_2} - \frac{R_4}{R_3 + R_4}\right)$$

$$\Delta V_o = (V_o + \Delta V_o) - V_o$$

$$\Delta V_o = V_i\left(\frac{R_2 + \Delta R_2}{R_1 + R_2 + \Delta R_2} - \frac{R_2}{R_1 + R_2}\right)$$

$$\text{if } \Delta R_2 \ll R_2, \quad \Delta V_o = V_i\left(\frac{R_2}{R_1 + R_2 + \Delta R_2} + \frac{\Delta R_2}{R_1 + R_2 + \Delta R_2} - \frac{R_2}{R_1 + R_2}\right)$$

$$\Delta V_o = V_i\left(\frac{R_2}{R_1+R_2} + \frac{\Delta R_2}{R_1+R_2} - \frac{R_2}{R_1+R_2}\right)$$

$$\Delta V_o = V_i\left(\frac{\Delta R_2}{R_1+R_2}\right)$$

The change in the output voltage drop is directly proportional to change in the resistance of the sensor.

(b) $\Delta R = R_T - R_o = R_o(1 + \alpha \cdot T - 1)$
$= R_o \cdot \alpha \cdot T = 100 \times 0.0039 \times 1 = 0.39\,\Omega/\mathrm{K}$
Since this resistance change is much smaller than the resistance itself,

$$\Delta V_o = V_i\left(\frac{\Delta R_2}{R_1+R_2}\right) = 12\left(\frac{0.39}{100+100}\right) = 24\,\mathrm{mV/°C}$$

Output voltage at 20 °C will be

$$\Delta V_o = 20 \times 24 = 480\,\mathrm{mV}$$

Problem 1.3.24

(a) Describe briefly the operating principle of a Coulter (blood cells) counter.
(b) In the bridge circuit shown in Fig. 1.46, $R_1 = R_3 = R_4 = R$. A cell counter orifice resting resistance is $R_2 = R$ ohms. An erythrocyte increases its resistance by ΔR.

Determine the output voltage of the circuit each time an erythrocyte passes through the orifice.

(c) Calculate V_{ab} if $R = 500\,\Omega$, $\Delta R = 20\,\Omega$, $U = 15\,\mathrm{V}$.

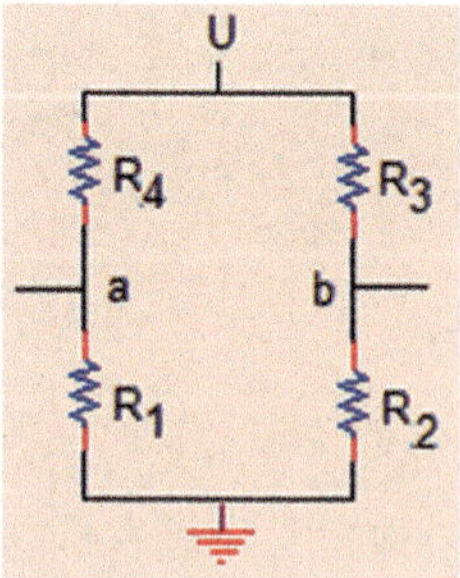

Fig. 1.46 Wheatstone bridge (for blood cells counting)

Solution

(a) In a Coulter counter (blood cell counting) aperture technology instrument, a tube with a small aperture on its wall is immersed into a beaker that contains particles suspended in a low concentration electrolyte. One electrode is placed inside the aperture tube and another electrode is dipped inside the beaker, and a current path is provided by the electrolyte when a potential difference is applied between the electrodes.

As a particle passes through the aperture, a volume of electrolyte equivalent to the volume of the particle is displaced from the sensing zone, causing a swift variation in the impedance between the electrodes. The pulse height is proportional to the volume of the particle. It is also proportional to the particle mass. The measured particle volume (or size) can be used to obtain particle size distribution.

A typical measurement takes less than a minute [14], as counting and sizing rates of up to 10,000 particles per second are possible. Aperture size typically ranges from 20 to 2000 μm.

(b)
$$\begin{aligned} V_{ab} &= U\left(\frac{R}{R+R} - \frac{R_x}{R+R_x}\right) = U\left(\frac{1}{2} - \frac{R+\Delta R}{R+R+\Delta R}\right) \\ &= U\left[\frac{2R+\Delta R - 2R - 2\Delta R}{2(2R+\Delta R)}\right] = U\left(\frac{-\Delta R}{4R+2\Delta R}\right) \end{aligned} \tag{1.45}$$

Since $\Delta R \ll R$, Eq. (1.45) can be simplified,

$$V_{ab} \cong -U\frac{\Delta R}{4R} \quad \rightarrow \quad V_{ba} = U\frac{\Delta R}{4R} \tag{1.46}$$

(c) Substituting given values into Eq. (1.46),

$$V_{ba} = 15 \times \frac{20}{4 \times 500} = \frac{300}{2000} = 0.15 = 150\,\text{mV}$$

If no simplification is made, one uses Eq. (1.45),

$$\left|V'_{ba}\right| = 15 \times \left(\frac{20}{4 \times 500 + 2 \times 20}\right) = \frac{300}{2040} = 0.147\,\text{V} = 147\,\text{mV}$$

Error due to simplification is 3 mV, or 2%.

Problem 1.3.25

(a) Determine V_{ab}, the Wheatstone bridge output voltage in the circuit shown in Fig. 1.46.
(b) If R_4 is a thermistor, state the maximum sensitivity condition with respect to the thermistor resistance.

(c) If R_4 is a thermistor with β = 4000 K and R_{25} = 100 Ω, calculate its value at T = 40 °C.
(d) Calculate V_{ab} at 40 °C with U = 10 V, $R_1 = R_2 = R_3$ = 100 Ω.
(e) Compute the numerical sensitivity value, $S = \Delta V_{ab}/\Delta T$.
(f) Compute temperature versus thermistor resistance, as well as temperature versus V_{ab} voltage for $35\,^o\mathrm{C} \le T \le 45\,°\mathrm{C}$; show these results in a single table and plot the graph of temperature versus Vab voltage, also demonstrating its linearity. Print the MATLAB script used in these computations (wheatstone_thermistor1.m).

Solution

(a) By voltage division rule application to full bridge circuit,

$$V_{ab} = U\left(\frac{R_1}{R_1+R_4} - \frac{R_2}{R_2+R_3}\right)$$

(b) $$\frac{dV_{ab}}{dR_1} = U\cdot\frac{d}{dR_1}\left(-\frac{R_1}{R_1+R_4}\right) = U\frac{1\times(R_1+R_4)-R_1(1)}{(R_1+R_4)^2} = U\frac{R_4}{(R_1+R_4)^2};$$

$$\frac{d}{dR_4}\frac{dV_{ab}}{dR_1} = \frac{d}{dR_4}\left(U\frac{R_4}{(R_1+R_4)^2}\right) = U\frac{R_4(R_1+R_4)^2 - 2R_4(R_1+R_4)}{(R_1+R_4)^4} = 0$$

or,

$$(R_1+R_4)(R_4+R_1-2R_4) = 0$$

R_1 $-R_4$ cannot be a solution (content of first parenthesis operation)
R_1 R_4 is the solution (content of the second parenthesis operation)

$$\text{Therefore, } \frac{dV_{ab}}{dR_1} \text{ is maximum when } R_1 = R_4.$$

This is also seen in the following graph of d (V_{ab})/dR_1 versus R_4. Maximum occurs at the value of $R_4 = R_1$ = 100 Ω (Figs. 1.47, 1.48 and 1.49).

(c) $$R_4 = R_{25}\cdot e^{\beta\left(\frac{1}{T}-\frac{1}{T_{25}}\right)} = 100\times e^{4000\left(\frac{1}{273+40}-\frac{1}{273+25}\right)} = 100\times e^{4000\left(\frac{1}{313}-\frac{1}{298}\right)}$$
$$R_4 = 100\times 0.5256 = 52.56\,\Omega$$

(d) $$V_{ab} = 10\left(\frac{100}{100+52.56} - \frac{100}{100+100}\right) = 10\left(\frac{100}{152.56} - 0.5\right) = 1.555\,\mathrm{V}$$

(e) $$S = \frac{\Delta V_{ab}}{\Delta T} = \frac{1.555}{15} = 0.104\,\mathrm{V/°C}$$

(g) Following is the MATLAB script for computing and plotting graphs requested in this part of the problem. (wheatstone_thermistor1.m)

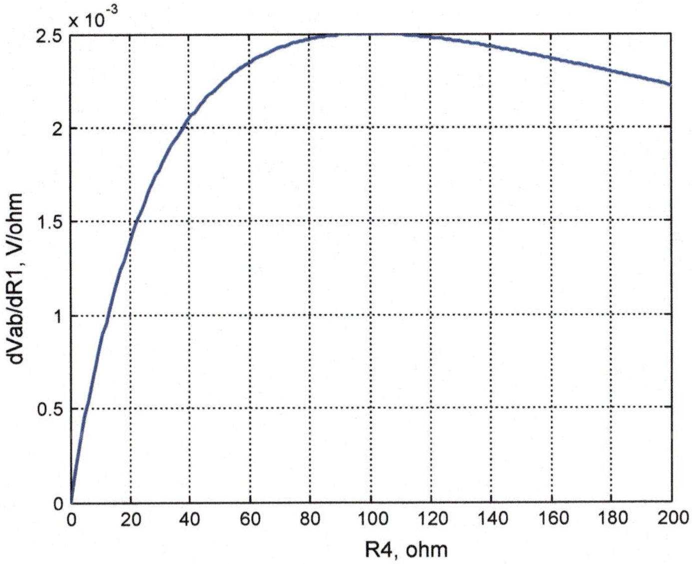

Fig. 1.47 The graph of d(Vab)/dR1 versus R4

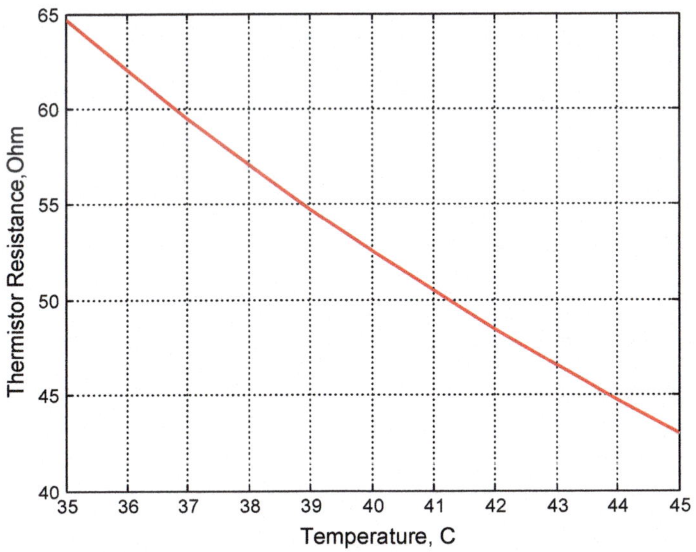

Fig. 1.48 NTC Thermistor resistance against body temperature

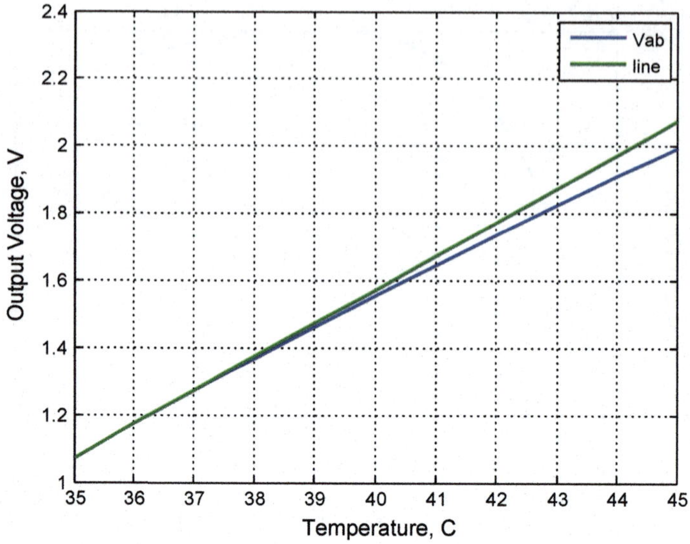

Fig. 1.49 The graph of output voltage deviation as a function of temperature

```
%Wheatstone bridge Thermistor
clc;clear;
U=10;
R1=100;R2=100;R3=100;
B=4000;R25=100;
%Determine max sensitivity condition for R4
R4=0:200;
plot (R4,R4./(R4+R1).^2,'linewidth',2);grid on;
xlabel('R4, ohm');ylabel('dVab/dR1, V/ohm');
figure;
T=35:45;
R4=R25*exp(B*(1./(T+273)-1/298));R4=R4';
plot(T,R4,'r','linewidth',2);grid on;
xlabel('Temperature, C');ylabel('Thermistor Resistance,Ohm');figure;
for k=1:length(R4);
Vab(k)=U*(R1/(R1+R4(k))-R2/(R2+R3));
end
fprintf('  Temp(C)     R4 (ohm)    Vab (V)')
[T'  R4  Vab'] %print of values in a table
k=1:length(Vab);
%nonlinearity:Vab(1)-k/10=0.9726
plot(k,Vab,'-',k,k/10+0.9726,'linewidth',2);grid on;
xlim([min(k) max(k)]);
set(gca,'XTick',k );set(gca,'XTickLabel',T );
xlabel('Temperature, C');
ylabel('Output Voltage, V');legend('Vab','line')
```

Temp(C)	R4(ohm)	Vab(V)
35.0000	64.6743	1.0726
36.0000	62.0125	1.1724
37.0000	59.4763	1.2705
38.0000	57.0591	1.3670
39.0000	54.7548	1.4618
40.0000	52.5573	1.5549
41.0000	50.4612	1.6462
42.0000	48.4612	1.7358
43.0000	46.5524	1.8235
44.0000	44.7302	1.9094
45.0000	42.9900	1.9935

Problem 1.3.26 In a Wheatstone bridge circuit shown in Fig. 1.50,

(a) Prove that

$$R_D = \frac{R_B R_C}{R_A} \times \left[\frac{1 - \left(\frac{\Delta V}{U}\right)\left(\frac{R_A + R_C}{R_C}\right)}{1 + \left(\frac{\Delta V}{U}\right)\left(\frac{R_A + R_C}{R_A}\right)} \right], \quad \Delta V = V_A - V_B$$

(b) What is the form of equation under balanced bridge condition?
(c) What will be the value of R_B, in the same form of equation given in part a?

Solution

(a) Voltage difference between A and B nodes:

$$\Delta V = V_A - V_B = \frac{R_C}{R_A + R_C} U - \frac{R_D}{R_B + R_D} U = U\left(\frac{R_C}{R_A + R_C} - \frac{R_D}{R_B + R_D}\right) \quad (1.47)$$

$$\frac{\Delta V}{U} = \frac{R_C}{R_A + R_C} - \frac{R_D}{R_B + R_D} \quad \rightarrow \quad \frac{R_D}{R_B + R_D} = \frac{R_C}{R_A + R_C} - \frac{\Delta V}{U} \quad (1.48)$$

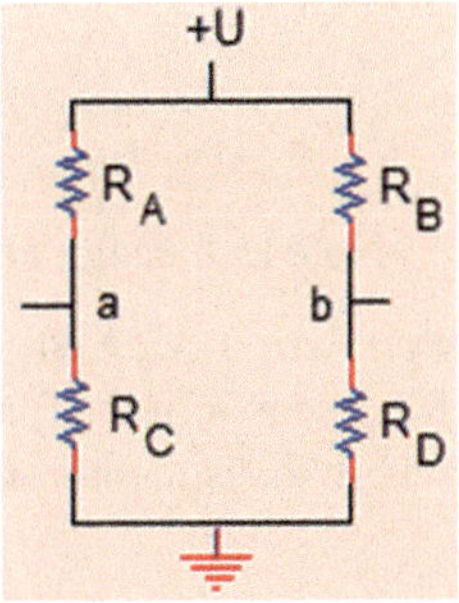

Fig. 1.50 Wheatstone bridge circuit for Problem 1.3.26

$$\begin{aligned} R_D &= \frac{R_C}{R_A + R_C}(R_B + R_D) - \frac{\Delta V}{U}(R_B + R_D) \\ &= \frac{R_B R_C}{R_A + R_C} + \frac{R_C R_D}{R_A + R_C} - \frac{\Delta V}{U} R_B - \frac{\Delta V}{U} R_D \end{aligned} \tag{1.49}$$

$$R_D = R_D\left(\frac{R_C}{R_A + R_C} - \frac{\Delta V}{U}\right) - \frac{\Delta V}{U} R_B + \frac{R_B R_C}{R_A + R_C} \tag{1.50}$$

$$\begin{aligned} R_D - R_D\left(\frac{R_C}{R_A + R_C} - \frac{\Delta V}{U}\right) &= R_D\left(1 - \frac{R_C}{R_A + R_C} - \frac{\Delta V}{U}\right) \\ &= -\frac{\Delta V}{U} R_B + \frac{R_B R_C}{R_A + R_C} \end{aligned} \tag{1.51}$$

$$R_D = \frac{-\frac{\Delta V}{U} R_B + \frac{R_B R_C}{R_A + R_C}}{\left(1 - \frac{R_C}{R_A + R_C} - \frac{\Delta V}{U}\right)} = \frac{\frac{R_B R_C}{R_A + R_C}\left(-\frac{\Delta V}{U}\frac{R_A + R_C}{R_C} + 1\right)}{\left(1 - \frac{R_C}{R_A + R_C} - \frac{\Delta V}{U}\right)} \tag{1.52}$$

$$= \frac{R_B R_C\left(1 - \frac{\Delta V}{U}\frac{R_A + R_C}{R_C}\right)}{R_A + R_C - R_C + \frac{\Delta V}{U}(R_A + R_C)} = \frac{R_B R_C\left(1 - \frac{\Delta V}{U}\frac{R_A + R_C}{R_C}\right)}{R_A + \frac{\Delta V}{U}(R_A + R_C)} \tag{1.53}$$

$$R_D = \frac{R_B R_C}{R_A}\frac{\left(1 - \frac{\Delta V}{U}\frac{R_A + R_C}{R_C}\right)}{\left(1 + \frac{\Delta V}{U}\frac{R_A + R_C}{R_A}\right)} \tag{1.54}$$

(b) The balanced bridge has $\Delta V = 0$ volt, therefore,

$$R_D = \frac{R_B R_C}{R_A} \tag{1.55}$$

(c) If all other resistor values are known, (except floating unknown resistor value, R_B),

$$R_B = \frac{R_A R_D}{R_C}\frac{1 + \frac{\Delta V}{U}\frac{R_A + R_C}{R_C}}{1 - \frac{\Delta V}{U}\frac{R_A + R_C}{R_C}} \tag{1.56}$$

Note to the sign changes in numerator and denominator of Eqs. (1.54) and (1.56).

Problem 1.3.27 Show that the output of a Wheatstone Bridge for a single resistive RTD sensor of the form $R(x) = R_N(1 + ax)$ is nonlinear function of the measurand x, where R_N is the nominal resistance of the sensor and a is a material constant, ($a < 1$).

Solution

Let the upper two arms of a Wheatstone bridge have resistor values of R_M while the lower two arms have the resistance values of R_N at nominal measurand value of x (temperature, displacement, pressure, etc.). If the resistance sensor is described as $R_S = R_N(1+ax)$, (Fig. 1.51)

$$V_{ab} = U\left[\frac{R_N}{R_N+R_M} - \frac{R_S}{R_S+R_M}\right] = U\left[\frac{R_N}{R_N+R_M} - \frac{R_N(1+ax)}{R_N(1+ax)+R_M}\right], \quad \text{let}\, k = \frac{R_M}{R_N}$$

$$V_{ab} = U\left[\frac{1}{1+k} - \frac{1(1+ax)}{(1+ax)+k}\right] = U\left[\frac{1+ax+k-(1+ax)(1+k)}{(1+k)[(1+ax)+k]}\right]$$

$$= U\frac{(1+ax)[1-(1+k)]+k}{(1+k)[1+ax+k]} = U\frac{(1+ax)(-k)+k}{(1+k)[1+ax+k]} = U\frac{-akx}{(1+k)(1+ax+k)}$$

If it is assumed that $V_b > V_a$, then $V_{ba} = -V_{ab}$,

$$V_{ba} = U\frac{akx}{(1+k)(1+ax+k)} = U\frac{akx}{(1+k)(1+k)(1+\frac{ax}{1+k})}$$

$$V_{ba} = U\left[\frac{k}{(1+k)^2}\right] \times \left[\frac{ax}{1+\frac{ax}{1+k}}\right]$$

This equation shows that Wheatstone bridge output voltage is a nonlinear function of measurand x of the form

$$V_{ba} = \frac{\propto x}{1+\beta x}, \quad \alpha = U\left[\frac{ka}{(1+k)^2}\right], \quad \beta = \frac{a}{1+k}$$

Problem 1.3.28 The value of maximum resistance of a Resistive Temperature Detector (RTD) is expected to be 150 Ω. Select the values of components in a Wheatstone bridge configuration.

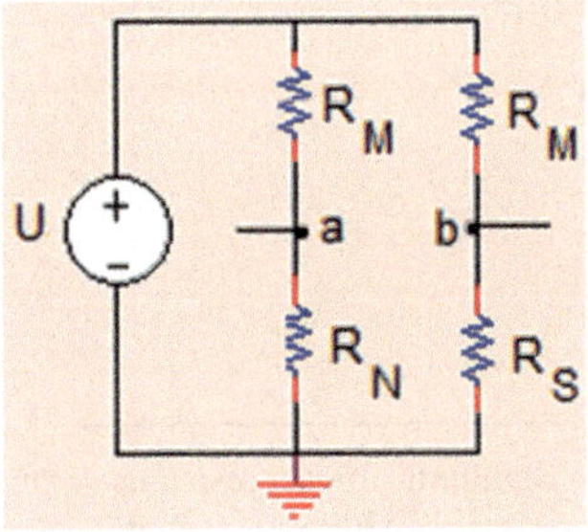

Fig. 1.51 Wheatstone bridge for a single resistive RTD sensor

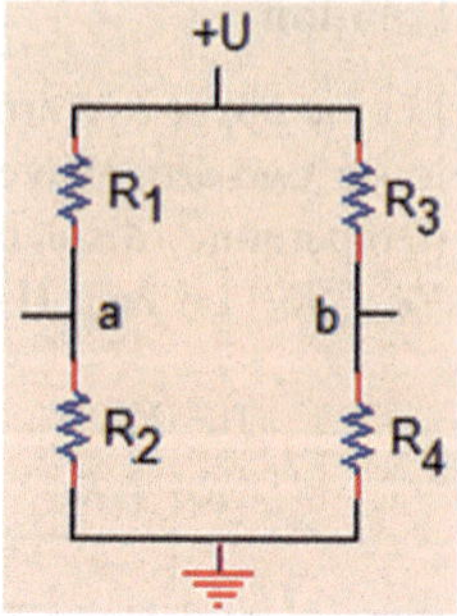

Fig. 1.52 Circuit for Problem 1.3.28

Solution

Assume that R_4 represents RTD in Fig. 1.52,

$$R_1 \cdot R_4 = R_2 \cdot R_3,$$

Let $R_1 = R_3$,

$$R \cdot R_4 = R_2 \cdot R \quad \rightarrow \quad R_4 = R_2$$

Select R_2 so that it covers 150 Ω. A standard variable resistor of 220 Ω is possible with a spindle-operated potentiometer.[1]

Problem 1.3.29

(a) In a Wheatstone Bridge circuit shown in Fig. 1.53, *Pt*-100 type RTD element is used for temperature measurement. $R_1 = R_3 = 100\ \Omega$. R_2 is a variable resistor, $R_{max} = 220\ \Omega$, with a linear taper. At some temperature, RTD resistance is measured to be 120 Ω. What fraction of the full slider range of the adjustable resistor travel for this (balanced) condition?

(b) What will be this fraction at 0 °C?

[1]Standard linear resistance range for spindle operated potentiometers is 100–2.2 MΩ, with resistance values 1, 2.2, 4.7 in each decade; See, for example, Tyco electronics, variable resistors literature No: 1773442-9, 2006.

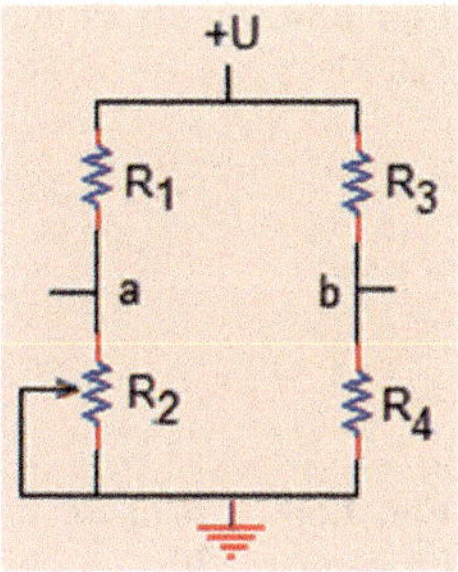

Fig. 1.53 Wheatstone Bridge for Problem 1.3.29

Solution

(a) In balanced case,

$$R_1 \cdot R_4 = R_2 \cdot R_3 \quad \rightarrow \quad R_4 = \frac{R_2 \times R_3}{R_1} = \frac{100 \times R_2}{100} = R_2 = 120\,\Omega \quad \rightarrow \quad R_2 = 120\,\Omega$$

$$\text{This is } \frac{120}{220} = 0.545 \quad \rightarrow \quad 54.5\%$$

in reference to ground potential.

(b) At 0 °C, RTD resistance will be 100 Ω. Then,

$$R_4 = \frac{100}{100} \cdot R_2 = 100\,\Omega \quad \rightarrow \quad R_2 = 100\,\Omega.\ \text{This is } \frac{100}{220} = 0.455 \quad \rightarrow \quad 45.5\%$$

in reference to ground potential.

Problem 1.3.30 A pressure transducer used in a patient lung ventilator device is made by using two strain gauges (SGs) glued on each side of a flexible thin plate, see, Fig. 1.54.

When a pressure P is exerted on the plate, it bends. This results in a strain value of k,

$$k = \frac{\Delta l_1}{l_1} = -\frac{\Delta l_2}{l_2}$$

Gauge factor for both SGs are defined as

$$G = G_1 = G_2 = \frac{\frac{\Delta R}{R_0}}{\frac{\Delta l}{l}} = \frac{\Delta R}{\Delta l}\frac{l}{R_0}$$

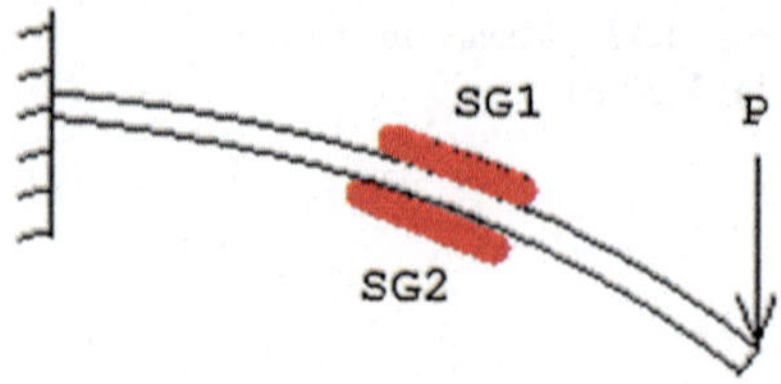

Fig. 1.54 Figure Problem 1.3.30

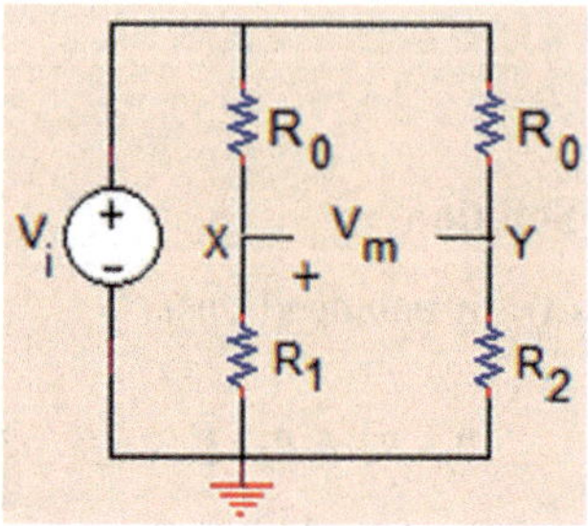

Fig. 1.55 Circuit for Problem 1.3.30

R_0 unstrained value of SGs,
R_1 strained resistance of SG_1,
R_2 strained resistance of SG_2,
l length of SGs
ΔR resistance change of SGs
Δl length change of SGs

The circuit of Fig. 1.55 is used to measure the voltage V_m.

(a) Determine V_m.
(b) For $G = 2$, $k = 0.25$, $V_i = 10$ V, $V_m = ?$

Solution

(a) Using the definition of gauge factor,

$$G = \frac{\frac{\Delta R}{R_0}}{\frac{\Delta l}{l}} = \frac{\Delta R}{\Delta l}\frac{l}{R_0}$$

Change in SG resistance,

$$\Delta R = R_0 G \frac{\Delta l}{l} = R_0 G k$$

SG resistances,

$$R_1 = R_0 + \Delta R = R_0 + R_0 Gk = R_0(1 + Gk)$$
$$R_2 = R_0 - \Delta R = R_0 - R_0 Gk = R_0(1 - Gk)$$

Bridge voltage,

$$V_m = V_X - V_Y = \frac{R_1}{R_0 + R_1} V_i - \frac{R_2}{R_0 + R_2} V_i = V_i \left(\frac{R_1}{R_0 + R_1} - \frac{R_2}{R_0 + R_2} \right)$$
$$V_m = V_i \left(\frac{R_0(1 + Gk)}{R_0(1 + Gk) + R_0} - \frac{R_0(1 - Gk)}{R_0(1 - Gk) + R_0} \right) = V_i \left(\frac{1 + Gk}{2 + Gk} - \frac{1 - Gk}{2 - Gk} \right)$$
$$V_m = V_i \left[\frac{(2 - Gk)(1 + Gk) - (2 + Gk)(1 - Gk)}{(2 + Gk)(2 - Gk)} \right]$$
$$= V_i \left[\frac{2 + 2Gk - Gk - (Gk)^2 - (2 - 2Gk + Gk - (Gk)^2}{(2 + Gk)(2 - Gk)} \right]$$

$$V_m = V_i \left(\frac{2Gk}{4 - (Gk)^2} \right) \tag{1.57}$$

(b) Using given data in Eq. (1.57),

$$V_{\mathrm{m}} = V_i \left(\frac{2 \times 2 \times 0.25}{4 - (2 \times 0.25)^2} \right) = \frac{10}{4 - (0.5)^2} = 2.667\,\mathrm{V}$$

Problem 1.3.31

(a) Four sensors each with resting resistance R are placed in a Wheatstone bride circuit so that two of them increase their resistance while the others decrease their resistance by the same amount ($= \Delta R$), see, Fig. 1.56. Determine the output voltage to supply voltage ratio in terms of sensor (resting) resistance R and ΔR.
(b) Give a practical example of this type of sensor circuit application topology in the area of measuring magnetic fields employing magnetoresistivity. Indicate referenced literature.

Solution

(a) $$V_{\mathrm{ab}} = U \left(\frac{R - \Delta R}{R - \Delta R + R + \Delta R} - \frac{R + \Delta R}{R + \Delta R + R - \Delta R} \right)$$

$$\frac{V_{\mathrm{ab}}}{U} = \frac{R - \Delta R}{2R} - \frac{R + \Delta R}{2R} = \frac{R - R - 2\Delta R}{2R} = -\frac{2\Delta R}{2R} = \frac{\Delta R}{R}$$

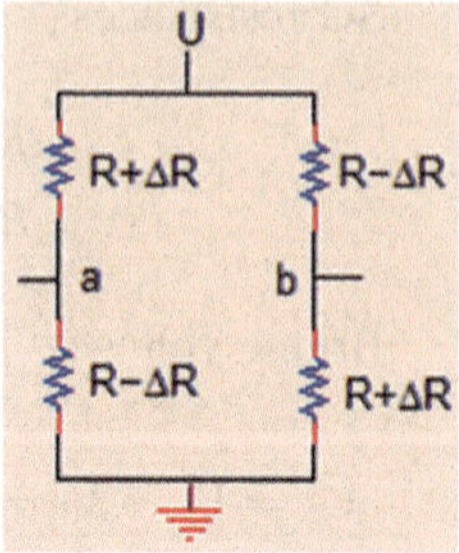

Fig. 1.56 Wheatstone bridge for Problem 1.3.31

(b) A current carrying magnetic material changes its resistivity in the presence of an external magnetic field. Various types of magnetoresistive sensors are reported in the literature. Anisotropic (AMR), spin valve (SV), semiconductor (InSb), multilayer giant (GMR), colossal (CMR) sensors are different types of magnetoresistive sensor structures. Among these, semiconductor types have strong temperature dependance and high nonlinearity.
Permalloy (19% Fe, 81% Ni) is a ferromagnetic material which shows relatively strong magnetoresistive effect [15].
The thin permalloy film strip resistance R depends on the angle α (the angle between the direction of current and the magnetic field which is parallel to the strip plane and perpendicular to the preferred direction), (Fig. 1.57)

$$R = R_0 + \Delta R \times \cos^2(\alpha) = R_0 + \Delta R\left[1 - \left(\frac{H_y}{H_o}\right)^2\right]$$

$$H_o = \text{constant for material}$$

$$\alpha = 0^\circ, \quad R_{max} = R$$

$$\alpha = 90^\circ, \quad R_{min} = R$$

$$\frac{\Delta R}{R} \cong 0.03 \text{ for permalloy}$$

By depositing aluminum stripes (barber poles) on top of the permalloy strip at 45° angle to the strip axis (see Fig. 1.58), current direction is rotated by 45°. This changes the angle between the magnetization vector and electrical current from α to $(\alpha - 45^\circ)$. This increases the sensitivity for weak magnetic fields and allows to detect the sign of (H_y).

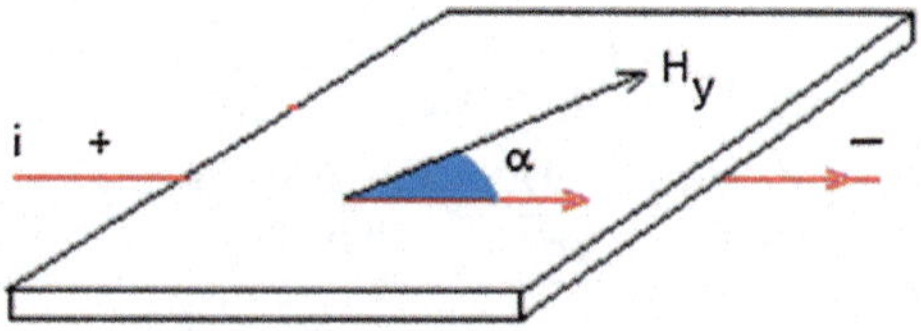

Fig. 1.57 Measuring magnetic fields using permalloy film strip

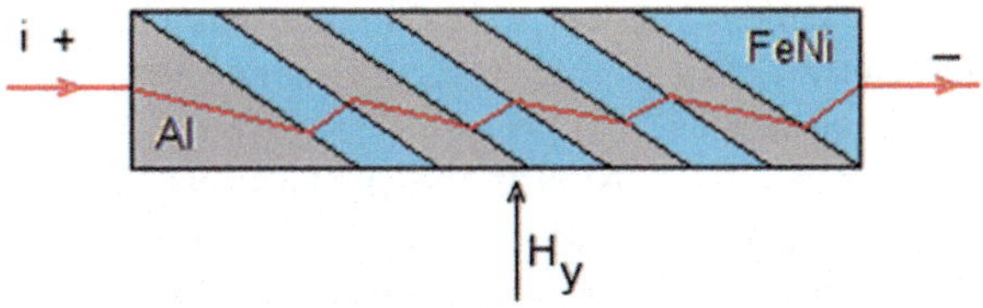

Fig. 1.58 Measuring magnetic fields employing magnetoresistivity

In Wheatstone bridge configuration, one diagonal pair is oriented 45° to the strip axis while the other pair is oriented at -45°. Thus, ΔR due to magnetic field is converted linearly into variation of the differential output voltage. Moreover, the inherent temperature coefficients of the four bridge resistances are mutually compensated [16].

Problem 1.3.32

(a) Four sensors each with resting resistance $R = 2\ \text{k}\Omega$ are placed in a Wheatstone bride circuit so that two of them increase their resistance while the others decrease their resistance by the same amount ($= \Delta R = 80\,\Omega$), see, Fig. 1.56. Calculate the output voltage if supply voltage is $U = 5$ V.

(b) Give a practical example of this type of sensor circuit application topology in the area of measuring respiratory pressure by employing piezoresistive effect. Compare them with strain gauges. Indicate the referenced literature.

Solution

(a) $V_{\text{ab}} = U\left(\dfrac{R-\Delta R}{R-\Delta R+R+\Delta R} - \dfrac{R+\Delta R}{R+\Delta R+R-\Delta R}\right) = \frac{\Delta R}{R}U = \frac{80}{2000} \times 5 = 200\,\text{mV}$

(b) Pressure Sensors based on piezoresistivity:

Pressure sensors convert the physical quantity "pressure" into an electrical signal. Measuring cell consists of a chip with an etched silicon diaphragm in which resistance paths are formed and a carrier chip is also formed on silicon base. The deflection of the diaphragm leads to changes in the implanted resistances when pressure is applied, The thickness of the diaphragm, its surface area and the geometric design of the resistors determine the permissible pressure range.

The Piezoresistive Effect

Metallic strain gauges change their resistance due to changes in their geometry. On the other hand, in piezoresistivity, the primary effect is the change in conductivity, ρ, dependent on the mechanical stresses within the crystal [17, 18]. This dependence is defined by the proportionality constant,

$$\frac{\delta\rho}{\rho} = \pi \cdot \sigma$$

Mechanical tension σ within the semiconductor crystal is used to produce resistance changes in resistors connected as a bridge circuit (π is the piezo-constant).

Compressive tension in the crystal causes an increase in the number of charge carriers in the direction of the compressive force. This increase in conductivity decreases its resistance. Conversely, dilatory tension causes an increase in resistance. The resistors are located on a flexible diaphragm to correspond with the maximum compressive and dilatory tensions. With reference to the relative directions of the components of mechanical tension σ, the electric field, and the current density, the following effects can be distinguished; longitudinal effect, transverse effect, shear tension effect.

While the gauge factor is the proportionality between the relative change in constant resistance and the relative change in length for a strain gauge sensor,

$$\frac{\delta R}{R} = K \cdot \frac{\delta L}{L}$$

the gauge factor for a piezoresistive element is $K = \pi E$.

Here, π is the piezo-constant and E the modulus of elasticity. The K-factor is approximately 2 for metals and up to 200 for semiconductors.

The main advantages of a pressure sensor using semiconductor technology as compared to conventional resistance strain gauges are greater sensitivity, high linearity, and fast response.

One disadvantage is the dependence on temperature, but it can be compensated by suitable circuitry.

Bridge Circuit

To achieve maximum measuring accuracy, four piezoresistors are connected to form a bridge circuit. The position of the individual resistors is chosen such that on deflection of the diaphragm, two oppositely situated resistors increase in resistance, while the other two decrease.

The equation for the output voltage can then be given as

$$V_{\text{out}} = V_{\text{in}} \cdot \frac{\Delta R}{R} = V_{\text{in}} \cdot K \cdot \varepsilon(p)$$

Here, $\varepsilon(p)$ is the mechanical deflection due to pressure that is linear for small diaphragm deflections [18].

Problem 1.3.33 If a Wheatstone bridge with $R_2 = R_3 = R + \Delta R$, and $R_1 = R_4 = R$ is driven by a constant current source, $I(A)$, (see, Fig. 1.59) determine the expression for the output voltage, V_{ab}.

What is the advantage of using such a circuit?

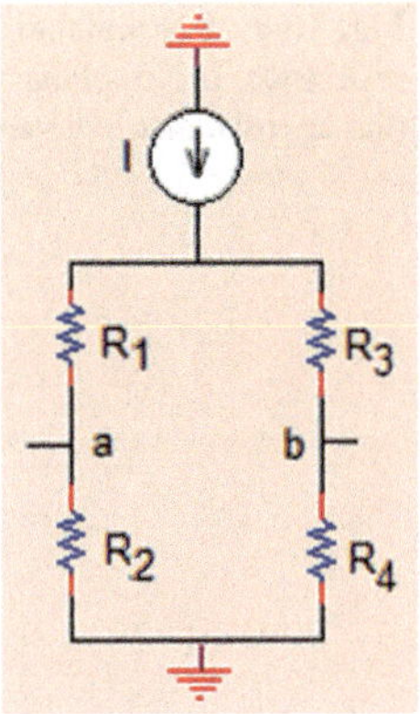

Fig. 1.59 Wheatstone bridge driven by a constant current source, for Problem 1.3.33

Solution

Since the resistances of both arms are equal, current I divides into two equal halves,

$$V_{ab} = V_a - V_b = \frac{I}{2} \cdot (R + \Delta R) - \frac{I}{2} R = \frac{I}{2} \cdot (R - R + \Delta R) = \frac{I}{2} \cdot \Delta R$$

Apparent zero linearity error is the advantage of using this circuit.

Problem 1.3.34 Let the resistance of each edge of a wire cube be R, (see Fig. 1.60a). Find the resistance between the vertices on the space diagonal of a wire cube with equal resistance values (i.e., $R_{ah} = R_{cf} = R_{ed} = R_{bg}$). Calculate this resistance if $R = 6$ kΩ.

Solution

Consider the resistance between vertices *a* and *h*. The edges ac, ab, and ae are equivalent; they carry the same current. Due to symmetry, the vertices *a*, *c*, and *e* have the same potential. The same is true for vertices *f*, *d*, and *h*.

Since the whole cube resistance does not change by merging these vertices into one, merging these vertices *a*, *c* and *e*, (then *f*, *d* and *h*) into one junction yields a planar equivalent circuit shown below.

$$R_{abce} = R_{dfgh} = \frac{R}{3}, \quad R_{bcedfg} = \frac{R}{6}, \quad R_{ah} = 2R_{abce} + R_{bcedfg} = \frac{2R}{3} + \frac{R}{6} = \frac{5}{6}R$$

For $R = 5\,\mathrm{k\Omega}$,

$$R_{ah} = \frac{5}{6} \times 6 = 5\,\mathrm{k\Omega}$$

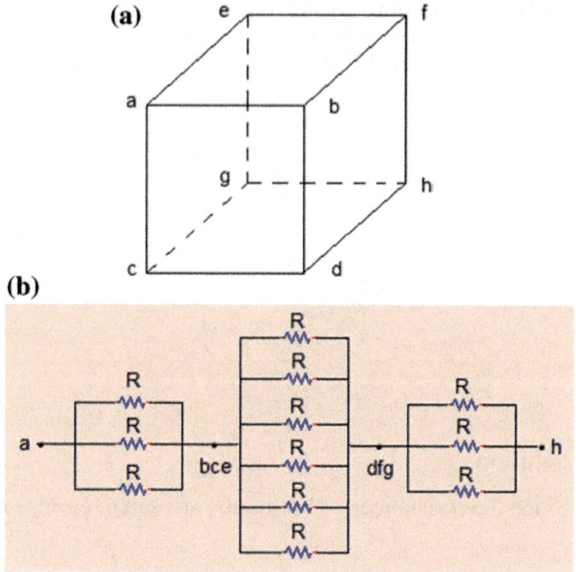

Fig. 1.60 **a** Nonplanar resistance. **b** Nonplanar to planar resistance conversion

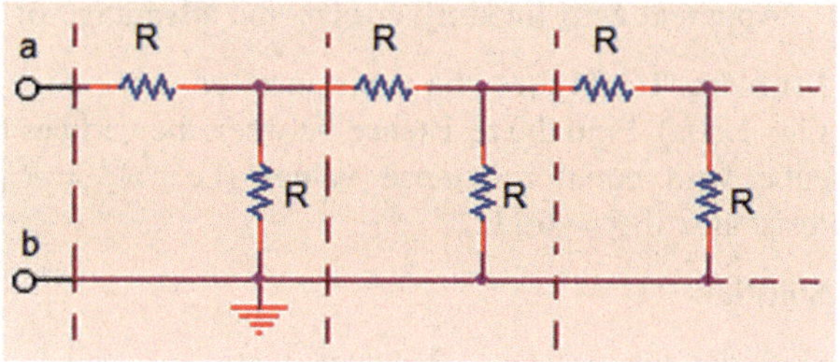

Fig. 1.61 Infinite ladder circuit

Problem 1.3.35 Compute the input resistance (equivalent resistance) of an infinite ladder circuit of the form shown in Fig. 1.61, and then find the value of R if the equivalent circuit resistance is 50 Ω.

Solution

Let input resistance = equivalent resistance = R_{in}

Adding one more section in front of it does not change the resistance of the overall circuit.

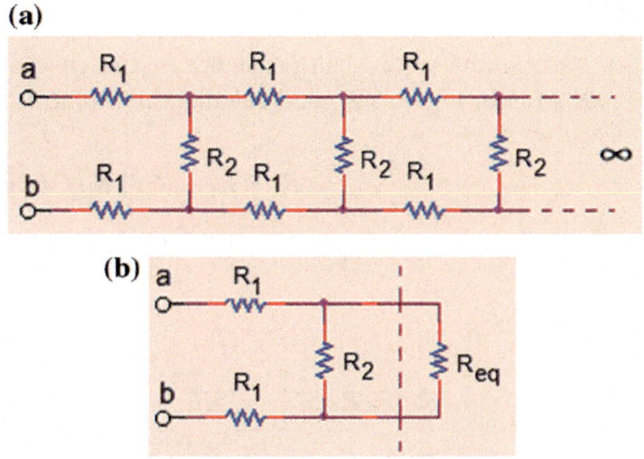

Fig. 1.62 **a**, **b** Another form of infinite ladder circuit

$$R_{\text{in}} = R + R \parallel R_{\text{in}} = R + \frac{R \cdot R_{\text{in}}}{R + R_{\text{in}}}$$
$$R \cdot R_{\text{in}} + R_{\text{in}}^2 = R + R \cdot R_{\text{in}}$$
$$R_{\text{in}}^2 - R \cdot R_{\text{in}} - R^2 = 0$$
$$x^2 - Rx - R^2 = 0$$
$$R_{\text{in}} = -\frac{-R}{2} \mp \frac{\sqrt{(-R)^2 - 4(-R^2)}}{2} = \frac{R}{2} \mp \frac{\sqrt{5R^2}}{2} = R\left(\frac{1}{2} \mp \frac{\sqrt{5}}{2}\right)$$
$$R_{\text{in}} = R \cdot \left(\frac{1 \mp \sqrt{5}}{2}\right), \text{only} \quad R_{\text{in}} > 0 \,\text{is meaningful},$$
$$R_{\text{in}} = 1.618\,R$$

If input resistance is 50 Ω, $50 = 1.618\,R, \quad R = 30.9\,\Omega$

Problem 1.3.36

(a) Find the equivalent (input) resistance of the infinite ladder circuit shown in Fig. 1.62a.
(b) Find the value of equivalent resistance if $R_2 = 10\ \Omega$, $R_1 = 20\ \Omega$.
(c) If the equivalent resistance is 50 Ω, $R_1 = 2R_2$, find the values of resistors.

Solution

(a) Assume that equivalent resistance is known. Add one more section in front of R_{eq} (as in Fig. 1.62b) and calculate the final R_{eq}:

$$R_{eq} = R_1 + R_1 + R_2 \parallel R_{eq} = 2R_1 + \frac{R_1 \cdot R_2}{R_1 + R_2}$$
$$R_{eq} \cdot R_2 + R_{eq}^2 = 2R_1R_2 + 2R_1R_{eq} + R_2R_{eq}$$

simplify,

$$R_{eq} = R_1 \pm \frac{1}{2} \cdot \sqrt{4R_1^2 + 8R_1R_2} = R_1 \pm \sqrt{R_1^2 + 2R_1R_2}$$
$$R_{eq} = R_1 \pm \sqrt{R_1(R_1 + 2R_2)}$$

Neglecting the negative result;

$$R_{eq} = R_1 + \sqrt{R_1(R_1 + 2R_2)}$$

(b) $R_2 = 10\,\Omega,\ R_1 = 12.5\,\Omega$

$$R_{eq} = 20 + \sqrt{(20)^2 + 2 \cdot 20 \cdot 12.5} = 20 + \sqrt{400 + 500}$$
$$= 20 + \sqrt{900} = 20 + 30 = 50\,\Omega$$

(c) $50 = 2R_2 + \sqrt{(2R_2)^2 + 2 \cdot 2R_2 \cdot R_2}$

$$= 2R_2 + \sqrt{4R_2^2 + 4R_2^2} = 2R_2 + \sqrt{8R_2^2} = 2R_2 + R_2 \cdot \sqrt{8} = R_2\left(2 + \sqrt{8}\right)$$

$$R_2 = \frac{50}{2 + \sqrt{8}} = 10.355\,\Omega, \quad R_1 = 2R_2 = 20.710\,\Omega$$

Problem 1.3.37 Compute the equivalent resistance (input resistance) of the infinite ladder circuit of the form in Fig. 1.63, and then find the value of R if the equivalent circuit resistance is 50 Ω.

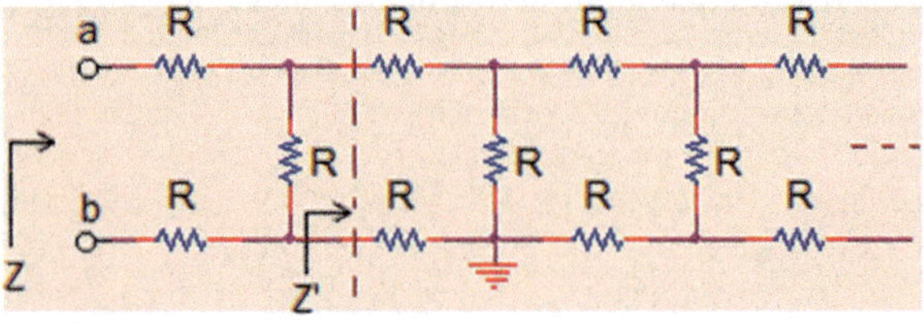

Fig. 1.63 The infinite ladder circuit of Problem 1.3.37

Solution

Let the input resistance be Z'. Adding one more section in front of it does not change the equivalent resistance of the overall circuit,

$$Z = Z' \text{ then,}$$

$$Z = 2R + Z \parallel R = 2R + \frac{ZR}{Z+R}$$

$$ZR + Z^2 = 2R^2 + 2ZR + 2Z \quad \rightarrow \quad Z^2 - 2ZR - 2R^2 = 0$$

$$Z_{1,2} = \frac{2R}{2} \mp \frac{\sqrt{(-2R)^2 - 4(-2R^2)}}{2} = R \mp \frac{\sqrt{12}}{2} R = R \cdot \left(1 + \frac{\sqrt{12}}{2}\right)$$

Only positive values are physically meaningful, $Z = 2.732R$ is the general result.

If the input resistance is 50 Ω then,

$$50 = 2.732R, \quad R = 50/2.732 = 18.3\,\Omega$$

Problem 1.3.38

(a) Compute the input resistance (equivalent resistance) of a ladder circuit which consists of m = 5 "L" sections, as shown in Fig. 1.61.
(b) If R = 100 Ω, R_{ab} = ? ($R_{ab} = R_{eq}$).
(c) Plot m versus $R(m)$, then find $R(\infty)$ = ?

Solution

(a)
$$R_{eq}(1) = R + R = 2R \tag{1.58}$$

$$R_{eq}(2) = R + R \parallel R_{eq}(1) = R + R \parallel 2R = R + \frac{R \cdot (2R)}{3R} = R + \frac{2R}{3}$$
$$= R\left(1 + \frac{2}{3}\right) = \frac{5}{3}R \tag{1.59}$$

$$R_{eq}(3) = R + R \parallel R_{eq}(2) = R + R \parallel \frac{5R}{3} = R + \frac{R \cdot \frac{5R}{3}}{R + \frac{5R}{3}} = R + \frac{5R^2}{8R}$$
$$= R\left(1 + \frac{5}{8}\right) = \frac{13}{8}R \tag{1.60}$$

Table 1.2 Section count versus multiplier

$R(m)$	Multiplier	Multiplier (decimal)
s	$1+\frac{1}{1}$	2.000000
$R(2)$	$1+\frac{2}{3}$	1.666667
$R(3)$	$1+\frac{5}{8}$	1.625000
$R(4)$	$1+\frac{13}{21}$	1.619047
$R(5)$	$1+\frac{34}{55}$	1.618181
$R(6)$	$1+\frac{89}{144}$	1.618055
$R(n)$	$1+\frac{x(n)}{y(n)}$	

Table 1.3 Section count versus resistance

m	$R(\Omega)$
6	1.618056
7	1.618037
8	1.618034
9	1.618034
10	1.618034

$$R_{\mathrm{eq}}(4) = R + R \parallel R_{\mathrm{eq}}(3) = R + \frac{R \cdot \dfrac{13R}{8}}{R + \dfrac{13R}{8}} = R + \frac{13R^2}{21R} = \left(1 + \frac{13}{21}\right)R = \frac{34}{21}R \tag{1.61}$$

$$R_{\mathrm{eq}}(5) = R + R \parallel R_{\mathrm{eq}}(4) = R + \frac{R \cdot \dfrac{34R}{21}}{R + \dfrac{34R}{21}} = R + \frac{34R^2}{55R} = \left(1 + \frac{34}{55}\right)R = \frac{89}{55}R \tag{1.62}$$

(b) $R_{\mathrm{ab}}(5) = R_{\mathrm{eq}} = \dfrac{89}{55}.100\,\Omega = 1618\,\Omega$

(c) Recursive formulation: (Table 1.2)

$$\begin{aligned} x(1) = y(1) = 1,\ x(m) &= x(m-1) + y(m-1), y(m) \\ &= x(m-1) + 2 \cdot y(m-1). \end{aligned}$$

The resulting $R(m)$ values for $m \geq 6$ are given in Table 1.3 as obtained using EXCEL or MATLAB.

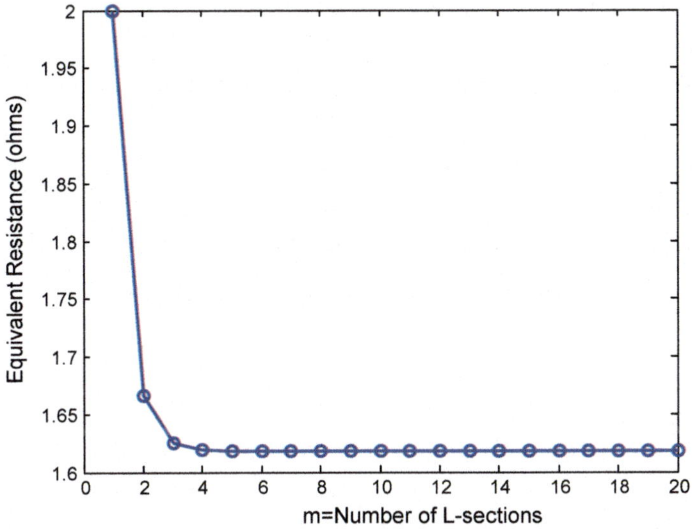

Fig. 1.64 The graph of input resistance (equivalent resistance) of a ladder circuit as a function of "L" sections

Note that For $m \geq 8$ one may approximate $R(\infty) = 1.618034$. A plot of first $m = 10$ values versus $R(m)$ is shown in Fig. 1.64.

References

1. Wang J (2000) Analytical electrochemistry, 2nd edn. Wiley
2. Compton RG and Banks CE (2011) Understanding voltammetry. World Scientific
3. Enderle J, Blanchard S, Bronzino J (2005) Introduction to biomedical engineering, 2nd edn. Elsevier Academic Press, pp 527–530
4. Janata J (2010) Principles of chemical sensors. Springer Science & Business Media (Chapter 7)
5. Blum LJ, Coulet PR (eds) (1991) Biosensor Principles and applications. CRC Press (Chapter 2)
6. American Public Health Association, American Water Works Association, Water Environment Federation (1999) Standard methods for the examination of water and wastewater, 20th edn
7. Synkera Technologies Inc., P/N 714, Longmont, CO, USA
8. Sevostianov I et al (2006) Elastic and electric properties of closed-cell aluminum foams. Mater Sci Eng A 420:87–99
9. Zhou D, Arbabi S, Stenby EH (1997) A percolation study of wettability effect on the electrical properties of reservoir rocks. Transp Porous Media 29:85–98
10. Mori T, Tanaka K (1973) Average stress in the matrix and average elastic energy of materials with misfitting inclusions. Acta Metall 21:571–574
11. Zhu K, Li C-F, Zhu Z-G (2007) Measurement of electrical conductivity of porous titanium and Ti6Al4 V prepared by the powder metallurgy method. Chin Phys Lett 24:187–190

12. Hanai T (1968) Electrical properties of emulsions. Chapter 5 of: Sherman (Ed.), Emulsion Science, Academic Press
13. Lorenzo AD, Andreoli A, Matthie J, Withers P (1997) Predicting body cell mass with bioimpedance by using theoretical methods: a technological review. J Appl Phys 82: 1542–1558
14. www.beckmancoulter.com, Accessed 09 Dec 2016
15. Tumanski S (2001) Thin film magnetoresistive sensors. Institute of Physics Publishing, ISBN 0750307021
16. Stork T (2000) Application note, AN00022 Philips Semiconductor
17. Doll JC, Pruitt BL (2013) Piezoresistor design and applications. Springer Science & Business Media
18. Siemens Datenbuch (1995) Silizium-Temperatur-und Drucksensoren

Chapter 2
Analysis Methods

2.1 Nodal Analysis

Problem 2.1.1 Two current sources with equal internal resistances feed a load as shown in Fig. 2.1.

$$I_a = 200\,\text{A}, \quad I_b = 100\,\text{A}, \quad R = 200\,\Omega, \quad R_L = 100\,\Omega.$$

(a) Find the current through the load resistor R_L.
(b) Find the node voltage value.

Solution

(a) Parallel-connected current sources, KCL applies, $200\,\text{A}||(-100\,\text{A}) \rightarrow 100\,\text{A}$
By current division,

$$I_{RL} = 100 \times \frac{200||200}{100 + 200||200} = 100 \times \frac{100}{100 + 100} = 50\,\text{A}.$$

(b) Applying Ohm's law,

$$V_x = 100\,\Omega \times 50\,\text{A} = 5000\,\text{V}.$$

Problem 2.1.2 Find the values of currents and voltages in the circuit shown in Fig. 2.2, for $R = 2\ \Omega$.

A.Ü. Keskin, *Electrical Circuits in Biomedical Engineering*,
DOI 10.1007/978-3-319-55101-2_2

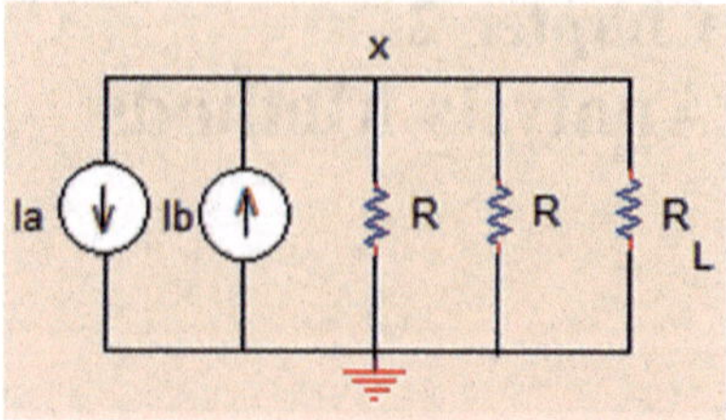

Fig. 2.1 The circuit for Problem 2.1.1

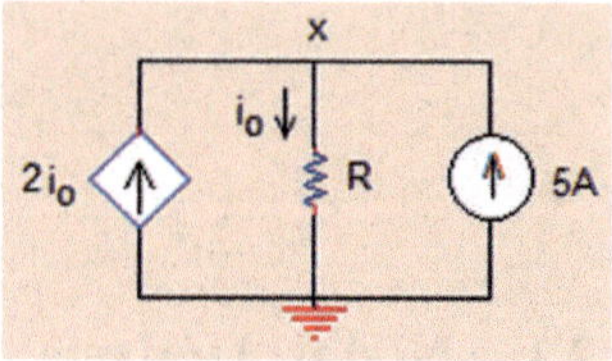

Fig. 2.2 The circuit for Problem 2.1.2

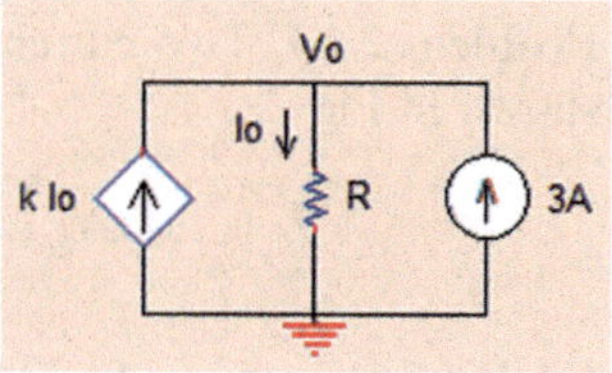

Fig. 2.3 The circuit for Problem 2.1.3

Solution

By Kirchhoff's current law, $2i_0 + 5 - i_0 = 0 \to i_0 = -5\,\text{A}$

$$v_x = i_0 R = -5 \times 2 = -10\,\text{V}.$$

Problem 2.1.3

(a) Find the value of V_0 in the circuit shown in Fig. 2.3.
(b) If the gain constant of dependent source is k, what are the limiting values of k, if I_0 has always a positive value? Resistor value is 4 Ω.

Solution

(a) KCL: $3 + 0.2I_0 - I_0 = 0 \to I_0 = 3.75\,\text{A}$

$$V_0 = 4I_0 = 4 \times 3.75 = 15\,\text{V}$$

(b) $3 + kI_0 - I_0 = 0 \to I_0 = -3/(k-1)\,\text{A},\ \ 0 \le k < 1.$

Problem 2.1.4 In the circuit shown in Fig. 2.4, the coefficient of current-controlled current source is 2 A/A. If the node voltage is 1 V, find the value of the voltage source, the current through $R_1 = 1\ \Omega$, and current through $R_2 = 2\ \Omega$.

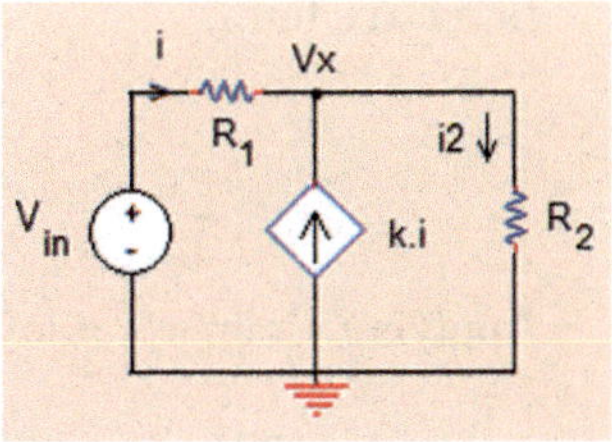

Fig. 2.4 The circuit for Problem 2.1.4

Solution
Applying KCL at the node, and using given component values,

$$\frac{V_{in} - V_x}{R_1} + ki - \frac{V_x}{R_2} = 0$$
$$V_{in} - V_x + 2(V_{in} - V_x) - \frac{V_x}{2} = 0 \rightarrow V_{in} - 1 + 2(V_{in} - 1) - \frac{1}{2} = 0$$
$$V_{in} = 1.167\,\text{V}$$
$$i_2 = \frac{V_x}{2} = \frac{1}{2} = 0.5\,\text{A}$$
$$i = \frac{V_{in} - V_x}{1} = V_{in} - V_x = 1.167 - 0.5 = 0.667\,\text{A}$$

Problem 2.1.5 Determine the node voltages in the circuit shown in Fig. 2.5. Use Cramer's rule, if necessary.

(a) For $I_1 = 1\,\text{A}, I_2 = 2\,\text{A}, \quad R_1 = 1/2\,\Omega, \quad R_2 = 1/8\,\Omega, \quad R_3 = 1/4\,\Omega$
(b) For $I_1 = 2\,\text{A}, I_2 = 4\,\text{A}, R_1 = 5\,\Omega, R_2 = 2\,\Omega, R_3 = 10\,\Omega$.

Solution

(a) For $I_1 = 1\,\text{A}, I_2 = 2\,\text{A}, R_1 = 1/2\,\Omega, R_2 = 1/8\,\Omega, R_3 = 1/4\,\Omega$

Node equations are

$$1 - 2V_1 - 4(V_1 - V_2) = 0$$
$$2 - 8V_2 + 4(V_1 - V_2) = 0$$

or

$$6V_1 - 4V_2 = 1$$
$$-4V_1 + 12V_2 = 2^{\cdot}$$

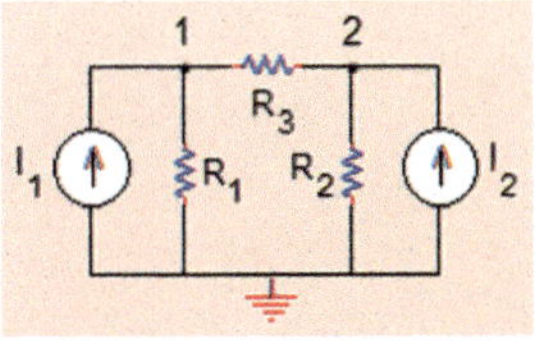

Fig. 2.5 The circuit for Problem 2.1.5

In matrix form,

$$\begin{bmatrix} 6 & -4 \\ -4 & 12 \end{bmatrix}\begin{bmatrix} V_1 \\ V_2 \end{bmatrix} = \begin{bmatrix} 1 \\ 2 \end{bmatrix}.$$

Applying Cramer's rule to solve this matrix equation yields

$$\Delta = 72 - 16 = 56, \quad \Delta_1 = 12 + 8 = 20, \quad \Delta_2 = 12 + 4 = 16$$
$$V_1 = \frac{\Delta_1}{\Delta} = \frac{5}{14} = 0.357\,\text{V}, \quad V_2 = \frac{\Delta_2}{\Delta} = \frac{4}{14} = 0.286\,\text{V}$$

(b) For $I_1 = 2\,\text{A}, I_2 = 4\,\text{A}, R_1 = 5\,\Omega, \quad R_2 = 2\,\Omega, \quad R_3 = 10\,\Omega.$

Nodal equations in matrix form can be formed using "analysis by inspection";

$$\begin{bmatrix} \frac{1}{5} + \frac{1}{7} & -\frac{1}{10} \\ -\frac{1}{10} & \frac{1}{10} + \frac{1}{2} \end{bmatrix}\begin{bmatrix} V_1 \\ V_2 \end{bmatrix} = \begin{bmatrix} 2 \\ 4 \end{bmatrix} \rightarrow \begin{bmatrix} \frac{3}{10} & -\frac{1}{10} \\ -\frac{1}{10} & \frac{6}{10} \end{bmatrix}\begin{bmatrix} V_1 \\ V_2 \end{bmatrix} = \begin{bmatrix} 2 \\ 4 \end{bmatrix}.$$

Applying Cramer's rule to solve this matrix equation yields

$$\Delta = \frac{3}{10} \cdot \frac{6}{10} - \frac{1}{100} = \frac{18}{100} - \frac{1}{100} = \frac{17}{100},$$
$$\Delta_1 = 2 \cdot \frac{6}{10} + 4 \cdot \frac{1}{10} = \frac{12}{10} + \frac{4}{10} = \frac{16}{10}$$
$$\Delta_2 = \frac{3}{10} \cdot 4 + \frac{1}{10} \cdot 2 = \frac{12}{10} + \frac{2}{10} = \frac{14}{10}$$

$$V_1 = \frac{\Delta_1}{\Delta} = \frac{\frac{16}{10}}{\frac{17}{100}} = \frac{160}{17} = 9.411\,\text{V}, \quad V_2 = \frac{\Delta_2}{\Delta} = \frac{\left(\frac{14}{10}\right)}{\frac{17}{100}} = \frac{140}{17} = 8.235\,\text{V}.$$

Problem 2.1.6 Determine the value of current I in the circuit shown in Fig. 2.6. Use Cramer's rule, when necessary $R_1 = R_2 = R_3 = R_4 = 1/2\,\Omega$, $R_5 = 1/4\,\Omega$, $I_1 = 1\,\text{A}, I_2 = 2\,\text{A}$.

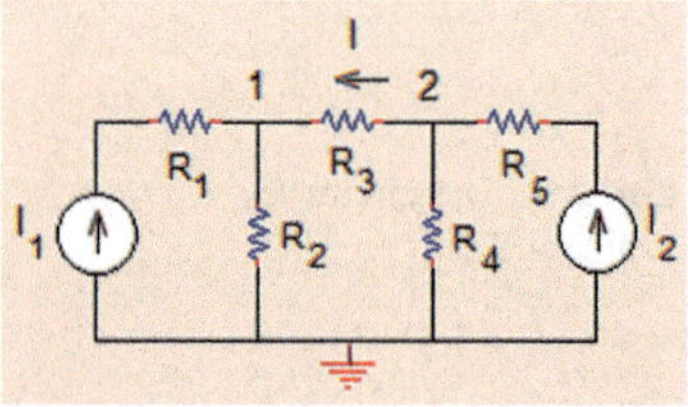

Fig. 2.6 The circuit of Problem 2.1.6

Solution

Nodal matrix equation of the circuit is obtained by applying "analysis by inspection" method,

$$\begin{bmatrix} G_1+G_2+G_3 & -G_3 \\ -G_3 & G_3+G_4+G_5 \end{bmatrix}\begin{bmatrix} V_1 \\ V_2 \end{bmatrix} = \begin{bmatrix} I_1 \\ I_2 \end{bmatrix} \rightarrow \begin{bmatrix} 2+2+2 & -2 \\ -2 & 2+2+4 \end{bmatrix}\begin{bmatrix} V_1 \\ V_2 \end{bmatrix} = \begin{bmatrix} 1 \\ 2 \end{bmatrix}$$

$$\begin{bmatrix} 6 & -2 \\ -2 & 8 \end{bmatrix}\begin{bmatrix} V_1 \\ V_2 \end{bmatrix} = \begin{bmatrix} 1 \\ 2 \end{bmatrix}.$$

Applying Cramer's rule to solve this matrix equation yields

$$\Delta = \begin{vmatrix} 6 & -2 \\ -2 & 8 \end{vmatrix} = 44, \quad \Delta_1 = \begin{vmatrix} 1 & -2 \\ 2 & 8 \end{vmatrix} = 12, \quad \Delta_2 = \begin{vmatrix} 6 & 1 \\ -2 & 2 \end{vmatrix} = 14$$

$$V_1 = \frac{\Delta_1}{\Delta} = \frac{12}{44} = 0.273\,\text{V}, \quad V_2 = \frac{\Delta_2}{\Delta} = \frac{14}{44} = 0.318\,\text{V}$$

$$I = \frac{V_2 - V_1}{R} = \frac{0.318 - 0.273}{\frac{1}{2}} = 0.09\,\text{A}$$

Problem 2.1.7 Find the values of voltages at the nodes of the circuit shown in Fig. 2.7.

$$G_1 = 0.5\,\text{S}, \quad G_2 = \frac{1}{4}\,\text{S}, \quad G_3 = 0.4\,\text{S}, \quad G_4 = \frac{1}{5}\,\text{S}, \quad G_5 = 1\,\text{S}, \quad I_1 = 5\,\text{A}, \quad I_2 = 4\,\text{A}.$$

Solution

$$G_A = G_1 + G_2 = 0.5 + 0.25 = 0.75\,\text{S}, \quad G_B = G_4 + G_5 = 0.2 + 1 = 1.2\,\text{S}$$

$$\begin{bmatrix} G_A+G_3 & -G_3 \\ -G_3 & G_B+G_3 \end{bmatrix}\begin{bmatrix} V_1 \\ V_2 \end{bmatrix} = \begin{bmatrix} I_1 \\ I_2 \end{bmatrix} = \begin{bmatrix} 5 \\ -4 \end{bmatrix} \rightarrow \begin{bmatrix} 1.15 & -0.4 \\ -0.4 & 1.6 \end{bmatrix}\begin{bmatrix} V_1 \\ V_2 \end{bmatrix} = \begin{bmatrix} 5 \\ -4 \end{bmatrix}.$$

Solution of this matrix equation for unknown voltages yields

$$V_1 = 3.8095\,\text{V}, \quad V_2 = -1.5476\,\text{V}.$$

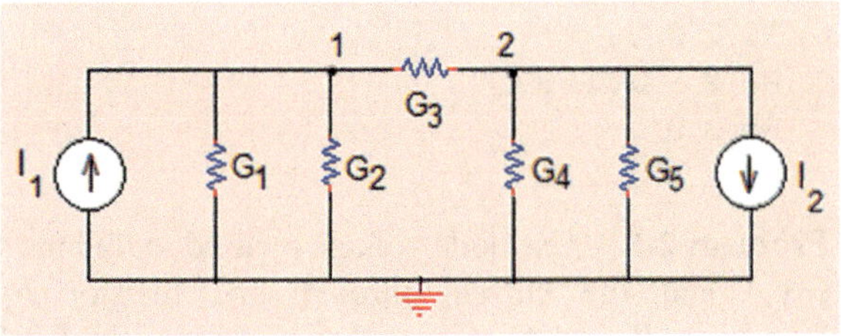

Fig. 2.7 The circuit of Problem 2.1.7

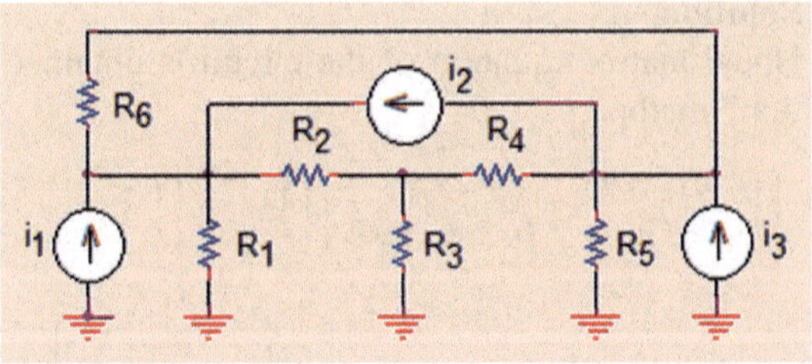

Fig. 2.8 The circuit of Problem 2.1.8

Problem 2.1.8 What is the voltage across resistor R_3 (in mV)? Use analysis by inspection and Cramer's rule if necessary ($i_1 = i_2 = i_3 = 1$ A, $R_1 = R_3 = R_5 = 1\ \Omega$, $R_2 = ½\ \Omega$, $R_4 = R_6 = 1/6\ \Omega$). Check the result using SPICE (Nodal 2.cir) (Fig. 2.8).

Solution

$$[G][V] = [I],$$

$$\begin{bmatrix} G_1+G_2+G_6 & -G_2 & -G_6 \\ -G_2 & G_2+G_3+G_4 & -G_4 \\ -G_6 & -G_4 & G_4+G_5+G_6 \end{bmatrix} \begin{bmatrix} V_1 \\ V_2 \\ V_3 \end{bmatrix} = \begin{bmatrix} 1+1 \\ 0 \\ 1-1 \end{bmatrix} = \begin{bmatrix} 2 \\ 0 \\ 0 \end{bmatrix}$$

$$\begin{bmatrix} 1+2+6 & -2 & -6 \\ -2 & 2+1+6 & -6 \\ -6 & -6 & 6+1+6 \end{bmatrix} \begin{bmatrix} V_1 \\ V_2 \\ V_3 \end{bmatrix} = \begin{bmatrix} 2 \\ 0 \\ 0 \end{bmatrix} \rightarrow \begin{bmatrix} 9 & -2 & -6 \\ -2 & 9 & -6 \\ -6 & -6 & 13 \end{bmatrix} \begin{bmatrix} V_1 \\ V_2 \\ V_3 \end{bmatrix} = \begin{bmatrix} 2 \\ 0 \\ 0 \end{bmatrix}.$$

$$\Delta = 1053 + (-72) + (-72) - [324 + 324 + 52] = 906 - 700 = 209$$

$$\Delta_2 = \begin{vmatrix} 9 & 2 & -6 \\ -2 & 0 & -6 \\ -6 & 0 & 13 \end{vmatrix} = (2)(-6)(-6) - [(13)(-2)(2)] = 72 - (-52) = 124$$

$$V_2 = \frac{\Delta_2}{\Delta} = \frac{124}{209} = 0.593\ \text{V} = 593\ \text{mV}$$

SPICE file:

```
*Operating point anaysis Nodal 2.cir
I1 0 1 1
I2 3 1 1
I3 0 3 1
R1 1 0 1
R2 1 2 0.5
R3 2 0 1
R4 2 3 0.1667
R5 3 0 1
R6 1 3 0.1667
```

Problem 2.1.9 Use node voltage method to find the values for the voltage at node C ($=V_c$) and the current through the resistor R_6 ($=i$). ($R_1 = R_2 = R_3 = 1\ \Omega$, $R_4 = R_5 = R_6 = 4\ \Omega$, $i_S = 2$ A) (Sim_Lin_Eq_Solve.m, matrix_solve.xlsx) (Fig. 2.9).

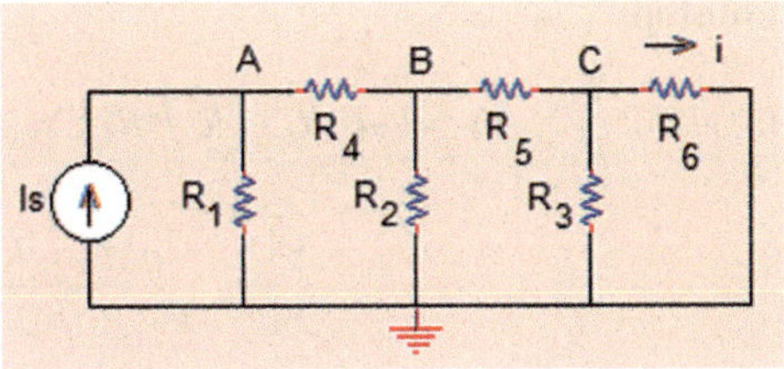

Fig. 2.9 The circuit of Problem 2.1.9

Solution

$$\begin{bmatrix} G1+G4 & -G4 & 0 \\ -G4 & G2+G4+G5 & -G5 \\ 0 & -G5 & G3+G5+G6 \end{bmatrix} \begin{bmatrix} V_A \\ V_B \\ V_C \end{bmatrix} = \begin{bmatrix} 15 \\ 0 \\ 0 \end{bmatrix}$$

$$\begin{bmatrix} 1+\frac{1}{4} & -\frac{1}{4} & 0 \\ -\frac{1}{4} & 1+\frac{1}{4}+\frac{1}{4} & -\frac{1}{4} \\ 0 & -\frac{1}{4} & 1+\frac{1}{4}+\frac{1}{4} \end{bmatrix} \begin{bmatrix} V_A \\ V_B \\ V_C \end{bmatrix} = \begin{bmatrix} 2 \\ 0 \\ 0 \end{bmatrix}$$

$$\begin{bmatrix} 1.25 & -0.25 & 0 \\ -0.25 & 1.50 & -0.25 \\ 0 & -0.25 & 1.50 \end{bmatrix} \begin{bmatrix} V_A \\ V_B \\ V_C \end{bmatrix} = \begin{bmatrix} 2 \\ 0 \\ 0 \end{bmatrix}.$$

Solution of this matrix equation by either manually using Cramer's rule or by employing available software (see, MATLAB m file or EXCEL file) yields

$$V_A = 1.657\,\text{V}, \quad V_B = 0.284\,\text{V}, V_C = 0.04734\,\text{V},$$

$$i = \frac{V_C}{R_6} = \frac{0.04734}{4} = 0.011834\,\text{A}.$$

Problem 2.1.10

(a) In the circuit shown in Fig. 2.10, find the voltage gain, i.e., $V_2/V_1 = ?$

(b) If $R_1 = 2.5$ kΩ, $R_o = 10$ kΩ, $R_L = 10$ kΩ, $k = 50$, find the numerical value of (V_2/V_1). Check the result using SPICE (cccs8.cir).

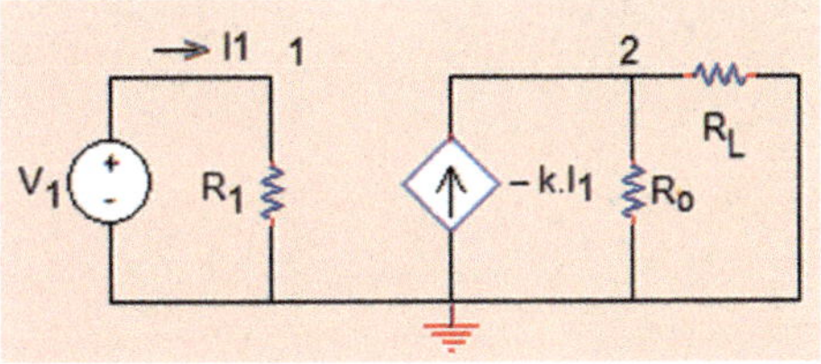

Fig. 2.10 The circuit of Problem 2.1.10

Solution

(a) $I_1 = \frac{V_1}{R_1}, \quad V_2 = -k \cdot I_1 \cdot (R_o//R_L) = -k \cdot \frac{V_1}{R_1} \cdot (R_o//R_L)$

$$\frac{V_2}{V_1} = -k \cdot \frac{R_o//R_L}{R_L} = k \cdot \frac{R_o//R_L}{R_o + R_L} \cdot R_1$$

(b) $$\frac{V_2}{V_1} = -50 \cdot \frac{10//10}{2.5} = -50 \cdot \left(\frac{5}{2.5}\right) = -100\,\mathrm{V/V}$$

SPICE netlist (cccr8.cir);

```
*DC Operating point analysis
*current controlled current source-nodal analysis
*fx N+ N- Vy Value
*Parameters:
*x  Name of the source
*N+ : Name of positive node
*N- : Name of negative node. Current flows from the + node
*   through the source to the - node
*Vr : Name of the voltage source where the controlling current flows.
*     The direction of positive control current is
*     from + node through the source to the - node of Vr=0
*Value: Current gain
v1 1 0 1m
f1 2 0 Vr 50
Vr 1 3 0
R1 3 0 2.5k
RO 2 0 10k
RL 2 0 10k
```

Problem 2.1.11 In the circuit shown in Fig. 2.11, $f = 2$, $R_1 = R_2 = R_3 = 1\,\Omega$, $i_S = 1\,\mathrm{A}$. $V_1 = ?, V_2 = ?, i = ?$

(Use node voltages method.) Check the result using SPICE (cccs7.cir).

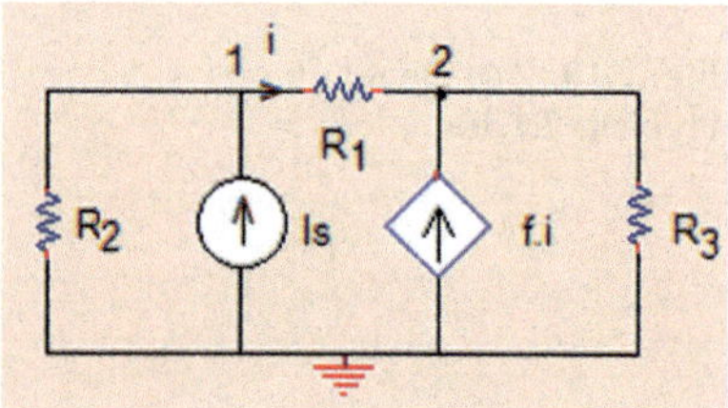

Fig. 2.11 The circuit of Problem 2.1.11

Solution

$$i_S - \frac{V_1}{R_2} - \frac{V_1 - V_2}{R_3} = 0$$

$$\frac{V_1 - V_2}{R_1} - \frac{V_2}{R_3} + \frac{2(V_1 - V_2)}{R_1} = 0.$$

Substituting given values of components into these equation yields

$$\begin{aligned} 1 - 2V_1 + V_2 &= 0 \\ 3V_1 - 4V_2 &= 0 \end{aligned}.$$

Simplifying,

$$\begin{aligned} 2V_1 - V_2 &= 1 \\ 3V_1 - 4V_2 &= 0 \end{aligned}.$$

Solving this simultaneous set of linear equations for unknown voltage values yields

$$\begin{aligned} V_1 &= 0.8V, \quad V_2 = 0.6V, \\ i &= \frac{V_1 - V_2}{R_1} = 0.8 - 0.6 = 0.2\,\text{A} \end{aligned}.$$

```
SPICE Netlist,cccs7.cir
Analysis: DC Operating Point
*fx N+ N- Vy Val
* x Name of the source, N+ : positive node
*N- : Name of negative node. Current flows from the + node
*  through the source to the - node
*Vy : Name of the voltage source
*The direction of positive control current is
*from + node through the source to the - node of Vy=0
*Val: Current gain
i1 0 1 1
f1 0 2 Vy 2
Vy 3 2 0
R1 1 3 1
R2 1 0 1
R3 2 0 1
```

Problem 2.1.12 In the circuit shown in Fig. 2.12, I_S = 1 A, $R_1 = R_2 = R_3 = R_4 =$ 1 Ω, f = 4 A/A. Find the values of currents flowing through resistors R_3 and R_4.

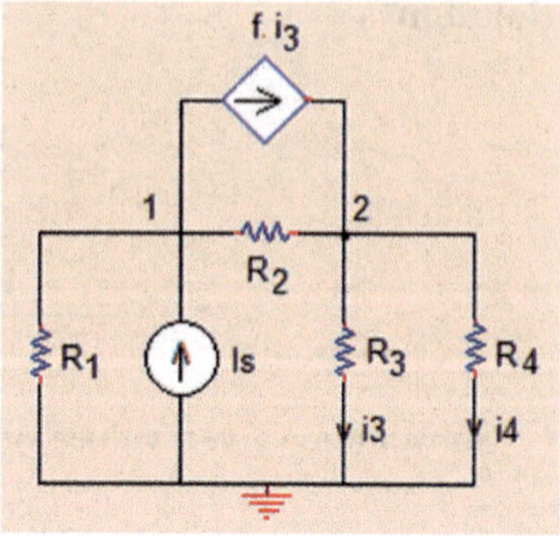

Fig. 2.12 The circuit of Problem 2.1.12

Solution

$$i = \frac{V_2}{R_3} = \frac{V_2}{1} = V_2.$$

KCL at nodes 1 and 2:

$$1 - 4V_2 - V_1 - (V_1 - V_2) = 0 \tag{2.1}$$

$$4V_2 + (V_1 - V_2) - V_2 - V_2 = (0). \tag{2.2}$$

Simplifying (2.1) and (2.2),

$$2V_1 + 3V_2 = 1 \tag{2.3}$$

$$V_1 + V_2 = 0. \tag{2.4}$$

Solving these equations for unknowns yields $V_1 = -1\,\text{V}, V_2 = 1\,\text{V}$,

$$i_4 = \frac{V_2}{R_4} = 1\,\text{A}, \quad i_3 = \frac{V_2}{R_3} = 1\,\text{A}.$$

Problem 2.1.13 In the circuit shown in Fig. 2.13, $R_1 = R_2 = R_3 = R_4 = 1\ \Omega$, $Is = 1$ A, and $f = 4$ A/A. $V_1 = ?$, $V_2 = ?$, $V_3 = ?$, $i = ?$ Use node voltages method. Check your results using SPICE (cccs4).

Solution
KCL at nodes 1 and 2 with $i = V_2/R_3$:

$$-\frac{V_1}{R_1} - f \cdot \frac{V_2}{R_3} - \frac{V_1 - V_2}{R_2} = 0 \tag{2.5}$$

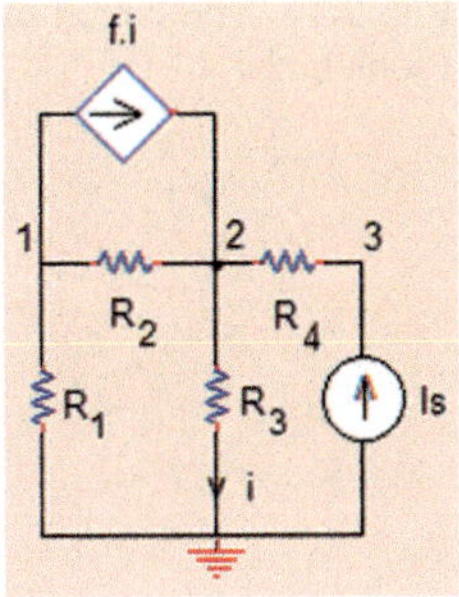

Fig. 2.13 The circuit of Problem 2.1.13

$$f \cdot \frac{V_2}{R_3} + \frac{V_1 - V_2}{R_2} - \frac{V_2}{R_3} = -1. \tag{2.6}$$

Using given values, these equations become

$$-V_1 - 4V_2 - V_1 + V_2 = 0 \tag{2.7}$$

$$4V_2 + V_1 - V_2 - V_2 = -1. \tag{2.8}$$

Simplifying,

$$2V_1 + 3V_2 = 0 \tag{2.9}$$

$$V_1 + 2V_2 = -1. \tag{2.10}$$

Solution of this set of simultaneous equations yields V_1 = 3 V, V_2 = −2 V.

$$V_{R4} = (-1\,\text{A}) \cdot R_4 = (-1) \cdot (1) = -1\,\text{V}$$
$$V_3 = V_2 - V_{R4} = -2 - (-1)(1) = -2 + 1 = -1\,\text{V}$$
$$i = \frac{V_2}{R_3} = -\frac{2}{1} = -2\,\text{A}$$

Netlist for SPICE check:

```
*cccs4.cir
*Analysis: DC Operating Point
   current controlled current source-nodal analysis
   i1 0 3 1
   f1 1 2 vref 4
   vref 4 0 0
   R1 1 0 1
   R2 1 2 1
   R3 2 4 1
   R4 2 3 1
```

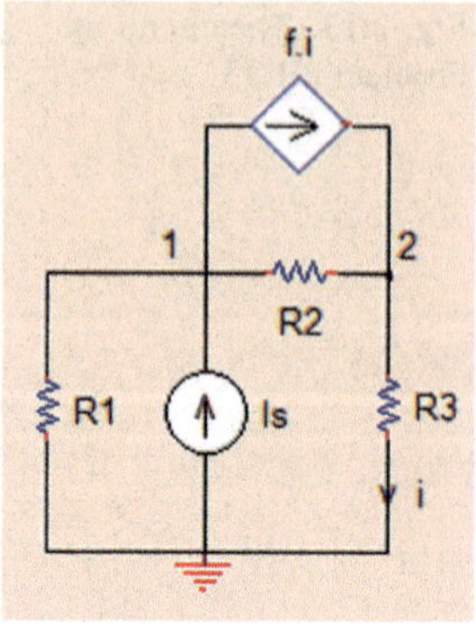

Fig. 2.14 The circuit of Problem 2.1.14

Problem 2.1.14 In the circuit shown in Fig. 2.14, $R_1 = R_2 = R_3 = 1\ \Omega$, $I_S = 1$ A, $f = 4$ A/A. $V_1 = ?$, $V_2 = ?$, $i = ?$, $i_{R2} = ?$ Use nodal analysis method.

Solution

$$i_s - f \cdot \frac{V_2}{R_3} - \frac{V_1}{R_1} - \frac{V_1 - V_2}{R_2} = 0 \tag{2.11}$$

$$f \cdot \frac{V_2}{R_3} + \frac{V_1 - V_2}{R_2} - \frac{V_2}{R_3} = 0. \tag{2.12}$$

Substituting the values,

$$1 - 4V_2 - V_1 - V_1 + V_2 = 0 \tag{2.13}$$

$$4V_2 + V_1 - V_2 - V_2 = 0. \tag{2.14}$$

Simplifying,

$$1 - 2V_1 - 3V_2 = 0 \tag{2.15}$$

$$V_1 + 2V_2 = 0. \tag{2.16}$$

In matrix form,

$$\begin{bmatrix} 1 \\ 0 \end{bmatrix} = \begin{bmatrix} 2 & 3 \\ 1 & 2 \end{bmatrix} \begin{bmatrix} V_1 \\ V_2 \end{bmatrix}. \tag{2.17}$$

Solution of this matrix equation yields

$$V_1 = 2\,\text{V}, \quad V_2 = -1\,\text{V}, \quad i = \frac{V_2}{R_3} = -\frac{1}{1} = -1\,\text{A}$$

$$i_{R2} = \frac{V_1 - V_2}{R_2} = \frac{2 - (-1)}{1} = 3\,\text{A}.$$

Problem 2.1.15

(a) Determine the voltage-to-current ratio (the input resistance) in the circuit shown in Fig. 2.15:

$$R_x = \frac{V_x}{I_x} = ?$$

(b) Determine the value of current through the voltage source, if

$$R_1 = R_2 = 1\,\Omega, \quad R_3 = 2\,\Omega, \quad R_4 = \frac{1}{2}\,\Omega, \quad V_x = \frac{1}{4}\,\text{V}, \quad k = 0.5\,\text{S}.$$

Solution

(a)
$$V_1 = I_x \cdot \left(\frac{(R_1 + R_2) \cdot R_3}{R_1 + R_2 + R_3}\right) \tag{2.18}$$

$$V_x = V_1 - V_2. \tag{2.19}$$

Node 2:

$$-I_x - kV_1 - \frac{V_2}{R_4} = 0 \tag{2.20}$$

$$V_2 = -R_4 \cdot I_x - kV_1 R_4 \tag{2.21}$$

Equation (2.21) → (2.19)

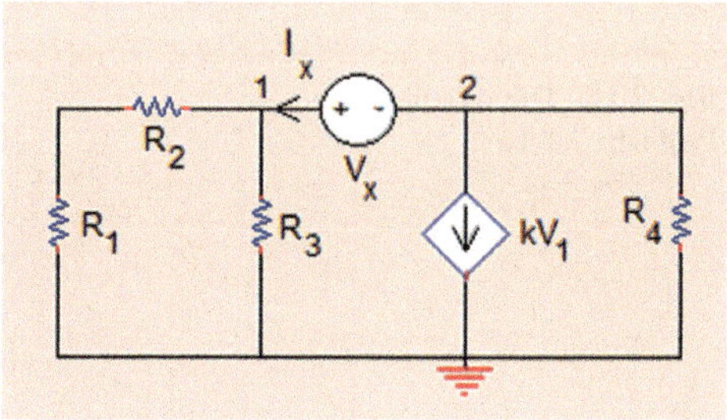

Fig. 2.15 The circuit of Problem 2.1.15

$$V_x = V_1 + R_4 \cdot I_x + kV_1R_4 = V_1(1 + kR_4) + R_4 \cdot I_x \tag{2.22}$$

divide all terms of (2.22) by I_x

$$\frac{V_x}{I_x} = R_x = \frac{V_1}{I_x}(1 + kR_4) + R_4. \tag{2.23}$$

Replace V_1/I_x by (2.18):

$$R_x = \left(\frac{(R_1 + R_2) \cdot R_3}{R_1 + R_2 + R_3}\right)(1 + kR_4) + R_4. \tag{2.24}$$

(b) Substituting given component values into (2.24) yields

$$R_x = \frac{(1+1) \cdot 2}{1+1+2}(1 + 0.5 \times 0.5) + 0.5 = 1.75\,\Omega$$

$$I_x = \frac{V_x}{R_x} = \frac{0.25}{1.75} = \frac{1}{7}A = 0.142857\,\text{A}.$$

Problem 2.1.16

(a) Use node voltage method and find the voltage drop across R_2 (in mV).
(b) Verify the solution using SPICE and print SPICE netlist (vccs2.cir).
(I_s = 1 A, R_1 = 1 Ω, R_2 = 1/2 Ω, R_3 = 1/4 Ω) (Fig. 2.16).

Solution
KCL at 1:

$$1 - V_1 - 2(V_1 - V_2) = 0$$

KCL at 2:

$$-\frac{V_1}{2} - 4V_2 + 2(V_1 - V_2) = 0$$

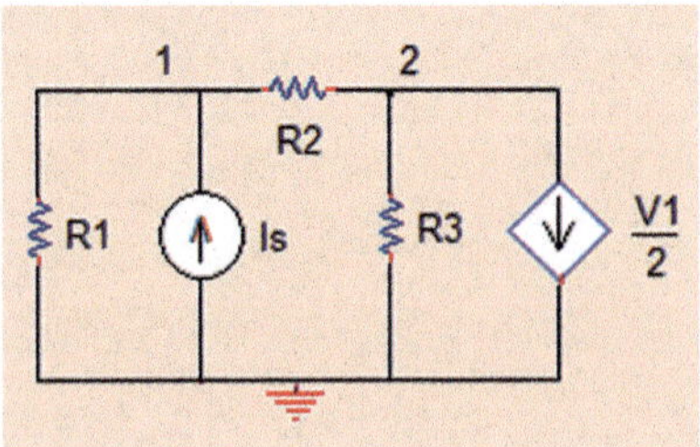

Fig. 2.16 The circuit of Problem 2.1.16

$$-3V_1 + 2V_2 = -1$$

$$-\frac{V_1}{2} + 2V_1 - 4V_2 - 2V_2 = 0$$

$$3V_1 - 2V_2 = 1 \tag{2.25}$$

$$3V_1 - 12V_2 = 0. \tag{2.26}$$

From (2.26) $3V_1$ + $12V_2$ $\rightarrow$ put into (2.25)

$$12V_2 - 2V_2 = 1 \quad \rightarrow \quad V_2 = 0.1\,\text{V}$$

$$3V_1 = 1.2 \quad \rightarrow \quad V_1 = 0.4\,\text{V}$$

$$V_1 - V_2 = 0.4 - 0.1 = 0.3\,\text{V} = 300\,\text{mV}.$$

(b) SPICE netlist: Operating point analysis, vccs2.cir

```
*gx  N+ N- NC+ NC- VALUE
*x      Name of the source
*N+     Name of positive node
*N-     Name of negative node
*NC+    Name of positive controlling node
*NC-    Name of negative controlling node
*VALUE  Transconductance in S
I1  0 1 1
G1  2 0 1 0  0.5
*Specifies that the current through G1 flowing from node 2 to ground
*is 0.5 times the potential difference between node 1 and ground.
R1  1 0 1
R2  1 2 0.5
R3  2 0 0.25
```

Problem 2.1.17 A *R*–2*R* ladder circuit is shown in Fig. 2.17.

(a) Find the node voltages and shunt branch currents.
(b) Find the current supplied by the voltage source.
(c) Compute numerical values if V_i = 5 V, R = 1 kΩ.

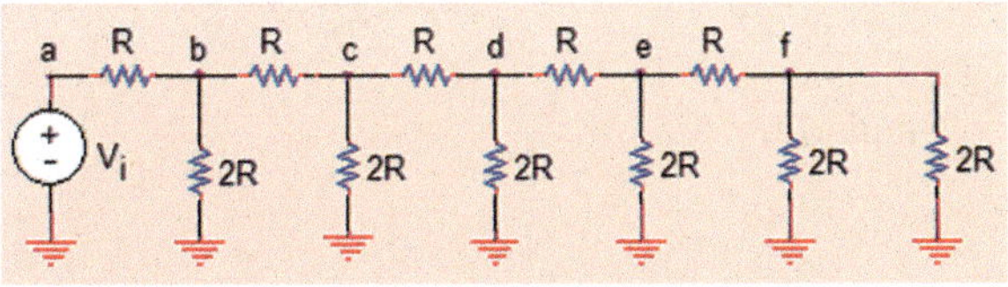

Fig. 2.17 The circuit for Problem 2.1.17

Solution

Start at the rightmost node of the circuit (node f), looking to the right of each node,

$$R_{\mathrm{f}} = 2R//2R = R, \quad R_{\mathrm{e}} = 2R//(R+R) = R, \quad R_{\mathrm{d}} = 2R//(R+R) = R$$

$$R_{\mathrm{c}} = 2R//(R+R) = R, R_{\mathrm{b}} = 2R//(R+R) = R.$$

Node voltages:

$$\begin{aligned}
V_{\mathrm{a}} &= V \\
V_{\mathrm{b}} &= \frac{R_{\mathrm{b}} \cdot V}{R+R_{\mathrm{b}}} = \frac{R \cdot V}{2R} = \frac{V}{2} V \\
V_{\mathrm{c}} &= \frac{R_{\mathrm{c}}}{R+R_{\mathrm{c}}} \cdot \frac{V}{2} = \frac{R}{2R} \cdot \frac{V}{2} = \frac{V}{4} V \\
V_{\mathrm{d}} &= \frac{R_{\mathrm{d}}}{R+R_{\mathrm{d}}} \cdot \frac{V}{4} = \frac{R}{2R} \cdot \frac{U}{4} = \frac{V}{8} V \\
V_{\mathrm{e}} &= \frac{R_{\mathrm{e}}}{R+R_{\mathrm{e}}} \cdot \frac{V}{8} = \frac{R}{2R} \cdot \frac{V}{8} = \frac{V}{16} V \\
V_{\mathrm{f}} &= \frac{R_{\mathrm{f}}}{R+R_{\mathrm{f}}} \cdot \frac{V}{16} = \frac{R}{2R} \cdot \frac{V}{16} = \frac{V}{32} V
\end{aligned}$$

The shunt branch currents are calculated by Ohm's law:
The right part branch departing from (f) carries a current of

$$I_{\mathrm{o}} = \frac{V_{\mathrm{f}}}{2R} = \frac{\frac{V}{32}}{2R} = \frac{V}{64R}.$$

This is the same current through the left branch departing from (f).
The shunt branch current departing from node (e) is

$$I_{\mathrm{eo}} = \frac{V_{\mathrm{e}}}{2R} = \frac{\frac{V}{16}}{2R} = \frac{V}{32} = 2I_{\mathrm{o}}.$$

The shunt branch current departing from node (d) is

$$I_{\mathrm{do}} = \frac{V_{\mathrm{d}}}{2R} = \frac{\frac{V}{8}}{2R} = \frac{V}{16R} = 2(2I_{\mathrm{o}}) = 4I_{\mathrm{o}}.$$

The shunt branch current departing from node (c) is

$$I_{\mathrm{co}} = \frac{V_{\mathrm{c}}}{2R} = \frac{\frac{V}{4}}{2R} = \frac{V}{8R} = 8I_{\mathrm{o}}.$$

Similarly,

$$I_{bo} = \frac{V_b}{2R} = \frac{\frac{V}{2}}{2R} = \frac{V}{4R} = 16I_o.$$

Overall current supplied by the voltage source is

$$I = I_{bo} + I_{co} + I_{do} + I_{eo} + I_{fo} + I_{fo} = I_o(16 + 8 + 4 + 2 + 1 + 1) = 32 \cdot I_o$$

$$I = 32 \cdot \frac{V}{64R} = \frac{V}{2R}$$

$$V = 5\,\text{V}, \quad R = 1\,\text{k}\Omega; \quad V_b = 2.5\,\text{V}, \quad V_c = 1.25\,\text{V},$$

$$V_d = 0.625\,\text{V}, \quad V_e = 0.3125\,\text{V}, \quad V_f = 0.15625\,\text{V},$$

$$I_o = \frac{V}{64R} = \frac{5}{64}\,\text{mA} = 0.078125\,\text{mA}$$

$$I = \frac{V}{2R} = \frac{5}{2}\,\text{mA} = 2.5\,\text{mA}$$

Problem 2.1.18 In the circuit shown in Fig. 2.18, find the voltage at node 1 (=V_1). Use node voltage method and Cramer's rule for the solution of matrix equations. $R_1 = R_2 = R_3 = R_4 = 1\,\text{k}\Omega$, $I_1 = I_2 = I_3 = I_4 = 1\,\text{mA}$.

Solution

$$[I] = [G][V], \quad G = R^{-1}$$

$$G_1 = G_2 = G_3 = G_4 = 10^{-3}\,\text{S}$$

$$10^{-3} \times \begin{bmatrix} I_3 \\ I_1 - (I_3 + I_4) \\ I_2 + I_4 \end{bmatrix} = \begin{bmatrix} G_1 + G_2 & -G_2 & 0 \\ -G_2 & G_2 + G_3 & -G_3 \\ 0 & -G_3 & G_3 + G_4 \end{bmatrix} \cdot \begin{bmatrix} V_1 \\ V_2 \\ V_3 \end{bmatrix} \times 10^{-3}$$

$$\begin{bmatrix} 1 \\ 1 - (1+1) \\ 1+1 \end{bmatrix} = \begin{bmatrix} 1+1 & -1 & 0 \\ -1 & 1+1 & -1 \\ 0 & -1 & 1+1 \end{bmatrix} \begin{bmatrix} V_1 \\ V_2 \\ V_3 \end{bmatrix} \rightarrow \begin{bmatrix} 1 \\ -1 \\ 2 \end{bmatrix} = \begin{bmatrix} 2 & -1 & 0 \\ -1 & 2 & -1 \\ 0 & -1 & 2 \end{bmatrix} \cdot \begin{bmatrix} V_1 \\ V_2 \\ V_3 \end{bmatrix}.$$

Using Cramer's rule for the solution of matrix equation,

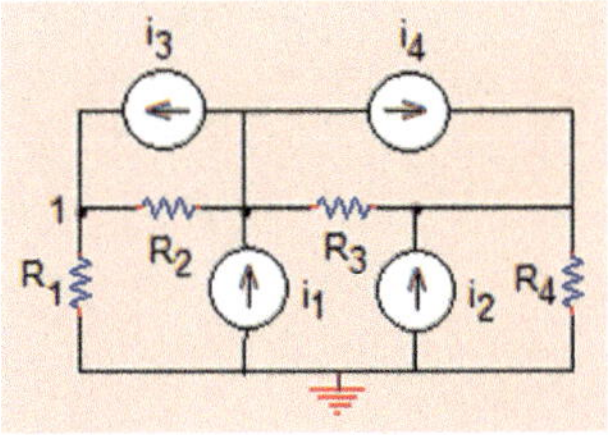

Fig. 2.18 The circuit for Problem 2.1.18

$$V_1 = \frac{\Delta_1}{\Delta} = \frac{\begin{vmatrix} 1 & -1 & 0 & \vdots & 1 & -1 \\ -1 & 2 & -1 & \vdots & -1 & 2 \\ 2 & -1 & 2 & & 2 & -1 \end{vmatrix}}{\begin{bmatrix} 2 & -1 & 0 & & 2 & -1 \\ -1 & 2 & -1 & \vdots & -1 & 2 \\ 0 & -1 & 2 & & 0 & -1 \end{bmatrix}} = \frac{4+2+0-(0+1+2)}{8+0+0-(0+2+2)} = \frac{3}{4}$$
$$= 0.75\,\text{V}.$$

Problem 2.1.19 In the circuit shown in Fig. 2.19, find the values of node voltages V_1 and V_2. Use Cramer's rule when necessary $R_1 = R_3 = 2\,\Omega, R_2 = 4\,\Omega, I_1 = I_3 = 1\,\text{A}, I_2 = I_4 = I_5 = 2\,\text{A}$.

Solution

$$\frac{V_1}{2} + \frac{V_1 - V_2}{4} - 5 = 0 \quad \rightarrow \quad \frac{3}{4}V_1 - \frac{1}{4}V_2 = 5$$

$$\frac{V_2 - V_1}{4} + \frac{V_2}{2} + 1 = 0 \quad \rightarrow \quad -\frac{1}{4}V_1 + \frac{3}{4}V_2 = -1$$

$$\begin{bmatrix} \frac{3}{4} & -\frac{1}{4} \\ -\frac{1}{4} & \frac{3}{4} \end{bmatrix} \cdot \begin{bmatrix} V_1 \\ V_2 \end{bmatrix} = \begin{bmatrix} 5 \\ -1 \end{bmatrix}$$

$$\Delta_1 = \begin{vmatrix} 5 & -\frac{1}{4} \\ -1 & \frac{3}{4} \end{vmatrix} = \frac{14}{4}, \quad \Delta_2 = \begin{vmatrix} \frac{3}{4} & 5 \\ -\frac{1}{4} & -1 \end{vmatrix} = \frac{1}{2}, \quad \Delta = \begin{vmatrix} \frac{3}{4} & -\frac{1}{4} \\ -\frac{1}{4} & \frac{3}{4} \end{vmatrix} = \frac{1}{2}$$

$$V_1 = \frac{\Delta_1}{\Delta} = 7\,\text{V}, V_2 = \frac{\Delta_2}{\Delta} = 1\,\text{V}.$$

Problem 2.1.20 In the circuit shown in Fig. 2.20, $I_1 = 1$ A, $I_2 = 1/2$ A, $R_1 = 1/2\ \Omega$, $R_2 = 1/4\ \Omega$, $R_3 = 1/8\ \Omega$.

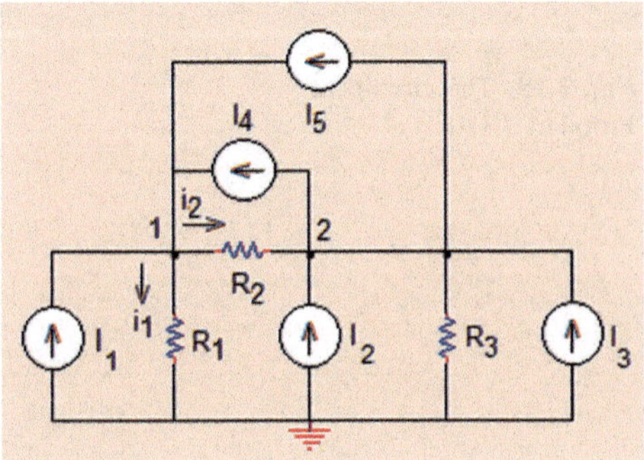

Fig. 2.19 The circuit for Problem 2.1.19

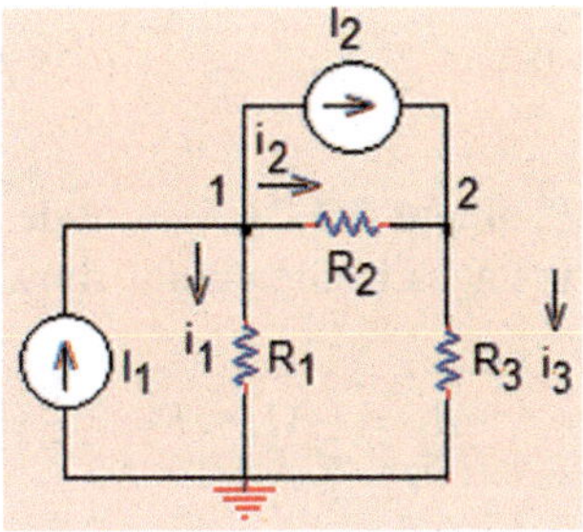

Fig. 2.20 The circuit for Problem 2.1.20

(a) Find the node voltages,
(b) Find the currents flowing in the circuit (Sim_Lin_Eq_Solve.m, matrix_solve.xlsx).

Solution

(a)
$$\frac{V_1}{R_1} + \frac{V_1 - V_2}{R_2} + I_2 - I_1 = 0 \tag{2.27}$$

$$\frac{V_2 - V_1}{R_2} + \frac{V_2}{R_3} - I_2 = 0 \tag{2.28}$$

$$V_1\left(\frac{1}{R_1} + \frac{1}{R_2}\right) - \frac{V_2}{R_2} = I_1 - I_2 \tag{2.29}$$

$$-\frac{V_1}{R_2} + V_2\left(\frac{1}{R_2} + \frac{1}{R_3}\right) = I_2. \tag{2.30}$$

Using last two equations,

$$\begin{bmatrix} \frac{1}{R_1} + \frac{1}{R_2} & -\frac{1}{R_2} \\ -\frac{1}{R_2} & \frac{1}{R_2} + \frac{1}{R_3} \end{bmatrix} \cdot \begin{bmatrix} V_1 \\ V_2 \end{bmatrix} = \begin{bmatrix} I_1 - I_2 \\ I_2 \end{bmatrix}$$

$$\begin{bmatrix} 6 & -4 \\ -4 & 12 \end{bmatrix} \cdot \begin{bmatrix} V_1 \\ V_2 \end{bmatrix} = \begin{bmatrix} 0.5 \\ 0.5 \end{bmatrix}.$$

Solution of this set of simultaneous linear equations by either manually using Cramer's rule or by substitution methods or by employing available software (see, MATLAB m file or EXCEL file referenced in the statement) yields

$$V_1 = 0.14\,\text{V}, \quad V_2 = 0.09\,\text{V}.$$

(b) $$i_1 = \frac{V_1}{R_1} = 0.28\,\text{A}, \quad i_2 = \frac{V_1 - V_2}{R_2} = 0.2\,\text{A}, \quad i_3 = \frac{V_2}{R_3} = 0.72\,\text{A}.$$

Problem 2.1.21 Use node voltages method and find the values of currents and voltages in the circuit shown in Fig. 2.21.

$$R_1 = \frac{1}{2}\,\Omega, \quad R_2 = \frac{1}{4}\,\Omega, \quad R_3 = R_4 = 1\,\Omega, \quad I_1 = I_2 = 1\,\text{A}, \quad I_3 = 2\,\text{A}.$$

Solution
Applying KCL at node 1,

$$I_1 + I_2 - i_1 - i_2 = 0 \quad \rightarrow \quad 1 + 1 - \frac{v_1}{R_1} - \frac{v_1 - v_2}{R_2} = 0$$

$$2 - \frac{v_1}{\frac{1}{2}} - \frac{v_1 - v_2}{\frac{1}{4}} = 0 \quad \rightarrow \quad 2 - 6v_1 + 4v_2 = 0$$

$$-6v_1 + 4v_2 = -2. \tag{2.31}$$

Applying KCL at node 2,

$$i_2 - i_3 - i_4 = 0 \quad \rightarrow \quad \frac{v_1 - v_2}{R_2} - \frac{v_2}{R_3} - \frac{v_2 - v_3}{R_4} = 0 \quad \rightarrow$$
$$\frac{v_1 - v_2}{\frac{1}{4}} - \frac{v_2}{1} - \frac{v_2 - v_3}{1} = 0$$

$$4v_1 - 6v_2 + v_3 = 0. \tag{2.32}$$

Applying KCL at node 3,

$$I_3 - I_2 + i_4 = 0 \quad \rightarrow \quad 2 - 1 + \frac{v_2 - v_3}{R_4} = 0 \quad \rightarrow \quad 1 + \frac{v_2 - v_3}{1} = 0$$

$$v_2 - v_3 = -1. \tag{2.33}$$

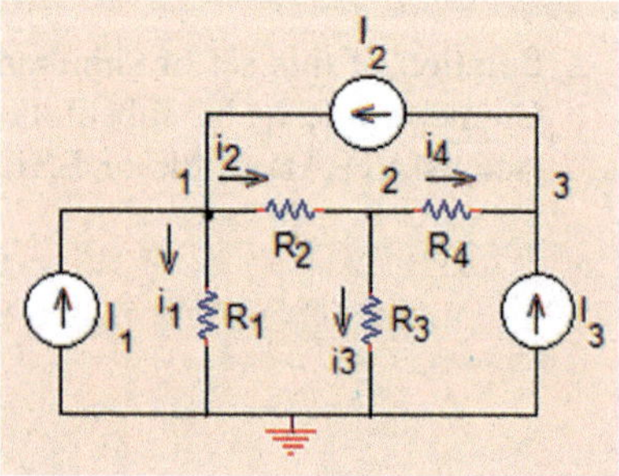

Fig. 2.21 The circuit for Problem 2.1.21

Combining Eqs. (2.31)–(2.33) into matrix form,

$$\begin{bmatrix} -6 & 4 & 0 \\ 4 & -6 & 1 \\ 0 & 1 & -1 \end{bmatrix} \begin{bmatrix} v_1 \\ v_2 \\ v_3 \end{bmatrix} = \begin{bmatrix} -2 \\ 0 \\ -1 \end{bmatrix}.$$

The solution of this matrix equation yields

$$v_1 = 1\,\text{V}, \quad v_2 = 1\,\text{V}, \quad v_3 = 2\,\text{V}$$

$$i_1 = \frac{v_1}{R_1} = 2\,\text{A}, \quad i_2 = \frac{v_1 - v_2}{R_2} = 0\,\text{A}, \quad i_3 = \frac{v_2}{R_3} = 1\,\text{A}, \quad i_4 = \frac{v_2 - v_3}{R_4} = -1\,\text{A}.$$

Problem 2.1.22 Determine the node voltages in the circuit shown in Fig. 2.22. $R = 1\Omega, \quad I_1 = I_2 = I_3 = 1\,\text{A}$.

Solution
Analysis by inspection,

$$G = \frac{1}{R} = 1\,\text{S}$$

$$\begin{bmatrix} G+G & -G & 0 \\ -G & G+G+G & G \\ 0 & -G & G+G \end{bmatrix} \begin{bmatrix} V_1 \\ V_2 \\ V_3 \end{bmatrix} = \begin{bmatrix} I_1 + I_2 + I_3 \\ -I_2 \\ -I_3 \end{bmatrix} \rightarrow \begin{bmatrix} 2 & -1 & 0 \\ -1 & 3 & -1 \\ 0 & -1 & 2 \end{bmatrix} \begin{bmatrix} V_1 \\ V_2 \\ V_3 \end{bmatrix} = \begin{bmatrix} 3 \\ -1 \\ -1 \end{bmatrix}.$$

This matrix equation is solved for unknown voltages and yields the following voltage values:

$$V_1 = 1.5\,\text{V}, \quad V_2 = 0\,\text{V}, \quad V_3 = -0.5\,\text{V}.$$

Problem 2.1.23

(a) Use node voltages and Cramer's methods to find the values of currents and voltages in the circuit shown in Fig. 2.23. Use SPICE for checking the results. Print the SPICE netlist (cccs5.cir).

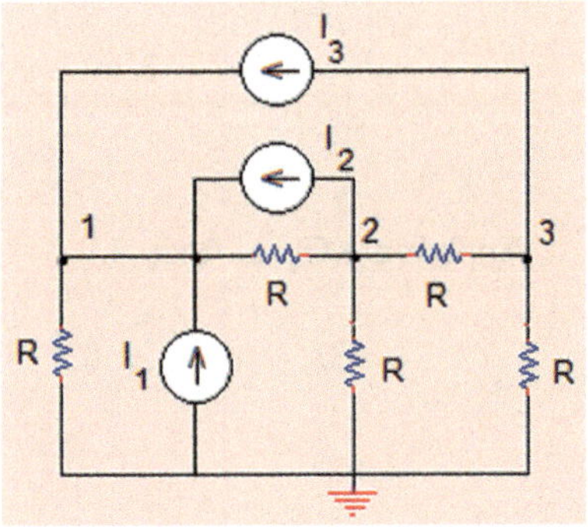

Fig. 2.22 The circuit for Problem 2.1.22

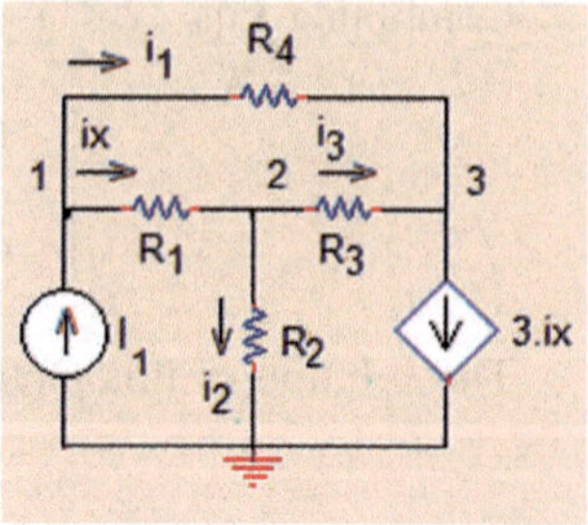

Fig. 2.23 The circuit for Problem 2.1.23

$$R_1 = R_4 = \frac{1}{2}\,\Omega, \quad R_2 = R_3 = 1\,\Omega, \quad I_1 = 2\,\text{A}, \quad f = 3\,\text{A/A}.$$

(b) Determine the node voltages using the following component values. Use SPICE for checking the results. Print the new SPICE netlist.

$$R_1 = 1\,\Omega, \quad R_2 = 4\,\Omega, \quad R_3 = 8\,\Omega, \quad R_4 = 4\,\Omega \quad I_1 = 3\,\text{A}, \quad f = 2\,\text{A/A}.$$

Solution

(a) Applying KCL at node 1,

$$I_1 - i_1 - i_x = 0 \rightarrow 2 - \frac{v_1 - v_3}{R_4} - \frac{v_1 - v_2}{R_1} = 0 \rightarrow 2 - \frac{v_1 - v_3}{\frac{1}{2}} - \frac{v_1 - v_2}{\frac{1}{2}} = 0$$

$$2 - 2(v_1 - v_3) - 2(v_1 - v_2) = 0$$

$$-4v_1 + 2v_2 + 2v_3 = -2. \tag{2.34}$$

Applying KCL at node 2,

$$i_x - i_2 - i_3 = 0 \rightarrow \frac{v_1 - v_2}{R_1} - \frac{v_2}{R_2} - \frac{v_2 - v_3}{R_3} = 0 \rightarrow \frac{v_1 - v_2}{\frac{1}{2}} - \frac{v_2}{1} - \frac{v_2 - v_3}{1} = 0$$

$$2(v_1 - v_2) - v_2 - (v_2 - v_3) = 0$$

$$2v_1 - 4v_2 + v_3 = 0. \tag{2.35}$$

Applying KCL at node 3,

$$-3i_x + i_1 + i_3 = 0 \rightarrow -3\frac{v_1 - v_2}{R_1} + \frac{v_1 - v_3}{R_4} + \frac{v_2 - v_3}{R_3} = 0$$

$$-3\frac{v_1 - v_2}{\frac{1}{2}} + \frac{v_1 - v_3}{\frac{1}{2}} + \frac{v_2 - v_3}{1} = 0 \rightarrow -3[2(v_1 - v_2)] + 2(v_1 - v_3) + (v_2 - v_3) = 0$$

$$-4v_1 + 7v_2 - 3v_3 = 0. \tag{2.36}$$

Combining Eqs. (2.34)–(2.36) into matrix form,

$$\begin{bmatrix} -4 & 2 & 2 \\ 2 & -4 & 1 \\ -4 & 7 & -3 \end{bmatrix} \begin{bmatrix} v_1 \\ v_2 \\ v_3 \end{bmatrix} = \begin{bmatrix} -2 \\ 0 \\ 0 \end{bmatrix}.$$

The solution of this matrix equation yields

$$v_1 = 0.5\,\text{V}, \quad v_2 = 0.2\,\text{V}, \quad v_3 = -0.2\,\text{V}$$

$$i_1 = \frac{v_1 - v_3}{R_4} = 1.4\,\text{A}, \quad i_2 = \frac{v_2}{R_2} = 0.2\,\text{A},$$

$$i_3 = \frac{v_2 - v_3}{R_3} = 0.4\,\text{A}, \quad i_x = \frac{v_1 - v_2}{R_1} = 0.6\,\text{A}.$$

```
*SPICE Netlist cccs5.cir
*Analysis: DC Operating Point
   current controlled current source-nodal analysis 5
   i1 0 1 2
   f1 3 0 vref 3
   vref 4 2 0
   R1 1 4 0.5
   R2 2 0 1
   R3 2 3 1
   R4 1 3 0.5
```

(b) Using new data set,

KCL at node 1, $12 = 5v_1 - 4v_2 - v_3$
KCL at node 2, $0 = 8v_1 - 11v_2 + v_3$
KCL at node 3, $0 = 14v_1 - 17v_2 + 3v_3$.

Using three nodal equations, one obtains the following matrix equation of the circuit:

$$\begin{bmatrix} 5 & -4 & -1 \\ 8 & -11 & 1 \\ 14 & -17 & 3 \end{bmatrix} \begin{bmatrix} v_1 \\ v_2 \\ v_3 \end{bmatrix} = \begin{bmatrix} 12 \\ 0 \\ 0 \end{bmatrix}$$

$$\Delta = \begin{vmatrix} 5 & -4 & -1 \\ 8 & -11 & 1 \\ 14 & -17 & 3 \end{vmatrix} = -58, \quad \Delta_1 = \begin{vmatrix} 12 & -4 & -1 \\ 0 & -11 & 1 \\ 0 & -17 & 3 \end{vmatrix} = -192$$

$$\Delta_2 = \begin{vmatrix} 5 & 12 & -1 \\ 8 & 0 & 1 \\ 14 & 0 & 3 \end{vmatrix} = -120, \quad \Delta_3 = \begin{vmatrix} 5 & -4 & 12 \\ 8 & -11 & 0 \\ 14 & -17 & 0 \end{vmatrix} = 216$$

$$v_1 = \frac{\Delta_1}{\Delta} \cong 3.31\text{V}, \quad v_2 = \frac{\Delta_2}{\Delta} \cong 2.06\,\text{V}, \quad v_3 = \frac{\Delta_3}{\Delta} \cong -3.72\,\text{V}.$$

Renewed SPICE Netlist;

```
* DC operating point analysis cccs5.cir
  i1 0 1 3
  f1 3 0 vref 2
  vref 4 2 0
  R1 1 4 1
  R2 2 0 4
  R3 2 3 8
  R4 1 3 4
```

Advantages of using SPICE are apparent here. It can be easily used for many different component variations of a circuit, rather than performing tedious calculations.

Problem 2.1.24 Determine the ratio of node voltages V_1/V_2 in the circuit shown in Fig. 2.24. Use Cramer's rule when necessary.

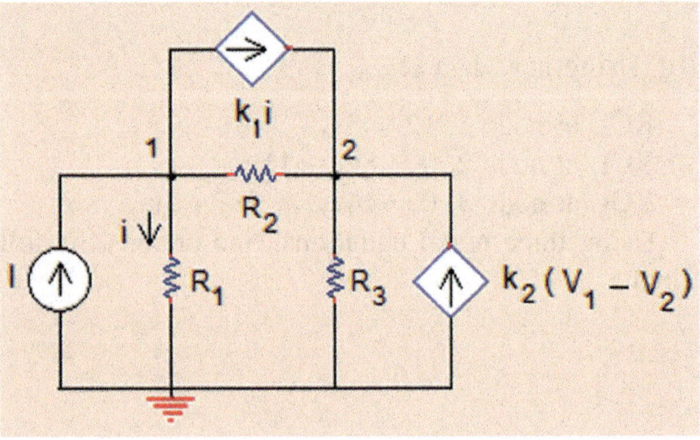

Fig. 2.24 The circuit for Problem 2.1.24

$$I = 1\,\text{A}, \quad R_1 = 10\,\Omega, \quad R_2 = 1\,\Omega, \quad R_3 = 5\,\Omega, \quad k_1 = 5\,\text{A/A}, \quad k_2 = 2\,\text{A/V}.$$

Solution
KCL at node 1:

$$I - i - 5i - i_{R_2} = I - \frac{V_1}{R_1} - 5\frac{V_1}{R_1} - \frac{V_1 - V_2}{R_2} = I - 6\frac{V_1}{R_1} - \frac{V_1}{R_2} + \frac{V_2}{R_2} = 0$$

$$I - 7V_1\left(\frac{1}{R_1} + \frac{1}{R_2}\right) + \frac{V_2}{R_2} = 0. \tag{2.37}$$

KCL at node 2:

$$5i + i_{R_2} - i_{R_3} + 2(V_1 - V_2) = 5\frac{V_1}{R_1} + \frac{V_1 - V_2}{R_2} - \frac{V_2}{R_3} - 2(V_1 - V_2) = 0$$

$$5\frac{V_1}{R_1} + \frac{V_1}{R_2} - \frac{V_2}{R_2} - \frac{V_2}{R_3} + 2V_1 - 2V_2 = 0. \tag{2.38}$$

From Eqs. (2.37) and (2.38),

$$\begin{bmatrix} 7\left(\frac{1}{R_1} + \frac{1}{R_2}\right) & -\frac{1}{R_2} \\ \frac{5}{R_1} + \frac{1}{R_2} + 2 & -\left(\frac{1}{R_2} + \frac{1}{R_3} + 2\right) \end{bmatrix} \begin{bmatrix} V_1 \\ V_2 \end{bmatrix} = \begin{bmatrix} 1 \\ 0 \end{bmatrix}$$

$$R_1 = 10\,\Omega, \quad R_2 = 1\,\Omega, \quad R_3 = 5\,\Omega$$

$$\begin{bmatrix} 7(0.1 + 1) & -1 \\ 0.5 + 1 + 2 & -(1 + 0.2 + 2) \end{bmatrix} \begin{bmatrix} V_1 \\ V_2 \end{bmatrix} = \begin{bmatrix} 1 \\ 0 \end{bmatrix}$$

$$\begin{bmatrix} 7.7 & -1 \\ 3.5 & -3.2 \end{bmatrix} \begin{bmatrix} V_1 \\ V_2 \end{bmatrix} = \begin{bmatrix} 1 \\ 0 \end{bmatrix}$$

$$V_1 = \frac{\Delta_1}{\Delta}, \quad V_2 = \frac{\Delta_2}{\Delta}, \quad \frac{V_1}{V_2} = \frac{\frac{\Delta_1}{\Delta}}{\frac{\Delta_2}{\Delta}} = \frac{\Delta_1}{\Delta_2}$$

$$\Delta_1 = \begin{vmatrix} 1 & -1 \\ 0 & -3.2 \end{vmatrix} = -3.2, \quad \Delta_2 = \begin{vmatrix} 7.7 & 1 \\ 3.5 & 0 \end{vmatrix} = -3.5$$

$$\frac{V_1}{V_2} = \frac{-3.2}{-3.5} = 0.9143(\text{V/V}).$$

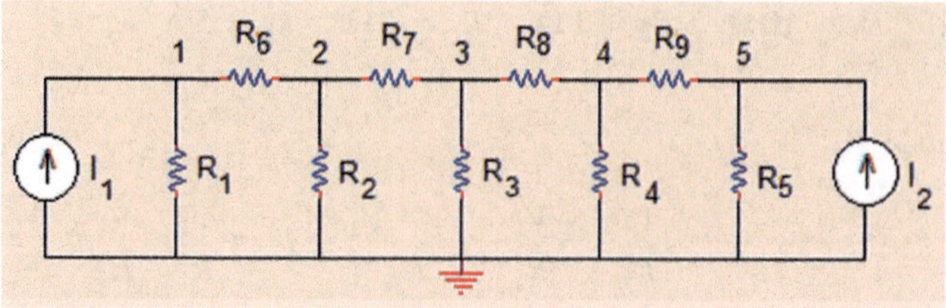

Fig. 2.25 The circuit for Problem 2.1.25

Problem 2.1.25 Determine currents flowing through each resistor in the circuit shown in Fig. 2.25 (ladder_node.xlsx).

$R_1 = R_2 = R_3 = R_4 = R_5 = 10\,\Omega, R_6 = R_7 = 5\,\Omega, R_8 = R_9 = 20\,\Omega, I_1 = I_2 = 2\,\text{A}.$

Solution

$$[G][V] = [I] = [2\,0\,0\,0\,2]^{\mathrm{T}} \tag{2.39}$$

where T stands for transpose operation.

$$G_{11} = \frac{1}{10} + \frac{1}{5}, \quad G_{22} = \frac{1}{10} + \frac{2}{5}, \quad G_{33} = \frac{1}{10} + \frac{1}{5} + \frac{1}{20}, \quad G_{44} = \frac{2}{20} + \frac{1}{10},$$

$$G_{55} = \frac{1}{10} + \frac{1}{20}, \quad G_{12} = G_{21} = -\frac{1}{5}, \quad G_{32} = G_{23} = -\frac{1}{5}, \quad G_{43} = G_{34} = -\frac{1}{20},$$

$$G_{45} = G_{54} = -\frac{1}{20},$$

$$[G] = \begin{bmatrix} 0.3 & -0.2 & 0 & 0 & 0 \\ -0.2 & 0.5 & -0.2 & 0 & 0 \\ 0 & -0.2 & 0.35 & -0.05 & 0 \\ 0 & 0 & -0.05 & 0.2 & -0.05 \\ 0 & 0 & 0 & -0.05 & 0.15 \end{bmatrix}.$$

Solution of Eq. (2.39) using these numerical values yields the node voltages:

$$V_1 = 10.545\,\text{V}, \quad V_2 = 5.818\,\text{V}, \quad V_3 = 4\,\text{V}, \quad V_4 = 4.727\,\text{V}, \quad V_5 = 14.909\,\text{V}.$$

Current values through resistors are obtained using Ohm's Law:

$$i_1 = \frac{V_1}{R_1} = 1.054545\,\text{A}, i_2 = \frac{V_2}{R_2} = 0.581818\,\text{A}$$
$$i_3 = \frac{V_3}{R_3} = 0.4\,\text{A}, i_4 = \frac{V_4}{R_4} = 0.472727\,\text{A}$$
$$i_5 = \frac{V_5}{R_5} = 1.490909\,\text{A}, i_6 = \frac{V_1 - V_2}{R_6} = 0.945459\,\text{A}$$
$$i_7 = \frac{V_2 - V_3}{R_7} = 0.363636\,\text{A}, i_8 = \frac{V_3 - V_4}{R_8} = -0.03636\,\text{A},$$
$$i_9 = \frac{V_4 - V_5}{R_9} = -0.50909\,\text{A}$$

Problem 2.1.26 Find the node voltage values for the circuit shown in Fig. 2.26. $I_1 = I_3 = 1$ A, $I_2 = 2$ A, $I_4 = -1$ A, $R_1 = R_2 = R_3 = R_5 = R_6 = R_7 = R_8 = 1\ \Omega$, $R_9 = 0.1\ \Omega$ (Sim_Lin_Eq_Solve.m, matrix_solve.xlsx).

Solution
The conductance values are

$$G1 = G2 = G3 = G4 = G5 = G6 = G7 = G8 = 1\,\text{S}, G9 = 10\,\text{S}$$

$$[G][V] = [I].$$

Solution of this matrix equation in MATLAB or EXCEL platform yields

$$V_1 = 1.795\,\text{V}, \quad V_2 = 0.590\,\text{V}, \quad V_3 = -1.081\,\text{V},$$

$$V_4 = -0.024\,\text{V}, \quad V_5 = -0.581\,\text{V}, \quad V_6 = -0.139\,\text{V}.$$

Problem 2.1.27

(a) Find the node voltage values in terms of current gain of the CCCS for the circuit shown in Fig. 2.27.
(b) Verify the solution using SPICE and print SPICE netlist (cccs1.cir).

$$I_S = 1\,\text{A}, \quad R_1 = R_2 = R_3 = 1\,\Omega.$$

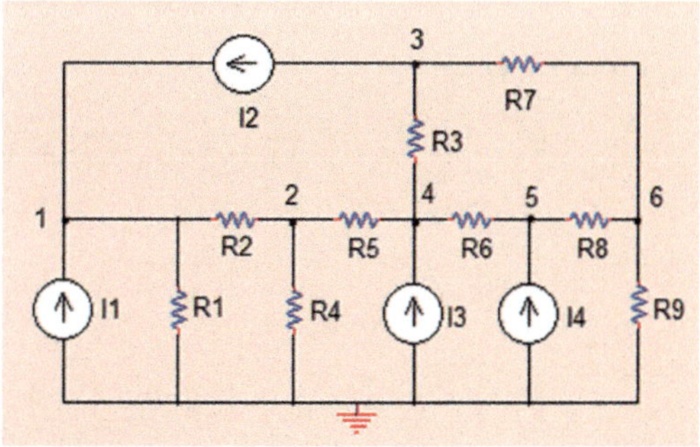

Fig. 2.26 The circuit for Problem 2.1.26

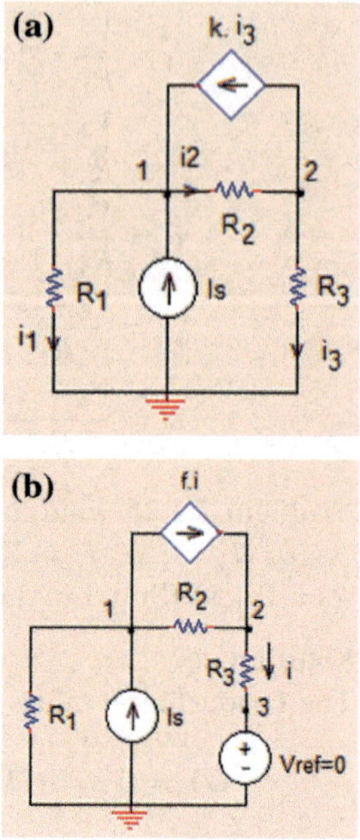

Fig. 2.27 The circuit for Problem 2.1.27

Solution

KCL at node 1: ($G_1 = G_2 = G_3 = 1$ S)

$$I_S - V_1 - (V_1 - V_2) - fi = 0 \quad \rightarrow \quad 1 - V_1 - V_1 + V_2 - f \cdot V_2 = 0$$

$$1 - 2V_1 + V_2(1 - f) = 0$$

$$2V_1 - (1 - f) \cdot V_2 = 1. \tag{2.40}$$

KCL at node 2:

$$fi + (V_1 - V_2) - i = 0 \quad \rightarrow \quad fV_2 + V_1 - V_2 - V_2 = 0$$

$$fV_2 + V_1 - 2V_2 = 0$$

$$V_1 + V_2(f - 2) = 0. \tag{2.41}$$

From (2.41),

$$V_1 = V_2(f - 2) = (2 - f) \cdot V_2. \tag{2.42}$$

Table 2.1 Circuit voltages and current as a function of current gain

f	V1 (V)	V2 (V)	*i* (A)
0.5	0.6	0.4	0.4
1	0.5	0.5	0.5
4	2	−1	−1

Substitute into (2.40):

$$2V_2(2-f) - (1-f)V_2 = 1 \rightarrow V_2[2(2-f) - (1-f)] = 1$$

$$V_2(4 - 2f - 1 + f) = 1 \rightarrow V_2(3-f) = 1$$

$$V_2 = \frac{1}{3-f}. \tag{2.43}$$

From (2.42),

$$V_1 = \frac{2-f}{3-f}. \tag{2.44}$$

Note that $f \neq 3$. As a check in SPICE, Table 2.1 displays the results.
Note that $f = 3$ yields a SPICE error message.
The circuit used in SPICE including the CCCS is shown in Fig. 2.27b.
SPICE Netlist (cccs1.cir):

```
*fx N+ N- Vy Value
*x  Name of the source
*N+ : Name of positive node
*N- : Name of negative node. Current flows from the + node
*   through the source to the - node
*Vref : Name of the voltage source through the controlling
*current flows.
* The direction of positive control current is
* from + node through the source to the - node of Vref=0
*Value: Current gain
i1 0 1 1
f1 1 2 Vref 4
Vref 3 0 0
R1 1 0 1
R2 1 2 1
R3 2 3 1
*.op
```

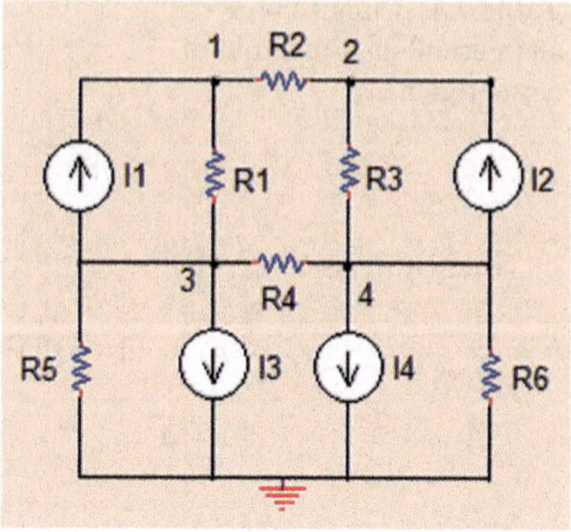

Fig. 2.28 The circuit for Problem 2.1.28

Problem 2.1.28 Find the node voltage values in the circuit shown in Fig. 2.28. All resistors are 1 Ω and $I_1 = 4$ A, $I_2 = 1$ A, $I_3 = 1$ A, $I_4 = 4$ A (Sim_Lin_Eq_Solve.m, matrix_solve.xlsx).

Solution

$$I_1 = \frac{V_1 - V_2}{R_2} + \frac{V_1 - V_3}{R_1} \quad \rightarrow \quad 2V_1 - V_2 - V_3 = 4$$

$$I_2 = \frac{V_2 - V_1}{R_2} + \frac{V_2 - V_4}{R_3} \quad \rightarrow \quad -V_1 + 2V_2 - V_4 = 1$$

$$I_1 + I_3 + \frac{V_3}{R_5} + \frac{V_3 - V_4}{R_4} + \frac{V_3 - V_1}{R_1} = 0 \quad \rightarrow \quad -V_1 + 3V_3 - V_4 = -5$$

$$I_2 + I_4 + \frac{V_4}{R_6} + \frac{V_4 - V_2}{R_3} + \frac{V_4 - V_3}{R_4} = 0 \quad \rightarrow \quad -V_2 - V_3 + 3V_4 = -5.$$

From these four equations, the following matrix equation is obtained:

$$[G] \cdot [V] = [I] \quad \rightarrow \quad \begin{bmatrix} 2 & -1 & -1 & 0 \\ -1 & 2 & 0 & -1 \\ -1 & 0 & 3 & -1 \\ 0 & -1 & -1 & 3 \end{bmatrix} \cdot \begin{bmatrix} V_1 \\ V_2 \\ V_3 \\ V_3 \end{bmatrix} = \begin{bmatrix} 4 \\ 1 \\ -1 \\ -5 \end{bmatrix}.$$

Solution of this matrix equation by employing any available software yields the voltage values as

$$V_1 = 2.727\,\text{V}, \quad V_2 = 1.273\,\text{V}, \quad V_3 = 0.182\,\text{V}, \quad V_4 = -1.182\,\text{V}.$$

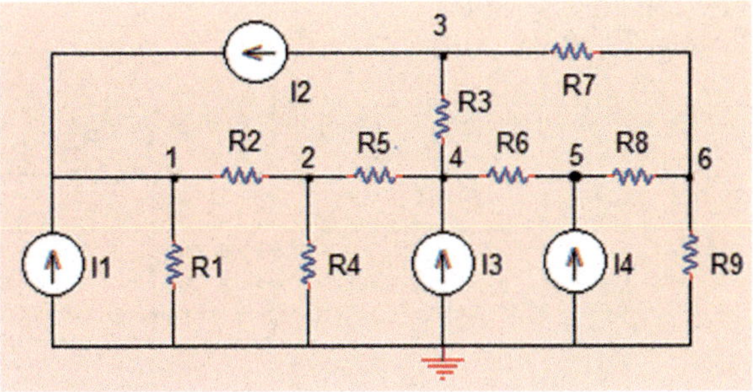

Fig. 2.29 The circuit for Problem 2.1.29

Problem 2.1.29 In the circuit shown in Fig. 2.29,

$$R_1 = R_2 = R_3 = R_4 = R_5 = R_6 = R_7 = R_8 = 1\,\mathrm{k\Omega}, \quad R_9 = 100\,\Omega,$$

$$I_1 = I_2 = I_3 = I_4 = 1\,\mathrm{mA}.$$

(a) Find the conductance matrix for the circuit.
(b) Compute the node voltages (Sim_Lin_Eq_Solve.m, matrix_solve.xlsx).

Solution

(a) $[G][V] = [I]$

$$G_1 = G_2 = G_3 = G_4 = G_5 = G_6 = G_7 = G_8 = 1\,\mathrm{mS}, \quad G_9 = 1 \times 10^{-2}\,\mathrm{mS}.$$

Then, the conductance matrix is

$$[G] = \begin{bmatrix} 2 & -1 & 0 & 0 & 0 & 0 \\ -1 & 3 & 0 & -1 & 0 & 0 \\ 0 & 0 & 2 & -1 & 0 & -1 \\ 0 & -1 & -1 & 3 & -1 & 0 \\ 0 & 0 & 0 & -1 & 2 & -1 \\ 0 & 0 & -1 & 0 & -1 & 12 \end{bmatrix} \times 10^{-3}\,\mathrm{S} \quad (= \mathrm{mS})$$

$$G_7 + G_8 + G_9 = 10^{-3} + 10^{-3} + 10^{-2} = \left(2 \times 10^{-3} + 10^{-2}\right)\mathrm{S} = 12 \times 10^{-3}\,\mathrm{S}.$$

(b) $I = [2 \quad 0 \quad -1 \quad 1 \quad 1 \quad 0]^{\mathrm{T}}\mathrm{mA}$, (T: transpose operator), $[G][V] = [I]$. Solution of this matrix equation for voltage vector (e.g., using MATLAB or EXCEL) yields

$$[V] = [\,1385.5 \quad 771.1 \quad 6.0 \quad 927.7 \quad 1006.0 \quad 84.3\,]^{\mathrm{T}}\,\mathrm{mV}.$$

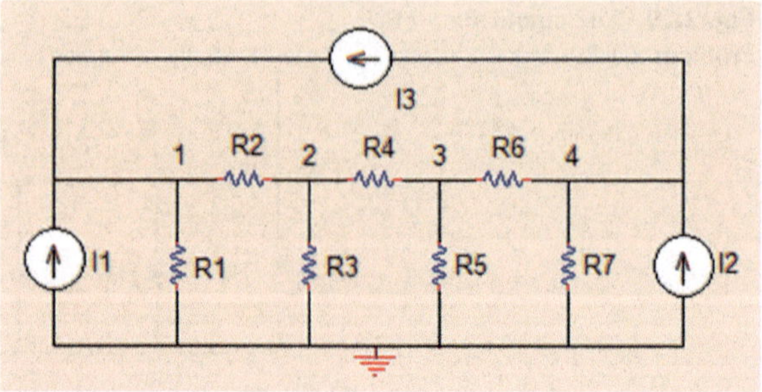

Fig. 2.30 The circuit for Problem 2.1.30

Problem 2.1.30 Find the node voltage values in the circuit shown in Fig. 2.30. $R1 = R6 = R7 = 1\ \Omega$, $R2 = R3 = R5 = 2\ \Omega$, $R4 = 4\ \Omega$, $I1 = 2$ A, $I2 = 1$ A, $I3 = 3$ A (Sim_Lin_Eq_Solve.m, matrix_solve.xlsx).

Solution
Analysis by inspection,

$$G_1 = 1\,\text{S}, \quad G_2 = 0.5\,\text{S}, \quad G_3 = 0.5\,\text{S}, \quad G_4 = 0.25\,\text{S}, \quad G_5 = 0.5\,\text{S},$$

$$G_6 = 1\,\text{S}, \quad G_7 = 1\,\text{S}$$

$$[G][V] = [I], \quad \begin{bmatrix} 1.5 & -0.5 & 0 & 0 \\ -0.5 & 1.25 & -0.25 & 0 \\ 0 & -0.25 & 1.75 & -1 \\ 0 & 0 & -1 & 2 \end{bmatrix} \begin{bmatrix} v_1 \\ v_2 \\ v_3 \\ v_4 \end{bmatrix} = \begin{bmatrix} 3+2 \\ 0 \\ 0 \\ -3+1 \end{bmatrix}.$$

Solution of this matrix equation (e.g., using EXCEL or MATLAB tools) gives node voltages:

$$v_1 = 3.806\,\text{V}, \quad v_2 = 1.419\,\text{V}, \quad v_3 = -0.516\,\text{V}, \quad v_4 = -1.258\,\text{V}.$$

Problem 2.1.31 In the circuit shown in Fig. 2.31a, $R_1 = R_2 = R_3 = 1\ \Omega$, $i_S = 1$ A.

(a) Use SPICE to find the values of the node voltages V_1, V_2, and the current i for current-controlled current source constants of $f = 2$, $f = 4$, and $f = 8$.
(b) Plot i versus i_S curve, $-1 \le i_S \le 1$ A, if the CCCS constant is 8 A/A. Include net list.

Solution

(a) The circuit used in SPICE analysis is shown in Fig. 2.31b: (Table 2.2)
(b) Fig. 2.31c displays the current sweep.

Fig. 2.31 **a** The circuit for Problem 2.1.31, **b** The circuit for Problem 2.1.31 for SPICE analysis, **c** SPICE analysis result for the Circuit of Problem 2.1.31. The current sweep

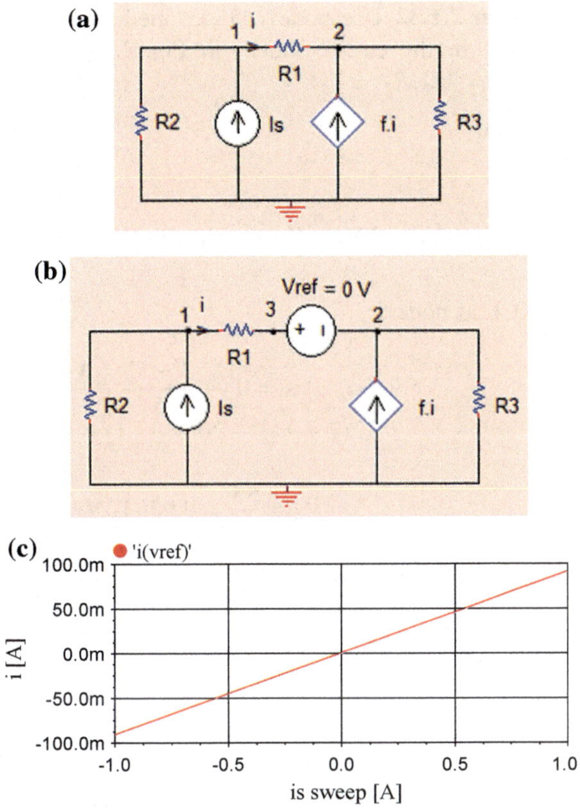

Table 2.2 Voltage and current values against current gain

f	V_1 (V)	V_2 (V)	i (A)
2	0.800	0.600	0.200
4	0.857	0.714	0.143
8	0.910	0.820	0.091

```
*SPICE Netlist for current sweep with f=8:
*Analysis: DC Transfer Curves
cccs2
is 0 1 dc 1
R1 1 3 1
R2 1 0 1
R3 2 0 1
f1 0 2 vref 8
vref 3 2 0
.dc is -1 1 .1
```

Problem 2.1.32 Use node voltages method and determine all currents (mA) and V_2(mV) in the circuit shown in Fig. 2.32. $V_1 = 2\,\text{V}, V_3 = 1\,\text{V}, R_1 = 5\,\Omega, R_2 = 3\,\Omega, R_3 = 4\,\Omega, R_4 = 2\,\Omega$.

Solution

$$V_1 = 2\,\text{V}, \quad V_3 = 1\,\text{V}, \quad i_3 = \frac{V_1 - V_3}{R_4} = \frac{2-1}{2} = 500\,\text{mA}.$$

KCL at node 2:

$$i_1 - i_2 - i_4 = 0 \rightarrow \frac{2 - V_2}{5} - \frac{V_2 - 1}{3} - \frac{V_2}{4} = 0 \quad \rightarrow$$
$$12(2 - V_2) - 20(V_2 - 1) - 15V_2 = 0$$

$$V_2 = \frac{44}{47} = 0.93617\,\text{V} = 936.17\,\text{mV}$$

$$i_1 = \frac{2 - 0.93617}{5} = 212.766\,\text{mA}, \quad i_2 = \frac{0.93617 - 1}{3} = -21.277\,\text{mA}$$

$$i_4 = \frac{0.93617}{4} = 234.043\,\text{mA}, \quad i_{V_1} = i_1 + i_3 = 212.766 + 500 = 712.766\,\text{mA}$$

$$i_{V_2} = -(i_2 + i_3) = -(-21.277 + 500) = -478.723\,\text{mA}.$$

Problem 2.1.33 Determine the node voltages in the circuit shown in Fig. 2.33.

$$V_1 = 12\,\text{V}, \quad V_2 = 6\,\text{V}, \quad R_1 = 4\,\Omega, \quad R_2 = 2\,\Omega, \quad R_3 = 2\,\Omega, \quad R_4 = 6\,\Omega.$$

Solution
There is a voltage source (V_2) connected between two nonreference nodes (2,3). These nodes form a supernode. KCL and KVL can be applied to obtain the node voltages in this circuit.

Fig. 2.32 The circuit for Problem 2.1.32

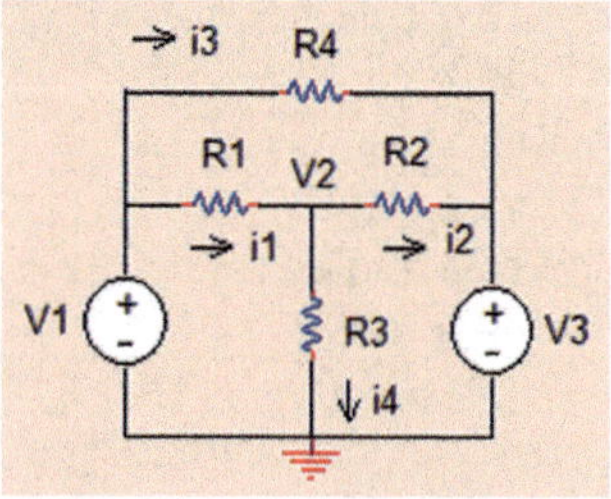

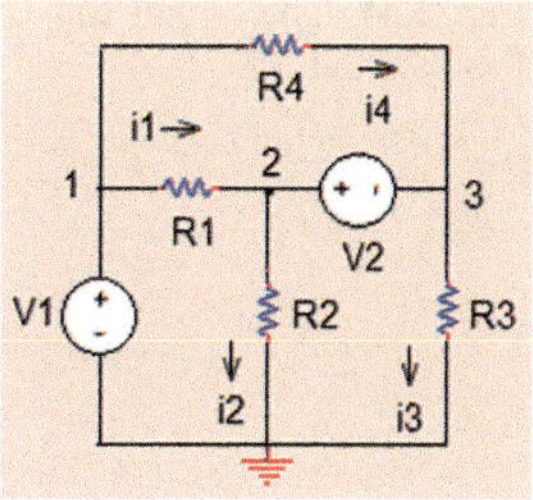

Fig. 2.33 The circuit for Problem 2.1.33

On the other hand, V_1 is connected between node 1 and ground. Thus, the voltage at node 1 equals to $v_1 = V_1 = 12\,\text{V}$.

At the supernode,

$$i_1 + i_4 - i_2 - i_3 = 0 \tag{2.45}$$

$$\frac{v_1 - v_2}{R_1} + \frac{v_1 - v_3}{R_4} - \frac{v_2}{R_2} - \frac{v_3}{R_3} = 0. \tag{2.46}$$

But constraint equation is

$$v_2 - v_3 = V_2 = 6\,\text{V} \quad \rightarrow \quad v_2 = V_2 + v_3 = 6 + v_3 \tag{2.47}$$

$$\frac{v_1 - v_2}{4} + \frac{v_1 - v_3}{6} = \frac{1}{2}(v_2 + v_3) = \frac{1}{2}(6 + v_3 + v_3) = \frac{6 + 2v_3}{2}. \tag{2.48}$$

Since $v_1 = V_1 = 12\,\text{V}$,

$$\frac{12 - (6 + v_3)}{4} + \frac{12 - v_3}{6} = \frac{6 + 2v_3}{2} \tag{2.49}$$

$$36 - 16 - 3v_3 + 24 - 2v_3 = 36 + 12v_3 \tag{2.50}$$

$$-17v_3 = -8 \quad \rightarrow \quad v_3 = \frac{8}{17}\,\text{V}.$$

From (2.47),

$$v_2 = 6 + \frac{8}{17} = \frac{110}{17}\,\text{V}.$$

Summary: $v_1 = 12\,\text{V}, \quad v_2 = 6.471\,\text{V}, \quad v_3 = 0.471\,\text{V}.$

Problem 2.1.34 Determine the voltage at node 2 and the current flowing through the voltage source in the circuit shown in Fig. 2.34. Prove the latter result by applying KCL at node 3.

$I_1 = 2\,\text{A}, V_1 = 2\,\text{V}, R_1 = 4\,\Omega, R_2 = 8\,\Omega, R_3 = 8\,\Omega, R_4 = 2\,\Omega$ (supernode1.cir).

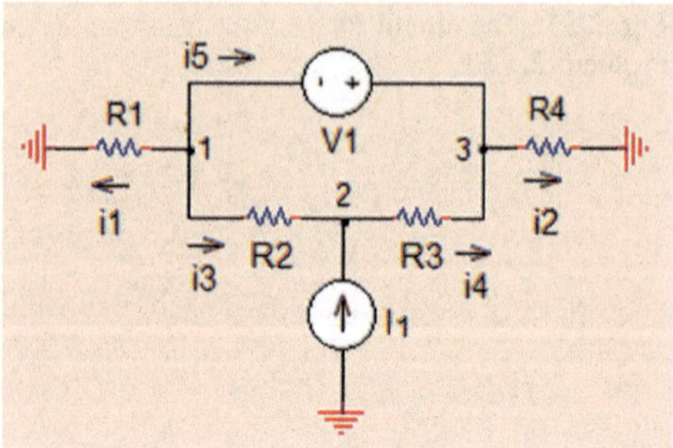

Fig. 2.34 The circuit for Problem 2.1.34

Solution

Since independent voltage source is connected between nodes (1, 3), these nodes form a supernode. Node 2 is included in this supernode. Thus,

$$-i_1 - i_2 + 5 = 0 \rightarrow -\frac{v_1}{R_1} - \frac{v_3}{R_4} + 2 = 0 \rightarrow -\frac{v_1}{4} - \frac{v_3}{2} + 2 = 0$$

$$v_1 + 2v_3 = 8. \tag{2.51}$$

The constraint is

$$v_3 - v_1 = 2\,\text{V} \quad \rightarrow \quad v_3 = 2 + v_1. \tag{2.52}$$

Substituting (2.52) into (2.51),

$$v_1 + 2(2 + v_1) = 8 \quad \rightarrow \quad v_1 + 4 + 2v_1 = 8 \quad \rightarrow \quad 3v_1 = 4$$

$$v_1 = \frac{4}{3} = 1.3333\,\text{V}.$$

From (2.52)

$$v_3 = 2 + 1.3333 = 3.3333\,\text{V}.$$

At node 2:

$$i_3 + I_1 - i_4 = 0 \rightarrow \frac{v_1 - v_2}{R_2} + I_1 - \frac{v_2 - v_3}{R_3} = 0. \tag{2.53}$$

Substituting the values for v_1 and v_2 into (2.53),

$$\frac{1.3333 - v_2}{8} + 2 - \frac{v_2 - 3.3333}{8} = 0 \quad \rightarrow \quad 4.6667 - 2v_2 = -16$$

$$-v_2 = \frac{-16 - 4.6667}{2} \quad \rightarrow \quad v_2 = \frac{20.6667}{2} = 10.3333\,\text{V}$$

$$i_3 = \frac{v_1 - v_2}{R_2} = \frac{1.3333 - 10.3333}{4} = -2.25\,\text{A}.$$

The current flowing through the voltage source is calculated by applying KCL at node 1:

$$-i_1 - i_5 - i_3 = 0 \quad \rightarrow \quad -\frac{v_1}{R_1} - i_5 - \frac{v_1 - v_2}{R_2} = 0$$

$$-\frac{1.3333}{4} - i_5 - \frac{1.3333 - 10.3333}{8} = 0 \quad \rightarrow \quad -0.3333 - i_5 + 1.125 = 0$$

$$i_5 = 0.79167\,\text{A}.$$

Proof KCL at node 3,

$$i_5 + i_4 - i_2 = 0 \quad \rightarrow \quad 0.79167 + \frac{v_2 - v_3}{R_3} - \frac{v_3}{R_4} = 0$$

$$0.79167 + \frac{10.3333 - 3.3333}{8} - \frac{3.3333}{2} = 0$$

$$0.79167 + \frac{7}{8} - \frac{3.3333}{2} = 0 \quad \rightarrow \quad 0.79167 + 0.875 - 1.66667 = 0 \quad \text{Q.E.D.}$$

SPICE netlist:

```
supernode1
*OP analysis
R1 1 0 4
R2 1 2 8
R3 2 3 8
R4 4 0 2
V1 3 1 2
I1 0 2 2
V3 3 4 0
```

Problem 2.1.35 Determine the current through dependent source and current though independent voltage source in the circuit shown in Fig. 2.35. Here, v_2 is the node voltage at node 2.

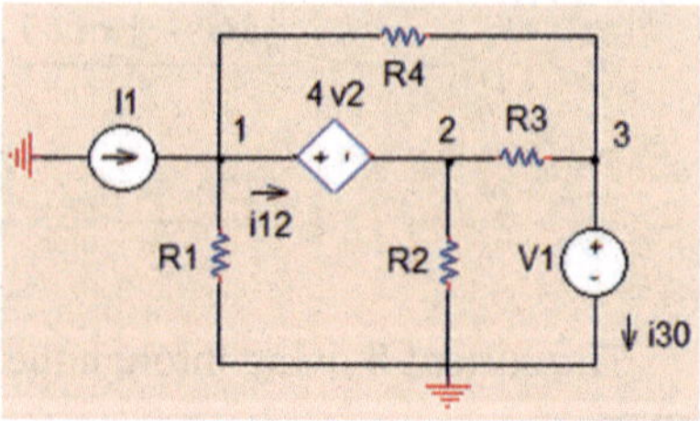

Fig. 2.35 The circuit for Problem 2.1.35

$$I_1 = 4\,\text{A}, \quad V_1 = 5\,\text{V}, \quad R_1 = R_4 = 2\,\Omega, \quad R_2 = 4\Omega, \quad R_3 = 1\,\Omega.$$

Solution

Consider the supernode consisting of nodes (1,2). Applying KCL,

$$I_1 - \frac{v_1}{R_1} - \frac{v_1 - v_3}{R_4} - \frac{v_2 - v_3}{R_3} - \frac{v_2}{R_2} = 0$$

$$4 - \frac{v_1}{2} - \frac{v_1 - v_3}{2} - \frac{v_2 - v_3}{1} - \frac{v_2}{4} = 0$$

$$16 - 2v_1 - 2(v_1 - v_3) - 4(v_2 - v_3) - v_2 = 0$$

$$4v_1 + 5v_2 - 6v_3 = 16 \tag{2.54}$$

But,

$$v_1 = v_2 + 4v_2 = 5v_2 \tag{2.55}$$

$$v_3 = V_1 = 5\,\text{V}. \tag{2.56}$$

Substituting (2.55), (2.56), into (2.54),

$$4(5v_2) + 5v_2 - 6 \times 5 = 16 \quad \rightarrow \quad 20v_2 + 5v_2 - 30 = 16 \quad \rightarrow \quad v_2 = 1.84\,\text{V}.$$

From (2.55),

$$v_1 = 5 \times 1.84 = 9.2\,\text{V}.$$

KCL at node 1:

$$i_{12} = I_1 - \frac{v_1}{R_1} - \frac{v_1 - v_3}{R_4} = 4 - \frac{9.2}{2} - \frac{9.2 - 5}{2} = -2.7\,\text{A}.$$

KCL at node 3:

$$i_{30} = \frac{v_2 - v_3}{R_3} + \frac{v_1 - v_3}{R_4} = \frac{1.84 - 5}{1} + \frac{9.2 - 5}{2} = -1.06\,\text{A}.$$

Following is the SPICE netlist (supernode2.cir) for the operating point analysis of this circuit:

```
supernode2
*OP analysis
R1 1 0 2
R2 2 0 4
R3 2 3 1
R4 1 3 2
V1 3 0 5
I1 0 1 4
*VCVS: e{name} {+node} {-node} {+cntrl} {-cntrl} {gain}
e1 1 2 2 0 4
```

2.2 Mesh Analysis

Problem 2.2.1 Find the values of V_x, V_0 in the circuit shown in Fig. 2.36. $U = 35$ V.

Solution
KVL around the loop, $U - V_x - 2V_x + V_0 = 0$

$$V_x = 10i, \quad V_0 = -5i$$

$$35 - 10i - 20i - 5i = 0$$

$$i = 1\,\text{A}, \quad V_0 = -5i = -5\,\text{V}, \quad V_x = 10i = 10\,\text{V}.$$

SPICE netlist (mesh01):

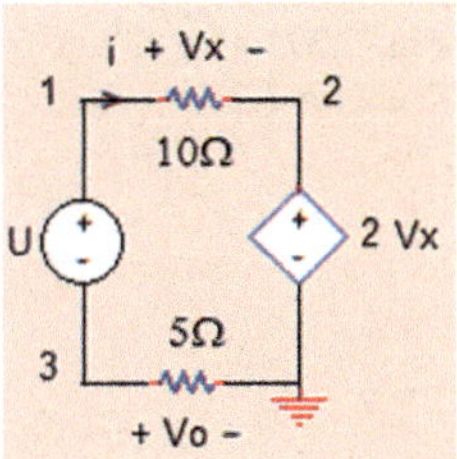

Fig. 2.36 The circuit for Problem 2.2.1

```
mesh01
*OP ANALYSIS
VU 1 3 35
R1 1 2 10
R2 3 0 5
*VCVS: Ex N+ N- NC+ NC- VALUE
E1 2 0 1 2 2
```

Problem 2.2.2

(a) Determine the current i_{ab} in the circuit shown in Fig. 2.37.
(b) If $U1 = 10$ V, $U2 = 6$ V, $R = 1$ kΩ what is the value of i_{ab}? (mA)
(c) If $U1 = U2 = 10$ V, $R = 1$ kΩ what is the value of i_{ab}? (mA)
(d) If $U1 = U2/2 = 10$ V, $R = 1$ kΩ what is the value of i_{ab}? (mA)

Solution

(a) Current through the left mesh (in CW direction),

$$i_1 = \frac{U1}{R}$$

Current through the right mesh, (in CW direction)

$$i_2 = -\frac{U2}{2R}$$

$$i_{ab} = i_1 + i_2 = \frac{U1}{R} - \frac{U2}{2R} = \frac{1}{R}\left(U1 - \frac{U2}{2}\right)$$

(b)

$$i_{ab} = \frac{\left(10 - \frac{6}{2}\right)}{10^3} = \frac{7}{10^3} = 7\,\text{mA}$$

(c) If $U1 = U2 = 10$ V,

$$i_{ab} = \frac{5}{10^3} = 5\,\text{mA}$$

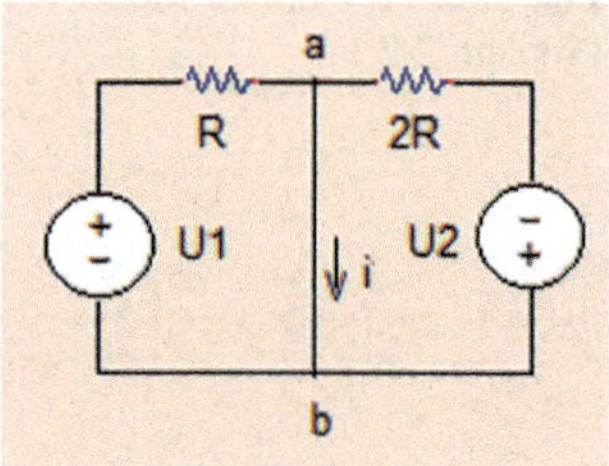

Fig. 2.37 The circuit for Problem 2.2.2

(d) If $U1 = U2/2 = 10$ V, R = 1 kΩ,

$$i_{\text{ab}} = \frac{\left(10 - \frac{20}{2}\right)}{10^3} = 0\,\text{mA}.$$

Problem 2.2.3 In the circuit shown in Fig. 2.38, use mesh currents method and find the value of voltage V_x. What is the voltage drop across R_2?. $R_1 = R_3 = 2R_2 = 4\,\Omega, V_1 = 3\,\text{V}, V_2 = 5\,\text{V}$ (Sim_Lin_Eq_Solve.m, matrix_solve.xlsx).

Solution
By applying mesh currents and analysis by inspection, the governing equation of the circuit is

$$\begin{bmatrix} R_1 + R_2 & -R_2 \\ -R_2 & R_2 + R_3 \end{bmatrix} \begin{bmatrix} \text{i}_1 \\ \text{i}_2 \end{bmatrix} = \begin{bmatrix} V_1 - V_2 \\ V_2 \end{bmatrix}$$

or

$$\begin{bmatrix} 6 & -2 \\ -2 & 6 \end{bmatrix} \begin{bmatrix} i_1 \\ i_2 \end{bmatrix} = \begin{bmatrix} -2 \\ 5 \end{bmatrix}.$$

From this matrix equation, i_1 and i_2 can be obtained as $i_1 = -0.0625\,\text{A}$; $i_2 = 0.8125\,\text{A}$.

The voltage V_x is calculated as $V_x = i_2 \cdot R_3 = 0.8125 \cdot 4 = 3.25\,\text{V}$.

The voltage drop across R_2, $V_{R2} = V_x - V_2 = 3.25 - 5 = -1.75\,\text{V}$.

Problem 2.2.4 Apply mesh analysis method to find the values of currents i_1 and i_2 and the node voltage in the circuit shown in Fig. 2.39 (V_1 = 2 V, V_2 = 1 V, R_1 = 1 Ω, R_2 = 1 Ω, R_3 = 2 Ω) (Sim_Lin_Eq_Solve.m, matrix_solve.xlsx).

Solution
Using mesh analysis for the circuit,

$$-V_1 + i_1 \cdot R_1 + (i_1 - i_2) \cdot R_3 = 0$$

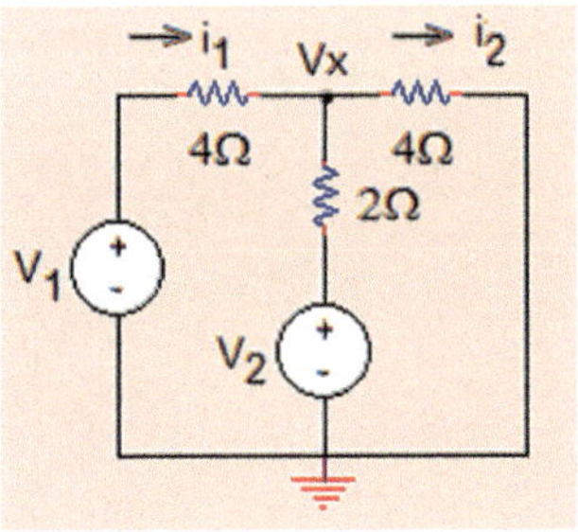

Fig. 2.38 The circuit for Problem 2.2.3

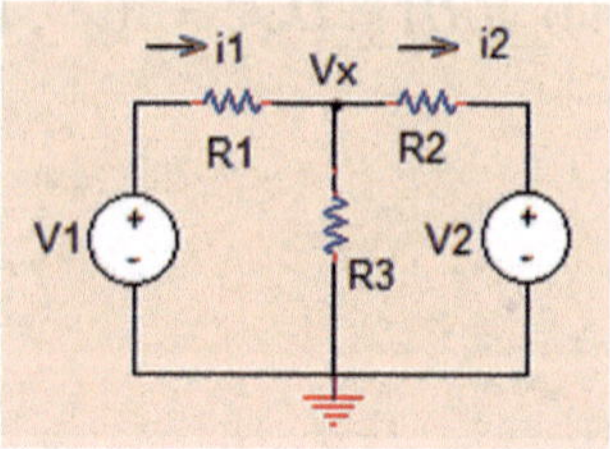

Fig. 2.39 The circuit for Problems 2.2.4 and 2.2.5

$$(i_2 - i_1) \cdot R_3 + i_2 \cdot R_2 + V_2 = 0.$$

Substituting V_1, V_2, R_1, and R_2 values in these equations,

$$-2 + 3i_1 - 2i_2 = 0$$

$$3i_2 - 2i_1 - 1 = 0.$$

Solution of this set of simultaneous linear equations for unknown current values yields

$$i_1 = 0.8\,\text{A}, i_2 = 0.2\,\text{A}$$

$$V_x = (i_1 - i_2).R_3 = 0.6 \times 2 = 1.2\,\text{V}.$$

Problem 2.2.5 In the circuit shown in Fig. 2.39, use mesh currents method and Cramer's rule to find the values for V_x, V_{R1}, V_{R2}. ($R_1 = R_2 = 2\ \Omega$, $R_3 = 1\ \Omega$, $V_1 = 2$ V, $V_2 = 1$ V) (Sim_Lin_Eq_Solve.m, matrix_solve.xlsx).

Solution

Using analysis by inspection and mesh currents method,

$$\begin{bmatrix} R_1 + R_3 & -R_3 \\ -R_3 & R_2 + R_3 \end{bmatrix} \begin{bmatrix} i_1 \\ i_2 \end{bmatrix} = \begin{bmatrix} V_1 \\ -V_2 \end{bmatrix}.$$

Substituting given component values,

$$\begin{bmatrix} 3 & -1 \\ -1 & 3 \end{bmatrix} \begin{bmatrix} i_1 \\ i_2 \end{bmatrix} = \begin{bmatrix} 2 \\ -1 \end{bmatrix}$$

$$\Delta = 3 \cdot 3 - 1 = 8, \quad \Delta_1 = \begin{vmatrix} 2 & -1 \\ -1 & 3 \end{vmatrix} = 6 - 1 = 5, \quad i_1 = \frac{\Delta_1}{\Delta} = \frac{5}{8}\,\text{A}$$

$$V_{R1} = i_1 R_1 = \frac{5}{8} \cdot 2 = \frac{10}{8} = 1.25\,\text{V}$$

$$V_x = V_1 - V_{R1} = 2 - 1.25 = 0.75\,\text{V}$$

$$V_{R2} = V_x - V_2 = 0.75 - 1 = 0.25\,\text{V}.$$

Problem 2.2.6 Find the values of mesh currents and the node voltage in the circuit shown in Fig. 2.40. $R1 = 10\ \Omega$, $R2 = 2\ \Omega$, $R3 = 1\ \Omega$, $V1 = 4$ V, $V2 = 2$ V, $V3 = 1$ V (Sim_Lin_Eq_Solve.m, matrix_solve.xlsx).

Solution
KVL in mesh1:

$$-4 + 10i_1 + 2(i_1 - i_2) + 2 = 0$$

$$12i_1 - 2i_2 = 2. \tag{2.57}$$

KVL in mesh2:

$$-2 + 2(i_2 - i_1) + i_2 + 1 = 0$$

$$-2i_1 + 3i_2 = 1. \tag{2.58}$$

Using Eqs. (2.57)–(2.58), one obtains the following matrix equation:

$$\begin{bmatrix} 12 & -2 \\ -2 & 3 \end{bmatrix} \times \begin{bmatrix} i_1 \\ i_2 \end{bmatrix} = \begin{bmatrix} 2 \\ 1 \end{bmatrix}.$$

The solution of this matrix equation gives

$$i_1 = 0.25\,\text{A}$$

$$i_2 = 0.5\,\text{A}$$

$$v_x = V_2 + (i_1 - i_2) \times R_2 = 2 - 0.25 \times 2 = 1.5\,\text{V}.$$

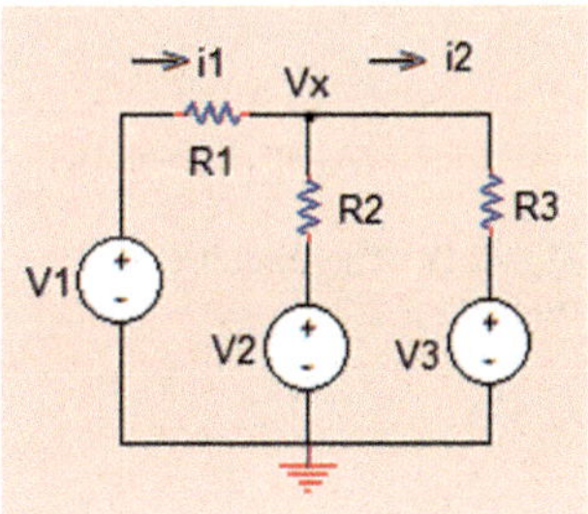

Fig. 2.40 The circuit for Problem 2.2.6

Alternatively,

$$v_x = V_3 + i_2R_3 = 1 + 0.5 \times 1 = 1.5\,\text{V}.$$

Problem 2.2.7 In the circuit shown in Fig. 2.41, determine the voltage drop across *R*2 using mesh analysis.

*R*1 = 3 Ω, *R*2 = 5 Ω, *R*3 = 4 Ω, *v*1 = 2 V, *I* = 2 A.

Solution

Since a current source exists in the second mesh,

$$i_2 = -I = -2\,\text{A}.$$

The mesh equation for the other mesh,

$$-2 + 3i_1 + (i_1 - i_2) \times 5 = 0.$$

Solving this equation for the unknown current,

$$3i_1 + 5i_1 + 10 - 2 = 0 \quad \rightarrow \quad 8i_1 + 8 = 0 \quad \rightarrow \quad i_1 = -1\,\text{A}.$$

The voltage drop across R_2 is

$$V_{R_3} = 5(i_1 - i_2) = 5(-1 + 2) = 5\,\text{V}.$$

Problem 2.2.8 Find the values of numbered (clockwise flowing) mesh currents in the circuit shown in Fig. 2.42. Use Cramer's rule when necessary $U = 4\,\text{V}, I = 2\,\text{A}, R = R_1 = R_2 = R_3 = R_4 = 1\,\Omega$ (mesh1.cir).

Solution

Assume clockwise rotation for mesh currents.

Since $i_2 = I$, it is not necessary to write down KVL equation associated with second mesh.

KVL in mesh 1:

$$-U = (i - i_3)R_1 + (i_1 - i_2)R_3 = 0$$

$$i_1(R_1 + R_3) - i_3R_1 = U + IR_3. \tag{2.59}$$

Fig. 2.41 The circuit for Problem 2.2.7

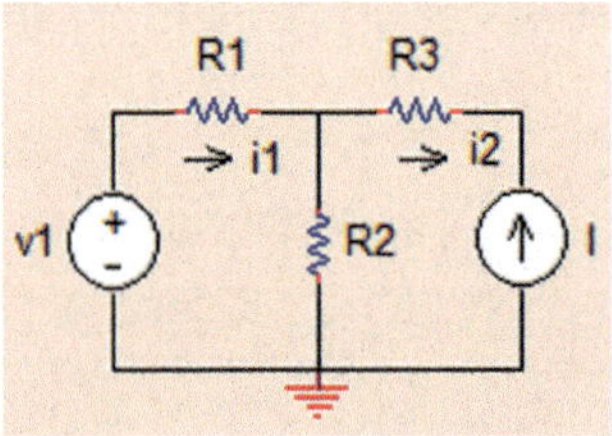

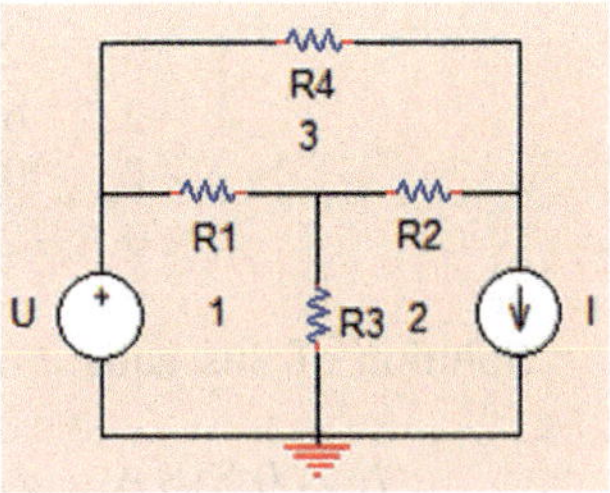

Fig. 2.42 The circuit for Problem 2.2.8

KVL in mesh 3:

$$i_3R_4 + (i_3 - i_1)R_1 + (i_3 - i_2)R_2 = 0$$

$$i_3(R_1 + R_3 + R_4) - i_1R_1 = U + IR_2. \tag{2.60}$$

From (2.59) and (2.60),

$$\begin{bmatrix} R_1 + R_3 & -R_1 \\ -R_1 & R_1 + R_2 + R_4 \end{bmatrix} \begin{bmatrix} i_1 \\ i_3 \end{bmatrix} = \begin{bmatrix} U + IR_3 \\ IR_2 \end{bmatrix}$$

$$\begin{bmatrix} 2 & -1 \\ -1 & 3 \end{bmatrix} \begin{bmatrix} i_1 \\ i_3 \end{bmatrix} = \begin{bmatrix} 6 \\ 2 \end{bmatrix}, \quad \Delta = 6 - 1 = 5, \quad \Delta_1 = \begin{vmatrix} 6 & -1 \\ 2 & 3 \end{vmatrix} = 20,$$
$$\Delta_2 = \begin{vmatrix} 2 & 6 \\ -1 & 2 \end{vmatrix} = 10$$

$$i_1 = \frac{\Delta_1}{\Delta} = \frac{20}{5} = 4\,\text{A}, i_2 = \frac{\Delta_2}{\Delta} = \frac{10}{5} = 2\,\text{A}.$$

Problem 2.2.9 For the circuit shown in Fig. 2.43, write down the circuit equation in matrix form and solve for mesh currents. $R = 2\ \Omega$, $V1 = V2 = V3 = 1$ V (Sim_Lin_Eq_Solve.m, matrix_solve.xlsx).

Solution
Analysis by inspection yields

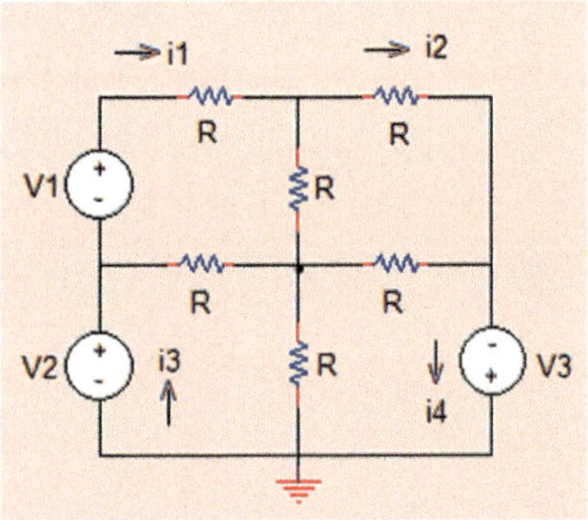

Fig. 2.43 The circuit for Problem 2.2.9

$$\begin{bmatrix} 6 & -2 & -2 & 0 \\ -2 & 6 & 0 & -2 \\ -2 & 0 & 4 & -2 \\ 0 & -2 & -2 & 4 \end{bmatrix} \cdot \begin{bmatrix} I_1 \\ I_2 \\ I_3 \\ I_4 \end{bmatrix} = \begin{bmatrix} 1 \\ 0 \\ 1 \\ 1 \end{bmatrix} \text{(V)}.$$

Solution of this equation (using given EXCEL or MATLAB tools) gives

$$I_1 = 0.818\,\text{A}, \quad I_2 = 0.682\,\text{A}, \quad I_3 = 1.273\,\text{A}, \quad I_4 = 1.227\,\text{A}.$$

Problem 2.2.10 In the circuit shown in Fig. 2.44, find the value of current i_{AB} through 3 Ω resistor.($i_{AB} = -i_{BA}$). Use Cramer's rule, when necessary.
$V1$ = 7 V, $V2$ = 6 V, $R1$ = $R5$ = 1 Ω, $R2$ = $R4$ = 2 Ω, $R3$ = 3 Ω.

Solution

$$i_{AB} = i_3 - i_2.$$

The mesh current equations yield the following matrix equation:

$$\begin{bmatrix} 3 & -1 & -2 \\ -1 & 6 & -3 \\ -2 & -3 & 6 \end{bmatrix} \begin{bmatrix} i_1 \\ i_2 \\ i_3 \end{bmatrix} = \begin{bmatrix} 7-6 \\ 0 \\ 6 \end{bmatrix} = \begin{bmatrix} 1 \\ 0 \\ 6 \end{bmatrix}$$

$$\Delta = \begin{vmatrix} 3 & -1 & -2 & & 3 & -1 \\ -1 & 6 & -3 & \vdots & -1 & 6 \\ -2 & -3 & 6 & & -2 & -3 \end{vmatrix} = 3 \times 36 + (-6) + (-6) - [24 + 27 + 6] = 39$$

$$\Delta_2 = \begin{vmatrix} 3 & 1 & -2 & & 3 & 1 \\ -1 & 0 & -3 & \vdots & -1 & 0 \\ -2 & 6 & 6 & & -2 & 6 \end{vmatrix} = 0 + 6 + 12 - (0 - 54 - 6) = 18 - (-60) = 78$$

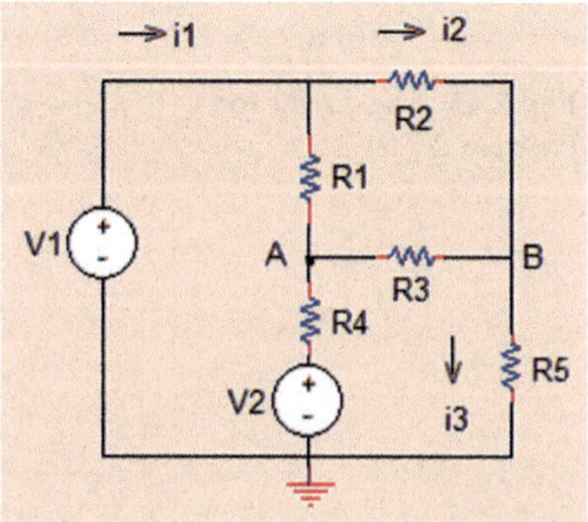

Fig. 2.44 The circuit for Problem 2.2.10

$$\Delta_3 = \begin{vmatrix} 3 & -1 & 1 & \vdots & 3 & -1 \\ -1 & 6 & 0 & \vdots & -1 & 6 \\ -2 & -3 & 6 & & -2 & -3 \end{vmatrix} = 108 + 0 + 3 - (-12 + 0 + 6)$$
$$= 111 - (-6) = 117$$

$$i_2 = \frac{\Delta_2}{\Delta} = \frac{78}{39} = 2\,\text{A}, \quad i_3 = \frac{\Delta_3}{\Delta} = \frac{117}{39} = 3\,\text{A}, \quad i_{AB} = i_3 - i_2 = 3 - 2 = 1\,\text{A}.$$

Problem 2.2.11 For the circuit shown in Fig. 2.45, determine the ratio of currents, $r = \frac{i_{R1}}{i_{R3}}$, for $k = -1, 0, 1, \infty$.

Solution

Assume mesh currents(i_1, i_2) flow clockwise in the left and right meshes, respectively,

$$v = (i_2 - i_1)R_2, \quad i_1 = i_{R1}, \quad i_2 = i_{R3}.$$

KVL at mesh 1:

$$kv + (R_1 + R_2)i_1 - i_2R_2 = 0 \quad \rightarrow \quad k(i_2 - i_1)R_2 + (R_1 + R_2)i_1 - i_2R_2 = 0$$

$$i_1(R_1 + R_2 - kR_2) + i_2(kR_2 - R_2) = 0. \tag{2.61}$$

KVL at mesh 2:

$$-R_2i_1 + (R_2 + R_3)i_2 = U. \tag{2.62}$$

From Eqs. (2.61) and (2.62),

$$\begin{bmatrix} R_1 + R_2(1-k) & -R_2(1-k) \\ -R_2 & R_2 + R_3 \end{bmatrix} \begin{bmatrix} i_1 \\ i_2 \end{bmatrix} = \begin{bmatrix} U \\ 0 \end{bmatrix}$$

$$\Delta_1 = \begin{vmatrix} 0 & -R_2(1-k) \\ U & R_2 + R_3 \end{vmatrix} = UR_2(1-k)$$

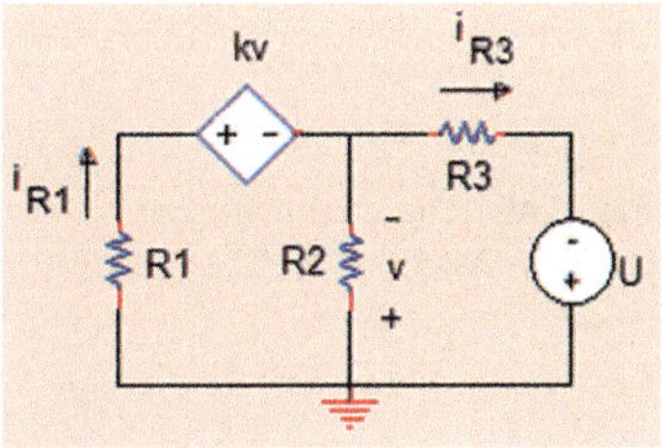

Fig. 2.45 The circuit for Problem 2.2.11

$$\Delta_2 = \begin{vmatrix} R_1 + R_2(1-k) & 0 \\ -R_2 & U \end{vmatrix} = U[R_1 + R_2(1-k)]$$

$$i_1 = \frac{\Delta_1}{\Delta}, i_2 = \frac{\Delta_2}{\Delta}, r = \frac{i_1}{i_2} = \frac{i_{R1}}{i_{R3}} = \frac{\Delta_1}{\Delta_2} = \frac{R_2(1-k)}{R_1 + R_2(1-k)} = \frac{R_2}{\frac{R_1}{1-k} + R_2}.$$

Values of r for different k parameters are shown in Table 2.3.

Problem 2.2.12 In the circuit shown in Fig. 2.46, use mesh analysis and calculate the value of current through 10 Ω internal resistance of the 24 V voltage source, currents through $R_2 = 12\ \Omega$ and $R_3 = 4\ \Omega$. Find the node voltage.

Solution

Applying KVL in the left mesh, taking clockwise current directions, $(i = i_1)$

$$10i_1 + 12i_1 - 12i_2 = 24$$

$$11i_1 - 6i_2 = 12. \tag{2.63}$$

KVL 2:

$$\begin{aligned} 12i_2 - 12i_1 + i_2 &= -4V_x \\ &= -4(12(i_1 - i_2)) \\ &= 48i_1 + 48i_2 \\ &\quad - 12i_1 + 48i_1 + 16i_2 - 48i_2 = 0 \end{aligned}$$

$$36i_1 - 32i_2 = 0$$

$$9i_1 - 8i_2 = 0 \tag{2.64}$$

$$9i_1 = 8i_2 \quad \rightarrow \quad i_1 = \frac{8}{9}i_2 \tag{2.65}$$

Table 2.3 Values of r for different k parameters

k	-1	0	1	∞
r	$\frac{R_2}{\frac{R_1}{2} + R_2}$	$\frac{R_2}{R_1 + R_2}$	0	1

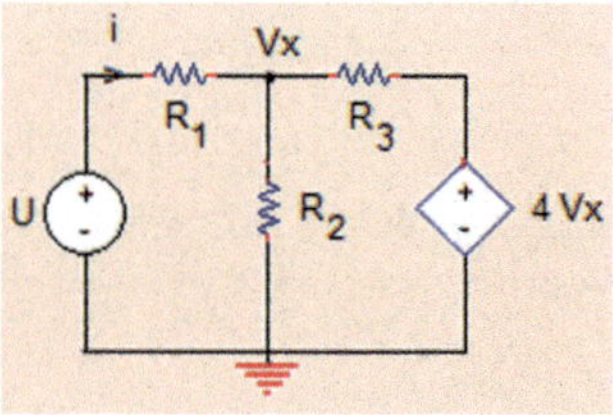

Fig. 2.46 The circuit for Problem 2.2.12

$$11\left(\frac{8}{9}i_2\right) - 6i_2 = 12 \quad \rightarrow \quad i_2\left(\frac{8}{9} - 6\right) = 12 \quad \rightarrow \quad i_2 = \frac{12}{\frac{34}{9}} = \frac{108}{34} = 3.176\,\text{A},$$

$$i = i_1 = \frac{8}{9} \times (3.176) = 2.824\,\text{A}$$

$$i_{R2} = i_1 - i_2 = 2.824 - 3.176 = -0.352\,\text{A}$$

$$V_x = 12 \times i_{R2} = -4.224\,\text{V}.$$

Problem 2.2.13 Find the values of currents i_1, i_2, i_3, i_{AB} in the circuit shown in Fig. 2.47. Use Cramer's rule, when necessary. $V1 = 4$ V, $R1 = 10\ \Omega$, $R2 = R3 = 4\ \Omega$, $R4 = 6\ \Omega$ (Sim_Lin_Eq_Solve.m, matrix_solve.xlsx).

Solution

$$i_0 = i_1 - i_2$$

$$\text{KVL at mesh1:} \; -4 + 10(i_1 - i_2) + 6(i_1 - i_3) = 0 \quad \rightarrow \quad 16i_1 - 10i_2 - 6i_3 = 4$$

$$\text{KVL at mesh2:} \; 4i_2 + 4(i_2 - i_3) + 10(i_2 - i_1) = 0 \quad \rightarrow \quad -10i_1 + 18i_2 - 4i_3 = 0$$

$$\text{KVL at mesh3:} \; 2(i_1 - i_2) + 6(i_3 - i_1) + 4(i_3 - i_2) = 0 \quad \rightarrow \quad -4i_1 - 6i_2 + 10i_3 = 0.$$

Collecting three equations in a matrix form,

$$\begin{bmatrix} 16 & -10 & -6 \\ -10 & 18 & -4 \\ -4 & -6 & 10 \end{bmatrix} \begin{bmatrix} i_1 \\ i_2 \\ i_3 \end{bmatrix} = \begin{bmatrix} 4 \\ 0 \\ 0 \end{bmatrix}$$

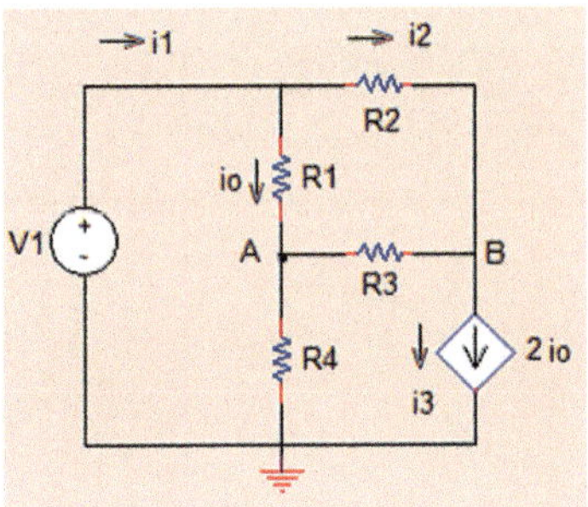

Fig. 2.47 The circuit for Problem 2.2.13

$$\Delta = \begin{vmatrix} 16 & -10 & -6 \\ -10 & 18 & -4 \\ -4 & -6 & 10 \end{vmatrix} = 544, \quad \Delta_1 = \begin{vmatrix} 4 & -10 & -6 \\ 0 & 18 & -4 \\ 0 & -6 & 10 \end{vmatrix} = 624$$

$$\Delta_2 = \begin{vmatrix} 16 & 4 & -6 \\ -10 & 0 & -4 \\ -4 & 0 & 10 \end{vmatrix} = 464, \quad \Delta_3 = \begin{vmatrix} 16 & -10 & 4 \\ -10 & 18 & 0 \\ -4 & -6 & 0 \end{vmatrix} = 528$$

$$i_1 = \frac{\Delta_1}{\Delta} \cong 1.15\,\text{A}, \quad i_2 = \frac{\Delta_2}{\Delta} \cong 0.85\,\text{A}, \quad i_3 = \frac{\Delta_3}{\Delta} \cong 0.97\,\text{A}, \quad i_{\text{ab}} = i_3 - i_2 \cong 0.12\,\text{A}.$$

Problem 2.2.14 In the circuit shown in Fig. 2.48, $R = 1\ \Omega$, $V1 = 2$ V, $V2 = 1$ V. Find the value of current i_X (Sim_Lin_Eq_Solve.m, matrix_solve.xlsx).

Solution
KVL at mesh 1:

$$-2 + (i_1 - i_2) + (i_1 + i_3) + (i_1 - i_5) + i_1 = 0.$$

Simplifying,

$$2 = 4i_1 - i_2 - i_3 - i_5$$

Similarly,

$$\text{KVL at mesh2} : i_2 + (i_2 - i_3) + (i_2 - i_1) = 0 \quad \rightarrow \quad 0 = -i_1 + 3i_2 - i_3$$

$$\text{KVL at mesh3} : 1i_3 + (i_3 - i_1) + (i_3 - i_2) = 0 \quad \rightarrow \quad 0 = -i_1 - i_2 + 4i_3 - i_4$$

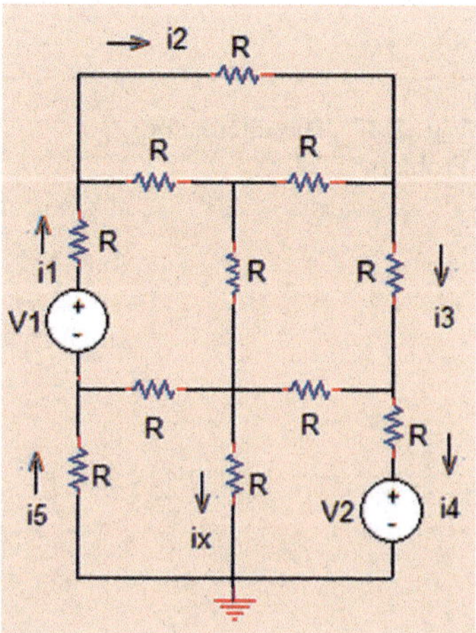

Fig. 2.48 The circuit for Problem 2.2.14

$$\text{KVL at mesh4}: (i_4 - i_3) + i_4 + 1 + (i_4 - i_5) = 0 \quad \rightarrow \quad 0 = -i_3 + 3i_4 - i_5$$

$$\text{KVL at mesh5}: i_5 + (i_5 - i_1) + (i_5 - i_4) = 0 \quad \rightarrow \quad 0 = -i_1 - i_4 + 3i_5.$$

Collecting these equations in a matrix form,

$$\begin{bmatrix} 4 & -1 & -1 & 0 & -1 \\ -1 & 3 & -1 & 0 & 0 \\ -1 & -1 & 4 & -1 & 0 \\ 0 & 0 & -1 & 3 & -1 \\ -1 & 0 & 0 & -1 & 3 \end{bmatrix} \begin{bmatrix} i_1 \\ i_2 \\ i_3 \\ i_4 \\ i_5 \end{bmatrix} = \begin{bmatrix} 2 \\ 0 \\ 0 \\ -1 \\ 0 \end{bmatrix}.$$

Solution of this matrix equation by employing available software (see, MATLAB m file or EXCEL file) yields

$$i_1 = 0.646\,\text{A}, \quad i_2 = 0.273\,\text{A}, \quad i_3 = 0.172\,\text{A}, \quad i_4 = -0.23\,\text{A}, \quad i_5 = 0.139\,\text{A}$$

$$i_y = i_2 = 0.273\,\text{A}$$

$$i_x = i_5 - i_4 = 0.139\,\text{A} + 0.23\,\text{A} = 0.369\,\text{A}.$$

Problem 2.2.15 Find the values of mesh currents and the node voltage V_X in the circuit shown in Fig. 2.49.

R = 1 kΩ (matrix_solve.xlsx).

Solution

$$-10 + i_1 + i_1 - i_2 + i_1 - i_4 = 0$$

$$3i_1 - i_2 - i_4 = 10 \tag{2.66}$$

$$i_2 + i_2 - i_3 + i_2 - i_5 + i_2 - i_1 = 0$$

$$-i_1 + 4i_2 - i_3 - i_5 = 0 \tag{2.67}$$

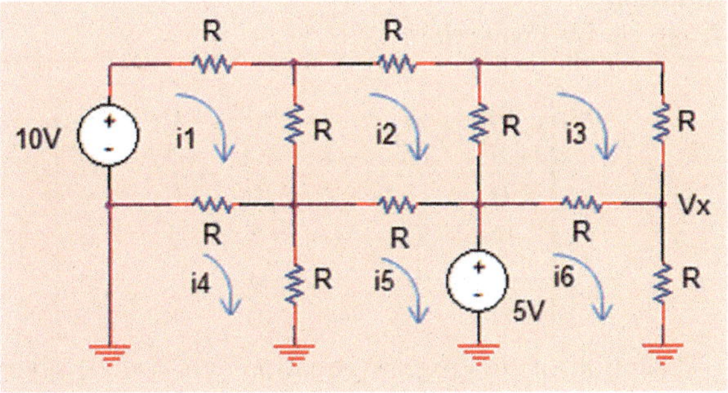

Fig. 2.49 The circuit for Problem 2.2.15

$$i_3 + i_3 - i_6 + i_3 - i_2 = 0$$

$$-i_2 + 3i_3 - i_6 = 0 \tag{2.68}$$

$$i_4 - i_1 + i_4 - i_5 = 0$$

$$-i_1 + 2i_4 - i_5 = 0 \tag{2.69}$$

$$i_5 - i_4 + i_5 - i_2 + 5 = 0$$

$$-i_2 - i_4 + 2i_5 = -5 \tag{2.70}$$

$$-5 + i_6 - i_3 + i_6 = 0$$

$$-i_3 + 2i_6 = 5. \tag{2.71}$$

Using Eqs. (2.66)–(2.71), one obtains the following matrix equation:

$$\begin{bmatrix} 3 & -1 & 0 & -1 & 0 & 0 \\ -1 & 4 & -1 & 0 & -1 & 0 \\ 0 & -1 & 3 & 0 & 0 & -1 \\ -1 & 0 & 0 & 2 & -1 & 0 \\ 0 & -1 & 0 & -1 & 2 & 0 \\ 0 & 0 & -1 & 0 & 0 & 2 \end{bmatrix} \times \begin{bmatrix} i_1 \\ i_2 \\ i_3 \\ i_4 \\ i_5 \\ i_6 \end{bmatrix} = \begin{bmatrix} 10 \\ 0 \\ 0 \\ 0 \\ -5 \\ 5 \end{bmatrix}.$$

The solution of this matrix equation gives

$$i_1 = 4.211\,\text{mA}, \quad i_2 = 1.118\,\text{mA}, \quad i_3 = 1.447\,\text{mA}$$

$$i_4 = 1.513\,\text{mA}, \quad i_5 = -1.184\,\text{mA}, \quad i_6 = 3.224\,\text{mA}$$

$$v_x = 5 - (i_6 - i_3) \times R = 5 - 1.777 = 3.223\,\text{V}.$$

Problem 2.2.16 In the circuit shown in Fig. 2.50, use mesh currents method to determine currents flowing through each resistor (mesh6.xlsx) $R_1 = R_2 = R_3 = 2\,\Omega$, $R_4 = R_5 = R_6 = 4\,\Omega, R_7 = R_{10} = 8\,\Omega, R_8 = R_9 = 1\,\Omega, V_1 = V_2 = 10\,\text{V}$.

Solution

Analysis by inspection, $[R][I] = [V]$,

$$\begin{bmatrix} 14 & -8 & 0 & -4 & 0 & 0 \\ -8 & 15 & -1 & 0 & -4 & 0 \\ 0 & -1 & 7 & 0 & 0 & -4 \\ -4 & 0 & 0 & 5 & -1 & 0 \\ 0 & -4 & 0 & -1 & 13 & -8 \\ 0 & 0 & -4 & 0 & -8 & 12 \end{bmatrix} = \begin{bmatrix} I_1 \\ I_2 \\ I_3 \\ I_4 \\ I_5 \\ I_6 \end{bmatrix} = \begin{bmatrix} 0 \\ 0 \\ 0 \\ 10 \\ 0 \\ -10 \end{bmatrix}.$$

Solution of this matrix equation (via Excel) gives mesh currents as

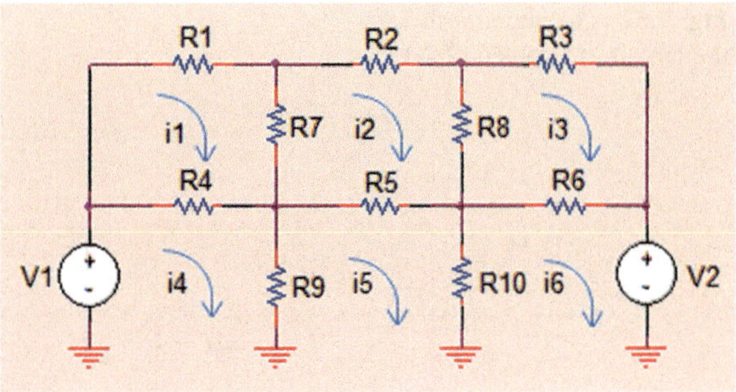

Fig. 2.50 The circuit for Problem 2.2.16

$$I_1 = 0.790\,\text{A}, \quad I_2 = 0.147\,\text{A}, \quad I_3 = -0.937\text{A}$$

$$I_4 = 2.473\,\text{A}, \quad I_5 = -0.796\,\text{A}, \quad I_6 = -1.677\,\text{A}.$$

Individual currents flowing through each resistor are calculated as follows:

$$I_{R1} = I_1 = 0.790\,\text{A}, \quad I_{R2} = I_2 = 0.147\,\text{A}, \quad I_{R3} = I_3 = -0.937$$

$$I_{R4} = I_1 - I_4 = -1.683\,\text{A}, \quad I_{R5} = I_2 - I_5 = 0.943\,\text{A}, \quad I_{R6} = I_3 - I_6 = 0.740\,\text{A}$$

$$I_{R7} = I_1 - I_2 = 0.644\,\text{A}, \quad I_{R8} = I_2 - I_3 = 3.269\,\text{A}, \quad I_{R9} = I_4 - I_5 = 3.269\,\text{A}$$

$$I_{R10} = I_5 - I_6 = 0.880\,\text{A}.$$

Problem 2.2.17 Use mesh analysis and determine the node voltages at nodes 2 and 4 in the circuit shown in Fig. 2.51. $V_a = 12\,\text{V}, V_b = 6\,\text{V}, I = 8\,\text{A}, R_1 = 2R_2 = 3R_3 = 12\,\Omega$ (Supermesh1.cir).

Solution

Since there is a current source between two meshes, a supermesh results by excluding the current source and resistor connected in series with it. Thus KVL around supermesh (Fig. 2.52):

$$-V_a + i_1R_1 + i_2R_2 + V_b = 0 \rightarrow V_b - V_a = -i_1R_1 - i_2R_2$$

$$V_a - V_b = i_1R_1 + i_2R_2 \rightarrow 6 = 12i_1 + 6i_2$$

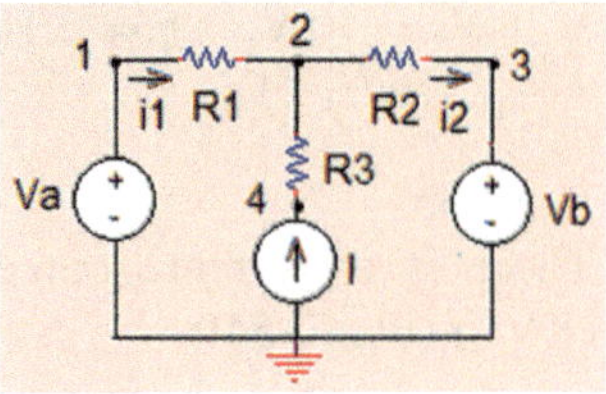

Fig. 2.51 The circuit for Problem 2.2.17

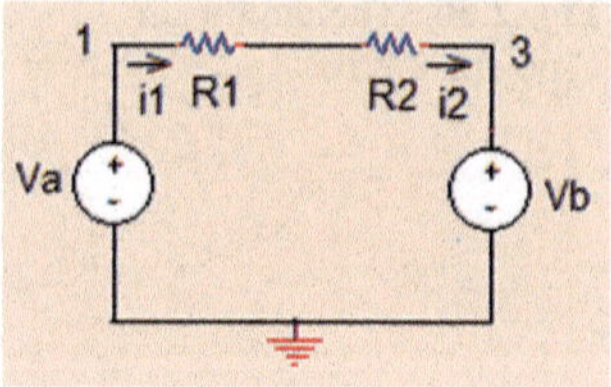

Fig. 2.52 The supermesh for the circuit of Problem 2.2.17

$$2i_1 + i_2 = 1. \tag{2.72}$$

Constraint equation is obtained by applying KCL at node,

$$i_1 - i_2 + 8 = 0 \rightarrow i_1 - i_2 = -8. \tag{2.73}$$

From (2.72) and (2.73),

$$-i_2 = -8 - i_1 \rightarrow i_2 = 8i_1 \tag{2.74}$$

$$2i_1 + 8 + i_1 = 1 \rightarrow 3i_1 = -7 \rightarrow i_1 = -\frac{7}{3}A = -2.333\,\mathrm{A}$$

$$i_2 = 8 - \frac{7}{3} = \frac{24-7}{3} = \frac{17}{3}A = 5.667\,\mathrm{A}$$

$$v_2 = V_b + i_2 \times R_2 = 6 + (5.667)(6) = 40\,\mathrm{V}, v_4 = 40 - 4 \times 8 = 8\,\mathrm{V}.$$

Following is the SPICE netlist for the operating point analysis of this circuit:

```
supermesh1
*OP Analysis
R1 1 2 12
R2 2 3 6
R3 2 4 4
Va 1 0 12
Vb 3 0 6
I1 0 4 8
```

Problem 2.2.18 In the circuit shown in Fig. 2.53, determine voltages at nodes 2, 3, and 4 using mesh current analysis (Supermesh2.cir).

$$V_a = 10\,\mathrm{V}, \quad g = 0.1(\mathrm{A/V}), \quad h = 10(\mathrm{V/A}), \quad R_1 = 5\,\Omega, \quad R_2 = 4\,\Omega, \quad R_3 = 3\,\Omega.$$

Solution
There is a current source between two meshes. For the supermesh circuit, (KVL) (Fig. 2.54),

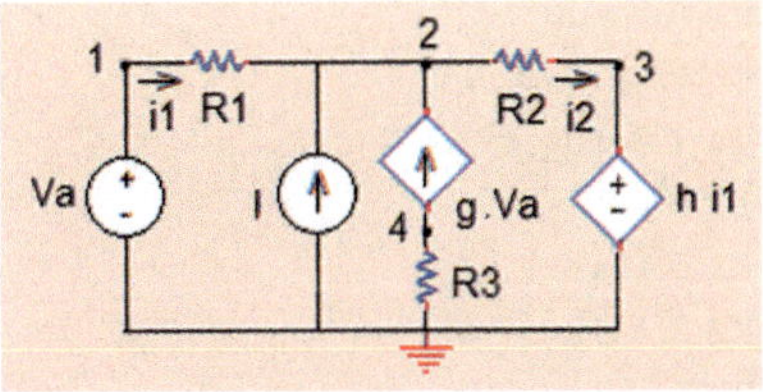

Fig. 2.53 The circuit for Problem 2.2.18

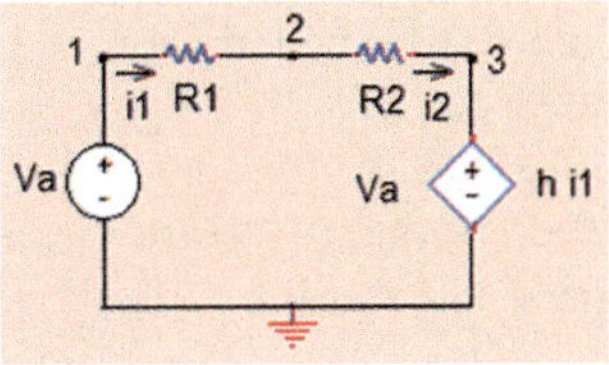

Fig. 2.54 The supermesh for the circuit of Problem 2.2.18

$$-V_a = R_1 i_1 + R_2 i_2 + h i_1 = 0$$

$$-10 = 5i_1 + 4i_2 + 10i_1 = 0 \rightarrow 15i_1 + 4i_2 = 10. \tag{2.75}$$

KCL at node 2 (constraint equation):

$$1 + i_1 - i_2 + \frac{V_a}{10} = 0 \rightarrow 1 + i_1 - i_2 + \frac{10}{10} = 0$$

$$i_1 - i_2 = -2\,\text{A}. \tag{2.76}$$

From (2.71) and (2.72),

$$i_2 = \frac{40}{19} = 2.1053\,\text{A}, \quad i_1 = \frac{2}{19} = 0.1053\,\text{A}$$

$$V_2 = i_2 R_2 + 10i_1 = \frac{40}{19}4 + 10\frac{2}{19} = \frac{180}{19} = 9.474\,\text{V}.$$

Alternatively,

$$V_2 = V_a - i_1 R_1 = 10 - \frac{2}{19} \times 5 = 9.474\,\text{V}$$

$$V_4 = 9V_a \times R_3 = 0.1 \times 10 \times 3 = -3\,\text{V}$$

$$V_3 = hi_1 = 10 \times 0.1053 = 1.053\,\text{V}.$$

Following is the SPICE netlist for the operating point analysis of this circuit:

```
supermesh2
*OP Analysis
Va 1 0 10
I1 0 2 1
R1 5 2 5
R2 2 3 4
R3 4 0 3
*CCVS: hxx N+ N- VNAME VALUE
*Controlling current is through a zero volt voltage source
VREF 1 5 0
h1 3 0 VREF 10
*VCCS: gxx N+ N- NC+ NC- VALUE
g1 4 2 1 0 0.1
```

2.3 Linearity and Superposition

Problem 2.3.1 In the circuit shown in Fig. 2.55, find the value of the current flowing through $R_2 = 9\,\Omega$ resistor using superposition. $R_1 = 6\Omega, U = 3\,\text{V}, I = 2\,\text{A}$.

Solution
Current due to voltage source alone is

$$i' = \frac{3}{9+6} = \frac{1}{5}\,\text{A}.$$

Current due to current source alone is

$$i'' = \frac{6}{9+6} \times 2 = \frac{4}{5}\,\text{A}.$$

The sum:

$$i = i' + i'' = \frac{1}{5} + \frac{4}{5} = 1\,\text{A}.$$

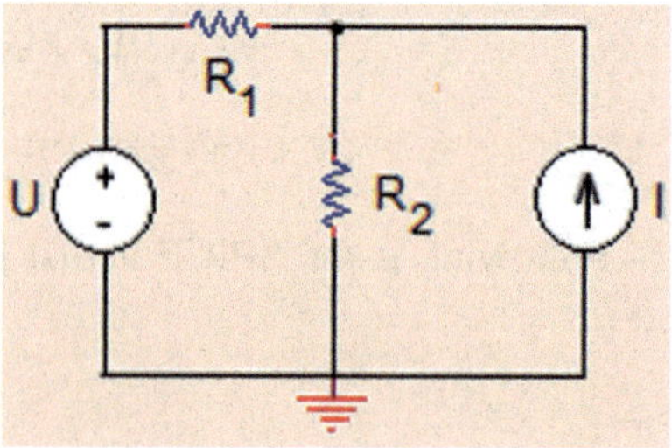

Fig. 2.55 The circuit for Problem 2.3.1

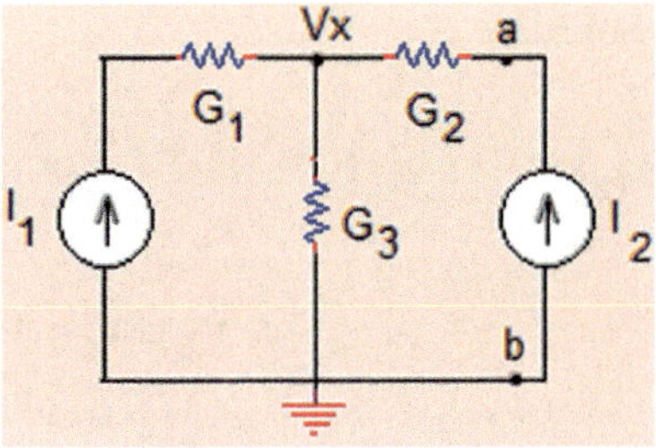

Fig. 2.56 The circuit for Problem 2.3.2

Problem 2.3.2

(a) Determine the node voltage in the circuit shown in Fig. 2.56 (use superposition).
(b) Calculate the node voltage if all conductances are 2 S, and current source values are both 1 A.

Solution

(a) By superposition,
I_1 off, I_2 on; $V_x' = I_2/G_3$; I_1 on, I_2 off; $V_x'' = I_1/G_3$;

$$V_x = V_x' + V_x'' = \frac{I_2}{G_3} + \frac{I_1}{G_3} = \frac{1}{G_3}(I_1 + I_2).$$

(b)
$$G_1 = G_2 = G_3 = 2\,\mathrm{S}, \quad I_1 = I_2 = 1\,\mathrm{A}$$

$$V_x = \frac{1}{2}(1+1) = 1\,\mathrm{V}.$$

Problem 2.3.3

(a) In the circuit shown in Fig. 2.57, determine the voltage at node a, if $V1$ = 1 V, $V2$ = 2 V, R = 1 kΩ.
(b) i = ? (Use superposition theorem).

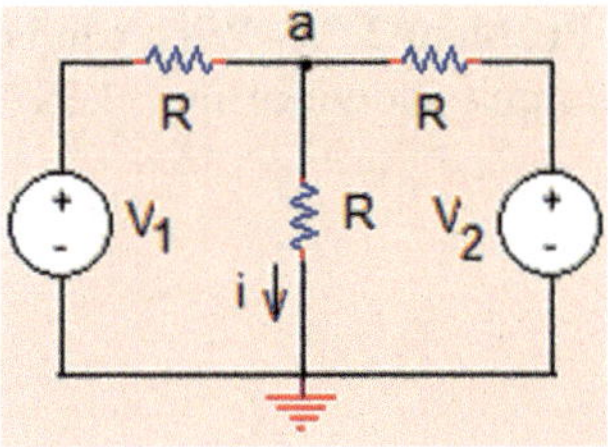

Fig. 2.57 The circuit for Problem 2.3.3

Solution

(a)
$$V_a' = \frac{V_1}{3}\,\text{V}, \quad V_a'' = \frac{V_2}{3}\,\text{V},$$

$$V_a = V_a' + V_a'' = \frac{V_1}{3} + \frac{V_2}{3} = \frac{1}{3} + \frac{2}{3} = 1\,\text{V}$$

(b)
$$i = \frac{V_a}{R} = \frac{1\,\text{V}}{1\,\text{k}\Omega} = 1\,\text{mA}.$$

Problem 2.3.4

(a) Assuming a single-input and single-output (SISO) system, state criteria to determine the linearity of such a system.
(b) If y is the output and x is the input of a system of the form $y = mx + n$, what can be said about its linearity?

Solution

(a) Assuming y is the output and x is the input of a system, three criteria to determine the linearity of such a system are as follows:

1. Homogeneity: if $y = f(x)$ then $k.y = f(k.x)$ where k is a constant factor (more generally stated, k is any real number for real systems and it is any complex number for complex-valued signals and systems).
2. Additivity: If $y1 = f(x1)$ and $y2 = f(x2)$, then $y1 + y2 = f(x1 + x2)$.
3. For $x = 0$, then $y = f(0) = 0$.

If a system satisfies all of these criteria stated above, it is a linear system.

(b) Let $m = 2$, $n = 1$, then $y = 2x + 1$, then

x	$x1 = 0$	$x2 = 1$	$x3 = 2$	$x4 = 3$
y	$y1 = 1$	$y2 = 3$	$y3 = 5$	$y4 = 7$

All criteria are violated. For example, $y(x2 + x3) \neq y(x2) + y(x3)$.

Therefore, this system is "incrementally linear" so that the output is a scaled reproduction of the input except for a fixed offset in the output.

Problem 2.3.5 What can be said about the linearity of the modified voltage divider circuit shown in Fig. 2.58?

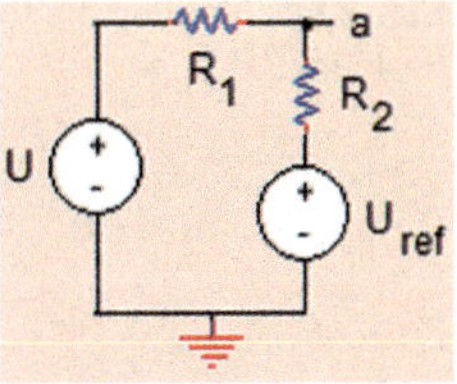

Fig. 2.58 The circuit for Problem 2.3.5

Solution

$$V_a(U) = \frac{R_2}{R_1+R_2}U + \frac{R_1}{R_1+R_2}U_{\mathrm{ref}} \quad \rightarrow \quad y = mx+n, \quad \text{where } x = U.$$

This circuit is "incrementally linear" so that the output voltage is a scaled reproduction of the input voltage except for a fixed offset in the output voltage.

Problem 2.3.6 In the circuit shown in Fig. 2.59, use linearity principle to find the values for the voltage at node C (=V_C) and the current i through the resistor R_6 (R_1 = R_2 = R_3 = 1 Ω, R_4 = R_5 = R_6 = 4 Ω, i_S = 2 A).

Solution
Let i = 1 A

$$V_{\mathrm{C}} = 4 \cdot i = 4 \cdot 1 = 4\,\mathrm{V}$$

$$i_{R3} = \frac{V_{\mathrm{C}}}{R_3} = \frac{4}{1} = 4\,\mathrm{A}$$

$$i_{\mathrm{BC}} = i_{R3} + i = 4 + 1 = 5\,\mathrm{A}$$

$$V_{\mathrm{B}} = i_{\mathrm{BC}} \cdot 4 + V_c = 5 \cdot 4 + 4 = 24\,\mathrm{V}$$

$$i_{R2} = \frac{V_{\mathrm{B}}}{R_2} = \frac{24}{1} = 24\,\mathrm{A}$$

$$i_{\mathrm{AB}} = i_{R2} + i_{\mathrm{BC}} = 24 + 5 = 29\,\mathrm{A}$$

$$V_{\mathrm{A}} = 4 \cdot i_{\mathrm{AB}} + V_{\mathrm{B}} = 4 \cdot 29 + 24 = 116 + 24 = 140\,\mathrm{V}$$

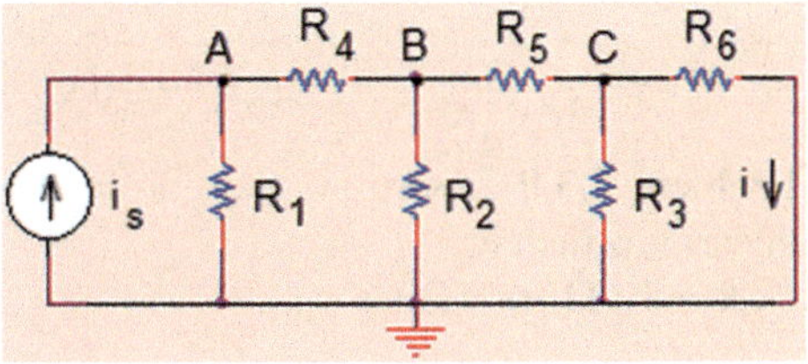

Fig. 2.59 The circuit for Problem 2.3.6

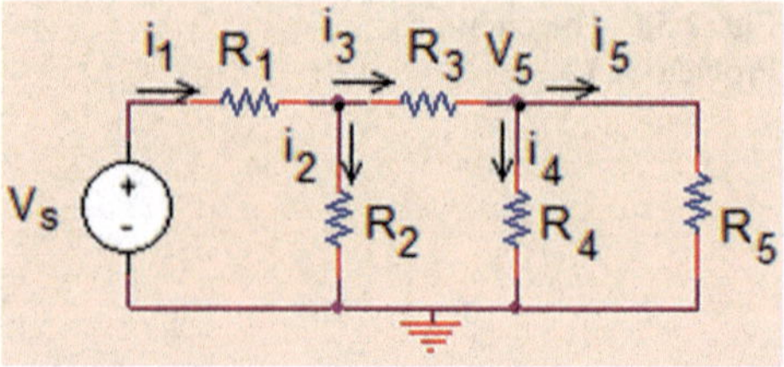

Fig. 2.60 The circuit for Problem 2.3.7

$$i_{R1} = \frac{V_A}{R_1} = 140\,\text{A}$$

$$i'_s = i_{R1} + i_{AB} = 140 + 29 = 169\,\text{A} \quad \rightarrow \quad \frac{i'_s}{i_s} = \frac{1}{i} \quad \rightarrow \quad i = \frac{2}{169} = 0.011834\,\text{A}$$

$$V_C = i \cdot 4 = 0.04734\,\text{V}.$$

Problem 2.3.7 Find the current through resistor R_5 in the circuit shown in Fig. 2.60 (use linearity principle). $V_s = 10\,\text{V}, R_1 = 0.5\,\Omega, R_2 = 8\,\Omega, R_3 = 2\,\Omega, R_4 = 2\,\Omega, R_5 = 1\,\Omega.$

Solution

$$\text{Let } i_5 = 1\,\text{A}, v_5 = i_5 \cdot R_5 = 1\,\text{V}$$

$$i_4 = \frac{v_5}{R_4} = 0.5\,\text{A}$$

$$i_3 = i_4 + i_5 = 1.5\,\text{A}$$

$$v_3 = i_3 R_3 + v_5 = (1.5)(2) + 1 = 4\,\text{V}$$

$$i_2 = \frac{v_3}{R_2} = \frac{4}{8} = 0.5\,\text{V}$$

$$i_1 = i_2 + i_3 = 0.5 + 1.5 = 2\,\text{A}$$

$$V_{sx} = i_1 R_1 + v_3 = (2)(0.5) + 4 = 5\,\text{V}$$

$$\text{When } V_s = 5\,\text{V}, \quad i_5 = 1\,\text{A}$$

$$\text{But } v_s = 10\,\text{V}, \quad \text{then } i_5 = 2\,\text{A}.$$

Problem 2.3.8 Determine the current (I_X) in the circuit shown in Fig. 2.61. Use linearity principle.

$R = 10\ \Omega$, $U = 20$ V.

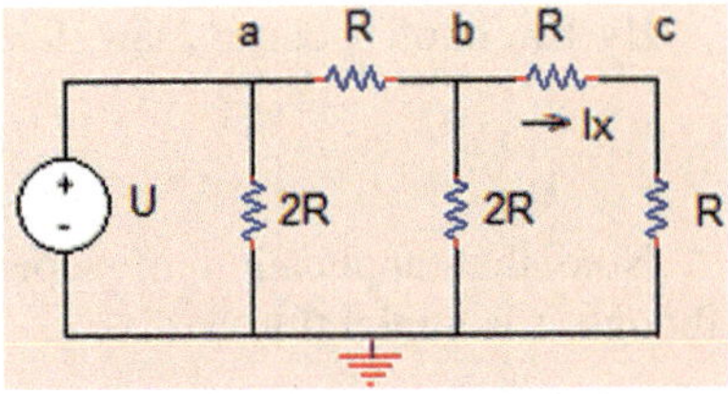

Fig. 2.61 The circuit for Problem 2.3.8

Solution
Let $I_x = 1\,\mathrm{A}$,

$$V_b = 2R \times 1 = 20\,\mathrm{V}$$

$$V_a = U = i_{\mathrm{ab}}R + V_b = (1+1)R + 20 = 2 \times 10 + 20 = 40\,\mathrm{V}.$$

Since given value of U = 20 V (which is half the calculated value), $I_x = 0.5\,\mathrm{A}$.

Problem 2.3.9 Calculate the value of currents through R_3 and R_1 in the circuit shown in Fig. 2.62. Use superposition. $R_1 = R_2 = 1\,\mathrm{k\Omega}, R_3 = 2\,\mathrm{k\Omega}, \mathrm{I}_1 = 9I_2 = 9\,\mathrm{mA}$.

Solution
By current division rule due to I_1

$$i_{R31} = I_1 \cdot \frac{R_1}{R_1 + R_2 + R_3}.$$

By current division rule due to I_2,

$$i_{R32} = -I_2 \cdot \frac{R_2}{R_1 + R_2 + R_3}.$$

The sum of the currents:

$$i_{R3} = i_{R31} + i_{R32} = \frac{1}{R_1 + R_2 + R_3}(I_1R_1 - I_2R_2) = \frac{1}{1+1+2}(9 \times 1 - 1 \times 1) = 2\,\mathrm{mA}.$$

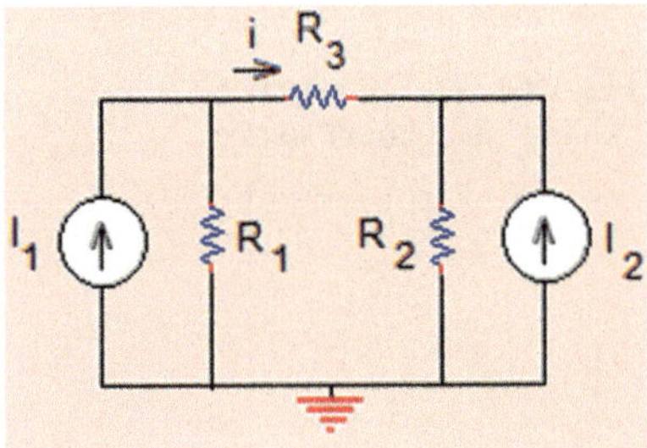

Fig. 2.62 The circuit for Problem 2.3.9

By Kirchhoff's current law, $9 = i_{R1} + i_{R3} = i_{R1} + 2$ or,

$$i_{R3} = 7\,\text{mA}.$$

Note that application of superposition principle is somewhat lengthy even though it is straightforward.

Problem 2.3.10 Using superposition theorem in the circuit shown in Fig. 2.63, find the value of

(a) Vx.
(b) Vx, if $R_1 = 0\,\Omega$.
(c) Vx, if $R_2 = 0\,\Omega$.

Solution

Step 1. (Fig. 2.64), $i_0 = 0\,\text{A} \rightarrow -\text{V}_1 + i\text{R}_1 + i\text{R}_2 + \text{kV}_1 = 0$

$$i(R_1 + R_2) = V_1 - kV_1 = V_1(1-k), \quad \rightarrow \quad i = \frac{V_1(1-k)}{R_1 + R_2}$$

$$V_{x1} = i \cdot R_2 + kV_1 = \frac{V_1(1-k)}{R_1 + R_2} \cdot R_2 + kV_1 = V_1\left[\frac{(1-k)R_2}{R_1 + R_2} + k\right].$$

Step 2. $V_1 = 0\,\text{V}$,

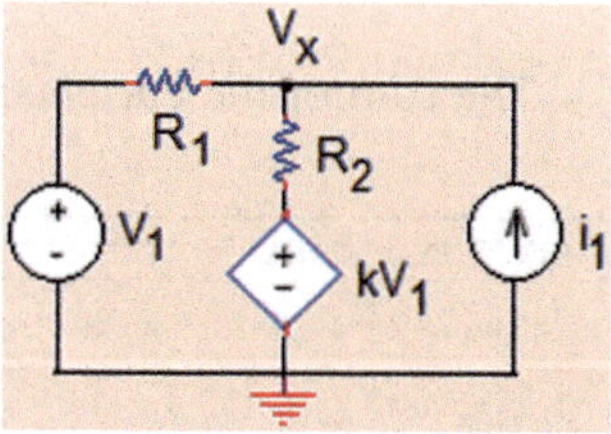

Fig. 2.63 The circuit for Problem 2.3.10

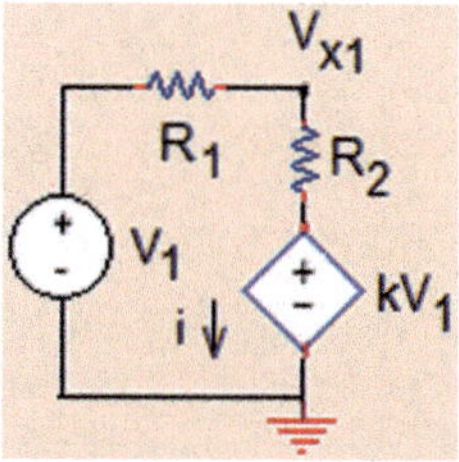

Fig. 2.64 The circuit after "killing" the current source

$$V_{x2} = i_1 \cdot \frac{R_1 \cdot R_2}{R_2 + R_2}$$

a. $V_x = V_{x1} + V_{x2} = V_1 \left[\frac{(1-k)R_2}{R_1 + R_2} + k\right] + i_1 \cdot \frac{R_1 R_2}{R_1 + R_2} \quad [\mathrm{V}]$

b. $R_1 = 0\,\Omega, \quad V_x = V_1 \quad [\mathrm{V}]$

c. $R_2 = 0\,\Omega, \quad V_x = kV_1 \quad [\mathrm{V}].$

Problem 2.3.11 In the circuit shown in Fig. 2.65, $R_1 = R_2 = R_3 = 2\,\Omega$. Find the values of $V_a, V_b, I_1 = I_{R1}, I_2 = I_{R2}, I_3 = I_{R3}$.

Solution

According to superposition theorem,

$$V_a = V_{a1} + V_{a2} + V_{a3} \text{ and } V_b = V_{b1} + V_{b2} + V_{b3}.$$

When 3 A current source is closed, I_2 = 1 A, I_1 = I_3 = 2 A.

$$V_{a1} = 2 \cdot I_2 = 2 \cdot 1 = 2\,\mathrm{V}$$

$$V_{b1} = 2 \cdot I_3 = 2 \cdot 2 = 4\,\mathrm{V}.$$

When 1 A current source is closed, I_2 = 5 A, I_1 = 1 A and I_3 = 5 A.

$$V_{a2} = 2 \cdot I_2 = 2 \cdot 5 = 10\,\mathrm{V}$$

$$V_{b2} = 2 \cdot I_3 = 2 \cdot 5 = 10\,\mathrm{V}$$

When 2 A current source is closed, I_2 = 4 A, I_1 = 2 A and I_3 = 4 A.

$$V_{a3} = 2 \cdot I_2 = 2 \cdot 4 = 8\,\mathrm{V}, V_{b3} = 2 \cdot I_3 = 2 \cdot 4 = 8\,\mathrm{V}$$

$$V_a = V_{a1} + V_{a2} + V_{a3} = 2 + 10 + 8 = 20\,\mathrm{V}$$

$$V_b = V_{b1} + V_{b2} + V_{b3} = 4 + 10 + 8 = 22\,\mathrm{V}.$$

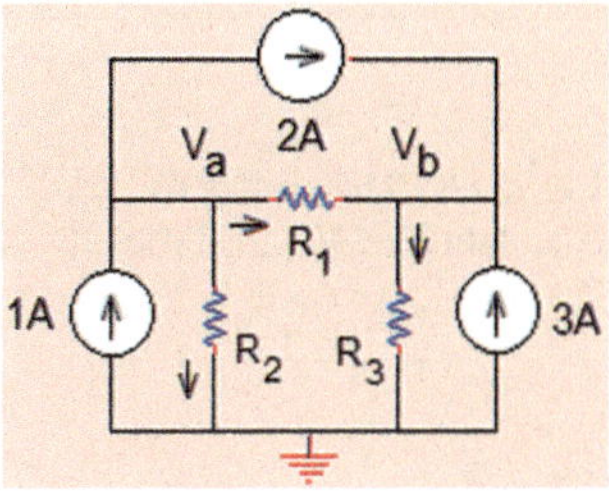

Fig. 2.65 The circuit for Problem 2.3.11

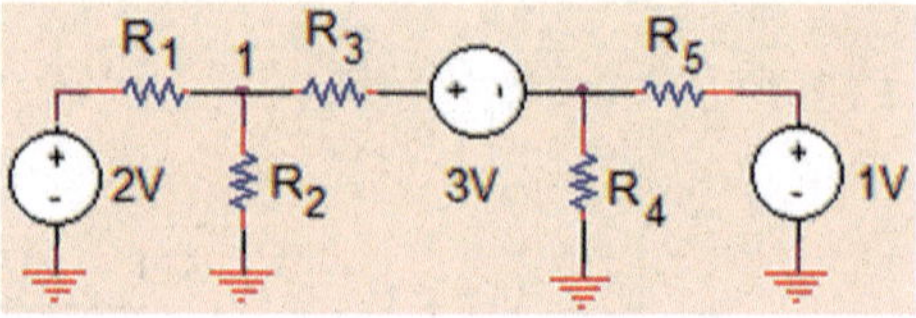

Fig. 2.66 The circuit for Problem 2.3.12

Problem 2.3.12

(a) Use superposition theorem and find the value of voltage at node 1 of the circuit shown in Fig. 2.66 ($R_1 = R_2 = R_4 = R_5 = 2\ \Omega$, $R_3 = 1\ \Omega$).
(b) Check your result using SPICE analysis. Print netlist (superposition check1.cir).

Solution
(a)

(i) See Fig. 2.67.

$$V_{11} = 2 \cdot \frac{(R_4//R_5 + R_3)//R_2}{(R_4//R_5 + R_3)//R_2 + R_1} = 2 \cdot \frac{1}{1+2} = \frac{2}{3}\ \text{V}.$$

(ii) See Fig. 2.68.

$$i = \frac{3}{1+1+1} = 1\ \text{A}, \quad V_{12} = 1 \cdot 1 = 1\ \text{V}.$$

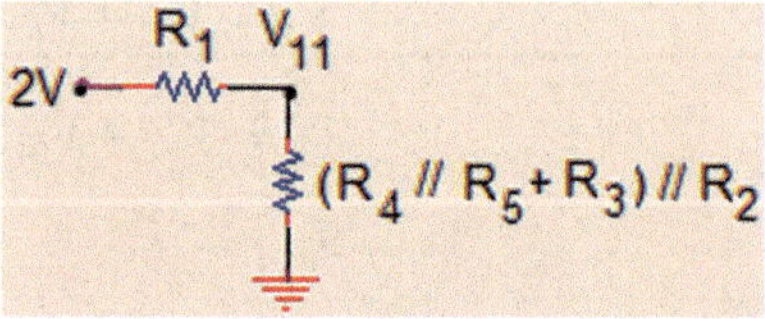

Fig. 2.67 The circuit for the calculation of V_{11} for Problem 2.3.12

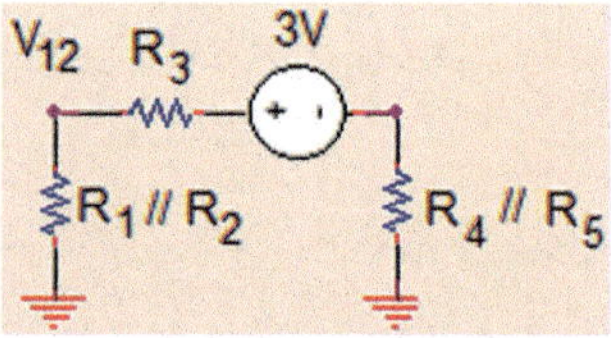

Fig. 2.68 The circuit for the calculation of V_{12} for Problem 2.3.12

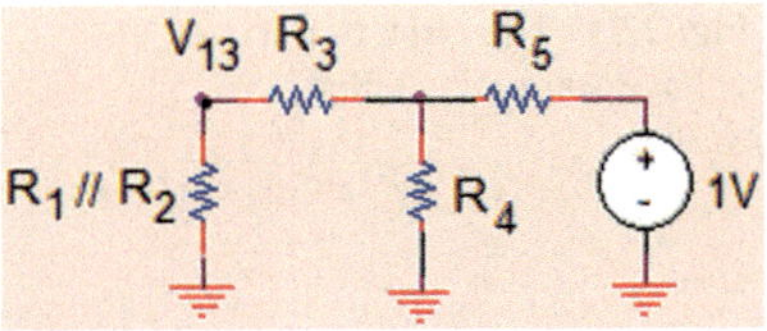

Fig. 2.69 The circuit after "killing" the current source

(iii) See Fig. 2.69.

$$V_{13} = \frac{(R_3 + R_1//R_2)//R_4}{(R_3 + R_1//R_2) + R_4} = \frac{(1+1)//2}{[(1+1)//2] + 2} = \frac{2//2}{2//2 + 2} = \frac{1}{1+2} = \frac{1}{3}\,\text{V}$$

(iv) $$V_{11} + V_{12} + V_{13} = \frac{2}{3} + 1 + \frac{1}{6} = \frac{4+6+1}{6} = \frac{11}{6} = 1833\,\text{V}.$$

(b) SPICE netlist,

```
*OP analysis,superposition check1
  v1 1 0 2
  v2 3 4 3
  v3 5 0 1
  R1 1 2 2
  R2 2 0 2
  R3 2 3 1
  R4 4 0 2
  R5 4 5 2
```

Problem 2.3.13 Use superposition theorem and find the value of voltage Vx in the circuit shown in Fig. 2.70.

Solution

First, voltage source is short circuited (Fig. 2.71).

The current flow in 4 Ω branch is calculated by current division,

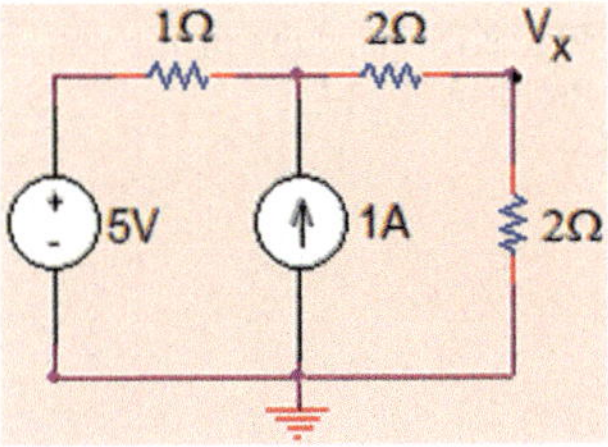

Fig. 2.70 The circuit for Problem 2.3.13

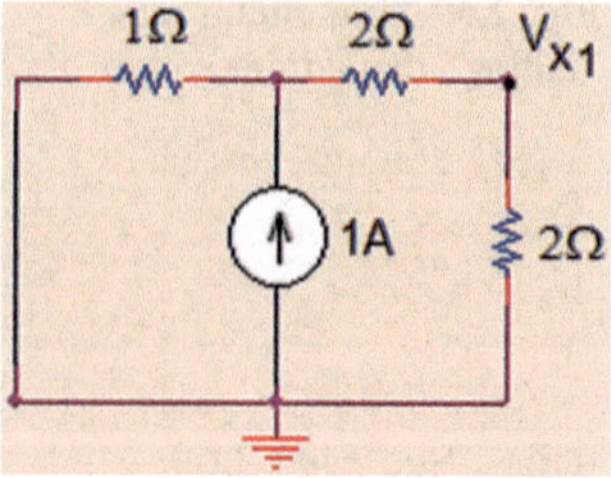

Fig. 2.71 The circuit after voltage source is "killed"

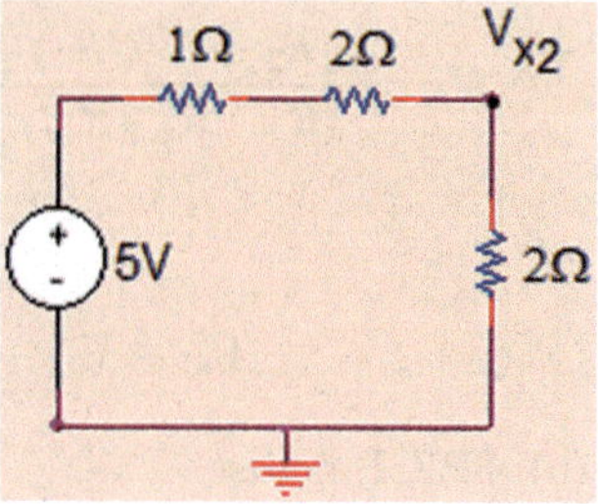

Fig. 2.72 The circuit after current source is "killed"

$$i_1 = 1 \times \frac{1}{1+4} = 0.2\,\text{A} \quad \rightarrow \quad V_{x1} = 0.2 \times 2 = 0.4\,\text{V}.$$

Second, current source is open circuited (Fig. 2.72).

In this circuit, one may use the voltage division rule and obtain the unknown voltage as

$$V_2 = 5 \times \frac{2}{5} = 2\,\text{V}.$$

Finally, superposition results are collected together, $V_x = V_{x1} + V_{x2} = 2.4\,\text{V}$.

Problem 2.3.14 Using superposition theorem, find the values of currents and voltages in the circuit shown in Fig. 2.73 (i_1 = 1 A, V = 10 V, R_1 = 1 kΩ, R_2 = 2 kΩ).

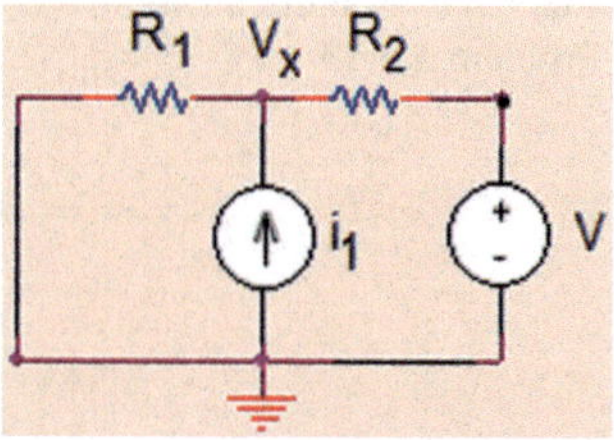

Fig. 2.73 The circuit for Problem 2.3.14

Solution
First, the voltage source is ignored (short circuited, Fig. 2.74).

$$R_{eq} = (R_1 // R_2) = 0.667\,\mathrm{kV}$$

$$V_{x1} = I \cdot R_{eq} = 1 \cdot 0.667 = 0.667\,\mathrm{kV}.$$

Then, the current source is ignored (open circuited, Fig. 2.75). By voltage division,

$$V_{x2} = \frac{R_1}{R_1 + R_2} \cdot V = \frac{1}{1+2} \cdot 10 = 3.333\,\mathrm{V}$$

$$V_x = V_{x1} + V_{x2} = 667\,\mathrm{V} + 3.33\,\mathrm{V} = 670.33\,\mathrm{V}$$

$$I_{R_1} = \frac{V_x}{R_1} = \frac{670.33\,\mathrm{V}}{1\,\mathrm{k}\Omega} = 0.67033\,\mathrm{A}$$

$$I_{R_2} = I - I_{R_1} = 1 - 0.67033 = 0.32967\,\mathrm{A}.$$

Problem 2.3.15 In the circuit shown in Fig. 2.76, find the value of i (in mA) by using superposition theorem.

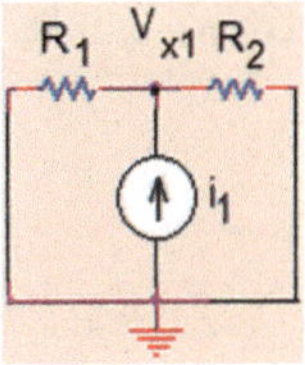

Fig. 2.74 The circuit after voltage source is "killed"

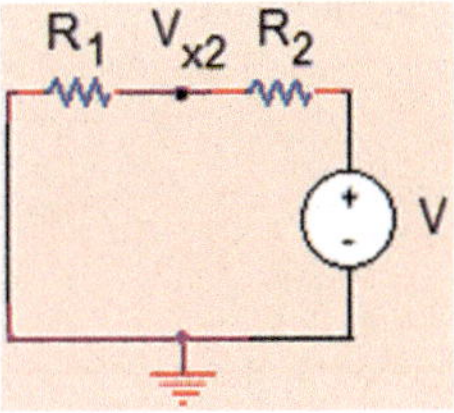

Fig. 2.75 The circuit after current source is "killed" (open circuited)

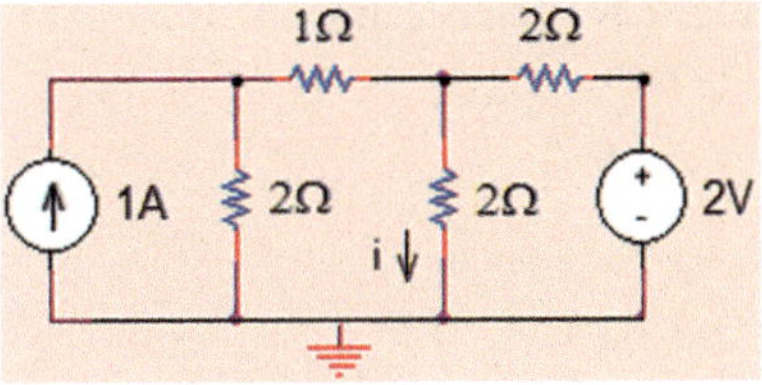

Fig. 2.76 The circuit for Problem 2.3.15

Solution

Deactivated voltage source: (Fig. 2.77). Applying current division,

$$i' = \frac{1}{2}\,\text{A}$$

$$V' = \frac{1}{2}\,\text{A} \cdot 1\,\text{V} = \frac{1}{2}\,\text{V}.$$

Deactivated current source (Fig. 2.78): By voltage division,

$$V'' = 2 \cdot \frac{3//2}{3//2+2} = 2 \cdot \frac{\frac{3 \cdot 2}{3+2}}{\frac{3 \cdot 2}{3+2}+2} = 2 \cdot \frac{\frac{6}{5}}{\frac{6}{5}+2} = 2 \cdot \frac{6}{16} = \frac{12}{16} = \frac{3}{4}\,\text{V}$$

$$V = V' + V' = \frac{1}{2} + \frac{3}{4} = 1.25\,\text{V}$$

$$i = \frac{V}{2} = \frac{1.25}{2} = 0.625\,\text{A} = 625\,\text{mA}.$$

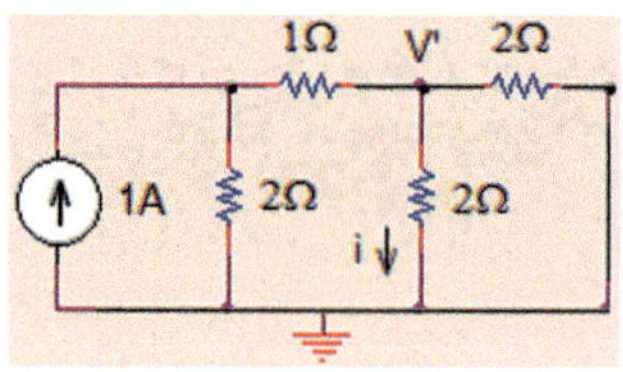

Fig. 2.77 Deactivated voltage source

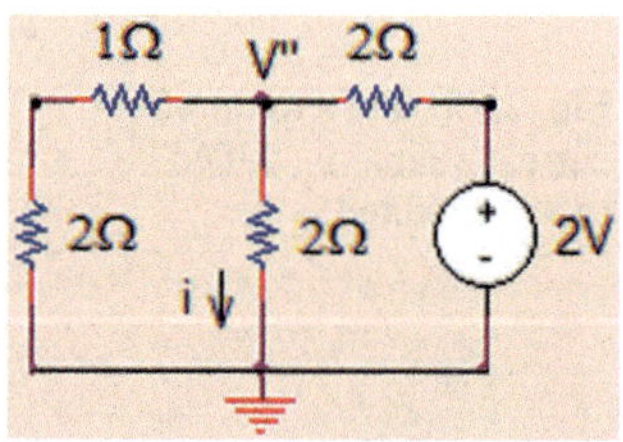

Fig. 2.78 Deactivated current source

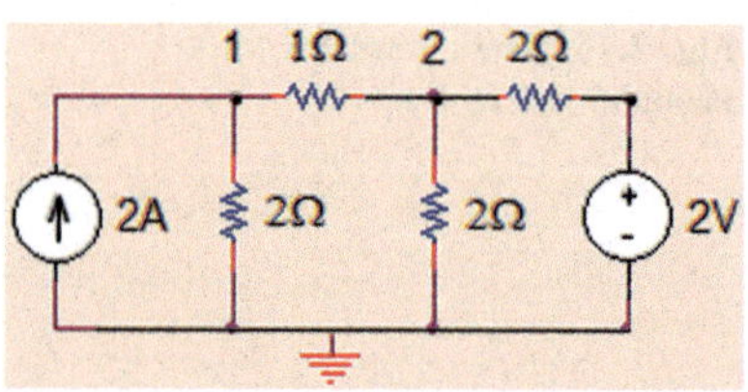

Fig. 2.79 The circuit for Problem 2.3.16

Problem 2.3.16 Use superposition theorem to find the values of voltages at nodes 1 and 2 in the circuit shown in Fig. 2.79.

Solution

(a) Kill the voltage source as shown in Fig. 2.80:

$$i_{1\Omega} = 1\,\text{A}, \quad V_2' = 1\,\Omega \times 1\,\text{A} = 1\,\text{V}$$

$$V_1' = 2[2//(1+1)] = (2//2) \times 2 = 1\,\Omega \cdot 2\,\text{A} = 2\,\text{V}.$$

(b) Kill the current source as shown in Fig. 2.81.

By voltage division,

$$R_p = \frac{(2+1)2}{(2+1)+2} = \frac{6}{5}\,\Omega$$

$$V_2'' = 2 \cdot \frac{R_p}{R_p+2} = 2 \cdot \frac{\frac{6}{5}}{\frac{6}{5}+2} = 2 \cdot \frac{\frac{6}{5}}{\frac{16}{5}} = \frac{12}{16}V = \frac{3}{4}\,\text{V}.$$

Voltage division:

$$V_1'' = V_2'' \cdot \frac{2}{2+1} = \frac{3}{4} \cdot \frac{2}{3} = \frac{1}{2}\,\text{V}$$

$$V_1 = V_1' + V_1'' = 2 + 0.5 = 2.5\,\text{V}$$

$$V_2 = V_2' + V_2'' = 1 + \frac{3}{4} = 1.75\,\text{V}.$$

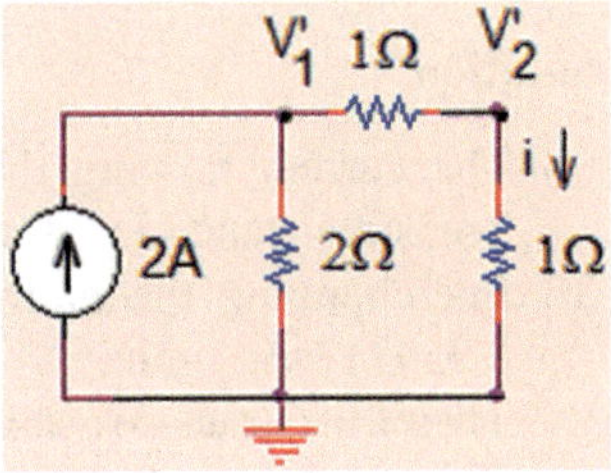

Fig. 2.80 Deactivated voltage source (short circuited)

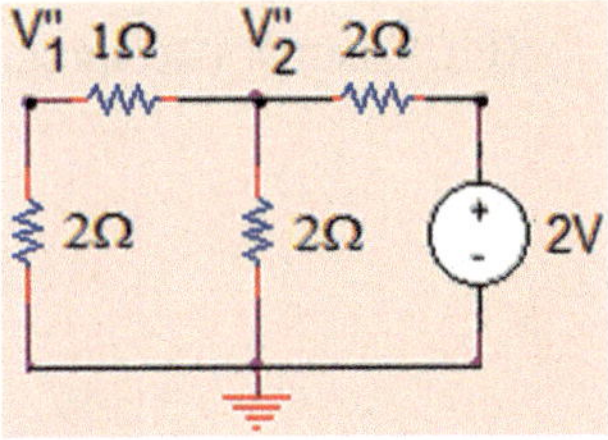

Fig. 2.81 Deactivated current source (open circuited)

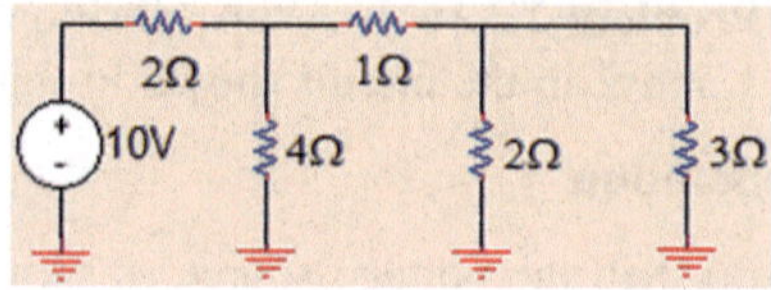

Fig. 2.82 The circuit for Problem 2.3.17

Problem 2.3.17 The supply voltage v and output current i are mutually transferable in a linear passive circuit. A circuit composed of linear bilateral elements (e.g., R, L, C) is reciprocal.

The ratio of v and i is called the transfer resistance (trans-resistance).

This means that if the positions of a voltage source and an ammeter are interchanged, the reading of ammeter remains the same, assuming ideal situation (i.e., internal resistance of both the voltage source and ammeter are null).

Alternatively, interchanging a current source and a voltmeter in a linear bilateral circuit does not change the voltmeter reading.

Reciprocity is based on the symmetry property of nodal conductance (mesh resistance) matrix. Thus, even a circuit containing dependent sources can be reciprocal for some specific dependent source coefficients, provided that its conductance or resistance matrix is symmetric.

Application of reciprocity theorem is limited only to circuits containing a single independent source.

(a) Use SPICE and determine the current flowing through 3 Ω resistor in the circuit shown in Fig. 2.82, assuming that an ammeter is placed in that branch. What is trans-resistance value?
(b) Interchange the ammeter and the voltage source and determine the new ammeter reading, again. What is new transresistance value?
(c) If the voltage is 50 V in part (b), determine the new ammeter reading.

Solution

(a) The current flowing through 3 Ω resistor (ammeter reading) is 0.754717 A. Trans-resistance is 13.24999966 Ω.
(b) Interchanging the ammeter and the voltage source, the ammeter reading is 0.754717 A, again.
    Therefore, trans-resistance is 13.24999966 Ω, as well.
(c) If the voltage is 100 V in part (b), (due to linearity) the new ammeter reading is 7.54717 A. This is also verified by SPICE analysis.

SPICE netlist (Reciprocity1.cir) is given below.

```
Reciprocity
*v1 1 0 10
V1 4 0 0
R1 1 2 2
R2 2 0 4
R3 2 3 1
R4 3 0 2
R5 3 4 3
*VX 4 0 0
VX 1 0 10
```

2.4 Source Transformation

Problem 2.4.1 In the circuit shown in Fig. 2.83, find the value of node voltage *Vx*, if $V_1 = 3$ V, $I = 9$ A, $R = 1\ \Omega$ using source transformation.

Solution
Applying source transformation to the given circuit gives the circuit shown in Fig. 2.84; then,

$$\frac{V_1}{R} - I + \frac{V_1}{R} + I = 3\frac{V_x}{R} = 3V_x$$

$$2V_1 = 3V_x$$

$$V_x = \frac{2}{3}V_1 = \frac{2}{3} \times 3 = 2\,\text{V}.$$

Problem 2.4.2 In the circuit shown in Fig. 2.85, use source transformation method and determine the current through resistor R_2.

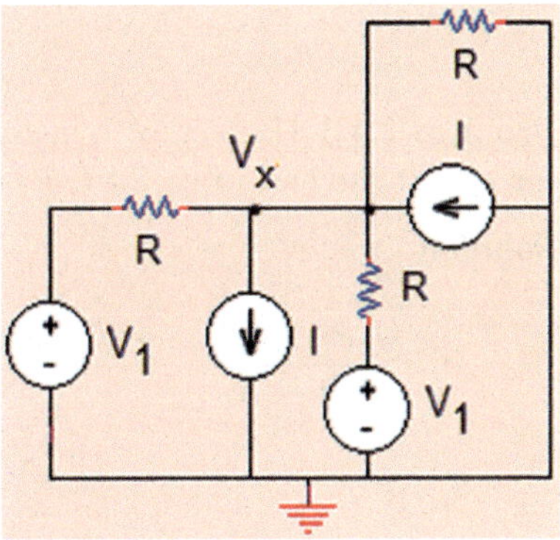

Fig. 2.83 The circuit for Problem 2.4.1

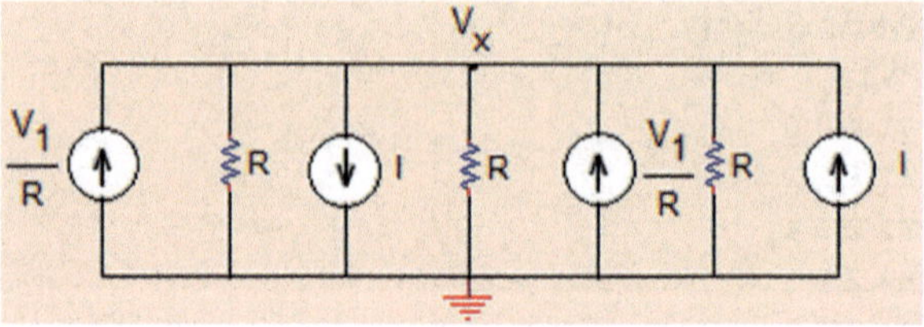

Fig. 2.84 Source transformed circuit of Problem 2.4.1

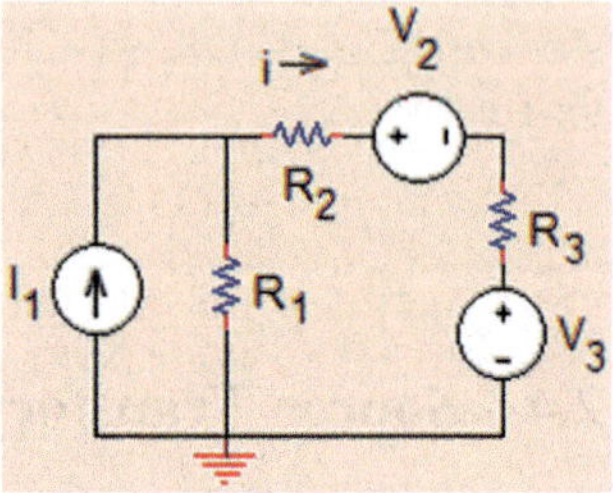

Fig. 2.85 The circuit for Problem 2.4.2

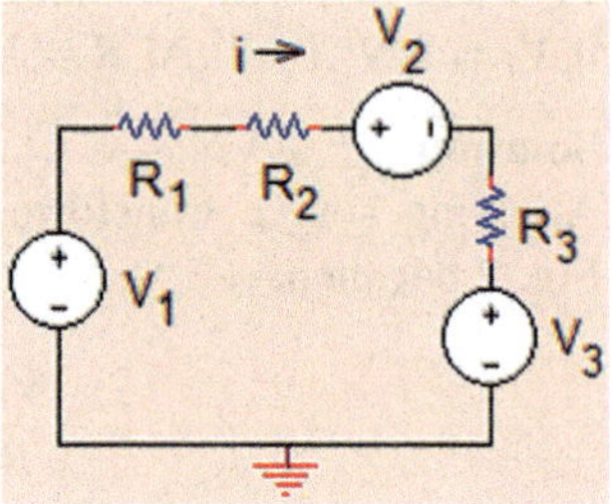

Fig. 2.86 Source transformation applied to circuit of Problem 2.4.2

Solution

By source transformation and KVL, (see Fig. 2.86),

$$V_1 - V_2 - V_3 = i(R_1 + R_2 + R_3)$$

$$I_1 R_1 - V_2 - V_3 = i(R_1 + R_2 + R_3)$$

$$i = \frac{I_1 R_1 - V_2 - V_3}{R_1 + R_2 + R_3}.$$

Problem 2.4.3 Use source transform to calculate the value of node voltage V_a in the circuit shown in Fig. 2.87. $R_1 = R_2 = 2R_3 = 8\ \Omega$, $I_1 = I_2 = 1$ A.

Solution

$$i = \frac{I_1 R_1 - I_2 R_3 - 4V_a}{R_1 + R_2 + R_3} = \frac{8 - 4 - 4V_a}{8 + 4 + 4} = \frac{4}{20}(1 - V_a) = \frac{1 - V_a}{5} \tag{2.77}$$

$$V_a = iR_3 + I_2 R_3. \tag{2.78}$$

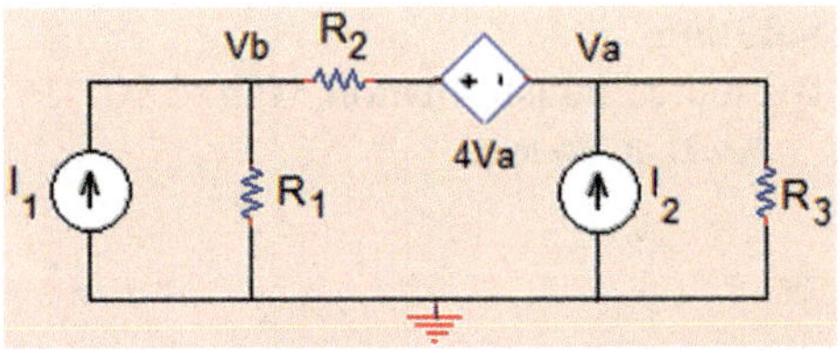

Fig. 2.87 The circuit for Problem 2.4.3

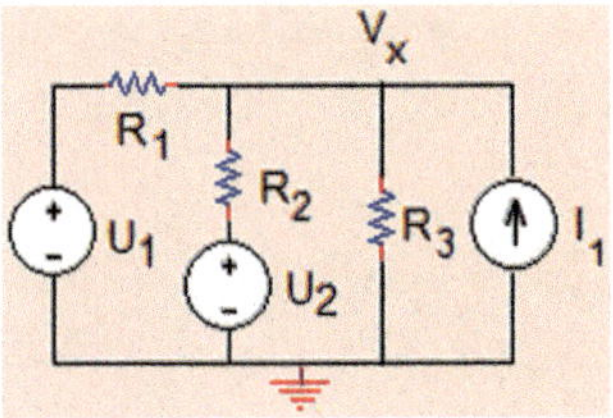

Fig. 2.88 The circuit for Problem 2.4.4

Substitute (2.77) in (2.78), use given data,

$$V_a = \frac{1 - V_a}{5} \times 4 + 1 \times 4 = \frac{4}{5} - \frac{4V_a}{5} + 4$$

$$\frac{9}{5}V_a = \frac{24}{5} \quad \rightarrow \quad V_a = \frac{24}{9} = \frac{8}{3} = 2.667\,\mathrm{V}.$$

Problem 2.4.4 In the circuit shown in Fig. 2.88, find the value of voltage V_x using source transformation.

$U_1 = 3$ V, $U_2 = 5$ V, $I_1 = 2$ A, $R_1 = R_3 = 2R_2 = 4\ \Omega$.

Solution

Application of source transformation to voltage sources results in with the following circuit equation:

$$\frac{3}{4} + \frac{5}{2} + 2 = \frac{V_x}{4}.$$

Solving for the unknown voltage yields $V_x = 21$ V.

Problem 2.4.5 Using source transformation, find the node voltage V_X in the circuit shown in Fig. 2.89.

$R_1 = R_2 = 4\ \Omega$.

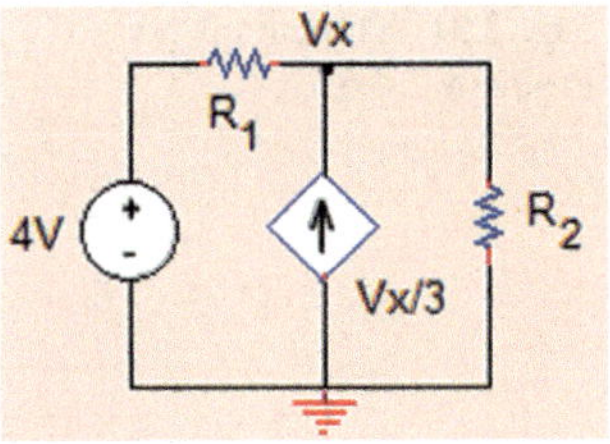

Fig. 2.89 The circuit for Problem 2.4.5

Solution

By source transformation, (Fig. 2.90), $R_1 \| R_2 = 2\Omega$.

KCL at node x:

$$1 + \frac{V_x}{3} - \frac{V_x}{2} = 0$$

$$V_x\left(\frac{1}{3} - \frac{1}{2}\right) = -1$$

$$V_x = -\frac{1}{\left(-\frac{1}{6}\right)} = 6\,\text{V}.$$

Problem 2.4.6 In the circuit shown in Fig. 2.91, find the value of node voltage using superposition theorem, and source transformation (E_1 = 20 V, R_2 = 10 Ω, R_1 = 10 Ω, I_1 = 2 A).

Solution

First, the current source is deactivated, and voltage source is transformed to current source. By KCL,

$$\frac{E_1}{R_2} + \frac{V_{x1}}{10} - \frac{V_{x1}}{R_p} = 0, \quad R_p = R_1//R_2 = 10//10 = 5\,\Omega$$

$$\frac{E_1}{R_2} + V_{x1}\left(\frac{1}{10} - \frac{1}{R_p}\right) = 0 \quad \rightarrow \quad V_{x1} = -\frac{\frac{E_1}{R_2}}{\frac{1}{10} - \frac{1}{R_p}} = \frac{-\frac{20}{10}}{\frac{1}{10} - \frac{1}{5}} = 20\,\text{V}.$$

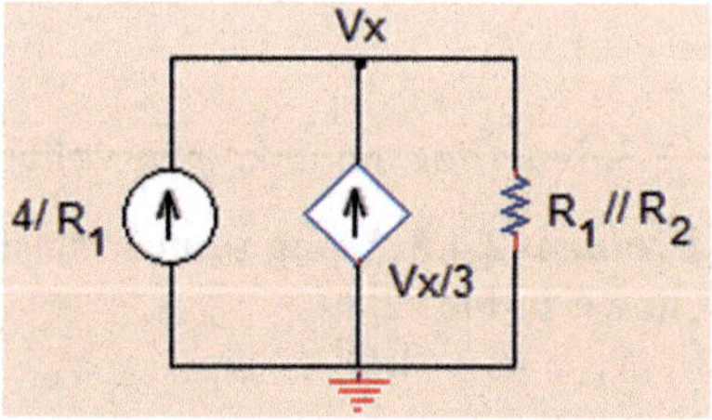

Fig. 2.90 Source transformation applied to the circuit of Problem 2.4.5

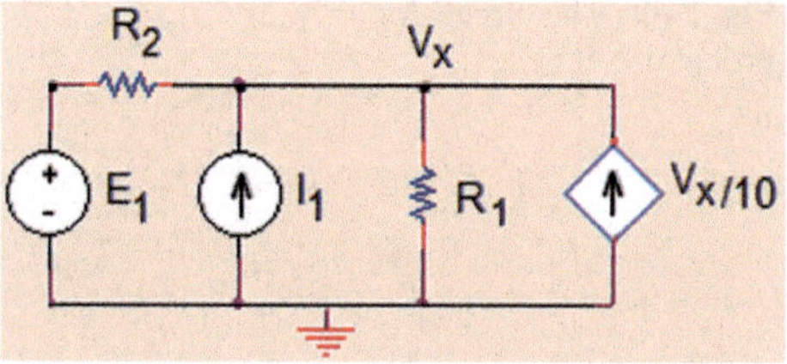

Fig. 2.91 The circuit for Problem 2.4.6

Fig. 2.92 Deactivated voltage source (short circuited)

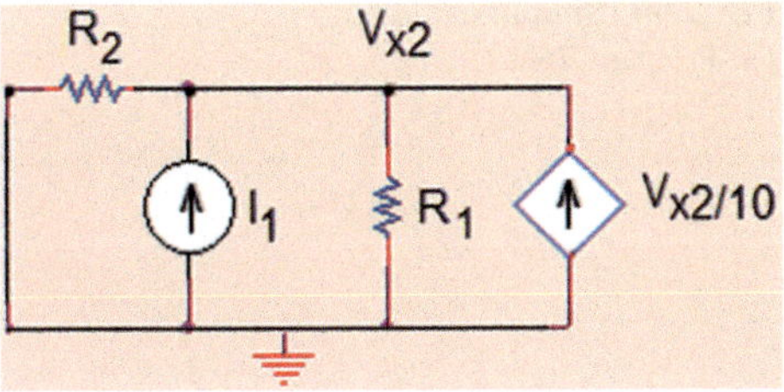

Then, the voltage source is deactivated by short circuiting it; and using KCL (Fig. 2.92),

$$I_1 + \frac{V_{x2}}{10} - \frac{V_{x2}}{R_p} = 0 \quad \rightarrow \quad I_1 + V_{x2}\left(\frac{1}{10} - \frac{1}{R_p}\right) = 0 \quad \rightarrow \quad V_{x2} = -\frac{I_1}{\frac{1}{10} - \frac{1}{R_p}} = 20\,\text{V}.$$

Finally, adding these superposition results, $V_x = V_{x_1} + V_{x_2} = 20 + 20 = 40\,\text{V}$.

Problem 2.4.7 In the circuit shown in Fig. 2.93, find the values of V_X by using source transformation and Kirchhoff's current law. Use Cramer's rule when necessary.

$$R_1 = R_2 = 2R_3 = R_4 = R_5 = 2R_2 = 2\,\Omega, \quad U_1 = U_2 = U_3 = 2\,\text{V}, \quad I_1 = 1\,\text{A}.$$

Solution

The voltage sources are transformed into 1 A current sources (Fig. 2.94), and simplified as shown in Fig. 2.95.

Fig. 2.93 The circuit for Problem 2.4.7

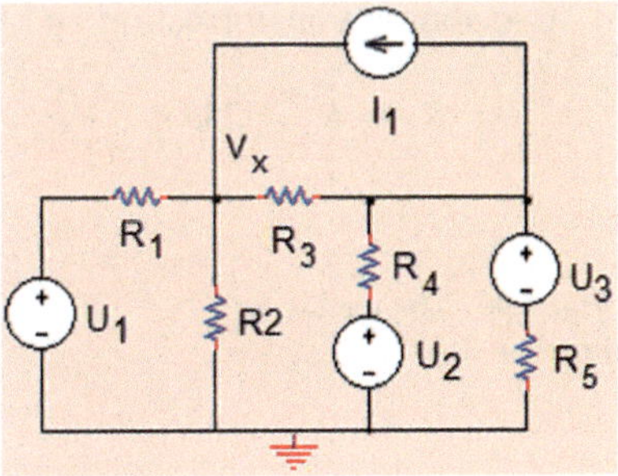

Fig. 2.94 The voltage sources are transformed into current sources

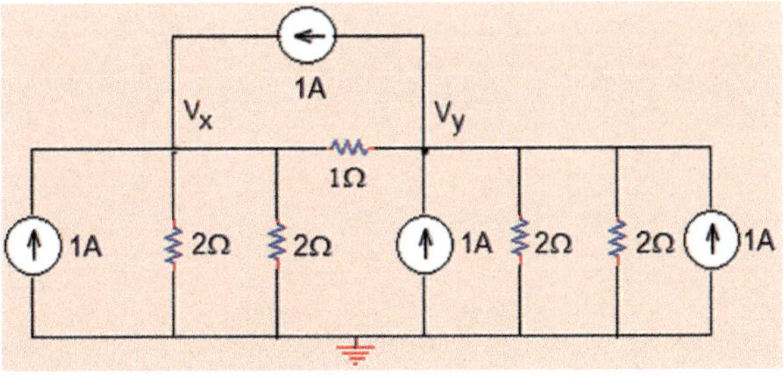

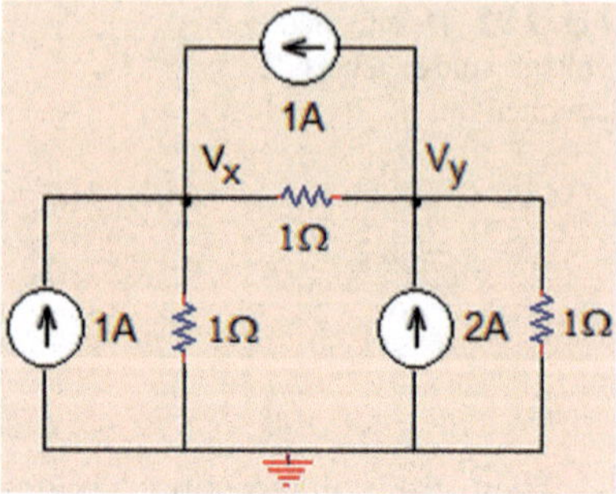

Fig. 2.95 Simplified circuit for Problem 2.4.7

Equivalent resistance of 2 Ω parallel resistors is calculated and the current sources are added:

$$[I] = [G][V]$$

$$\begin{bmatrix} 1+1 \\ 2-1 \end{bmatrix} = \begin{bmatrix} 1+1 & -1 \\ -1 & 1+1 \end{bmatrix} \begin{bmatrix} V_x \\ V_y \end{bmatrix}$$

$$\Delta = 2 \cdot 2 - 1 \cdot 1 = 4 - 1 = 3$$

$$\Delta_x = 2 \cdot 2 - (-1)(1) = 5$$

$$V_x = \frac{\Delta_x}{\Delta} = \frac{5}{3}\,\text{V} = 1.667\,\text{V}.$$

Problem 2.4.8 Determine the value of voltage at node 2 in the circuit shown in Fig. 2.96.

Use source transformation and Cramer's rule, when necessary.

$$R_1 = R_2 = R_3 = 2\,\text{k}\Omega, \quad R_4 = R_5 = 4\,\text{k}\Omega, \quad I_1 = 4I_2 = 4\,\text{mA}, \quad U = 2\,\text{V}.$$

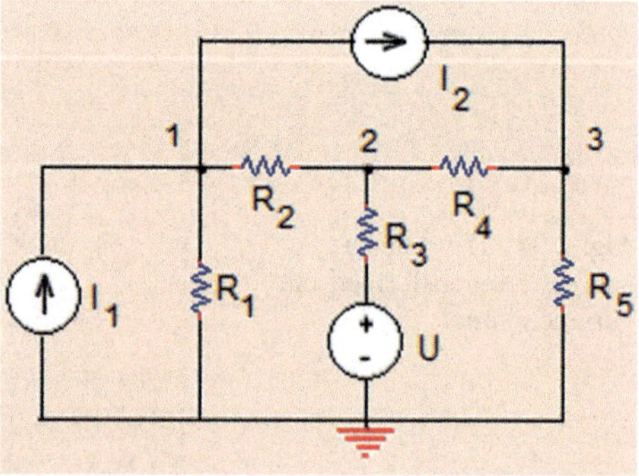

Fig. 2.96 The circuit for Problem 2.4.8

Solution

$$GV = I$$

$$\begin{bmatrix} 1/R_1 + 1/R_2 & -1/R_2 & 0 \\ -1/R_2 & 1/R_2 + 1/R_3 + 1/R_4 & -1/R_4 \\ 0 & -1/R_4 & 1/R_4 + 1/R_5 \end{bmatrix} \begin{bmatrix} V_1 \\ V_2 \\ V_3 \end{bmatrix} = \begin{bmatrix} I_1 - I_2 \\ \dfrac{U}{R_3} \\ I_2 \end{bmatrix}$$

$$10^{-3} \times \begin{bmatrix} 1/2 + 1/2 & -1/2 & 0 \\ -1/2 & 1/2 + 1/2 + 1/4 & -1/4 \\ 0 & -1/4 & 1/4 + 1/4 \end{bmatrix} \begin{bmatrix} V_1 \\ V_2 \\ V_3 \end{bmatrix} = \begin{bmatrix} (4-1) \times 10^{-3} \\ 2 \times 10^{-3} \\ 10^{-3} \end{bmatrix} = \begin{bmatrix} 3 \\ 2 \\ 1 \end{bmatrix} \times 10^{-3}$$

or

$$10^{-3} \times \begin{bmatrix} 1 & -\frac{1}{2} & 0 \\ -\frac{1}{2} & \frac{5}{4} & -\frac{1}{4} \\ 0 & -\frac{1}{4} & \frac{1}{2} \end{bmatrix} \begin{bmatrix} V_1 \\ V_2 \\ V_3 \end{bmatrix} = \begin{bmatrix} 3 \\ 2 \\ 1 \end{bmatrix} \times 10^{-3}$$

10^{-3} terms cancel out;

$$V_2 = \frac{\Delta_2}{\Delta},$$

$$\Delta_2 = \begin{vmatrix} 1 & 3 & 0 \\ -\frac{1}{2} & 2 & -\frac{1}{4} \\ 0 & 1 & \frac{1}{2} \end{vmatrix} = 2, \quad \Delta = \begin{vmatrix} 1 & -\frac{1}{2} & 0 \\ -\frac{1}{2} & \frac{5}{4} & -\frac{1}{4} \\ 0 & -\frac{1}{4} & \frac{1}{2} \end{vmatrix} = \frac{5}{8} - \left(\frac{1}{16} + \frac{1}{8}\right) = \frac{4}{8} - \frac{1}{16} = \frac{7}{16}$$

$$V_2 = \frac{\Delta_2}{\Delta} = \frac{2}{\frac{7}{16}} = \frac{32}{7}\,\text{V} = 4.571\,\text{V}.$$

Problem 2.4.9 In the circuit shown in Fig. 2.97, find the value of current I using source transformation method $2R_1 = 2R_2 = R_3 = R_4 = 2\,\Omega, I_1 = 1\,\text{A}, U = 2\,\text{V}$ (matrix_solve.xlsx).

Solution

2 V voltage source is transformed to 1 A independent current source, VCVS is transformed and circuit is simplified by taking only equivalent resistance of parallel resistors into consideration, see Fig. 2.98.

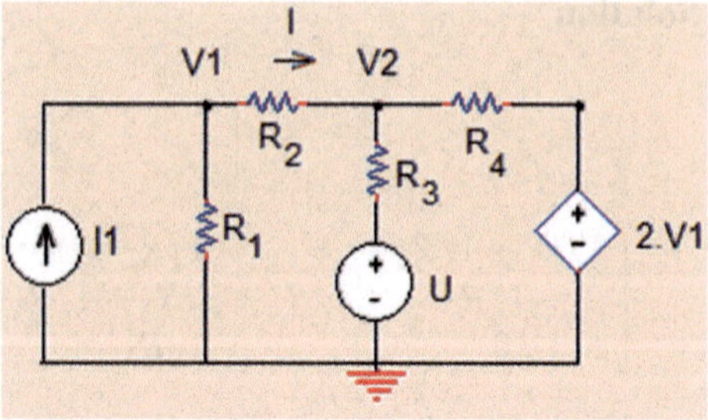

Fig. 2.97 The circuit for Problem 2.4.9

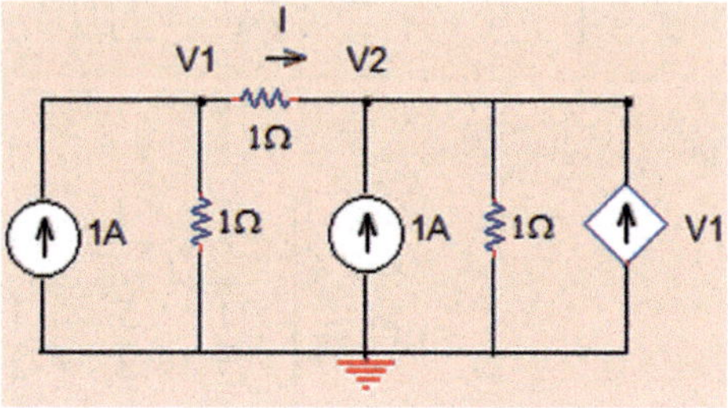

Fig. 2.98 Simplified circuit for Problem 2.4.9

Apply Kirchhoff's Current Law:
At node 1: $1 = V_1 + V_1 - V_2$

$$2V_1 - V_2 = 1. \tag{2.79}$$

At node 2: $V_1 + 1 + V_1 - V_2 - V_2 = 0$

$$V_1 - V_2 = -\frac{1}{2}. \tag{2.80}$$

Put (2.80) and (2.81) in matrix form:

$$\begin{bmatrix} 2 & -1 \\ 1 & -1 \end{bmatrix} \cdot \begin{bmatrix} V_1 \\ V_2 \end{bmatrix} = \begin{bmatrix} 1 \\ -0.5 \end{bmatrix}.$$

Solving the matrix equation yields

$$V_1 = 1.5\,\text{V}, \quad V_2 = 2\,\text{V},$$

$$I = (V_1 - V_2)/R = 1.5 - 2 = -0.5\,\text{A}.$$

Problem 2.4.10 In the circuit shown in Fig. 2.99, use source transformation to find the value of node voltage V_x, if $V_1 = V_2 = 10$ V, $I_1 = 1$ A, $R_1 = R_2 = R_3 = 1$ kΩ.

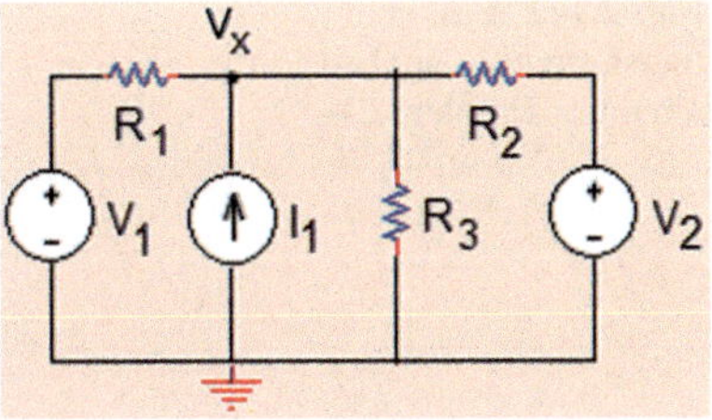

Fig. 2.99 The circuit for Problem 2.4.10

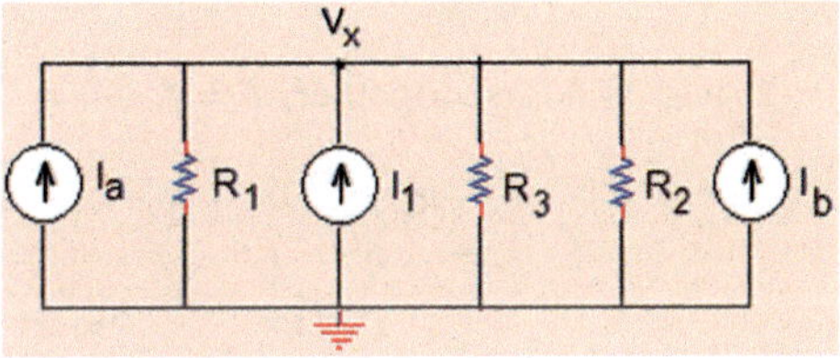

Fig. 2.100 Source transformation applied to the circuit of Problem 2.4.10

Solution
If source transformation is used for the circuit (Fig. 2.100),
V_1 is transformed into I_a and V_2 transformed into I_b

$$I_a = \frac{10\,\text{V}}{10^3\,\Omega} = 10^{-2}\,\text{A}$$

$$I_b = \frac{10\,\text{V}}{10^3\,\Omega} = 10^{-2}\text{A}, \quad I_1 + I_a + I_b = 1.02\,\text{A}$$

$$\frac{1}{R_t} = \frac{1}{10^3} + \frac{1}{10^3} + \frac{1}{10^3} = \frac{3}{10^3} \quad \rightarrow \quad R_t = 333.\bar{3}\,\Omega,$$

$$V_x = I \cdot R_t = 1.02 \times 333.\bar{3} = 340\,\text{V}.$$

Problem 2.4.11 In the circuit shown in Fig. 2.101, find the value of current through resistor R_1 using source transformation method (E_1 = 20 V, R_1 = 2 Ω, R_2 = 3 Ω, R_3 = 5 Ω, I_1 = 8 A).

Solution
Source transformation is applied on E_1 and R_3 (see, Fig. 2.102).

$$I_2 = \frac{E_1}{R_3} = \frac{20\,\text{V}}{5\,\Omega} = 4\,\text{A}$$

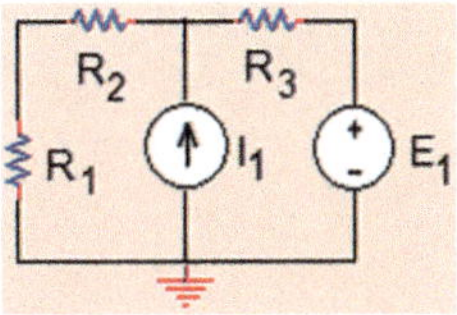

Fig. 2.101 The circuit for Problem 2.4.11

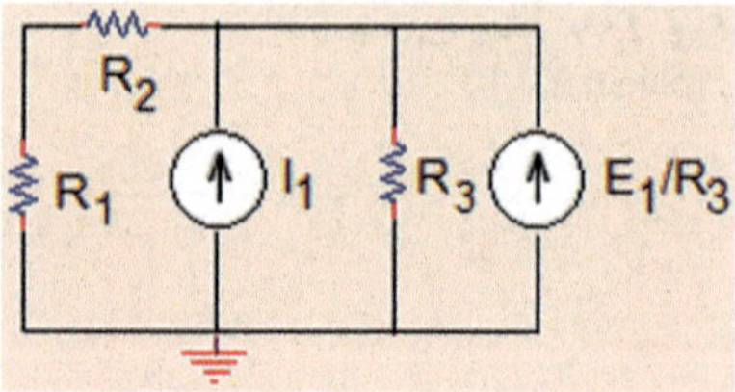

Fig. 2.102 Source transformation applied to the circuit of Problem 2.4.11

I_1 and I_2 added together, $I = 8 + 4 = 12\,\text{A}$

$$R_{\text{eq}} = (R_1 + R_2)//R_3 = (2+3)//5 = 2.5\,\Omega$$

$$V_{R_{\text{eq}}} = I \cdot R_{\text{eq}} = 12 \cdot 2.5 = 30\,\text{V}$$

$$I_{R1} = \frac{V_{R_{\text{eq}}}}{R_1 + R_2} = \frac{30\,V}{5\,\Omega} = 6\,\text{A}.$$

Problem 2.4.12 In the circuit shown in Fig. 2.103, $V_S = 1$ V, $R_1 = R_2 = R_3 = R_4 = 1\ \Omega$, $f = 4$ A/A.

V_1 = ?, V_a = ?, i = ? Use source transformation and node voltage method.

Solution

Use source transformation and note that $V_a = V_2$ (see, Fig. 2.104),

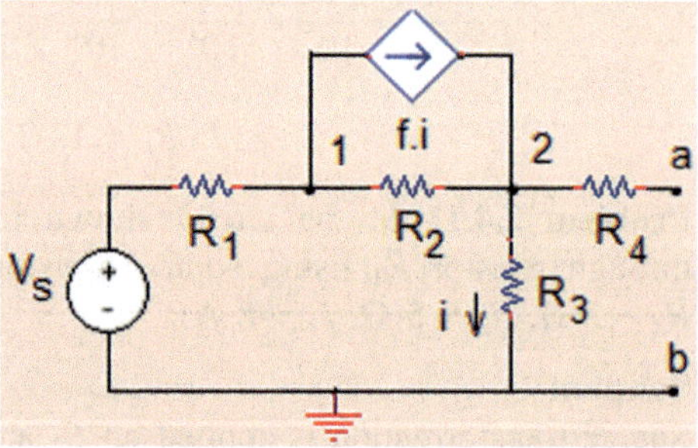

Fig. 2.103 The circuit for Problem 2.4.12

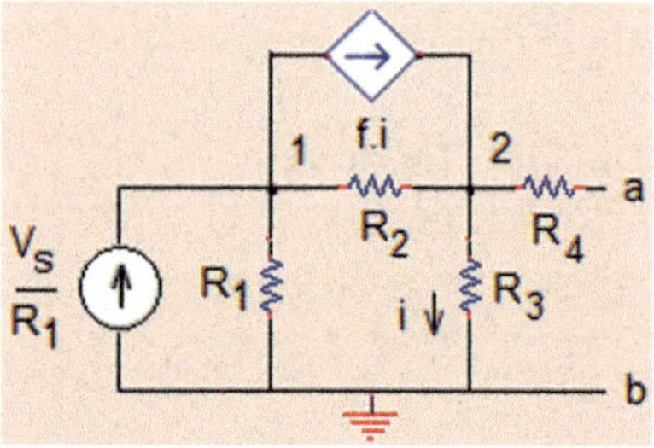

Fig. 2.104 Source transformation applied to the circuit of Problem 2.4.12

$$\frac{V_s}{R_1} - f \cdot \frac{V_2}{R_3} - \frac{V_1}{R_1} - \frac{V_1 - V_2}{R_2} = 0 \tag{2.81}$$

$$f \cdot \frac{V_2}{R_3} + \frac{V_1 - V_2}{R_2} - \frac{V_2}{R_3} = 0. \tag{2.82}$$

Substituting numerical values and rearranging above equations,

$$\begin{bmatrix} 2 & 3 \\ 1 & 2 \end{bmatrix} \begin{bmatrix} V_1 \\ V_2 \end{bmatrix} = \begin{bmatrix} 1 \\ 0 \end{bmatrix}$$

Solution of this matrix equation gives

$$V_1 = 2\,\text{V}, \quad V_2 = V_a = -1\,\text{V}, \quad i = \frac{V_2}{R_3} = -\frac{1}{1} = 1\,\text{A}.$$

2.5 Thévenin–Norton Equivalent Circuits and Maximum Power Transfer

Problem 2.5.1 A signal source has an open-circuit voltage of 1 mV and a short-circuit current of 100 nA. What is the source resistance?

Solution
Open-circuit voltage = Thévenin voltage, Short-circuit current = Norton current

$$R_s = \frac{V_\text{T}}{I_\text{N}} = \frac{10^{-3}}{100 \cdot 10^{-9}} = \frac{10^{-3}}{10^{-7}} = 10^4\,\Omega = 10\,\text{k}\Omega.$$

Problem 2.5.2 For which one of the following circuits in Fig. 2.105, Thévenin's theorem cannot be applied?

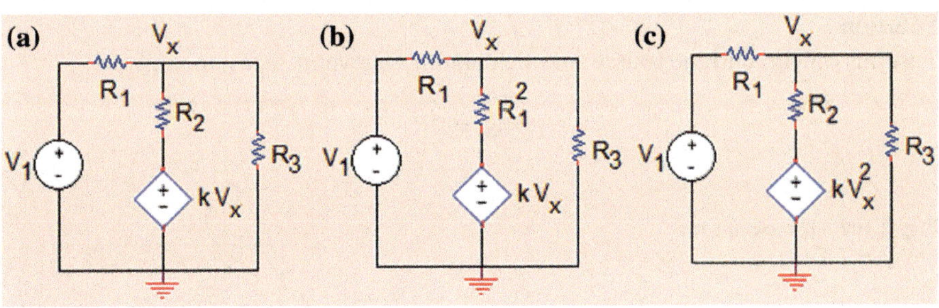

Fig. 2.105 The circuit for Problem 2.5.2

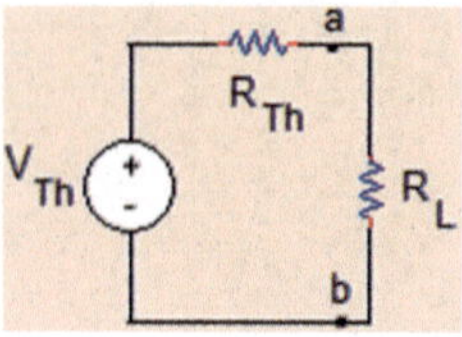

Fig. 2.106 The circuit for Problem 2.5.3

Solution
Thévenin's theorem helps to reduce any one-port linear electrical network to a single-voltage source and a single impedance. The circuit in Fig. 2.105c contains a nonlinear-dependent source. Therefore, it does not suit for the application of this theorem.

Problem 2.5.3 A carbon–zinc battery can be thought of its Thévenin's equivalent circuit with Thévenin resistance being the internal battery resistance, see Fig. 2.106. In an experiment, the open-circuit voltage of a battery is measured as 1.596 V. When a resistor of R = 33.0 Ω is connected across its terminals, the load voltage is measured as 1.580 V. What is the internal battery resistance?

Solution
Assuming that measuring equipment probes and battery terminals have no contact resistances, internal resistance of the battery is serially connected to the load. The load current is

$$I_L = \frac{V_{\text{Th}}}{R_{\text{Th}} + R_L} = \frac{V_L}{R_L}$$

$$I_L = \frac{1.596}{R_{\text{Th}} + 33} = \frac{1.580}{33} = 0.47878\,\text{A}$$

$$R_{\text{Th}} = (28.886)(1.596) - 33$$

$$R_{\text{Th}} = 0.334\,\Omega.$$

Problem 2.5.4 In the circuit shown in Fig. 2.107, determine the current through R_L at maximum power transfer condition if I = 4 A.

Solution
Turning off the current source and calculating Thévenin resistance gives

$$R_{\text{Th}} = 2R.$$

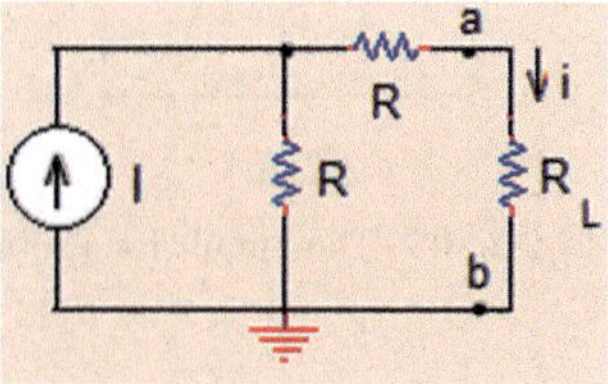

Fig. 2.107 The circuit for Problem 2.5.4

Thévenin voltage is the voltage drop across the grounded resistor,

$$V_{\text{Th}} = V_{\text{oc}} = IR$$

$$i_{\text{RLpmax}} = \frac{V_{\text{Th}}}{R_{\text{Th}} + R_L} = \frac{IR}{2R + 2R} = \frac{I}{4} = 1\,\text{A}.$$

Problem 2.5.5

(a) Determine the Thévenin and Norton equivalents for the circuit shown in Fig. 2.108, between a and b terminals.
(b) Find the power delivered to a load resistance, if $R_L = 5\,\Omega$.
(c) Determine the value of load resistor for maximum power transfer. $R_1 = R_2 = 6\,\Omega,\quad R_3 = 12\,\Omega,\quad U = 10\,\text{V}.$

Solution

(a) Thévenin equivalent circuit values are

$$R_{\text{Th}} = 12 + 6\|6 = 12 + 3 = 15\,\Omega$$

$$V_{\text{Th}} = V_{\text{ab}} = V_{\text{oc}} = \frac{6}{6+6} \cdot 10 = 5\,\text{V}.$$

Note that 12Ω resistance has no effect here.
Norton equivalent circuit values are

$$R_{\text{N}} = R_{\text{Th}} = 15\,\Omega$$

$$I_{\text{N}} = \frac{V_{\text{Th}}}{R_{\text{Th}}} = \frac{5}{15} = \frac{1}{3} = 0.333\,\text{A}$$

(b) $I = \dfrac{V_{\text{Th}}}{R_{\text{Th}} + R_L} = \dfrac{5}{15+5} = \dfrac{1}{4}\,\text{A}$

$$P = I^2 R_L = \left(\frac{1}{4}\right)^2 \times 5 = \frac{5}{16} = 0.3125\,\text{W}$$

(c) For maximum power transfer; $R_L = R_{\text{Th}} = 5\,\Omega$.

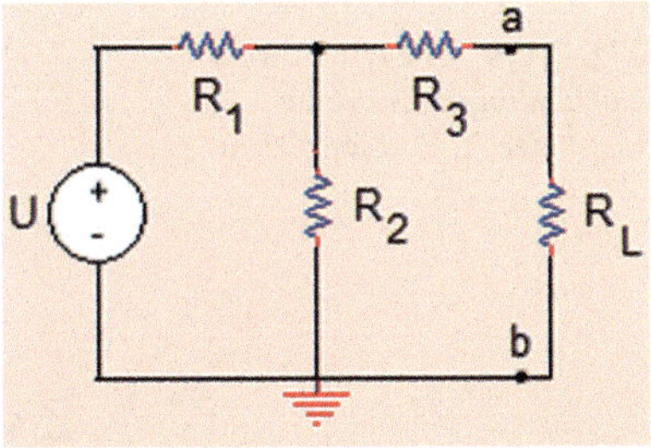

Fig. 2.108 The circuit for Problem 2.5.5

Problem 2.5.6 Use Thévenin's theorem to find the value of current, i_{R6} (Fig. 2.109).

$$(R_1 = 50\,\Omega,\quad R_2 = 5\,\Omega,\quad R_3 = R_4 = 10\,\Omega,\quad R_5 = 4\,\Omega,\quad R_6 = 2\,\Omega,\quad V_1 = 20\,V).$$

Solution

Remove R_6 from circuit: $R_3//R_4 = 5\,\Omega$, $V_{ab} = V_{Th}$

Voltage division:

$$V_{ab} = \frac{R_3//R_4}{R_2 + R_3//R_4} \cdot 20$$

$$V_{Th} = V_{ab} = \frac{5}{5+5} \cdot 20 = 10\,\text{V}.$$

Thévenin resistance: When the voltage source is short circuited, $V_1 = 0$ V, R_1 is shorted (see, Fig. 2.110):

$$R_{Th} = R_5 + (R_2 \| R_3 \| R_4) = 4 + (5\|10\|10)$$

$$R_{Th} = 4 + (5\|5) = 4 + 2.5 = 6.5\,\Omega.$$

The value of the current flowing through resistor R_6

$$i_{R_6} = \frac{V_{Th}}{R_{Th} + R_6} = \frac{10}{6.5+2} = \frac{10}{8.5} = 1.177\,\text{A}.$$

Problem 2.5.7 In the circuit shown in Fig. 2.111, use Thévenin's Theorem and source transformation method to determine the current through the resistor R_L.

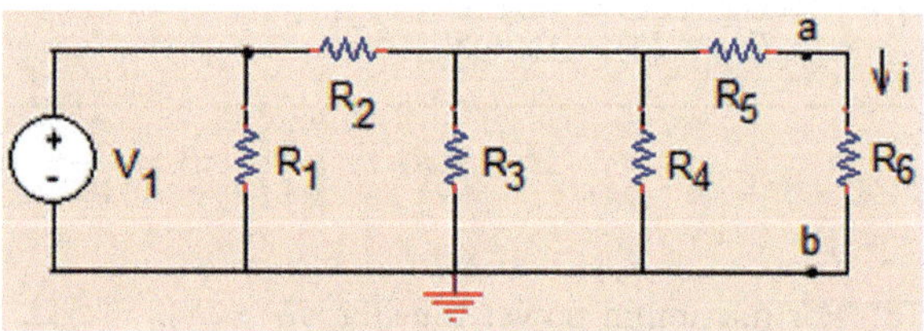

Fig. 2.109 The circuit for Problem 2.5.6

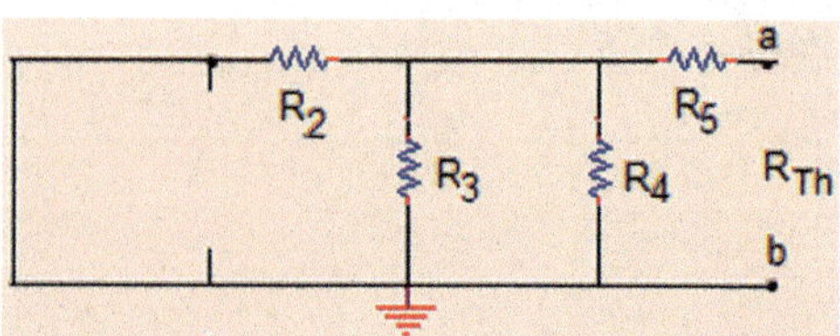

Fig. 2.110 The circuit for the calculation of Thévenin resistance in Problem 2.5.6

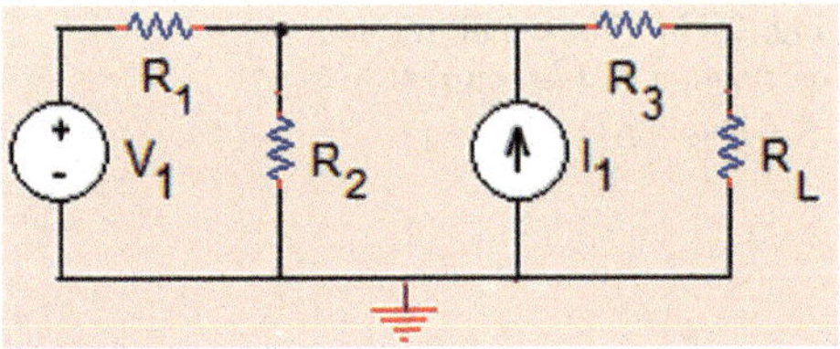

Fig. 2.111 The circuit for Problem 2.5.7

$$R_1 = R_2 = 2\,\Omega,\quad R_3 = R_L = 1\,\Omega,\quad V_1 = 1\,\mathrm{V},\quad I_1 = 0.5\,\mathrm{A}.$$

Solution

First calculate Thévenin resistance for the circuit to the left of R_L (Fig. 2.112):

$$R_{\mathrm{Th}} = (R_1 \| R_2) + R_3.$$

Then, determine Thévenin voltage, noting that R_3 has no current flow at the output terminals (a-ground), as shown in Fig. 2.113.

The node voltage V_x becomes the open-circuit voltage.

Using KCL at this node,

$$\frac{V_1}{R_1} + I_1 = \frac{V_x}{R_1} + \frac{V_x}{R_2} \quad \rightarrow \quad V_x = V_{\mathrm{Th}} = \frac{\frac{V_1}{R_1} + I_1}{\frac{1}{R_1} + \frac{1}{R_2}}.$$

The load current through the resistor R_L is found using Fig. 2.114:

$$i_{RL} = \frac{V_{\mathrm{Th}}}{R_{\mathrm{Th}} + R_L} = \frac{\dfrac{\frac{V_1}{R_1} + I_1}{\frac{1}{R_1} + \frac{1}{R_2}}}{(R_1 \| R_2) + R_3 + R_L} = \frac{\dfrac{\frac{1}{2} + \frac{1}{2}}{\frac{1}{2} + \frac{1}{2}}}{(2 \| 2) + 1 + 1} = \frac{1}{3}\,\mathrm{A}.$$

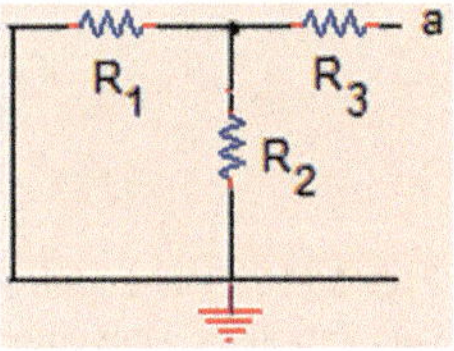

Fig. 2.112 The circuit for the calculation of Thévenin resistance in Problem 2.5.7

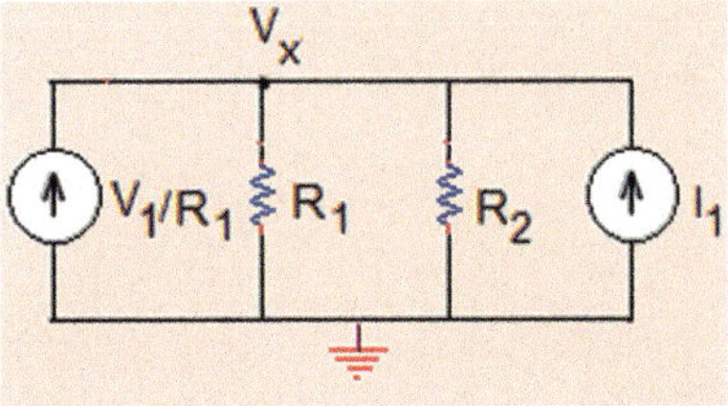

Fig. 2.113 The circuit for the calculation of Thévenin voltage in Problem 2.5.7

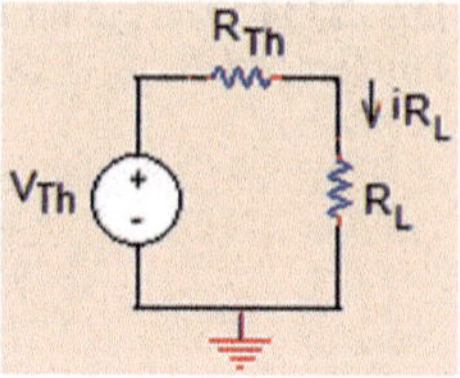

Fig. 2.114 The circuit for the calculation of load current in Problem 2.5.7

Problem 2.5.8 Determine Thévenin equivalent parameters between a and b terminals of the circuit shown in Fig. 2.115. $R = 2\ \Omega$. (Hint: Apply source transformation for the current source in calculating Thévenin equivalent voltage.)

Solution

Thévenin equivalent resistance is found by eliminating independent sources (i.e., short-circuit voltage source and open-circuit current source) in the circuit and finding the resistance between a and b terminals.

$$R_{\mathrm{Th}} = R_{\mathrm{ab}} = [(R+R)\|2R] + R = 2R = 4\,\Omega.$$

Thévenin equivalent voltage between (a) and (b) terminals of the circuit can be found by applying source transformation to the current source, and then determining the voltage at node (c). This is due to the fact that the voltage at terminal (a) equals to voltage at node (c), $V_{\mathrm{cb}} = V_{\mathrm{ab}}$ (Fig. 2.116).

By KVL, $i = (2-1)/(R+R+2R) = 1/8 = 0.125\ \mathrm{A}$

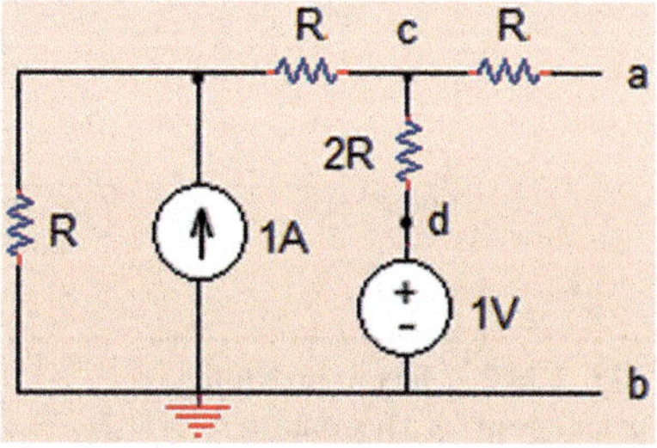

Fig. 2.115 The circuit for Problem 2.5.8

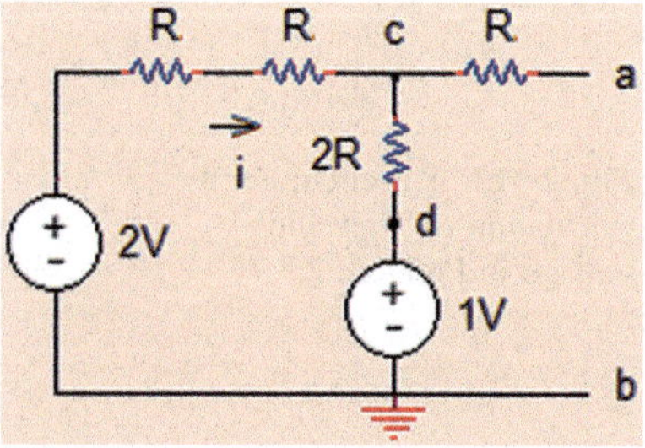

Fig. 2.116 The circuit for the calculation of Thévenin voltage in Problem 2.5.8

$$V_{Th} = V_{ab} = V_{cb} = 2R \cdot i + 1 = 4 \times 0.125 + 1 = 1.5\,V.$$

In summary, $V_{Th} = 1.5\,V, R_{Th} = 4\,\Omega$.

Problem 2.5.9 For the circuit shown in Fig. 2.117,

(a) Find Thévenin's equivalent to the left of terminals a and b.
(b) If R_1 = 1 kΩ, R_2 = 1 kΩ, R_3 = 100 Ω, k = 0.1, R_L = 1 kΩ, what is V_{ab}?

Solution
For Thévenin's equivalent circuit to the left of terminals a and b, the voltage source is removed, the first circuit becomes a short circuit, so the R_{Th} only depends on R_3 (Fig. 2.118).

$$R_{Th} = R_3 = 100\,\Omega.$$

The current generated with current-controlled current source, CCCS becomes

$$i_{Th} = k \times i_{R_2} = k \times \frac{V_1}{R_1 + R_2} = 0.1 \times \frac{V_1}{2000} = \frac{V_1}{2} \times 10^{-4}\,A$$

$$V_{Th} = R_{Th} \times i_{Th}$$

$$V_{Th} = 100 \times \frac{V_1}{2} \times 10^{-4} = 50 \times 10^{-4} \times V_1$$

$$i_L = \frac{V_{Th}}{R_3 + R_L} = \frac{50 \times 10^{-4} \times V_1}{100 + 1000} = 4.6 \times 10^{-6} \times V_1\,A$$

$$V_{R_L} = R_L \times i_L = 1000 \times 4.6 \times 10^{-6} \times V_1 = 4.6 \times 10^{-3} \times V_1.$$

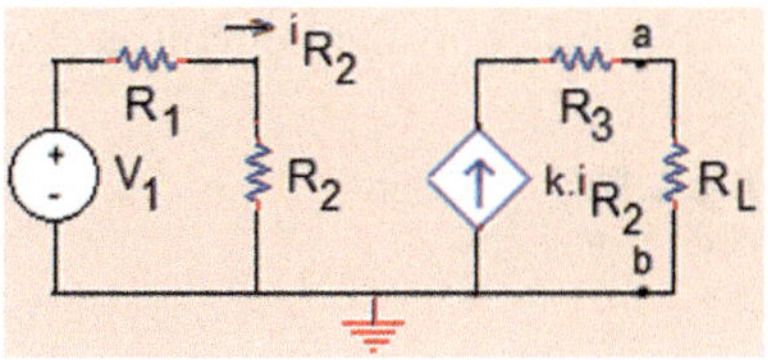

Fig. 2.117 The circuit for Problem 2.5.9

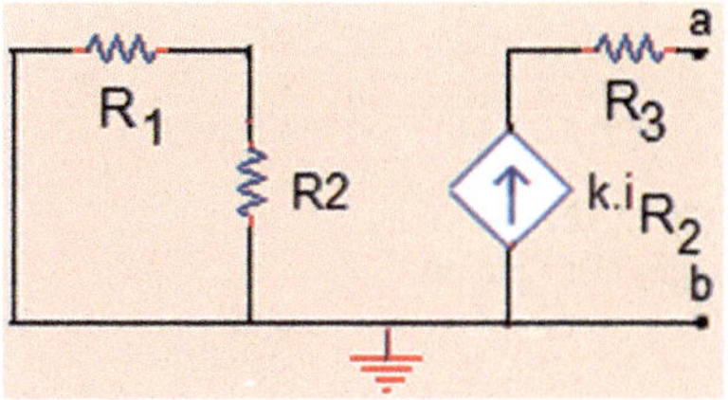

Fig. 2.118 R_{Th} only depends on R_3

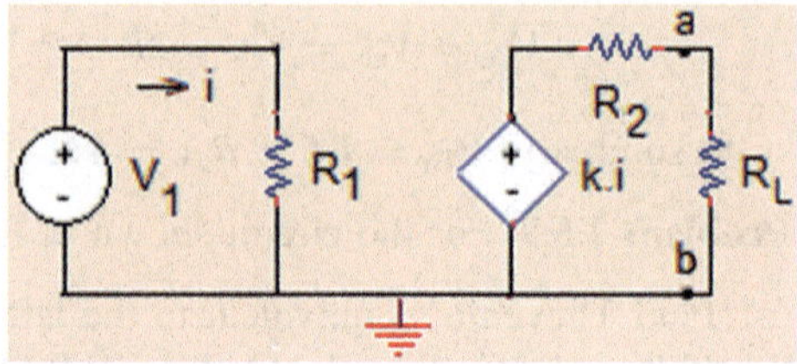

Fig. 2.119 The circuit for Problem 2.5.10

Problem 2.5.10 For the circuit shown in Fig. 2.119,

a. Find the current through R_L, and voltage across R_L, using Thévenin's method (as function of k, V_1, R_1, R_2, R_L)
b. What is the Norton's equivalent circuit to the left of a–b if $V_1 = 2$ V, $k = 2$ V/V, $R_1 = 10\ \Omega$, $R_2 = 5\ \Omega$.

Solution

(a) $i = \dfrac{V_1}{R_1}, \quad V_{oc} = V_T = k \cdot i = k \cdot \dfrac{V_1}{R_1}$

$$R_T = R_2,$$

$$I_{R_L} = \frac{V_T}{R_2 + R_L} = \frac{kV_1}{R_1(R_2 + R_L)}, \quad V_{R_L} = I \cdot R_L = \frac{k \cdot V_1 R_L}{R_1(R_2 + R_L)}.$$

b. Thévenin equivalent circuit is shown in Fig. 2.120.

$$V_T = k \cdot \frac{V_1}{R_1} = 2 \cdot \frac{2}{10} = 0.4\,\text{V}, \quad R_T = R_N = R_2 = 5\,\Omega.$$

Norton equivalent circuit is shown in Fig. 2.121.

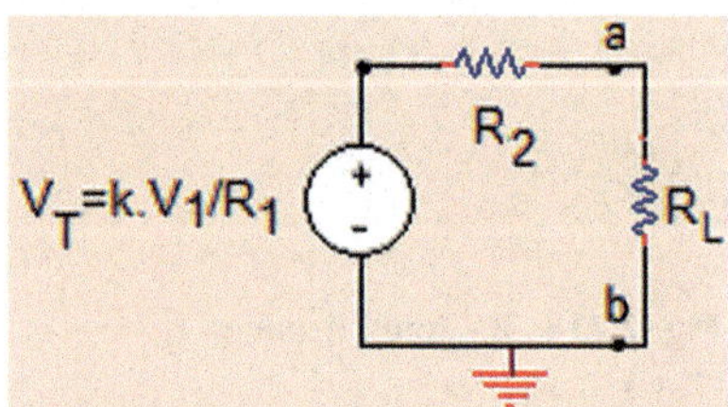

Fig. 2.120 Thévenin equivalent circuit

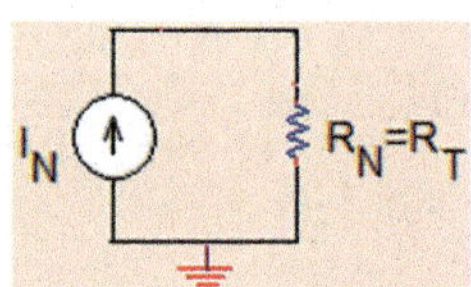

Fig. 2.121 Norton equivalent circuit

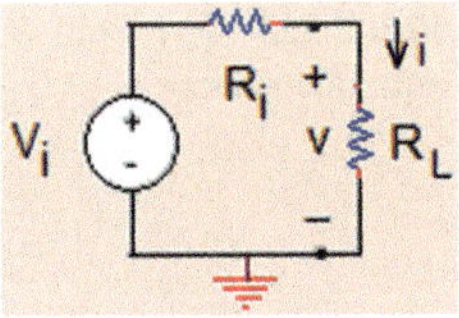

Fig. 2.122 The circuit described in Problem 2.5.11

Problem 2.5.11 A DC voltage source with internal resistance of 10 Ω and with an open-circuit voltage of 12 V feeds a resistive load, R_L. Determine the range of load resistance values so that the circuit operates for $V_L \geq 5\,\text{V}$ and $i_L \geq 0.5\,\text{A}$.

Solution

Figure 2.122 shows the equivalent circuit.

Using voltage constraint, $R_{\text{Th}} = R_i = 10\,\Omega, V_{\text{oc}} = V_{\text{Th}} = 12\,\text{V}$

$$\frac{R_L}{R_{\text{Th}} + R_L} V_{\text{Th}} \geq 5 \quad \rightarrow \quad \frac{R_L}{10 + R_L} \times 12 \geq 5 \quad \rightarrow \quad 7R_L \geq 50 \quad \rightarrow \quad R_L \geq 7.14\,\Omega.$$

Using current constraint,

$$i = \frac{V_{\text{Th}}}{R_{\text{Th}} + R_L} \geq 0.5\,\text{A} \quad \rightarrow \quad \frac{12}{10 + R_L} \geq 0.5 \quad \rightarrow \quad 12 \geq 5 + 0.5R_L \quad \rightarrow \quad R_L \leq 14\,\Omega.$$

Therefore, $7.14\,\Omega \leq R_L \leq 14\,\Omega$.

A proper value of load resistance is to choose the arithmetic mean of limiting values. This allows for component variations in the circuit. In that case, $R_L = 10\,\Omega$ can be a suitable value of the load resistance.

Problem 2.5.12 Maximum power delivered by a DC circuit to a resistor of $R = 20.25\ \Omega$ is 1 W. Find the open-circuit voltage at the output of the circuit. Draw its Thévenin's equivalent circuit.

Solution

$$P_{\text{max}} = \frac{V_{\text{th}}^2}{4R_L}, \quad V_{\text{Th}} = V_{\text{oc}}, \quad R = R_{\text{Th}}$$

$$1 = \frac{V_{\text{oc}}^2}{4 \cdot (20.25)} = \frac{V_{\text{oc}}^2}{81} \quad \rightarrow \quad V_{\text{oc}}^2 = 81 \quad \rightarrow \quad V_{\text{oc}} = 9\,\text{V}.$$

Figure 2.123 describes the Thévenin equivalent circuit. $V_{\text{Th}} = V_{\text{oc}} = 9\,\text{V}, R_{\text{Th}} = R = 20.25\,\Omega$.

Problem 2.5.13 Determine Thévenin and Norton equivalent circuit parameters for the circuit between terminal b and ground, as shown in Fig. 2.124.

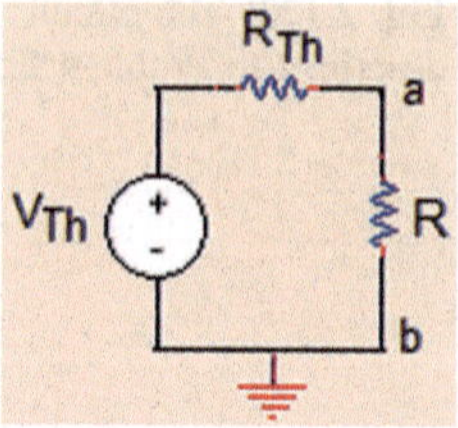

Fig. 2.123 Thévenin equivalent circuit

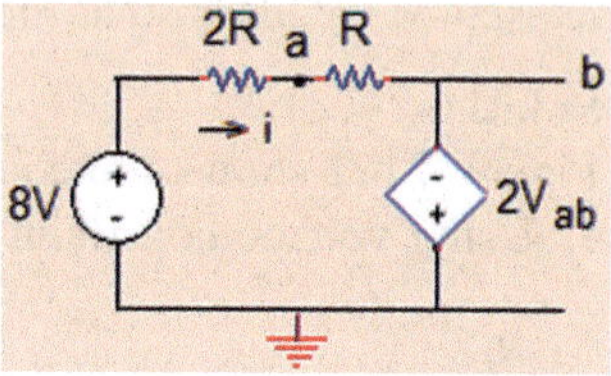

Fig. 2.124 The circuit described in Problem 2.5.13

Solution

First, determine the loop current under open-circuit conditions (i.e., no load is connected between terminal b and ground). By KVL and assuming clockwise current flow in the loop,

$$8-(4+2)i+V_{\mathrm{ab}}=8-6i+2(2i)=8-2i=0 \quad \rightarrow \quad i=4\,\mathrm{A}$$

$$V_{\mathrm{oc}}=V_{\mathrm{th}}=2V_{\mathrm{ab}}=2(2i)=4i=4\times 4=16\,\mathrm{V}.$$

When dependent source is shorted to ground, one can determine the short-circuit current at the output port, as

$$i_{\mathrm{sc}}=\frac{8}{4+2}=\frac{8}{6}\,\mathrm{A}.$$

Then,

$$R_{\mathrm{th}}=R_{\mathrm{N}}=\frac{V_{\mathrm{oc}}}{i_{\mathrm{sc}}}=\frac{16}{\frac{8}{6}}=12\,\Omega, \quad I_{\mathrm{N}}=\frac{V_{\mathrm{th}}}{R_{\mathrm{th}}}=\frac{16}{12}=1.333\,\mathrm{A}.$$

Problem 2.5.14 For the circuit shown in Fig. 2.125, find

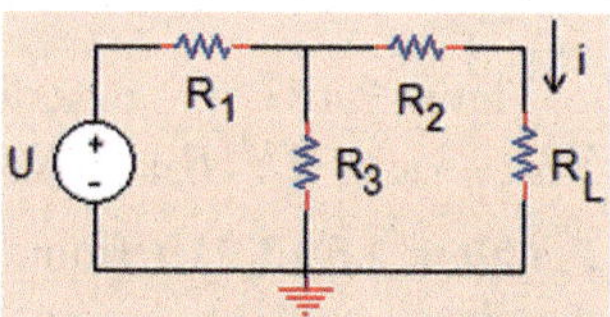

Fig. 2.125 The circuit described in Problem 2.5.14

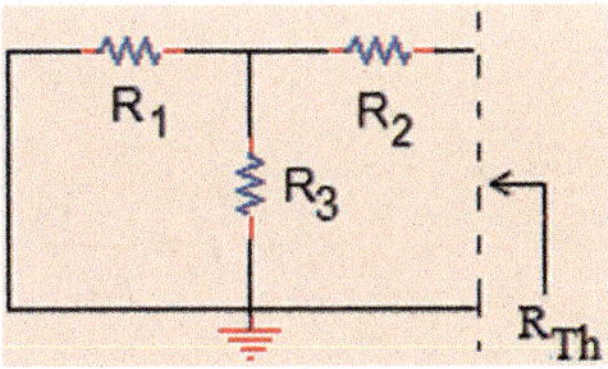

Fig. 2.126 The circuit for the calculation of Thévenin resistance in Problem 2.5.14

(a) the value of load at maximum power transfer.
(b) the value of current through load resistor at maximum power.
(c) the value of maximum power transferred to the load $(U = 2\,\text{V}, R1 = 6\,\Omega, R2 = 8\,\Omega, R3 = 4\,\Omega)$.

Solution

(a) Thévenin resistance (see Fig. 2.126)

$$R_{\text{Th}} = (6//4) + 8 = \frac{24}{10} + 8 = \frac{104}{10}$$

$$R_L = R_{\text{Th}} = 10.4\,\Omega.$$

(b) The value of current at maximum power is calculated from Thévenin's equivalent circuit, see Figs. 2.127 and 2.128.

By voltage division,

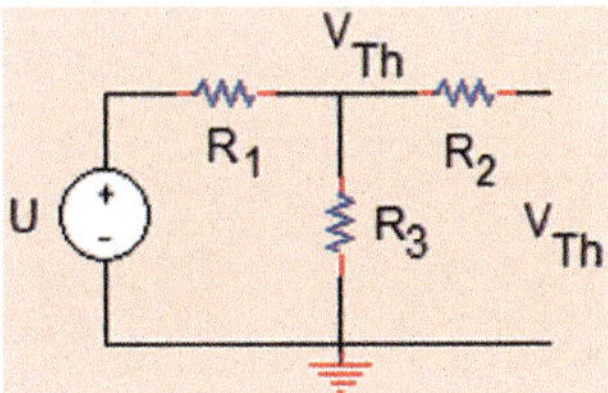

Fig. 2.127 The circuit for the calculation of Thévenin voltage in Problem 2.5.14

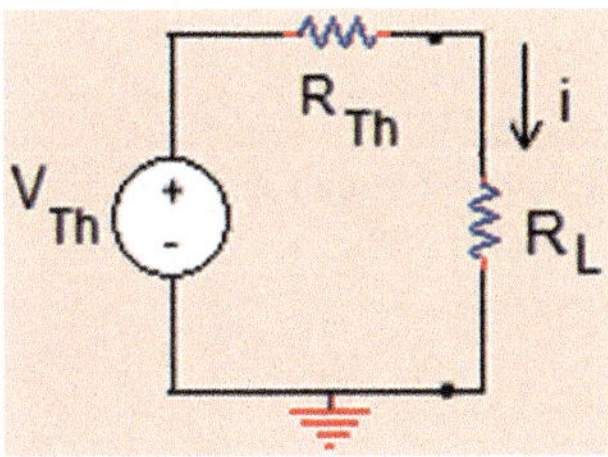

Fig. 2.128 Thévenin equivalent circuit

$$V_{Th} = V_{oc} = 2 \cdot \frac{4}{4+6} = \frac{8}{10} = 0.8\,\text{V}$$

$$i = \frac{V_{Th}}{R_{Th} + R_L} = \frac{0.8}{10.4 + 10.4} = \frac{0.8}{20.8} = 0.038\,\text{A}.$$

(c) $P_{max} = i^2 \cdot R_L = (0.038)^2 \cdot (10.4) = 0.015\,\text{W}.$

Problem 2.5.15 In the circuit shown in Fig. 2.129, all resistors (except R_x) have the same resistance of R = 10 Ω, and V_1 = 10 V. Find the value of R_x for maximum power transfer (matrix_solve.xlsx) (Fig. 2.130).

Solution
Thévenin's equivalent circuit

$$R_x = R_{Th} = \frac{v_{oc}}{i_{sc}}.$$

Open-circuit voltage (Fig. 2.131),

$$\begin{bmatrix} 3R & -R & 0 & -R & 0 \\ -R & 4R & -R & 0 & -R \\ 0 & -R & 4R & 0 & 0 \\ -R & 0 & 0 & 2R & -R \\ 0 & -R & 0 & -R & 3R \end{bmatrix} \cdot \begin{bmatrix} i_1 \\ i_2 \\ i_3 \\ i_4 \\ i_5 \end{bmatrix} = \begin{bmatrix} V_1 \\ 0 \\ 0 \\ 0 \\ 0 \end{bmatrix}.$$

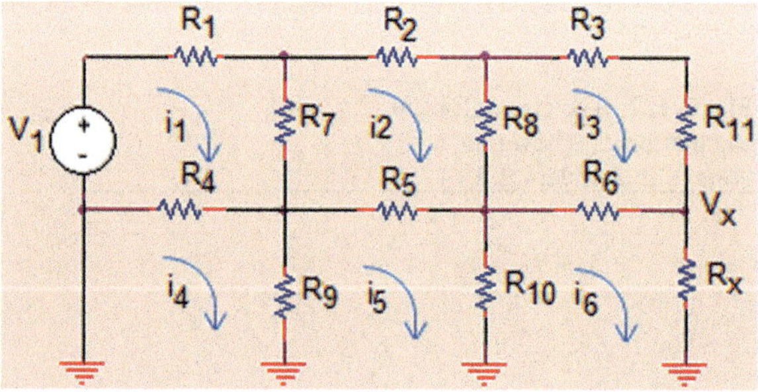

Fig. 2.129 The circuit of Problem 2.5.15

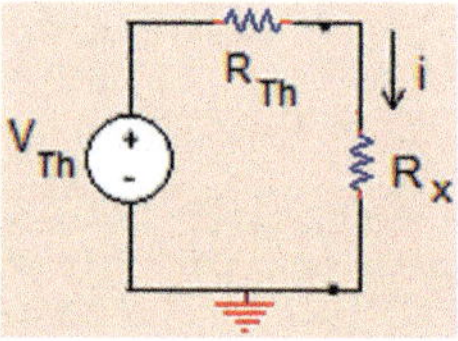

Fig. 2.130 Thévenin's equivalent circuit in Problem 2.5.15

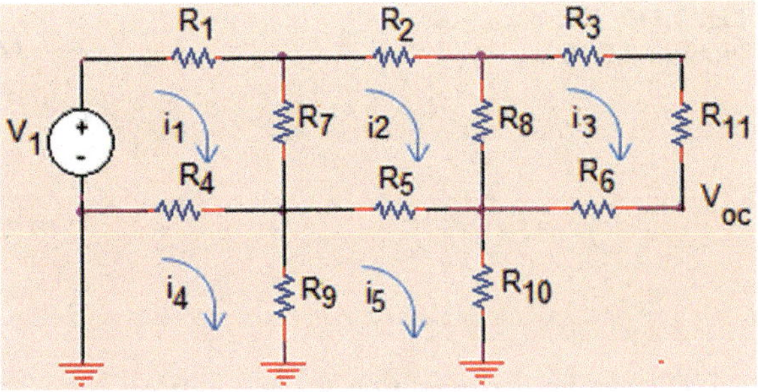

Fig. 2.131 Circuit for Thévenin voltage calculation in Problem 2.5.15

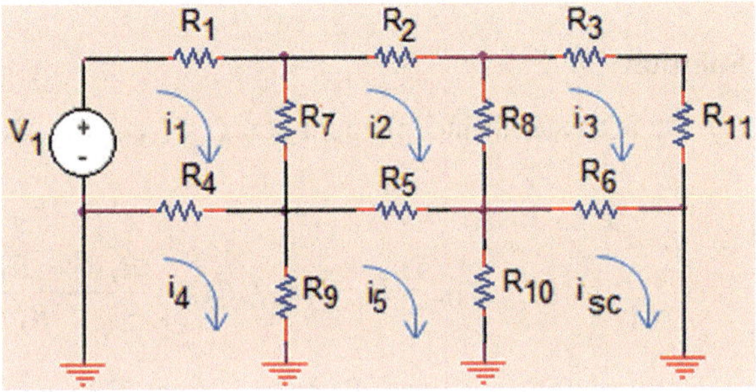

Fig. 2.132 Circuit for short circuit current calculation in Problem 2.5.15

Mesh currents can be found here using an EXCEL spreadsheet as

$$i_1 = 0.509\,\text{A},\quad i_2 = 0.185\,\text{A},\quad i_3 = 0.046\,\text{A},\quad i_4 = 0.343\,\text{A},\quad i_5 = 0.176\,\text{A}$$

$$V_{\text{oc}} = V_{\text{Th}} = i_5 \cdot R + i_3 \cdot R = 10(0.046 + 0.176) = 10 \times 0.222 = 2.22\,\text{V}.$$

Short-circuit current (Fig. 2.132):

$$\begin{bmatrix} 3R & -R & 0 & -R & 0 & 0 \\ -R & 4R & -R & 0 & -R & 0 \\ 0 & -R & 4R & 0 & 0 & -R \\ -R & 0 & 0 & 2R & -R & 0 \\ 0 & -R & 0 & -R & 3R & -R \\ 0 & 0 & -R & 0 & -R & 2R \end{bmatrix} \begin{bmatrix} i_1 \\ i_2 \\ i_3 \\ i_4 \\ i_5 \\ i_{\text{sc}} \end{bmatrix} = \begin{bmatrix} V \\ 0 \\ 0 \\ 0 \\ 0 \\ 0 \end{bmatrix}$$

$$i_{\text{sc}} = 0.195\,\text{A}$$

$$R_x = R_{\text{Th}} = \frac{V_{\text{oc}}}{i_{\text{sc}}} = \frac{2.2}{0.195} = 11.282\,\Omega.$$

Problem 2.5.16 In the circuit shown in Fig. 2.133,

(a) Determine Thévenin's equivalent circuit parameters.
(b) If $R_1 = R_2 = R_3 = 1\ \Omega$, $V_1 = 2$ V, $I_1 = 1$ A, $R_{\text{Th}} = ?$, $V_{\text{Th}} = ?$

Fig. 2.133 The circuit of Problem 2.5.16

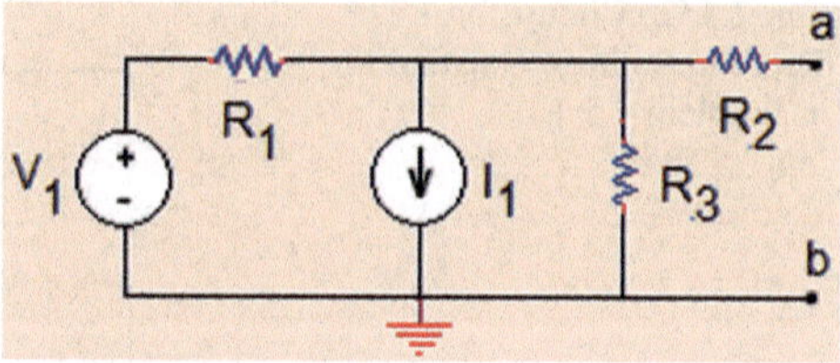

(c) Determine the condition for $V_{ab} < 0$.
(d) Norton equivalent circuit parameters in (b) ?

Solution

(a) De-activate all independent sources (see, Fig. 2.134):

$$R_{\mathrm{Th}} = R_2 + (R_1 // R_3) = \frac{R_1R_2 + R_1R_3 + R_2R_3}{R_1 + R_3}.$$

Because node a is open, R_2 has no effect. By source transformation of the voltage source, following circuit is obtained. Then, applying KCL at node a (Fig. 2.135),

$$\frac{V_1}{R_1} - \frac{V_a}{R_1} - I_1 - \frac{V_a}{R_3} = 0 \quad \rightarrow \quad \frac{V_1}{R_1} - I_1 = V_a\left(\frac{1}{R_1} + \frac{1}{R_3}\right) \quad \rightarrow$$

$$V_a = \frac{\left(\frac{V_1}{R_1} - I_1\right)}{\left(\frac{1}{R_1} + \frac{1}{R_3}\right)} = V_{\mathrm{Th}}.$$

Fig. 2.134 The circuit for the calculation of Thévenin resistance in Problem 2.5.16

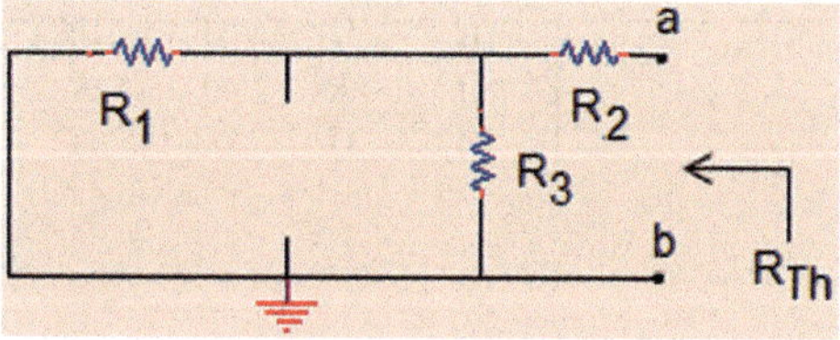

Fig. 2.135 The circuit for the calculation of Thévenin voltage in Problem 2.5.16

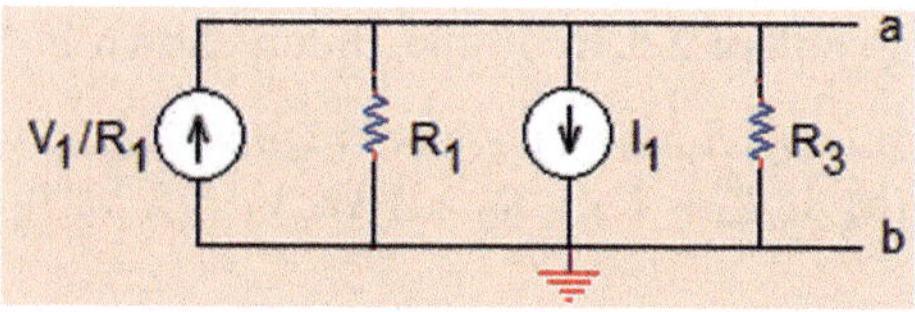

(b) $R_1 = R_2 = R_3 = 1\ \Omega$, $V_1 = 2$ V, $I_1 = 1$ A, substituting the values,

$$R_{\mathrm{Th}} = \frac{1+1+1}{1+1} = 1.5\,\Omega$$

$$V_{\mathrm{Th}} = V_a = \frac{\frac{2}{1} - 1}{1+1} = 0.5\,\mathrm{V}$$

(c) If $\frac{V_1}{R_1} < I_1,\quad V_{\mathrm{Th}} = V_a < 0.$

(d) Norton equivalent circuit parameters are

$$I_{\mathrm{Th}} = \frac{V_{\mathrm{Th}}}{R_{\mathrm{Th}}} = \frac{0.5}{1.5} = 0.333\,\mathrm{A}$$

$$R_{\mathrm{N}} = R_{\mathrm{Th}} = 1.5\,\Omega.$$

Problem 2.5.17

(a) Find Thévenin equivalent circuit for the circuit shown in Fig. 2.136.
(b) Find the limiting value of k if $R_1 = R_2 = 1\ \Omega$.
(c) Norton equivalent circuit?

Solution

(a) The circuit has no independent sources. Apply source transform to dependent source and 1 A current at terminals a, b (Fig. 2.137).

KCL at terminal (a), with $i = V_a/R_2$

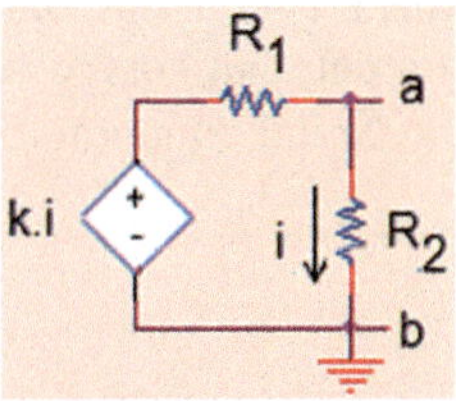

Fig. 2.136 The circuit of Problem 2.5.17

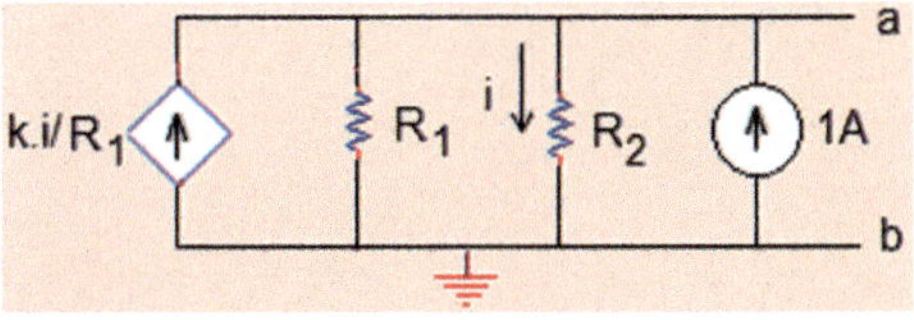

Fig. 2.137 Source transformation of dependent source and application of 1 A current at terminals a, b of the circuit of Problem 2.5.17

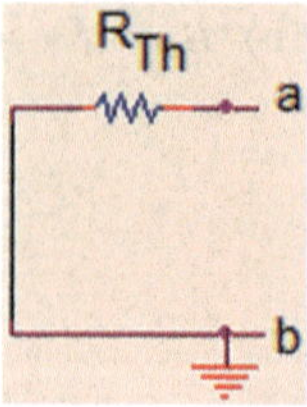

Fig. 2.138 Thévenin's (=Norton's) equivalent of the circuit in Problem 2.5.17

$$\frac{kV_a}{R_1 \cdot R_2} + 1 - \frac{V_a}{R_1} - \frac{V_a}{R_2} = 0 \quad \rightarrow \quad V_a\left(\frac{k}{R_1R_2} - \frac{1}{R_1} - \frac{1}{R_2}\right) = -1$$

$$V_{\text{Th}} = V_a = \frac{-1}{\dfrac{k}{R_1R_2} - \left(\dfrac{1}{R_1} + \dfrac{1}{R_2}\right)}$$

$$R_{\text{Th}} = \frac{V_a}{1\,\text{A}} = \frac{1}{\dfrac{1}{R_1} + \dfrac{1}{R_2} - \dfrac{k}{R_1R_2}}.$$

(b) Denominator of Thévenin resistance must be different than zero,

$$\frac{k}{R_1R_2} \neq \frac{1}{R_1} + \frac{1}{R_2}.$$

If $R_1 = R_2 = 1\ \Omega$, then $k \neq 2$

(c) Norton equivalent circuit = Thévenin equivalent circuit (Fig. 2.138).

Problem 2.5.18 In the circuit shown in Fig. 2.139, determine the inequality condition on parameter C in terms of known quantities of the circuit so that, $R_{ab} < 0\ \Omega$.

Solution

The circuit does not contain an independent source, therefore its Thévenin equivalent circuit has only a Thévenin resistance. We assign a current source at the output, and source transform-dependent source (Fig. 2.140a, b),

Nodal equation:

$$I_0 + I_x + \frac{C}{2}I_x - \frac{V_a}{2} = 0, \quad (V_b = 0\,\text{V}).$$

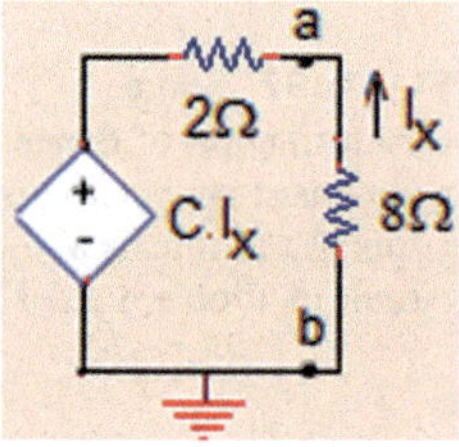

Fig. 2.139 The circuit of Problem 2.5.18

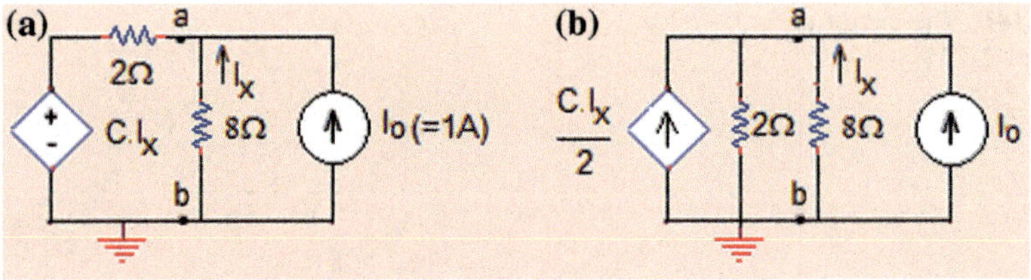

Fig. 2.140 Assign a current source at the output, and source transform-dependent source

But,

$$V_a = -8I_x \quad \rightarrow \quad I_x = -\frac{V_a}{8}$$

$$I_0 - \frac{V_a}{8} + \frac{C}{2}\left(-\frac{V_a}{8}\right) - \frac{V_a}{2} = 0,$$

$$I_0 = \frac{CV_a}{16} + \frac{V_a}{8} + \frac{V_a}{2} = V_a\left(\frac{C}{16} + \frac{1}{8} + \frac{1}{2}\right) = V_a\left(\frac{C+10}{16}\right)$$

$$V_a = \frac{16I_0}{10+C}.$$

Thévenin equivalent of the circuit,

$$R_{\mathrm{Th}} = \frac{V_a}{I_0} = \frac{\left(\frac{16I_0}{10+C}\right)}{I_0} = \frac{16}{10+C}.$$

Therefore, for $R_{\mathrm{ab}} = R_{\mathrm{Th}} < 0$, $C < -10$.

Problem 2.5.19 Find Thévenin equivalent circuit for the circuit shown in Fig. 2.141 (Vs = 1 V, f = 4, $R_1 = R_2 = R_3 = R_4 = 1$ Ω) (matrix_solve.xlsx).

Solution

R_4 has no influence since node a is open. $V_{\mathrm{Th}} = V_{\mathrm{oc}} = V_2$. Apply source transform to voltage source and write nodal equations at 1–2, with $i = V_2/R_3$ (see, Fig. 2.142)

$$\frac{V_s}{R_1} - f \cdot \frac{V_2}{R_3} - \frac{V_1}{R_1} - \frac{V_1 - V_2}{R_2} = 0 \tag{2.83}$$

$$f \cdot \frac{V_2}{R_3} + \frac{V_1 - V_2}{R_2} - \frac{V_2}{R_3} = 0. \tag{2.84}$$

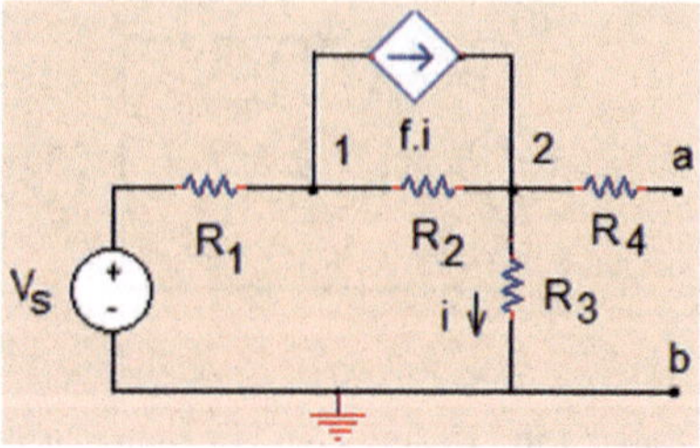

Fig. 2.141 The circuit of Problem 2.5.19

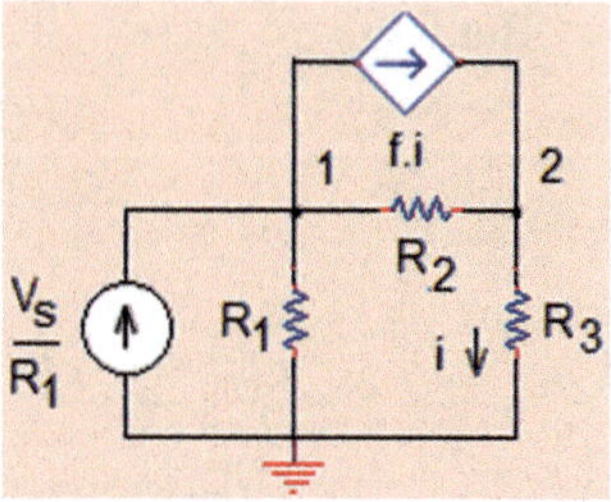

Fig. 2.142 The circuit of Problem 2.5.19 after independent source transformation

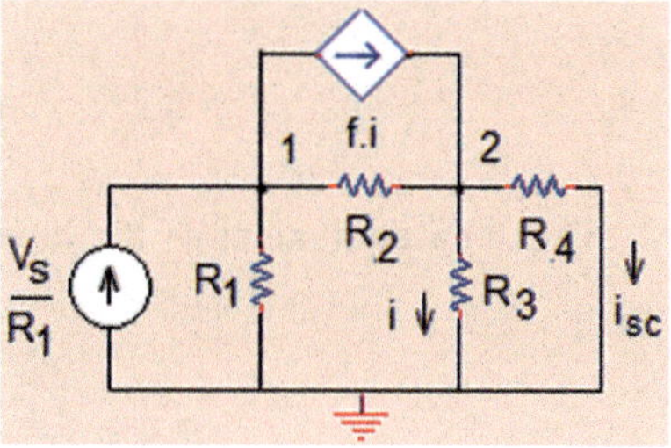

Fig. 2.143 Determining short-circuit current

Substituting numerical values and rearranging equations,

$$\begin{bmatrix} 2 & 3 \\ 1 & 2 \end{bmatrix} \begin{bmatrix} V_1 \\ V_2 \end{bmatrix} = \begin{bmatrix} 1 \\ 0 \end{bmatrix} \tag{2.85}$$

Solving for V_2,

$$V_2 = V_{\mathrm{Th}} = -1\,\mathrm{V} = V_{\mathrm{oc}}$$

$$R_{\mathrm{Th}} = \frac{V_{\mathrm{oc}}}{I_{\mathrm{sc}}}.$$

KCL at nodes 1, 2; with $i = V_2/R_3$, (Fig. 2.143)

$$1 - 4V_2 - V_1 - (V_1 - V_2) = 0 \tag{2.86}$$

$$4V_2 + (V_1 - V_2) - V_2 - V_2 = 0 \tag{2.87}$$

Simplify (2.86) and (2.87),

$$2V_1 + 3V_2 = 1 \tag{2.89}$$

$$V_1 + V_2 = 0. \tag{2.89}$$

Solving these two simultaneous equations for V_2 yields

$$V_2 = 1\,\mathrm{V}, \quad I_{sc} = \frac{V_2}{R_4} = \frac{1}{1} = 1\,\mathrm{A}.$$

Therefore, Thévenin equivalent circuit consists of a negative resistor (Fig. 2.144), with

$$R_{\mathrm{Th}} = \frac{V_{oc}}{I_{sc}} = -\frac{1}{1} = -1\,\Omega.$$

Problem 2.5.20 In the circuit shown in Fig. 2.145, determine the maximum power (in Watts) that can be transferred to load resistance R_L. Given that, when R_L is removed from the circuit, the voltage at node 4 is measured as 4.25 V:

$$R_1 = R_5 = R_6 = R_7 = 1\,\Omega, \quad R_2 = R_3 = R_4 = 2\,\Omega, \quad I_1 = I_2/2 = 2\,\mathrm{A}.$$

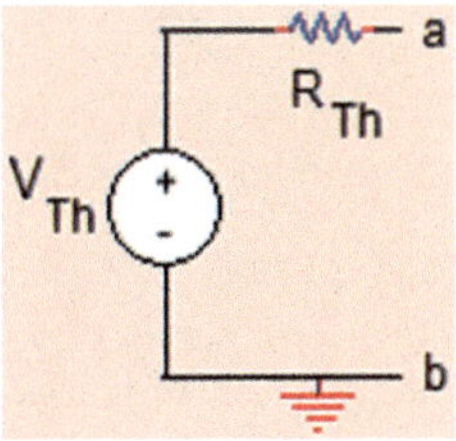

Fig. 2.144 Thévenin equivalent circuit consists of a negative resistor

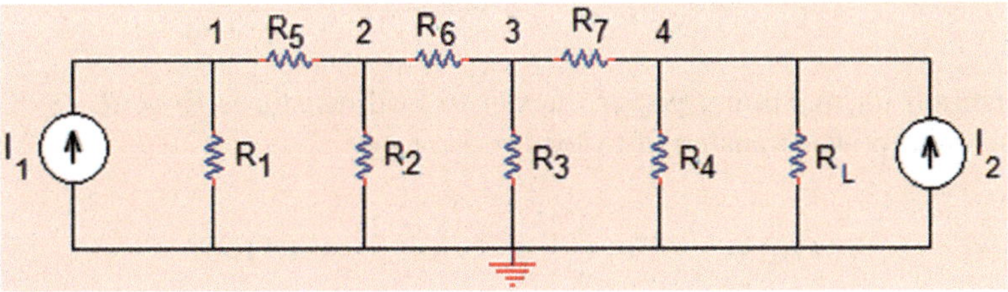

Fig. 2.145 The circuit of Problem 2.5.21

Solution

$$P_{\max} = \frac{V_{\mathrm{Th}}^2}{4R_{\mathrm{Th}}} = \frac{4.25^2}{4 \times R_{\mathrm{Th}}}$$

$$R_{\mathrm{th}} = \{\{[(R_1 + R_5)\|R_2] + R_6\}\|R_3 + R_7\}\|R_4 = 1\,\Omega$$

$$P_{\max} = \frac{4.25^2}{4 \times 1} = \frac{18.0625}{4} = 4.516\,\mathrm{W}.$$

Problem 2.5.21 In the circuit shown in Fig. 2.145, determine the maximum power transferred to R_L :

$R_1 = R_5 = R_6 = R_7 = 1\,\Omega, R_2 = R_3 = R_4 = 2\Omega, i_1 = 2\,\mathrm{A}, i_2 = 4\,\mathrm{A}$ (matrix_-solve.xlsx).

Solution

$$R_{\mathrm{th}} = \{\{[(R_1 + R_5)\|R_2] + R_6\}\|R_3 + R_7\}\|R_4$$

$$R_{\mathrm{th}} = 1\,\Omega$$

$$V_{\mathrm{oc}} = V_4, \quad (R_L = \infty).$$

Using node voltages method, finding V_4 yields Th $= V_{\mathrm{oc}} = V_4$.

$$GV = I$$

$$G = \begin{bmatrix} 1/R_1 + 1/R_5 & -1/R_5 & 0 & 0 \\ -1/R_5 & -1/R_5 + 1/R_2 + 1/R_6 & -1/R_6 & 0 \\ 0 & -1/R_6 & 1/R_3 + 1/R_6 + 1/R_7 & -1/R_7 \\ 0 & 0 & -1/R_7 & 1/R_4 + 1/R_7 \end{bmatrix}$$

$$\begin{bmatrix} 2 & -1 & 0 & 0 \\ -1 & 2.5 & -1 & 0 \\ 0 & -1 & 2.5 & -1 \\ 0 & 0 & -1 & 1.5 \end{bmatrix} \begin{bmatrix} V_1 \\ V_2 \\ V_3 \\ V_4 \end{bmatrix} = \begin{bmatrix} 2 \\ 0 \\ 0 \\ 4 \end{bmatrix}.$$

Solution of this matrix equation for V_4 yields, $V_4 = 4.25\,\mathrm{V}$

$$(V_1 = 1.844\,\mathrm{V}, V_2 = 1.588\,\mathrm{V}, V_3 = 2.375\,\mathrm{V}).$$

Condition for maximum power transfer to load resistor is $R_L = R_{\mathrm{Th}} = 1\,\Omega$. Maximum power transferred to load resistor,

$$P_{\max} = \frac{V_{\mathrm{Th}}^2}{4R_{\mathrm{Th}}} = \frac{4.25^2}{4 \times 1} = \frac{18.0625}{4} = 4.516\,\mathrm{W}.$$

Chapter 3
Capacitors and First-Order RC Circuits

3.1 Capacitors, Permittivity, Capacitance

Problem 3.1.1

(a) In the circuit shown in Fig. 3.1, $C_2 = C_3 = 4C, C_1 = 2C(\text{F})$, determine the voltage across each capacitor, in terms of U.
(b) If $U = 100$ V, calculate the voltage across each capacitor.

Solution

(a) Q = Total charge = $C_{\text{eq}}U$

$$\frac{1}{C_{\text{eq}}} = \frac{1}{C_1} + \frac{1}{C_2} + \frac{1}{C_3} = \frac{1}{2C} + \frac{1}{4C} + \frac{1}{4C} = \frac{1}{C}, \quad C_{\text{eq}} = C(\text{F})$$

$$V_1 = \frac{Q}{C_1} = \frac{CU}{2C} = \frac{U}{2}\text{V}, \quad V_2 = V_3 = \frac{Q}{C_2} = \frac{CU}{4C} = \frac{U}{4}\text{V}$$

(b) $V_1 = 100/2 = 50\,\text{V}, \quad V_2 = V_3 = 100/4 = 25\,\text{V}$

Problem 3.1.2 In the circuit shown in Fig. 3.2, series-connected four capacitors are fed by 12 V DC voltage source. Voltage readings across each capacitor are: $\Delta V_1 = 1\,\text{V}, \Delta V_2 = 2\,\text{V}, \Delta V_3 = 5\,\text{V}, C_{\text{eq}} = 0.1\,\text{F}$.

(a) ΔV_4 = ?
(b) How much charge is stored in each capacitor?
(c) What is the capacitance of each capacitor?
(d) V_x = ?

A.Ü. Keskin, *Electrical Circuits in Biomedical Engineering*,
DOI 10.1007/978-3-319-55101-2_3

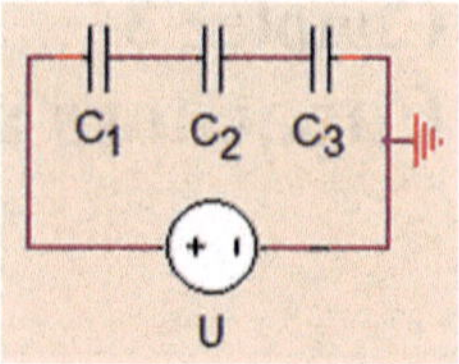

Fig. 3.1 The circuit for Problem 3.1.1

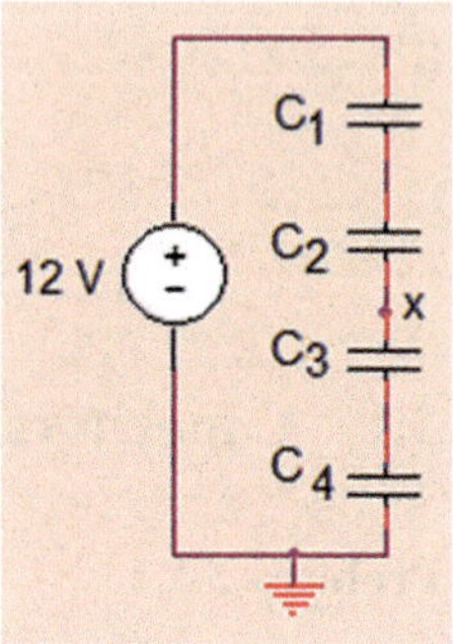

Fig. 3.2 The circuit for Problem 3.1.2

Solution

(a) $\Delta V_4 = 12 - 1 - 2 - 5 = 4\,\text{V}$

(b) $\Delta V = V_{\text{dc}} = C_{\text{eq}}Q \quad \rightarrow \quad 12\,\text{V} = 0.1FQ \quad \rightarrow \quad Q = \dfrac{12}{0.1} = 120\,\text{C}$

(c) $\Delta V \cdot C_1 \pm \Delta V_2 \cdot C_2 \pm \Delta V_3 \cdot C_3 \pm \Delta V_4 \cdot C_4 = Q$

$$C_1 = \frac{\Delta V_1}{Q} = \frac{1\,\text{V}}{120\,\text{C}} = \frac{1}{120}\,\text{F}, \quad C_2 = \frac{\Delta V_2}{Q} = \frac{2\,\text{V}}{120\,\text{C}} = \frac{1}{60}\,\text{F}$$
$$C_3 = \frac{\Delta V_3}{Q} = \frac{5}{120\,\text{C}} = \frac{1}{24}\,\text{F}, \quad C_4 = \tfrac{\Delta V_4}{Q} = \tfrac{4}{120\,\text{C}} = \tfrac{1}{30}\,\text{F}$$

(d) $V_x = \Delta V_3 + \Delta V_4 = 5 + 4 = 9\,\text{V}$

Problem 3.1.3

(a) In the circuit shown in Fig. 3.3, find the relationship between the capacitor values in order that the equivalent capacitance has the value of capacitance of capacitor C_2.
(b) If C_2 = 1 nF, C_1 = ?

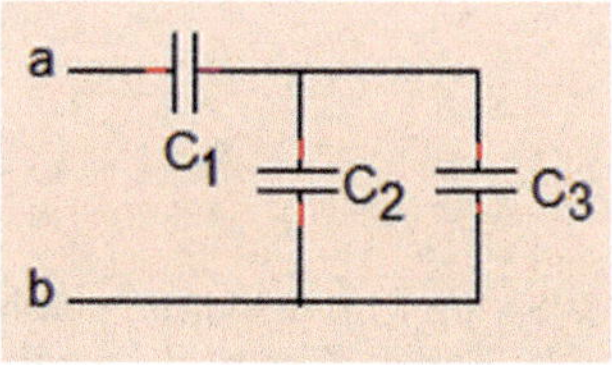

Fig. 3.3 The circuit for Problem 3.1.3

Solution

(a) $$C_{ab} = C_{\text{eq}} = \frac{C_1(C_1 + C_2)}{C_1 + (C_1 + C_2)} \quad \rightarrow \quad C_1^2 + C_1C_2 = 2C_1C_2 + C_2^2$$

$$C_2^2 + C_1C_2 - C_1^2 = 0 \quad \rightarrow \quad x^2 + xy - y^2 = 0$$

Solving last equation for x ($=C_2$) yields,

$$x = -\frac{y}{2} \pm \frac{\sqrt{y^2 - 4(-y^2)}}{2} = -\frac{y}{2} \pm \frac{\sqrt{5} \cdot y}{2} = \frac{1}{2}\left(\sqrt{5} - 1\right)y = 0.618y$$

Therefore, if $C_2 = 0.618C_1$ the equivalent capacitance value is the same as that of the capacitor C_2.

(b) $1\,\text{nF} = 0.618 \cdot C_1$

$$C_1 = \frac{1}{0.618}\,\text{nF} = 1.618\,\text{nF}$$

Problem 3.1.4

(a) Find the equivalent capacitance of the infinite capacitor ladder network shown in Fig. 3.4.
(b) If $C_{\text{eq}} = 10$ pF, C = ?

Solution

(a) If the ladder equivalent capacitance is C_{eq} and one more L section is added to the left of it, equivalent capacitance does not change, see Fig. 3.5.

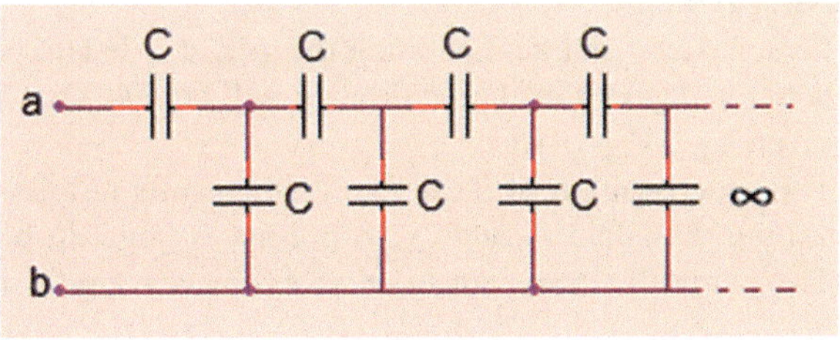

Fig. 3.4 The circuit for Problem 3.1.4

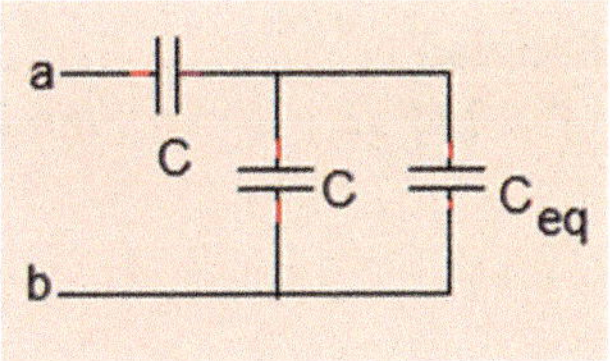

Fig. 3.5 The equivalent circuit for Problem 3.1.4

$$C_{eq} = C \parallel (C + C_{eq}) = \frac{C(C + C_{eq})}{2C + C_{eq}}$$

$$2C_{eq} \cdot C + C_{eq}^2 = C^2 + C_{eq} \cdot C$$

$$C_{eq}^2 + C_{eq} \cdot C - C^2 = 0$$

Solving for C_{eq} and dropping negative options,

$$C_{eq} = \frac{1}{2}\left(\sqrt{5} - 1\right) \cdot C = 0.618\,\mathrm{C}$$

(b) $10\,\mathrm{pF} = 0.618\,\mathrm{C}$

$$C = \frac{10}{0.618}\,\mathrm{pF} = 16.18\,\mathrm{pF}$$

Problem 3.1.5 An infinite number of identical capacitors each having a capacitance C are connected to form an infinite square grid. Find the capacitance between adjacent vertical or horizontal neighboring nodes.

Solution

Case 1: Let a charge q enter the grid at a node x and let it come out of the grid at a distant point. Then, the charge flowing through each of the four capacitors connected to the node will be equal. Therefore each one of them will carry a charge of $q/4$.

Thus, the resulting voltage drop between node x and an adjacent node y will be $\Delta V = q/4$ C.

Case 2: Let a charge q entering the grid at a distant point and exiting at the adjacent node, y. Again, the charge flowing will be $q/4$. The voltage drop from x to y will be given by $\Delta V = q/4$ C.

The superposition of these two cases results in a charge q entering the node x and exiting the adjacent node y with a net voltage drop equals to $\Delta V = q/2$ C.

Therefore, the capacitance between adjacent nodes is $C_{xy} = 2$ C.

Problem 3.1.6 Find the capacitance (in pF) if the areas of parallel plates are 10 cm^2 and the separation distance is (a) 1 cm. (b) 1 mm. (c) 0.1 mm. ($\epsilon_r = 5$)

Solution

$\epsilon_0 = 8.8542 \times 10^{-12}$ [F/m] is the permittivity of vacuum.

$$A = 10\,\mathrm{cm}^2 = 10^{-3}\,\mathrm{m}^2$$

$$C = \epsilon_r.\epsilon_0 \frac{A}{d} = 5 \times 8.8542 \times 10^{-12} \times \frac{A}{d} = 44.271 \times 10^{-12} \times \frac{A}{d}$$
$$= \frac{44.271 \times 10^{-15}}{d} \text{ F}$$

(a) $d = 1\,\text{cm} = 10^{-2}\,\text{m}, \quad C = \dfrac{44.271 \times 10^{-15}}{10^{-2}} \cong 4.427\,\text{pF}$

(b) $d = 1\,\text{mm} = 10^{-3}\,\text{m}, \quad C = \dfrac{44.271 \times 10^{-15}}{10^{-3}} \cong 44.27\,\text{pF}$

(c) $d = 0.1\,\text{mm} = 10^{-4}\,\text{m}, \quad C = \dfrac{44.271 \times 10^{-15}}{10^{-4}} \cong 442.7\,\text{pF}$

Problem 3.1.7 A capacitor consists of a 0.1 mm thick polyester film sandwiched between two rectangular aluminum foils (1 cm by 100 cm, with negligible thickness), and it is rolled into cylindrical form. Relative dielectric constant of polyester is 3. Find the capacitance of this element.

Solution
This is equivalent to a parallel plate system with A = 100 cm^2, d = 0.1 mm;

$$C = \epsilon_r \epsilon_o \frac{A}{d},$$
$$C = 3 \times \left(8.85 \times 10^{-12}\right) \times \frac{100 \times 10^{-4}\,\text{m}^2}{10^{-4}\,\text{m}}$$
$$= 26.55 \times 10^{-10}\,\text{F} = 2.655\,\text{nF}$$

Problem 3.1.8 One plate of a parallel plate capacitor (in air) is connected to a spring with the spring constant of k. The system is initially at rest. If the capacitor is charged by q, neglecting any friction, what is the expression for the expansion of the spring? (Fig. 3.6)

Solution
Force on plate a due to spring is

$$F_s = -kx \tag{3.1}$$

If the charges on plates a, b are $q+$ and $q-$, respectively, the electrostatic force due to electric field generated by b is given as:

$$F_e = qE \tag{3.2}$$

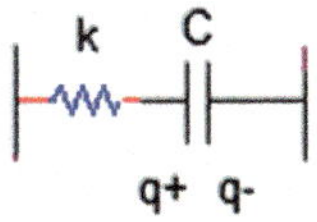

Fig. 3.6 The spring and capacitor configuration

The electric field is

$$E = \frac{\sigma}{2\epsilon_0} \tag{3.3}$$

Here, σ is the surface charge density (not conductivity!),

$$\sigma = \frac{q}{A} \tag{3.4}$$

A is the area of plates. Therefore

$$F_e = q\frac{\frac{q}{A}}{2\epsilon_0} = \frac{q^2}{2A\epsilon_0} \tag{3.5}$$

Plate a is pulled toward plate b. At equilibrium,

$$F_e = F_s \quad \rightarrow \quad \frac{q^2}{2A\epsilon_0} = kx \quad \rightarrow \quad x = \frac{q^2}{2k\epsilon_0 A}$$

Problem 3.1.9 Determine the resistance of material of permittivity ϵ and conductivity σ placed between two concentric spheres (a = radius of inner sphere, b = radius of outer sphere), Fig. 3.7.

Solution
By Ohm's law, $J = \epsilon E$, and assuming that a potential difference U is applied between two spheres,

$$I = \int_s J\mathrm{d}S = \int_s \sigma E\mathrm{d}S = \sigma \int_s E\mathrm{d}S = \sigma\frac{q}{\epsilon} = \sigma\frac{CU}{\epsilon}$$

The capacitance of concentric spheres is known to be

$$C = \frac{4\pi\epsilon}{\frac{1}{a} - \frac{1}{b}}$$

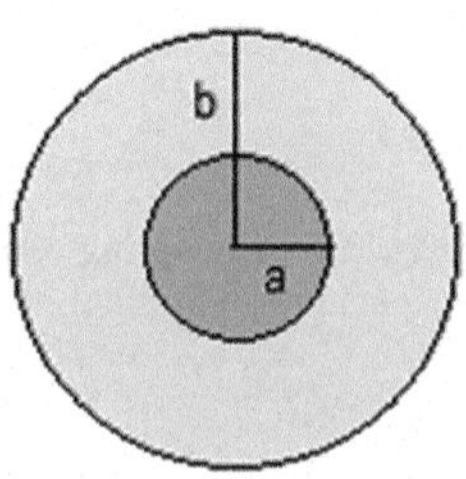

Fig. 3.7 The concentric spheres of Problem 3.1.9

Substituting these values into current equation,

$$I = \frac{\sigma}{\epsilon}\frac{4\pi\epsilon U}{\frac{1}{a}-\frac{1}{b}} = \frac{4\pi\sigma U}{\frac{1}{a}-\frac{1}{b}}$$

The resistance is

$$R = \frac{U}{I} = \frac{U}{\frac{4\pi\sigma U}{\frac{1}{a}-\frac{1}{b}}} = \frac{\frac{1}{a}-\frac{1}{b}}{4\pi\sigma}\ [\Omega]$$

Problem 3.1.10 Determine the capacitance of spherical capacitor with two concentric dielectric materials with relative permittivities of ϵ_{r_1} and ϵ_{r_2}.

Solution
Let

r_1 internal (conducting) spherical radius,
r_2 radius including the innermost sphere and first dielectric material,
r_3 radius including both dielectric of outermost sphere,

$$\text{Electric field} = E = \frac{V}{r} \quad \rightarrow \quad V = Er \quad \rightarrow \quad dV = E\mathrm{d}r$$

$$E_1 = \frac{1}{4\pi\epsilon_0\epsilon_{r_1}}\frac{q}{r^2}, \quad E_2 = \frac{1}{4\pi\epsilon_0\epsilon_{r_2}}\frac{q}{r^2}$$

Potential at radius r_2,

$$V_1 = \int \mathrm{d}V_1 = -\int_{r_2}^{r_1} \frac{1}{4\pi\epsilon_0\epsilon_{r_1}}\frac{q}{r^2}\mathrm{d}r = -\frac{q}{4\pi\epsilon_0\epsilon_{r_1}}\int_{r_2}^{r_1}\frac{\mathrm{d}r}{r^2}$$

$$V_1 = -\frac{q}{4\pi\epsilon_0\epsilon_{r_1}}\left(-\frac{1}{r}\right)\Bigg|_{r_2}^{r_1} = \frac{q}{4\pi\epsilon_0\epsilon_{r_1}}\left(\frac{1}{r_1}-\frac{1}{r_2}\right)$$

Potential at radius r_3

$$V_2 = \frac{q}{4\pi\epsilon_0\epsilon_{r_2}}\left(\frac{1}{r}\right)\Bigg|_{r_3}^{r_2} = \frac{q}{4\pi\epsilon_0\epsilon_{r_2}}\left(\frac{1}{r_2}-\frac{1}{r_3}\right)$$

Total capacitance:

$$C = \frac{q}{V_1 + V_2} = \frac{q}{\frac{q}{4\pi\epsilon_0}\left[\frac{1}{\epsilon_{r_1}}\left(\frac{1}{r_1} - \frac{1}{r_2}\right) + \frac{1}{\epsilon_{r_2}}\left(\frac{1}{r_2} - \frac{1}{r_3}\right)\right]} = \frac{4\pi\epsilon_0 r_2}{\frac{1}{\epsilon_{r_1}}\left(\frac{r_2}{r_1} - 1\right) + \frac{1}{\epsilon_{r_2}}\left(1 - \frac{r_2}{r_3}\right)}$$

$$C = \frac{4\pi\epsilon_0 r_2}{\frac{1}{\epsilon_{r_1}}\left(\frac{r_2}{r_1} - 1\right) + \frac{1}{\epsilon_{r_2}}\left(1 - \frac{r_2}{r_3}\right)}$$

Problem 3.1.11 Two conducting spheres of radius r are separated by a distance $L \gg r$, and a constant voltage source U is connected to the spheres. Assuming the charge is distributed uniformly on each sphere, determine;

(a) the capacitance of each sphere,
(b) the charge on each sphere,
(c) the resistance of each sphere,
(d) time constant of the circuit, if the voltage source is disconnected (shorted) (Fig. 3.8).

Solution

(a) The capacitance of each sphere is $C = 4\pi\epsilon r$
(b) Due to symmetry, the changes are $\mp q$ and potentials are $\mp U/2$,

Therefore, the charge of each sphere is

$$|q| = C\frac{U}{2} = 4\pi\epsilon r\frac{U}{2} = 2\pi\epsilon r U$$

(c) By Ohm's law, $\boldsymbol{J} = \epsilon \mathbf{E}$ and the current I is

$$I = \int_S \mathbf{J}\mathrm{d}S = \int_S \sigma\mathbf{E}\mathrm{d}S = \sigma\int_S \mathbf{E}\mathrm{d}S = \sigma\frac{q}{\epsilon} = \sigma\frac{2\pi\epsilon r U}{\epsilon} = 2\pi\sigma r U$$

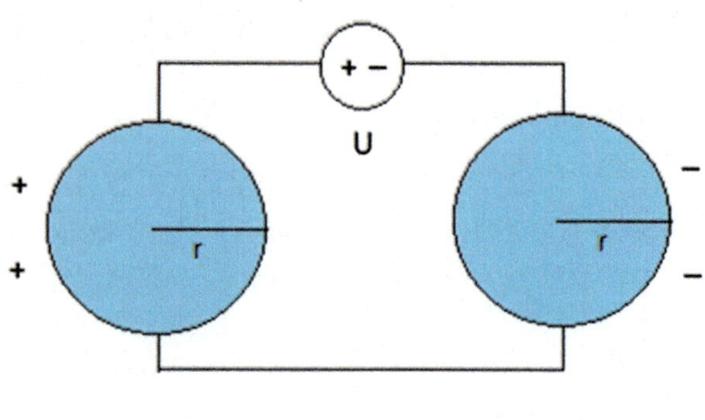

Fig. 3.8 Two conducting spheres of Problem 3.1.11

The resistance of circuit is

$$R = \frac{U}{I} = \frac{U}{2\pi\sigma r U} = \frac{1}{2\pi\sigma r}$$

Each sphere has a resistance of

$$R_s = \frac{1}{4\pi\epsilon r}$$

(d) Time constant of the circuit is

$$\tau = CR = (4\pi\epsilon r)\frac{1}{4\pi\sigma r} = \frac{\epsilon}{r}$$

Problem 3.1.12 Determine the resistance of a circular disc with Radius r, conductivity σ and relative permittivity ϵ_r.

Determine the capacitance and resistance of a thin circular graphite disc with diameter of 2 cm. ($\epsilon_r = 10$, $\sigma = 1350\,\mu\Omega\,\text{cm}$), see Fig. 3.9.

Solution

Assuming that a potential difference U is applied, the capacitance of disc is $C = 8\epsilon_o\epsilon_r r$.

$$I = \sigma\frac{q}{\epsilon_o\epsilon_r} = \sigma\frac{CU}{\epsilon_o\epsilon_r} = \sigma\frac{(8\epsilon_o\epsilon_r r)U}{\epsilon_o\epsilon_r} = 8\sigma r U$$

$$R = \frac{U}{I} = \frac{U}{8\sigma r U} = \frac{1}{8\sigma r}$$

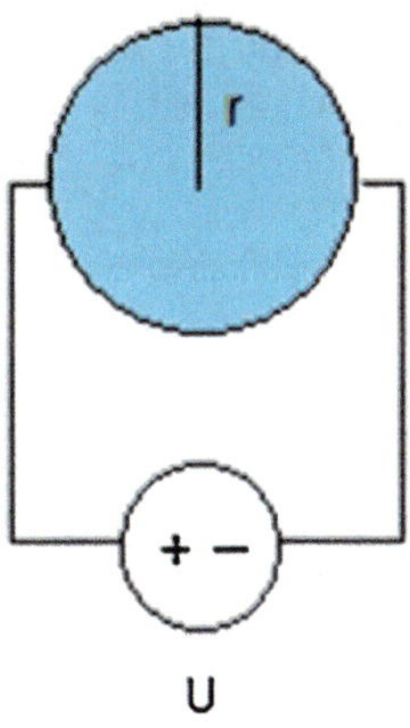

Fig. 3.9 The circular disc of Problem 3.1.12

For graphite, $\epsilon_r = 10$, $\sigma = 1350\,\mu\Omega\,\text{cm} = 1350 \times 10^{-8}\,\Omega\,\text{m}$

$$C = 8\epsilon_o\,\epsilon_r r = 8 \times 8.85 \times 10^{-12} \times 10 \times 10^{-2} = 7.08 \times 10^{-12} = 7.08\,\text{pF}$$

$$R = \frac{1}{8\sigma r} = \frac{1}{8 \times 1350 \times 10^{-8} \times 10^{-2}} = 9.26 \times 10^5\,\Omega = 926\,\text{k}\Omega$$

Problem 3.1.13

(a) Find the resistance of a spherical cell membrane if the radius of the cell is $r = 25$ microns, membrane thickness, $d = 6$ nm, and resistivity of 35 MΩ m (see, Fig. 3.10).
(b) Determine the electric field inside the resting cell, if the potential difference across the membrane is $U = -60$ mV,
(c) Determine the capacitance of the membrane if its relative dielectric constant is 9.
(d) Determine the electrical charge over the cell.
(e) Calculate the electric charge density on the membrane.
(f) What is the membrane capacitance/cm^2?
(g) Determine the time constant of cell membrane (ms) and comment on this value.
(h) How much leakage current flows through the membrane? What is the direction of this current flow?

Solution

(a) The cell area,

$$A = 4\pi r^2 = 4 \times \pi \times (25 \times 10^{-6})^2 = 7853.98 \times 10^{-12}\,\text{m}^2$$

$$R = \frac{\rho d}{A} = \frac{(35 \times 10^6) \times (6 \times 10^{-9})}{7853.98 \times 10^{-12}} = 0.026738 \times 10^9 = 26.738\,\text{M}\Omega$$

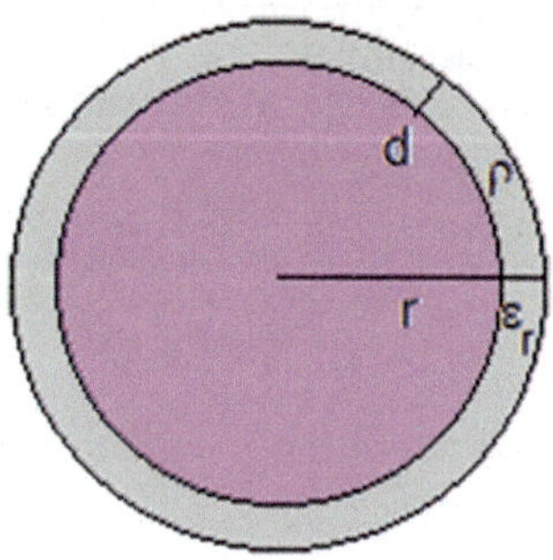

Fig. 3.10 The spherical model of a cell

(b) The electric field inside resting cell is

$$E = \frac{U}{d} = \frac{-60 \times 10^{-3}}{6 \times 10^{-9}} = -10 \times 10^{6}\,\text{V/m} = -10\,\text{MV/m}$$

The electric field points conventionally from positive to negative; therefore in this case electric field points inwards.

(c) Capacitance of the membrane is

$$C = \frac{\varepsilon_o \varepsilon_r A}{d} = \frac{(8.85 \times 10^{-12}) \times 9 \times (7853.98 \times 10^{-12})}{6 \times 10^{-9}}$$
$$= 104261.6 \times 10^{-15}\,\text{F}$$
$$C = 104.3\,\text{pF}$$

(d) Electric charge over the cell,

$$Q = CU = 104.3 \times 10^{-12} \times (-60) \times 10^{-3}$$
$$= -6258 \times 10^{-15}\,\text{C} = -6.258\,\text{pC}$$

(e) Electric charge density on the membrane of the cell,

$$\sigma = \frac{Q}{A} = \frac{-6258 \times 10^{-15}}{7853.98 \times 10^{-12}} = -0.7968 \times 10^{-3}\,\text{C/m}^2 = -79.68\,\text{nC/cm}^2$$

(f) Capacitance per centimeter squared is

$$C = \frac{104.3 \times 10^{-12}\,\text{F}}{7853.98 \times 10^{-12}\,\text{m}^2} = \frac{104.3 \times 10^{-12}\,\text{F}}{7853.98 \times 10^{-12} \times 10^{4}\,\text{cm}^2}$$
$$= 0.0133 \times 10^{-4}\,\text{F/cm}^2$$
$$C = 0.00133 \times 10^{-3}\,\text{F/cm}^2 = 1.33\,\mu\text{F/cm}^2$$

(g) Time constant of cell membrane is

$$\tau = RC = 26.738 \times 10^{6} \times 104.3 \times 10^{-12} = 2788.8 \times 10^{-6} = 2.79\,\text{ms}$$

If membrane voltage is raised large enough to be measured, but not sufficient to create an action potential, this voltage decays to its initial resting value with a time constant of 2.79 ms.

(h) Leakage current (dc) is calculated as

$$I_L = \frac{U}{R} = \frac{-60 \times 10^{-3}}{0.026738 \times 10^{9}} = -2244 \times 10^{-12}\,\text{A} = -2.244\,\text{nA}$$

It flows from outside of the cell membrane to inside of the cell.

Problem 3.1.14 Relative dielectric constant of pure water is temperature dependent and it is described as $\epsilon(T) = Be^{-bT}$, $B = 87.85306$, $b = 0.00456992$, $T =$ temperature, °C [1]

(a) Determine and plot dielectric constant of water and the capacitance of a parallel plate capacitor (plates' area = 1000 cm^2, distance between the plates = 0.5 cm, water filled) between 0 and 100 °C, at 1 °C steps, on the same graph.
(b) Calculate temperature at which the capacitance of a parallel plate capacitor (each of Aluminum plates' area = 0.1 m^2, distance between the plates = 0.005 m) becomes larger than 13 nF? (water_eps_T.m)

Solution

(a) Dielectric constant of water and the capacitance of a parallel plate capacitor (plates' area = 1000 cm^2, distance between the plates = 0.5 cm, water filled) between 0 and 100 °C are displayed in Fig. 3.11.
(b) Using capacitance formula for parallel plate capacitor,

$$C = \epsilon_0 \epsilon_r \frac{A}{d} = \epsilon_0 \frac{A}{d} B e^{-bT}$$

$$\frac{Cd}{\epsilon_0 AB} = e^{-bT}$$

Taking logarithm of both sides,

$$\ln\left(\frac{Cd}{\epsilon_0 AB}\right) = -bT \quad \rightarrow \quad T = \frac{\ln\left(\frac{Cd}{\epsilon_0 AB}\right)}{-b} = \frac{\ln\left(\frac{13000 \times 0.005}{8.85 \times 0.1 \times 87.85306}\right)}{-0.00456992} = 39.19\,°\mathrm{C}$$

When temperature of pure water (dielectric material between the plates) is less than 39.19 °C, capacitance of the capacitor is larger than 13 nF.

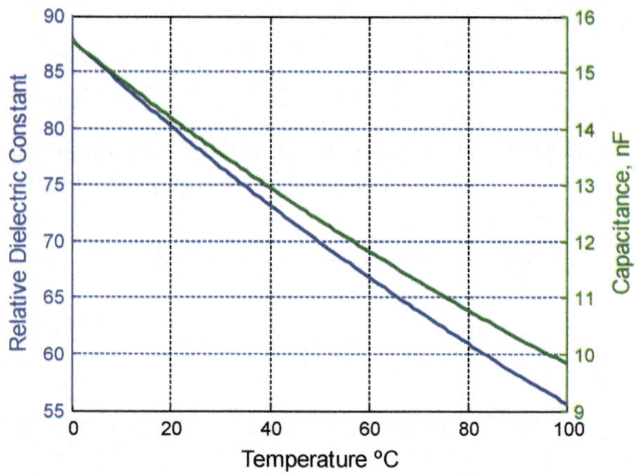

Fig. 3.11 Dielectric constant and the capacitance of a parallel plate capacitor (water filled) between 0 and 100 °C, in Problem 3.1.14

Problem 3.1.15 A parallel plate capacitor with $A = 50\,\text{cm} \times 10\,\text{cm}, d = 2\,\text{mm}$ is immersed into an ethanol–water mixture at $T = 25\,°\text{C}$ and the capacitance is measured as 0.91 nF. Determine the concentration of ethanol in solution. Neglect fringe field effects.

Solution

$$C_x = \epsilon_0 \epsilon_r \frac{A}{d} = 8.85 \times 10^{-12} \times \epsilon_r \times \frac{5 \times 10^{-12} \times 10^{-2}}{2 \times 10^{-3}} = 22.125 \times \epsilon_r = 910 \ \ [\text{pF}]$$

Static dielectric constant values for ethanol–water mixtures at 25 °C can be obtained from chemistry hand books [2]. Then, Table 3.1 is prepared.

Since the measured capacitance is $C_x = 910\,\text{pF} = 22.125 \times \epsilon_r$,

$$\epsilon_r = \frac{910}{22.125} = 41.13$$

This value corresponds to about 70% ethanol concentration.

Problem 3.1.16 Maxwell–Garnett (MG) model for effective permittivity of dielectric mixtures is valid for dilute conductive phases. This model assumes that the particles are solid homogeneous objects in an isotropic medium.

f The volume fraction of inclusions $= \frac{V_s}{V_s + V_m}$
ϵ_{eff} relative permittivity of mixture (composite),
ϵ_s relative permittivity of spherical inclusions (grains dispersed phase),
ϵ_m relative permittivity of matrix (background-host-medium).
V_s Volume of inclusions
V_m Volume of matrix

For one type of embedded spherical inclusions, (two-phase formulation),

$$\epsilon_{\text{eff}} = \epsilon + \frac{\dfrac{3f\epsilon(\epsilon_s - \epsilon)}{\epsilon_s + 2\epsilon}}{1 - \dfrac{f(\epsilon_s - \epsilon)}{\epsilon_s + 2\epsilon}}$$

Compute and plot the variation of the permittivity of a mixture containing polystyrene beads in water as a function of the volume fraction of the beads using MG model (emix1.m). Polystyrene beads: $\epsilon_s = 2.55$; Water: $\epsilon_m = 78.5$.

Table 3.1 Dielectric constant versus ethanol concentration in water

eps	79.5	73.73	70.23	66.08	59.67	55.07	47.56	41.15	35.42	29.27	24.95
Conc.	0	10	20	30	40	50	60	70	80	90	100

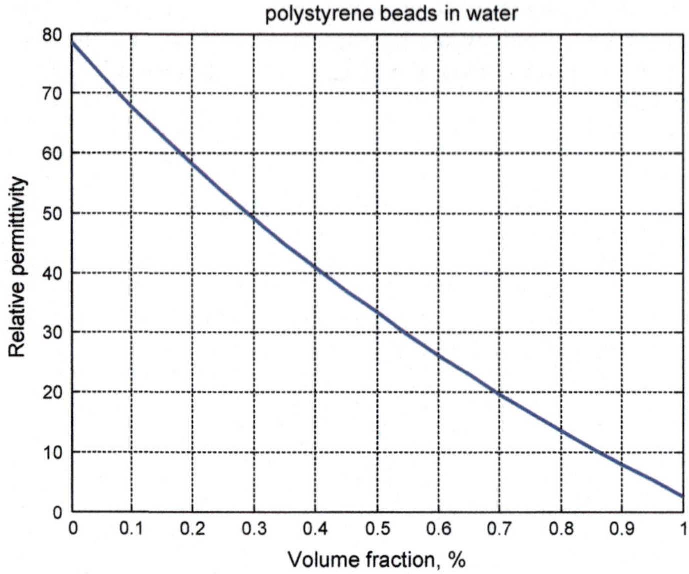

Fig. 3.12 Relative permittivity of polystyrene beads (simulating cells) in water

Solution

The permittivity for the mixture of polystyrene beads in water as a function of the volume fraction of the beads using MG model is displayed in Fig. 3.12.

Problem 3.1.17 Voigt model for the permittivity of dielectric mixtures (two-phase formulation at a constant temperature and pressure) can be stated as

$$\epsilon_{\text{eff}} = \epsilon_s f + \epsilon_m (1 - f)$$

where,

f volume fraction of inclusion,
ϵ_{eff} relative permittivity of mixture (composite),
ϵ_s relative permittivity of inclusion (grain),
ϵ_m relative permittivity of matrix (background-host-medium).

(a) Derive the preceding equation (Voigt model for mixtures) using the concept of parallel-connected capacitors.
(b) Plot the variation of composite static relative permittivity as a function of volume fraction of air bubbles in water ($\epsilon_s = 1, \epsilon_m = 78.5$), using Voigt model and also Maxwell–Garnett (MG) model.

Display composite static relative permittivity as predicted by both models as a function of volume fraction of air bubbles in water, for different air volume fraction values ($0 \leq f \leq 1$) (emix1.m, emix2.m, emix3.m).

Solution

(a) Consider two parallel-connected capacitors with different permittivities, and having the same distance (=d) between the parallel plates with a composite area of $A = xw$, while electric field vector **E** is aligned in parallel to inclusions. See, Fig. 3.13.

Effective capacitance is

$$C_{\text{eff}} = C_s + C_m$$

$$\frac{\epsilon_0 \epsilon_{\text{eff}} A}{d} = \frac{\epsilon_0 \epsilon_s A_s}{d} + \frac{\epsilon_0\ \epsilon_m A_m}{d},$$

where A_s and A_m are the area of inclusion and matrix (background). This further simplifies to

$$\epsilon_{\text{eff}} x = \epsilon_s t + \epsilon_m (x - t)$$

Divide both sides of equation by edge dimension value (=x),

$$\epsilon_{\text{eff}} = \epsilon_s \frac{t}{x} + \epsilon_m \frac{(x - t)}{x}$$

Since volume fraction of inclusion is $f = t/x$,

$$\epsilon_{\text{eff}} = \epsilon_s f + \epsilon_m (1 - f)$$

This gives the permittivity of dielectric mixtures in ROM for Voigt Model.

(b) Static relative permittivity in air–water mixture as a function of volume fraction of air bubbles in water, using "Maxwell Garnett (MG) model for one type of embedded spherical inclusions", (two-phase formulation at a constant temperature and pressure) is calculated using following equation:

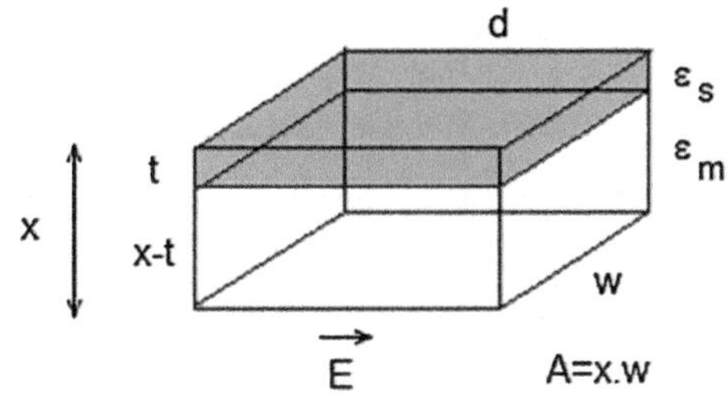

Fig. 3.13 Two parallel-connected capacitors with different permittivities with electric field vector **E** is aligned in parallel to inclusions

$$\epsilon_{\text{eff}} = \epsilon_m + \frac{\dfrac{3f\epsilon_m(\epsilon_s - \epsilon_m)}{\epsilon_s + 2\epsilon_m}}{1 - \dfrac{f(\epsilon_s - \epsilon_m)}{\epsilon_s + 2\epsilon_m}}$$

The meanings of the terms in this equation are given in part (a). This model assumes that the particles are solid homogeneous objects in an isotropic medium. Substituting static permittivity values of water and air into this formula, and plotting it by changing volume fraction of air bubbles parameter (*f*) yields the graph shown in Fig. 3.14.

Problem 3.1.18 Permittivity of a composite material (mixture) can be described using "the rule of mixtures-ROM-" and with the following assumptions: (1) Inclusions in the composite material are continuous and uniformly distributed throughout the base matrix. (2) Perfect bonding between inclusions and matrix exists. (3) Matrix is void free. (4) Linearity applies to both inclusions and the matrix.

Based on the Wiener theory for bounds of effective dielectric permittivity of two-phase composite materials, a general permittivity equation is given as:

$$\epsilon_{\text{eff}}^n = f\epsilon_s^n + (1 - f)\epsilon_m^n \tag{3.6}$$

The parameter n varies within the [1, −1] range, and describes a specific topology of a composite,

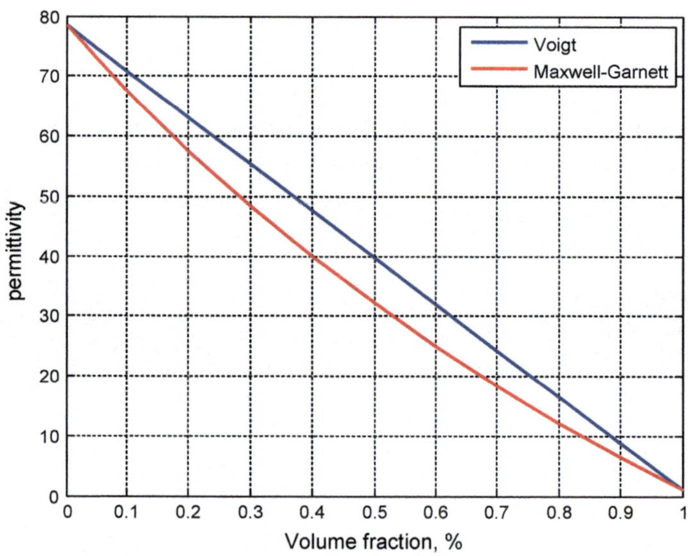

Fig. 3.14 Static relative permittivity as a function of volume fraction of air bubbles in water

f The volume fraction of fibers $= \frac{V_f}{V_f + V_m}$
ϵ_{eff} relative permittivity of mixture (composite),
ϵ_s relative permittivity of spherical inclusions (grains dispersed phase),
ϵ_m relative permittivity of matrix (background-host-medium).
V_s Volume of inclusions
V_m Volume of matrix

Here, $n = 1$ corresponds to electric field parallel to the inclusion (Voigt model). When $n = -1$, the lower bound of permittivity for the composite material, which corresponds to electric field perpendicular to the inclusion (Reuss model or inverse ROM model) is

$$\epsilon_{\text{eff}} = \frac{1}{\frac{f}{\epsilon_s} + \frac{1-f}{\epsilon_m}} \tag{3.7}$$

Therefore,

$$\frac{1}{\frac{f}{\epsilon_s} + \frac{1-f}{\epsilon_m}} \leq \epsilon_{\text{eff}} \leq f\epsilon_s + (1-f)\epsilon_m \tag{3.8}$$

Krawzenski model has $n = 1/2$ [3] while Looyenga model has parameter $n = 1/3$ in Eq. (3.6) [4].

(a) Derive Eq. (3.7) [i.e., Eq. (3.6) for $n = -1$] for composite permittivity using the concept of series-connected capacitors.
(b) Plot the variation of composite static relative permittivity as a function of volume fraction of air bubbles in water ($\epsilon_s = 1, \epsilon_m = 78.5$), using four different mixture models mentioned above (emix4.m).

Solution

(a) Assume that capacitance of the composite material is made out of two capacitors connected in series (see, Fig. 3.15). In this case, direction of electric field is perpendicular to the plane of inclusions (Reuss model of composites).

$$\frac{1}{C_{\text{eff}}} = \frac{1}{C_s} + \frac{1}{C_m} \tag{3.9}$$

$$f = \frac{V_s}{V_s + V_m} = \frac{d_1}{D} \tag{3.10}$$

$\epsilon = \epsilon_o \epsilon_r, \quad \epsilon_o = 8.85 \times 10^{-12}\,\text{F m}^{-1}, \quad \epsilon_r =$ Relative permittivity of material

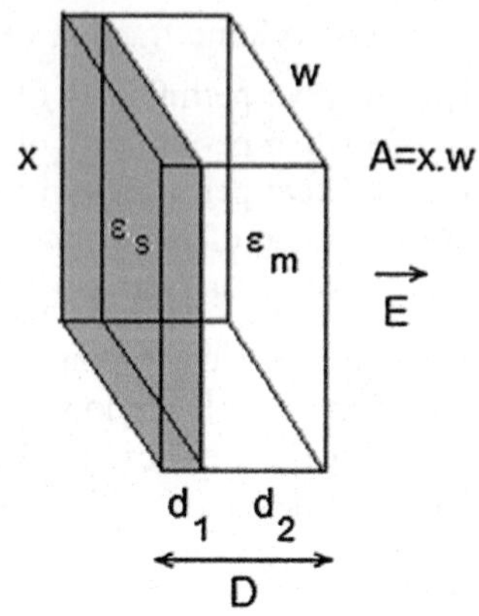

Fig. 3.15 Two capacitors connected in series with direction of electric field perpendicular to the plane of inclusions (Reuss model of composites)

$$C_{\text{eff}} = \frac{C_s C_m}{C_s + C_m} = \epsilon_o \frac{\frac{\epsilon_s A}{d_1} \cdot \frac{\epsilon_m A}{D - d_1}}{\frac{\epsilon_s A}{d_1} + \frac{\epsilon_m A}{D - d_1}} = \epsilon_o \frac{\epsilon_s \epsilon_m A^2}{A[\epsilon_s (D - d_1) + \epsilon_m d_1]}$$

$$C_{\text{eff}} = \epsilon_o \frac{\epsilon_s \epsilon_m A}{\epsilon_s (D - d_1) + \epsilon_m d_1} \tag{3.11}$$

Since effective capacitance is

$$C_{\text{eff}} = \epsilon_o \frac{\epsilon_{\text{eff}} A}{D} \tag{3.12}$$

From equivalence of (3.11) and (3.12),

$$\epsilon_o \frac{\epsilon_{\text{eff}} A}{D} = \epsilon_o \frac{\epsilon_s \epsilon_m A}{\epsilon_s (D - d_1) + \epsilon_m d_1}$$

$$\frac{\epsilon_{\text{eff}}}{D} = \frac{\epsilon_s \epsilon_m}{\epsilon_s (D - d_1) + \epsilon_m d_1}$$

Inverting both sides of the equation,

$$\frac{1}{\epsilon_{\text{eff}}} = \frac{\epsilon_s (D - d_1) + \epsilon_m d_1}{\epsilon_s \epsilon_m D} = \frac{d_1}{\epsilon_s D} + \frac{(D - d_1)}{\epsilon_m D}$$

Using Eq. (3.7),

$$\frac{1}{\epsilon_{\text{eff}}} = \frac{f}{\epsilon_s} + \frac{(1 - f)}{\epsilon_m}$$

which is the expected result.

(b) Variations of composite static relative permittivity as a function of volume fraction of air bubbles in water ($\epsilon_s = 1, \epsilon_m = 78.5$), using four different mixture models (Voigt, Reuss, Krawzenski, and Looyenga) are displayed in Fig. 3.16.

```
Rel. permittivity of inclusions = 1.000, Rel. permmittivity of Matrix=78.5
          f     Voigt  Krawzenski Looyenga  Reuss
          0   78.5000  78.5000  78.5000  78.5000
     0.1000   70.7500  65.1898  61.7982   8.9714
     0.2000   63.0000  53.1152  47.6511   4.7576
     0.3000   55.2500  42.2762  35.8468   3.2371
     0.4000   47.5000  32.6728  26.1732   2.4531
     0.5000   39.7500  24.3050  18.4182   1.9748
     0.6000   32.0000  17.1728  12.3698   1.6526
     0.7000   24.2500  11.2762   7.8158   1.4208
     0.8000   16.5000   6.6152   4.5442   1.2460
     0.9000    8.7500   3.1898   2.3430   1.1095
     1.0000    1.0000   1.0000   1.0000   1.0000
```

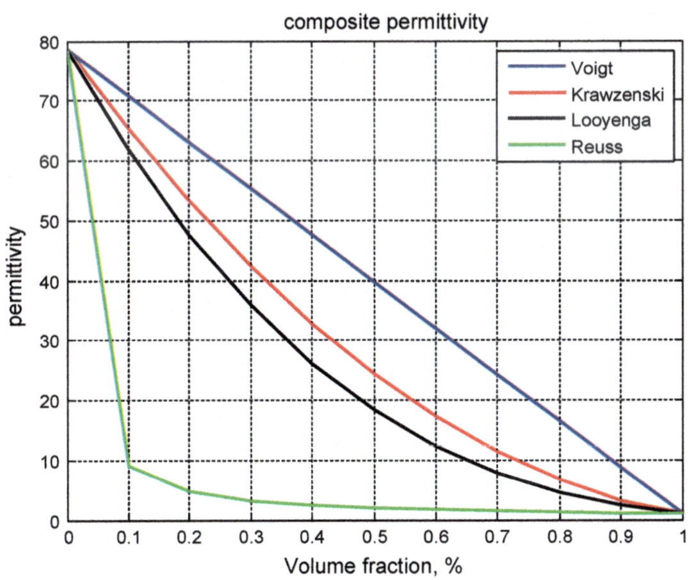

Fig. 3.16 Variations of composite static relative permittivity as a function of volume fraction of air bubbles in water ($\epsilon_s = 1, \epsilon_m = 78.5$), using four different mixture models

Problem 3.1.19 For air bubbles in water, determine and plot the capacitance of the sensor and capacitance difference to the initial capacitance as a function of f = air volume fraction, with 1% increments up to f = 10%. Assume that Voigt model is valid for the mixture. (emix5.m)

Area of the parallel plates = 1 cm^2, distance between parallel plates = 1 mm.

$$\epsilon_0 = 8.85 \times 10^{-12}\,\mathrm{Fm}^{-1}, \quad \epsilon_m = \epsilon_{\mathrm{water}} = 78, \quad \epsilon_{\mathrm{air}} = 1$$

Solution

Capacitance of the air free sensor is

$$C = \epsilon_0 \epsilon_m \frac{A}{d} = 8.85 \times 10^{-12} \times 78 \times \frac{10^{-4}}{10^{-3}} = 69.03\,\mathrm{pF}$$

Mixture capacitance and difference in capacitance,

$$C_{\mathrm{eff}} = \epsilon_0 \epsilon_{\mathrm{eff}} \frac{A}{d}, \quad \Delta C = C - C_{\mathrm{eff}},$$

Voigt model describes permittivity of the mixture,

$$\epsilon_{\mathrm{eff}} = f\epsilon_{\mathrm{air}} + (1 - f)\epsilon_{\mathrm{water}}$$

Following is a list of "f", relative permittivity, mixture capacitance and sensitivity $k = \Delta C/C$ (Figs. 3.17 and 3.18).

f	rel.eps	Cap(pF)	dC/C(%)
0	78.0000	69.0300	0
0.0100	77.2300	68.3486	0.9970
0.0200	76.4600	67.6671	2.0141
0.0300	75.6900	66.9857	3.0519
0.0400	74.9200	66.3042	4.1111
0.0500	74.1500	65.6227	5.1922
0.0600	73.3800	64.9413	6.2960
0.0700	72.6100	64.2598	7.4232
0.0800	71.8400	63.5784	8.5746
0.0900	71.0700	62.8970	9.7509
0.1000	70.3000	62.2155	10.9531

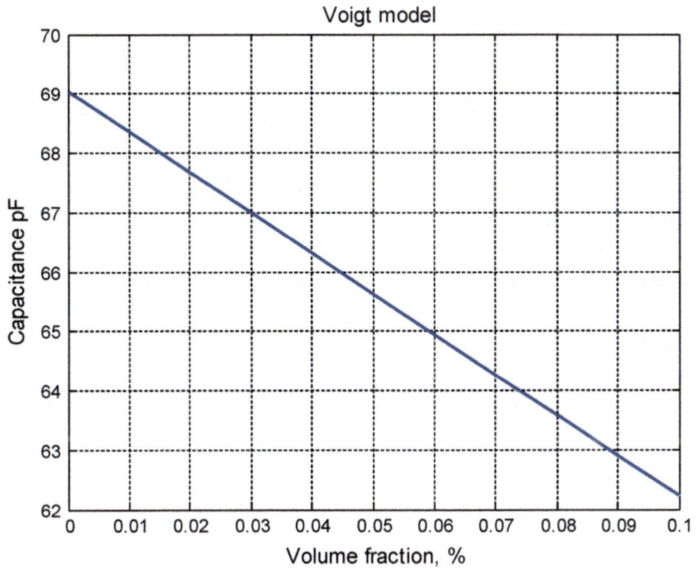

Fig. 3.17 Capacitance for air bubbles in water versus volume fraction

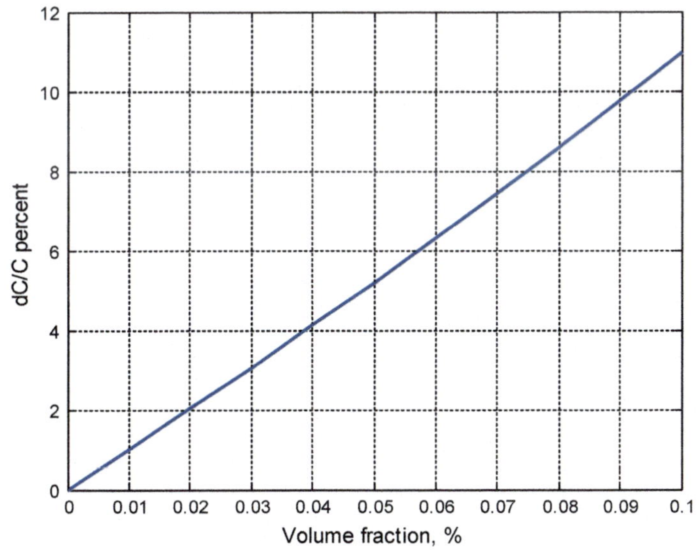

Fig. 3.18 Capacitance sensitivity to volume fraction of air bubbles in water

Problem 3.1.20 Lichtenecher model for the effective permittivity of mixtures is of the form

$$\log(\epsilon_{\text{eff}}) = f \,.\, \log(\epsilon_s) + (1 - f) \,.\, \log(\epsilon_m)$$

(a) For air bubbles in water, plot effective mixture permittivity as a function of inclusion volume fraction for full range. (emix_Lichtenecker.m, emix_Lichtenecker2.m)

(b) Determine the capacitance of the sensor and the ratio of capacitance difference to the initial capacitance as a function of f = air volume fraction, with 1% increments up to f = 10%. Assume that Lichtenecker model is valid for the mixture.
Area of the parallel plates = 1 cm^2, distance between parallel plates = 1 mm.

$$\epsilon_0 = 8.85 \times 10^{-12}\,\text{Fm}^{-1}, \quad \epsilon_m = \epsilon_{\text{water}} = 78, \quad \epsilon_{\text{air}} = 1$$

Solution

(a) Applying the formulation for the full range of (f) values, yields the result as displayed in Fig. 3.19;

(b) Capacitance of the air free sensor is

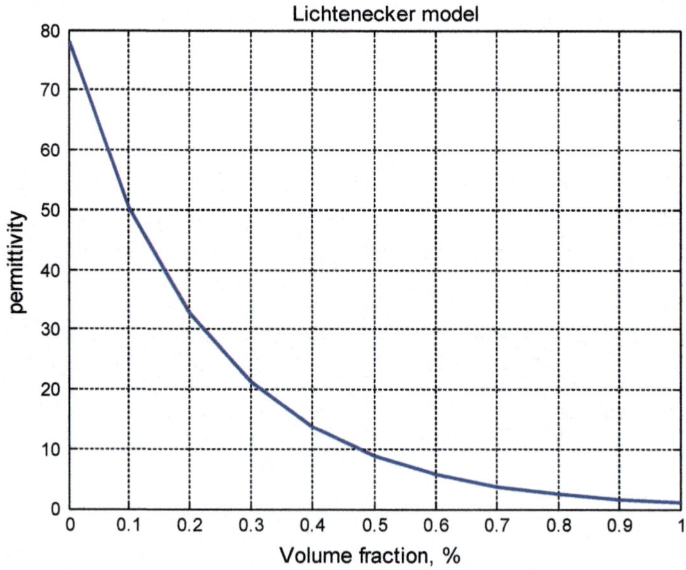

Fig. 3.19 Effective mixture permittivity as a function of inclusion volume fraction (air bubbles in water)

$$C = \epsilon_0 \epsilon_m \frac{A}{d} = 8.85 \times 10^{-12} \times 78 \times \frac{10^{-4}}{10^{-3}} = 69.03\,\text{pF}$$

Mixture capacitance and difference in capacitance,

$$C_{\text{eff}} = \epsilon_0 \epsilon_{\text{eff}} \frac{A}{d}, \quad \Delta C = C - C_{\text{eff}},$$

Lichtenecker model describes permittivity of the mixture,

$$\log(\epsilon_{\text{eff}}) = f.\log(\epsilon_{\text{air}}) + (1 - f).\log(\epsilon_{\text{water}})$$

Following is a list of (*f*), relative permittivity, mixture capacitance and sensitivity $k = \Delta C/C$.

f	eps	C	k
0	78.0000	69.0300	0
0.0100	74.6747	66.0871	4.4530
0.0200	71.4912	63.2697	9.1043
0.0300	68.4434	60.5724	13.9627
0.0400	65.5256	57.9901	19.0375
0.0500	62.7321	55.5179	24.3382
0.0600	60.0577	53.1511	29.8750
0.0700	57.4974	50.8852	35.6584
0.0800	55.0462	48.7159	41.6993
0.0900	52.6995	46.6390	48.0091
0.1000	50.4528	44.6507	54.6000

Problem 3.1.21 Show that for an RC circuit, the unit for time constant τ is in seconds.

Solution
Since time constant expression is given as

$$\tau = R.C = \frac{V}{I} \cdot C = \frac{VC}{I} = \frac{Q}{I} = t; \quad \tau = t$$

Therefore, the unit for time constant τ is in seconds.

Problem 3.1.22 Capacitance per unit length (F/m) of a coaxial cylindrical capacitor is calculated as

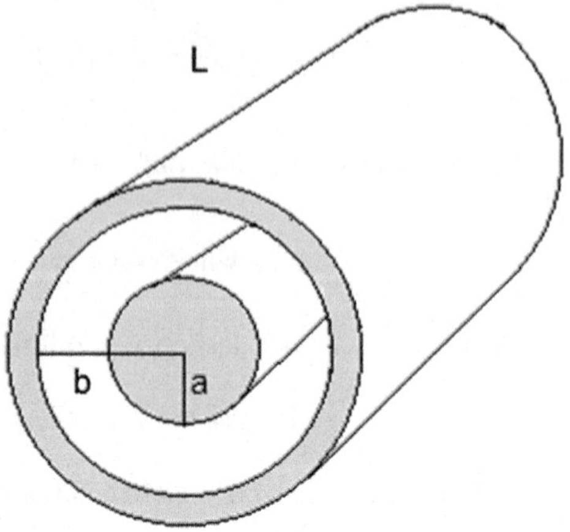

Fig. 3.20 Coaxial cylindrical capacitor

$$\frac{C}{L} = \frac{2\pi\epsilon_r\epsilon_0}{\ln\left(\frac{b}{a}\right)},$$

where, a = inner radius, b = outer radius of cylinder, L = length of the cylinder, ϵ_r = relative permittivity of liquid, $\epsilon_0 = 8.85 \times 10^{-12}$ F m^{-1}.

(a) What is the value of capacitance of the sensor (pF), if a = 2 mm, b = 5 mm, L = 1.5 m, and it is mounted in an empty vertical cylindrical container? (see, Fig. 3.20).
(b) Determine the level of liquid (pure) glycerol, $C_3H_8O_3$, (with relative permittivity of 42.5) in the container, if measured capacitance is C = 2.21 nF. What is the percentage value of vertical cylindrical container filling, neglecting the effect of air filled section?
(c) Repeat part (b), taking the effect of air filled section of the sensor into consideration. What is the percent error in making approximate analysis?
(d) What is the minimum time for a stable level measurement in a sampled pulsed mode of operation, if measuring circuit has an equivalent resistance of 100 kΩ, and the minimum read period of the microcontroller is 1.8 ms? (Coax_C1.m)

Solution

(a) Capacitance (in air) is,

$$C = \frac{2\pi\epsilon_r\epsilon_0 L}{\ln\left(\frac{b}{a}\right)} = \frac{2 \times \pi \times 8.85 \times 10^{-12} \times 1.5}{\ln\left(\frac{5}{2}\right)}$$
$$= 9.103 \times 10^{-11}\,\text{Fm}^{-1} = 91.03\,\text{pF}$$

(b) Glycerol is a covalently linked molecule, it does not ionize and conduct electricity. Since measured capacitance value is 2.21 nF, and the relative permittivity of glycerol is much higher than that of the air, one may neglect the effect of air capacitance, as a simplification. In this case,

$$2.21 \times 10^{-9} = \frac{2 \times \pi \times 42.5 \times 8.85 \times 10^{-12} \times L}{\ln\left(\frac{5}{2}\right)}$$

Solving above equation for L,

$$L = 0.857\,\text{m} = 857\,\text{mm}$$

This means that 857/1500 = 57.13, i.e., 57.13% of the vertical cylindrical container is filled with glycerol.

(c) Equivalent circuit of the sensor can be taken as parallel-connected capacitors,

$$C_{\text{measured}} = C_{\text{glycerol}} + C_{\text{air}} = \frac{2\pi\epsilon_r\epsilon_0 f}{\ln\left(\frac{b}{a}\right)} + \frac{2\pi\epsilon_r\epsilon_0(1-f)}{\ln\left(\frac{b}{a}\right)}$$

$$= \frac{2\pi\epsilon_0}{\ln\left(\frac{b}{a}\right)}(\epsilon_r f + 1 - f)$$

Because volume fraction of materials are equivalent to their (height) levels,

$$C_{\text{measured}} = \frac{2\pi\epsilon_0}{\ln\left(\frac{b}{a}\right)}\left(\epsilon_r L_{\text{glycerol}} + 1 - L_{\text{glycerol}}\right)$$

Substitution of values into this equation and solving for glycerol level,

$$2.21 \times 10^{-9} = \frac{2 \times \pi \times 8.85 \times 10^{-12}}{\ln\left(\frac{5}{2}\right)}\left(42.5 \times L_{\text{glycerol}} + 1 - L_{\text{glycerol}}\right)$$

$$L_{\text{glycerol}} = \frac{\dfrac{2.21 \times 10^{-9} \times \ln\left(\frac{5}{2}\right)}{2 \times \pi \times 8.85 \times 10^{-12}} - 1}{41.5} = 0.853\,\text{m}$$

$$\Delta C = C_{\text{actual}} - C_{\text{approx}} = 853 - 857 = -4\,\text{mm}$$

Absolute error, in percent,

$$\text{err}(\%) = \frac{\Delta C}{C_{\text{actual}}} = \frac{4}{853} = 0.47\%$$

(d) Given $C = 2.21$ nF, $R = 100$ kΩ; Settling time of the measuring pulse waveform is

$$T_r = 5 \times \tau = 5 \times 2.21 \times 10^{-9} \times 10^5 = 11.05 \times 10^{-4}\,\text{s} = 1.105\,\text{ms}$$
$$T = T_r + T_{\text{Comp}} = 1.105 + 1.8 = 2.905\,\text{ms}$$

This value is the minimum time for a stable level measurement in a sampled pulsed mode of operation.

Problem 3.1.23 Determine the time constant (in ms) of two concentric conducting cylinders with outer and inner diameters 64 and 50 mm, respectively, and height of each 90 mm, inner spacing is fully filled with distilled water at 20 °C (conductivity = 1 μs/m). Bottom of each cylinder is nonconducting.

Solution
For water, $\epsilon_r \cong 80$ at 20 °C (Data is found in handbooks and on the web), and noting that the ratio of diameters equals to the ratio of radii,

$$C = \frac{2\pi\epsilon_0\epsilon_r L}{\ln\left(\frac{r_b}{r_a}\right)} = 2\pi \times 8.85 \times 80 \times \frac{9 \times 10^{-2}}{\ln\left(\frac{64}{50}\right)} = 1621.83\,\text{pF}$$

On the other hand, considering the resistance between conducting cylinders filled with distilled water, since the current passing through shells of radius r and thickness dr, with the resistance of each shell being dR,

$$\text{d}R = \rho\frac{\text{d}r}{2\pi rL} \quad \rightarrow \quad R = \int \text{d}R = \frac{\rho}{2\pi rL}\int_{r_a}^{r_b}\frac{\text{d}r}{r} = \frac{\rho}{2\pi rL}\ln\left(\frac{r_b}{r_a}\right)$$

$$R = \frac{\frac{1}{10^{-6}}}{2\pi \times 9 \times 10^{-2}}\ln\left(\frac{64}{50}\right) = 0.4365\,\text{M}\Omega$$

Time constant of the system,

$$\tau = RC = 0.4365 \times 10^6 \times 1621 \times 10^{-12} = 708 \times 10^{-6}\,\text{s} = 0.708\,\text{ms}$$

Problem 3.1.24 Current–voltage relations on a capacitor of unknown capacitance are as follows:

$$i_C(t) = \begin{cases} 2\,\text{A} & 0 \le t < 1 \\ -2\,\text{A} & 0 \le t < 2 \end{cases}, \quad v_C(t) = \begin{cases} 4t & 0 \le t < 1 \\ -4t + 4 & 0 \le t < 2 \end{cases} \quad (t \text{ in seconds})$$

Determine the capacitance. Plot current and voltage waveforms (civ.cir).

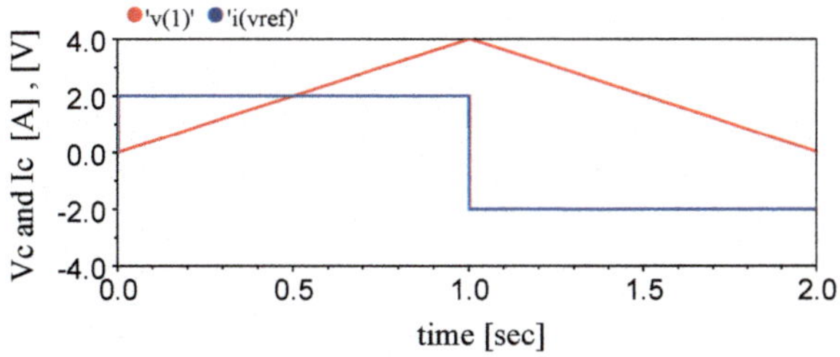

Fig. 3.21 The voltage (*red line*) and current (*blue line*) waveforms in a capacitor

Solution
Current and voltage equations on a capacitor are

$$i_C(t) = C\frac{\mathrm{d}v_C(t)}{\mathrm{d}t}, \quad v_C(t) = \frac{1}{C}\int\limits_{t_0}^{t} i_C(t')\mathrm{d}t' + v_C(0)$$

Therefore for $0 \le t < 1$,

$$4t = \frac{1}{C}\int\limits_{0}^{t} 2\mathrm{d}t' = \frac{1}{C}2t \quad \rightarrow \quad C = \frac{1}{2}F$$

As a check for this result, use data for $1 \le t < 2$, and find the voltage across the capacitor,

$$v_C(t) = \frac{1}{0.5}\int (-2)\mathrm{d}t' + 4\mathrm{V} = 2(-2t) + 4 = -4t + 4\,\mathrm{V}$$

In agreement with given $v_c(t)$, within the time interval of $1 \le t < 2$.

Following is a SPICE netlist (civ.cir) that provides a display of current and voltage waveforms in Fig. 3.21.

```
voltage-current on a capacitor
*transient analysis
I1 0 1 PWL( 0 2 1 2 1.001 -2 1.999 -2 2 -2)
vref 1 2 0
C 2 0 0.5
```

Problem 3.1.25 If the voltage across a 10,000 μF capacitor is given as

$$V_c(t) = \begin{cases} 0 & t = 0 \\ 5t & 0 \leq t \leq 1 \\ 10 - 5t & 1 \leq t \leq 2 \\ 0 & t \geq 2 \end{cases}$$

Determine the capacitor current (t in seconds).

Solution

$$i_c(t) = C \cdot \frac{dv}{dt}, \quad C = 10\,\text{mF} = 10^{-2}\,\text{F},$$
$$i_c(t) = 0 \quad \text{for } t<0 \quad \text{and} \quad t > 2.$$

For $0 \leq t \leq 1$,

$$i_c(t) = 10^{-2} \times \frac{d}{dt}(5t) = 0.05\,\text{A} = 50\,\text{mA}.$$

For $1 \leq t \leq 2$,

$$i_c(t) = 10^{-2} \times \frac{d}{dt}(10 - 5t) = -0.05 = -50\,\text{mA}$$

in summary,

$$i(t) = \begin{cases} 0 & t = 0 \\ 50\,\text{mA} & 0 \leq t \leq 1 \\ -50\,\text{mA} & 1 \leq t \leq 2 \\ 0 & t \geq 2 \end{cases}$$

Problem 3.1.26 Current through a 1000 μF capacitor is 1 mA for $0 \leq t \leq 1$ s and zero for the outside of this time range. Initial voltage across the capacitor is $v_C(t) = 5\,\text{V}$. Obtain the voltage across the capacitor as a function of time. What is the maximum voltage across the capacitor?

Solution

$$i_c(t) = \begin{cases} 0 & t<0 \\ 1\,\text{mA} & 0 \leq t \leq 1 \\ 0 & t > 1 \end{cases}$$

$$v_C(t) = \frac{1}{C} \int_{t_0}^{t} i_c(t')dt' + V_c(t_0), \quad t > t_0$$

For $0 \leq t \leq 1\,\text{s}$,

$$v_C(t) = \frac{10^{-3}}{C}t + 5 = \frac{10^{-3}}{10^{-3}}t + 5 = t + 5 = 1 + 5 = 6\,\text{V}$$

in summary,

$$V_c(t) = \begin{cases} t + 5\,\text{V} & 0 \leq t \leq 1 \\ 6\,\text{V} & t > 1 \end{cases}$$

Final capacitor voltage value after the current has stopped charging it depends on the initial value of the capacitor voltage (besides the current history). The voltage across the capacitor then remains at the maximum value of 6 V.

Problem 3.1.27 Current through a capacitor with capacitance C is given by $i_c(t) = I_0 \cdot e^{-t/\tau} \cdot u(t)$, where $u(t)$ (switching) unit step function, and τ has a unit of time.

Determine the final value of voltage and stored charge on the capacitor, then find the capacitor's energy. Assume that there is no energy stored initially on the capacitor.

Solution

$$v_C(t) = v_c(0) + \frac{1}{C}\int_0^t i_c(t')\mathrm{d}t' = \frac{1}{C}\int_0^t i_c(t')\mathrm{d}t' = \frac{I_0 \cdot \tau}{C}\left(-e^{-t/\tau}\right)\Big|_0^t$$
$$= \frac{I_0 \cdot \tau}{C}\left(1 - e^{-t/\tau}\right)$$

Therefore,

$$v_C(\infty) = \frac{I_0 \cdot \tau}{C}\,\text{V}$$

The charge stored is

$$q = C \cdot v_C(\infty) = C \cdot \frac{I_0 \cdot \tau}{C} = I_0 \cdot \tau C$$

Capacitor's energy is

$$W = \frac{1}{2}C \cdot v_C^2(\infty) = \frac{1}{2}C \cdot \left(\frac{I_0 \cdot \tau}{C}\right)^2 = \frac{I_0^2 \cdot \tau^2}{2C}\,\text{J}.$$

Problem 3.1.28 Voltage across a capacitor with $C = 2.2\,\mu\text{F}$ at time t_0 is -4 V. Determine the energy stored on this capacitor at $t = t_0$. Can stored energy on a capacitor have a negative sign?

Solution

$$V_C(t_0) = -4\,\text{V},$$
$$W(t_0) = \frac{1}{2}C \cdot V_C^2(t_0) = \frac{1}{2} \times 2.2 \times 10^{-6} \times (-4)^2 = 17.6\,\mu\text{J}$$

Stored energy on a capacitor cannot be negative.

Problem 3.1.29 The voltage across a $C = 1\,\text{mF}$ capacitor is shown in Fig. 3.22. Determine the energy stored on this capacitor at $t = 1.3\,\text{s}$. (plot_time_function1.m)

Solution

In Fig. 3.22, the voltage equation is given as (graph title),

$$v_C(t) = \left(-e^{-t} + 2e^{-2t} - 2e^{-4t} + 2e^{-5t}\right)u(t)$$

A plot of this function in MATLAB verifies the graph and numerical result. Evaluating it at $t = 1.3\,\text{s}$ yields $v_C(1.3) = 0.132\,\text{V}$.
Energy stored at $t = 1.3\,\text{s}$ is

$$W(1.3) = \frac{1}{2}CV_C^2(1.3) = \frac{1}{2} \times 10^{-3} \times (-0.132)^2 = 8.7\,\mu\text{J}$$

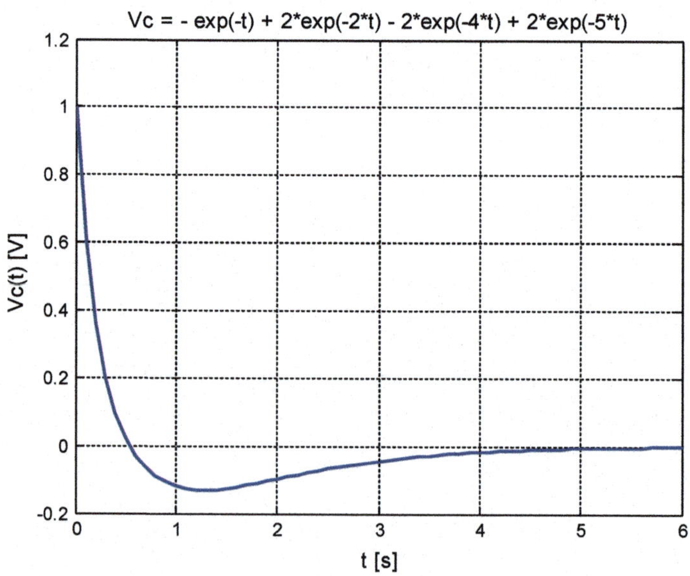

Fig. 3.22 The voltage across a 1 mF capacitor

A MATLAB script (plot_time_function1.m) is listed below.

```
C = 1e-3; t = 1.3;
Vc = -exp(-t)+2*exp(-2*t)-2*exp(-4*t)+2*exp(-5*t)
W = 0.5 * C * Vc^2
```

3.2 First-Order RC Circuits

Problem 3.2.1 In a first-order source free RC circuit, find the power dissipated (mW) in 1 kΩ resistor 0.5 s after discharge begins, if the initial voltage over the 1 mF capacitor is $V_0 = 10$ V.

Solution

$$p(t) = i(t)^2 \cdot R = \left(\frac{V_0}{R} \cdot e^{-\frac{t}{\tau}}\right)^2 . R = \frac{V_0^2}{R} \cdot e^{-\frac{2t}{\tau}} = \frac{100}{1000} \cdot e^{-1}$$
$$= 0.1 \times 0.03679 = 36.79\,\text{mW}$$

Problem 3.2.2 Find the energy dissipated up to time t in a first-order source-free RC circuit of Fig. 3.23, if the initial voltage over the capacitor is $V_c = V_0$ V. Comment on this result.

Solution

$$p(t) = i(t)^2 \cdot R = \left(\frac{V_0}{R} \cdot e^{-\frac{t}{\tau}}\right)^2 .R = \frac{V_0^2}{R} \cdot e^{-\frac{2t}{\tau}}$$

Initial energy stored on the capacitor,

$$W = \frac{1}{2} Cv(0)^2 = \frac{1}{2} CV_0^2$$

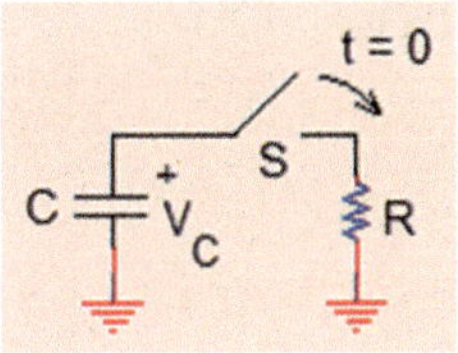

Fig. 3.23 The first-order source-free RC circuit

The energy dissipated on the resistor up to time t,

$$W_t = \int_0^t \frac{v(t)^2}{R} \mathrm{d}t = \int_0^t i(t)^2 R \mathrm{d}t = \int_0^t \frac{V_0^2}{R} \cdot \mathrm{e}^{-\frac{2t}{\tau}} \mathrm{d}t = -\frac{\tau V_0^2}{2R} \cdot \mathrm{e}^{-\frac{2t}{\tau}}\Big|_0^t$$

$$W_t = \frac{1}{2} C V_0^2 \left(1 - \mathrm{e}^{-\frac{2t}{\tau}}\right) = W\left(1 - \mathrm{e}^{-\frac{2t}{\tau}}\right)$$

This shows that

$$\frac{W_t}{W} = \left(1 - \mathrm{e}^{-\frac{2t}{\tau}}\right)$$

which means $\left[100\left(1 - \mathrm{e}^{-\frac{2t}{\tau}}\right)\right]$ percentage of the initial energy stored on the capacitor is dissipated on the resistor at the end of t seconds following the closure of switch S.

Problem 3.2.3 In the circuit shown in Fig. 3.24, power supply voltage value is 10 V, R = 1 MΩ, C = 1 pF. Determine the current (μA) flowing through resistor for $t < 0$ and $t \geq 0$.

Solution

$$t < 0; \quad I_0 = \frac{V_{\mathrm{dc}}}{R} = \frac{10}{10^6} = 10^{-5}\ \mathrm{A}$$

$$t \geq 0; \quad i_R(t) = I_0 \mathrm{e}^{-t/RC}$$

$$\tau = RC = 10^6 \times 10^{-12} = 10^{-6}\ \mathrm{s}.$$

$$i_R(t) = 10^{-5} \mathrm{e}^{-t/10^{-6}} = 10^{-5} \mathrm{e}^{-10^6 t}\ \mathrm{A} = 10 \mathrm{e}^{-10^6 t}\ \mu\mathrm{A}$$

Problem 3.2.4 In the source-free circuit shown in Fig. 3.25, $v_{C_1}(0^-) = 9.5\ \mathrm{V}$, C_1 = 10 F, C_2 = 5 F, C_3 = 3 F.

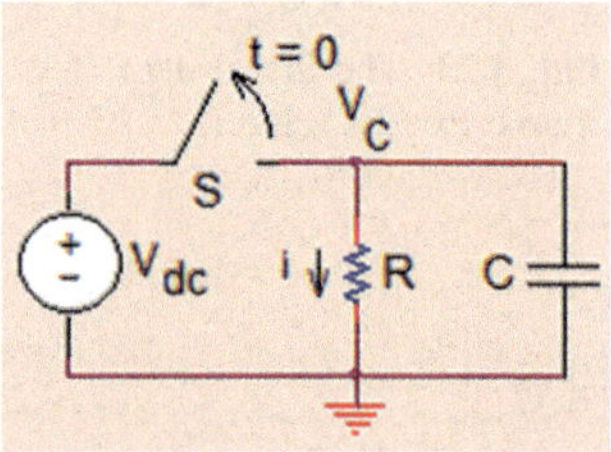

Fig. 3.24 The circuit of Problem 3.2.3

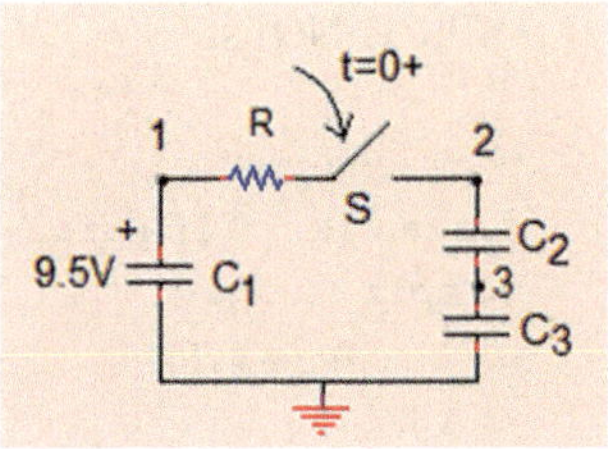

Fig. 3.25 The circuit of Problem 3.2.4

(a) For $R = 0$, find the values of voltage at 1, 2, and 3 for $t = \infty$.
(b) Repeat (a) for $R = 0.5\ \Omega$ using SPICE to plot the waveforms, print netlist. (C_dischargeR.cir)
(c) Determine the current through the resistor at $t = 0^+$, $t = 1$ s, and $t = \infty$.
(d) Check the results found in (c) using SPICE, plot the current waveform and print netlist. (CdischargeR2.cir)

Solution

(a)

$$\text{for } t<0, \quad V_2 = V_3 = 0\,\text{V}, \quad V_1 = 9.5\ \text{V}$$

$$\text{for } t\geq 0, \quad C_{\text{eq}} = \frac{C_1C_2C_3}{C_1C_2 + C_1C_3 + C_2C_3} = \frac{10\cdot 5\cdot 3}{50+30+15} = \frac{150}{95}\ \text{F}$$

$$q_{c10} = C_1\cdot V_{c10} = 10\cdot 9.5 = 95\,\text{C}$$

$$q_\infty = C_{\text{eq}}\cdot V_{c10} = \frac{150}{95}\cdot 95 = 15\,\text{C}$$

$$\Delta q = q_{c10} - q_\infty = 95 - 15 = 80\,\text{C}$$

$$V_{c1\infty} = \frac{\Delta q}{C_1} = \frac{80}{10} = 8\,\text{V}$$

$$V_1 = V_{c1\infty} = V_2 = 8\,\text{V}$$

$$V_3 = V_{c3\infty} = V_{(c1\infty)}\cdot\frac{C_2}{C_1+C_2} = 8\cdot\frac{5}{5+3} = 5\,\text{V}$$

(b) Figure 3.26 displays circuit voltages against time.

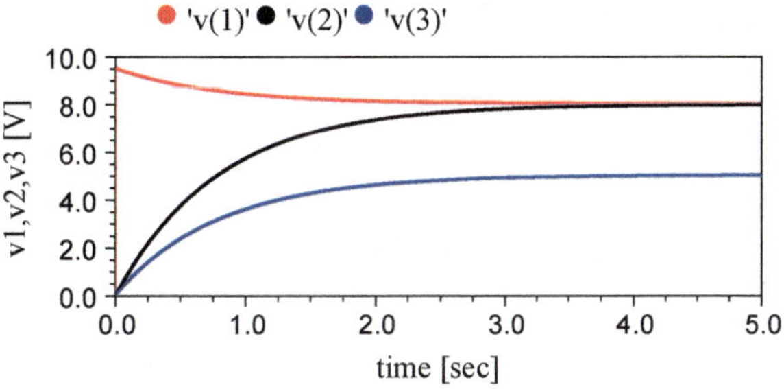

Fig. 3.26 The voltage waveforms in the circuit of Fig. 3.25

SPICE Netlist

```
C_dischargeR.cir
*.tran 1M 5 0 10M uic
*C split at dc
C1 1 0 10 ic 9.5v
R1 1 2 0.5
C2 2 3 5
C3 3 0 3
```

(c) The current through the resistor is

$$i(t) = I_0 e^{-\frac{t}{\tau}} = \frac{V_{C_1}(0)}{R} e^{-\frac{t}{\tau}}$$

$$\tau = RC_{eq} = 0.5\left(\frac{150}{95}\right) = \frac{75}{95} = 0.78947 = 789.47\,\text{ms}$$

$$i(t) = \left(\frac{9.5}{0.5}\right) e^{-\frac{t}{0.78947}} = 19e^{-1.267t}$$

$$i(0) = 19e^{-1.267\times 0} = 19\,\text{A}$$

$$i(1) = 19e^{-1.267\times 1} = 15e^{-1.267} = 15 \times 0.28177 = 5.354\,\text{A}$$

$$i(\infty) = 19e^{-1.267\times\infty} = 0\,\text{A}$$

(d) The current waveform in the circuit is plotted in Fig. 3.27, using SPICE analysis.

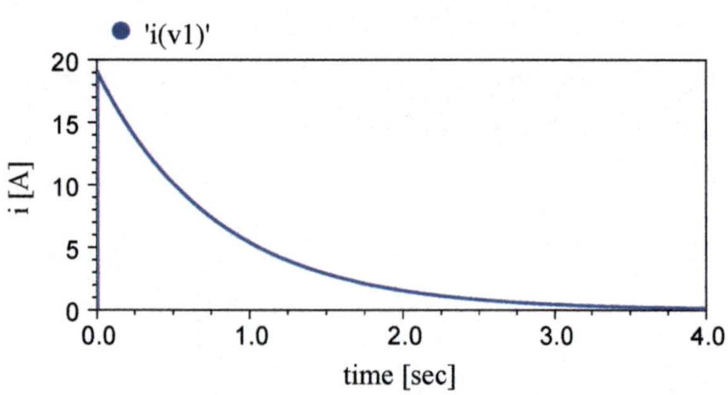

Fig. 3.27 The current waveform in the circuit of Fig. 3.25

SPICE Netlist,

```
C_dischargeR2.cir
Analysis: Transient
Current waveform in C split circuit
C1 1 0 10 ic 9.5v
R1 4 2 0.5
C2 2 3 5
C3 3 0 3
v1 1 4 0
*.tran 1M 5 0 10M uic
```

Problem 3.2.5 For the circuit shown in Fig. 3.28, (a) $V_c(0^-)$ = ? (b) $V_c(0^+)$ = ? (c) $V_c(t)$ = ?, $t > 0$ (d) $V_c(\infty)$ = ?

($R_1 = 1\ \Omega$, $R_2 = 2\ \Omega$, $C = 2$ F, $V_{dc} = 10$ V)

Solution

(a) $V_c(0^+) = \dfrac{R_2}{R_1 + R_2} \cdot V_{dc} = \dfrac{2}{1+2} \cdot 10 = \dfrac{20}{3} = 6.667\,\mathrm{V}$

(b) $V_c(0^+) = V_c(0^-) = 6.667\,\mathrm{V}$

(c) $V_c(t) = V_c(0^-) \cdot \mathrm{e}^{-\frac{t}{\tau}} = V_c(0^-) \cdot \mathrm{e}^{-\frac{t}{RC}} = 6.667 \times \mathrm{e}^{-\frac{t}{2\times 2}} = 6.667\mathrm{e}^{-0.25t}$

(d) $V_c(\infty) = 0\,\mathrm{V}$

Problem 3.2.6 In the circuit of Fig. 3.29, switch (S_0) has been in the position (01) for long time. At $t = 0$, it is brought to position (02).

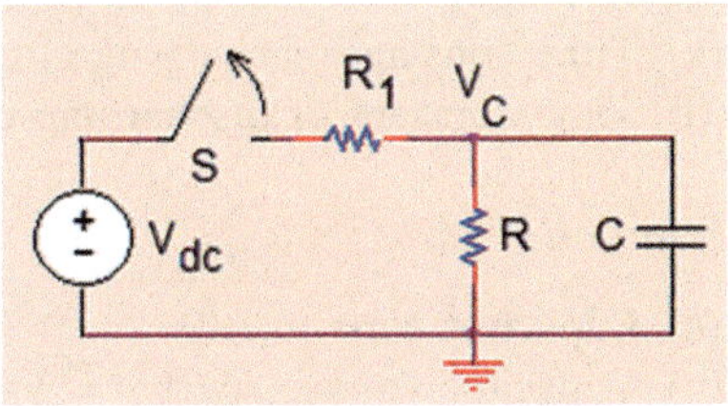

Fig. 3.28 The circuit of Problem 3.2.5

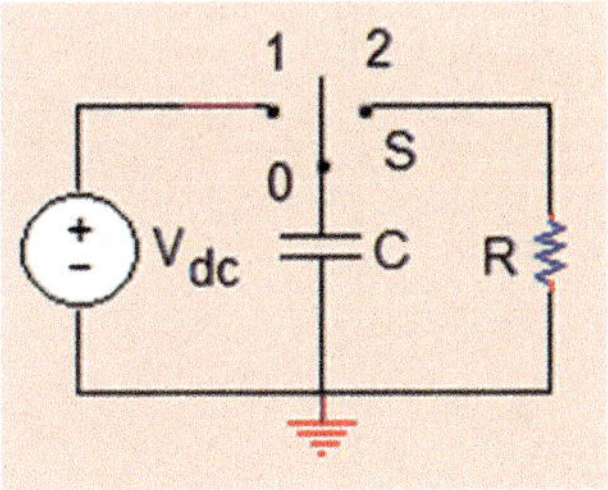

Fig. 3.29 The circuit of Problem 3.2.6

For $t \geq 0$,

(a) What is the time constant of the circuit?
(b) What is the equation of voltage over resistance?
(c) What is the value of current through resistor at $t = 1$ ms? (in mA)
(d) What is the value of initial current flowing through resistor? (in mA)

Solution

(a) $\tau = RC = 10^3 \times 10^{-6} = 10^{-3}\,\text{s} = 1\,\text{ms}$
(b) $v_R(t) = 10 \cdot \text{e}^{-\frac{t}{\tau}} = 10 \cdot \text{e}^{-\frac{t}{10^{-3}}} = 10 \cdot \text{e}^{-10^3 t} = 10 \cdot \text{e}^{-1000t}$
(c) $$i_R(t) = \left.\frac{v_R(t)}{10^3}\right|_{t=10^{-3}\,\text{s}} = \frac{10 \cdot \text{e}^{10^3 \times 10^{-3}}}{10^3} \cong 10^{-2} \times (0.37)$$
$$i_R(10^{-3}) = 3.7 \times 10^{-3}\,\text{A} = 3.7\,\text{mA}$$
(d) $i_R(0) = 10 \cdot \frac{\text{e}^{-1000.(0)}}{10^3} = \frac{10}{10^3} = 10^{-2}\,\text{A} = 10\,\text{mA}$

Problem 3.2.7 In Fig. 3.30, which represents a monophasic defibrillator circuit, the switch S was in position A for a long time. It is thrown into position B at $t = 0$ s. Find;

(a) Capacitor current, $i_c(0-) = ?$
(b) Capacitor voltage, $v_c(0-) = ?$
(c) Capacitor current, $i_c(0+) = ?$
(d) Capacitor voltage, $v_c(0+) = ?$
(e) Capacitor voltage, $v_c(\infty) = ?$
(f) Maximum magnitude of charging current, $i_c(-\infty) = ?$
(g) Maximum magnitude of discharging current, $i_c(\infty) = ?$
(h) Energy stored on capacitor at $t = 0$?
(i) Energy dissipated on resistor R_2 at $t = \infty$?
(j) Average power dissipated on resistor R_2, T seconds after switch S is closed?
(k) Time constant of charging circuit?
(l) Time constant of discharging circuit?

Solution

(a) Capacitor current, $i_c(0-) = 0$ A
(b) Capacitor voltage, $v_c(0-) = V_{dc}$ V

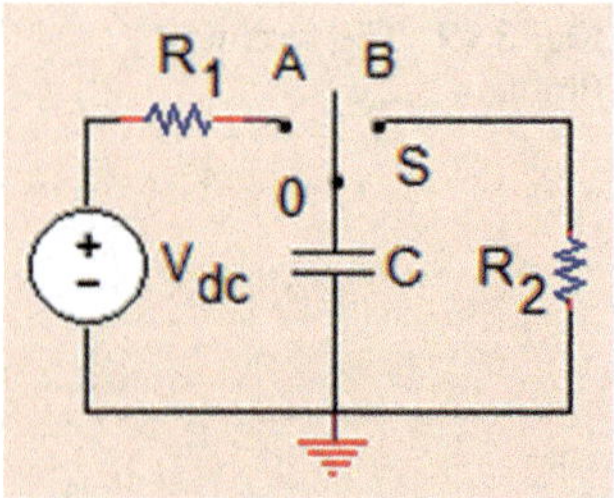

Fig. 3.30 The circuit of Problem 3.2.7

(c) Capacitor current,

$$i_C(0+) = \frac{V_{dc}}{R_2} \text{ A}$$

(d) Capacitor voltage, $v_C(0+) = V_{dc}$ V
(e) Capacitor voltage, $v_c(\infty) = 0$ V
(f) Capacitor acts as a short circuit if it has no initial charge on it,

$$i_C(-\infty) = \frac{V_{dc}}{R_1} \text{ A}$$

Maximum magnitude of charging current is $i_c(-\infty)$, found in part (f).
(g) Maximum magnitude of discharging current is $i_c(0+)$, found in part (c).
(h) Energy stored on capacitor at $t = 0+$ is

$$W_C(0+) = \frac{CV_{dc}^2}{2}$$

(i) Energy dissipated on resistor R_2 at $t = \infty$?

$$W_{R_2}(\infty) = W_C(0+) = \frac{CV_{dc}^2}{2}$$

(j) Average power dissipated on resistor R_2, T seconds after switch S is closed?

$$P_{R_2}(T) = \frac{W_C(0+)}{T}$$

(k) Time constant of charging circuit is $\tau_{charge} = (R_1)C$ s
(l) Time constant of discharging circuit is $\tau_{discharge} = (R_2)C$ s

Problem 3.2.8 In the circuit shown in Fig. 3.31a, the switch has been closed for a long time. It is opened at $t = 0$. Find the voltage variation across the capacitor, $t \geq 0$. ($V_{dc} = 20$ V, $C = 20$ mF).

Solution
For $t < 0$, the capacitor is open circuit to dc (Fig. 3.31b),
By voltage division rule; for $t < 0$,

$$V_c(t) = \frac{9}{3+9} \cdot 20 = 15 \text{ V}$$

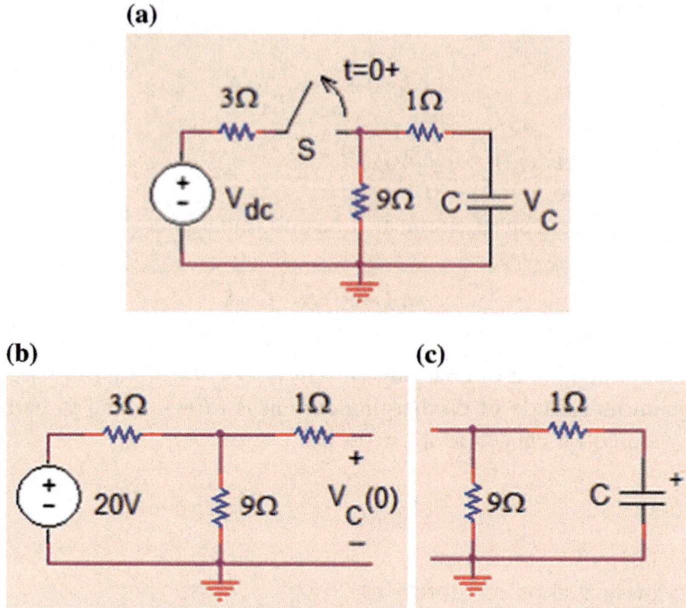

Fig. 3.31 a–c The circuit of Problem 3.2.8

Since the voltage across the capacitor cannot change instantaneously, $V_c(0) = V_0 = 15\,\text{V}$. At $t = 0$, the switch is opened: (Fig. 3.31c), $R_{\text{eq}} = 1 + 9 = 10\,\Omega$

$$\tau = R_{\text{eq}} \cdot C = 10 \cdot 20 \times 10^{-3} = 0.2\,\text{s}$$

$$v(t) = V_c(0) \cdot \mathrm{e}^{-t/\tau}$$
$$v(t) = 15 \cdot \mathrm{e}^{-t/0.2} = 15 \cdot \mathrm{e}^{-5t}\,\text{V}$$

Problem 3.2.9 A capacitor is initially charged at V_c = 10 V. It discharges through a resistor of R = 10 kΩ. If V(10 μs) = 3.7 V,

(a) C = ?
(b) $i_c(0)$ = ?

Solution

(a) $\tau = RC \cong 10\,\mu\text{s}, \quad \rightarrow \quad C = \dfrac{10^{-5}}{10^4} = 10^{-9} = 1\,\text{nF}$

(b) $i_c(t) = \dfrac{v_c(t)}{R} \quad \rightarrow \quad i_c(0) = \dfrac{v_c(0)}{R} = \dfrac{10}{10^4} = 10^{-3}\,\text{A} = 1\,\text{mA}$

Problem 3.2.10 In the circuit shown in Fig. 3.32, for $t \geq 0$, $v(t)$ = ? $R_2 = R_3$ = 2 Ω, R_2 = 10 Ω, V_{dc} = 12 V, C = 1 uF.

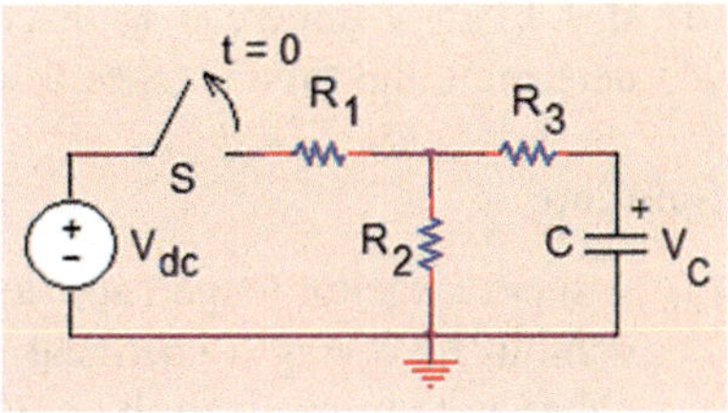

Fig. 3.32 The circuit of Problem 3.2.10

Solution

For $t < 0$, the capacitor is open circuit, therefore no current flows through R_3. Voltage drop across resistor R_2 is equal to the voltage across the capacitor, see, Fig. 3.33.

$$V_c = \frac{R_2}{R_2 + R_1}.V_{\text{dc}} \quad t<0 \quad \rightarrow \quad V_c(0) = V_0 = V_{\text{dc}}$$

$$V_c = \frac{10}{10+2}.12 = 10\,\text{V} \quad t<0 \quad \rightarrow \quad V_c(0) = V_0 = 10\,\text{V}$$

For $t > 0$: The switch is opened (Fig. 3.34).

$$R_{\text{eq}} = R_2 + R_3 = 10 + 2 = 12\,\Omega, \quad \rightarrow \quad \tau = R_{\text{eq}}.C = 12 \times 10^{-6} = 1.2 \times 10^{-5}\ \text{s}$$

$$V(t) = V_c(0).\text{e}^{-t/\tau} = 10.\text{e}^{-83333,\bar{3}t}\ \text{V}$$

Problem 3.2.11

(a) What is a supercapacitor? Draw its equivalent circuit.
(b) Compare the energy storing performance of a supercap with $C = 1$ F, nominal voltage of 5.5 V with that of a tantalum capacitor, $C = 1500$ μF rated at 6.3 V.
(c) Compare the charge storage performance of the two capacitors, given in (b).

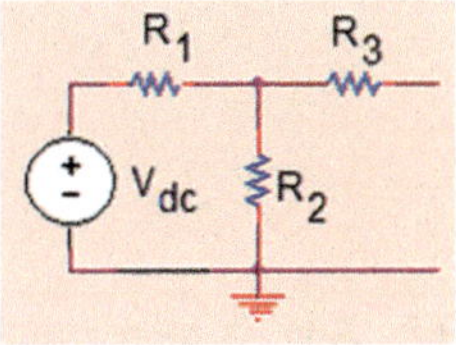

Fig. 3.33 The circuit of Problem 3.2.10 for $t < 0$

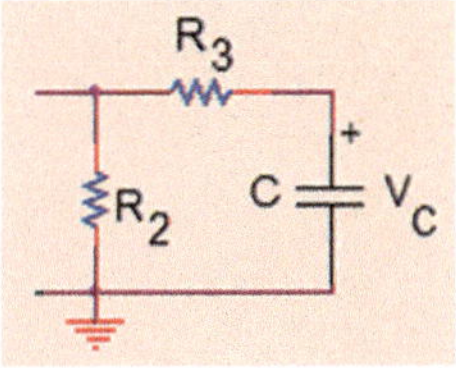

Fig. 3.34 The circuit of Problem 3.2.10 for $t > 0$

(d) If 1 F/5.5 V supercap is discharged from 4.2 V at 10 mA constant current, determine the time it takes to complete discharge (in minutes).

Solution

(a) A supercapacitor (supercap) stores energy by polarizing an electrolytic solution, without involving a chemical reaction [5–8]. It can be viewed as two porous plates within an electrolyte. The applied voltage on the (+) plate attracts the negative ions in the electrolyte. The positive ions are attracted by the voltage on the (−) plate. This creates two layers; one where the positive and negative charges are separated on positive plate and the other at the negative plate. The porous carbon-based electrodes provide a larger area per mass (typically, 2000 m^2/g), while the charge separation distance depends on the size of the ions in the electrolyte, (about 10 Å). Simplified electrical equivalent model of a supercap is shown in Fig. 3.35.
Here, R_s = series resistance, R_i = insulation resistance.
A detailed equivalent circuit model has multiple series resistors and capacitors, as shown in Fig. 3.36. Such capacitors can be used in defibrillators.

(b) $E_{sc} = \frac{1}{2}CV^2 = \frac{1}{2} \times 1 \times (5.5)^2 = 15.125\,\text{J}$

$$E_t = \frac{1}{2} \times \left(1500 \times 10^{-6}\right) \times (6.3)^2 = 19.845 \times 1.5 \times 10^{-3} \cong 0.0298\,\text{J}$$

$$\frac{E_{sc}}{E_t} = \frac{15.125}{0.0298} \cong 507.6$$

Energy storing capacity of 1 F/5.5 V supercap is 507.6 times higher than the energy storing capacity of a 1.5 mF/6.3 V tantalum capacitor.

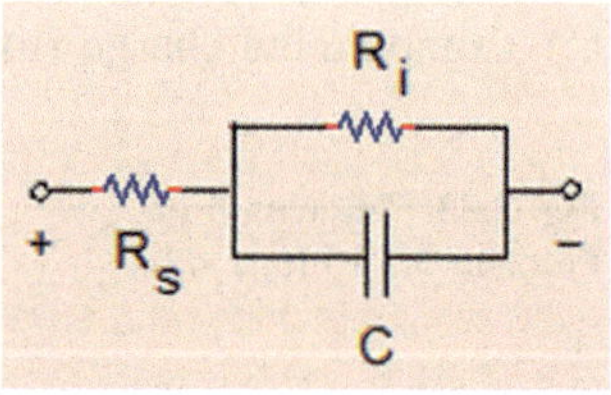

Fig. 3.35 Simplified electrical equivalent model of a supercapacitor

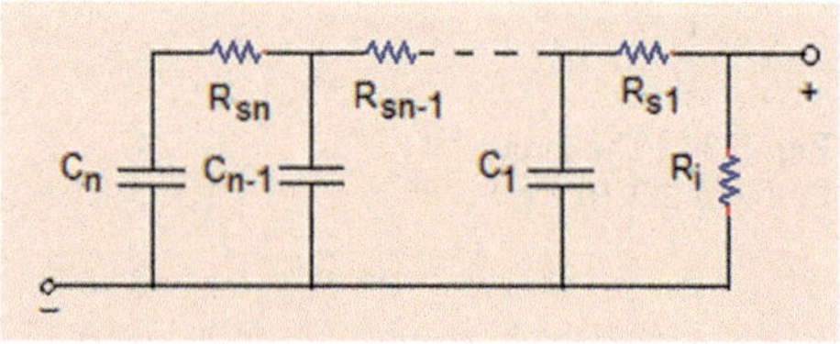

Fig. 3.36 A more detailed equivalent circuit model of a supercapacitor

(c) $Q_{sc} = CV = 1 \times 5.5 = 5.5\,\mathrm{C}$

$$Q_t = \left(1.5 \times 10^{-3}\right) \times 6.3 = 0.00945\,\mathrm{C}$$
$$\frac{Q_{sc}}{Q_t} = \frac{5.5}{0.00945} \cong 582$$

The charge storing performance of 1 F/5.5 V supercap is 582 times better than the charge storing performance of a 1.5 mF/6.3 V tantalum capacitor.

(d) $I = C\frac{\mathrm{d}V}{\mathrm{d}t} \quad \rightarrow \quad \left|\frac{\Delta V}{\Delta t}\right| \approx \frac{I}{C} \quad \rightarrow \quad \Delta t = \frac{\Delta \mathrm{V}}{\mathrm{I}} C$

$$\Delta t = \frac{4.2}{0.01} \times 1 = 420\,\mathrm{s} = 7\,\mathrm{min}$$

Problem 3.2.12 A capacitor $C = 10\,\mathrm{mF}$ is charged to 36 V. It discharges through a variable electronically controlled resistor so that a constant current of 1 mA flows through the source-free circuit.

(a) How many minutes does it take to discharge it?
(b) Determine the initial resistance of the circuit.
(c) Determine the rate of voltage change on the capacitor.

Solution

(a) Capacitor accumulated charge: $Q = CV = 10 \times 10^{-3} \times 35 = 0.35\,\mathrm{C}$

$$Q = It \rightarrow 0.35 = 10^{-3} \times t \quad \rightarrow \quad t = 360\,\mathrm{s} = 6\,\mathrm{min}$$

(b) Initial resistance of circuit,

$$R(0) = \frac{V(0)}{10^{-3}} = \frac{36}{10^{-3}} = 36\,\mathrm{k\Omega}$$

(c) $i(t) = C\frac{\mathrm{d}V(t)}{\mathrm{d}t} = I = 10^{-3}\,\mathrm{A} \quad \rightarrow \quad \frac{\mathrm{d}V(t)}{\mathrm{d}t} = \frac{I}{C} = \frac{10^{-3}}{10^{-2}} = 0.1\,\mathrm{V\,s^{-1}}$

In other words, the voltage across the capacitor varies with a (negative) linear slope of 0.1 V/s.

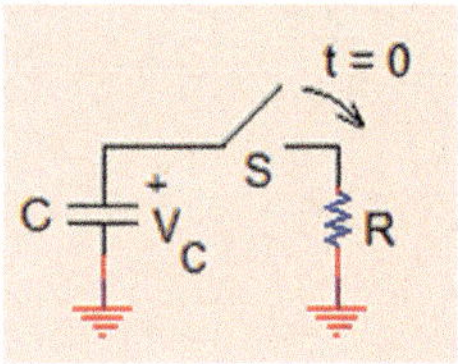

Fig. 3.37 The pulsed switching circuit of Problem 3.2.5

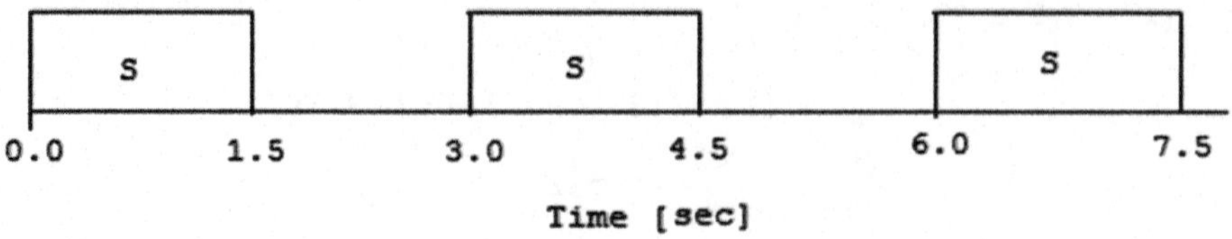

Fig. 3.38 The switching action timing diagram

Problem 3.2.13 A capacitor $C = 1$ F with initial voltage of 5 V is discharged via a switch through a resistor of $R = 1\ \Omega$. The turn on time of the switch is 1.5 s with a period of 3 s (see, Figs. 3.37 and 3.38).

(a) Calculate the voltage across the capacitor at $t = 0$ s, $t = 1.5$ s, $t = 3$ s and $t = 4.5$ s.
(b) What is the initial energy stored on the capacitor?
(c) What is the energy dissipated within the first 1.5 s?
(d) Plot the voltage across the capacitor and show the switching time diagram.
(e) Check the results found in parts (a) and (d) by using SPICE analysis and print SPICE netlist. (SW3.cir)

Solution

(a) Since voltage across a capacitor cannot change abruptly, $v_C(0) = v_R(0) = 5\,\mathrm{V}$
Time constant of the circuit is $\tau = RC = (1)(1) = 1\,\mathrm{s}$.
Using given data,

$$v_C(t) = v_C(0)\mathrm{e}^{-\frac{t}{\tau}} = 5\mathrm{e}^{-t}$$

$$v_C(1.5) = 5\mathrm{e}^{-1.5} = 1.116\,\mathrm{V}$$

$$v_C(3) = v_C(1.5) = 1.116\,\mathrm{V}$$

$$v_C(4.5) = v_C(1.5)\mathrm{e}^{-\frac{(t-3)}{\tau}}$$

$$v_C(4.5) = 1.116\mathrm{e}^{-1.5}$$

$$v_C(4.5) = (1.116)(0.223) = 0.249\,\mathrm{V}$$

(b) Initial energy stored on the capacitor,

$$W = \frac{1}{2}Cv(0)^2 = 0.5(1)\left(5^2\right) = 12.5\,\mathrm{J}$$

(c) Energy dissipated within the first 1.5 s,

$$W = \int_0^{1.5} \frac{v(t)^2}{R} \mathrm{d}t = \int_0^{1.5} v(t)^2 \mathrm{d}t = \int_0^{1.5} (5\mathrm{e}^{-t})^2 \mathrm{d}t = \int_0^{1.5} 25\mathrm{e}^{-2t} \mathrm{d}t = 25 \int_0^{1.5} \mathrm{e}^{-2t} \mathrm{d}t$$

$$W = (25)\left(\frac{1}{-2}\right)\left[\mathrm{e}^{-2(1.5)} - 1\right] = -12.5\left(\mathrm{e}^{-3} - 1\right) = 12.5\left(1 - \mathrm{e}^{-3}\right) = 12.5(0.95)$$

$$W = 11.875\,\mathrm{J}$$

This means that 95% of the initial energy stored on the capacitor is dissipated at the end of first 1.5 s following the closure of switch *S*.

(d) The voltage variation across the capacitor (red) and the switching time (green) are shown in Fig. 3.39.

(e)

```
SPICE net list SW3.cir
       Analysis: Transient
   truncated exponential monophasic waveform
   C1 1 0 1 IC=5v
   s2 1 2 3 0 sm2
   *v5 3 0 dc 1
   r1 2 0 1
   *pulse(V1 V2 TD TR TF PW PER)
   v5 3 0 pulse(0 2 0 0 0 1.5 3)
   .MODEL sm2 SW(Ron=1u)
   .tran .01 10 0 .01 uic
   Analysis Parameters
        *Stepsize ...........        : .01
        *Final time .........        : 10
        *Display start time ..       : 0
        *Maximum step size ...       : .01
        *Use Initial Conditions: On
```

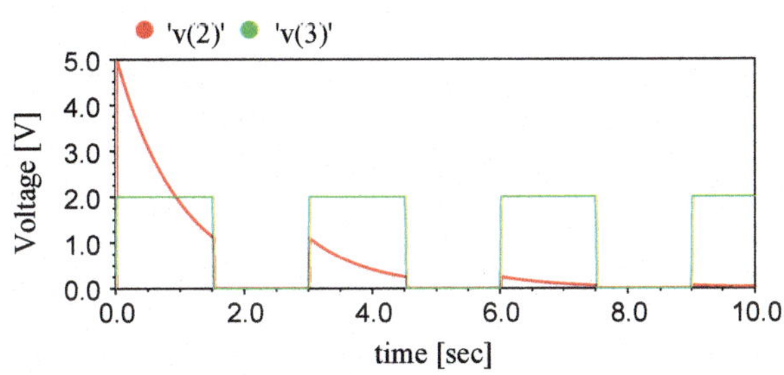

Fig. 3.39 The voltage waveforms on the circuit of Problem 3.2.13

Problem 3.2.14 A capacitor with $C = 1$ F and initial voltage of 5 V is discharged through a resistor $R_1 = 1\ \Omega$ and a switch S_1. Electronic switching is realized as a periodic on-off switch S_1 that is "on" for 1.5 s in a period of 3 s. Another switch S_2 is turned on so long as switch S_1 is in "off" position (otherwise S_2 remains turned off) and quickly discharges residual charge on the capacitor through a resistor with small resistance, $R_2 = 0.1\ \Omega$. Assume that the electronic switches are ideal, i.e., no energy is lost through these components (see, Fig. 3.40).

(a) Draw the circuit,
(b) Plot "timing signals" of electronic switches S_1, S_2,
(c) What is the initial energy stored?
(d) Determine the maximum current.
(e) What is the equation of current within the first 1.5 s?
(f) Determine the current at time $t = 1.5$ s.
(g) What is the percentage value of current flowing through the resistor R_1 at $t = 1.5$ s to the initial current?
(h) What is the amount of dissipated energy within the first 1.5 s?
(i) How much energy is dissipated by resistor R_2?
(j) Plot the capacitor voltage across the resistor R_1, R_2 and switching waveforms on the same graph and check the results using SPICE analysis. (`SW1.cir`)

Solution

(a) See Fig. 3.40
(b) Timing diagram for switches is shown in Fig. 3.41;
(c) $W = \dfrac{Cv(0)^2}{2} = \dfrac{1 \times 5^2}{2} = 12.5\,\text{J}$

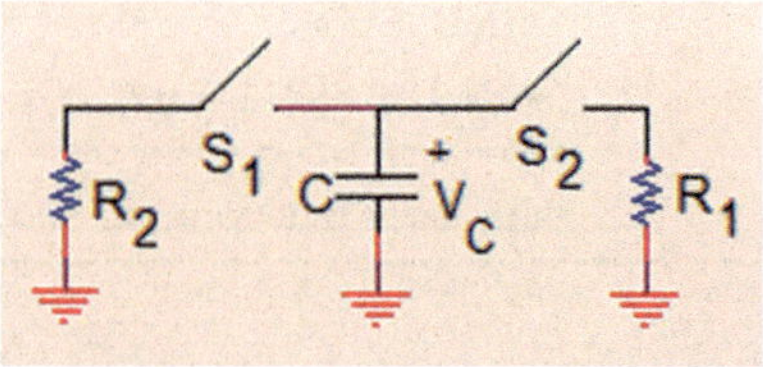

Fig. 3.40 The circuit of Problem 3.2.14

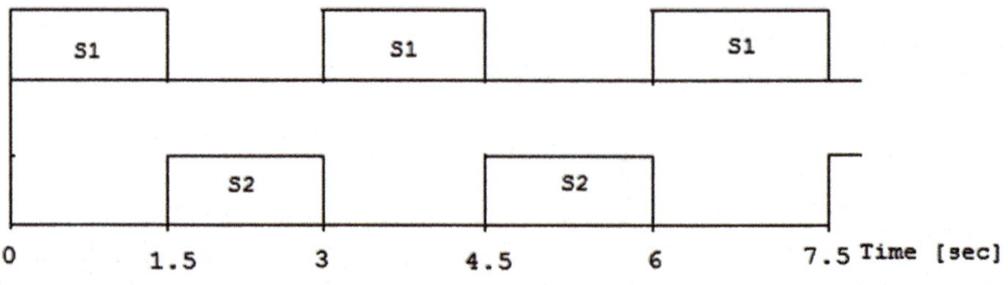

Fig. 3.41 Timing diagram for switches in Fig. 3.40

(d) Maximum current flows at time $t = 0+$;

$$i(0) = \frac{v(0)}{R_1} = \frac{5}{1} = 5\,\text{A}$$

(e) Time constant for the main circuit is $\tau = R_1 C = 1 \times 1 = 1\,\text{s}$
The current is described by the following equation:

$$i(t) = \frac{v(t)}{R_1} = \frac{v(0)e^{-\frac{t}{\tau}}}{R_1} = \frac{5e^{-t}}{1} = 5e^{-t}$$

(f) The current at time $t = 1.5$ s is calculated from the following equation:

$$i(1.5) = 5e^{-1.5} = (5)(0.2231) = 1.116\,\text{A}$$

(g) The percentage value of current flowing through the resistor R_1 at $t = 1.5$ s to the initial current is

$$p = \frac{i(1.5)}{i(0)} = \frac{1.116}{5} = 22.31\%$$

(h) Energy dissipated within the first 1.5 s,

$$W_{1.5} = \int_0^{1.5} i(t)^2 R\mathrm{d}t = \int_0^{1.5} (5e^{-t})^2 \mathrm{d}t$$

$$W_{1.5} = W\left(1 - e^{-2(1.5)}\right) = 12.5\left(1 - e^{-3}\right) = 12.5(0.95)\,\text{J} = 11.875\,\text{J}$$

This means that 95% of the initial energy stored on the capacitor is dissipated at the end of first 1.5 s following the closure of switch S.

(i) The energy dissipated by resistor R_2 can be calculated as

$$W_{R_2} = W - W_{1.5} = 12.5 - 11.875 = 0.625\,\text{J}$$

(j) The voltage variation across the resistor R_1(red) R_2 (blue) and switching waveform for S_1 are shown in Fig. 3.42.

```
SPICE netlist;
    *Transient analysis
truncated exponential monophasic waveform SW1.cir
C1 1 0 1 IC=5v
R1 2 0 1
s1 1 2 3 0 sm2
*PULSE(V1 V2 TD TR TF PW PER)
```

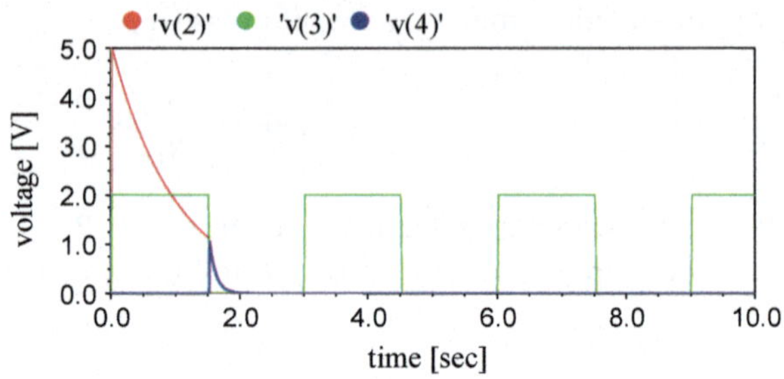

Fig. 3.42 The voltage variations across the resistors and the switching waveform

```
vx 3 0 pulse(0 2 0 0 0 1.5 3)
*s2 totally discharges C1 (residual discharging)
R2 4 0 .1
s2 1 4 0 3 sm2
.MODEL sm2 SW(Ron = 1u)
   .tran .01 10 0 .01 uic
   *Stepsize ...........         : .01
   *Final time .........         : 10
   *Display start time ..        : 0
   *Maximum step size ...        : .01
   *Use Initial Conditions: On
```

Problem 3.2.15 Discharge section of a truncated monophasic exponential defibrillator circuit is to be designed using a capacitor with $C = 48$ μF, and some electronic switching components. Patient's trunk is represented as resistor $R_1 = 50\ \Omega$.

An initially charged capacitor discharges through a switch S_1 and "resistor". Electronic switching is realized as a periodic on-off switch S_1 that is "on" for 10 ms in a period of 20 ms. Another switch S_2 is turned on so long as switch S_1 is in "off" position (otherwise S_2 remains turned off) and quickly discharges residual charge on the capacitor through a resistor with small resistance, $R_2 = 0.1\ \Omega$. Assume that the electronic switches are ideal, i.e., no energy is lost through these components.

(a) Draw the circuit,
(b) Plot "timing signals" of electronic switches S_1, S_2,
(c) What must be the initial voltage of the capacitor, if the energy stored on it is to be 360 J?
(d) Determine the maximum current.
(e) What is the equation of current within the first 10 ms?
(f) Determine the current at time $t = 10$ ms.
(g) What is the percentage value of current flowing through the patient trunk at $t = 10$ ms to the initial current?

(h) What is the amount of dissipated energy on patient's trunk within the first 10 ms?
(i) How much energy is dissipated by resistor R_2?

Solution

(a) The circuit is shown in Fig. 3.43.
(b) Timing diagram for switches is shown in Fig. 3.44.
(c) $W = \frac{Cv(0)^2}{2} = \frac{48 \times 10^{-6} \times v(0)^2}{2} = 360\,\text{J}$

$$v(0)^2 = \frac{360}{24 \times 10^{-6}} = \frac{15}{10^{-6}} = 15 \times 10^6\,\text{J}$$
$$v(0) = 3873\,\text{V}$$

(d) Maximum current flows at time t = 0+;

$$i(0) = \frac{v(0)}{R_1} = \frac{3873}{50} = 77.46\,\text{A}$$

(e) Time constant for "discharge through the patient" circuit is

$$\tau = R_1 C = 50 \times 48 \times 10^{-6} = 2400 \times 10^{-6} = 2.4\,\text{ms}$$

The current is described by the following equation:

$$i(t) = \frac{v(t)}{R_1} = \frac{v(0)e^{-\frac{t}{\tau}}}{R_1} = \frac{3873e^{-\left(\frac{t}{2.4}\right)10^3}}{50} = 77.46e^{-416.7t}$$

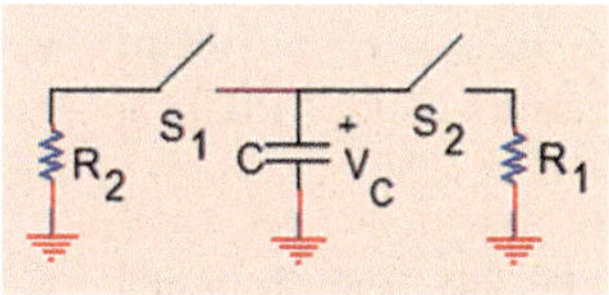

Fig. 3.43 The charging and discharging of a capacitor

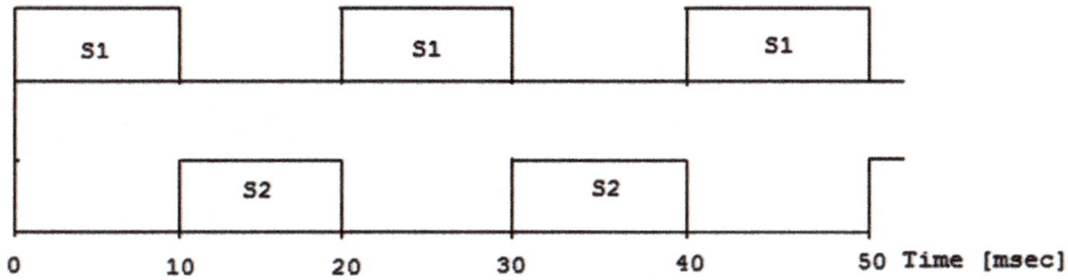

Fig. 3.44 Timing diagram for switches

(f) The current at time t = 10 ms is calculated from the following equation:

$$i(10\times10^{-3}) = 77.46\mathrm{e}^{-416.7\times10^{-2}} = (77.46)(0.01549) = 1.200\,\mathrm{A}$$

(g) The percentage value of current flowing through the patient trunk at t = 10 ms to the initial current is

$$p = \frac{i(10\times10^{-3})}{i(0)} = \frac{1.2}{77.46} = 1.55\%$$

(h) Energy dissipated within the first 10 ms,

$$W_{10} = \int_0^{10\,\mathrm{m}} i(t)^2 R\mathrm{d}t = R\int_0^{10\,\mathrm{m}} (77.46\mathrm{e}^{-416.7t})^2\mathrm{d}t = 50(6000.0516)\int_0^{10\,\mathrm{m}} \mathrm{e}^{-2\times416.7t}\mathrm{d}t$$

$$W_{10} = (300002.58)\left(\frac{1}{-833.3}\right)\left[\mathrm{e}^{-833.3\times0.01} - 1\right] = -360(\mathrm{e}^{-8.333} - 1)$$

$$W_{10} = 360(1 - \mathrm{e}^{-8.333}) = 360(0.99976) = 359.91\,\mathrm{J}$$

This means that 99.98% of the initial energy stored on the capacitor is dissipated at the end of first 10 ms following the closure of switch S.

(i) Energy dissipated by resistor R_2

$$W - W_{10} = 360 - 359.91 = 0.09\,\mathrm{J}$$

This means that major electrical energy is already dissipated by R_1.

Problem 3.2.16 In the circuit of Fig. 3.45, find the voltages $v_{C_1}(t), v_{C_2}(t)$ across both of the capacitors after the switch is closed. $R = R_1 = R_2 = 10$ kΩ, $C = C_1 = C_2 = 100$ μF, $V_i = 1000$ V, $V_{C_1}(0) = 0\,\mathrm{V}$, $V_{C_2}(0) = 400\,\mathrm{V}$. Check the results using SPICE analysis, print the netlist. (RCx2.cir)

Fig. 3.45 The circuit of Problem 3.2.16

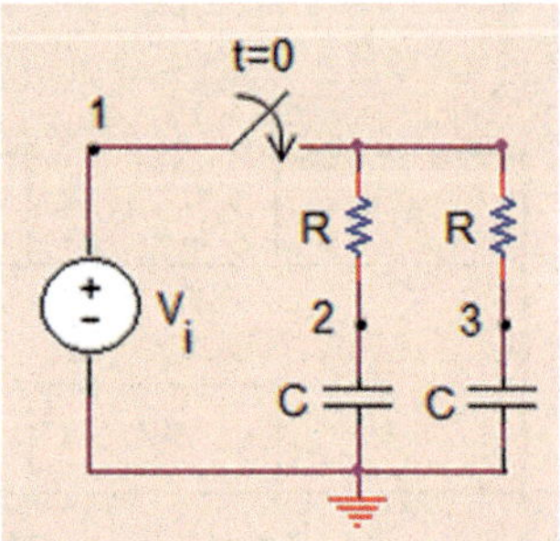

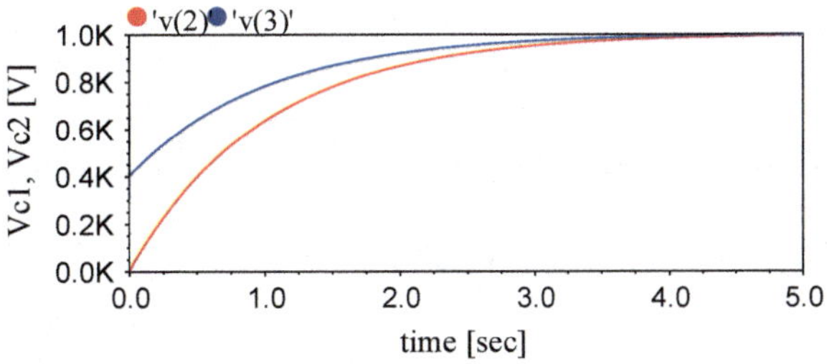

Fig. 3.46 The voltage variations in the circuit of Fig. 3.45

Solution

$$\tau = RC = 10^4 \times 100 \times 10^{-6} = 1\,\text{s}$$

$$v_C = v(\infty) + [v(0) - v(\infty)]e^{-\frac{t}{\tau}}$$

$$v_{C_1} = 1000 + (0 - 1000)e^{-t} = 1000(1 - e^{-t})$$

$$v_{C_2} = 1000 + (400 - 1000)e^{-t} = 1000 - 600e^{-t}$$

These voltages are plotted in Fig. 3.46.

```
Following is the SPICE Netlist (RCx2.cir);
   RCx2
   Vi 1 0 1000
   R1 1 2 10k
   R2 1 3 10k
   C1 2 0 100u
   C2 3 0 100u IC 400
   *.tran .1m 5 0 .1m uic
```

Problem 3.2.17 A simple RC charging circuit is used to test a supercapacitor. In test circuit, the current through the capacitor is measured to be 1 mA 30 min after the switch is closed. If

$$U = 10\,\text{V}, \quad R = 100\,\Omega, \quad C = ?$$

Solution
During charging of a capacitor, the current through the capacitor decreases exponentially,

$$i_c = \frac{U}{R}e^{-\frac{t}{\tau}} \quad \rightarrow \quad 10^{-3} = \frac{10}{100}e^{-\frac{30\times 60}{100\times C}} = 10^{-1} \times e^{-\frac{18}{C}} \quad \rightarrow \quad 10^{-2} = e^{-\frac{18}{C}}$$

$$\ln(10^{-2}) = -\frac{18}{C} \quad \rightarrow \quad C = \frac{-18}{\ln(0.01)} = 3.909\,\mathrm{F}$$

Problem 3.2.18 In the circuit of Fig. 3.47a, switch S opens at $t = 0$. Determine $v_X(t),\ t > 0$.

Comment on the result.

Solution

Initial capacitor voltage is determined by applying voltage division rule, and because voltage across a capacitor cannot change abruptly,

$$v_X(0^-) = v_C(0^-) = U\frac{R}{R+R} = \frac{U}{2}\,\mathrm{V}$$

After the switch S opens,

$$v_X(t) = v_X(\infty) + [v_X(0) - v_X(\infty)]\mathrm{e}^{-t/RC} = U + \left(\frac{U}{2} - U\right)\mathrm{e}^{-t/RC}$$
$$= U\left(1 - \frac{\mathrm{e}^{-t/RC}}{2}\right)u(t)$$

This means that the capacitor voltage is initially at half the value of input voltage, (=U/2) and after the switch is closed, it exponentially charges up to U volts, within approximate duration of 5RC seconds. For example, if U = 10 V, R = 1 Ω, C = 0.1 F, Fig. 3.47b displays $v_X(t),\ t > 0$.

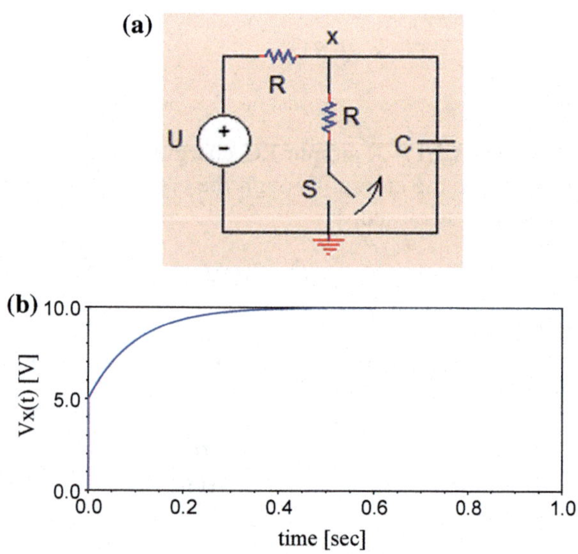

Fig. 3.47 **a** The circuit of Problem 3.2.18, **b** the voltage across charging capacitor

Problem 3.2.19 A capacitor $C_1 = 1$ F and initial voltage of 5 V is discharged through a resistor $R = 1\ \Omega$ and a switch S_1. Another capacitor $C_2 = 1$ F, having initial voltage of 5 V but opposite polarity is discharged through the same resistor and a switch S_2. Electronic switching is realized by alternate on-off action of S_1 and S_2. S_1 is "on" for 1.5 s while S_2 is off, and switch S_2 is turned on while S_1 is in "off" position for the next 1.5 s duration. Both of the switches are turned off at the end of 3 s. Switches are ideal.

(a) Draw the circuit,
(b) Plot "timing diagram" of electronic switches S_1 and S_2,
(c) What is the initial energy stored on C_1? What is the initial energy stored on C_2?
(d) Determine the magnitude(s) of maximum current(s).
(e) What is the equation of current within the first switching period of 1.5 s?
(f) What is the equation of current within the second switching period of 1.5 s?
(g) Determine the currents at time $t = 1.5$ s and $t = 3$ s.
(h) What is the percentage value of current flowing through the resistor R at $t = 1.5^-$ s to the initial current at $t = 0^+$?
(i) What is the amount of dissipated energy within the first 1.5 s?
(j) What is the amount of dissipated energy within the second 1.5 s period due to S_2?
(k) Plot the voltage across the resistor R and switching waveforms on the same graph and print SPICE netlist.

Plot the voltage across the resistor R and switching waveforms on the same graph and print SPICE netlist if the beginning of the second switching (negative) period has a short delay, e.g., 1.55 s in stead of 1.5 s. Comment on this case. (`SW2.cir`).

Solution

(a) The circuit is shown in Fig. 3.48a. The current flow reverses its direction at the end of first 1.5 s, then it ceases flowing in the resistor at the end of 3 s.
(b) Timing of electronic switches S_1 and S_2 are shown in Fig. 3.48b.
(c) Initial energy stored on C_1,

$$W = \frac{C_1 v(0)^2}{2} = \frac{1 \times 5^2}{2} = 12.5\,\mathrm{J}$$

Since $C_2 = C_1$ and initial charges are the same, initial energy stored on C_2 is the same as that on C_1.
(d) Maximum magnitudes of currents flow at time $t = (0+) = (1.5+)$ s

$$i(0) = \frac{v(0+)}{R} = \frac{5}{1} = 5\,\mathrm{A}$$

$$i(1.5+) = \frac{v(1.5+)}{R} = \frac{-5}{1} = -5\,\mathrm{A}$$

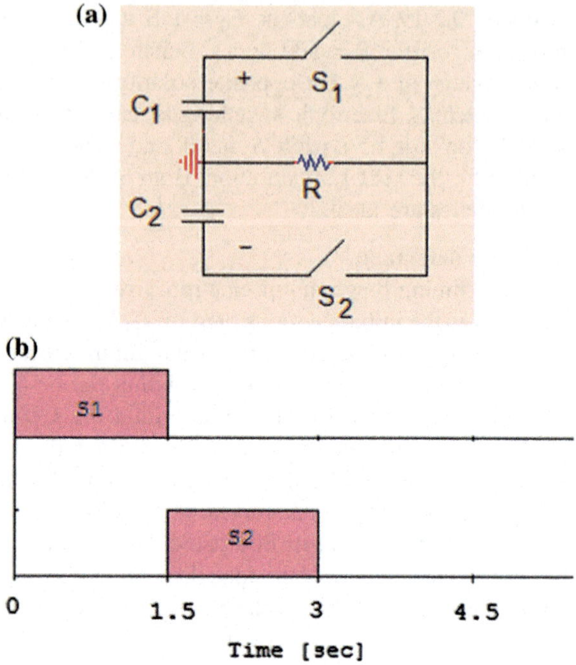

Fig. 3.48 **a** The circuit of Problem 3.2.19, **b** timing of electronic switches in the circuit of Fig. 3.47a

(e) Time constants for the main circuit are

$$\tau_1 = RC_1 = 1 \times 1 = 1\,\mathrm{s}$$
$$\tau_2 = RC_2 = 1 \times 1 = 1\,\mathrm{s}$$

The currents are described by the following equations:

$$i(t) = \frac{v(t)}{R} = \frac{v(0)\mathrm{e}^{-\frac{t}{\tau 1}}}{R} = \frac{5\mathrm{e}^{-t}}{1} = 5\mathrm{e}^{-t} \quad 0 < t < 1.5\,\mathrm{s}$$

(f) $i(t) = \frac{v(t)}{R} = \frac{v(1.5)\mathrm{e}^{-\frac{t-1.5}{\tau 2}}}{R} = \frac{-5\mathrm{e}^{-t+1.5}}{1} = -5\mathrm{e}^{-t+1.5} \quad 1.5 < t < 3\,\mathrm{s}$

(g) The current at time t = 1.5 s is calculated from the following equation:

$$i(t = 1.5-) = 5\mathrm{e}^{-1.5} = (5)(0.2231) = 1.116\,\mathrm{A}$$
$$i(t = 1.5+) = -5\mathrm{e}^{-t+1.5} = (-5)(1) = -5\,\mathrm{A}$$
$$i(t = 3) = -5\mathrm{e}^{-t+1.5} = (-5)(0.2231) = -1.116\,\mathrm{A}$$

(h) The percentage value of current flowing through the resistor R at t = 1.5− s to the initial current is

$$p = \frac{i(1.5-)}{i(0+)} = \frac{1.116}{5} = 22.31\%$$

(i) Energy dissipated within the first 1.5 s,

$$W_{1.5} = \int_0^{1.5} i(t)^2 R \mathrm{d}t = \int_0^{1.5} (5\mathrm{e}^{-t})^2 \mathrm{d}t = W\left(1 - \mathrm{e}^{-2(1.5)}\right)$$
$$= 12.5\left(1 - \mathrm{e}^{-3}\right) = 11.875\,\mathrm{J}$$

This means that 95% of the initial energy stored on the capacitor is dissipated at the end of first 1.5 s following the closure of switch *S*.

(j) The amount of dissipated energy within the second 1.5 s switching period (due to S_2) is the same as that of the dissipated energy within the first 1.5 s switching period, although the direction of current for $t > 1.5$ s is opposite that of the current for $0 < t < 1.5$ s.

(k) Truncated biphasic exponential voltage variation across resistor *R* and switching waveforms are shown in Fig. 3.49.

```
SPICE netlist (SW4.cir)
    *Analysis: Transient
    Truncated exponential biphasic waveform
    C1 1 0 1 IC=5v
    s1 1 2 3 0 sm2
    R1 2 0 1
    *    PULSE(V1  V2  TD  TR  TF  PW  PER)
    v5 3 0 pulse(0  2  0  0  0  1.5  0 )
    v6 6 0 pulse(0  2  1.5 0  0 1.5  0 )
    C2 4 0 1 IC=-5v
```

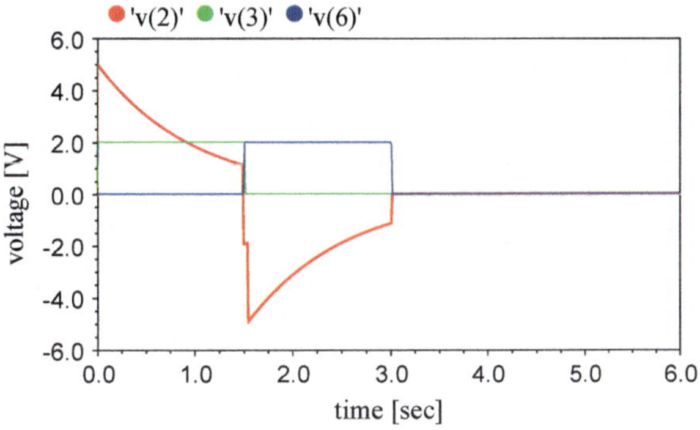

Fig. 3.49 Truncated biphasic exponential voltage variation across resistor *R* and switching waveforms in the circuit of Fig. 3.47a

```
s2 4 2 6 0 sm2
.MODEL sm2 SW(Ron=1u)
*Analysis Parameters
* .tran .01 10 0 .01 uic
```

(l) Negative leading edge of the voltage waveform is improved by introducing a short delay of 50 ms. Truncated biphasic exponential voltage variation across resistor *R* and switching waveforms are shown for this improved performance case in Fig. 3.50.

```
Truncated exponential biphasic waveform
C1 1 0 1 IC=5v
s1 1 2 3 0 sm2
R1 2 0 1
*    PULSE(V1  V2  TD  TR  TF  PW  PER)
v5 3 0 pulse(0  2  0  0  0  1.5  0 )
v6 6 0 pulse(0  2  1.55 0  0 1.5  0 )
C2 4 0 1 IC=-5v
s2 4 2 6 0 sm2
.MODEL sm2 SW(Ron=1u)
*.tran .01 10 0 .01 uic
```

Problem 3.2.20 In the RC circuit shown in Fig. 3.51, the capacitor is initially uncharged.

$R = 1\ \text{k}\Omega$, $C = 1\ \text{mF}$, $V_{dc} = 12\ \text{V}$

(a) What is the equation of current in the circuit for $t \geq 0$?
(b) How long does it take approximately to charge the capacitor?
(c) Determine the voltages over resistor and capacitor for $t \geq 0$.

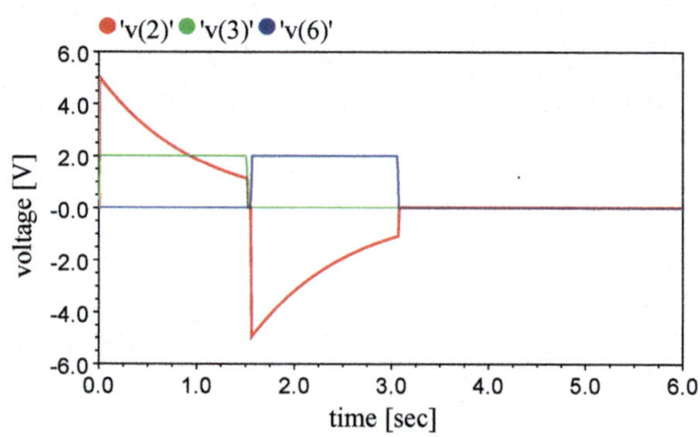

Fig. 3.50 Truncated biphasic exponential voltage variation across resistor *R* and switching waveforms for the improved performance case in the circuit of Fig. 3.47a

Fig. 3.51 The circuit of Problem 3.2.20

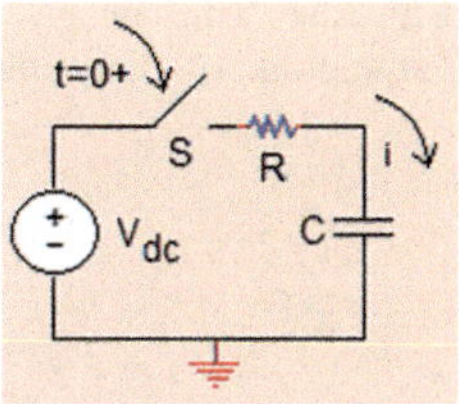

(d) Determine the voltage over capacitor for $t \geq 0$, if the initial voltage over the capacitor is 4 V.
(e) Plot the voltage variation across the capacitor in part (c) using SPICE analysis and print the netlist. (`SW4.cir`)

Solution

(a) $i = \left(\frac{V_{\text{dc}}}{R}\right)\text{e}^{-\frac{t}{\tau}}$

$$\tau = RC = 10^3 10^{-3} = 1\text{ s}$$

$$i = \frac{12}{10^3}\text{e}^{-t} = 12\text{e}^{-t}\text{ mA}$$

(b) Time to charge the capacitor $\approx 5\tau = 5\text{ s}$. Note that this is approximate time to charge the capacitor. Actually this is 99.3% of time to charge the capacitor, but for practical purposes it is a satisfactory assumption.

(c) $v_R = iR = \left(\frac{V_{\text{dc}}}{R}\right)\text{e}^{-\frac{t}{\tau}}.\ R = V_{\text{dc}}\text{e}^{-\frac{t}{\tau}} = 12\text{e}^{-t}\text{ V}$

$$v_C = V_{\text{dc}} - v_R = V_{\text{dc}} - V_{\text{dc}}\text{e}^{-\frac{t}{\tau}} = V_{\text{dc}}\left(1 - \text{e}^{-\frac{t}{\tau}}\right) = 12(1 - \text{e}^{-t})\text{ V}$$

(d) General equation of capacitor charging voltage in an RC circuit is

$$v_C = V_{\text{Cf}} + (V_{\text{Ci}} - V_{\text{Cf}})\text{e}^{-\frac{t}{\tau}}$$

$$V_{\text{Cf}} = \text{final capacitor voltage}, \quad V_{\text{Ci}} = \text{initial capacitor voltage}$$

Substituting the given values of the components yields

$$v_C = 12 + (4 - 12)\text{e}^{-\frac{t}{\tau}} = 12 - 8\text{e}^{-t}\text{ V}$$

(e) Voltage variation across the capacitor when $V_C(0) = 0$ V is shown in Fig. 3.52.

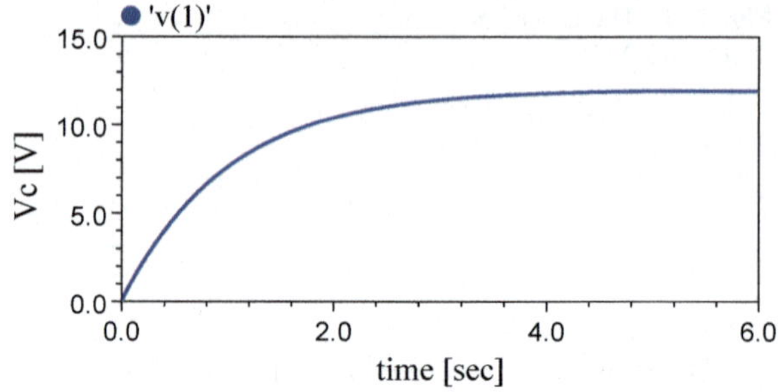

Fig. 3.52 The voltage variation across the capacitor

SPICE Netlist:

```
SW4.cir
 Analysis: Transient
capacitor charging
C 1 0 1m IC=0v
*    pulse(V1 V2 TD  TR TF PW  PER)
v5 3 0 pulse(0  2 0 0 0  5  0 )
Vdc 10 0 12
S 10 11 3 0 sm2
R 11 1 1 k
.MODEL sm2 SW(Ron=1n) *.tran 0.01 6 0 0.01 uic
```

Problem 3.2.21 A device[1] measures the prothrombin time (PT) in blood (plasma), based on the conductivity variation principle as a function of time. Determine PT using the concept and definition of RC time constant from the registered conductance versus time graph, shown in Fig. 3.53. Is this a "normal" value? Comment.

Solution
Prothrombin time, PT, is commonly used parameter to screen bleeding disorders as well as to monitor the accuracy of blood thinning treatment and the coagulation with drugs such as warfarin [9]. It measures the time needed to clot formation after the calcium, phosoholipids and thromboplastin are added to citrated blood plasma. A reagent includes various activators for causing plasma to experience thrombosis. Such activators that include thromboplastin and calcium reagents are available

[1]Such a device is hypothetical, since there is not a physical equivalence between an RC circuit behavior and the variation of conductivity as a function of time, as given here. Most existing recent laboratory instrumentation techniques are either based on optical or rheological (viscosity) measurements.

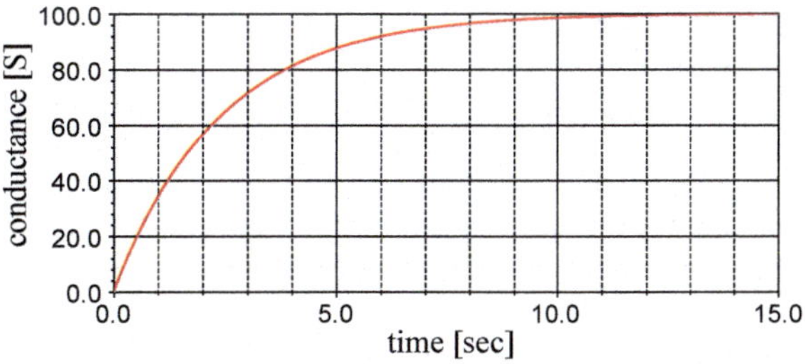

Fig. 3.53 Conductance versus time curve in blood plasma

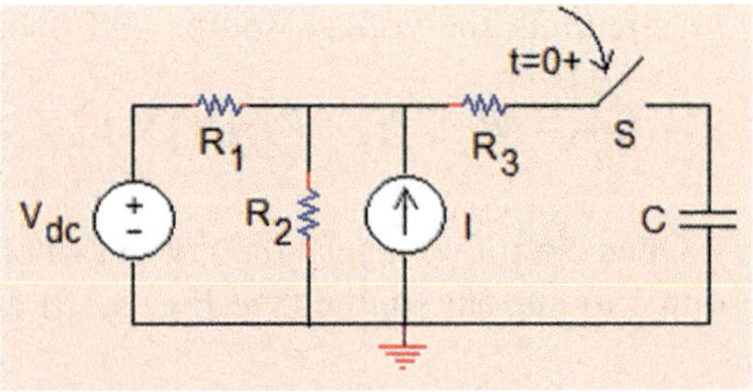

Fig. 3.54 The circuit of Problem 3.2.22

commercially. Plasma (at the standard operating temperature) is then dispensed in one of the tubes containing reagent, and its PT is measured.[2]

Therefore, injection time of the reagent is the starting time of this measurement. For a charging capacitor circuit, time constant of an RC circuit is the time required to charge the capacitor, through the resistor, by 63% of the final value of the voltage across the capacitor. Tracking y-axis at 63% corresponds to time value of 2.4 s. Multiplication of this value by 5 equals 12 s. Therefore, PT time is 12 s. Normal values of PT time are between 11 and 14 s (for adults). However, the range has some variation because of the differences in the thromboplastin produced by various commercial vendors, method of the end point detection implemented by different device manufacturers, as well as storage and incubation conditions of the samples. Therefore, a derived parameter, so-called INR (International Normalized Ratio) is used worldwide to account for this variation.

A test result may be abnormal due to a clotting factor deficiency or existence of a factor inhibitor. Such a condition is differentiated by a mixing study. Here, patient's plasma is mixed with a pooled normal plasma at 1:1 ratio and then the test is run on this sample again. Normalization of the result denotes a factor deficiency. Else, patient has a factor inhibitor.

[2]In systems for whole blood testing, adding calcium ions in the commercial thromboplastin mixture is immaterial, since the blood calcium is sufficient for the clotting action, in vitro. However, if citrated venous blood is tested, calcium chloride is added to the thromboplastin mixture.

Problem 3.2.22 In the RC circuit shown in Fig. 3.54, the capacitor is initially uncharged.

$R_1 = R_2 = 2.R_3 = 1\ \text{k}\Omega,\ C = 1\text{mF},\ V_{\text{dc}} = 12\ \text{V},\ I = 8\ \text{mA}$

(a) What is the equation of current through the capacitor for $t \geq 0$?
(b) How long does it take approximately to charge the capacitor?
(c) Determine the voltage over capacitor for $t \geq 0$.

Solution

(a) Thévenin's equivalent resistance of the circuit to the left of switch *S* (after short circuiting the voltage source and open circuiting the current source),

$$R_{\text{Th}} = (R_1 \parallel R_2) + R_3 = (1\,\text{k}\Omega \parallel 1\,\text{k}\Omega) + 0.5\,\text{k}\Omega = 0.5\,\text{k}\Omega + 0.5\,\text{k}\Omega = 1\,\text{k}\Omega$$

Open circuit voltage at the left of switch *S* is calculated by converting the voltage source to current source (see Fig. 3.55) and noting that

$$R = (R_1 \parallel R_2) = (1\,\text{k}\Omega \parallel 1\,\text{k}\Omega) = 0.5\,\text{k}\Omega$$

$$V_{\text{Th}} = V_X = V_{OC}$$

By KCL at node for V_X,

$$\frac{V_{\text{dc}}}{R_1} + I - \frac{V_X}{R} = 0$$

$$\frac{V_X}{R} = \frac{V_{\text{dc}}}{R_1} + I$$

$$V_X = R\left(\frac{V_{\text{dc}}}{R_1} + I\right) = (R_1 \parallel R_2)\left(\frac{V_{\text{dc}}}{R_1} + I\right)$$

$$= (0.5\,\text{k}\Omega)\left(\frac{12}{10^3} + 8.10^{-3}\right) = (0.5\,\text{k}\Omega)(20\,\text{mA})$$

$$V_X = V_{\text{Th}} = 10\,\text{V}$$

Fig. 3.55 Circuit for the calculation of open circuit voltage at the left of switch S

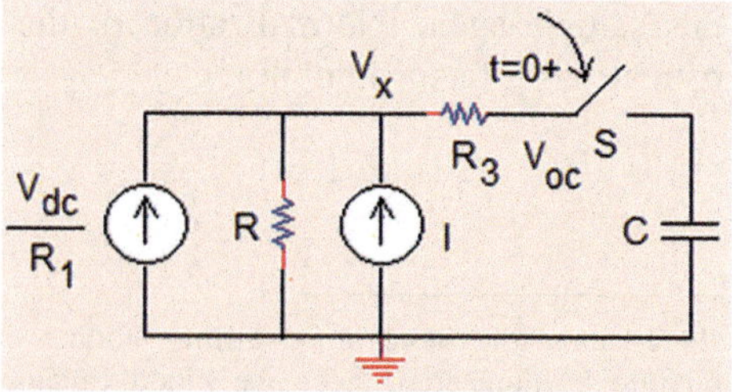

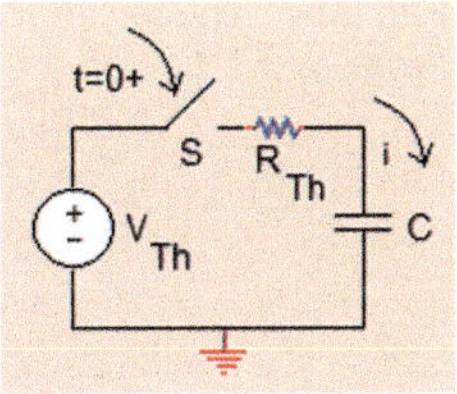

Fig. 3.56 Thévenin equivalent for the circuit of Problem 3.2.22

Note that although Thévenin resistance is placed to the right of the switch in the equivalent circuit of Fig. 3.56, actual physical placement of this resistor is at the left side of the switch. Nevertheless, for the calculations, it does not matter which side of the switch Thévenin's equivalent resistor is placed.

$$i = \left(\frac{V_{\mathrm{Th}}}{R_{\mathrm{Th}}}\right)\mathrm{e}^{-\frac{t}{\tau}}$$

$$\tau = R_{\mathrm{Th}}C = 10^3 10^{-3} = 1\,\mathrm{s}$$

$$i = \frac{10}{10^3}\mathrm{e}^{-t} = 10\mathrm{e}^{-t}\,\mathrm{mA}$$

(b) Time to charge the capacitor $\approx 5\tau = (5)(1) = 5\,\mathrm{s}$
(c) The voltage over capacitor for $t \geq 0$,

$$v_{R_{\mathrm{Th}}} = iR_{\mathrm{Th}} = \left(\frac{V_{\mathrm{Th}}}{R_{\mathrm{Th}}}\right)\mathrm{e}^{-\frac{t}{\tau}}.R_{\mathrm{Th}} = V_{\mathrm{Th}}\mathrm{e}^{-\frac{t}{\tau}} = 10\mathrm{e}^{-t}\,\mathrm{V}$$

$$v_C = V_{\mathrm{Th}} - v_{R_{\mathrm{Th}}} = V_{\mathrm{Th}} - V_{\mathrm{Th}}\mathrm{e}^{-\frac{t}{\tau}}$$

$$v_C = V_{\mathrm{Th}}\left(1 - \mathrm{e}^{-\frac{t}{\tau}}\right)$$

$$v_C = 10(1 - \mathrm{e}^{-t})\,\mathrm{V}$$

Problem 3.2.23 In the circuit shown in Fig. 3.57, the switch was in position 1 for a long time. After the switch is brought into position 2 at $t = 0$ s, at what time the current through R_2 reaches 1 mA? ($U = 10$ V, C = 1 mF, $R_2 = 10\ R_2 = 10$ kΩ)

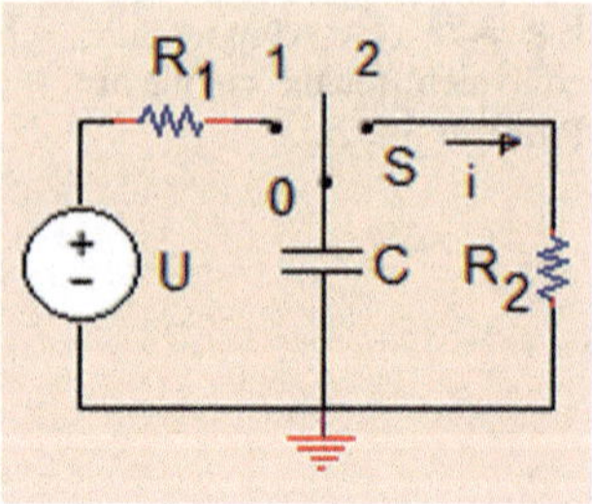

Fig. 3.57 The circuit of Problem 3.2.23

Solution

$$v_c(0^-) = v_c(0^+) = U = 10\,V$$

$$t \geq 0:$$

$$v_c(t) = V_0 e^{-\frac{t}{\tau}} = 10e^{\frac{-t}{R_2 C}} = 10e^{-\frac{t}{10^3 \times 10^{-3}}} = 10e^{-t}$$

$$i_{R_2}(t) = \frac{v_c(t)}{R_2} = \frac{U}{R_2}e^{-t} = \frac{10}{1}e^{-t}\,(\text{mA})$$

$$10^{-3} = 10^{-2}e^{-t}\,(\text{A})$$

$$\text{e}^{-t} = \frac{1}{10}$$

$$\ln(0.1) = -t \quad \rightarrow \quad -2.3 = -t \quad \rightarrow \quad t = 2.3\,\text{s}.$$

Problem 3.2.24 In the RC circuit shown in Fig. 3.58, the capacitor is initially uncharged. The switch S is in position (b) for a long time. At time $t = 0$ it is thrown to position

(a) What is the equation of current in the circuit for $t \geq 0$?
(b) How long does it take (approximately) to charge the capacitor?
(c) What is the value of current in the capacitor at $t = 60$ s?
(d) Determine the voltage over capacitor for $t \geq 0$.
(e) What is the value of voltage on the capacitor at $t = 60$ s?

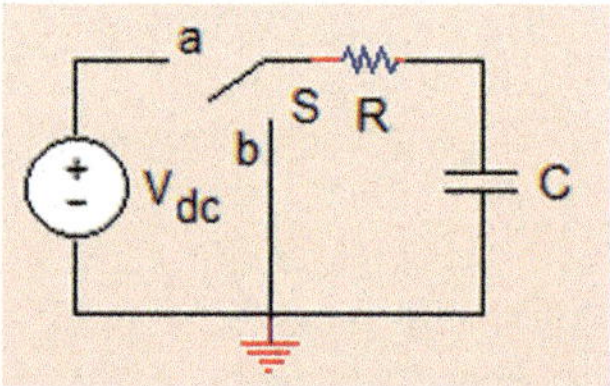

Fig. 3.58 The circuit of Problem 3.2.24

(f) The switch S is thrown from position (a) to (b) at t = 70 s. What is the equation of voltage across the capacitor?
(h) How long does it take (approximately) to discharge the capacitor?

R = 1 kΩ, C = 1 mF, V_{dc} = 12 V.

Solution

(a) $i = \left(\frac{V_{dc}}{R}\right)e^{-\frac{t}{\tau}} \quad \rightarrow \quad \tau = RC = 10^3 10^{-3} = 1\,\mathrm{s}$

$$i = \frac{12}{10^3}e^{-t} = 12e^{-t}\,\mathrm{mA}$$

(b) Time to charge the capacitor $\approx 5\tau = (5)(1) = 5\,\mathrm{s}$
(c) The current through capacitor, at t = 60 s is i = 0 A
(d) $v_C = V_{dc}\left(1 - e^{-\frac{t}{\tau}}\right) = 12(1 - e^{-t})$
(e) The voltage on the capacitor at t = 60 s is 12 V.
(f) The capacitor discharges through resistor R for $t \geq$ 70 s.

$$v_C = V_{dc}e^{-\frac{t-70}{\tau}} = 12e^{-(t-70)}\,\mathrm{V}$$

(g) Time to discharge the capacitor $\approx 5\tau = (5)(1) = 5\,\mathrm{s}$

Problem 3.2.25 In the circuit shown in Fig. 3.59, the capacitor is initially uncharged. Both of the switches are in open position for a long time. $R_1 = R_2$ = 1 kΩ, C = 1 mF, V_{dc} = 12 V. At time t = 0, S_1 is closed, while S_2 remains open.

(a) What is the equation of current in the circuit for $t \geq 0$?
(b) How long does it take (approximately) to charge the capacitor?
(c) What is the value of current in the capacitor at t = 6 s?
(d) Determine the voltage over capacitor for $t \geq 0$.
(e) What is the value of voltage on the capacitor at t = 6 s?
(f) The switch S_1 is opened and simultaneously switch S_2 is closed at t = 6 s. What is the equation of voltage across the capacitor?
(g) How long does it take (approximately) to discharge the capacitor?
(h) Plot the voltage variation across the capacitor (to check the results found in parts d, e, f, and g) using SPICE analysis and print the netlist. (`SW5.cir`)

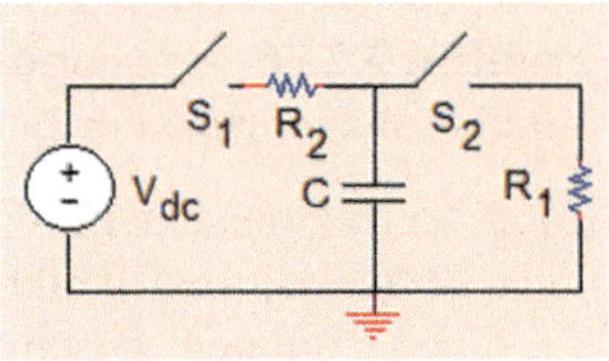

Fig. 3.59 The circuit of problem 3.2.25

Solution

(a) $i = \left(\frac{V_{dc}}{R_2}\right)e^{-\frac{t}{\tau}}$

$$\tau = R_2C = 10^3 10^{-3} = 1\text{ s}$$

$$i = \frac{12}{10^3}e^{-t} = 12e^{-t}\text{ mA}$$

(b) Time to charge the capacitor $\approx 5\tau = (5)(1) = 5\text{ s}$
(c) The current through capacitor, at $t = 6$ s is approximately 0 A
(d) $v_C = V_{dc}\left(1 - e^{-\frac{t}{\tau}}\right) = 12(1 - e^{-t})$
(e) The voltage on the capacitor at $t = 6$ s is 12 V.
(f) The capacitor discharges through resistor R_1 for $t \geq 6$ s.

$$v_C = V_{dc}e^{-\frac{t-6}{\tau}} = 12e^{-(t-6)}\text{ V}$$

(g) Time to discharge the capacitor $\approx 5\tau = (5)(1) = 5\text{ s}$
(h) Voltage variation across the capacitor is shown in Fig. 3.60, along with the switching time diagram.

SPICE Netlist (SW5.cir);

```
capacitor charging and discharging circuit
C 1 0 1m IC=0v
*    pulse(V1 V2 TD  TR TF PW  PER)
v5 3 0 pulse(0  2 0 0 0  6  0 )
*charging section
Vdc 10 0 12
S1 10 11 3 0 sm2
R2 11 1 1 k
*discharging section
S2 1 2 0 3 sm2
R1 2 0 1 k
*.MODEL sm2 SW(Ron=1u)
*Analysis Parameters
*.tran .01 12 0 .01 uic
```

Problem 3.2.26 A major design condition for a defibrillator is that it would charge a high-voltage capacitor quickly and efficiently.

(a) A battery rated at 12 V, 2.3 Ah feeds a charging unit to charge the capacitor of a 360 J adjusted defibrillator device. What is the shortest capacitor charging time? Comment on this result.

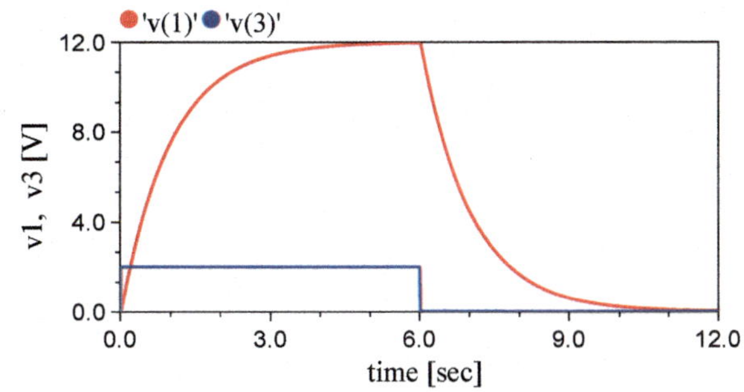

Fig. 3.60 Voltage across the capacitor and the switching time for the circuit of Fig. 3.59

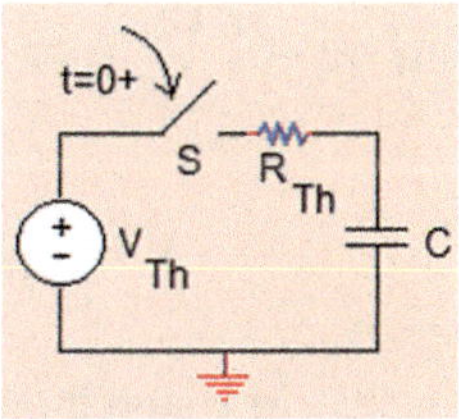

Fig. 3.61 The circuit of Problem 3.2.26b

(b) Determine Thévenin's equivalent circuit of the charging unit for C = 100 uF (see, Fig. 3.61). What is the maximum charge stored on the capacitor? Plot the charging (voltage) waveform using SPICE analysis and print SPICE netlist. (SW6.cir)
(c) Study and give a brief description about construction of high-voltage capacitor charging section for a defibrillator. Describe earlier design principles for these sections and state their disadvantages. Describe newer design basics. Give a short list of reference literature.

Solution

(a) For the battery, in one hour operation available energy capacity is $12 \times 2.3 = 27.6\,\text{Wh}$
Assuming that there is a lossless front-end circuitry and power limit for the battery is less than half of this value, P = 13 W,

$$t = \frac{360}{13} = 27.7\ \text{s}$$

Note that this is a relatively long-time duration, and it becomes a serious issue for emergency patient treatment due to limitations for the number of multiple shocks that can be administered to a patient in a short time. Most modern external defibrillator units have a charging time of less than 10 s at 360 J setting, or 5 s or less at 200 J setting.

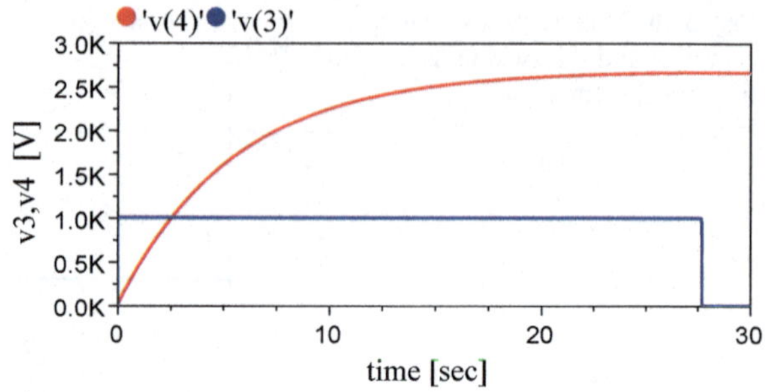

Fig. 3.62 Charging voltage waveform and switching time

(b) $W = \dfrac{CV^2}{2} \quad \rightarrow \quad 360\,\text{J} = \dfrac{100 \times 10^{-6} \times V^2}{2}$

$$V = V_{\text{Th}} = \sqrt{\frac{2 \times 360}{100 \times 10^{-6}}} = \sqrt{\frac{720}{10^{-4}}} = 10^2\sqrt{720} = 2683.3\,\text{V}$$

Charge stored on the capacitor, $Q = CV = 100 \times 10^{-6} \times 2683.3 = 0.26833\,\text{C}$

$$t = \frac{360}{13} = 27.7 = 5\tau = 5RC = R(500 \times 10^{-6})$$

$$R = R_{\text{Th}} = \frac{27.7}{500 \times 10^{-6}} = 0.2 \times 27.7 \times 10^4 = 55.4\,\text{k}\Omega$$

Charging voltage waveform is shown in Fig. 3.62.

SPICE Netlist (SW6.cir) is given below;

```
*Transient Analysis
Charging capacitor equivalent circuit
C 4 0 100u IC=0v
R 2 4 55.4k
s1 1 2 3 0 sm2
Vdc 1 0 2683
*PULSE(V1 V2 TD TR TF PW PER)
vx 3 0 pulse(0 1k 0 0 0 27.7 0)
.MODEL sm2 SW(Ron=1u)
*   Transient Analysis Parameters
    *Stepsize ........... : .01
    *Final time ......... : 30
    *Display start time .. : 0
    *Maximum step size ... : .01
    *Use Initial Conditions: On
```

(c) Some earlier designs of defibrillator charging sections use auto-transformers to manually adjust the energy level of discharge. Another step-up transformer along with a high-voltage rectifier and a high voltage change over switch which are connected in series to step-up transformer's secondary winding follows the auto-transformer, so that desired dc voltage is set to charge the capacitor.

When the shock is to be delivered to the patient, a switch mounted on the handle of the electrode is used to change the position of the high voltage change over switch. Then, the capacitor is discharged across the torso between two electrodes. After shock delivery, the capacitor is again charged for succeeding defibrillation procedure.

Disadvantages of such a design are that: (1) These defibrillators have to be permanently connected to ac power plug precluding the equipment mobility. (2) They consume valuable time for preparation to provide the therapy which limits the number of multiple shocks that can be administered to a patient in a short time available. (3) Equipment becomes heavy due to the use of bulky auto-transformers, step-up transformers and high-voltage vacuum switches.

Some newer designs are based on charging a high-voltage capacitor through the application of a current, the magnitude of which has a fixed frequency waveform [10]. During a charging sequence in which the current is applied repeatedly to the capacitor, the duty cycle of the fixed frequency current waveform is controlled dynamically based on the capacitor voltage. Specifically, the rate at which the energy is transferred to the capacitor is modified according to the efficiency at which the energy can be delivered to the capacitor. This increases the speed at which the high-voltage capacitor is charged.

Recent defibrillators employ high frequency solid state converter technology and switching power supplies (e.g., flyback topology) [11–14] in generating the voltage for charging a high-voltage capacitor.

Problem 3.2.27

(a) Concentration of potassium ions inside of a mammalian excitable cell is 14 mM, while its concentration outside the cell is 1 mM. What is the equilibrium potential between two sides of this cell's membrane at body temperature of 37 °C?
(b) Concentration of sodium ions inside of a mammalian excitable cell is 1 mM, while its concentration outside the cell is 14 mM. What is the equilibrium potential between two sides of this cell's membrane at body temperature of 37 °C?
(c) Concentration of calcium ions inside of a mammalian excitable cell is 1 mM, while its concentration outside the cell is 14 mM. What is the equilibrium potential between two sides of this cell's membrane at body temperature of 37 °C?
(d) Plot the Nernst voltage (in mV) for sodium ions in part (b), as a function of temperature between 35 and 42 °C. (Nernst1.m)

(e) Plot Nernst voltage change (in mV) for sodium ions in part (b) at 37 °C, as a function of extracellular ionic concentration between 1 and 100 mM, while intracellular ionic concentration is 1 mM. Make a survey of literature for common ionic concentrations of an excitable human cell (Nernst2.m).
(Note: Solutes in the body are measured in milligrams and millimoles. When converting to milliequivalents: 1 meq = 10^{-3} eq, for monovalent ions, 1 meq = 1 mmol, for divalent ions, 1 meq = 0.5 mmol, for trivalent ions, 1 meq = 0.333 mmol)

Solution

(a) Using Nernst equation;

$$V_{eq} = \frac{RT}{zF} \ln\left(\frac{[X]_{out}}{[X]_{in}}\right)$$

Faraday's constant and universal gas constants are given as

F 96485 C mol^{-1} (Coulombs per mole)
R 8.314 J K^{-1} mol^{-1} (Joules per Kelvin per mole)
z +1 (valance of Potassium)
F 96475 C mol^{-1},
R 8.314 J K^{-1}

One may use the ion symbol as a subscript for the equilibrium potential of that ion (e.g., V_{Na}, V_K, V_{Ca}, etc.).

$V_{Eq.}$ will also be the membrane potential (V_m) if there is only one ionic species present in the system.

Substituting given potassium data into Nernst equation,

$$V_K = \frac{(8.314)(273+37)}{(1)(96485)} \ln\left(\frac{1}{14}\right)$$
$$= 0.0267 \times \ln(0.0714) = -0.070495\,\text{V} = -70.5\,\text{mV}$$

(b) Substituting given data for sodium ions into Nernst equation,

$$V_{Na} = \frac{(8.314)(273+37)}{(1)(96485)} \ln\left(\frac{14}{1}\right)$$
$$= 0.0267 \times \ln(14) = +0.070495\,\text{V} = +70.5\,\text{mV}$$

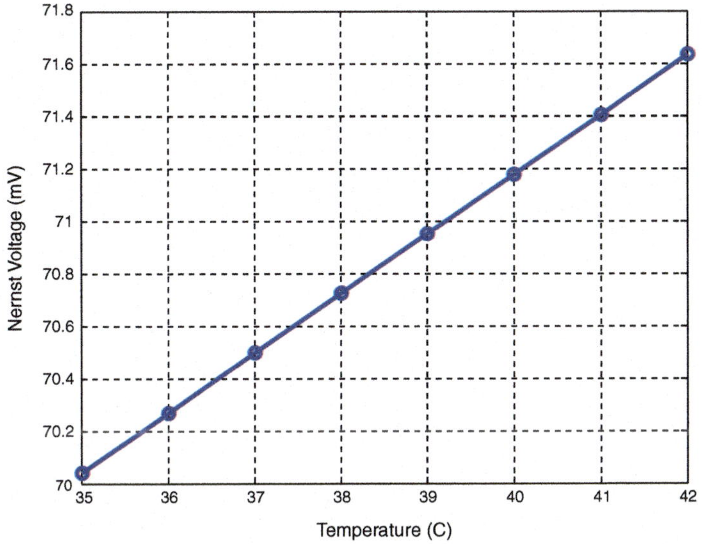

Fig. 3.63 The graph of Nernst voltage as a function of temperature

(c) Substituting given data for calcium ions into Nernst equation,

$$V_{\mathrm{Ca}} = \frac{(8.314)(273+37)}{(2)(96485)} \ln\left(\frac{14}{1}\right)$$
$$= 0.01335 \times \ln(14) = +0.03524769\,\mathrm{V} = +35.25\,\mathrm{mV}$$

(d) Nernst voltage plot (in mV) for sodium ions in part (a), as a function of temperature between 35 and 42 °C is shown in Fig. 3.63.

(e) Figure 3.64 shows Nernst voltage change (in mV) for sodium ions in part (b) at 37 °C, as a function of extracellular ionic concentration between 1 and 100 mM while intracellular ionic concentration is 1 mM. Sodium is the most common extracellular cation in physiological systems.

A typical extracellular Na^{+1} concentration is around 135–147 mM, while an intracellular concentration is in 10–15 mM range. This means that extracellular to intracellular sodium ion concentration ratio is around 10.

As for the potassium cations, typical extracellular K^{+1} concentration is around 3.5–5.5 mM, while an intracellular concentration is about 140 mM. This means that intracellular to extracellular sodium ion concentration ratio is around 140/4 = 35.

Table 3.2 lists these concentration levels for common ions in human cellular environment. (Mean values may differ according to gender, age of the subject, test methodology and conditions of assay.)

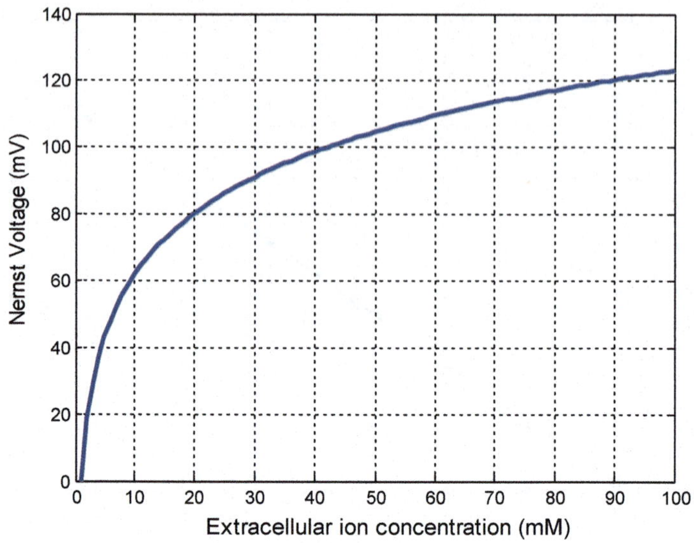

Fig. 3.64 The graph of Nernst voltage as a function of extracellular ion concentration

Table 3.2 Intracellular and extracellular ion concentrations

Ion	Intracellular conc. (mM) [15]	Extracellular conc. (mM) [16]
Sodium	10–15	135–147
Potassium	139	3.5–5.5
Calcium	<0.0002	4.5–5.5
Chloride	4	98–106
Magnesium	0.8	1.5–3
Bicarbonate	12	18–30

Problem 3.2.28 A biological neuron behavior can be modeled as a "leaky integrator" of its input current excitation, $i(t)$, which is called leaky integrate-and-fire (LIF) neuron model. Here, $v(t)$ represents the membrane potential at time t, R, and C are the membrane resistance and capacitance, respectively (Fig. 3.65).

(a) Write the governing equation of this equivalent circuit.
(b) Assuming constant current excitation $i(t) = I$ and $v(0) = V_o$ initial voltage across the cell membrane (over the capacitor), find the expression for the voltage change across the membrane of the neuron.
(c) Let constant current excitation be $I = 10$ mA, $C = 1$ mF, $R = 10\ \Omega$, and initial voltage across the cell membrane $v(0) = -70$ mV,

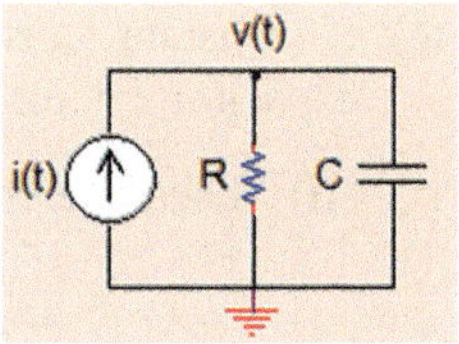

Fig. 3.65 LIF neuron model

(i) Find the time constant of the circuit.
(ii) Find the expression for the voltage change across the membrane of the neuron. Check this result using SPICE analysis, and print SPICE netlist. (LIF1.cir)

(d) The neuron spiking behavior cannot be explicitly modeled by this circuit, alone, in the LIF model algorithm. As soon as the membrane potential $v(t)$ reaches a threshold voltage (V_{Th}), a Dirac delta (impulse) is added, then the membrane potential is reset to a lower voltage value (reset voltage = V_{R}) and the integration process starts again with the reset voltage being the initial value of this recharging period.
Assume that the threshold voltage, V_{Th}, is −30 mV; Reset voltage = V_{R} = −70 mV. Determine the period of this voltage variation across the cell membrane. Plot this voltage across the cell membrane versus time, without showing added impulses.

(e) What will be the period of spiking if the excitation current is doubled, all else remaining the same? Plot a generalized "period versus input excitation current" curve. (LIF_IvsT.m)

Solution

(a) The governing equation of this equivalent circuit is obtained by applying KCL at the node,

$$i - C\frac{\mathrm{d}v}{\mathrm{d}t} - \frac{v}{R} = 0 \quad \rightarrow \quad i = C\frac{\mathrm{d}v}{\mathrm{d}t} + \frac{v}{R} \rightarrow \quad \frac{\mathrm{d}v}{\mathrm{d}t} + \frac{v}{RC} = \frac{i}{C}$$

(b) Assuming constant current excitation and no initial voltage over the capacitor, the expression for the voltage change across the membrane of the neuron is

$$\frac{\mathrm{d}v}{\mathrm{d}t} + \frac{v}{RC} = \frac{I}{C}$$

The solution of this differential equation is

$$v(t) = v(\infty) + [v(0) - v(\infty)]\mathrm{e}^{-\frac{t}{\tau}}$$

$$v(\infty) = IR, \quad v(0) = V_0, \quad \tau = RC$$

(c) If constant current excitation is $I = 10$ mA, and initial voltage over the capacitor (membrane) $v(0) = -70$ mV, with $C = 1$ mF, $R = 10\ \Omega$,

(i) The time constant of the circuit is

$$\tau = RC = 10 \times 10^{-3} = 10\,\text{ms}$$

(ii) The expression for the voltage change (across the membrane of the neuron) is

$$v(t) = v(\infty) + [v(0) - v(\infty)]e^{-\frac{t}{\tau}}$$
$$v(t) = 10 \times 10^{-3} \times 10 + [-0.07 - 10 \times 10^{-3} \times 10]e^{-\frac{t}{10}\times 1000}$$
$$v(t) = 100 + [-70 - 100]e^{-\frac{t}{10}\times 1000}\,\text{mV} = 100 - 170e^{-100t}\,\text{mV}$$

Figure 3.66 displays this voltage change as a function of time. SPICE netlist is also given.

```
LIF (Leakage Integrate and Fire) spiking neuron model
C1 1 0 1m IC=-70m
R1 1 0 10
Iin 0 1 10m
*tran 0.1m 100m 0 1m
```

(d) Since the membrane potential is reset to a voltage value of $V_R = -70$ mV and the integration process starts again with the reset voltage being the initial value of this recharging period when the membrane potential $v(t)$ reaches a threshold voltage of $V_{Th} = -26$ mV, the period of this voltage variation across the cell membrane is determined by equating the voltage function to its given threshold voltage and solving the equation for time $t = T$,

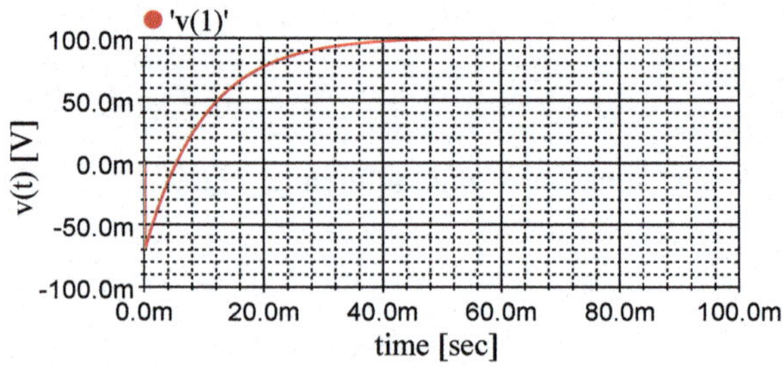

Fig. 3.66 The voltage change across the membrane of the neuron

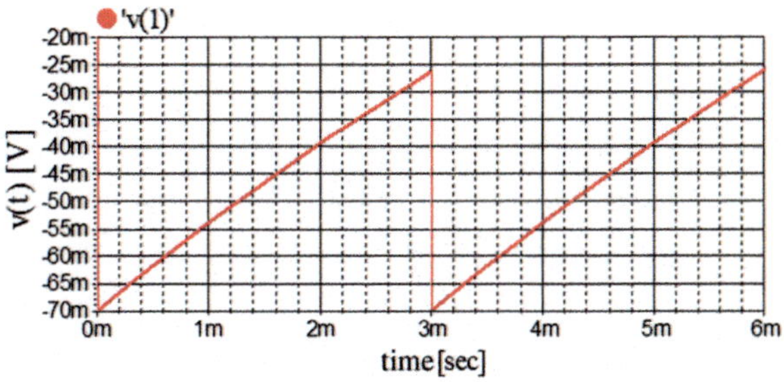

Fig. 3.67 Voltage variation across the cell membrane as a function of time

$$v(T) = 100 - 170e^{-100T} = -26\,\text{mV} \quad \rightarrow \quad 170e^{-100T} = 126$$

$$e^{-100T} = \frac{126}{170} = 0.741 \quad \rightarrow \quad -100T = \ln(0.741) = -0.2995$$

$$T = 3\,\text{ms}$$

Voltage variation across the cell membrane as a function of time is displayed in the Fig. 3.67.

(e) Rewriting the change of voltage as a function of time expression,

$$v(t) = v(\infty) + [v(0) - v(\infty)]e^{-\frac{t}{\tau}}$$

$$v(t) = IR + [V_0 - IR]e^{-\frac{t}{\tau}}$$

$$V_{\text{Th}} = IR + [V_0 - IR]e^{-\frac{T}{\tau}} \quad \rightarrow \quad V_{\text{Th}} - IR = [V_0 - IR]e^{-\frac{T}{\tau}}$$

$$\frac{V_{\text{Th}} - IR}{V_0 - IR} = e^{-\frac{T}{\tau}} \quad \rightarrow \quad \ln\left(\frac{V_{\text{Th}} - IR}{V_0 - IR}\right) = -\frac{T}{\tau}$$

$$T = -\tau.\ln\left(\frac{V_{\text{Th}} - IR}{V_0 - IR}\right) \tag{3.13}$$

Letting $I = 20$ mA in this equation,

$$T = -10 \times \ln\left(\frac{-26 - 20 \times 10}{-70 - 20 \times 10}\right) = -10 \times \ln\left(\frac{-226}{-270}\right)$$
$$= -10 \times \ln(0.837) = 1.78\,\text{ms}$$

This result shows that increasing the excitation current decreases the period of spiking, in other words, increasing the excitation current increases the frequency of spikes. In this case frequency of spikes is increased from 3.33 to 5.62 kHz.

A plot for Eq. (3.13) demonstrating this relationship in graphical form is given in Fig. 3.68.

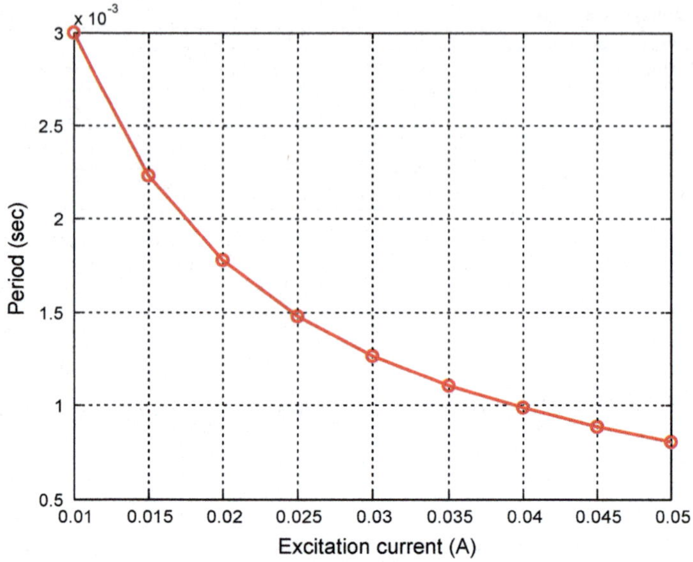

Fig. 3.68 Increased excitation current decreases the period of spikes

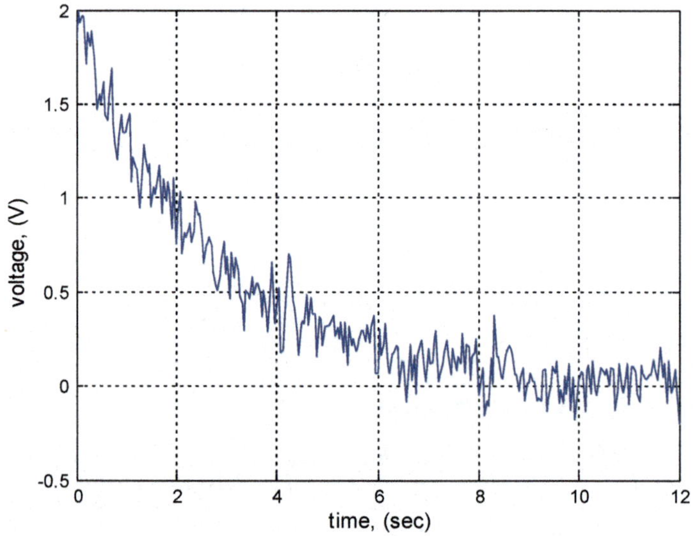

Fig. 3.69 Noisy electrical discharge signal of a biological cell

Problem 3.2.29 Amplified noisy electrical discharge signal of a biological cell recorded during an experiment is displayed in Fig. 3.69. (Signal is time scaled by a factor of 1000.)

(a) Determine the equation of discharging action of the cell as a function of time. (fit_exp1.m).
(b) Find the time constant of discharging signal.

Solution

(a) Assuming that observed biological action is a first-order process and applying curve fitting to recorded experimental data (using MATLAB) yields the following values;
General model for fitting; $f(x) = Ae^{bx}$
Coefficients (with 95% confidence): $A = 1.945$, $b = -0.3808$
Equation of fitted data is $v(t) = 1.945e^{-0.381t}$ V. It will be noted that the approximation $v(t) = 2e^{-0.4t}$ will remain within the confidence bounds. The time unit is in milliseconds due to given time scaling factor of 1000.
(b) The time constant of discharging signal is

$$\tau = \frac{1}{\sigma} = \frac{1}{0.4} = 2.5\,\text{ms}$$

Result of curve fitting as well as recorded noisy bioelectrical signal are displayed simultaneously in Fig. 3.70.

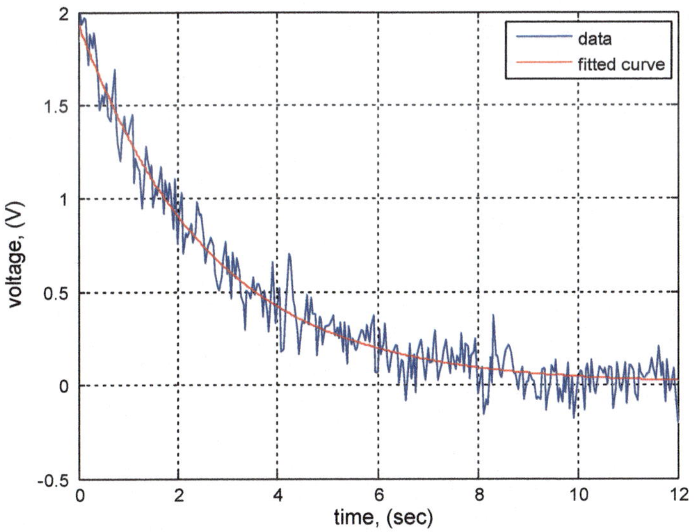

Fig. 3.70 The result of curve fitting and noisy bioelectrical signal

Following is the MATLAB script for the m file (fit_exp1.m);

```
x = (0:0.04:12)';
%Single exponential
y1 = 2*exp(-0.4*x) + 0.1*randn(size(x)); %Data generator
plot(x,y1);grid on;xlabel('time, (sec)');ylabel('voltage, (V)');figure;
f = fit(x,y1,'exp1')
plot(f,x,y1,'b');grid on; xlabel('time, (sec)'); ylabel('voltage, (V)')
```

Problem 3.2.30 A capacitor (with no initial voltage across its terminals) is charged through a resistor of $R = 2\Omega$ in an electrically noisy environment Fig. 3.71. The charging voltage is sampled and recorded in a storage oscilloscope, as displayed in Fig. 3.72. Determine dc voltage of the battery and the value of capacitor. (fit_exp2.m).

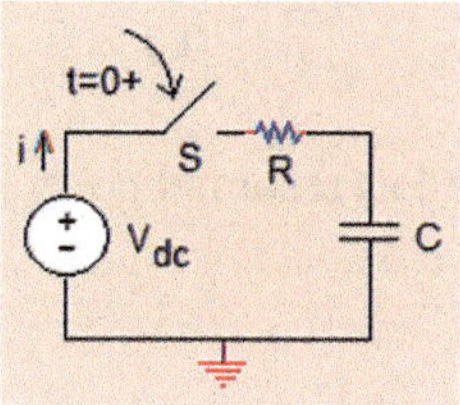

Fig. 3.71 The circuit of Problem 3.2.30

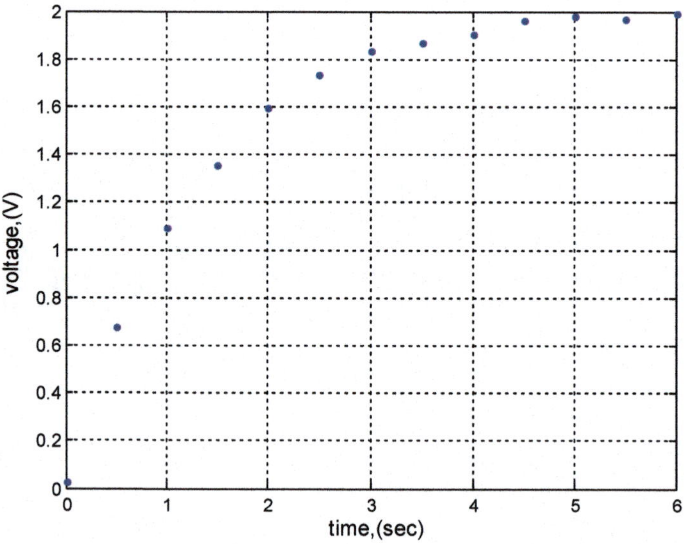

Fig. 3.72 Display of sampled data

Solution

Applying curve fitting to recorded experimental data (using MATLAB curve fitting toolbox) yields the following values;

General model: $v(t) = Ae^{Bt} + Ce^{Dt}$

Coefficients (with 95% confidence bounds):

$$A = 2.04(1.869, 2.212), \quad B = -0.002534(-0.01722, 0.01215)$$
$$C = -2.007(-2.171, -1.842), \quad D = -0.7551(-0.8432, -0.6669)$$

(SSE: 0.002671, R-square: 0.9994, RMSE: 0.01723)

Substituting these values in the general model gives

$$v(t) = 2.04e^{-0.002534t} + (-2.007)e^{-0.7551t}$$

Approximating the function coefficients within confidence bounds,

$$v(t) = 2e^{-0.00t} - 2e^{-0.75t} = 2(1 - e^{-0.75t})$$

The battery voltage is 2 V.

Since resistor value is 2 Ω, the capacitor value is calculated using time constant relationship as,

$$\tau = RC = \frac{1}{\sigma} \quad \rightarrow \quad 2 \times C = \frac{1}{0.75}$$

$$C = \frac{1}{2 \times 0.75} = \frac{2}{3} = 0.6667 \text{ F}$$

Result of curve fitting as well as recorded noisy capacitor voltage signal samples are displayed concurrently in the Fig. 3.73.

MATLAB script for the m file (fit_exp2.m);

```
% Curve fitting of two term exponential model (sum of two exponentials)
%Data generator
x = (0:0.5:6)';
%y2 = 2*exp(-.0*x) -2*exp(-1.0*x)+ 0.02*randn(size(x));
y2 = 2*(1-exp(-0.8*x))+ 0.02*randn(size(x));
plot(x,y2,'.');grid
on;xlabel('time, (sec)');ylabel('voltage, (V)');figure;
%
f = fit(x,y2,'exp2')
plot(f,x,y2);grid on; xlabel('time, (sec)'); ylabel('voltage, (V)')
```

Problem 3.2.31 A simplified Hodgin–Huxley neuron equivalent circuit for a single type of ion channel is shown in Fig. 3.74.

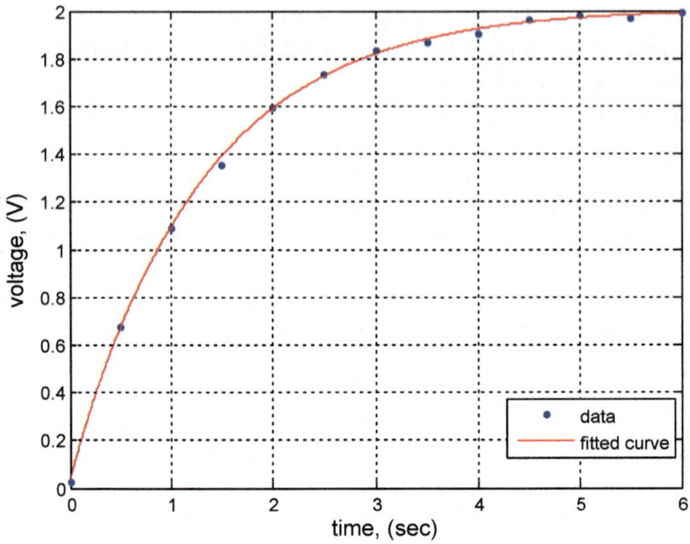

Fig. 3.73 Display for the result of curve fitting and noisy capacitor voltage signal samples

Fig. 3.74 Simplified Hodgin–Huxley neuron equivalent circuit

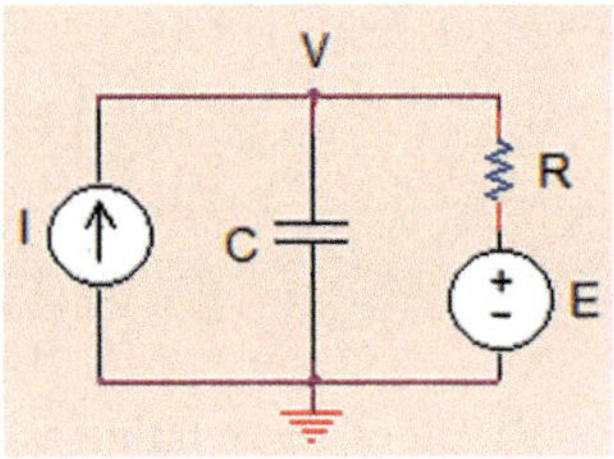

(a) What is the differential equation describing the behavior of the circuit (No initial conditions)?
(b) What will be the steady-state voltage at node 1, if $I = 0$ A?
(c) What is the time constant of the circuit?
(d) For $i(t) = 0.1e^{(-t/\tau)}$ (mA), $\tau = 100$ ms, $C = 10$ nF, $R = 1$ kΩ, $E = -72$ mV. Plot $v(t)$.

Solution

(a) By KCL,

$$I - C\frac{(\mathrm{d}v)}{\mathrm{d}t} - \frac{V-E}{R} = 0 \quad \rightarrow \quad C\frac{\mathrm{d}v}{\mathrm{d}t} + \frac{V-E}{R} = I$$

(b) $V = E$
(c) $\tau = RC$
(d) For $i(t) = (10^{-4})e^{-\frac{t}{\tau}}$, with τ = 100 ms, C = 10 nF, R = 1 kΩ, E = −72 mV, the plot of $v(t)$ is given in Fig. 3.75.

Problem 3.2.32 When there is no excitation current, the conductance values for sodium, potassium, and leakage (chloride) ion channels at a specific instant of a cell membrane are 100, 10, and 0.3 mS/cm^2, respectively. For this cell, Nernst potentials for sodium, potassium, and leakage (chloride) ions are 55, −72, and −49 mV, respectively.

If a steady state is reached under these circumstances, and considering a Hodgkin–Huxley model for this excitable cell (Fig. 3.76), determine the membrane potential and ionic currents at each channel. (Hodgkin–Huxley3.cir)

Solution

If a dc steady state is reached, KCL at the node yields,

$$\begin{aligned} 0 &= (U - 55) \cdot 100 + (U + 72) \cdot 10 + (U + 49) \cdot (0.3) \\ &= 110.3U - 5500 + 720 + 17.7 \\ 0 &= 110.3U - 4765.3 \end{aligned}$$

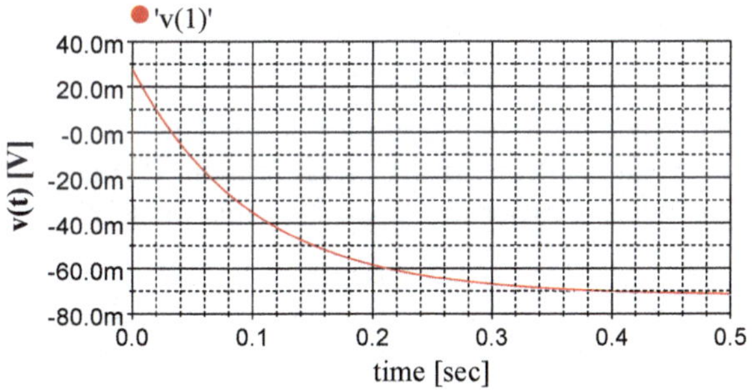

Fig. 3.75 Voltage signal in a HH neuron model

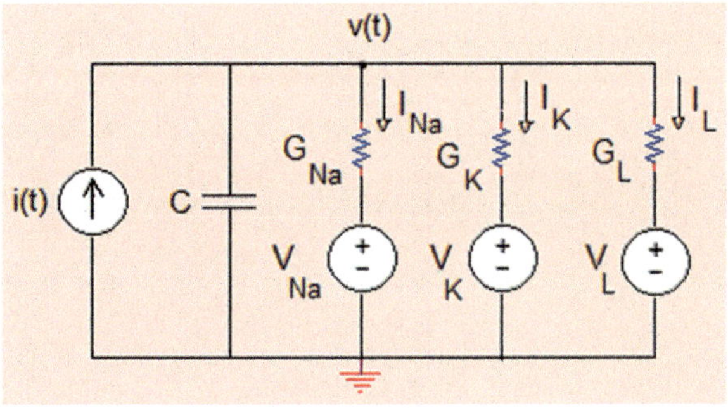

Fig. 3.76 Hodgkin–Huxley model of an excitable cell

Membrane potential is,

$$U = \frac{4765.3}{110.3} = 43.2\,\text{mV}$$

Ionic currents at each channel are,

$$\begin{aligned} i_K &= (U - V_K) \cdot G_K = (43.2 - (-72)) \cdot (0.01\ \text{S/cm}^2) \\ &= 115.2 \times 0.01 = 1.152\,\text{mA/cm}^2 \end{aligned}$$

$$\begin{aligned} i_{\text{Na}} &= (U - V_{\text{Na}}) \cdot G_{\text{Na}} = (43.2 - (55)) \cdot \left(0.1\ \text{S/cm}^2\right) \\ &= -11.8 \times 0.1 = -1.18\,\text{mA/cm}^2 \end{aligned}$$

$$\begin{aligned} i_L &= (U - V_L) \cdot G_L = (43.2 - (-49)) \cdot \left(0.0003\,\frac{\text{S}}{\text{cm}^2}\right) \\ &= 92.2 \times 0.0003 = 27.67\,\mu\text{A/cm}^2 \end{aligned}$$

Problem 3.2.33 Radiation detector is indispensable measuring/monitoring equipment in a Nuclear Medicine department.

In a Geiger–Müller radiation detector, a dc voltage source is applied to the center element of a Geiger tube through a current-limiting resistor. If an ionizing beam passes through the tube, it ionizes some of the gas inside. This causes the tube momentarily to conduct a current. A model for an over-simplified Geiger–Müller counter circuit is given here (Fig. 3.77). Perform a circuit simulation study and answer the following questions:

(a) What is the time constant of the decaying pulse observed at point 3?
(b) Print a SPICE netlist for the circuit. (Geiger1.cir)
(c) What is major discrepancy of this model?

V_1 = 500 V, R_1 = 1 MΩ, R_2 = 1 Ω, C_2 = 4 pf, C_1 = 1 μf, R_3 = 1 kΩ, R_1 is the current-limiting resistor, C_1 is the blocking capacitor.

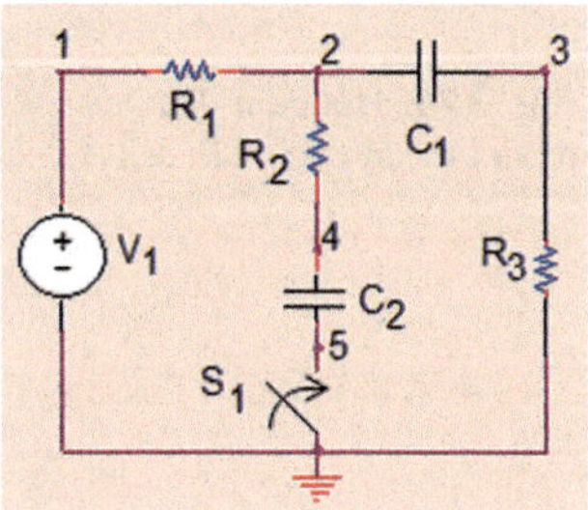

Fig. 3.77 Simplified Geiger–Müller counter principle

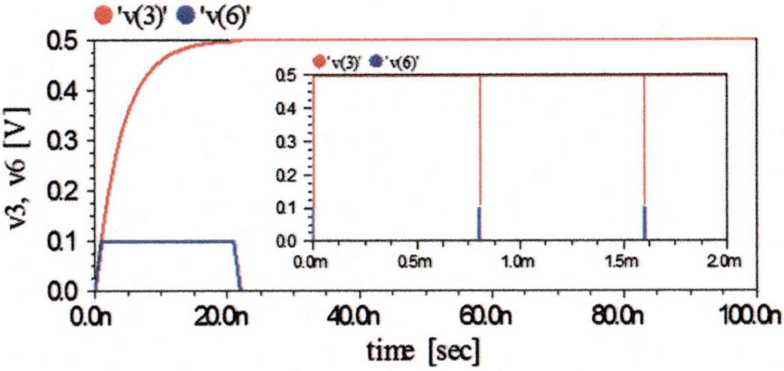

Fig. 3.78 Voltage signal due to radiation

Solution

(a) Assuming a 0.1 V amplitude and 20 ns duration of a pulse due to radiation (blue), one obtains the curves in Fig. 3.78.
Time constant of the circuit = $R_2.C_2$

(b)

```
The SPICE netlist:
   * plot v3,v6
   GEIGER1
   v1 1 0 dc 500
   R1 1 2 1MEG
   R2 2 4 1
   C2 4 5 4P
   C1 2 3 1U
   R3 3 0 1K
   S1 5 0 6 0 SW1 ON
   vin 6 0 pulse(0 0.1 0 1ns 1ns 20n 800u)
  *PULSE(V1 V2 TD TR TF PW PER)
  .MODEL SW1 SW
  * Analysis Parameters
    Stepsize ...........          : 0.01u
    Final time .........          : 4000u
    Display start time ..         : 0
    Maximum step size ...         : 10U
    Use Initial Conditions: On
```

(c) This crude GM counter electrical equivalent model does not take the Geiger plateau into account. A GM tube operates only at a constant slope of the plateau, for example, between 400 and 600 V. However, this model does not differentiate such a plateau and operates at much wider supply voltage range.

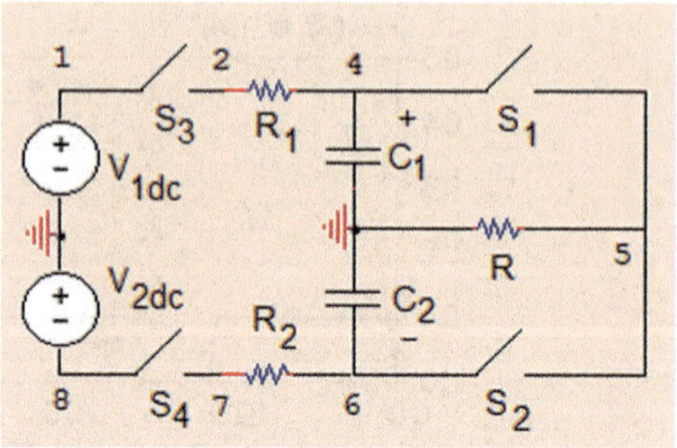

Fig. 3.79 Simplified ideal defibrillator circuit with biphasic output current waveform

Problem 3.2.34 A simplified ideal defibrillator circuit with biphasic output current waveform is shown in Fig. 3.79. This circuit includes both charging and discharging sections. Timing diagram for all switches are also given in Fig. 3.80.

$$C_1 = C_2 = 100\,\mu\text{F}, \quad R = R_1 = R_2 = 50\,\Omega, \quad V_{1\text{dc}} = -V_{2\text{dc}} = V_{\text{dc}} = 2\,\text{kV}$$

(a) What is the equation of current in the circuit after switch S_3 remains closed?
(b) How long does it take to charge the capacitors?
(c) Determine the voltages over resistor R_1 and capacitor C_1 for $0 \leq t < 30$ ms.
(d) Determine the voltage over capacitor after switch S_3 is closed for $0 \leq t < 30$ ms, if the initial voltage over the capacitor is 260 V.
(e) What is the initial energy stored on C_1? What is the initial energy stored on C_2?
(f) Determine the magnitude(s) of maximum current(s) flowing through R.

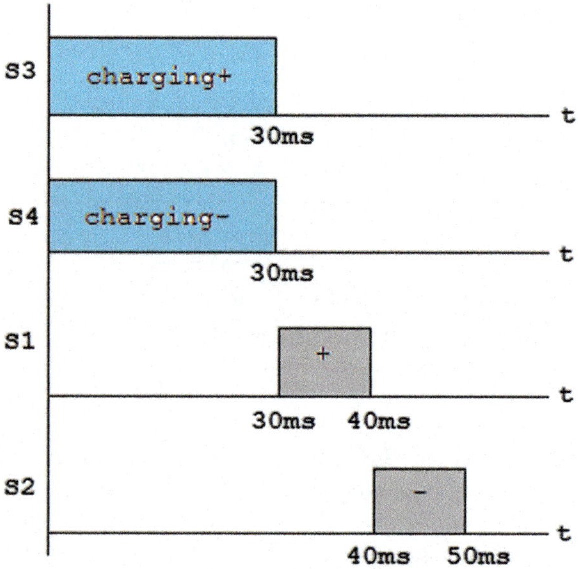

Fig. 3.80 Timing of a defibrillator

(g) What is the equation of current within the S_1 switching period of 10 ms?
(h) What is the equation of current within the S_2 switching period of 10 ms?
(i) Determine the currents at time $t = 0.04$ s and $t = 0.05$ s.
(j) What is the percentage value of current flowing through the resistor R at $t = 0.04^-$ s to the initial current at $t = 0^+$?
(k) What is the amount of dissipated energy within the first 1.5 s?
(l) What is the amount of dissipated energy within the second 1.5 s period due to S_2?
(m) Plot the voltage variation across the resistor which represents patient's torso, use SPICE analysis and print the SPICE netlist. Modify the timing circuit, if necessary. Explain the reasons for such modifications. (defib1.cir)
(n) What further timing modification is required to make this model be more realistic? Redraw timing diagram, if necessary. (defib3.cir)
(o) What is the value of residual voltage across the capacitor C_1 after it discharges over the patient's torso (resistor R)?
(p) What is the new equation of charging (+) voltage across the capacitor with the residual voltage, following its first discharge ?

Solution

(a) $i = \left(\frac{V_{dc}}{R_1}\right)e^{-\frac{t}{\tau 1}}$

$$\tau 1 = R_1 C = 50 \times 100 \times 10^{-6} = 5\,\text{ms}$$

$$i = \frac{2000}{50} e^{-\frac{t}{5}\times 1000} = 40e^{-200t}\,\text{A}$$

(b) Time to charge the capacitor $\approx 5\tau = 5 \times 5 \times 10^{-3} = 25\,\text{ms}$
Note that this is approximate time to charge the capacitor. Actually this is 99.3% of time to charge the capacitor, but for practical purposes it is a satisfactory assumption.
Since the switch remains closed for 30 ms, it reaches to its full charge of 2000 V at the end of 30 ms.

(c) $v_R = iR = \left(\frac{V_{dc}}{R}\right)e^{-\frac{t}{\tau}}.\ R = V_{dc}e^{-\frac{t}{\tau}} = 2000e^{-200t}\,\text{V}$

$$v_C = V_{dc} - v_R = V_{dc} - V_{dc}e^{-\frac{t}{\tau}} = V_{dc}\left(1 - e^{-\frac{t}{\tau}}\right) = 2000\left(1 - e^{-200t}\right)\,\text{V}$$

(d) General equation of capacitor charging voltage in an RC circuit is

$$v_C = V_{Cf} + (V_{Ci} - V_{Cf})e^{-\frac{t}{\tau}}$$
$$V_{Cf} = \text{final capacitor voltage}, \quad V_{Ci} = \text{initial capacitor voltage}$$

Substituting the given values of the components yields

$$v_C = 2000 + (260 - 2000)e^{-200t} = 2000 - 1740e^{-200t}\ \text{V}$$

(e) Energy stored on C_1

$$W = \frac{C_1 v(0.030)^2}{2} = \frac{100 \times 10^{-6} \times 2000^2}{2} = 200\ \text{J}$$

Since $C_2 = C_1$ and initial charges are the same, initial energy stored on C_2 is the same as that on C_1.

(f) Maximum magnitudes of currents flow at time $t = 0.030$ s, and at $t = 0.04$ s

$$i(0.03) = \frac{v(0.03)}{R} = \frac{2000}{50} = 40\ \text{A}$$
$$i(0.04) = \frac{v(0.04)}{R} = \frac{-2000}{50} = -40\ \text{A}$$

(g) Time constants for the main circuit are

$$\tau_1 = RC_1 = 50 \times 100 \times 10^{-6} = 5\ \text{ms}$$
$$\tau_2 = RC_2 = 50 \times 100 \times 10^{-6} = 5\ \text{ms}$$

The currents are described by the following equations:

$$i(t) = \frac{v(t)}{R} = \frac{2000e^{-200(t-0.03)}}{50} = 40e^{-200(t-0.03)} \quad 0.03\ \text{ms} < t < 0.04\ \text{ms}$$

(h) $i(t) = \frac{v(t)}{R} = \frac{-2000e^{-200(t-0.04)}}{50} = -40e^{-200(t-0.04)} \quad 0.04\ \text{ms} < t < 0.05\ \text{ms}$

(i) The current at time $t = 0.04$ s is calculated from the following equation:

$$i(t = 0.04-) = 40e^{-200(t-0.03)} = (40)e^{-200(0.04-0.03)} = 40 \times 0.1353 = 5.413\,\text{A}$$
$$i(t = 0.04+) = -40e^{-200(t-0.04)} = (t = 0.04-)(1) = -40\,\text{A}$$
$$i(t = 0.05-) = -40e^{-200(0.05-0.04)} = (-40)(0.1353) = -5.413\,\text{A}$$

(j) The percentage value of current flowing through the resistor R at $t = 0.04-$ s to the initial current is

$$p = \frac{i(0.04-)}{i(0+)} = \frac{5.413}{40} = 13.53\%$$

(k) Energy dissipated within the first discharge period of 10 ms,

$$W_{0.04} = \int_{0.03}^{0.04} i(t)^2 R\mathrm{d}t = \int_{0.03}^{0.04} (40e^{-200(t-0.03)})^2 \mathrm{d}t$$
$$W_{0.04} = W\left(1 - e^{-200(0.02)}\right) = 200\left(1 - e^{-4}\right) = 200(0.9817)\,\text{J} = 196.34\,\text{J}$$

This means that 98.17% of the initial energy stored on the capacitor is dissipated at the end of first 10 ms following the closure of switch S_1.

(l) The amount of dissipated energy within the second 10 ms switching period (due to S_2) is the same as that of the dissipated energy within the first 10 ms switching period, although the direction of current for $t > 0.04$ ms is opposite that of the current for $(0.03 < t < 0.04)$ ms.

(m) Truncated biphasic exponential voltage variation across resistor R is shown in Fig. 3.81. SPICE netlist is given for this waveform.

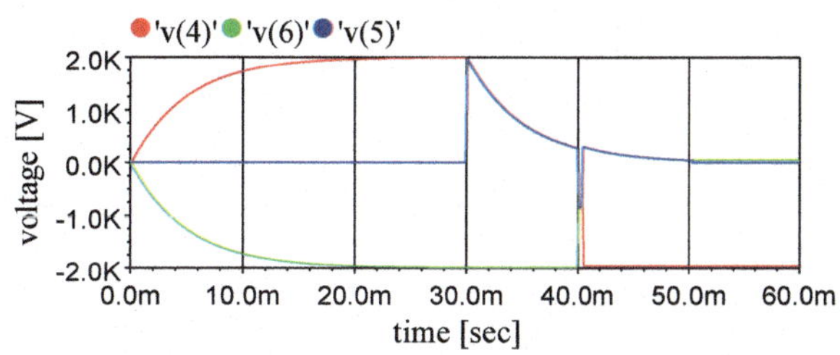

Fig. 3.81 Truncated biphasic exponential voltage variation across the load resistor

```
biphasic defibrillator complete equivalent circuit
    *Cap charging+
    C1 4 0 100u IC=0v
    R1 2 4 50
    s3 1 2 3 0 sm2
    V1dc 1 0 2k
    *       pulse(V1 V2 TD TR TF PW  PER)
    vx 3 0 pulse(0  1k 0  0  0  0.03 0)
        *Cap charging-
    C2 0 6 100u IC=0v
    R2 7 6 50
    s4 8 7 3 0 sm2
    V2dc 0 8 2k
        *Cap discharging+
    s1 4 5 9 0 sm2
    R  5 0 50
    *       pulse(V1 V2  TD    TR TF PW    PER)
    vy 9 0 pulse(0  1k  0.03  0  0  0.01  0 )
        *Cap discharging-
    s2 6 5 10 0 sm2
    *       pulse(V1  V2  TD    TR TF PW    PER)
    vz 10 0 pulse(0  1k  40.0m  0  0  0.01  0 )
    .MODEL sm2 SW(Ron=1u)
    *.tran 0.1m 60m 0 0.1m uic
```

Note that this waveform (in blue color) is not the expected one since the negative truncated discharge does not appear in the time diagram (Fig. 3.82). This is not corrected by SPICE transient parameter improvement, either (changes in SPICE netlist portions in red print).

```
*Cap discharging-
    s2 6 5 10 0 sm2
    *       pulse(V1  V2  TD    TR TF PW    PER)
    vz 10 0 pulse(0  1k  40.0m  0  0  0.01  0 )
    .MODEL sm2 SW(Ron=1u)
    *.tran 0.01m 60m 0 0.01m uic
```

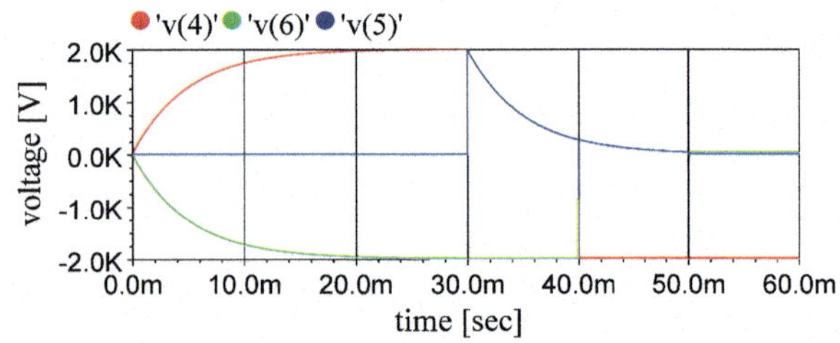

Fig. 3.82 Uncorrected transients of Problem 3.2.34

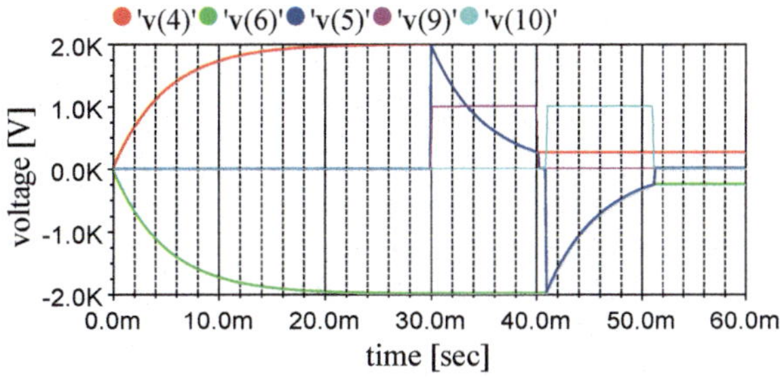

Fig. 3.83 Truncated biphasic exponential voltage variation across the load and switching waveforms are shown for the improved performance case

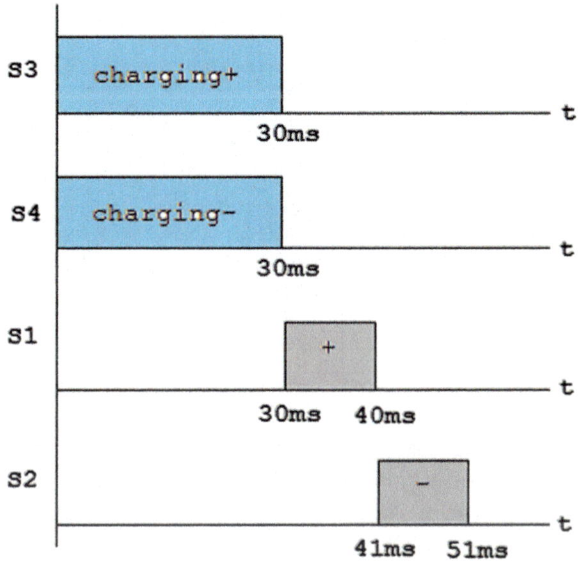

Fig. 3.84 Improved timing diagram for all switches

Negative leading edge of the voltage waveform is improved by introducing a short delay of 1 ms for S_4 switching time. Truncated biphasic exponential voltage variation across resistor R and switching waveforms are shown for this improved performance case in Fig. 3.83. Switching actions (pulses) for discharging are also included on the same graph. A separate improved timing diagram for all switches is also shown in Fig. 3.84.

```
Analysis: Transient defib1.cir
biphasic defibrillator complete equivalent circuit
*Cap charging+
C1 4 0 100u IC=0v
R1 2 4 50
s3 1 2 3 0 sm2
V1dc 1 0 2k
*       pulse(V1 V2 TD TR TF PW  PER)
vx 3 0 pulse(0  1k 0  0  0  0.03 0)
    *Cap charging-
C2 0 6 100u IC=0v
R2 7 6 50
s4 8 7 3 0 sm2
V2dc 0 8 2k
    *Cap discharging+
s1 4 5 9 0 sm2
R  5 0 50
*       pulse(V1 V2  TD    TR TF PW    PER)
vy 9 0 pulse(0  1k  0.03  0  0  0.01  0 )
     *Cap discharging-
s2 6 5 10 0 sm2
*       pulse(V1  V2  TD    TR TF PW    PER)
vz 10 0 pulse(0  1k  41m  0  0  0.01  0 )
.MODEL sm2 SW(Ron=1u)
*.tran 0.1m 60m 0 0.1m uic
```

(n) Timing diagram for the charging section of the model shows that it takes merely 30 ms to fully charge the capacitor with $C = 100$ microfarads to a voltage of 2000 V through a resistor of 50 Ω. This means that an electrical energy of 200 J is accumulating on the capacitor in that time duration which requires large equipment volume and electronic switching components with extreme ratings. Charging durations for external defibrillators recently available on the market are in the ranges of 5–10 s.

For a charging period of 5 s and using the same capacitor values,

$$5 = 5 \times \tau$$

$$\tau = R_1 C = 1\,\mathrm{s}$$

$$R_1 = \frac{\tau}{C} = 10\,\mathrm{k\Omega}$$

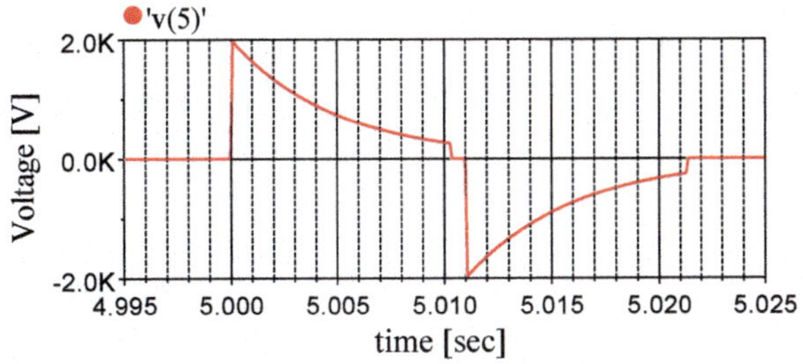

Fig. 3.85 The voltage waveform over the load in Problem 3.2.34n

The voltage waveform over the resistor *R* becomes as shown in Fig. 3.85. Modified SPICE netlist is also given with highlighted portions indicating the changed parameter values.

```
biphasic defibrillator equivalent circuit (modified) defib3.cir
    *Cap charging+
    C1 4 0 100u IC=0v
    R1 2 4 10k
    s3 1 2 3 0 sm2
    V1dc 1 0 2k
    *      pulse(V1 V2 TD TR TF PW  PER)
    vx 3 0 pulse(0  1k 0  0  0  5    0)
        *Cap charging-
    C2 0 6 100u IC=0v
    R2 7 6 10k
    s4 8 7 3 0 sm2
    V2dc 0 8 2k
        *Cap discharging+
    s1 4 5 9 0 sm2
    R  5 0 50
    *      pulse(V1 V2  TD  TR TF PW   PER)
    vy 9 0 pulse(0  1k  5   0  0  0.01  0 )
         *Cap discharging-
    s2 6 5 10 0 sm2
    *      pulse(V1  V2  TD     TR TF PW    PER)
    vz 10 0 pulse(0  1k  5.011  0  0  0.01  0 )
    .MODEL sm2 SW(Ron=1u)
    *.tran 0.1m 6 0 0.1m uic
```

Modified timing diagram (unscaled) is shown in Fig. 3.86.

Fig. 3.86 Modified timing diagram (unscaled)

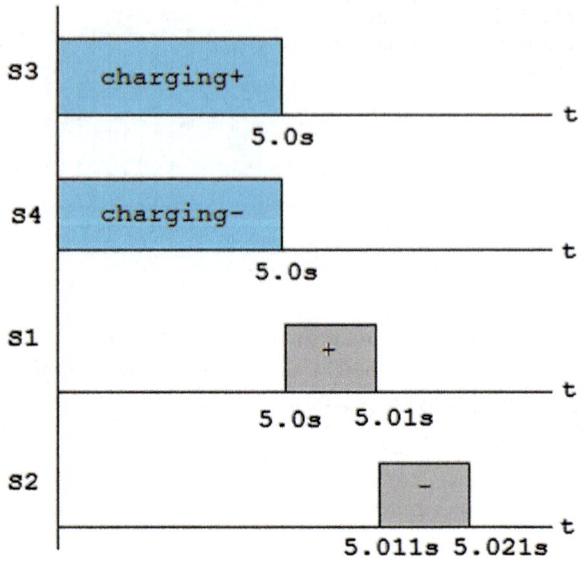

(o) The amount of residual voltage across the capacitor C_1 can be calculated using the equation for a discharging RC circuit,

$$v_C = V_{dc}e^{-\frac{t}{\tau}}$$

For a 50 O resistor (patient's torso),

$$\tau = RC = 50 \times 100 \times 10^{-6} = 5\,\text{ms}$$

At the end of discharging truncation at 10 ms, residual charge is

$$v_C = 2000e^{-200\times0.01} = 2000(e^{-2}) = 2000(0.1353) = 270.7\,\text{V}$$

(p) Following the first discharge, the next charging of the capacitor begins from 270.7 V and goes up to its final value of 2000 V. The new equation of charging (+) voltage across the capacitor is

$$\tau = R_1C = 10 \times 10^3 \times 100 \times 10^{-6} = 1\,\text{s}$$

$$v_C = v(\infty) + [v(0) - v(\infty)]e^{-\frac{t}{\tau}} = 2000 + (270.7 - 2000)e^{-t}$$
$$v_C = 2000 - 1729.3e^{-t}$$

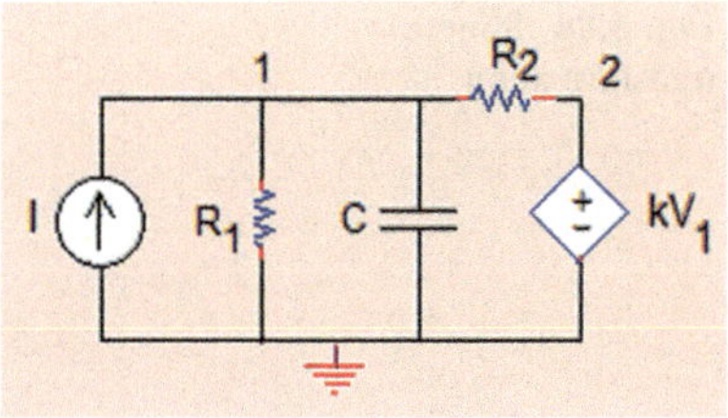

Fig. 3.87 The circuit of Problem 3.2.35

Problem 3.2.35 For the circuit shown in Fig. 3.87,

(a) Set up differential equation at node1;
(b) Find the analytical expression of the voltage at nodes 1 and 2 as functions of time;
(c) Evaluate these functions for $k = 3$, $C = 1$ F, $R_1 = 10\ \Omega$, $R_2 = 100\ \Omega$, $I = 1$ A at $t = 20$ s;
(d) Use SPICE analysis and plot the voltage variations at nodes 1 and 2, print netlist.

Solution

(a) KCL at node 1:

$$I - C\frac{\mathrm{d}v_1}{\mathrm{d}t} - \frac{v_1}{R_1} - \frac{(v_1 - v_2)}{R_2} = 0,$$

$$v_1 = kv_1$$

$$I - C\frac{\mathrm{d}v_1}{\mathrm{d}t} - v_1\left(\frac{1}{R_1} + \frac{1}{R_2} - \frac{k}{R_2}\right) = 0,$$

$$a = \frac{1}{R_1} + \frac{1}{R_2} - \frac{k}{R_2}, \quad \rightarrow \quad \frac{\mathrm{d}v_1}{\mathrm{d}t} + b \cdot v_1 = \frac{I}{C}$$

where

$$b = \frac{a}{C} = \frac{1}{\tau} \quad \rightarrow \quad \tau = \frac{C}{a}$$

(b) $v_1(t) = v_1(\infty)\left(1 - \mathrm{e}^{-\frac{t}{\tau}}\right)$

Fig. 3.88 Source transformation

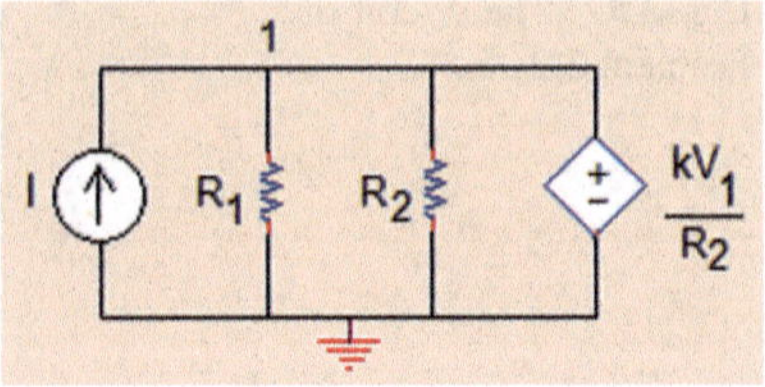

Performing dc analysis for the steady-state (final) value of the voltage at node 1, by source transforming dependent voltage source into dependent current source (Fig. 3.88),

$$I_1 - \frac{v_1}{R_1} - \frac{v_1}{R_2} + \frac{kv_1}{R_2} = 0 \quad \rightarrow \quad I_1 - v_1\left(\frac{1}{R_1} + \frac{1}{R_2} - \frac{k}{R_2}\right) = 0$$

$$I_1 - a \cdot v_1 = 0, \quad v_1(\infty) = \frac{I_1}{a}$$

$$v_1(t) = \frac{I_1}{a}\left(1 - \mathrm{e}^{-\frac{t}{\tau}}\right)$$

(c) Substitution of given numerical data into these equations yield the following results:

$$a = \frac{1}{R_1} + \frac{1}{R_2} - \frac{k}{R_2} = \frac{1}{10} + \frac{1}{100} - \frac{3}{100} = 0.1 + 0.01 - 0.03 = 0.08$$

$$v_1(\infty) = \frac{I_1}{a} = \frac{1}{0.08} = 12.5\,\mathrm{V}$$

$$\tau = \frac{C}{a} = \frac{1}{0.08} = 12.5\,\mathrm{s}$$

$$v_1(t) = 12.5\left(1 - \mathrm{e}^{-\frac{t}{12.5}}\right) = 12.5\left(1 - \mathrm{e}^{-0.08t}\right)$$

$$v_1(20) = 12.5\left(1 - \mathrm{e}^{-0.08\times 20}\right) = 9.976\,\mathrm{V}$$

$$v_2(t) = 3v_1(t) = 37.5\left(1 - \mathrm{e}^{-0.08t}\right)$$

$$v_2(20) = 29.928\,\mathrm{V}$$

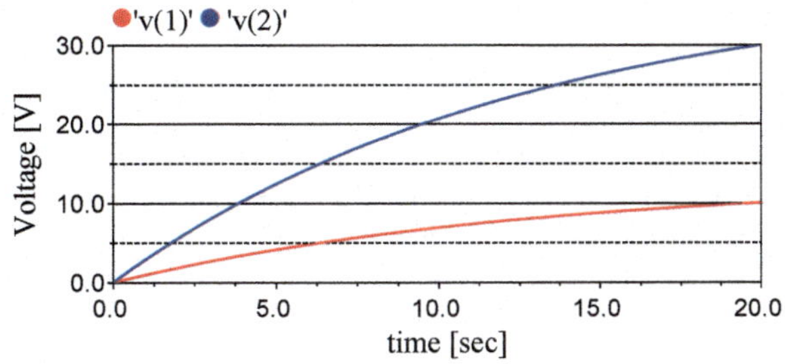

Fig. 3.89 Voltage variations at nodes 1 and 2

(d) A plot of the voltage variations at nodes 1 and 2 is displayed in Fig. 3.89, and SPICE netlist is given below.

```
nodal analysis
I1 0 1 1
C1 1 0 1
*voltage controlled voltage source
e1 2 0 1 0 3
* controlling voltage = 1 0
R1 1 0 10
R2 1 2 100
*tran 10m 20 0 10m UIC
```

References

1. Ellison WJ et al (1996) Water: a dielectric reference. J Mol Liq 68:171–279
2. Landdolt-Börnstein (2008) V-17, static dielectric constants of pure liquids and binary liquid mixtures. Springer, Berlin
3. Krawzenski A (1977) J. Microwave Power 12(3):215
4. Looyenga H (1965) Physica 31:401
5. Beguin F, Frackowiak E (2013) Supercapacitors: materials, systems and applications. Wiley, New York
6. Yu A, Chabot V, Zhang J (2013) Electrochemical supercapacitors for energy storage and delivery: fundamentals and applications. CRC Press, Boca Raton
7. Conway D (2014) Supercapacitor 108 success secrets—108 most asked questions on a supercapacitor—What you need to know. Emereo Pty Limited, Brisbane
8. Murata supercapacitor Technical Note No: C2M1CXS-053K
9. Marchant K, Davis B (2012) Laboratory hematology practice, Chapter 32. Wiley, New York
10. Brink GD, Freeman WC (2002) System and method for charging a capacitor using a variable frequency, variable duty cycle current waveform. US Patent 6,411,064
11. Sokal NO, Redl R (1997) Control algorithms and circuit designs for optimal flyback-charging of an energy-storage capacitor. IEEE Trans Power Electron 12(5)

12. Joseph S, Lawrence B, Richard N (2002) H-bridge circuit for generating a high-energy biphasic waveform in a external defibrillator. US Patent 6,477,413
13. Power DJ (2001) Apparatus for controlling delivery of defibrillation energy. US Patent 6,230,054
14. Pressman AI (1998) Switching power supply design, 2nd edn. McGraw-Hill, New York
15. FDA Investigations Operations Manual (2015), Appendix C, p 443
16. Lodish H, Berk A, Zipursky SL et al (2000) Molecular cell biology. W. H. Freeman, New York

Chapter 4
Inductors and First-Order RL Circuits

4.1 Inductors

Problem 4.1.1 In the system shown in Fig. 4.1, one of the two parallel conductors is attached to an electrically isolated spring with the spring constant of k and residing on the same plane in the air with the conducting wires. The system is initially at rest. If wires carry DC currents I_1 and I_2, neglecting any friction, what is the expression for the expansion of the spring in balanced condition?

Solution
Force on conductor 1 due to spring is

$$F_s = -kx. \tag{1}$$

Since the currents in the wires are I_1 and I_2, the magnetic field density on the second conductor due to first one is

$$B = \frac{\mu_o I_1}{2\pi d}, \tag{2}$$

where μ_o is permeability constant in vacuum (≅air). The force due to magnetic field is

$$F_B = b\, I_2 B = b\, I_2 \frac{\mu_o I_1}{2\pi d}. \tag{3}$$

Conductor 1 is pulled toward conductor 2. At equilibrium,

$$F_B = F_s \quad \rightarrow \quad b\, I_2 \frac{\mu_o I_1}{2\pi d} = kx \quad \rightarrow \quad x = \frac{\mu_o b I_1 I_2}{2\pi k d}.$$

A.Ü. Keskin, *Electrical Circuits in Biomedical Engineering*,
DOI 10.1007/978-3-319-55101-2_4

Fig. 4.1 Parallel conductors

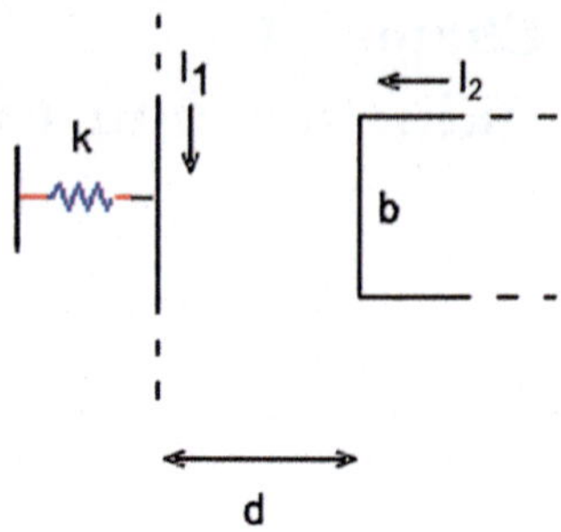

Problem 4.1.2 A resistive magnetic resonance unit is to be designed for small animal experiments. The field is generated using a solenoid of 20 cm length carrying $I = 1$ A current. The magnetic field in the mid-axis of this equipment is B. Determine the number of turns of solenoid if B is to be (a) 1 Gauss, (b) 10 Gauss, (c) 100 Gauss, and (d) 0.1 T.

Solution

$$B = \mu_0 nI = \mu_0 \frac{N}{L} I \quad \rightarrow \quad N = \frac{L}{\mu_0 I} B = kB$$

$$k = \frac{20 \times 10^{-2}}{4\pi \times 10^{-7} \times 1} = \frac{10^6}{2\pi} = 0.159 \times 10^6$$

$$N = 0.0001 \times 0.159 \times 10^6 = 0.159 \times 10^2 \cong 16.$$

Table 4.1 lists the results for different *B* values.

Problem 4.1.3 A coil of radius *a*, length *L*, number of turns *N*, moves with an axial velocity *v* toward a magnet. The distance from the coil top to the mid of the magnet (in *z*-axis) at equilibrium is *b*. The radial component of the field B_r at the axial distance *z* from the magnet is computed as

$$B_r = \frac{3\mu a z}{(a^2 + z^2)^{\frac{5}{2}}}.$$

Determine the induced voltage e_i on the coil.

Table 4.1 Results for different B values

B (Tesla)	N (turns)
0.0001	16
0.001	160
0.01	1600
0.1	16000

Solution
Induced electromotive force on a single-turn loop of radius a,

$$\mathrm{e}_i = \oint (v \times B)\mathrm{d}l = 2\pi a v B_r.$$

Since the radial component of the field at the axial distance z from the magnet is

$$B_r = \frac{3\mu a z}{(a^2 + z^2)^{\frac{5}{2}}}.$$

The induced emf is generated in the N-turns of coil of length L where the number of turns in a differential element of length dz and b is distance from the coil top to the mid-plane of the magnet at is equilibrium,

$$\mathrm{d}N = \left(\frac{N}{L}\right)\mathrm{d}z$$

$$\mathrm{e}_i = \int 2\pi a v B_r \mathrm{d}N = \int_b^{b+L} 2\pi a \frac{3azv}{(a^2+z^2)^{\frac{5}{2}}} \frac{N\mathrm{d}z}{L} = \frac{3N}{L} 2\pi a v \int_b^{b+L} \frac{z\mathrm{d}z}{(a^2+z^2)^{\frac{5}{2}}}$$

$$= \left.\frac{-2\pi a N \mu v}{L(a^2+z^2)^{\frac{3}{2}}}\right|_b^{b+L} = \frac{(2\pi a^2)N\mu v}{L}\left\{\frac{1}{(a^2+z^2)^{\frac{3}{2}}} - \frac{1}{\left(a^2+(b+L)^2\right)^{\frac{3}{2}}}\right\}.$$

Problem 4.1.4 A spring–magnet system with spring constant k_s and magnet mass M and dipole moment m oscillates while interacting with a non-magnetizable metallic plate of electrical conductivity σ, and thickness h as shown in Fig. 4.2. The vertical distance of the center of magnet to the plate is b, and $h \ll b$. The radial component of magnetic field is given as [1]

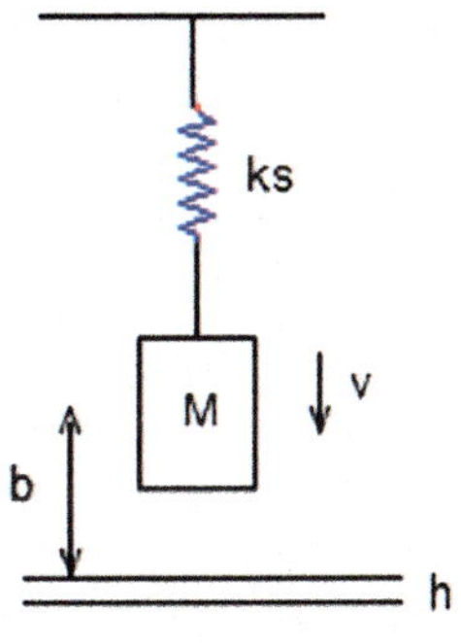

Fig. 4.2 A spring–magnet system with known spring constant and magnet mass

$$B_r = \frac{\mu_\rho m}{4\pi}\frac{3rb}{(r^2+b^2)^{\frac{5}{2}}}.$$

Determine the equation of motion of this system (without calculating numerical value of the phase) in time domain, if $M = 0.1\,\text{kg}$, $k_s = 2\,\text{Nm}^{-1}$, $m = 3\,\text{Am}^2$, $h = 1\,\text{mm}$, $b = 25\,\text{mm}$, $\sigma = 58.5\times10^7\,(\Omega\text{m})^{-1}$, initial displacement A = 10 mm.

Solution

Assume an infinitesimal ring section of the plate of width dr and radius r. Infinite small current di is induced by the magnetic field. Induced emf in the ring is

$$e_i = \oint (v\times B)\text{d}l = 2\pi r v B_r,$$

where m is the dipole moment of the magnet. The conductance of the ring

$$\text{d}\sigma = \sigma\left(\frac{h\text{d}r}{2\pi r}\right),$$

where σ is the conductivity of plate:

$$\text{d}i = \sigma_i\text{d}\sigma = (2\pi r B_r v)\frac{\sigma h\text{d}r}{2\pi r}.$$

Here, the inductance of the ring is ignored (since the frequency of oscillations is expected to be small). The vertical magnetic force dF_z on the ring,

$$\text{d}F_z = \int \text{d}i\left(\vec{\text{d}l}\times\vec{B}\right)_z = 2\pi B_r\text{d}i = 2\pi r(\sigma h v)B_r^2\text{d}r.$$

Since the plate has infinite radius,

$$F_z = \int_{r=0}^{\infty}\text{d}F_z = 2\pi\sigma h v\int_0^{\infty}B_r^2 r\text{d}r.$$

Let

$$u = \frac{r}{b},$$

$$F_z = 2\pi\sigma h v b^2\left(\frac{3\mu_\rho m}{4\pi b^3}\right)^2\int_0^{\infty}\frac{u^3}{(l+u^2)^5}\,\text{d}u.$$

Since

$$\int_0^\infty \frac{u^3}{(l+u^2)^5}\,\mathrm{d}u = \frac{1+4u^4}{24(1+u^2)^4}\Big|_0^\infty = \frac{1}{24}$$

$$F_z = \frac{3\pi\sigma h}{4b^4}\left(\frac{\mu_\rho m}{4\pi}\right)^2 v = kv$$

k is the damping (viscous) parameter,

$$k = \frac{3\pi\sigma h}{4b^4}\left(\frac{\mu_\rho m}{4\pi}\right)^2.$$

Since M is the mass of the magnet, k_s = spring constant, the equation of motion is

$$M\ddot{z} + k\dot{z} + k_s z = 0 \quad \rightarrow \quad \ddot{z} + \frac{k}{M}\dot{z} + \frac{k_s}{M}z = 0 \quad \rightarrow \quad \ddot{z} + 2\lambda\dot{z} + \omega_0^2 z = 0.$$

The solution of this differential equation is of the form

$$z(t) = A\mathrm{e}^{-\lambda t}\cos(\omega t + \varphi),$$

where

$$\lambda = \frac{k}{2M}, \quad \omega_0 = \sqrt{\frac{k_s}{M}}, \quad \omega = \sqrt{\omega_0^2 - \lambda^2}$$

A = 10 mm is the initial amplitude (displacement). Using given data,

$$k = \frac{3\pi\sigma h}{4b^4}\left(\frac{\mu_\rho m}{4\pi}\right)^2 = \frac{3\times\pi\times 58.5\times 10^7\times 10^{-3}}{4(25\times 10^{-3})^4}\times\left(\frac{4\pi\times 10^7\times 3}{4\pi}\right)^2 \cong 0.318$$

$$\lambda = \frac{k}{2M} = \frac{0.318}{2\times 0.1} = 1.59$$

$$\omega_0 = \sqrt{\frac{k_s}{M}} = \sqrt{\frac{2}{0.1}} = 4.47\ \text{rad s}^{-1}, \quad f_0 = \frac{\omega_0}{2\pi} = \frac{4.47}{2\pi} = 0.711\ \text{Hz}$$

$$\omega = \sqrt{\omega_0^2 - \lambda^2} = \sqrt{4.47^2 - 1.59^2} = 4.178\ \text{rad s}^{-1},$$

$$f = \frac{\omega}{2\pi} = \frac{4.178}{2\pi} = 0.665\ \text{Hz}.$$

Equation of the motion is

$$z(t) = 10e^{-1.59t}\cos(4.178t + \varphi) \quad [\text{mm}].$$

Problem 4.1.5 In 50-turn coil of radius 10 cm, the magnetic field is changing from 10 to 40 mT in 200 ms. Determine the induced voltage in the coil (The field is parallel to coil axis).

Solution

$$V = N\frac{d\varphi}{dt} \cong N\frac{\Delta\varphi}{\Delta t} = N\frac{\Delta(BA)}{\Delta t} = NA\frac{\Delta B}{\Delta t} = 50 \times 2\pi \times 0.1 \times \frac{(40 - 10) \times 10^{-3}}{2 \times 10^{-1}}$$
$$= \pi \times \frac{3 \times 10^{-2} \times 10}{2 \times 10^{-1}} = \pi \times \frac{3}{2} = 4.71\,\text{V}.$$

Problem 4.1.6 The current through an inductor changes linearly from 2 to 12 mA in 4 ms, inducing 240 mV across the inductor. Find the inductance.

Solution

$$v = L\frac{di}{dt} = L\frac{\Delta i}{\Delta t} = L\frac{(12 - 2) \times 10^{-3}}{4 \times 10^{-3}} = L\frac{10}{4} = L(2.5) = 240\,\text{mV}$$
$$L = \frac{240 \times 10^{-3}}{2.5} = 96\,\text{mH}.$$

Problem 4.1.7 Current through an inductor L = 1 H has a triangular waveform as shown in Fig. 4.3. Determine and plot the voltage across this inductor (L_v_i2.cir).

Solution

$$i(t) = \begin{cases} 3t, & 0 < t < 1\,\text{s} \\ -3t + 6, & 1 < t < 2\,\text{s} \\ 3t - 6, & 2 < t < 3\,\text{s} \\ -3t + 12 & 3 < t < 4\,\text{s} \end{cases}$$

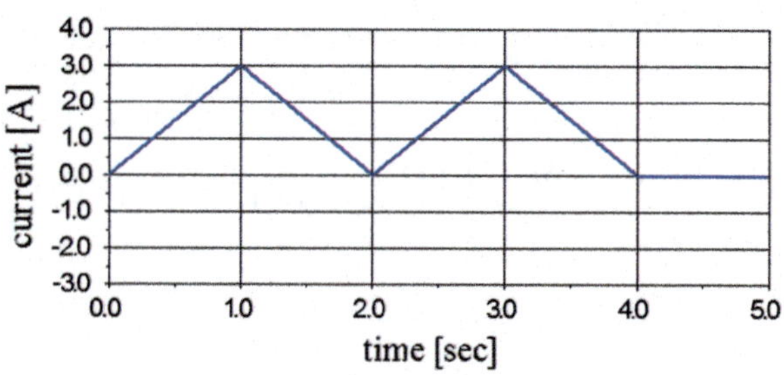

Fig. 4.3 Current flowing through an inductor with L = 1 H

$$v = L\frac{di}{dt} = 1 \times \frac{di}{dt}\ V$$

$$v = \begin{cases} 1 \times 3 = 3\,\text{V}, & 0 < t < 1\,\text{s} \\ 1 \times (-3) = -3\,\text{V}, & 1 < t < 2\,\text{s} \\ 1 \times (3) = 3\,\text{V}, & 2 < t < 3\,\text{s} \\ 1 \times (-3) = -3\,\text{V}, & 3 < t < 4\,\text{s} \end{cases}.$$

A plot of voltage across (in red color) and current through the inductor (in blue color) is shown in Fig. 4.4. A SPICE net list is also given below.

```
Current through L generates a voltage v
I1 0 1 PWL(0 0 1 3 2 0 3 3 4 0 5 0)
L1 1 0 1
R1 1 0 1 meg
*tran(0.1 5 0 0.1 uic)
```

Problem 4.1.8 The current through a L = 10 mH inductor is given in Fig. 4.5. Find the voltage values at (a) 1 ms, (b) 4 ms, and (c) 6 ms.

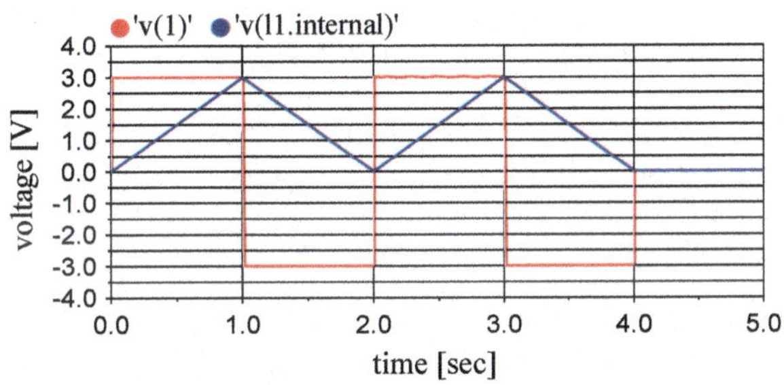

Fig. 4.4 Graph of voltage across (in red color) and current through the inductor (in blue color)

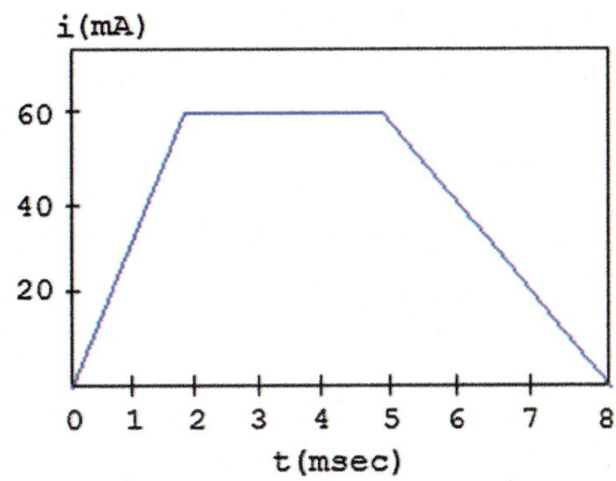

Fig. 4.5 The current flowing through a L = 10 mH inductor

Solution

$$i(t) = \begin{cases} 30\,t, & 0<t<2\,\text{ms} \\ 60, & 2<t<5\,\text{ms} \\ 160 - 20\,t, & 5<t<8\,\text{ms} \end{cases}$$

$$v = L\frac{\mathrm{d}i}{\mathrm{d}t} = 10 \times \frac{\mathrm{d}i}{\mathrm{d}t}\ \text{mV}$$

$$v = \begin{cases} 10 \times 30 = 300\,\text{mV}, & 0<t<2\,\text{ms} \\ 0\,\text{mV}, & 2<t<5\,\text{ms} \\ 10 \times (-20) = -200\,\text{mV}, & 5<t<8\,\text{ms} \end{cases}$$

(a) The voltage at 1 ms, $v = 300\,\text{mV}$
(b) The voltage at 4 ms, $v = 0\,\text{mV}$
(c) The voltage at 6 ms, $v = -200\,\text{mV}$

Figure 4.6 displays resulting voltage waveform and input current signal.

Problem 4.1.9 The voltage across an inductor with L = 100 mH is given as

$$v(t) = 5t^3 - 3t^2 + 4t - 2\ \text{mV}, \quad t \geq 0.$$

Find the current at t = 1 s if the initial current at t = 0 is 2 A.

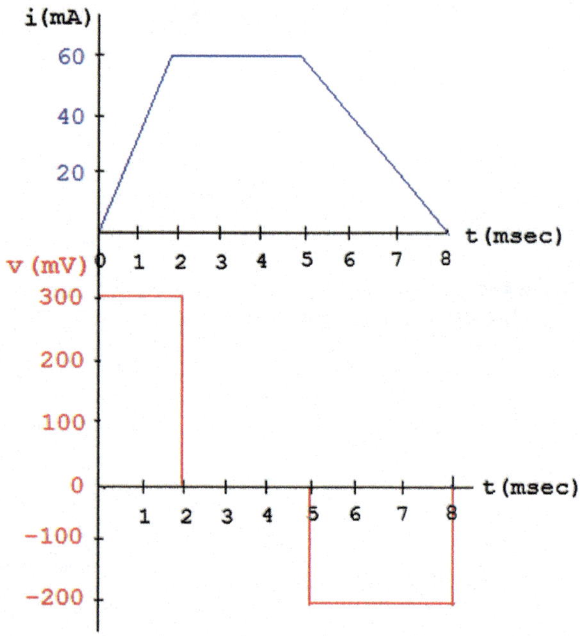

Fig. 4.6 Display of resulting voltage waveform and input current signal

Solution

$$i(t) = \frac{1}{L}\int_0^t v(\theta)\mathrm{d}\theta + i(0) = \frac{1}{100 \times 10^{-3}}\int_0^t (5\theta^3 - 3\theta^2 + 4\theta - 2)\mathrm{d}\theta + 2$$

$$i(t) = 10\left[\frac{5}{4}t^4 - \frac{3}{3}t^3 + \frac{4}{2}t^2 - 2\mathrm{t}\right] + 2 = 10\left[\frac{5}{4}t^4 - t^3 + 2t^2 - 2\mathrm{t}\right] + 2$$

$$i(t) = 12.5t^4 - 10t^3 + 20t^2 - 20\mathrm{t} + 2 \quad \rightarrow \quad i(1) = 12.5 - 10 + 20 - 20 + 2 = 4.5\mathrm{A}$$

Problem 4.1.10 Electrical charge variation through a cross section of a coil of $L = 2\mathrm{H}$ inductance is given as a function of time in the following form $q(t) = 3t^3 - 2t^2 + 2t + 5 \quad C.$

Find the voltage across the coil at $t = 1$ s.

Solution

$$i = \frac{\mathrm{d}q}{\mathrm{d}t}$$

$$v = L\frac{di}{\mathrm{d}t} = L\frac{\mathrm{d}^2q}{\mathrm{d}t^2} = 2\frac{\mathrm{d}^2q}{\mathrm{d}t^2} = 2 \times \frac{\mathrm{d}q}{\mathrm{d}t}\left(9t^2 - 4t + 2\right) = 2 \times (18t - 4) = 36t - 8$$

$$v(1) = 36(1) - 8 = 28 \text{ V}.$$

Problem 4.1.11

Voltage across an inductor with $L = 3\mathrm{H}$ is $v_L(t) = 6\mathrm{e}^{-2t}u(t)$.
Initial current through the inductor is 1 A.
Determine the current through this inductor element as a function of time.
Determine the current at time zero and infinity. Check the results using SPICE analysis. Print SPICE netlist. (inductor current1.cir)

Solution

$$\begin{aligned} i_L(t) &= \frac{1}{L}\int_0^t v(\alpha)\mathrm{d}\alpha + i(0) = \frac{1}{3}\int_0^t 6\mathrm{e}^{-2\alpha}\mathrm{d}\alpha + 1 \\ &= \frac{6}{3}\left(\frac{1}{-2}\mathrm{e}^{-2\alpha}\Big|_0^t\right) + 1 = 2 \times \frac{1}{-2} \times (\mathrm{e}^{-2t} - 1) + 1 \\ &= -(\mathrm{e}^{-2t} - 1) + 1 = -\mathrm{e}^{-2t} + 1 + 1 = (2 - \mathrm{e}^{-2t}) \cdot u(t)\, A \\ i_L(0) &= (2 - \mathrm{e}^0) \cdot u(t) = 2 - 1 = 1\, A \\ i_L(\infty) &= (2 - \mathrm{e}^{-\infty}) \cdot u(t) = 2 - 0 = 2\, A. \end{aligned}$$

Figure 4.7 displays the current through the inductor. SPICE netlist (inductor current1.cir) is given below.

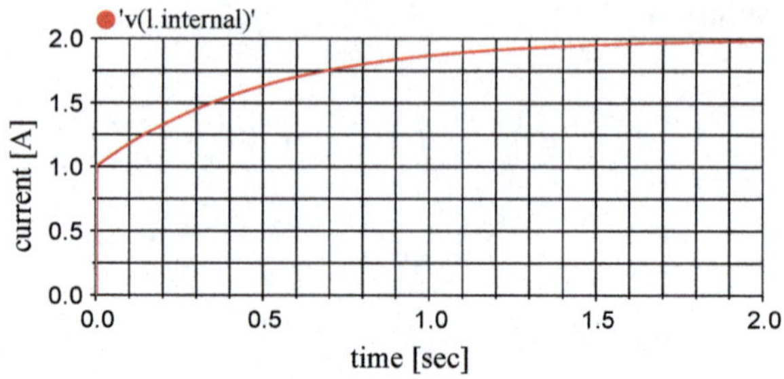

Fig. 4.7 The current flow through the inductor for problem 4.1.11

```
inductor current1
L 1 0 3 ic = 1
r 1 2 1m
*v1 2 0 EXP(V1 V2 TD1 TAU1 TD2 TAU2)
V1 2 0 EXP(0 6 0 0 0 0.5S)
```

Problem 4.1.12 Voltage and current periodic waveforms across an inductor are shown in Fig. 4.8. What is the inductance? Check the result using SPICE analysis, print the netlist. `(L_v_i1.cir)`

Solution
From current waveform,

$$i_L = \begin{cases} 4t - 2, & 0 < t < 1\,\text{s} \\ 6 - 4t, & 1 < t < 2\,\text{s} \end{cases}.$$

One can obtain the voltage across the inductor,

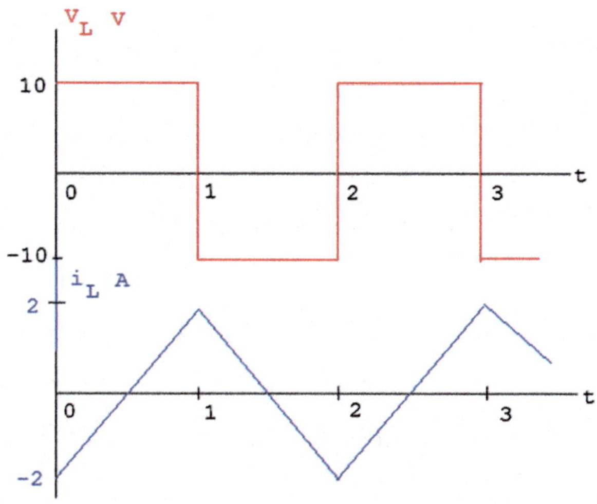

Fig. 4.8 Voltage and current periodic waveforms across an inductor

$$v_L = L\frac{\mathrm{d}i}{\mathrm{d}t} = \begin{cases} 4L, & 0<t<1\,\mathrm{s} \\ -4L, & 1<t<2\,\mathrm{s} \end{cases}.$$

The voltage variation is already given in Fig. 4.8,

$$v_L = \begin{cases} 10\,\mathrm{V}, & 0<t<1\,\mathrm{s} \\ -10\,\mathrm{V}, & 1<t<2\,\mathrm{s} \end{cases}.$$

This means, $10 = 4L$ or, $L = 2.5\,\mathrm{H}$
Following is a SPICE analysis (L_v_i1.cir) (Fig. 4.9).
SPICE netlist (L_v_i1.cir):

```
inductor current1
r  4 0 .1
L  1 4 2.5 ic=-2
*       pulse( V1  V2 TD TR TF  PW  PER)
v1 1 0 pulse(-10  10  0  0  0  1    2 )
*.tran 0.001 3.5 0 0.001 uic
```

Problem 4.1.13 A current signal is uniformly sampled as given in Table 4.2. It is passed through a L = 1 H inductor. Plot the current and voltage waveforms on this inductor, concurrently. (diff1.m)

Solution

$$v(t) = L\frac{\mathrm{d}i(t)}{\mathrm{d}t} = \frac{\mathrm{d}i(t)}{\mathrm{d}t}.$$

Numerical differentiation using MATLAB is performed and numerical results are plotted. Note that time and current vector lengths are reduced by 1 unit, due to numerical definition of the differentiation operation. These results are plotted as

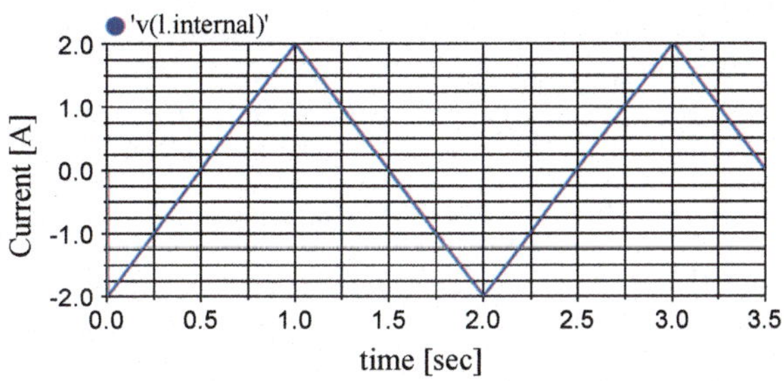

Fig. 4.9 Check for inductor current in Problem 4.1.12

Table 4.2 Current versus time

t (sec)	0	0.5	1.0	1.5	2.0	2.5	3.0	3.5	4.0	4.5	5.0	5.5
i (mA)	0	288	415	488	532	559	575	584	590	594	597	598

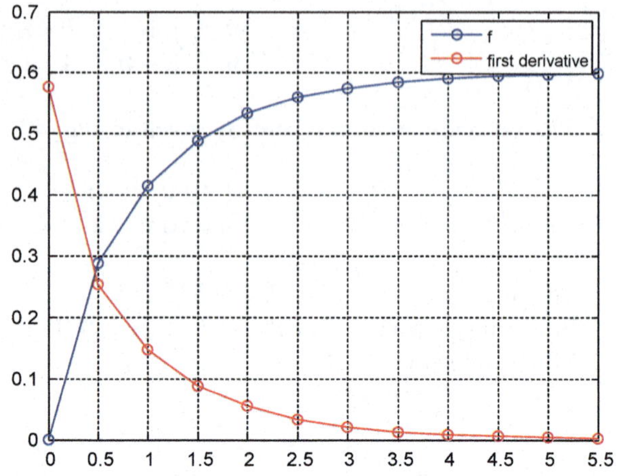

Fig. 4.10 Current and voltage waveforms in an inductor

shown in Fig. 4.10. Here, blue-colored "*f*" in legend indicates the current passing through the inductor, while the "first derivative" in red color denotes the voltage over the inductor.

MATLAB m file script (diff1.m) performing these operations is listed below.

```
h = 0.5;        % step size
x = (0:h:6);      % domain
f = 0.5*(1-exp(-x))+0.1*(1-exp(-5*x)) % input signal
Y = diff(f)/h    % first derivative
plot(x,f,'o-',x(:,1:length(Y)),Y,'ro-');grid on;
xlim([0 6-h]); legend('f','first derivative');
x(end)=[]; f(end)=[]; A
```

4.2 First-Order RL Circuits

Problem 4.2.1

(a) For the RL circuit of Fig. 4.11a with an initial current I_0, prove that

$$i(t) = I_0 e^{-\frac{t}{\tau}},\ \tau = \frac{L}{R}$$

(b) Find the voltage on the inductor in the circuit of Fig. 4.11b, for $t > 0 (R = 100\,\Omega)$.

Solution

(a) $i(0) = I_0, \quad v_L + v_R = 0, \quad v_L = L\frac{\mathrm{d}i}{\mathrm{d}t}, \quad v_R = iR$

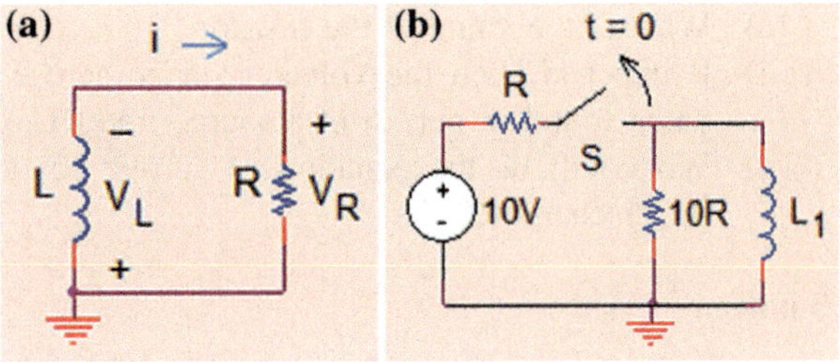

Fig. 4.11 RL circuits with initial currents

$$L\frac{\mathrm{d}i}{\mathrm{d}t} + Ri = 0 \quad \rightarrow \quad \frac{\mathrm{d}i}{\mathrm{d}t} + \frac{R}{L}i = 0$$
$$\ln\left(\frac{i(t)}{I_0}\right) = -\frac{R}{L}t \quad \rightarrow \quad i(t) = I_0\mathrm{e}^{-\frac{R}{L}t},\ \tau = \frac{L}{R},\ i(t) = I_0\mathrm{e}^{-\frac{t}{\tau}}$$

(b) $i_L(0^-) = I_0 = \dfrac{10\,V}{100\,\Omega} = 0.1\,\mathrm{A}$

$$i_L(t) = I_0\mathrm{e}^{-t/\tau}, \quad \tau = \frac{L}{R} = \frac{0.1}{1000} = 0.1\,\mathrm{ms}$$
$$v_L(0^+) = -i(0^+) \times 10R = -0.1 \times 10 \times 100 = -100\,V$$
$$v_L(t) = -100\mathrm{e}^{-t/\tau} = -100\mathrm{e}^{-10{,}000t}u(t).$$

Problem 4.2.2 In an RL circuit with $L = 1$ mH, $R = 1$ kΩ and the initial current through the inductor is 1 mA, find the current transient.

Solution

$$i(t) = I_0\mathrm{e}^{-\frac{t}{\tau}}, \quad \tau = \frac{L}{R} = \frac{10^{-3}\,\mathrm{H}}{1000\,\Omega} = 10^{-6}\,\mathrm{s}$$
$$i(t) = \mathrm{e}^{-10^6 t}\,\mathrm{mA}.$$

Problem 4.2.3 Voltage change across the 1 μF capacitor of a source-free RC circuit is shown in Fig. 4.12.

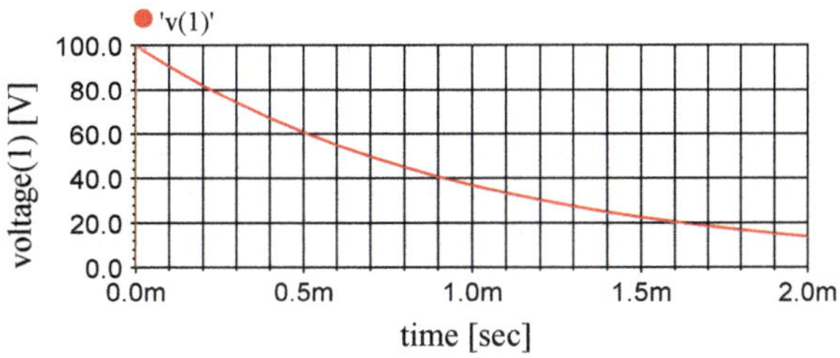

Fig. 4.12 Voltage change across the capacitor of a source-free RC circuit

(a) What is the value of the resistor?
(b-i) If this had been the voltage change across the resistor (having the same value found in part a) of a source-free RL circuit, what is the inductance?
(b-ii) What will be the equation of current for the RL circuit (Ampere unit)? (C_dischargeR1.cir)

Solution

(a) Time constant of a discharging RC circuit is the time required to discharge the capacitor, through the resistor, to its 37% of the difference between the initial and final values of the voltage across the capacitor. In other words, the voltage will decrease to 37% of its maximum value in one time constant period. Here, initial voltage is 100 V, final voltage tends to be 0 V.
Voltage level reaches 37% of its initial value at 1 ms,

$$\tau = RC = 10^{-3} = R \times 10^{-6} \quad \rightarrow \quad R = 10^3\,\Omega = 1\,\text{k}\Omega.$$

(b-i) If this graph was showing the voltage change across the resistor of a source-free RL circuit (with the same value found in part a), the inductance could be determined using time constant relationship of a source-free RL circuit,

$$\tau = \frac{L}{R} = \frac{L}{10^3} = 10^{-3} \quad \rightarrow \quad L = 10^3 \times 10^{-3} = 1\,\text{H}.$$

(b-ii) The equation of discharging current for the RL circuit (Ampere unit) is

$$i(t) = i(0) \times e^{-\frac{t}{\tau}} = \frac{v(0)}{R} \times e^{-\frac{t}{\tau}} = \frac{100}{1000} \times e^{-\frac{t}{10^{-3}}} = 0.1 \times e^{-1000t}\,\text{A}.$$

Following is the SPICE netlist (C_dischargeR1.cir) to prove the solution of this problem.

```
Current waveform in a source free RC circuit
C1 1 0 1u ic 100v
R1 1 0 1 k
*tran(0.1 m 2 m 0 0.1 m uic)
```

Problem 4.2.4 In a source-free RL circuit, an initial current of 1 A passes through an inductor and fades out through a resistor of R = 100 Ω, generating a voltage (natural response) across the resistor as shown in Fig. 4.13. What is the value of inductance?

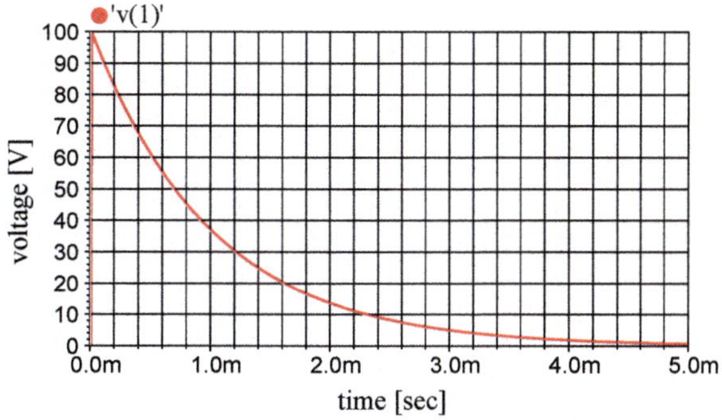

Fig. 4.13 Voltage across the resistor in a source-free RL circuit

Solution

For the source-free circuit with an initial current I_0

$$i(t) = I_0 \mathrm{e}^{-\frac{t}{\tau}}, \quad \tau = \frac{L}{R}.$$

Voltage generated across the resistor is $v(t) = I_0 R \mathrm{e}^{-\frac{t}{\tau}}$.

This voltage is reduced to its 37% value in 1 ms, and therefore time constant of the circuit is 1 ms:

$$\tau = 1 \times 10^{-3} = \frac{L}{100}$$
$$L = 1 \times 10^{-3} \times 100 = 0.1\,\mathrm{H}.$$

Problem 4.2.5

(a) In a first-order RL circuit, determine the energy–time relationship, if the initial current I_0 A is flowing through the inductor toward ground.
(b) Calculate the steady-state energy, i.e., w_R $(t = \infty) = ?$

Solution

(a) The current and voltage expressions for the resistor,

$$i(t) = I_0 \mathrm{e}^{-\frac{t}{\tau}}, \quad v(t) = I_0 R \mathrm{e}^{-\frac{t}{\tau}},$$

where $\tau = L/R$. The power dissipated in the resistor, $p = v(t).i(t) = I_0^2 R \mathrm{e}^{-\frac{2t}{\tau}}$.

The energy absorbed, $w_R(\mathrm{t}) = \int_0^t p.\mathrm{d}t = \int_0^t I_0^2 R \mathrm{e}^{-\frac{2t}{\tau}} \mathrm{d}t$

$$\text{When} \quad t \geq 0, \quad w_R(t) = \frac{1}{2} L I_0^2 \left[1 - \mathrm{e}^{-2\left(\frac{R}{L}\right)t}\right].$$

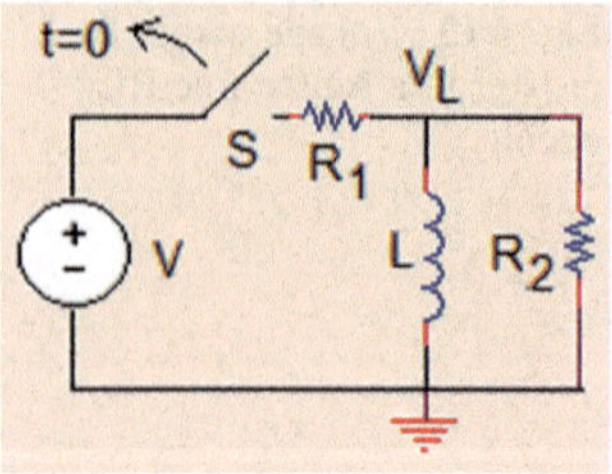

Fig. 4.14 A model of the strain relaxation for a muscle fiber

(b) As the final condition,

$$\text{when } t = \infty, \quad w_R(\infty) = \frac{1}{2} L I_0^2.$$

Problem 4.2.6 The electrical circuit shown in Fig. 4.14 is used to model the strain relaxation of a muscle fiber in an experiment ($V = 10$ V, $R_1 = 100\ \Omega$, $R_2 = 2\ \Omega$, $L = 1$ H). Find the voltage across the inductor at

(a) $t = 0^+$ s,
(b) $t = 0.5$ s.

Solution

$$i_L(0^-) = \frac{V}{R_1} = \frac{10}{100} = 0.1\ \text{A}$$
$$i_L(0^+) = i_L(0^-) = -i_R(0^+) = 0.1\ \text{A}$$
$$i_R(t) = (-0.1) \cdot \text{e}^{-\frac{t}{\tau}}, \quad t \geq 0$$
$$\tau = \frac{L}{R} = \frac{1}{2} = 0.5\ \text{s}$$
$$i_R(t) = -0.1\text{e}^{-\frac{t}{0.5}} = -0.1\text{e}^{-2t}\text{A}, \quad t \geq 0$$
$$v_L(t) = v_R(t) = i_{R_2}(t) \cdot R_2 = (-0.1 \cdot \text{e}^{-2t}) \cdot 2 = -0.2 \cdot \text{e}^{-2t}\ \text{V}, \quad t \geq 0$$

(a) $v_L(0^+) = -0.2\ \text{V}$
(b) $v_L(0.5) = (0.37)(-0.2) = -0.74\ \text{V}$.

Problem 4.2.7 In the circuit shown in Fig. 4.15a, the switch has been closed for a long time. At $t = 0$, it is opened. Find the current flowing through inductor with $L = 1.6$ H, $t > 0$.

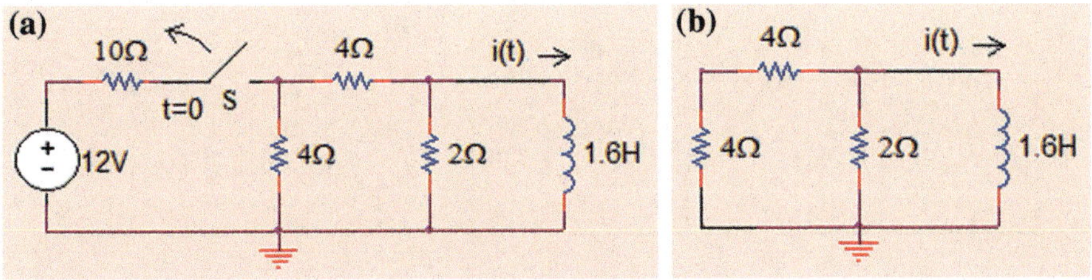

Fig. 4.15 The circuit of Problem 4.2.7 **a** for t < 0, **b** for t > 0

Solution

When $t<0$, *inductor* acts as a short circuit, shorting the 2 Ω resistor, the current supplied by 12 V voltage source is

$$i_1 = \frac{12}{10 + (4 \parallel 4)} = \frac{12}{10 + \left(\frac{4 \cdot 4}{4+4}\right)} = \frac{12}{12} = 1 \text{ A}.$$

The *inductor* attains its initial current and its value is calculated bycurrent division;

$$i(0^-) = \frac{4}{4+4} \cdot i_1 = \frac{4}{8} \cdot 1 = 0.5 \text{ A} \quad \text{at} \quad t<0.$$

When $t \geq 0$;

The current through L cannot change instantaneously; therefore $i(0) = i(0^-) = 0.5$ A.

$$R_{\text{eq}} = (4+4) \parallel 2 = 8 \parallel 2 = \frac{8.2}{8+2} = \frac{16}{10} = 1.6\,\Omega$$

$$\tau = \frac{L}{R_{\text{eq}}} = \frac{1.6}{1.6} = 1 \text{ s}.$$

$$i(t) = i(0) \cdot e^{-t/\tau} \quad \rightarrow \quad i(t) = 0.5 \cdot e^{-t} \text{ A}.$$

Problem 4.2.8 In the circuit shown in Fig. 4.16a, the switch S was closed for long time. It is opened at t = 0.

(a) $i(t)$ =? $t \geq 0$.
(b) What is the initial energy stored in L?

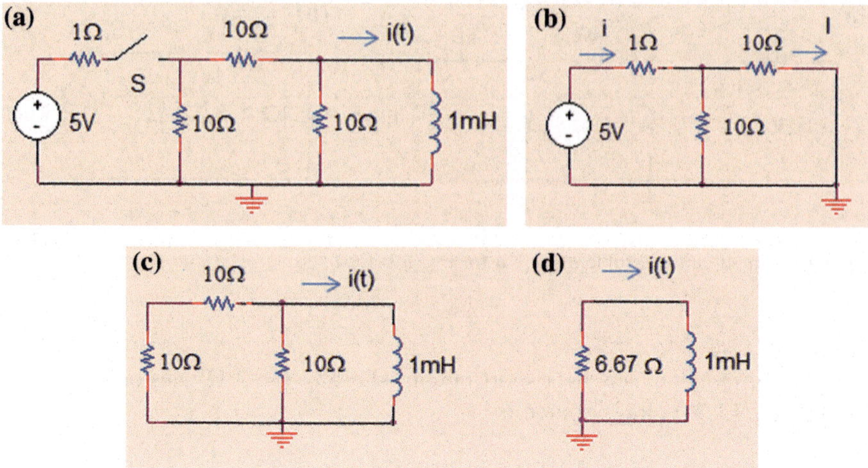

Fig. 4.16 The circuit of Problem 4.2.8 **a** for t < 0, **b** for t >> 0, **c** The circuit for the calculation of equivalent resistance, **d** RL equivalent circuit

Solution

(a) For $t < 0$,

$$i = \frac{5\,\text{V}}{1+5} = \frac{5}{6}\,\text{A},\ 10 \parallel 10 = 5\,\Omega,\ I = i \times \frac{10}{20} = \frac{5}{6} \times \frac{10}{20} = 0.4617\,\text{A}$$

$$i(t) = 0.41\bar{6}7 \cdot \text{e}^{-t/\tau}$$

$$R_{\text{eq}} = \frac{20 \times 10}{30} \cong 6.67\,\Omega.$$

For $t > 0$, current through the inductor cannot change abruptly,

$$1\,\text{mH} = 10^{-3}\,\text{H},\quad \tau = \frac{L}{R_{\text{eq}}} = \frac{10^{-3}}{6.67} \cong 1.5 \times 10^{-4}\,\text{s},\quad i(t) = 0.41\bar{6}7.\text{e}^{-6666.\bar{6}7t}\,\text{A}$$

(b) $W_l(0) = \frac{1}{2}L(i_0)^2 = \frac{1}{2} \times 10^{-3} \times (0.41\bar{6}7)^2 = 0.087 \times 10^{-3}\,\text{J} = 87\,\mu\text{J}.$

Problem 4.2.9 Isotonic (Normal) saline solution (0.9% NaCl) is used for injecting medicines via the veins (intravenous infusion). If the height of this solution is h in a cylindrical glass bottle container with area A, its volumetric flow rate q, pressure values $p1$ and $p2$, density of the solution, hydraulic resistance of the exit tube R_h, gravitational constant g:

(a) Set up differential equation relating the change of height of solution as a function of time.

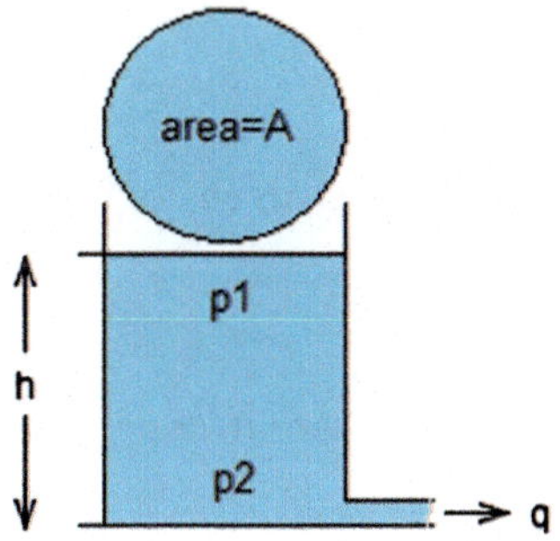

Fig. 4.17 Model for intravenous infusion set up

(b) What is the analogy with an electrical RL circuit?
(c) What is the analytical expression for h during infusion? (Fig. 4.17)

Solution

(a)
$$p_1 - p_2 = R_h q = h\rho g$$

$$q = -\frac{\mathrm{d}V}{\mathrm{d}t} = -\frac{\mathrm{d}(Ah)}{\mathrm{d}t} = -A\frac{\mathrm{d}h}{\mathrm{d}t}$$

$$h\rho g = -R_h A\frac{\mathrm{d}h}{\mathrm{d}t}$$

$$R_h A\frac{\mathrm{d}h}{\mathrm{d}t} + \rho g h = 0$$

$$\frac{\mathrm{d}h}{\mathrm{d}t} + \frac{\rho g}{R_h A}h = 0.$$

(b) Analogy between a source-free RL circuit and emptying saline bottle container is

$$\frac{\mathrm{d}i}{\mathrm{d}t} + \frac{R}{L}i = 0,$$

where R is the electrical resistor value and L is the inductance. Therefore, current change models the height change in the bottle.

(c) Assume that initial height of the solution is $h(0)$. Solving the differential equation

$$\frac{dh}{dt} + \frac{\rho g}{R_h A} h = 0$$

for h yields (in analogy with the solution of electrical circuit equation)

$$h(t) = h(0).e^{-\frac{t}{\tau}}$$

(in analogy with the solution of electrical circuit equation)

$$i(t) = i(0).e^{-\frac{t}{\tau}},$$

where the time constant for the hydraulic circuit is obtained as

$$\tau = \frac{R_h A}{\rho g}.$$

In contrast to the electrical RL circuit,

$$\tau = \frac{L}{R}.$$

Problem 4.2.10 In an RL circuit, $L = 2$ H, $R = 1$, $i(0) = 480$ A, $i(\infty) = 100$ A. Compute the current through the circuit for $0 \leq t \leq 10$ s and plot it.

Solution

$$\begin{aligned} i(t) &= i(\infty) + [i(0) - i(\infty)]e^{-\frac{t}{\tau}} \\ \tau &= \frac{L}{R} = \frac{2}{1} = 2\,\text{s} \\ i(t) &= 100 + [480 - 100]e^{-\frac{t}{2}} = 100 + 380e^{-0.5t}\,\text{A}. \end{aligned}$$

Figure 4.18 shows the MATLAB plot of current values computed at different times. The script is given below.

```
t=[0,1,2,3,4,5,6,7,8,9,10];
i=[480,330.5,239.8,184.8,151.4,131.2,118.9,111.5,106.9,104.2,102.6];
plot(t,i,'o-','linewidth',2);ylim([0 500]);
xlabel('time,[sec]');ylabel('i(t), A');grid on;
```

Problem 4.2.11 In the circuit shown in Fig. 4.19, the coil of a relay has inductance of 50 mH and resistance 100. It operates from a 12 V dc source. Determine the equation of current as a function of time, after switch is closed.

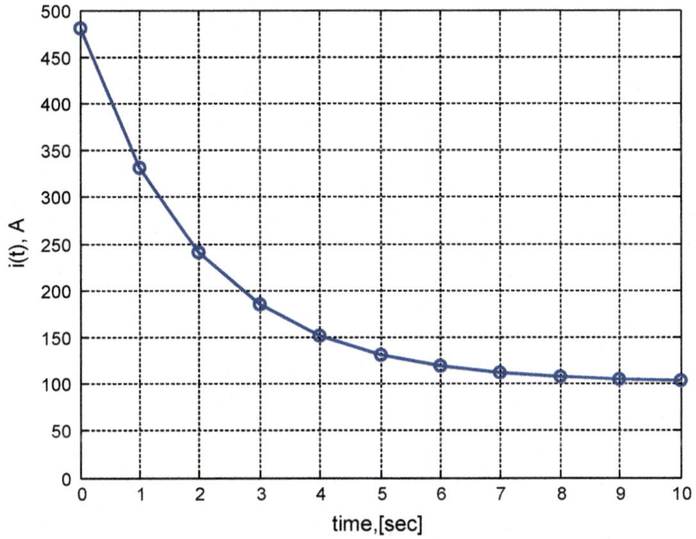

Fig. 4.18 Current variation through inductor as a function of time

Fig. 4.19 Excitation circuit for a relay coil

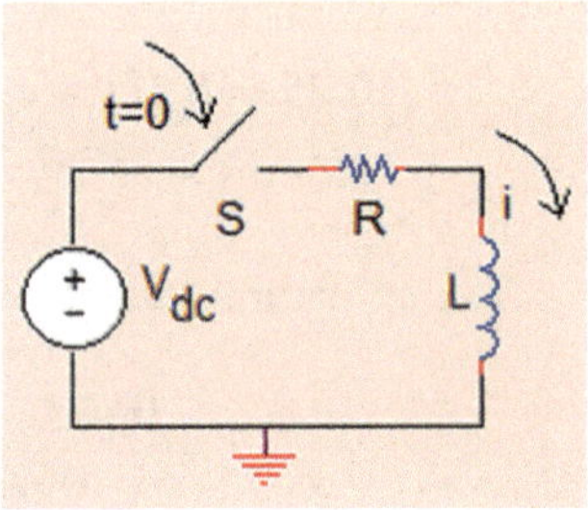

Solution

$$i(t) = i(\infty) + [i(0) - i(\infty)]\mathrm{e}^{-\frac{t}{\tau}}$$

$$i(\infty) = \frac{12\text{V}}{100\Omega} = 120\text{mA}, \quad i(0) = 0, \tau = \frac{L}{R} = \frac{50 \times 10^{-3}\text{H}}{100\Omega} = 0.5\text{ms}$$

$$i(t) = i(\infty)\left(1 - \mathrm{e}^{-\frac{t}{\tau}}\right) = (120\text{mA})\left(1 - \mathrm{e}^{-\frac{t}{0.5}\times 1000}\right) = (120)\left(1 - \mathrm{e}^{-2000t}\right)\text{mA}.$$

Problem 4.2.12 In the circuit shown in Fig. 4.20, $V_{dc} = 5$ V, $i(0^{-}) = 0$ A, $L = 50$ mH, $R = 10\ \Omega$ At what time (in milliseconds) the current reaches 250 mA?

Fig. 4.20 The circuit for Problem 4.2.12

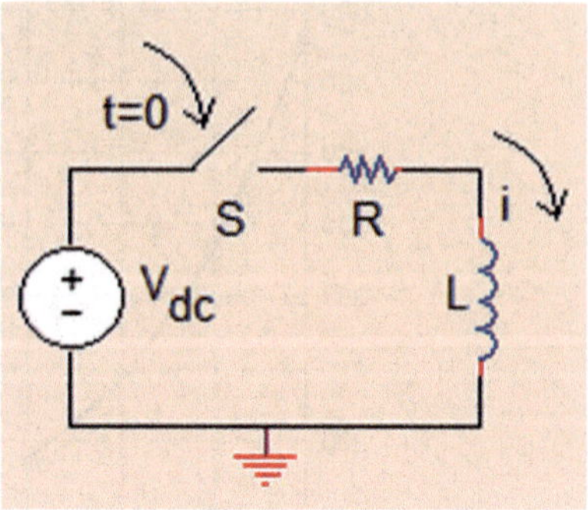

Solution

After the switch is closed,

$$i(t) = i(\infty) + [i(0) - i(\infty)]e^{-\frac{t}{\tau}}$$

$$i(\infty) = V_{dc}R = \frac{5}{10} = 0.5\,\text{A}, \quad \tau = \frac{L}{R} = \frac{50\,\text{mH}}{10\Omega} = 5\,\text{ms}$$

$$i(t) = 0.5\left(1 - e^{-\frac{t}{\tau}}\right) = 0.5\left(1 - e^{-200t}\right)\,\text{A}.$$

Since the current level is given as 250 mA,

$$0.25 = 0.5\left(1 - e^{-200t}\right) \qquad t \geq 0\,\text{s}$$

$$0.5 = 1 - e^{-200t} \quad \rightarrow \quad -0.5 = -e^{-200t} \quad \rightarrow \quad 0.5 = e^{-200t}.$$

Taking natural logarithm of both sides of last equation,

$$\ln(0.5) = -200t \quad \rightarrow \quad -0.693 = -200t$$

$$t = \frac{0.693}{200} = 0.00347\,\text{s} = 3.47\,\text{ms}.$$

Fig. 4.21 The coil of a relay

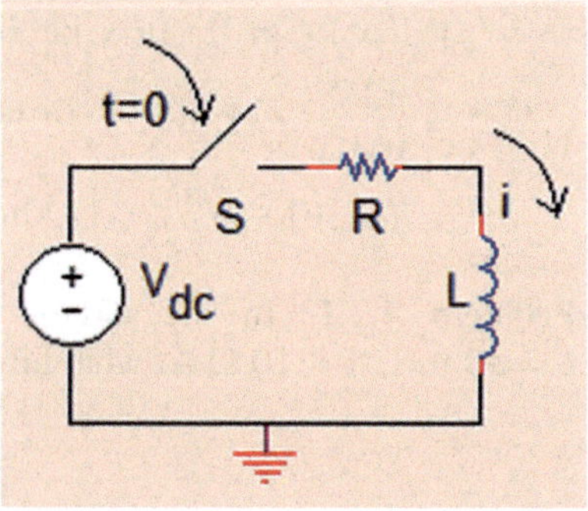

Problem 4.2.13 The coil of a relay has inductance of 25 mH and resistance 50 Ω (see, Fig. 4.21). It operates from a 5 V_{dc} source. Current required to pull the relay is 60 mA. What is the relay's delay time in milliseconds?

Solution

$$i(t) = i(\infty) + [i(0) - i(\infty)]e^{-\frac{t}{\tau}}$$

$$i(\infty) = \frac{5\,\text{V}}{50\,\Omega} = 100\,\text{mA}, \quad i(0) = 0, \quad \tau = \frac{L}{R} = \frac{25 \times 10^{-3}\,\text{H}}{50\,\Omega} = 0.5\,\text{ms}$$

$$i(t) = i(\infty)\left(1 - e^{-\frac{t_d}{\tau}}\right) = (100\,\text{mA})\left(1 - e^{-\frac{t_d}{0.5}\times 1000}\right) = 60\,\text{mA}$$

$$1 - e^{-2000t_d} = \frac{6}{10} = 0.6 \quad \rightarrow \quad e^{-2000t_d} = 0.4 \quad \rightarrow \quad -2000t_d = \ln(0.4)$$

$$t_d = -\frac{\ln(0.4)}{2000} = -\frac{(-0.9163)}{2000} = 0.46\,\text{ms}.$$

Problem 4.2.14 In the circuit shown in Fig. 4.22, how much time will pass until a quick blow 100 mA fuse to break the current flow after the switch is closed? R = 120 Ω, L = 60 mH, V_{dc} = 24 V.

Solution

$$\begin{aligned} i(t) &= i(\infty) + [i(0) - i(\infty)]e^{-\frac{t}{\tau}} \\ \tau &= \frac{L}{R} = \frac{60\,\text{mH}}{120\,\Omega} = 0.5\,\text{ms} \\ i(0) &= 0\,\text{A}, \quad i(\infty) = \frac{24\,\text{V}}{120\,\Omega} = 200\,\text{mA} \end{aligned} \tag{1}$$

Substitute into (1),

$$100 = 200 + [0 - 200]e^{-\frac{t}{0.5\times 10^{-3}}} = 200(1 - e^{-2000t})$$

$$0.5 = 1 - e^{-2000t} \quad \rightarrow \quad 0.5 - 1 = -e^{-2000t} \quad \rightarrow \quad 0.5 = e^{-2000t}.$$

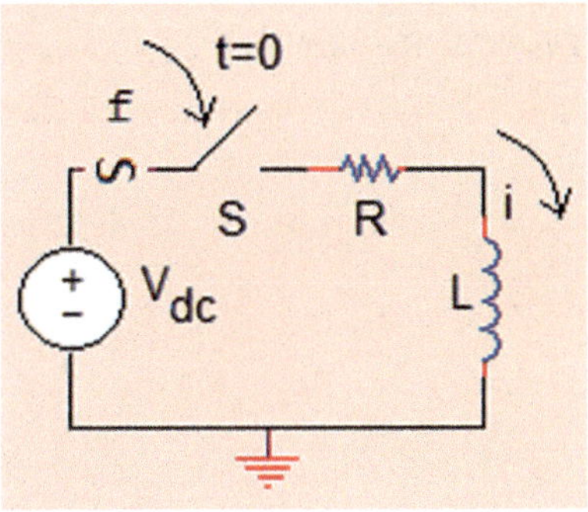

Fig. 4.22 The circuit for Problem 4.2.14

Take the natural logarithm of both sides of the last equation,

$$\ln(0.5) = -2000t \quad \rightarrow \quad -0.69315 = -2000t$$

$$t = \frac{0.69315}{2000} = 0.347\,\text{ms}.$$

Problem 4.2.15

(a) Determine the voltage at node 1 of the circuit of Fig. 4.23 as a function of time, for $t \geq 0$ s. $R = 10\,\Omega$, $L_1 = 1\,\text{H}$, $L_2 = 2\,\text{H}$, $R_1 = 0.1\,\Omega$, $R_2 = 0.2\,\Omega$.
(b) Use SPICE analysis to check the result found above. Print the SPICE net list (LD1.cir).

Solution

$$R_1 << R, \quad R_2 << R, \quad L_{eq} = \frac{L_1 L_2}{L_1 + L_2} = \frac{1 \times 2}{1 + 2} = \frac{2}{3}\text{H},$$

$$\tau = \frac{L_{eq}}{R} = \frac{\frac{2}{3}}{10} = 0.0667\,\text{s}$$

$$i(t) = I_0 e^{-\frac{t}{\tau}} = -e^{-\frac{t}{0.0667}}\,\text{A}$$

$$v(t) = RI_0 e^{-\frac{t}{\tau}} = -10e^{-\frac{t}{0.0667}}\,\text{V}.$$

SPICE netlist (LD1.cir): (Fig. 4.24)

```
Inductor discharges over another inductor
L1 2 0 1 IC = 1
R1 1 2 0.1
L2 3 0 2
R2 1 3 0.2
R 1 0 10
*.tran 1 m .5 0 1 m uic
```

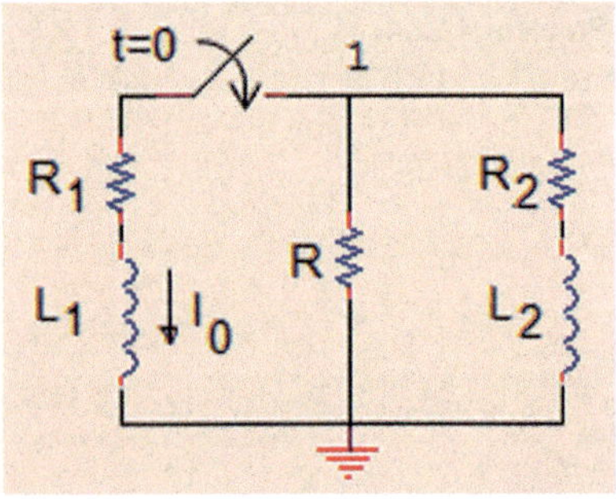

Fig. 4.23 The circuit for Problem 4.2.15

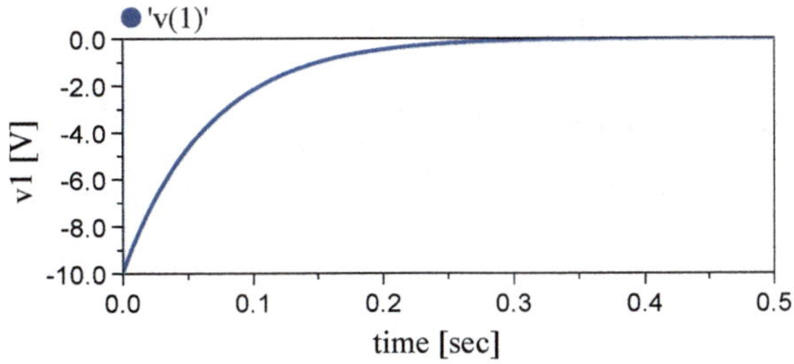

Fig. 4.24 The voltage at node1 of the circuit of Fig. 4.23

Problem 4.2.16 Shock wave generation system of an ESWL (Extracoporeal Shock Wave Lithotripter) is a device used to crash kidney stones. A spark gap device is the basic source of acoustic shock wave generator. In some systems, it is placed at the focus of an ellipsoid. Generated shock waves are concentrated on the other focus where the kidney stone is centered for destruction. The medium is assumed to be water equivalent.

The Spark gap device is parallel connected to a coil of inductance L and a DC power supply feeds current to this parallel combination through an electronic switch. A simplified circuit is shown in Fig. 4.25. Assuming that $L = 100$ mH, $R = 10\ \Omega$, $V_{dc} = 38$ V, the switching takes place (switch opens) in 20 ms, calculate the transient time until steady current level is reached after the switch is closed, steady current level, maximum voltage over the spark gap device, and energy stored in the inductor.

Solution

When the switch is in closed, the current reaches its maximum level in approximately at the end of

$$T_{\text{charge}} = 5.\tau = 5.\frac{L}{R} = 5.\frac{0.1}{10} = 0.05\ \text{s} = 50\ \text{ms}.$$

The steady current level,

$$I = \frac{V_{\text{dc}}}{R} = \frac{38}{10} = 3.8\ \text{A}$$

maximum voltage over the spark gap device is observed at the time when switch is opened,

$$V_{sg} = L\frac{\text{d}i}{\text{d}t} \cong L\frac{\Delta i}{\Delta t} = (0.1)\frac{3.8}{20.10^{-6}} = 19,000\ \text{V}$$

$$W = \frac{1}{2}LI^2 = \frac{1}{2}(0.1)(3.8)^2 = 722\ \text{mJ}.$$

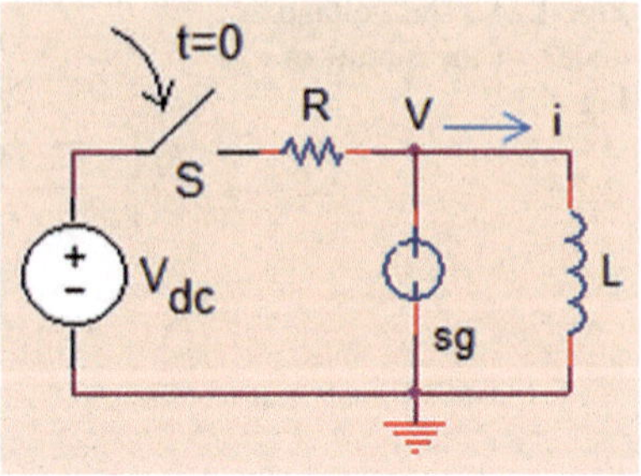

Fig. 4.25 Model for a shock wave generation system of an ESWL (Extracorporeal Shock Wave Lithotripter)

Problem 4.2.17

(a) What is the meaning of the "critical temperature" of a material? Which biomedical applications of superconductivity are known today?
(b) Assume that a direct current flows in a superconducting closed loop (coil). How this current can be measured?
(c) A closed loop with inductance L has an initial current Io at $t = 0$. Assuming that the circuit has a very small residual resistance R, what will be the time constant of the circuit? What does it imply in medical imaging area?
(d) What does "quenching" of an MRI system mean? What are the consequences of quenching a magnet in MRI?

Solution

(a) A material is said to be a superconductor if its resistivity is zero below certain temperature. This limiting value of temperature is called the critical temperature of material. For example, its value is 7.2 K for Pb, 4.1 K for Hg, 3 K for Cr, and 3.7 K for Sn, while some alloys have higher values of Tc (e.g., NbTi: Tc = 9.8 K, Nb3Sn: Tc = 18.1 K). For certain ceramics such as thallium-doped mercuric-cuprate, Hg0.8Tl0.2Ba2Ca2Cu3O8.33, Tc = 138 K. Superconducting electromagnets are employed in magnetic resonance imaging (MRI) systems for diagnostic imaging and in SQUID (superconducting quantum interference device) magnetometers to measure very small magnetic fields originating from the electro-physiological activity in the brain.
(b) If a current flows round a superconducting loop (coil), it cannot be measured by inserting an ammeter into the loop, since the current rapidly decays due to resistance of the ammeter. However, the magnetic field generated by the current in the coil can be used to measure this current value without consuming energy in the circuit.
(c) Time constant of the circuit will be $= L/R$. Since $R \cong 0$, $\tau \cong \infty$, and the current in a source-free RL circuit would exponentially decay with time, this implies that the current persists as long as the material remains superconducting. A further consequence of the persistent current in a superconducting coil is that the magnetic flux that passes through a closed loop of such a

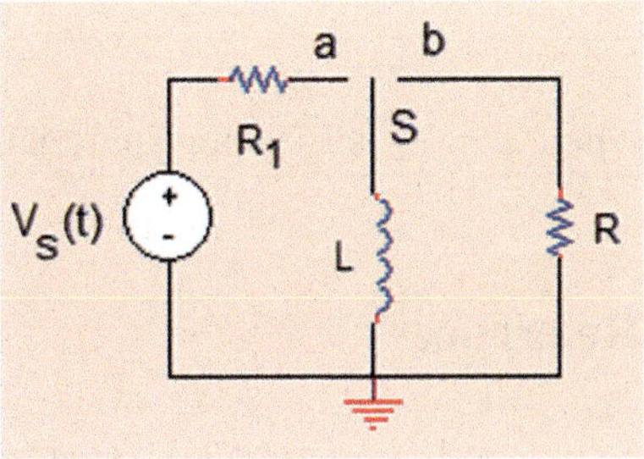

Fig. 4.26 Nuclear magnetic resonance (NMR) experiment

superconductive material remains constant. This is the principle of obtaining constant and high magnetic field densities in MRI [2–4].

(d) Quenching an MRI system happens by sudden discharging of the cooling liquefied gas (He) in the gantry. Then, superconductivity of the coil material disappears, and magnetic field intensity is dramatically reduced, within seconds.

This action may be required due to certain safety reasons in the patient (gantry) room. The gantry needs to be refilled by cryogenic (He) for a ramp-up operation. This is costly regarding the provision of cryogenic material as well as time lost during this procedure which may take several days to begin patient scans (including recalibration and alignment of complete imaging sequences).

Problem 4.2.18 In a simplified nuclear magnetic resonance (NMR) experiment, a coil which encloses a test tube filled with pure liquid water is placed vertical to a constant magnetic field density of 352.3 mT, as shown in Fig. 4.26. The switch S is thrown from position a to b at time $t = 0$. The source is a pure sinusoidal signal at a frequency of 15 MHz. Resistor R represents the input resistance of an ideal amplifier. Determine the form of the signal at b after the switch is thrown into this position.

Solution
Note that the source is not a dc voltage source, but a sinusoidal waveform generator,

$$v_s = V_s \cos(2\pi f_0 t), \quad f_0 = 15\,\text{MHz}.$$

On the other hand,

$$\gamma = \frac{f_0}{B_0} = \frac{15}{352.3} \times 10^9 = 42.577\,\text{MHz.T}^{-1}.$$

This is the gyromagnetic ratio of a proton.

Therefore, after the switch S is thrown into position b, the waveform at b is due to free induction decay (FID) which is of the form

$$v_b(t) = V_b e^{-t/\tau} \cos(2\pi f_0 t),$$

where $\tau = T_2$ is the spin–spin relaxation time for pure liquid water (=2.5 s [5]).

References

1. Ladera CL, Donoso G (2013) Oscillations of a spring-magnet system damped by a conductive plate. Eur J Phys 34:1187–1197
2. Weinstock H (2013) *Applications of superconductivity* (Chapters 3 and 7). Berlin: Springer Science & Business Media
3. Seidel P (ed) (2015) *Applied superconductivity: handbook on devices and applications* (vol 1, Chapter 4) New Jersey: Wiley
4. Ansorge R, Graves M (2016) *The physics and mathematics of MRI* (Chapter 2). San Rafael: Morgan & Claypool Publishers
5. Chary KVR, Govil G (2008) *NMR in biological systems: from molecules to human* (pp 401–404). Berlin: Springer Science & Business Media

Chapter 5
Second-Order Circuits

5.1 Second-Order RLC Circuits

Problem 5.1.1 Find the equation of current as a function of time in a source-free series RLC circuit with initial capacitor voltage V_0, assuming that a series connected switch discharges the capacitor through R, L by closing at time $t = 0$.

Solution

The circuit is excited by initial voltage over the capacitor, V_0.

$$v_c(t) = \frac{1}{C}\int_{-\infty}^{t} i(\tau)\,\mathrm{d}\tau + V_0 \tag{5.1}$$

By Kirchhoffs' voltage law (KVL),

$$R \cdot i + L\frac{\mathrm{d}t}{\mathrm{d}i} + \frac{1}{C}\int_{-\tau}^{t} + V_0 = 0 \tag{5.2}$$

Differentiation,

$$\frac{\mathrm{d}^2 i}{\mathrm{d}t^2} + \frac{R}{L}\frac{\mathrm{d}i}{\mathrm{d}t} + \frac{i}{LC} = 0 \tag{5.3}$$

The characteristic equation is

$$s^2 + \frac{R}{L}s + \frac{1}{LC} = 0 \tag{5.4}$$

A.Ü. Keskin, *Electrical Circuits in Biomedical Engineering*,
DOI 10.1007/978-3-319-55101-2_5

Its roots

$$s_1 = -\frac{R}{2L} \pm \sqrt{\left(\frac{R}{2L}\right)^2 - \frac{1}{LC}} \tag{5.5}$$

$$s_{1,2} = -\alpha \pm \sqrt{\alpha^2 - \omega_0^2} \tag{5.6}$$

$$\alpha = \frac{R}{2L} \tag{5.7}$$

$$\omega_0 = \frac{1}{\sqrt{LC}} \tag{5.8}$$

The natural response is

$$i(t) = A_1 \cdot e^{s_1 t} + A_2 \cdot e^{s_2 t} \tag{5.9}$$

Here A_1, A_2 are determined from initial values $i(0)$ and $di(0)/dt$.
Three types of solutions may exist:

1. Overdamped case, $\alpha > \omega_0$

$$i(t) = A_1 \cdot e^{s_1 t} + A_2 \cdot e^{s_2 t} \tag{5.10}$$

2. Critically damped case, $\alpha = \omega_0$

$$i(t) = (A_1 \cdot t + A_2) \cdot e^{-\alpha t} \tag{5.11}$$

3. Underdamped case, $\alpha < \omega_0$

$$i(t) = e^{-\alpha t}[(B_1 \cdot \cos(\omega_d t) + B_2 \cdot \sin(\omega_d t)] \tag{5.12}$$

The damped natural frequency is

$$\omega_d = \sqrt{\omega_0^2 - \alpha^2} \tag{5.13}$$

Problem 5.1.2 What is the range of resistor values for overdamped, critical, and underdamped responses in a series RLC circuit with $C = 200$ nF, $L = 50$ mH?

Solution

The roots of the characteristic equation of the series RLC circuit is

$$s_{1,2} = -\alpha \pm \sqrt{\alpha^2 - \omega_0^2}, \quad \omega_0 = \sqrt{\frac{1}{LC}}, \quad \alpha = \frac{R}{2L}$$

Critical resistance value in terms of known L and C is found by equating damping factor to undamped frequency, and solving the equation for resistance,

$$\sqrt{\frac{1}{LC}} = \frac{R}{2L} \quad \rightarrow \quad R = 2\sqrt{\frac{L}{C}} = 2\sqrt{\frac{50 \times 10^{-3}}{200 \times 10^{-9}}} = 2\sqrt{25 \times 10^4} = 1\,\text{k}\Omega$$

(a) Overdamped: $\alpha > \omega_0$. $R > 10^3\ \Omega$
(b) Critically damped: $\alpha = \omega_0$. $R = 1\ \text{k}\Omega$
(c) Underdamped: $\alpha < \omega_0$. $R < 1\ \text{k}\Omega$

Problem 5.1.3 Let $L = 10$ mH and $C = 1$ mF component values are kept fixed, initial voltage over the capacitor is 10 V, and the resistance value is changed from 10 to 1 Ω, gradually, following the recharging of the capacitor each time. Determine and plot the waveforms for the voltage over capacitor and the voltage drop across the resistor, for typical resistor values (RLC1.cir).

Solution

$$\alpha = \frac{R}{2L} = \frac{R}{2 \times 10 \times 10^{-3}} = 50R, \quad \omega_0 = \frac{1}{\sqrt{LC}} = \frac{1}{\sqrt{10 \times 10^{-3} \times 0^{-3}}} = \frac{1000}{\sqrt{10}}$$

$$50R = \frac{1000}{\sqrt{10}} \quad \rightarrow \quad R = \frac{1000}{50 \times \sqrt{10}} = \frac{20}{\sqrt{10}} = 6.32\,\Omega$$

If $R > 6.32\ \Omega$, circuit response is over-damped, if $R < 6.32\ \Omega$, the response is under-damped.

Figures 5.1, 5.2, and 5.3 display the variation of voltages across the capacitor and the resistor in the circuit.

Problem 5.1.4 Find the equation of current as a function of time in a source-free series underdamped RLC circuit with initial capacitor voltage V_0.

Solution

Underdamped case:

$\alpha < \omega_0$,

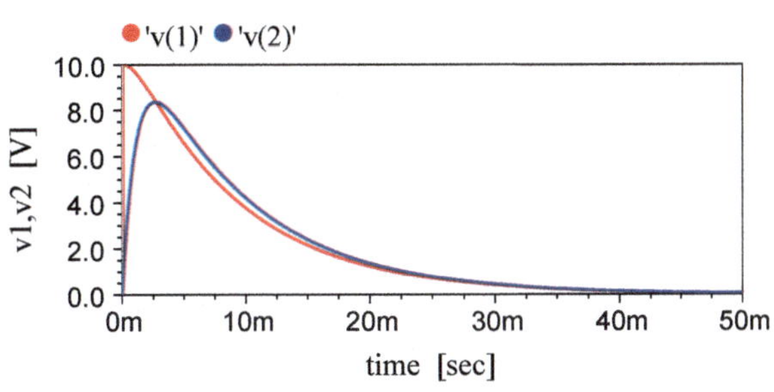

Fig. 5.1 The v_C and v_R variation for $R = 10\ \Omega$, $L = 10$ mH, $C = 1000\ \mu$F

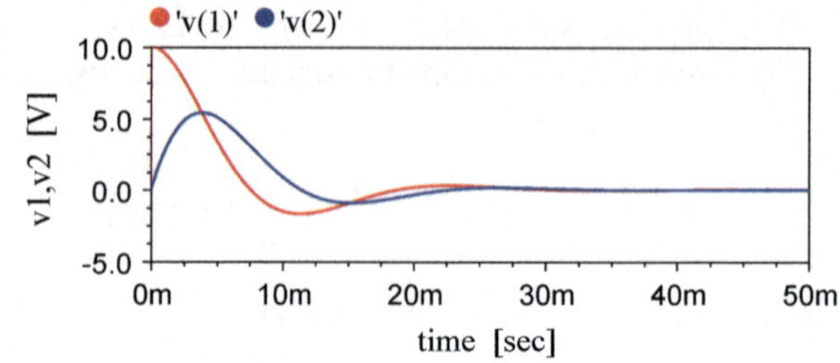

Fig. 5.2 The v_C and v_R variation for $R = 3.1\ \Omega$, $L = 10$ mH, $C = 1000\ \mu$F

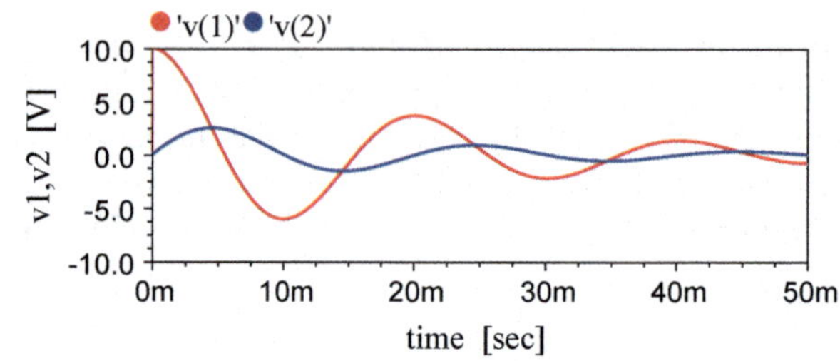

Fig. 5.3 The v_C and v_R variation for $R = 1\ \Omega$, $L = 10$ mH, $C = 1000\ \mu$F

$$i(t) = \mathrm{e}^{-\alpha t}(B_1 \cdot \cos(\omega_d t) + B_2 \cdot \sin(\omega_d t)) \tag{5.14}$$

With $v_c(0^-) = V_0$ volts,

$$i(t) = B_1 \cdot e^{-\alpha t} \cdot \cos(\omega_d t) + B_2 \cdot \mathrm{e}^{-\alpha t} \cdot \sin(\omega_d t) \tag{5.15}$$

Since $i(0^+) = 0$ A, $B_1 = 0$,

$$v_R(0^+) = 0\,\mathrm{V}, \quad v_c(0^+) = v_L(0^+) = L \cdot \mathrm{d}i(0^+)/\mathrm{d}t = V_0 \tag{5.16}$$

$$\frac{\mathrm{d}i(0^+)}{\mathrm{d}t} = \frac{V_0}{L} \tag{5.17}$$

Because of the derivative identity,

$$\left(\frac{\mathrm{d}}{\mathrm{d}t}[b \cdot \mathrm{e}^{ax}(\sin \omega \cdot x)]\right) = b \cdot \mathrm{e}^{ax} \cdot [a \cdot \sin(\omega \cdot x) + \omega \cdot \cos(\omega \cdot x)] \tag{5.18}$$

with $a = -\alpha$,

$$\left(\frac{\mathrm{d}}{\mathrm{d}t}[B_2 \cdot \mathrm{e}^{-\alpha t} \cdot \sin(\omega_d t)]\right) = (B_2 \cdot \mathrm{e}^{-\alpha t}[(-\alpha) \cdot \sin(\omega_d \cdot t) + \omega_d \cdot \cos(\omega_d \cdot t)]$$

$$\frac{\mathrm{d}i(0^+)}{\mathrm{d}t} = \frac{V_0}{L} = B_2 \cdot \mathrm{e}^0[(-\alpha) \cdot \sin(0) + \omega_d . \cos(0)] \tag{5.19}$$

$$\frac{V_0}{L} = B_2 \cdot \omega_d \tag{5.20}$$

$$B_2 = \frac{V_0}{\omega_d \cdot L} \tag{5.21}$$

Substituting this formula into the equation for current yields

$$i(t) = \left(\frac{V_0}{\omega_d \cdot L}\right) \cdot e^{-\alpha t} \cdot \sin(\omega_d . t) \tag{5.22}$$

Note that the capacitor voltage is

$$v_c(t) = L \cdot \frac{\mathrm{d}i}{\mathrm{d}t} + iR = -\frac{1}{C}\int_0^t i\,\mathrm{d}\tau + V_0 \tag{5.23}$$

Problem 5.1.5 How is dissipated energy in an underdamped series RLC circuit calculated using the current waveform?

Solution

$$\begin{aligned} W &= \int_0^t i^2(\tau) \cdot R\mathrm{d}\tau \\ &= R\int_0^t i^2(\tau)\mathrm{d}\tau = R\int_0^t [B \cdot \mathrm{e}^{-\alpha\tau} \cdot \sin(\omega_d . \tau)]^2 \mathrm{d}\tau = R \cdot B^2 \int_0^t \mathrm{e}^{-2\alpha\tau} \cdot \sin^2(w_d \cdot \tau)\mathrm{d}\tau \end{aligned} \tag{5.24}$$

The last integral equation is of the from

$$\int \mathrm{e}^{ax} \cdot \sin^2(bx)\mathrm{d}x = \frac{\mathrm{e}^{ax}[4b^2 + a^2 - a^2\cos(2bx) - 2ab \cdot \sin(2bx)]}{2(a^3 + 4ab^2)} \tag{5.25}$$

where $a = -2\alpha$, $b = \omega_d$.

Instead of computing Eq. (5.25), it is easier and more meaningful to perform numerical integration of Eq. (5.24). In addition, numerical methods of computation of cumulative integrals is rather intuitive to demonstrate instantaneous energy dissipation by a graphical construction of cumulative energy as function of time.

Problem 5.1.6 A series RLC circuit with initially charged capacitor has following component values: $C = 100$ nF, $L = 100$ mH, $R = 56\ \Omega$, $V_c(0) = 100$ V (see, Fig. 5.4a). Determine the equation of current flow in the circuit after the switch S is closed, and plot it in MATLAB platform (RLCp1.m).

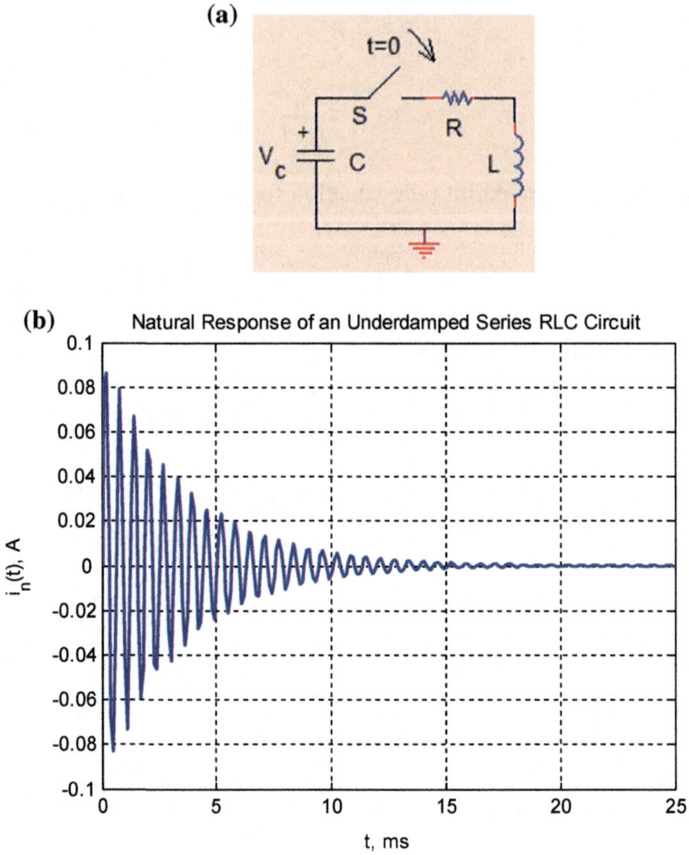

Fig. 5.4 **a** A series RLC circuit with initially charged capacitor and **b** C = 100 nF, L = 0.1 H, R = 56 Ω, V_0 = 100 V; $i(t)$ = 0.01 exp(−280t) sin(996t)

Solution

$$\omega_0 = \frac{1}{\sqrt{LC}} = 10^4\,\frac{\text{rad}}{\text{s}}, \qquad \alpha = \frac{R}{2L} = 280\,\text{s}^{-1}$$

$\alpha < \omega_0$, (underdamped case)

$$w_d = \sqrt{\omega_0^2 - \alpha^2} = \sqrt{(10)^8 - (280)^2} = 996\,\text{rad/s}$$
$$i(t) = \left(\frac{V_0}{\omega_d}\cdot L\right)\text{e}^{-\alpha t}\cdot\sin(\omega_d\cdot t) = \left(\frac{100}{996}\times 10^{-1}\right)\cdot \text{e}^{-280t}\cdot\sin(996\,t)$$
$$i(t) = (0.01)\text{e}^{-280t}\sin(996t)\,\text{A}$$

A plot of this waveform in MATLAB is given in Fig. 5.4b.

MATLAB script (RLCp1.m):

```
% Underdamped series RLC circuit (RLCp1.m)
L=0.1; C=100e-9; R=56; V0=100;
% a=damping coefficient, w0=natural freq., wd=damped resonance freq.
a=R/(2*L)
w0=1/sqrt(L*C)
wd=sqrt(w0*w0 - a*a)
% Set coefficients based on initial conditions
B1=0; B2= V0/(wd*C)
% Create a time base and calculate the response at those times
t=0:0.0001:0.025;
I=B1*exp(-a*t).*cos(wd*t) + B2*exp(-a*t).*sin(wd*t);
plot(1000*t,I,'b'); grid on; ylabel('i_n(t), A'); xlabel('t, ms');
title('Natural Response of an Underdamped Series RLC Circuit');
```

Problem 5.1.7 For an RLC circuit with $C = 0.1$ mF, $L = 100$ mH, $R = 56\ \Omega$, $V_c(0) = 100$ V, determine and plot the variation of current flowing in the circuit.

Solution

This is underdamped RLC case, because $\alpha < \omega_0$

$$\omega_0 = \frac{1}{\sqrt{LC}} = 316\,\mathrm{rad/s}, \quad \alpha = \frac{R}{2L} = 280\,\mathrm{s}^{-1}, \quad \omega_d = \sqrt{\omega_0 - \alpha^2} = 147\ \mathrm{rad/s}$$

$$i(t) = \left(\frac{V_0}{\omega_d} \cdot L\right) \mathrm{e}^{-\alpha t} \cdot \sin(\omega_d \cdot t) = (6.80)\mathrm{e}^{-280t} \sin(147t)\,\mathrm{A}$$

Figure 5.5 displays the variation of current flowing in the circuit.

Problem 5.1.8 A series RLC circuit with initially charged capacitor ($V_0 = 100$ V) has following component values: $C = 48\ \mu$F, $L = 0.1$ H, $R = 56\ \Omega$.

(a) Determine the equation of current and plot it.
(b) Is this waveform suitable for biphasic Lown defibrillator design?
(c) Modify the circuit for 360 J of energy delivering capability. What are the maximum and minimum current values under this condition?
(d) Plot instantaneous power dissipation and cumulative energy dissipation of the circuit as functions of time.
(e) Plot the voltage waveforms over the resistor and the capacitor during the overall dissipation of 360 J of energy.
(f) Print MATLAB m-file list to compute and plot the current flowing in the RLC series circuit.
(g) Print MATLAB m-file list to plot the cumulative energy dissipation in the circuit (RLCs2.m, RLCs3.m)(RLCs.cir).

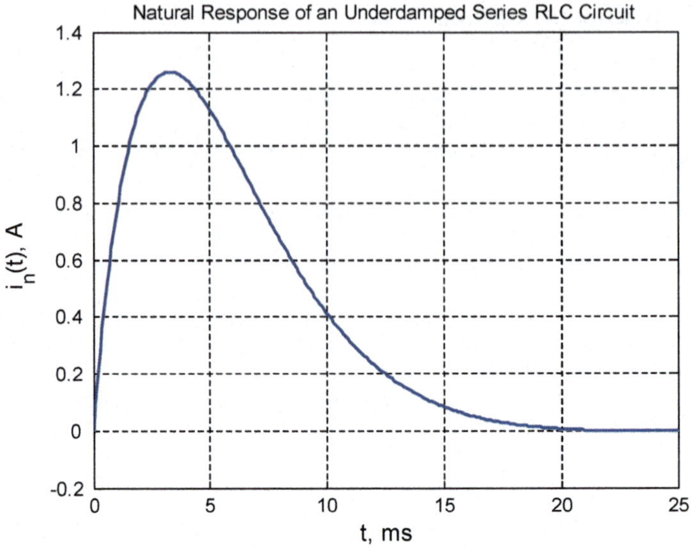

Fig. 5.5 $C = 100$ μF, $L = 0.1$ H, $R = 56$ Ω, $V_0 = 100$ V; $i(t) = 6.80 \exp(-280t) \sin(147t)$

Solution

(a) $\omega_0 = \dfrac{1}{\sqrt{LC}} = 316\,\text{rad/s}, \quad \alpha = \dfrac{R}{2L} = 280\,\text{s}^{-1},$

$\omega_d = \sqrt{\omega_0 - \alpha^2} = 360\,\text{rad/s}$

This is underdamped RLC case, because $\alpha < \omega_0$.

$$i(t) = \left(\frac{V_0}{\omega_d} \cdot L\right) e^{-\alpha t} \cdot \sin(\omega_d \cdot t) = (2.77)e^{-280t} \sin(360t)\,\text{A}$$

This current waveform is shown in Fig. 5.6.

(b) This current waveform shown in Fig. 5.6 is suitable for Lown type biphasic defibrillator design. However, it requires some energy improvement.

(c) Considering the availability of only certain rated values of capacitors with maximum voltages and, $C = 48$ μF, $L = 100$ mH; A maximum of 360 J of stored energy can be dissipated over the resistor ($R = 56$ Ω), if the initial voltage over the capacitor is $V_c(0) = V_0 = 3873$ V. In this case,

$$i(t) = \left(\frac{V_0}{\omega_d} \cdot L\right) e^{-\alpha t} \cdot \sin(\omega_d \cdot t) = (107.45)e^{-280t} \sin(360t)\,\text{A}$$

This current waveform is shown in Fig. 5.7.

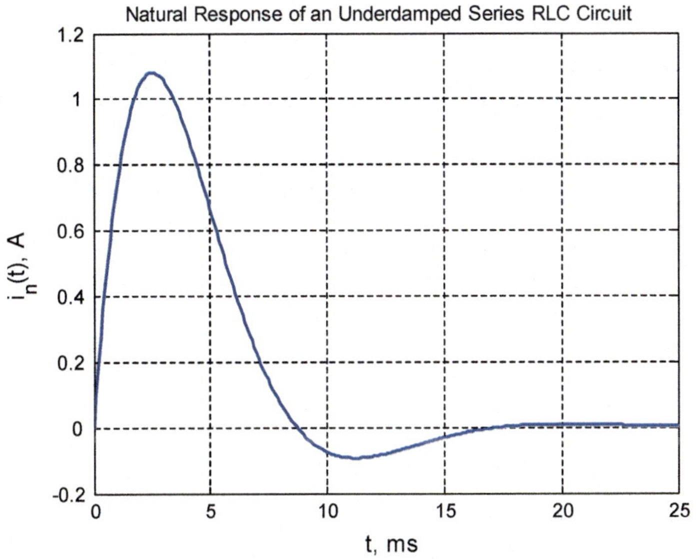

Fig. 5.6 $C = 48\ \mu F$, $L = 0.1$ H, $R = 56\ \Omega$, $V_0 = 100$ V; $i(t) = 2.77 \exp(-280t) \sin(360t)$

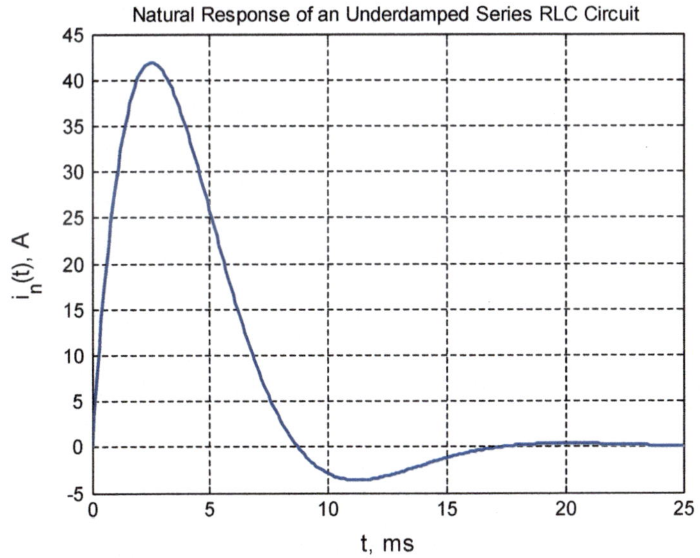

Fig. 5.7 Natural response of an underdamped series RLC circuit. Dissipation of 360 J of energy over the resistor $R = 56\ \Omega$, $C = 48\ \mu F$, $L = 0.1$ H, $V_0 = 3873$ V; $i(t) = 107.45 \exp(-280t) \sin(360t)$

Inspection of this waveform shows that maximum peak positive current i (2.5 ms) = 41.81 A flows through the circuit, while a negative peak current of 3645 A is reached at $t = 11.2$ s.

(d) Fig. 5.8 displays instantaneous power dissipation of the circuit.

Cumulative dissipation of 360 J of energy in the circuit is given in Fig. 5.9. It is to be noted that 97.5% of stored energy is dissipated across the R within 6.4 ms.

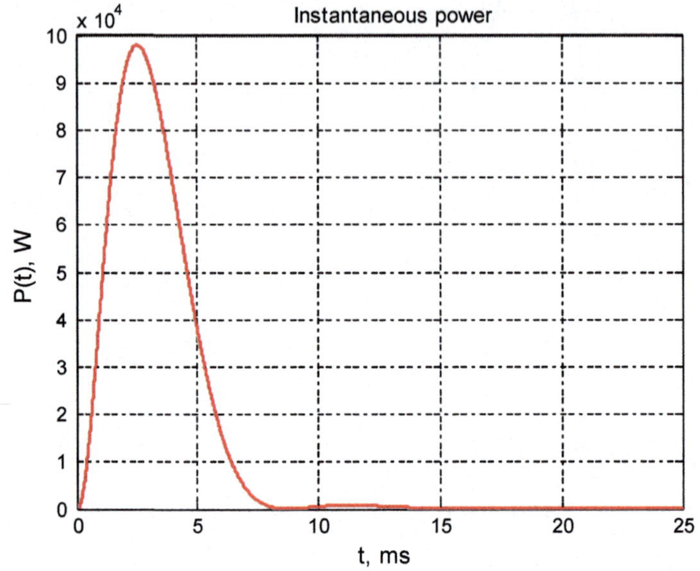

Fig. 5.8 Instantaneous power over the resistor (in dissipation of 360 J of energy)

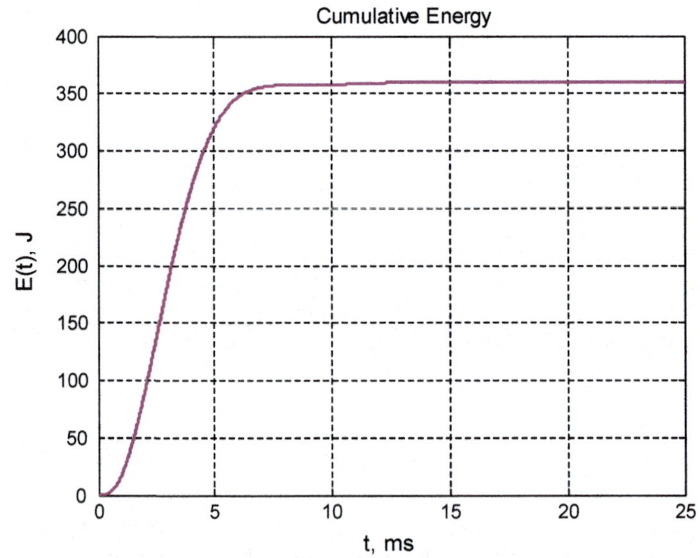

Fig. 5.9 Cumulated energy dissipation waveform over the resistor (during the overall dissipation of 360 J of energy over the resistor). Note that initial stored energy E_0 = 360.0031 J, is dissipated across the resistor element, E = 360.0029 J before 18 ms, while 97.5% of the stored energy in the capacitor is dissipated across the resistor element within the first 6.4 ms

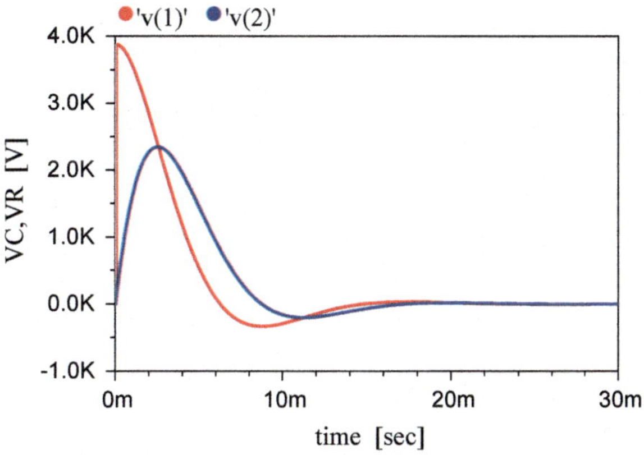

Fig. 5.10 The voltage waveforms over the resistor (*blue line*) and the capacitor (*red line*) during dissipation of 360 J of energy over the resistor. $R = 56\ \Omega$, $C = 48\ \mu F$, $L = 0.1$ H, $V_0 = 3873$ V, $i(t) = 107.45 \exp(-280) \sin(360t)$. Note that initial voltage diminishes and stored energy is dissipated across the resistor element, before 18 ms

(e) Simultaneous variation of voltage waveforms across the resistor and the capacitor elements are shown in Fig. 5.10.

(f) MATLAB m-file list to compute and plot the current flowing in the RLC series circuit:

```
% Underdamped series RLC circuit
L=0.1; C=48e-6; R=56; V0=100;
% a=damping coefficient, w0=natural frequency,
% wd=damped resonance frequency
a=R/(2*L)
w0=1/sqrt(L*C)
wd=sqrt(w0*w0 - a*a)
% Set coefficients based on initial conditions
B1=0; B2= V0/(wd*L)
% Create a time base and calculate the response
t=0:0.0001:0.025;
I=B1*exp(-a*t).*cos(wd*t) + B2*exp(-a*t).*sin(wd*t);
plot(1000*t,I,'b');grid on; ylabel('i_n(t), A'); xlabel('t, ms');
title('Natural Response of an Underdamped Series RLC Circuit');
```

(g) MATLAB m-file list to compute and plot the cumulative energy dissipation in the RLC series circuit:

```
%Energy dissipated in series RLC circuit using integration by
%trapezoids
Vc0=3873; L=0.1; C=48e-6; R=56;
E0=C*Vc0*Vc0/2 %initial energy stored on capacitor
a=R/(2*L)
w0=1/sqrt(L*C)
A=w0-a;
if A<=0. ' not underdamped case, w0<=a'
else
wd=sqrt(w0*w0 - a*a);
% Set coefficients based on initial conditions
B2= Vc0/(wd*L);
end;
% Create a time base and calculate the response at those times
t=0:0.00001:0.025;
I=B2*exp(-a*t).*sin(wd*t);
%hold off
plot(1000*t,I,'b','linewidth',2);grid on;figure;
%hold on
y=I.*I*R; % Array multiplication is used.
E=trapz(t,y) % energy dissipated on resistor.
plot(1000*t,y,'r','linewidth',2);grid on;ylabel('P(t), W');
xlabel('t, ms'); title('Instantaneous power');
figure;
Z=cumtrapz(t,y);% cumulative energy dissipated on resistor.
plot(1000*t,Z,'m','linewidth',2);grid on;
ylabel('E(t), J'); xlabel('t, ms'); title('Cumulative Energy');
```

Problem 5.1.9 Determine the Inductance for a flat spiral air core (defibrillator) coil if, A = cross-sectional area = 3 cm^2, l = length of the coil = 50 cm, N = number of turns = 45, μ_0 = permeability of air ($4\pi \times 10^{-7}$ H/m).

Discuss the influence of the coiled cables' inductance on discharge current in a Lown defibrillator circuit with $C = 48$ μF, $L = 0.1$ H, initial voltage across the capacitor $V_0 = 3873$ V, and an overall energy dissipation of 360 J of energy across the load resistor $R = 56\ \Omega$.

Solution

$$L = \frac{N^2 \mu_0 A}{l} = \frac{(45)^2 \times 4 \times \pi \times 10^{-7} \times 3 \times 10^{-4}}{0.5} = 152{,}681 \times 10^{-11}\,\text{H} = 1.52\,\mu\text{H}$$

If there is no coupling (i.e., mutual inductance Inductance, mutual is null) between the two paddles of the defibrillator, total inductance is approximately

3 μH. This value is too small as compared to 100 mH inductance of the Lown circuit. Therefore, its effect can be ignored.

Problem 5.1.10 Determine the maximum current in a Lown defibrillator circuit with $C = 48$ μF, $L = 0.1$ H, initial voltage across the capacitor $V_0 = 3873$ V, and an overall energy dissipation of 360 J of energy across the load resistor $R = 56\ \Omega$.

Solution

A Lown type defibrillator is a series connected resistor inductor and capacitor circuit where, the resistance is the sum of coil internal resistance and the transthoracic resistance between electrodes placed on sternum and apex of the patient.

With $C = 48$ μF, $L = 0.1$ H, initial voltage across the capacitor $V_0 = 3873$ V, and an overall energy dissipation of 360 J of energy across the resistor $R = 56\ \Omega$, the current waveform is defined as

$$i(t) = B_1 e^{-\alpha t}\cos(\omega_d t) + B_2 e^{-\alpha t}\sin(\omega_d t)\ \text{A}$$

$$\omega_0 = \frac{1}{\sqrt{LC}} = \frac{1}{\sqrt{(0.1\,\text{H})(48\,\mu\text{F})}} = 456.4\frac{\text{rad}}{\text{s}},\ \alpha = \frac{R}{2L} = 280\,\text{s}^{-1}$$

$$\omega_d = \sqrt{\omega_0^2 - \alpha^2} = \sqrt{456.4^2 - 280^2} = 360.5\,\text{rad/s}$$

Here, α, ω_d, ω_0 are Neper Frequency, damped resonance frequency of the circuit, and the resonance frequency, respectively.

Using given component values and initial condition yields $B_1 = 0$, and the equation of current waveform is

$$i(t) = \left(\frac{V_0}{\omega_d}\cdot L\right)e^{-\alpha t}\cdot\sin(\omega_d\cdot t) = (107.445)e^{-280t}\sin(360.5t)\ \text{A}$$

This is of the form $i(t) = Ae^{-\alpha t}\sin(\omega_d t)$

Maximum current level at $t_{\max}$ can be determined by differentiating the current with respect to time and then equating it to zero,

$$\frac{di(t)}{dt} = A\omega_d e^{-\alpha t}\cos(\omega_d t) - \alpha A e^{-\alpha t}\sin(\omega_d t) = Ae^{-\alpha t}[\omega_d\cos(\omega_d t) - \alpha\sin(\omega_d t)] = 0$$

$$\omega_d\cos(\omega_d t) = \alpha\sin(\omega_d t) \quad\rightarrow\quad \frac{\omega_d}{\alpha} = \frac{\sin(\omega_d t)}{\cos(\omega_d t)} = \tan(\omega_d t)$$

$$\omega_d t = \text{a}\tan\left(\frac{\omega_d}{\alpha}\right)$$

$$t = t_{\max} = \frac{1}{\omega_d}\text{a}\tan\left(\frac{\omega_d}{\alpha}\right)$$

Substituting given values in this equation gives $t = 2.5\,\text{ms}$. Plugging this value of time into the current yields

$$i(t_{max}) = Ae^{-\alpha t_{max}} \sin(\omega_d t_{max})$$
$$i(2.5\,\text{ms}) = 41.81\,\text{A}$$

In other words, the maximum positive peak current of 41.833 A occurs at t = 2.5 ms.

Problem 5.1.11 Determine the resistance value in a parallel RLC circuit so that its transient behavior is the same with that of a series RLC circuit having same values for inductance and capacitance.

Solution

Transient behavior of simple series or parallel three-element RLC circuits are same if their damping coefficients have the same values.

$$\alpha = \frac{1}{2RC} \quad (\text{parallel}\, RLC), \quad \alpha = \frac{R}{2L} \quad (\text{series}\, RLC)$$

Therefore, equating both expressions, one obtains

$$\frac{1}{2RC} = \frac{R}{2L} \quad \rightarrow \quad R = \sqrt{\frac{L}{C}}$$

All other characteristic equations are same, with

$$s_{1,2} = -\alpha \pm \sqrt{\alpha^2 - \omega_0^2} \quad , \quad \omega_0 = \frac{1}{\sqrt{LC}}$$

The natural response is

$$i(t) = A_1 \cdot e^{s_1 t} + A_2 \cdot e^{s_2 t}$$

Here A_1, A_2 are determined from inital values $i(0)$ and $di(0)/dt$. Three types of solutions may exist for both series and parallel RLC circuits Parallel RLC circuits:

1. Overdamped case, $\alpha > \omega_0$

$$i(t) = A_1 \cdot e^{s_1 t} + A_2 \cdot e^{s_2 t}$$

2. Critically damped case, $\alpha = \omega_0$

$$i(t) = (A_1 \cdot t + A_2) \cdot e^{-\alpha t}$$

3. Underdamped case, $\alpha < \omega_0$

$$i(t) = e^{-\alpha t}[(B_1 \cdot \cos(\omega_d t) + B_2 \cdot \sin(\omega_d t)]$$

The damped natural frequency Damped natural frequency is

$$\omega_d = \sqrt{\omega_0^2 - \alpha^2}$$

Problem 5.1.12 The switch in the dc circuit shown in Fig. 5.11 has been closed for a long time. $C = 0.1$ F.

It is opened at $t = 0$.

Find:

(a) $i_L(0^-)$, $v_C(0^-)$;
(b) $i_L(\infty)$, $v_C(\infty)$;

Solution

(a) Before switch is opened, L is " short," C is "open" (Fig. 5.12),

$$i(0^-) = \frac{12}{4+2} = 2\,\text{A}$$
$$v_C(0^-) = 2 \cdot i(0^-) = 2 \cdot 2 = 4\,\text{V}$$

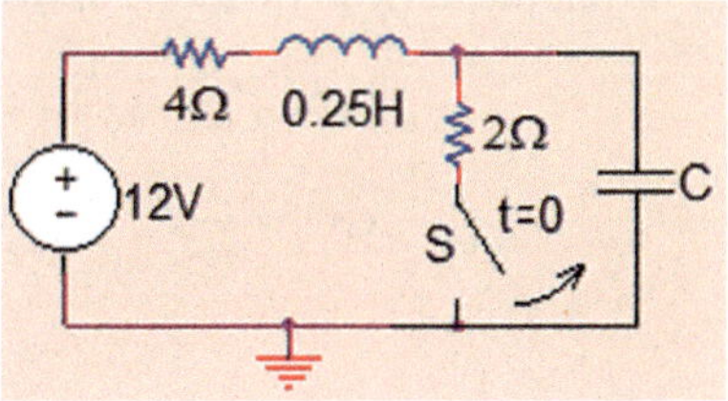

Fig. 5.11 The circuit of Problem 5.1.12

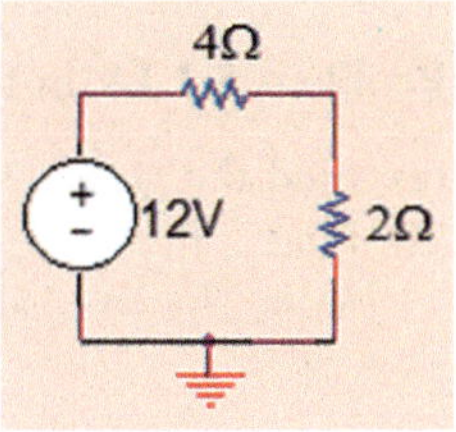

Fig. 5.12 The circuit of Problem 5.1.12 before switch is opened

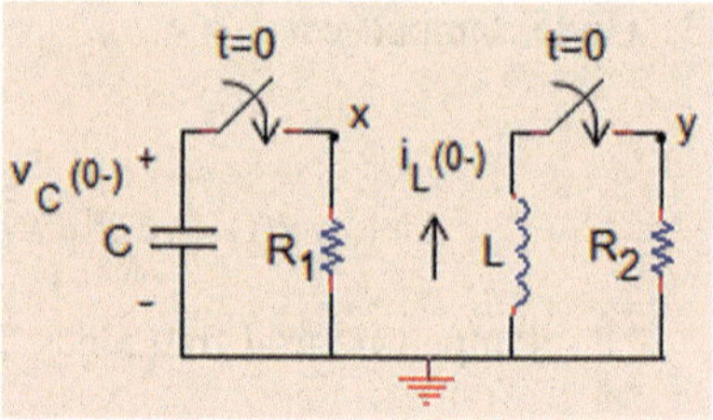

Fig. 5.13 The circuit of Problem 5.1.13

(b) For $t = \infty$, circuit reaches another steady state, L is "short," C is "open".

$$i_L(\infty) = 0\,\text{A}, \quad v_C(\infty) = 12\,\text{V}$$

Problem 5.1.13 In Fig. 5.13, voltage variations at points X and Y are found to be identical to each other after switches have been closed. $C = 2$ F, $L = 3$ H, $v_C(0^-) = V_0 = 10$ V, $i_L(0^-) = I_0 = 1$ A.

(a) Find the values of both resistors.
(b) What is the time constant of each circuit (in milliseconds)?

Solution

(a)
$$V_0 = v_C(0^-) = 10\text{ V}, \quad I_0 = i_L(0^-) = 1\,\text{A}$$

$$v_x = v_C(t) = V_0 \cdot \mathrm{e}^{-\frac{t}{\tau}}\,\text{V}, \quad v_y = v_L(t) = (R_2 \cdot I_0) \cdot \mathrm{e}^{-\frac{t}{\tau}}\,\text{V}$$

From these two equations above, one obtains $V_0 = R_2 \cdot I_0$

$$10 = R_2 \cdot 1 \quad \rightarrow \quad R_2 = 10\,\Omega, \quad \tau_{RC} = \tau_{RL} = \tau$$
$$R_1 C = \frac{L}{R_2} \quad \rightarrow \quad R_1 = \frac{L}{C \cdot R_2} = \frac{3}{2.10} = 0.15\,\Omega$$

(b)
$$\tau_{RC} = \tau_{RL} = \tau = R_1 C = \frac{L}{R_2} = 0.15 \times 2 = \frac{3}{10} = 0.3\,\text{s} = 300\text{ ms}$$

Problem 5.1.14 In the circuit shown in Fig. 5.14,

(a) Find the steady state current flowing through $L = 1.6$ H inductor,
(b) Find the dc steady state voltage across $C = 100$ μF capacitor. $R_1 = 2R_2 = 2$ kΩ, $I_{dc} = 100$ mA (Fig. 5.15).

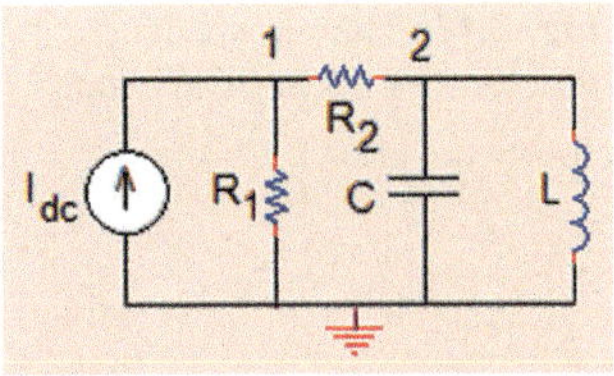

Fig. 5.14 The circuit of Problem 5.1.14

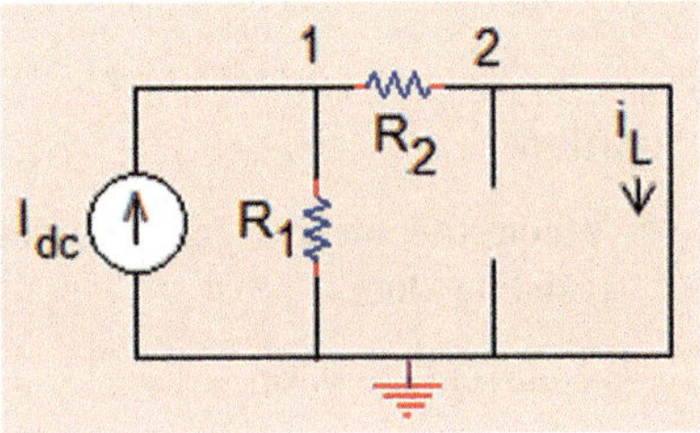

Fig. 5.15 The circuit of Problem 5.1.14 in steady state

Solution

(a) When the steady state condition is reached, the circuit becomes a purely resistive one,

By current divisionCurrent-division principle,

$$i_L(\infty) = I_{\mathrm{dc}} \frac{R_1}{R_1 + R_2} = 100\,\mathrm{mA} \times \frac{2}{2+1} = \frac{200}{3}\,\mathrm{mA} = 66.667\,\mathrm{mA}$$

(b) Because the capacitor becomes shorted by the inductor when steady state condition is reached,

$$v_C(\infty) = 0\,\mathrm{V}$$

Problem 5.1.15 In the circuit shown in Fig. 5.16

(a) Determine the steady state current flowing through L = 1 mH inductor,
(b) Find the steady state voltage across C = 10 μF capacitor. $R_1 = 2R_2$ = 2 kΩ, I_{dc} = 100 mA.

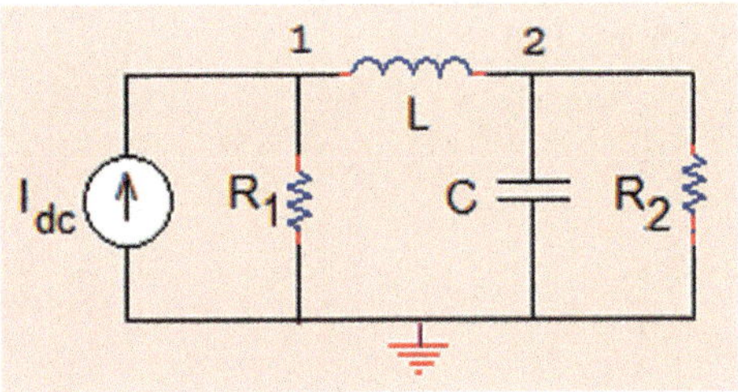

Fig. 5.16 The circuit of Problem 5.1.15

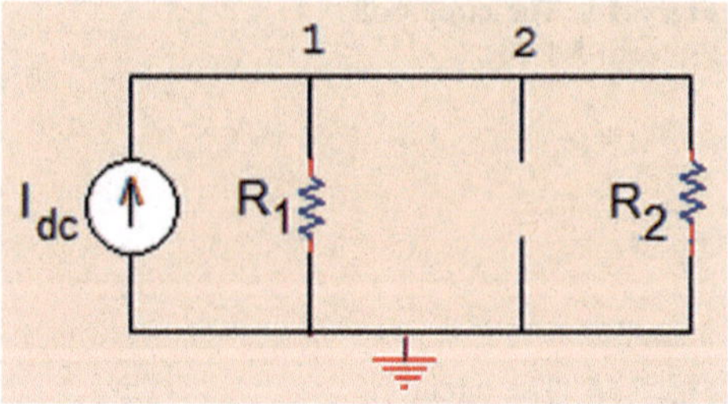

Fig. 5.17 The circuit of Problem 5.1.15 in steady state

Solution

(a) When the steady state condition is reached, the circuit becomes a purely resistive one,

By current division,

$$i_L(\infty) = I_{dc}\frac{R_1}{R_1+R_2} = 100\,\text{mA} \times \frac{2}{2+1} = \frac{200}{3}\,\text{mA} = 66.667\,\text{mA}$$

(b) Because the capacitor becomes opened when steady state condition is reached (see, Fig. 5.17), by Ohm's law,

$$v_C(\infty) = I_{dc}\frac{R_1.R_2}{R_1+R_2} = 100\,\text{mA} \times \frac{2\times 1}{2+1}\,\text{k}\Omega = \frac{200}{3}\,\text{V} = 66.667\,\text{V}$$

Alternatively,

$$v_C(\infty) = i_L(\infty) \times R_2 = 66.667\,\text{mA} \times 1\,\text{k}\Omega = 66.667\,\text{V}$$

Problem 5.1.16 In the circuit shown in Fig. 5.18, $v_c(t) = 0\,\text{V}, \quad i_L(t) = 0\,\text{A}, t<0$.

Determine $i_L(t), i_C(t), v_C(t), v_{R1}(t)$ and, $i_{R1}(t)$ (a) at $t = 0^+$, (b) at $t = \infty$

$$R_1 = R_2 = 1\,\text{k}\Omega, \quad C = 1\,\mu\text{F}, \quad L = 1\,\text{mH}, \quad U = 10\,\text{V}$$

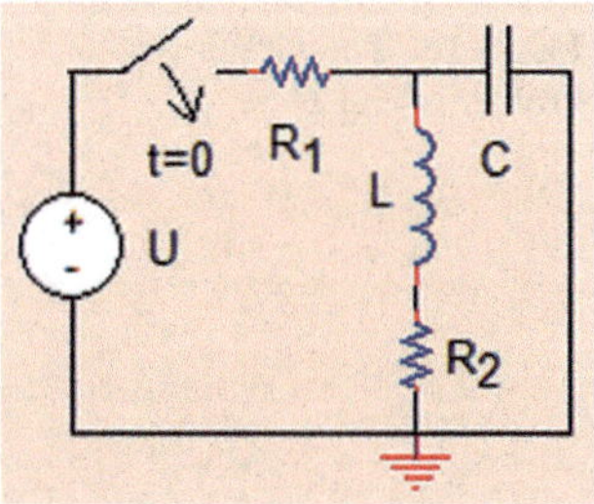

Fig. 5.18 The circuit of Problem 5.1.16

Solution

(a) $$t = 0^+; i_L(0^+) = 0\,\text{A}, i_C(0^+) = i_{R1}(0^+) = \frac{U}{R_1} = \frac{10}{1\text{k}} = 10\,\text{mA}$$
$$v_C(0^+) = 0\,\text{V}, v_{R1}(0^+) = U - v_C(0^+) = 10 - 0 = 10\,\text{V}$$

$$t = \infty; \quad i_C(\infty) = 0\,\text{A}, \quad v_C(\infty) = 10 \times \frac{R_2}{R_1 + R_2} = 10\frac{1}{2} = 5\,\text{V}$$

(b) $$i_L(\infty) = \frac{V_C(\infty)}{R_2} = \frac{5}{1} = 5\,\text{mA}, \quad v_{R1}(\infty) = U - v_C(\infty) = 10 - 5 = 5\,\text{V}$$
$$i_{R1}(\infty) = \frac{V_{R1}(\infty)}{R_1} = 5\,\text{mA}$$

Problem 5.1.17 In the circuit of Fig. 5.19, $i_{L1}(t) = i_{L2}(t) = 0\,\text{A}$ and $v_C(t) = 0\,\text{V}$ for $t < 0$.

Determine;

(a) $$i_{L1}(0^+), \quad v_x(0^+), \quad v_y(0^+), \quad v_c(0^+), \quad i_{R1}(0^+)$$

(b) $$v_x(\infty), \quad i_{L1}(\infty), \quad i_{L2}(\infty)$$

$$R_1 = R_2 = R_2 = 1\,\text{k}\Omega,\ L_1 = L_2 = 1\,\text{mH},\ C = 10\,\mu\text{F},\ U = 10\,\text{V}$$

Solution

Voltage across a capacitor (current through an inductor) can not change abruptly. On the other hand, inductor is a short and capacitor is an open circuit in steady state (DC).

(a) $i_{L1}(0^+) = i_{L2}(0^+) = 0\,\text{A}, v_c(0^+) = 0\,\text{V}, v_x(0^+) = v_y(0^+) = 10\,\text{V}, i_{R1}(0^+) = 0\,\text{A}$

(b) Final values are,

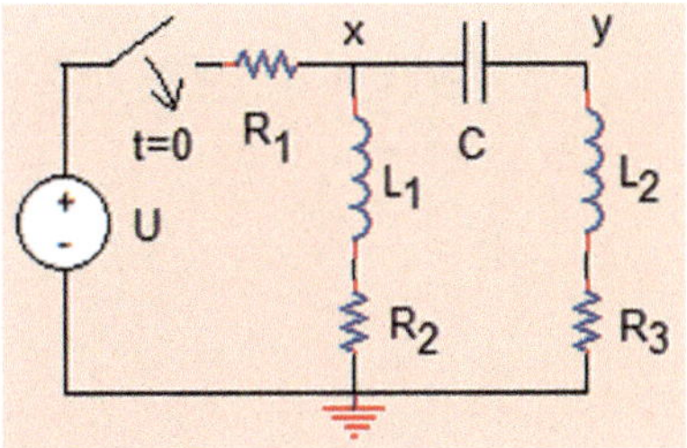

Fig. 5.19 The circuit of Problem 5.1.17

$$v_x(\infty) = U\frac{R_2}{R_1 + R_2} = 10\frac{1}{2} = 5\,\text{V}, \quad i_{L1}(\infty) = \frac{v_x(\infty)}{R_2} = \frac{5}{1\,\text{k}\Omega} = 5\,\text{mA},$$
$$i_{L2}(\infty) = 0\,\text{A}$$

Problem 5.1.18 In the circuit showing Fig. 5.20a, $R_2 = R_3 = R_4 = 2R_1 = 1$ kΩ, $C = 1$ mF, $L = 0.1$H, $V_{dc} = 10$ V.

Find the values of,

a) $v_2(\infty)$, b) $v_3(\infty)$, c) $v_4(\infty)$, d) i_C (∞), e) $i_{R2}(\infty)$, f) $i_{R1}(\infty)$, g) $v_3(0^+)$, h) $i_{R3}(0^+)$, i) $v_4(0^+)$, j) $v_2(0^+)$, k) $i_{R4}(0^+)$, l) $i_{R1}(0^+)$, m) $i_{R2}(0^+)$.

Solution

The circuit at $t = \infty$: ($v_2 = v_3 = v_4$)

If $V_{dc} = v_1 = 10$ V, $v_2 = v_1/2 = 5$ V, for $t = \infty$:

(a) $v_2 = 5$ V,
(b) $v_3 = 5$ V,
(c) $v_4 = 5$ V,
(d) $i_C = 0$ A,

(e) $i_{R2} = \dfrac{5\,\text{V}}{1\text{k}} = 5\,\text{mA},$

(f) $i_{R1} = \dfrac{(10 - 5)\,\text{V}}{0.5\text{k}} = \dfrac{5}{0.5\text{k}} = 10\,\text{mA}$

(g) $v_3(0^+) = 0$ V,
(h) $i_{R3}(0^+) = 0$ A,
(i) $v_C(0^+) = v_4(0^+) = 0$ V,

(j) $v_2(0^+) = \left[\dfrac{R_2//R_4}{(R_2//R_4) + R_1}\right] \cdot v_1(0^+) = 5\,\text{V},$

Fig. 5.20 (a) The circuit of Problem 5.1.18. (b) Steady state circuit

(a)

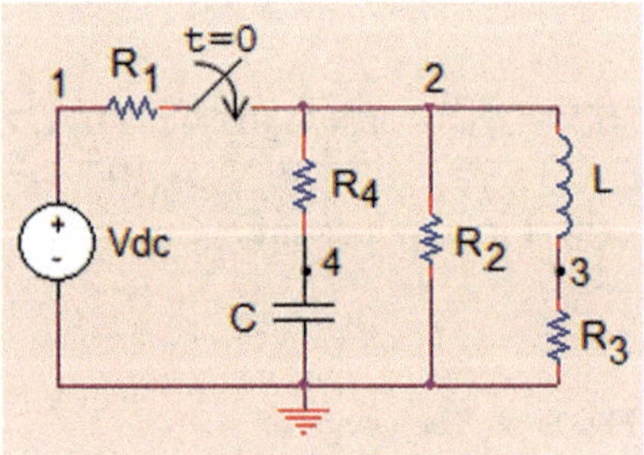

(b)

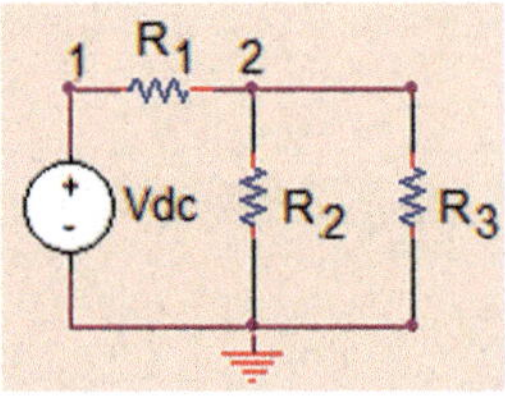

(k) $i_{R4}(0^+) = \frac{5}{1\text{k}} = 5\,\text{mA},$

(l) $i_{R1}(0^+) = \frac{5}{5\text{k}} = 10\,\text{mA},$

(m) $i_{R2}(0^+) = \frac{5V}{1\text{k}} = 5\,\text{mA}$

Problem 5.1.19 In the circuit shown in Fig. 5.21, $i_{R3}(0^+) = ?$, $i_{R1}(\infty) = ?$, $v_3(\infty) = ?$ ($V_{dc} = 10$ V, $R_1 = 500\ \Omega$, $R_2 = R_3 = 2R_1$, $C_1 = 1\ \mu\text{F}$, $L = 0.1$ H).

Solution

See Fig. 5.20b, it has the same circuit structure.

$$i_{R3}(0^+) = 0 \quad \text{(Current through an inductor cannot change abruptly)}$$

$$v_2(\infty) = v_3(\infty) = V_{dc}\frac{R_2//R_3}{R_2//R_3 + R_1} = 10 \cdot \frac{0.5\text{k}}{0.5\text{k} + 0.5\text{k}} = 5\text{ V}$$

$$i_{R1}(\infty) = \frac{V_{dc} - v_2(\infty)}{R_1} = \frac{10 - 5}{0.5\text{k}} = 10 \times 10^{-3}\text{ A} = 10\,\text{mA}$$

Problem 5.1.20 For the circuit shown in Fig. 5.22, $R_1 = R_2 = R_3 = 10\ \Omega$, $V_{dc} = 10$ V, $C = 1$ F, $L = 1$ H. Find;

$$v_1(0^+), v_2(0^+), i_{R2}(0^+), v_3(0^+), i_{R1}(0^+), i_C(0^+), i_L(0^+),$$
$$i_{R2}(\infty),\quad v_2(\infty),\quad v_3(\infty),\quad i_L(\infty),\quad i_{R1}(\infty),\quad i_C(\infty).$$

Use SPICE analysis and check the results found for currents through inductor and capacitor. Print SPICE netlist (inifin1.cir).

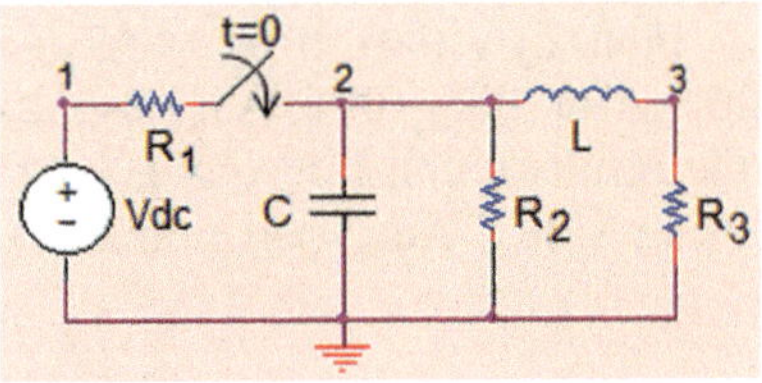

Fig. 5.21 The circuit of Problem 5.1.19

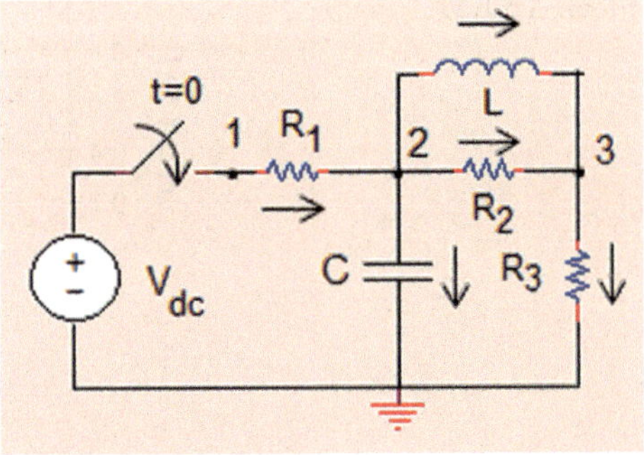

Fig. 5.22 The circuit of Problem 5.1.20

Solution

$$v_1(0^+) = V_{dc}V, \quad v_2(0^+) = 0\,\text{V}, \quad i_{R2}(0^+) = 0\,\text{A}, \quad v_3(0^+) = 0\,\text{V},$$

$$i_{R1}(0^+) = \frac{v_1}{R_1} = \frac{V_{dc}}{R_1} = \frac{10}{10} = 1\,\text{A}, \quad i_C(0^+) = i_{R1}(0^+) = 1\,\text{A}, \quad i_L(0^+) = 0\,\text{A},$$

$$i_{R2}(\infty) = 0\,\text{A}, \quad v_2(\infty) = V_{dc}\cdot\frac{R_3}{R_1+R_3} = 10\cdot\frac{10}{10+10} = 5\,\text{V}, \quad v_3(\infty) = v_2(\infty)$$

$$i_L(\infty) = \frac{V_{dc}}{R_1+R_3} = \frac{10}{10+10} = 0.5\,\text{A}, \quad i_{R1}(\infty) = i_L(\infty) = 0.5\,\text{A}, \quad i_C(\infty) = 0\,\text{A}$$

Figure 5.23 is obtained using SPICE analysis and proves the values of currents through inductor and capacitor.

SPICE Netlist (inifin1.cir) is given below.

```
inifin1
vdc 1 0 10
r1 1 2 10
c1 2 4 1
r2 2 3 10
L1 2 3 1
r3 3 0 10
vref 4 0 0 *tran 10 m 30 0 10 m uic
```

Problem 5.1.21 In the circuit shown, V_{dc} = 10 V, $R_1 = R_2$ = 10 Ω, R_3 = 20 Ω, L = 1 H, C = 1 F (Zero initial conditions) (Fig. 5.24).

Find; a) $v_x(\infty)$, b) $i_{R1}(\infty)$, c) $i_{R2}(\infty)$, d) $v_y(\infty)$, e) $i_L(\infty)$, f) $i(\infty)$, g) $i_L(0+)$, h) $v_y(0+)$, i) $|i_{R2}(0+)|$, j) $i_{R1}(0+)$, k) $v_x(0+)$, l) $i_{R3}(0+)$, m) $i(0+)$, n) $i_C(0+)$, p) check the results found for $v_x(0+)$, $v_x(\infty)$, $v_y(0+)$, $v_y(\infty)$, $i_L(0+)$, i_L (∞), $i_C(0+)$, $i_C(\infty)$ using SPICE Analysis (inifin2.cir).

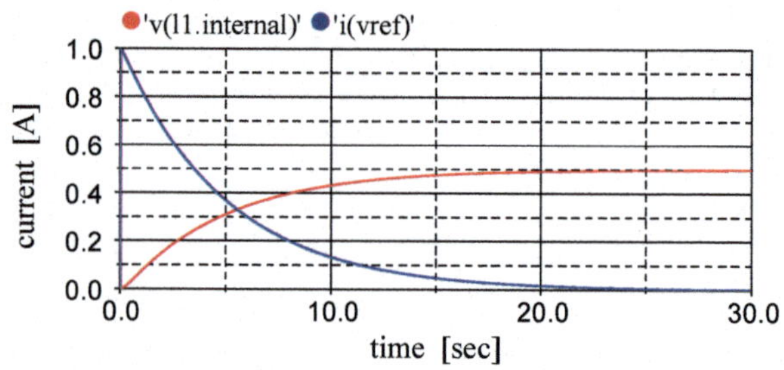

Fig. 5.23 SPICE check for Problem 5.1.20

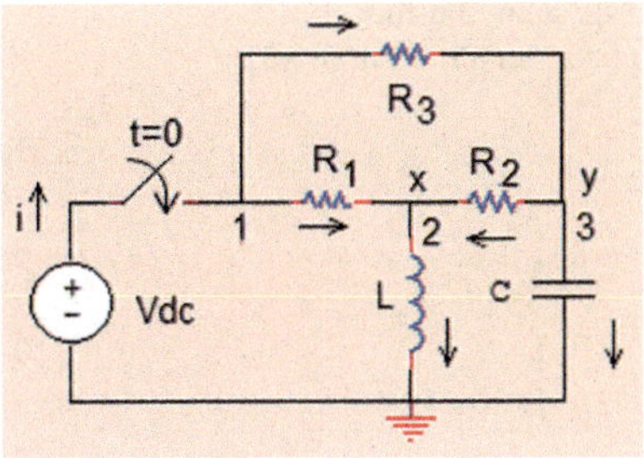

Fig. 5.24 The circuit of Problem 5.1.21

Solution

After the switch is closed at $t = 0$,

(a) $v_x(\infty) = 0\,\text{V}$,

(b) $i_{R1}(\infty) = \dfrac{V_{dc}}{R_1} = \dfrac{10}{10} = 1\,\text{A}$,

(c) $i_{R2}(\infty) = \dfrac{V_{dc}}{R_2 + R_3} = \dfrac{10}{30} = 0.333\,\text{A}$

(d) $v_y(\infty) = R_2 \cdot i_{R2}(\infty) = 10 \times 0.333 = 3.333\,\text{V}$,

(e) $i_L(\infty) = i_{R1}(\infty) + i_{R2}(\infty) = 1.333\,\text{A}$,

(f) $i(\infty) = i_L(\infty) = 1.333\,\text{A}$,

(g) $i_L(0^+) = i_L(0-) = 0\,\text{A}$,

(h) $v_y(0+) = v_C(0-) = 0\,\text{V}$,

(i) $|i_{R2}(0^+)| = \dfrac{V_{dc}}{R_2 + R_1} = \dfrac{10}{20} = 0.5\,\text{A}$,

(j) $i_{R1}(0^+) = |i_{R2}(0^+)| = 0.5\,\text{A}$

(k) $v_x(0+) = \dfrac{V_{dc} \times R_2}{R_2 + R_1} = \dfrac{10 \times 10}{10 + 10} = \dfrac{100}{20} = 5\,\text{V}$,

(l) $i_{R3}(0+) = \dfrac{V_{dc}}{R_3} = \dfrac{10}{20} = 0.5\,\text{A}$

(m) $i(0^+) = i_{R1}(0^+) + i_{R3}(0^+) = 0.5 + 0.5 = 1\,\text{A}$,

(n) $i_C(0^+) = i(0^+) = 1\,\text{A}$

(o) Checking the values of $v_x(0+)$, $v_x(\infty)$, $v_y(0+)$, $v_y(\infty)$ using SPICE Analysis is demonstrated in Fig. 5.25. Initial and final values of these voltages ($v_x = v_2$, $v_y = v_3$) proves the results found above; $v_x(0+) = 5$ V, $v_x(\infty) = 0$ V, $v_y(0+) = 0$ V, $v_y(\infty) = 3.333$ V.

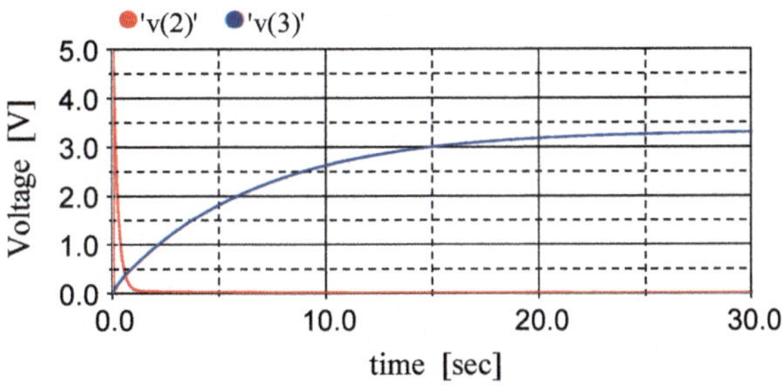

Fig. 5.25 SPICE check for Problem 5.1.21

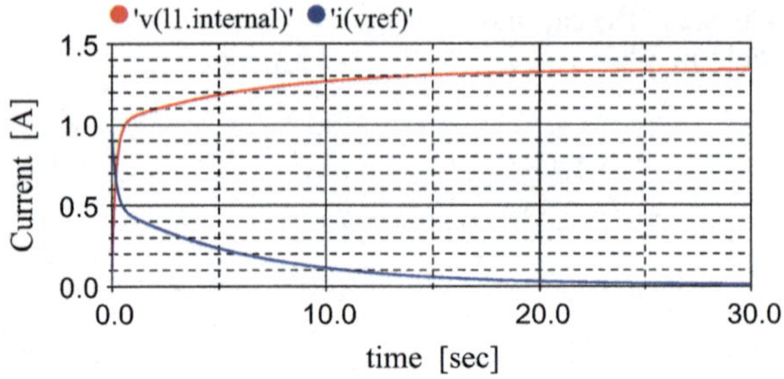

Fig. 5.26 Further SPICE check for Problem 5.1.21

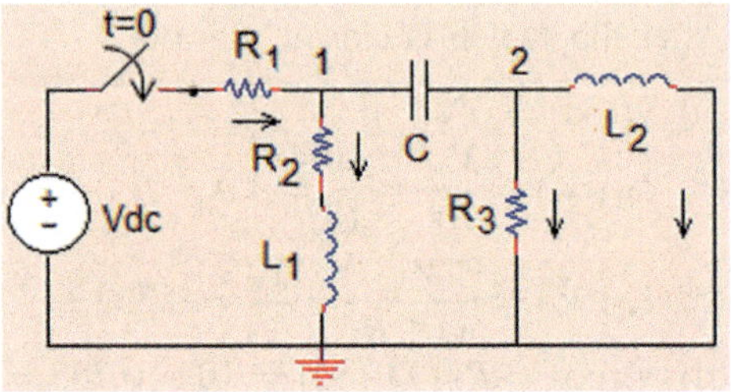

Fig. 5.27 The circuit of Problem 5.1.22

Checking the values of $i_L(0^+)$, $i_L(\infty)$, $i_C(0^+)$, $i_C(\infty)$ using SPICE Analysis is also demonstrated in Fig. 5.26. Initial and final values of these currents (i_L = l1.internal, $i_C = i_{\text{vref}}$) proves the results found in above operations; $i_L(0^+) = 0$ A, $i_L(\infty) = 1.333$ A, $i_C(0^+) = 1$ A, $i_C(\infty) = 0$ A.

Problem 5.1.22 In the circuit shown in Fig. 5.27, $v_1(0^+) = 8$ V, $v_1(\infty) = 5$ V.

Find the values of R_2 and R_3 if $R_1 = 6\ \Omega$, $L1 = 1$ H, $L2 = 2$ H, $C = 1$ F, $V_{\text{dc}} = 10$ V (no initial conditions).

Solution

$$v_1(0^+) = \frac{R_3}{R_1 + R_3} \cdot V_{\text{dc}} = \frac{R_3}{6 + R_3} \cdot 10 = 8\,\text{V}$$

$$10R_3 = 8(6 + R_3) = 48 + 8R_3 \quad \rightarrow \quad R_3 = 24\,\Omega$$

$$v_1(\infty) = \frac{R_2}{R_1 + R_2} \cdot V_{\text{dc}} = \frac{R_2}{6 + R_2} \cdot 10 = 5\,\text{V}$$

$$10R_2 = 5(6 + R_2) = 30 + 5R_2 \quad \rightarrow \quad R_2 = 6\,\Omega$$

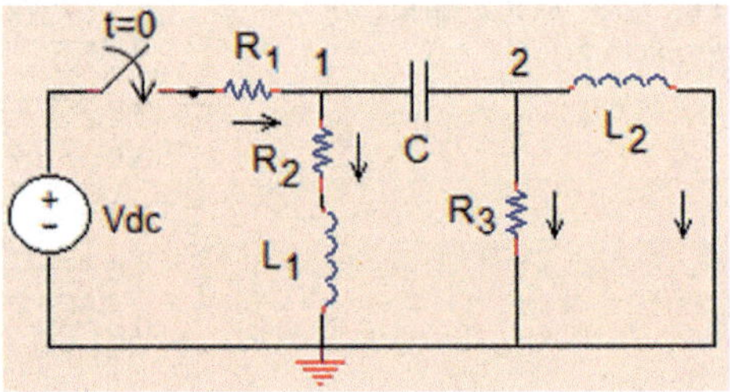

Fig. 5.28 The circuit of Problem 5.1.23

Problem 5.1.23 For the circuit shown in Fig. 5.28, $v_C(0^-) = 0$ V, $i_{L1}(0^-) = i_{L2}(0^-) = 0$ A; $R_1 = 6\ \Omega$, $R_2 = 2\ \Omega$, $R_3 = 4\ \Omega$, $L_1 = 1$ H, $L_2 = 2$ H, $V_{dc} = 10$ V.

Find the values of,

(a) $v_1(0^+)$, (b) $v_2(0^+)$, (c) $v_3(0^+)$, (d) $i_{R1}(0^+)$, (e) $i_C(0^+)$, (f) $v_1(\infty)$, (g) $v_2(\infty)$, (h) $v_3(\infty)$, (i) $i_C(\infty)$, (j) $i_{R1}(\infty)$, (k) check the results using SPICE analysis for initial and final values of voltages v_1, v_2 and v_3. (inifin3.cir)

Solution

Since voltage across a capacitor and current through an inductor cannot change abruptly, and a capacitor acts as an open circuit while an inductor behaves like a short circuit at steady state dc conditions,

(a) $v_1(0^+) = \dfrac{R_3}{R_1 + R_3} \cdot V_{dc} = \dfrac{4}{10} \cdot 10 = 4\,\text{V},$

(b) $v_2(0^+) = v_1(0^+) = 4\,\text{V},$

(c) $v_3(0^+) = v_1(0^+) = 4\,\text{V},$

(d) $i_{R1}(0^+) = \dfrac{V_{dc}}{R_1 + R_3} = \dfrac{10}{10} = 1\,\text{A},$

(e) $i_C(0^+) = i_{R1}(0^+) = 1\,\text{A},$

(f) $v_1(\infty) = \dfrac{R_2}{R_1 + R_2} \cdot V_{dc} = \dfrac{2}{8} \cdot 10 = 2.5\,\text{V}$

(g) $v_2(\infty) = 0\,\text{V},$

(h) $v_3(\infty) = 0\,\text{V},$

(i) $i_C(\infty) = 0\,\text{A},$

(j) $i_{R1}(\infty) = \dfrac{V_{dc}}{R_1 + R_2} = \dfrac{10}{8} = 1.25\,\text{A},$

(k) In Fig. 5.29, the results for initial and final values of voltages v_1, v_2, and v_3 are checked using SPICE analysis.

Problem 5.1.24 At the receiving end of a signal processor of one type of a data acquisition system in a CT Scanner (X-ray Computer Tomography unit), the received signal (which is a sampled and quantized analog voltage waveform) is applied to a digital-to-analog (D/A) converter whose output is constant at each time interval. It is smoothed by passing the output of DAC through a low-pass filter

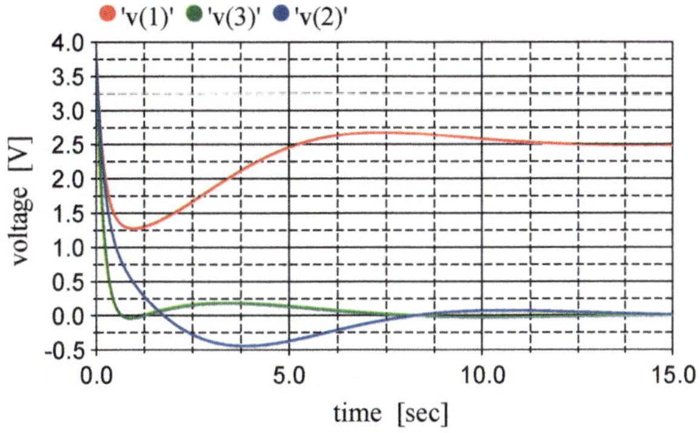

Fig. 5.29 Complete voltage waveforms to check the results of calculations for initial and final values of voltages

circuit to recover the transmitted analog signal. As a demonstrative example, let the output of a D/A converter (input to the smoothing filter) at a certain time interval be

$$v_i(t) = \begin{cases} 0\,\text{V} & t = 0 \\ 5\,\text{V} & 0 < t \leq 1 \\ 9\,\text{V} & 1 < t \leq 2 \\ -5\,\text{V} & 2 < t \leq 3 \\ 0\,\text{V} & 3 < t \leq 4 \end{cases}$$

Time (t) is in seconds. There is no initially stored energy in the circuit.

If this waveform is applied to the input of the RLC circuit in Fig. 5.30, Use SPICE to plot both input and output voltages on the same graph. $R = 1\ \Omega$, $L = 1$ H, $C = 1$ F.

Solution

Characteristic equation is $s^2 + 2\alpha s + \omega_0^2 = 0$. The RLC circuit will produce an underdamped response, since damping factor is less than the undamped natural frequencyNatural frequency:

$$\alpha = \frac{R}{2L} = \frac{1}{2}, \qquad \omega_0 = \frac{1}{\sqrt{LC}} = 1, \qquad \alpha < \omega_0$$

A plot for both input and output voltages on the same graph is shown in Fig. 5.31.

A SPICE netlist (RLC_smoothing.cir) is given below.

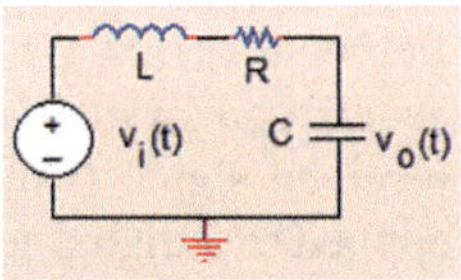

Fig. 5.30 RLC smoothing filter

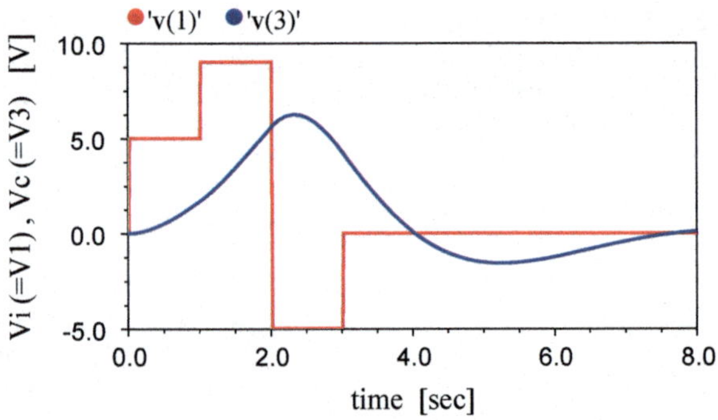

Fig. 5.31 Exemplary input and output voltages of a data smoothing filter

```
smoothing
Vi 1 0 PWL(0 0 0.01 5 1 5 1.01 9 2 9 2.01 -5 3 -5 3.01 0 4 0)
R1 1 2 1
L1 2 3 1
C1 3 0 1
*.tran .01 8 0 .1 uic
```

Problem 5.1.25

(a) A prosthetic limb device is simulated by a series RLC circuit as shown in Fig. 5.4a. Designer aims to realize an overdamped response with time constants of $\tau_1 = 50$ ms and $\tau_2 = 250$ ms using a 10 Ω resistor. Determine circuit component values, and undamped natural frequency (in Hz).
(b) If $v_C(0) = 10\,\text{V}$, Plot the current waveform for $t \geq 0$.
(c) If designer uses a parallel RLC circuit during his simulation of the same prosthetic limb device, what will be the component values and undamped natural frequency of the system, in Hz (R = 10 Ω).
(d) In the parallel RLC model, if initial conditions on the energy storage elements are −180 mA and 2 V, determine the voltage variation on each component.

Solution

(a) An overdamped RLC circuit has the capacitor voltage in the form of

$$v_C(t) = K_1 e^{s_1 t} + K_1 e^{s_2 t}, \quad t \geq 0$$

$$\tau_1 = \frac{1}{|s_1|} = 250 \times 10^{-3} \quad \rightarrow \quad s_1 = -\frac{1000}{250} = -4$$

$$\tau_2 = \frac{1}{|s_2|} = 50 \times 10^{-3} \quad \rightarrow \quad s_2 = -\frac{1000}{50} = -20$$

$$s_{1,2} = -\alpha \pm \sqrt{\alpha^2 - \omega_0^2} \quad \rightarrow \quad s_1 + s_2 = -2\alpha \quad \rightarrow \quad -4 - 20 = -2\alpha$$

$$\alpha = 12 = \frac{R}{2L} \quad \rightarrow \quad L = \frac{R}{24} = \frac{10}{24} = 0.417\,\text{H}$$

$$s_2 = -\alpha + \sqrt{\alpha^2 - \omega_0^2} \quad \rightarrow \quad s_2 + \alpha = \sqrt{\alpha^2 - \omega_0^2} \quad \rightarrow \quad \omega_0^2 = \alpha^2 - (s_2 + \alpha)^2$$

$$\omega_0 = \sqrt{\alpha^2 - (s_2 + \alpha)^2} = \sqrt{(12)^2 - (-20 + 12)^2} = 10.39\,\text{rad/s}$$

$$f_0 = \frac{\omega_0}{2\pi} = 1.65\,\text{Hz}$$

$$\omega_0 = \frac{1}{\sqrt{LC}} \quad \rightarrow \quad C = \frac{1}{\omega_0^2 L} = \frac{1}{(10.392)^2 (0.417)} = 22.2\,\text{mF}$$

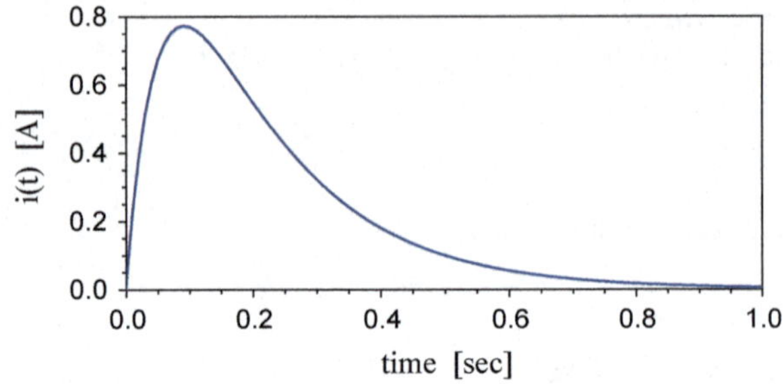

Fig. 5.32 The current waveform in prosthetic limb device simulation

(b) The current waveform is shown in Fig. 5.32.

A SPICE netlist (RLC1.cir) is also listed below.

```
RLC analysis
C1 1 0 22.2 m IC=10v
L 1 2 0.417
R1 2 0 10
*.tran .001 1 0 0.01 uic
```

(c) Since $s_1 = -4$, $s_2 = -20$,

$$\alpha = -\frac{s_1 + s_2}{2} = 12 = \frac{1}{RC}$$

It should be pointed out here that the expression of α in a parallel RLC circuit is different than the α for series RLC circuit. Solving this equation for C and substituting known values yields;

$$C = \frac{1}{2\alpha R} = \frac{1}{2 \times 12 \times 10} = 4.17\,\text{mF}$$

Undamped angular frequency for the parallel RLC circuit is the same as that of the series RLC circuit,

$$\omega_0 = \frac{1}{\sqrt{LC}} \quad \rightarrow \quad L = \frac{1}{\omega_0^2 C} = \frac{1}{(10.392)^2 \times 4.17 \times 10^{-3}} = 22.2\,\text{mH}$$

(d) Since overdamped response of the circuit is of the form

$$v(t) = K_1 e^{-4t} + K_2 e^{-20t}, \quad t \geq 0$$

Two equations are needed to determine the values of K_1, K_2 . The first one is

$$v_C(0) = v(0) = 2 = K_1 + K_2$$

The second equation to find out K_1 and K_2 is obtained from the node equation for the parallel RLC circuit:

$$i_L(t) + C\frac{dv(t)}{dt} + \frac{v(t)}{R} = 0 \rightarrow \frac{dv(t)}{dt} - \frac{1}{RC}v(t) - \frac{i_L(t)}{C}$$

Using initial values and R = 10 Ω, C = 4.17 mF,

$$\frac{dv(0)}{dt} = -\frac{1}{RC}v(0) - \frac{1}{C}i_L(0) = -\frac{10^3}{10 \times 4.17}(2) - \frac{10^3}{4.17}(-0.18) = -4.796$$

On the other hand, first derivative of the overdamped RLC circuit response is

$$\frac{dv(t)}{dt} = -4K_1e^{-4t} - 20K_2 - e^{-20t}$$

Evaluating this last equation at t = 0;

$$-4.796 = -4K_1 - 20K_2$$

Therefore, two equations for two unknown parameters are in hand,

$$K_1 + K_2 = 2$$
$$4K_1 + 20K_2 = 4.796$$

Solution of this simultaneous equations set yields $K_1 = 2.2, K_2 = -0.2$;
The node voltage equation becomes,

$$v(t) = v_R(t) = v_L(t) = v_C(t) = 2.2e^{-4t} - 0.2e^{-20t}, t \geq 0$$

As a check for this result, it is verified that $v(0) = 2\,\text{V}$.

Problem 5.1.26 What is ventricular fibrillation, VF? What is internal and external defibrillation? Which types of waveforms are used in defibrillators? Discuss historical development of defibrillating equipment. What are the specific properties of external defibrillator capacitors? What are the advantages and disadvantages of external defibrillator paddles?

Solution

Ventricular fibrillation (VF) results from asynchronous contraction of heart muscle, and it appears as an uncoordinated movement of ventricle walls of the heart. During VF, adjacent cells of heart muscle fibers are continuously stimulated so that the regular sequence of events that follow synchronized heart action deteriorates, causing the loss of cardiac output, which may finally be the reason for the patient's death.

ECG features of VF show that the cardiac rhythm is highly irregular, heart rate is not measurable, the P wave is absent, the PR interval is not measurable, baseline is wandering and the QRS waveform does not exist, see, Fig. 5.33.

An electric shock which can be delivered to the fibrillating heart externally or by the electrodes in direct contact with the heart during a surgical operation (internal defibrillation) may restart regular heart activity.

Early studies on defibrillation utilized capacitor discharge [1, 2], i.e., a decaying monophasic exponential current. Many studies investigated the effect of the value of the capacitor used and the voltage on the Capacitor . Tacker [3] showed that neither short nor long capacitor discharges defibrillated effectively. Short time constant discharges were thought to cause functional damage. However, discharges from long time constants failed to be effective, even though high energies were delivered.

Damped Sinusoidal Waveforms

The addition of an inductor in series with a series capacitor discharge, converts the waveform to a damped sinusoid. However, the energy required to defibrillate was about half that required for capacitor discharge circuits [4].

This discovery was further investigated by Edmark [5] and Lown et al. [6]. The damped sinusoidal waveform offered technical advantages as well as lower defibrillation thresholds. Capacitors rated at high voltage and high energy density could be used, and the circuits were more reliable. Additionally, small, lightweight, portable defibrillators using damped sinusoidal waveforms have been designed for use by paramedics and other emergency personnel outside the hospital.

Lown type defibrillator is a biphasic underdamped RLC circuit, in contrast with the monophasic overdamped Edmark circuit.

A technical disadvantage of using inductor is that a practical inductor has its internal resistance and dissipates part of the supplied energy during the discharge process. A significant clinical advantage for under-damped Lown type defibrillator was that it did not produce atrial fibrillation, as did the Alternating Current (AC) waveform in about 80% of shocks.

After extensive testing, it has been determined that biphasic waveforms are more efficacious than monophasic waveforms [7–11]. One hypothesis for the increased efficacy of biphasic waveforms over that of monophasic waveforms holds that first positive phase defibrillates the heart and second phase performs a stabilizing action

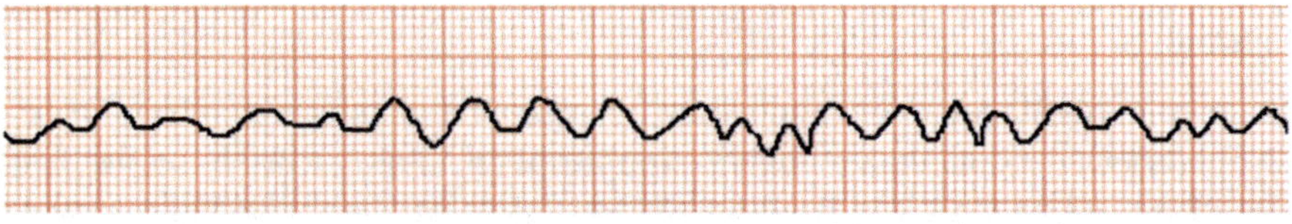

Fig. 5.33 ECG recording during ventricular fibrillation (VF)

that keeps the heart from re-fibrillating. Biphasic defibrillation waveforms are now the standard of care in clinical use for defibrillation with implantable cardioverter-defibrillators.

Some researchers [12] have developed cardiac cell response models of defibrillation to understand these significantly different outcomes. Waveform design criteria have been postulated from these studies and have been applied to monophasic and biphasic waveforms to optimize their parameters.

Kroll [13] proposed that the stabilizing action of the second phase removed the charge deposited by positive phase from those cells which were not stimulated in the first phase. This is known as "charge burping." Kroll supported his hypothesis with retrospective analysis of the studies by Dixon et al. [14], Tang et al. [15], and Freeser et al. [16] regarding single capacitor, biphasic waveform studies.

The charge burping hypothesis can be used to develop equations that describe the time course of a cell's membrane potential during a biphasic shock pulse. At the end of first phase, those cells that were not stimulated have a residual charge due to the action of first phase on the cell.

Babs and Whistler [17] proposed a method to determine values of circuit elements in an underdamped sine wave (Lown waveform) defibrillator solely from measurements of the outputPower measurement, using two or more power resistors and a storage oscilloscope.

Jones and Charbonnier [18] developed a technique to determine peak current, transthoracic impedance, and delivered energy during a damped sinusoidal defibrillation pulse. The discharge waveform information is generated from sampling the peak discharge current through a current transformer and measuring the voltage stored on the energy storage capacitor. For a given defibrillator circuit a unique relationship exists between the peak discharge current and the unknown external impedance presented to the defibrillator by the patient; hence, measurement of peak discharge current allows calculation of external impedance. A real time algorithm provides delivered energy information using known internal resistance, capacitance, and inductance parameters. The benefit of this method of delivered energy calculation is that the current and voltage waveforms need not be digitized and then integrated to provide the desired information. This method also keeps defibrillation circuitry ground isolated and simplifies operation through the high electromagnetic fields generated during the discharge.

Truncated Exponential Discharge

There are two kinds of truncated capacitive exponential discharge waveforms to be used with defibrillator systems: monophasic and biphasic waveforms. A monophasic waveform is a single exponentially decaying electrical pulse that is truncated before the capacitor system is fully discharged. A biphasic waveform is a pair of decaying electrical pulses (phases) that are of different (+/−) polarity [19].

Truncated Biphasic Exponential Discharge

One method to design a defibrillator with biphasic waveform is to discharge a positive pulse from the capacitor and then, at the point the positive pulse is truncated, a switching circuit connected to the electrodes is used to reverse the discharge polarity of the capacitor system as seen by the patient's trunk (resistor) between the electrodes. This action produces the second pulse of the biphasic waveform that is of the negative polarity. Solid-state H-bridge circuit structure is widely employed to generate a biphasic waveform in a defibrillator unit.

It has been demonstrated that [20, 21], truncated biphasic waveforms achieve shock therapies using relatively lower currents, voltages, and energies than monophasic waveforms of similar durations.

Biphasic waveform generating defibrillators are most common type of defibrillators available today (2016). Basically, the first (positive) phase of this biphasic waveform has a larger area under the curve than the second (negative) phase.

Most external defibrillator devices designed for adult defibrillation have energy settings of 50–360 J, while devices for internal or pediatric defibrillation or synchronized cardioversion of 5–50 J, and devices for neonatal applications have energy settings of 1–20 J. Units are able to charge to maximum energy in less than 15 s over their full range.

A basic external defibrillator is used as a crash-cart unit to provide defibrillation. These devices become suitable for transport applications and emergency medical services paramedic use besides their crash-cart applications by adding external pacing and advanced monitoring options.

There are two major types of biphasic defibrillators on the medical market. One group of manufacturers use the biphasic truncated exponential waveform using different energy settings. For example, Medtronic-Physio Control use the term ADAPTIV Biphasic™ in their commercial product, and the energy settings go up to 360 J. Additionally, they vary the voltage and automatically extend the period of the shock for patients having higher torso resistance. Philips Medical uses the biphasic truncated exponential waveform in their defibrillator product named as SMART™ Biphasic, distributing voltage, and current so that maximum current is delivered at 200 J (The SMART Biphasic waveform uses a 100 μF capacitor to store the energy; the Medtronic Physio Control ADAPTIV biphasic waveform uses a 200 μF capacitor). A Swiss Company Schiller uses pulsed high frequency biphasic waveform in their product Multipulse Biowave™.

Another original defibrillator equipment manufacturer, Zoll Medical, uses Rectilinear Biphasic™ waveform which is fixed at 10 ms duration. It adjusts the equipment-based resistance and gives a constant current during the first positive pulse application to patient's torso by adding–subtracting resistors in the equipment. Nihon Kohden's ActiBiphasic™ technology in their Cardiolife™ products provides biphasic defibrillation by maintaining constant second phase pulse width particularly for high impedance patients.

Defibrillator Capacitors

Capacitors create a limiting factor in defibrillator design in terms of cost, size, parasitic values, as well as their reliability and lifetime issues. Defibrillator capacitors are housed in round or oval metal, oil-filled cases; or in a dry, epoxy-filled plastic housing version. Plastic cases offer design flexibility, and can be customized to specific requirements. They are available in voltage ranges from less than 1000 VDC up to 6000 VDC, delivering in excess of 500 J at full charge. Defibrillator film capacitors are designed to meet the reliability demands of a Class III medical device. An inspection of defibrillator capacitor specifications shows that capacitance values as well as their rated voltage values are quite limited. This fact sets another constraint in defibrillator design.

External Defibrillator Paddles and Cables

External defibrillator hard paddles have been used for many years by healthcare staff.

Advantages are that the shock can be provided quickly with hard paddles, they are needed especially for patients who are diaphoretic due to hemodynamic instability, patients with body incisions or wounds, patients with a large amount of hair on their chest, or for patients having severe trauma.

As for the disadvantages; there is the problem of the paste smearing with paddles over the chest and the potential of electrical arcing, if there is not good skin contact. The person performing the actual defibrillation has to lean over the patient's body to apply the necessary pressure to the paddles. Additionally, hard paddles are costlier than disposable pad electrodes.

There are numerous types of commercially presented external defibrillator paddles and cables. Ideally, the paddles would have unrestricted movement, perform as ECG electrode as well, their cables would not tangle, and both the paddles and cables would detach easily from the defibrillator for storage. Each paddle is connected to a respective coiled cable having its other end attached to a single connector. Unfortunately, these two coiled cables can become tangled together during use, with potentially life-threatening results. Furthermore, due to the total length of cable required, storage of the paddles and cables can be difficult.

A uni-cable paddle system for use with a defibrillator comprises two paddles with contact surfaces and a main cable connected at one of its ends to one of the paddles, (the sternum paddle); its other end having a connector for attachment to the defibrillator, [22]. An inter-paddle cable extends between the sternum and apex paddles. The main cable includes two high-voltage (HV) wires. One of these HV wires is connected to the contact surface of the sternum paddle, and the other HV wire is connected to the contact surface of the apex paddle via a HV wire in the inter-paddle cable. The main and inter-paddle cables are identical, having two HV wires arranged in a shielded, twisted pair to minimize external electromagnetic interference. The "unused" HV wire in the inter-paddle cable may be connected to the sternum contact surface (but left unconnected at its other end) to minimize common mode effects in the cable and to match capacitances between the apex and sternum paddles and ground. An alternative "Y" type structure includes a main cable terminating in a connector for connection with a defibrillator. The main cable's other end terminates at a junction box.

This construction is claimed to minimize cable tangling during use and storage and also to provide a defibrillator cable and paddle system which produces a minimum of electromagnetic interference.

Major defibrillator paddle cabling available in the medical field today are of the latter type, with the exception that main coiled connection cable is placed by a rather short and relatively thick plug directly connecting to the socket on the defibrillator. This can be termed as "V type" cable connection to defibrillator paddles.

5.2 Duality Principle

Problem 5.2.1 Determine the dual of the circuit shown in Fig. 5.34, and write its characteristic equation in differential form.

Solution

Parallel → series, current source → voltage source, $C \to L$, $G \to R$,
Node voltage → mesh current
Characteristic equation:

$$C\frac{\mathrm{d}v}{\mathrm{d}t} + Gv = i$$

Drawing the dual (Fig. 5.35):
Characteristic equation:

$$L\frac{\mathrm{d}i}{\mathrm{d}t} + Ri = v$$

Fig. 5.34 The circuit of Problem 5.2.1

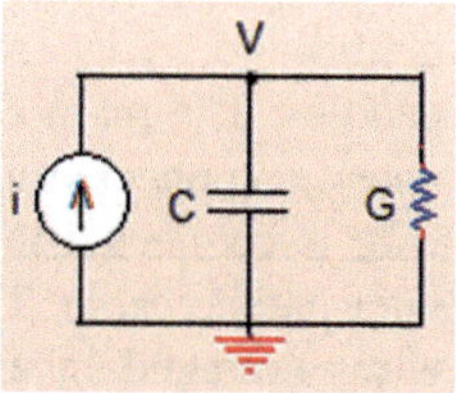

Fig. 5.35 Dual circuit

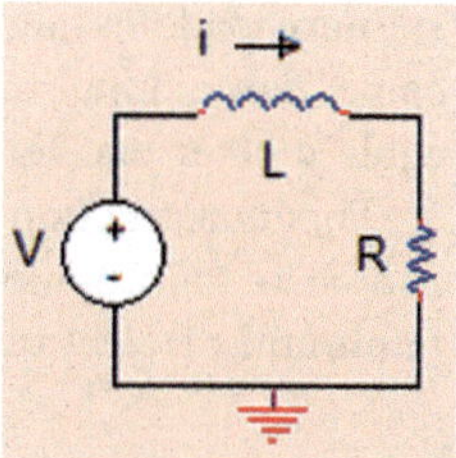

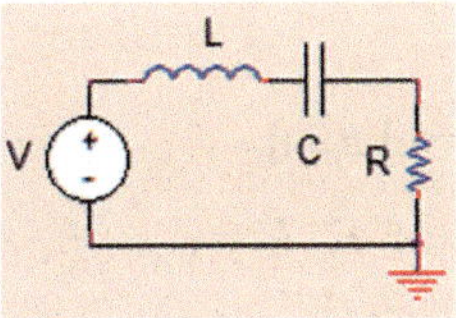

Fig. 5.36 The circuit of Problem 5.2.2

Problem 5.2.2 Construct the dual of the circuit of Fig. 5.36, and write defining equations for both circuits (L = 2 mH, C = 4 nF, R = 10 kΩ, v = 3 cos(t) [V].

Solution

Series–parallel duality principle (Fig. 5.37),

$$L' = 4\,\text{nH}, \quad C' = 2\,\text{mF}, \quad G' = 10\,\text{kS}, \quad R = \frac{1}{10^4} = 10^{-4}\,\Omega, \quad i = 3\cos(t)\,[\text{A}]$$

Since no initial conditions are specified,

$$v = L\frac{\mathrm{d}i}{\mathrm{d}t} + \frac{1}{C}\int i\,\mathrm{d}t + iR \quad \rightarrow \quad i = C'\frac{\mathrm{d}v}{\mathrm{d}t} + \frac{1}{L}\int v_L \mathrm{d}t + vG'$$

Problem 5.2.3

(a) Write the Mesh equation for the circuit of Fig. 5.38.
(b) Determine the current in the circuit.
(c) Find the voltage drop over R_3.
(d) Construct the dual of this circuit.
(e) Write down the governing (constraint) equation of the dual circuit.
(f) Calculate the branch currents of the dual circuit.

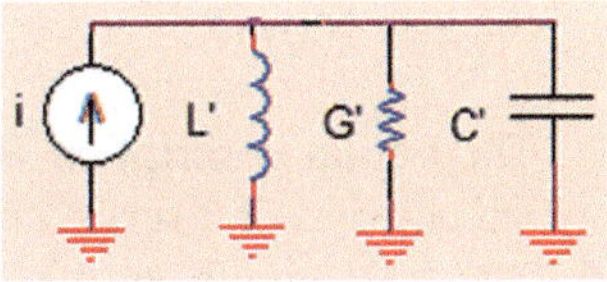

Fig. 5.37 Dual circuit

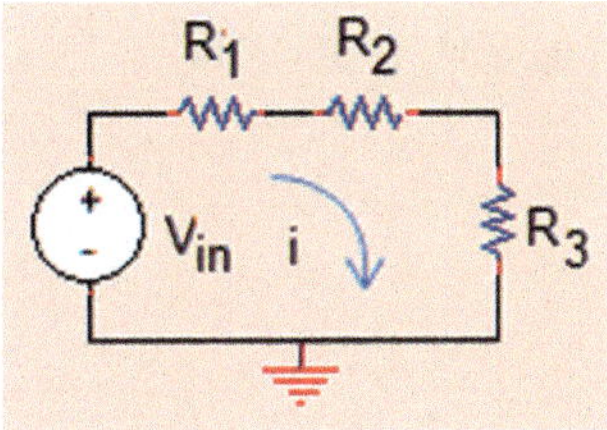

Fig. 5.38 The circuit of Problem 5.2.3

$$v_{in} = 10\cos(314t)\ (\text{V}),\quad R_1 = 2.2\,\Omega,\quad R_2 = 3.9\,\Omega,\quad R_3 = 5.1\,\Omega$$

Solution

(a) KVL: $v_{in} - iR_1 - iR_2 - iR_3 = 0 \quad \rightarrow \quad v_{in} = (R_1 + R_2 + R_3)\cdot i$

(b) $\mathrm{i} = \dfrac{\mathrm{v_{in}}}{\mathrm{R_1 + R_2 + R_3}} = \dfrac{10\cdot\cos(314\mathrm{t})\ [\mathrm{V}]}{11.2\ [\Omega]} = 0.893\ \cos\ 314\mathrm{t}\ [\mathrm{A}]$

(c) $\mathrm{v_{R3}} = \mathrm{i}\cdot\mathrm{R_3} = 0.893 \times 5.1\ \cos\ 314\,\mathrm{t} = 4.554\ \cos\ 314\mathrm{t}\ [\mathrm{V}]$

(d) Fig. 5.39 shows the dual circuit.

(e) $\mathrm{G_1} = 2.2\,\mathrm{S},\ \mathrm{G_2} = 3.9\,\mathrm{S},\mathrm{G_3} = 5.1\,\mathrm{S}$

$$i_{in} - VG_1 - VG_2 - VG_3 = 0 \quad (\text{from dualty})$$
$$i_{in} = (G_1 + G_2 + G_3)V$$

$$\mathrm{V} = \frac{\mathrm{i_{in}}}{\mathrm{G_1 + G_2 + G_3}} = \frac{(10\ \cos(314\mathrm{t})\ [\mathrm{A}])}{11.2\ [\mathrm{S}]} = 0.893\ \cos\ 314\mathrm{t}\ [\mathrm{V}]$$

(f) $\mathrm{i_1} = \mathrm{V}\cdot\mathrm{G_1} = 0.893 \times 2.2\ \cos\ 314\mathrm{t} = 1.965\ \cos\ 314\mathrm{t}$

$\mathrm{i_2} = \mathrm{V}\cdot\mathrm{G_2} = 0.893 \times 3.9\ \cos\ 314\mathrm{t} = 3.483\ \cos\ 314\mathrm{t}$

$\mathrm{i_3} = \mathrm{V.G_3} = 0.893 \times 5.1\ \cos 314\mathrm{t} = 4.554\ \cos\ 314\mathrm{t}\ [\mathrm{A}]$

The last equation verifies (c).

Problem 5.2.4 Determine the dual of the circuit shown in Fig. 5.40. V = 100 mV, L = 5 H, R = 0.5 Ω, C = 2 F.

Solution

$$i = 100\,\text{mA},\quad G' = 0.5\,\text{S} \quad \rightarrow \quad R' = 2\,\Omega,\quad L' = 2\,\text{H},\quad C' = 5\,\text{F}$$

The verbal description of the circuit is that the voltage source sees a network which consists of a series connected (*L*//*R*) and *C*. This translates into the statement for the dual circuit as the current source sees a network consisting of a parallel connected (*G*′–*C*′)//*L*′. Then, the circuit of Fig. 5.41 is drawn.

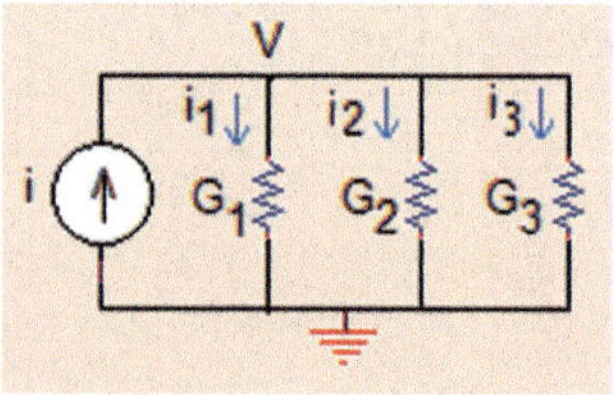

Fig. 5.39 Dual circuit

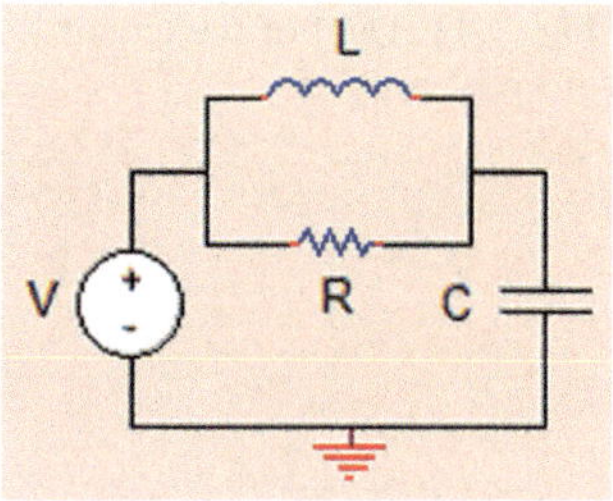

Fig. 5.40 The circuit of Problem 5.2.4

Problem 5.2.5 Construct the dual of circuit Shown in Fig. 5.42.

Solution
The verbal description of the circuit is that the current source sees a network which consists of a series connected two parallel branches, 2 H//4 S, 6 F//2 S.

This translates into the statement for the dual circuit as follows;

The voltage source sees a network consisting of a parallel connected two series branches, 2 F–4 Ω, 6 H–2 Ω.

Then, dual circuit is drawn as Shown in Fig. 5.43:

Problem 5.2.6 Obtain the dual of the circuit shown in Fig. 5.44, using parallel–series duality concept (i = 10 mA, V = 20 mV, C = 3F, L = 2 H, R = 10 Ω).

Solution

Start from the right most side of the circuit.

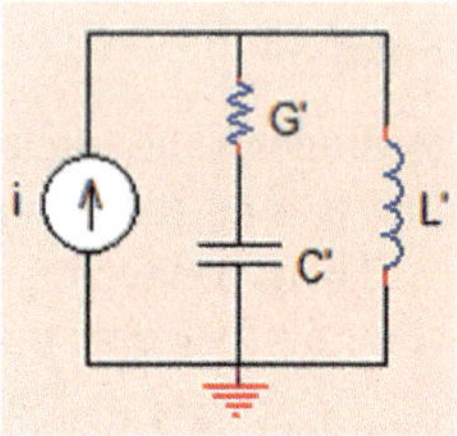

Fig. 5.41 Dual circuit

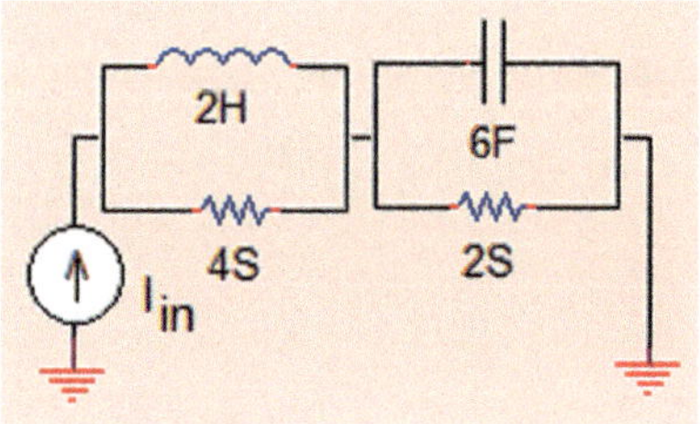

Fig. 5.42 The circuit of Problem 5.2.5

Fig. 5.43 Dual of the circuit of Problem 5.2.5

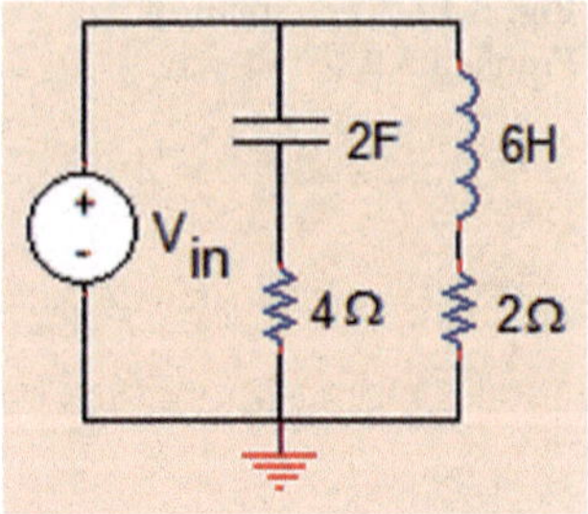

Fig. 5.44 The circuit of Problem 5.2.6

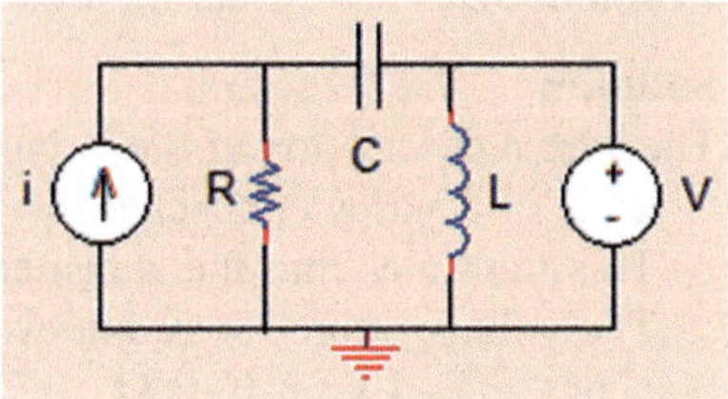

(*L*//*V*) is in series with *C*, then it connects to (*i*//*R*). This translates into the following statement: (*C*′//*i*′) is in parallel with *L*′ then it connects to (*V*′–*G*′). The circuit of Fig. 5.45 can be drawn;

$$i' = 20\,\text{mA},\quad V' = 10\,\text{mV},\quad C' = 2\,\text{F},\quad L' = 3\,\text{H},\quad G' = 10\,\text{S} \rightarrow \quad R' = 0.1\,\Omega$$

Problem 5.2.7 Obtain the dual for the circuit shown in Fig. 5.46.

Solution

Redrawing the circuit and placing nodes at the center of each mesh and a ground line around the circuit as shown in Fig. 5.47,

Then connecting nodes and writing duals as in Fig. 5.48, and reorganizing (Fig. 5.49) yields the dual circuit.

Problem 5.2.8 Draw the dual of the circuit shown in Fig. 5.50 ($i = 2$ A, $R = 1/8\ \Omega$, $L = 2$ H, $C = 10$ F).

Fig. 5.45 Dual circuit

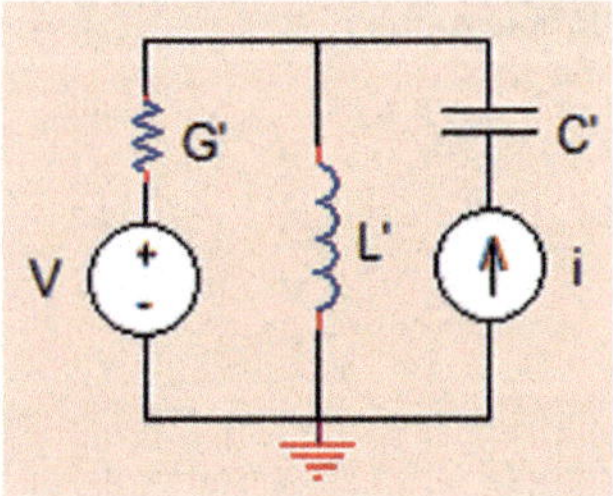

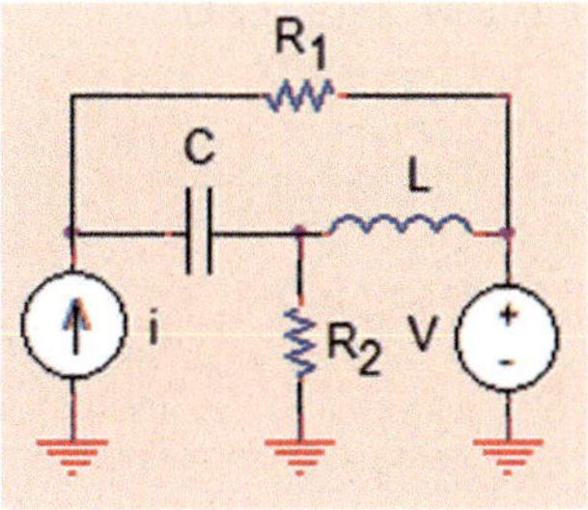

Fig. 5.46 The circuit of Problem 5.2.7

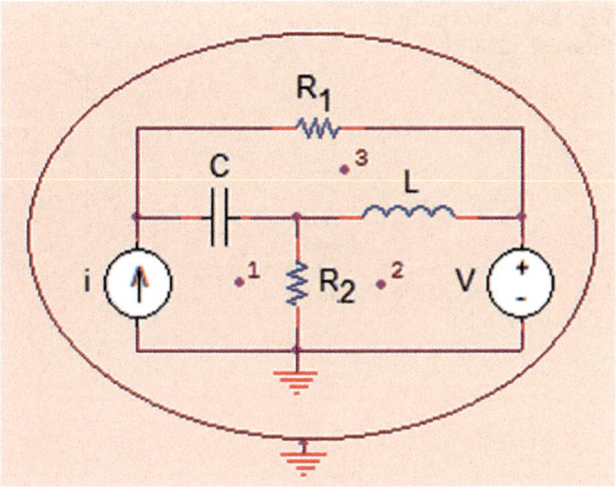

Fig. 5.47 Redrawing the circuit and placing nodes at the center of each mesh and a ground line around the circuit of Problem 5.2.7

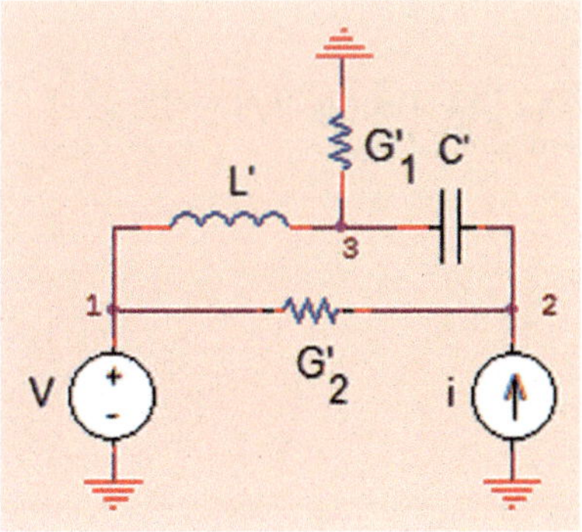

Fig. 5.48 Connecting nodes and writing duals

Solution

Three parallel branches to the left of node 1 are converted into series connected elements.

Series connected (L, C) elements are converted into parallel (C, L) braches.

All are being replaced with their duals.(Figure 5.51).

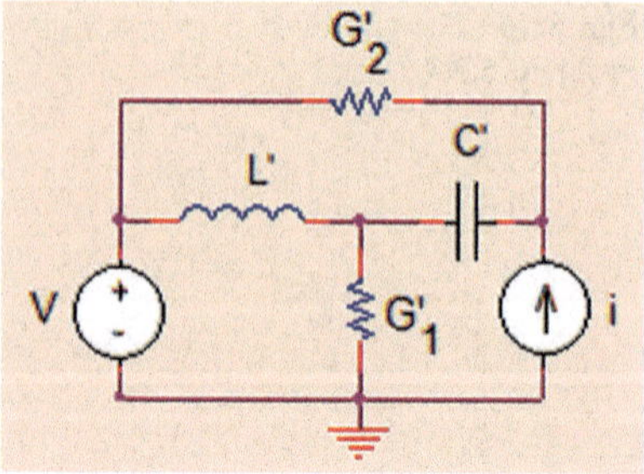

Fig. 5.49 Dual circuit

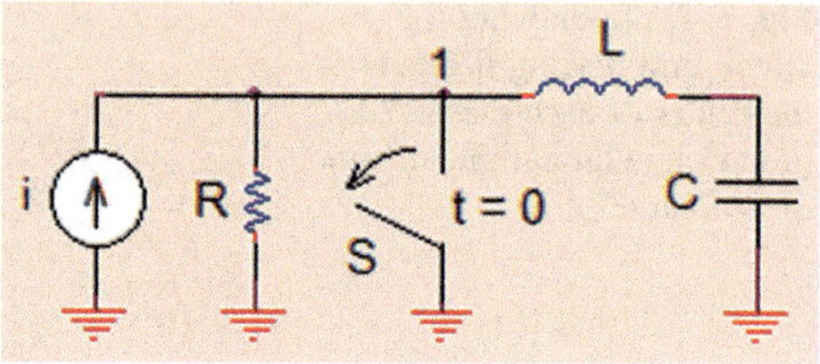

Fig. 5.50 The circuit of Problem 5.2.8

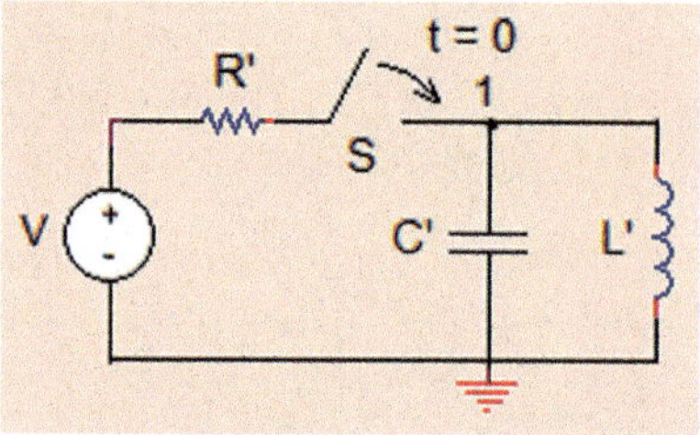

Fig. 5.51 Dual of the circuit of Problem 5.2.8

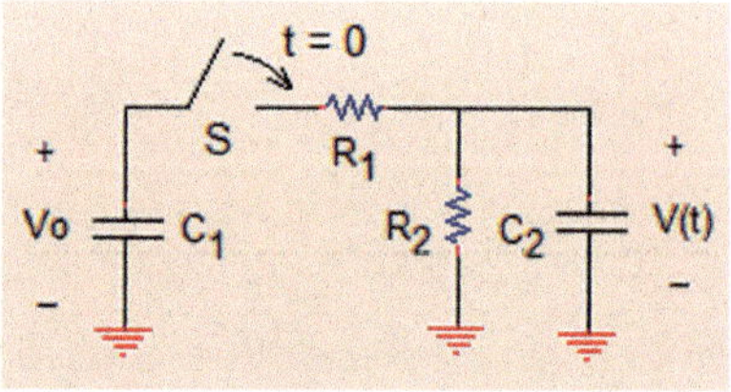

Fig. 5.52 The circuit of Problem 5.2.9

$$R' \equiv \frac{1}{R} = 8\,\Omega, \quad L' \equiv C = 10\,\text{H}, \quad C \equiv L = 2\,\text{F}, \quad v \equiv i = 2\,\text{V}$$

Problem 5.2.9 Draw the dual of the network in Fig. 5.52 (R_1 = 2 Ω, R_2 = 20 Ω, C_1 = 10 F, C_2 = 5 F).

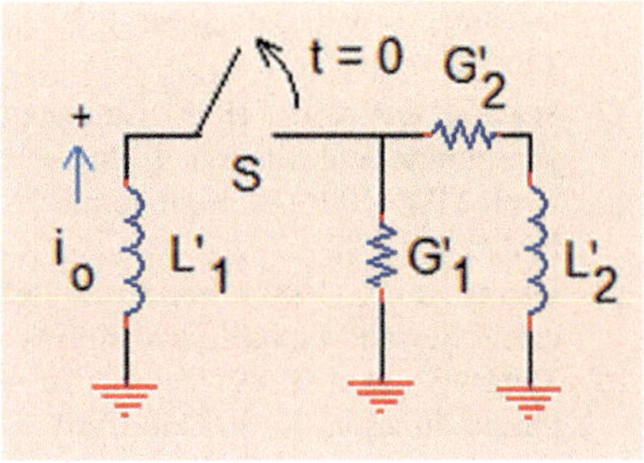

Fig. 5.53 Dual circuit

Solution

C_1 and R_1 are series connected through a switch that turns on at $t = 0$, ($R_2//C_2$). Initial voltage of C_1 is V_0^+. Dual case:

L_1' and G_1' are parallel connected through a switch that turns off at $t = 0$, (G_2–L_2). Initial current of L_1' is i_0'. Dual circuit is shown in Fig. 5.53, with

$$G_1' = 2\,\text{S},\quad R_1' = 0.5\,\Omega,\quad G_2' = 20\,\text{S},\quad R_2' = 0.05\,\Omega,\quad L_1' = 10\,\text{H},\quad L_2' = 5\,\text{H}$$

References

1. Beck CS, Pritchard WH, Feil HS (1947) Ventricular fibrillation of long duration abolished by electric shock. JAMA 135(15):985
2. Kouwenhoven WB, Milnor WR (1954) Treatment of ventricular fibrillation using a capacitor discharge. J Appl Physiol 7:253
3. Tacker WA, Geddes LA, McFarlane II, Milnor W, Gullet J et al (1969) Optimum current duration for capacitor-discharge defibrillation of canine ventricles. J Appl Physiol 27(4):480
4. Mackay RS, Leeds SE (1953) Physiological effects of condenser discharges. J Appl Physiol 667
5. Edmark KW (1963) Simultaneous voltage and current waveforms generated during internal and external direct current defibrillation. Surg Forum 262
6. Lown B, Neuman J, Amarasingham R, Berkovitz BV (1962) Comparison of alternating current with direct current countershock. Am J Cardiol 10:223
7. Kavanagh KM, Tang ASL, Rollins DL, Smith WM, Ideker RE (1989) A comparison of the internal defibrillation thresholds for monophasic, double and single capacitor biphasic waveforms. J Am Coll Cardiol 14(5):1343
8. Flaker GC, Schuder JC, McDaniel WC, Stoeckle H, Dbeis M (1989) Superiority of biphasic shocks in the defibrillation of dogs by epicardial patches and catheter electrodes. Am Heart J 118:228
9. White RD (2004) Waveforms for defibrillation and cardioversion: recent experimental and clinical studies. Curr Opin Crit Care 10(3):202–7
10. Geddes LA, Havel W (2000) Evolution of the optimum bidirectional (+/− biphasic) wave for defibrillation. Biomed Instrum Technol 34(1):39–54
11. Morrison LJ, Dorian P, Long J, Vermeulen M, Schwartz B, Sawadsky B, Frank J, Cameron B, Burgess R, Shield J, Bagley P, Mausz V, Brewer JE, Lerman BB (2005) Out-of-hospital cardiac arrest rectilinear biphasic to monophasic damped sine defibrillation

waveforms with advanced life support intervention trial (ORBIT). Recuscitation 66(2): 149–57
12. Walcott GP et al (1995) Choosing the optimal monophasic and biphasic waveforms for ventricular defibrillation. J Cardiovasc Electrophysiol 6:737–750
13. Kroll MW (1994) A minimal model of the single capacitor biphasic defibrillation waveform. PACE 17:1782–1792
14. Dixon et al (1987) Improved defibrillation thresholds With large contoured epicardial electrodes and biphasic waveforms. Circulation 76:1176–1184
15. Tang ASL et al (1989) Ventricular defibrillation using biphasic waveforms: the importance of phasic duration. J Am Coll Cardiol 13:207–214
16. Freeser SA et al (1990) Strength-duration and probability of success curves for defibrillation with biphasic waveforms. Circulation 82:2128–2141
17. Babbs CF, Whistler SJ (1978) Evaluation of the operating internal resistance, inductance, and capacitance of intact damped sine wave defibrillators. Med Instrum 12(1):34–37
18. Jones VC, Charbonnier FM, Long P (1981) Determining transthoracic impedance, delivered energy, and peak current during defibrillation episodes. Med Instrum 15(6):380–382
19. Adams TP, Kroll MW (1999) Apparatus for generating biphasic waveforms in an implantable defibrillator. U S Patent 005(871):505A
20. Bardy GH et al (1989) A prospective randomized evaluation of biphasic vs. monophasic waveform pulses on defibrillation efficiency in humans. J Am Coll Cardiol 14:728–733
21. Wyse DG et al (1993) Comparison of biphasic and monophasic shocks for defibrillation using a non-thoracotomy system. Am J Cardiol 71:197–202
22. Cameron DB (1993) Uni-cable defibrillator paddles. US Patent: 5,203,347

Chapter 6
Sinusoidal Steady-State Analysis

6.1 Average and RMS Values of Signals

Problem 6.1.1

(a) Show that rms value of the sinusoidal voltage, $v(t) = V_p$. $\sin(2\pi ft)$ is $V_p/\sqrt{2}$ volts.
(b) If $f = 1000$ Hz, $V_p = 10$ V, what is the rms value of $v(t)$?

Solution

(a)

$$\omega = 2\pi f$$

$$V_{rms} = \sqrt{\frac{1}{T}\int_0^T [V_p \cdot \sin(\omega t)]^2 dt} = V_p\sqrt{\frac{1}{T}\int_0^T \sin^2(\omega t)dt}$$

$$= V_p \cdot \sqrt{\frac{1}{T}\int_0^T \frac{1-\cos(2\omega t)}{2}dt}$$

$$V_{rms} = V_p\sqrt{\frac{1}{T}\left[\frac{t}{2} - \frac{\sin(2\omega t)}{4\omega}\right]\Bigg|_0^T} = V_p\sqrt{\frac{1}{T}\left(\frac{t}{2}\right)\Bigg|_0^T}$$

$$= V_p\sqrt{\frac{1}{T}\left(\frac{T}{2} - 0\right)} = V_p \cdot \sqrt{\frac{1}{2}} = \frac{V_p}{\sqrt{2}}$$

(b) The rms value is independent of signal frequency,

$$V_{rms} = \frac{V_p}{\sqrt{2}} = \frac{10}{\sqrt{2}} = 7.07\,\text{V}$$

A.Ü. Keskin, *Electrical Circuits in Biomedical Engineering*,
DOI 10.1007/978-3-319-55101-2_6

Problem 6.1.2 Show that the rms value of a sinusoidal voltage with maximum value A volts is the same for a pulse train with amplitude A (volts) and duty cycle = 1/2.

Solution

$$\text{rms value of a sinusoid} = A/\sqrt{2}\ \text{volts}$$
$$\text{rms value of pulse train} = A\cdot\sqrt{d/T}$$
$$d = \text{pulse width (s)},\ T = \ \text{period}$$

Since d/T = 1/2 = duty cycle, the rms value of pulse train is

$$A\cdot\sqrt{\frac{1}{2}} = \frac{A}{\sqrt{2}}$$

Therefore, rms values are the same for sinusoidal and pulse train signals if both have the same maximum levels and duty cycle = 1/2 for pulse train.

Problem 6.1.3

(a) Compute the average power (in mW) absorbed by 1 kΩ resistor connected to an independent voltage source supplying periodic triangular pulses at 50% duty cycle and maximum at Vm volts.
(b) Find the average power if the frequency of this waveform is increased 10 times.

Solution

(a)
$$P_{\text{av}} = \frac{1}{T}\int_0^T \frac{v^2(t)}{R}\,\mathrm{d}t = \frac{1}{T}\int_0^{\frac{T}{2}}\left(\frac{2V_{\text{m}}t}{T}\right)^2\cdot\frac{1}{R}\mathrm{d}t$$
$$P_{\text{av}} = \frac{1}{T}\int_0^{\frac{T}{2}}\frac{4V_{\text{m}}^2t^2}{T^2R}\mathrm{d}t = \frac{1}{T}\cdot\frac{4V_{\text{m}}^2}{T^2R}\int_0^{\frac{T}{2}}t^2\mathrm{d}t = \frac{4V_{\text{m}}^2}{T^3R}\cdot\frac{t^3}{3}\bigg|_0^{\frac{T}{2}} = \frac{V_{\text{m}}^2}{6R}$$

substituting $R = 1\,\text{k}\Omega,\ V_{\text{m}} = 10\,\text{V},\ P_{\text{av}} = 10^2/6000 = 16.667\,\text{mW}$

(b) No matter how much the frequency of this waveform is increased, average power absorbed remains the same, i.e., 16.667 mW.

Problem 6.1.4 In the sinusoidal state condition, the current and voltage relations on an element are

$$i(t) = 12\cos(6280t + 15^\circ)\ [\text{mA}],\quad v(t) = 3\cos(6280t + 75^\circ)\ [\text{V}]$$

respectively.

(a) Determine the average power consumed by this element, (mW).
(b) Is this an inductor or a capacitor?

Solution

$$P_{av} = \frac{1}{T}\int_{o}^{T} i(t) \cdot v(t)\mathrm{d}t = \frac{1}{2} V_m \cdot I_m \cos(\varphi_v - \varphi_i)$$

(a)

$$= \frac{1}{2} \times 3 \times 12 \times 10^{-3} \times \cos(75^\circ - 15^\circ)$$

$$= 18 \times 10^{-3} \times \cos 60^\circ = \frac{18 \times 10^{-3}}{2}$$

$$P_{av} = 9\,\mathrm{mW}$$

(b) This is neither an inductor nor a capacitor, since $\Delta\varphi \neq \pm 90^\circ$ (in other words, the difference between the phase angles is neither +90° nor −90°).

Problem 6.1.5 Compute the rms value of thesignal $x(t) = A(t) \cdot B(t)$, which is a modulated 50 Hz, and 10 V peak sinusoidal signal with another 4 Hz sinusoidal signal,

$$A(t) = 10 \cdot \sin(2\pi \cdot 50t), \quad B(t) = \sin(2\pi \cdot 4t).$$

Plot the modulated signal and its rms value on the same figure.

Solution
Using MATLAB, the rms value of the signal is computed as

$$\mathrm{Yrms} = 4.9009\,\mathrm{V}$$
$$\mathrm{Yaverage} = -0.1014\,\mathrm{V}$$

The signal and its rms value are displayed in Fig. 6.1.

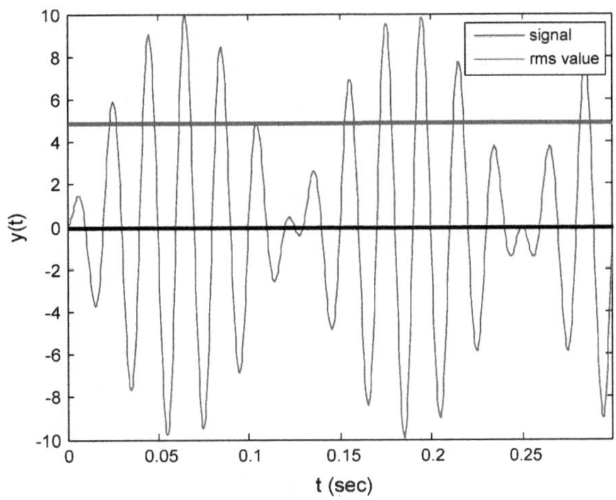

Fig. 6.1 The modulated signal and its rms value

A MATLAB script for the solution of this problem is given below.

```
%Compute the RMS level of a signal X, sampled at fs Hz.
fs=2000; ts=1/fs; tmax=0.3-ts;
t = 0:ts:tmax;
X = 10*sin(2*pi*50*t).*sin(2*pi*4*t);
%plot(t,X,'k-');grid on;xlim([0 tmax]);
Yrms = rms(X)
Yav = mean(X)
Y=0;
plot(t,X,t,Yrms,'r-',t,Y,'k-');xlim([0 tmax]);
xlabel('t (sec)');ylabel('y(t)');legend('signal','rms value');
```

Problem 6.1.6 A medical grade Carbon dioxide Laser unit produces 25 kW pulsed power in bursts tp = 100 ns long at a 400 Hz repetition rate. Determine the average power output of this laser unit.

Solution

$$f = \text{repetition rate} = 400 = \frac{1}{T} \quad \rightarrow \quad T = \frac{1}{400}\ \text{s}$$

$$P_{\text{av}} = \frac{1}{T}\int_0^T p(t)\text{d}t = \frac{1}{\frac{1}{400}}\int_0^{T_\text{p}} 25000\text{d}t = 400 \times 25000 \times t\big|_0^{T_\text{p}}$$
$$= 400 \times 25000 \times 100 \times 10^{-9} = 1\ \text{W}$$

Problem 6.1.7 Femtosecond lasers emit laser pulses of extremely short duration in the range of 300–500 fs. They do not transfer heat or shock to the material being operated. Therefore, surgical incisions are performed with much higher precision than conventional procedures.

A 1050 nm femtosecond laser unit produces 0.5 W average power in bursts of Tp = 500 fs duration at a 1 kHz repetition rate. Determine, (a) Frequency of the optical emission within a pulse package, (b) Single pulse power output of this laser unit (GW).

Solution

(a) Optical signal frequency within each pulse is

$$f\text{c} = \frac{c}{\lambda} = \frac{3 \times 10^8\,(\text{ms}^{-1})}{1050 \times 10^{-9}\,(\text{m})} = 0.002857 \times 10^{17}\ \text{Hz} = 285.7 \times 10^{12}\ \text{Hz}$$
$$= 285.7\ \text{THz}$$

(b) Let single pulse power be $P(W)$

$$f = \text{repetition rate} = 1000\,\text{Hz} = \frac{1}{T}$$

$$T = \frac{1}{1000}\,\text{s} = 1\,\text{ms}$$

$$P_{\text{av}} = 0.5\,\text{W} = \frac{1}{T}\int_0^T p(t)\text{d}t = \frac{1}{\frac{1}{1000}}\int_0^{T_\text{p}} P\text{d}t = 1000 \times P \times t|_0^{T_\text{p}}$$

$$0.5 = 1000 \times P \times 500 \times 10^{-15} = 5 \times 10^5 \times 10^{-15} \times P = 5 \times 10^{-10} \times P$$

$$P = \frac{0.5}{5 \times 10^{-10}} = 1 \times 10^9\,\text{W} = 1\,\text{GW}$$

Problem 6.1.8 A pacemaker is attached to the heart tissue at 250 Ω via its electrodes. The pulse generated by this unit is a rectangular pulse with 300 ms duration and 2.5 V peak, for 60 beats per minute. It draws 3 μA from the battery when the pulse is off. Pacemaker battery capacity is 2 A h at 2.5 V.

(a) Calculate the energy delivered to the heart during one period.
(b) Compute the lifetime of battery in years, if it operates only in this mode. Comment on this result.

Solution

(a) Period is 1 s, the energy delivered to the heart in 1 s

$$W = \frac{V^2}{R} \times T_\text{p} = \frac{2.5^2}{250} \times 300 \times 10^{-6} = 7.5\,\mu\text{J} + (2.5 \times 3 \times 1)\,\mu\text{J} = 15\,\mu\text{J}$$

(b) Total energy of the battery, $W_{\text{Bt}} = 2\,\text{A}\,\text{h} \times V = 2 \times 3600 \times 2.5 = 18{,}000\,\text{J}$

$$\text{Battery Life time} = \frac{18{,}000}{15} \times 10^6\,\text{s} = \frac{1200 \times 10^6}{365 \times 24 \times 3600} = 38\,\text{years}$$

This is an ideal case. By time, internal battery resistance increases, and this causes a serious reduction for the lifetime of pacemaker [1].

Problem 6.1.9 A pacemaker generates rectangular pulses of 1 ms duration and 3.5 V amplitude at 70 pulses/min, driving a load of 350 Ω.

(a) Determine total energy supplied to this load in 5 years.
(b) Determine the capacity of its battery in A h, (ampere-hours), if the battery is a 5.6 V Lithium type and 40% of its energy is spent for the pulses in 5 years period.

Solution

(a) $\text{Energy/pulse} = V_p \times I_p \times t_p = 3.5 \times \dfrac{3.5}{350} \times 1 \times 10^{-3} = 3.5 \times 10^{-5}\,\text{J/p}$

Number of pulses in 5 years:

$$70 \times 60 \times 24 \times 365 \times 5 = 183.96 \times 10^6\,\text{pulses}$$

Total energy:

$$E = 3.5 \times 10^{-5} \times 183.96 \times 10^6 = 641.76 \times 10 = 6417.6\,\text{J}$$

(b) Battery energy:

$$\begin{aligned} E_B &= 6417.6\,\text{J} = \text{VA s} \\ &= 0.4 \times 5.6 \times (\text{A s}), \quad \text{A s} = \text{capacity in seconds} \end{aligned}$$

$$\begin{aligned} &6417.6 = 2.24 \times \text{Capacity} \\ &\text{Capacity} = \frac{6417.6}{2.24}\,\text{A s} = 2865\,\text{A s} \\ &\text{Capacity in A h} = \frac{2865\,(\text{A s})}{3600\,(\text{s/h})} = 0.796\,\text{A h} \end{aligned}$$

6.2 The Phasor, Admittance, Impedance

Problem 6.2.1 Find the expression of the voltage signal in Fig. 6.2 (ac1.cir).

Solution

$$f = \frac{1}{10}\,\text{Hz}, \quad v(t) = 0.5 + \sin(2\pi f)t = 0.5 + \sin\left(628 \cdot \frac{1}{10}\right)t = 0.5 + \sin(62.8t)$$

Problem 6.2.2 Consider a sinusoidal voltage signal, $v(t) = 2.4\sin(628t)$.

(a) What is the amplitude of $v(t)$?
(b) What is the argument of $v(t)$?
(c) What is the angular frequency of $v(t)$?
(d) What is the period of $v(t)$?
(e) What is the frequency of $v(t)$, in Hz?

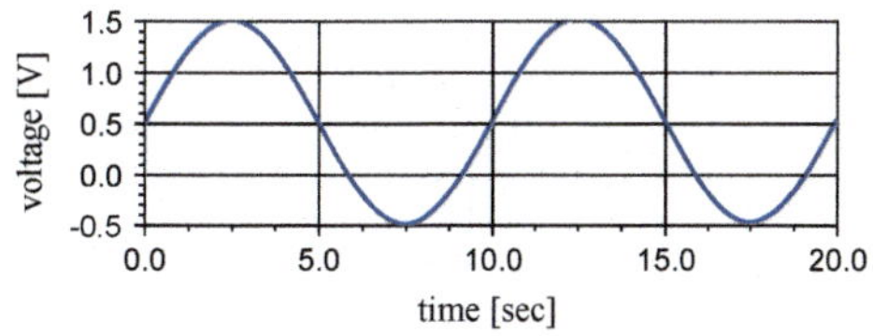

Fig. 6.2 The graph of the voltage signal

Solution

(a) Amplitude of $v(t)$ is 2.4 V
(b) Argument of $v(t)$ is $\omega t = 628t$
(c) Angular frequency of $v(t)$ is 628 rad/s.
(d) Period of $v(t)$ is

$$T = \frac{2\pi}{\omega} = \frac{6.28}{628} = \frac{1}{100} = 0.01\ \text{s}$$

(e) Frequency of $v(t)$ is $f = 1/T = 100\,\text{Hz}$
or, alternatively,

$$\omega = 2\pi f \quad \rightarrow \quad f = \frac{\omega}{2\pi} = \frac{628}{6.28} = 100\,\text{Hz}$$

Problem 6.2.3 Show that a sinusoidal voltage signal repeats itself at every T seconds.

Solution

$$v(t) = V_\text{m} \sin(\omega t)$$

Let, $(t + T)$ substitute t,

$$v(t+T) = V_\text{m} \sin(\omega(t+T)) = V_\text{m} \sin\left(\omega\left(t + \frac{2\pi}{\omega}\right)\right) = V_\text{m} \sin(\omega t + 2\pi)$$
$$= V_\text{m} \sin(\omega t)$$

which is $v(t)$. Therefore, $v(t+T) = v(t)$, in general, $v(t+nT) = v(t)$ for all integers n.

Problem 6.2.4 What is the amount of instantaneous power consumed over a 10 Ω resistor at t = 10 ms, if the sinusoidal current flowing through it has a peak value of 10 A and period of 20 ms?

Solution

$$\omega = \frac{2\pi}{T} = \frac{2\pi}{20 \times 10^{-3}} = 314\,\text{rad/s}$$
$$p(t) = i(t) \cdot v(t) = i^2(t) \cdot R = (10\cos(314t))^2 \cdot 10 = 1000 \cdot \cos^2(314t)$$
$$p\left(10 \times 10^{-3}\right) = 1000\cos^2\left(314 \times 10 \times 10^{-3}\right) = 1000\cos^2(3.14) = 1000\cos^2\pi$$
$$\cos\pi = -1,$$
$$p\left(10 \times 10^{-3}\right) = 1000 = 1\,\text{kW}$$

Problem 6.2.5 For the signals shown in Fig. 6.3, determine the equation of $v_1(t)$ and $v_2(t)$, if $v_2(t)$ crosses horizontal axis at $4\pi/9$ radians. Which one is leading the other? How many degrees? (plot_sine1.m)

Solution

$$v_1(t) = 1 \cdot \sin\omega t, \quad v_2(t) = 0.5 \times \sin(\omega t + \theta) = 0.5 \times \sin\left(\omega t + \frac{4\pi}{9}\right)$$

v_2 leads v_1 by

$$\frac{4 \times 180}{9} = 80°$$

Problem 6.2.6 Determine graphically the phase angle between

(a) $3\cos\omega t - 3\sin\omega t$,
(b) $3\cos\omega t + 4\sin\omega t$,
(c) $\cos\omega t - 2\sin\omega t$,

(Note: Positive direction of the sine function is down).

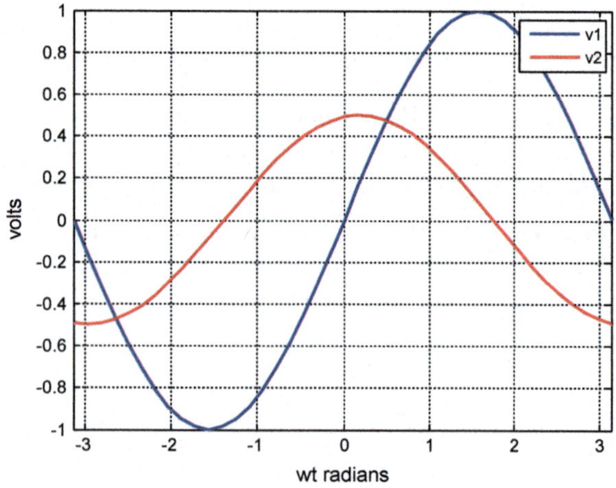

Fig. 6.3 The signals of Problem 6.2.5

Solution

(a) $\theta = \tan^{-1}(3/3) = 45°$
(b) $\theta = \tan^{-1}(4/3) = 53.1°$
(c) $\theta = \tan^{-1}(2/1) = 63.4°$.

Figure 6.4 shows these phasor angles.

Problem 6.2.7 Calculate the phase angle between $v_1(t) = -\cos(\omega t + 40°)$, $v_2(t) = 2\sin(\omega t - 5°)$. Which one leads the other?

Solution

$$\theta = 90 - 40 - 5 = 45°$$

Here, v_2 leads v_1 (phasors rotate counterclockwise). See, Fig. 6.5.

Problem 6.2.8 If $z = 3 + j4$, write z in polar and exponential forms.

Solution
Cartesian form is $z = 3 + j4$,
Polar form: $z = r\angle\theta = \sqrt{3^2 + 4^2}\angle\tan^{-1}(4/3) = 5\angle 53.1°$
Exponential form: $z = re^{j\emptyset} = 5e^{j53.1}$

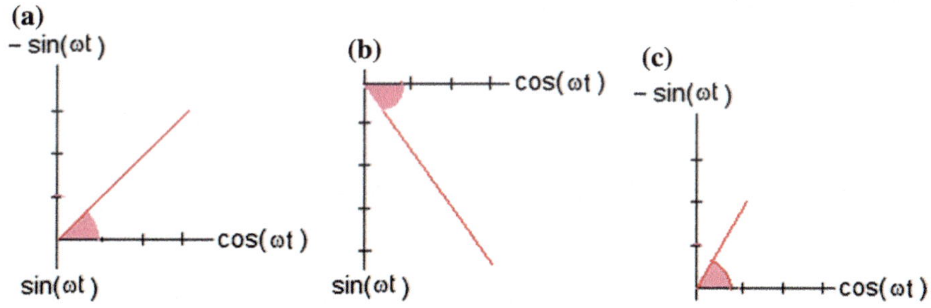

Fig. 6.4 The phasors of Problem 6.2.6

Fig. 6.5 The phasors of Problem 6.2.7

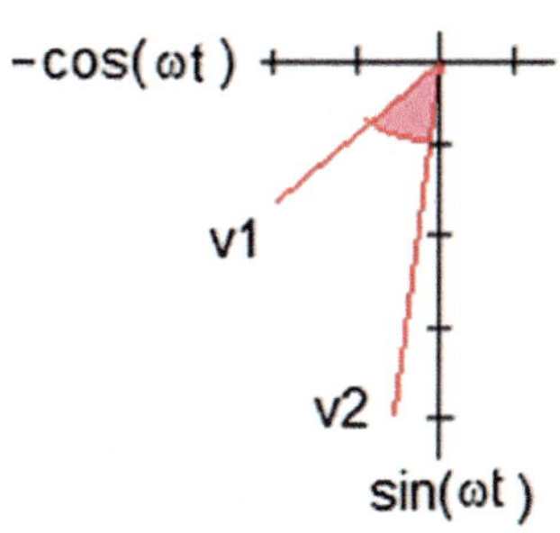

Problem 6.2.9 Using series expansions, prove the validity of Euler's theorem.

Solution
Euler's theorem: $\cos\theta + j\sin\theta = e^{j\theta}$

$$e^{\theta} = 1 + \theta + \frac{\theta^2}{2!} + \frac{\theta^3}{3!} + \frac{\theta^4}{4!} + \frac{\theta^5}{5!} + \cdots$$
$$\cos\theta = 1 - \frac{\theta^2}{2!} + \frac{\theta^4}{4!} - \cdots$$
$$\sin\theta = \theta - \frac{\theta^3}{3!} + \frac{\theta^5}{5!} - \frac{\theta^7}{7!} + \cdots$$
$$\cos\theta + j\sin\theta = 1 + j\theta - \frac{\theta^2}{2!} - j\frac{\theta^3}{3!} + \frac{\theta^4}{4!} + j\frac{\theta^5}{5!} - \cdots$$
$$\cos\theta + j\sin\theta = e^{j\theta}$$

Problem 6.2.10 If $Z = 2\angle 45°$, write Z in cartesian form, and in exponential form.

Solution

$$Z = 2\angle 45° = 2(\cos 45° + j\sin 45°) = 2\left(\frac{\sqrt{2}}{2} + j\frac{\sqrt{2}}{2}\right) = \sqrt{2} + j\sqrt{2}$$

In exponential form,

$$Z = re^{j\varphi} = 2e^{j45°}$$

Problem 6.2.11 A sinusoidal voltage reaches a positive maximum of 10 V at 30° before $\omega t = 0$.

Express this voltage waveform in phasor form.

Solution
The phase angle θ is 30°

$$30° = \frac{\pi}{6}\text{ rad}$$
$$v(t) = 10 \cdot \cos\left(\omega t + \frac{\pi}{6}\right) V = \text{Re}\left\{10\,e^{j\left(\omega t + \frac{\pi}{6}\right)}\right\} = \text{Re}\left\{10\,e^{j\left(\frac{\pi}{6}\right)}e^{j(\omega t)}\right\} V$$
$$V = 10\,e^{j\left(\frac{\pi}{6}\right)} = 10\angle\frac{\pi}{6}$$

Problem 6.2.12 Represent the voltage $v(t) = 310 \cdot \sin(\omega t + 30°)$ V by a phasor.

Solution

$$\cos\left(x - \frac{\pi}{2}\right) = \sin x$$
$$v(t) = 310 \cdot \cos\left(\omega t + \frac{\pi}{6} - \frac{\pi}{2}\right) = 310 \cdot \cos\left(\omega t - \frac{\pi}{3}\right)\, V$$

The voltage phasor is $\boldsymbol{V} = 310\, \mathrm{e}^{-j\left(\frac{\pi}{3}\right)}\, V = 310\angle - 60°\, V$

Problem 6.2.13 $s_1 = a_1 + jb_1, s_2 = a_2 + jb_2, \quad s_1/s_2 = ?$
(ln cartesian coordinates).

Solution

$$\frac{s_1}{s_2} = \frac{a_1 + jb_1}{a_2 + jb_2} = \frac{(a_1 + jb_1)(a_2 - jb_2)}{(a_2 + jb_2)(a_2 - jb_2)} = \frac{a_1 a_2 + b_1 b_2 + j(a_2 b_1 - a_1 b_2)}{a_2^2 + b_2^2}$$
$$\frac{s_1}{s_2} = \frac{a_1 a_2 + b_1 b_2}{a_2^2 + b_2^2} + j\frac{(a_2 b_1 - a_1 b_2)}{a_2^2 + b_2^2}$$

Problem 6.2.14 If $Z_1 = 3 + j4$, $Z_2 = 1 - j$. Evaluate the following:

(a) $Z_1 + Z_2$, (b) $Z_1 - Z_2$, (c) $Z_1 \,.\, Z_2$, (d) Z_1/Z_2, (e) $1/Z_2$, (f) $\sqrt{Z_2}$.

Solution

(a) $Z = Z_1 + Z_2 = 3 + j4 + 1 - j = 4 + j3$
(b) $Z = Z_1 - Z_2 = 3 + j4 - 1 + j = 2 + j5$
(c) $Z = Z_1 \cdot Z_2 = (3 + j4) \cdot (1 - j) = 3 - j3 + j4 + 4 = 7 + j$

$$Z = \sqrt{49 + 1}\angle\tan^{-1}\left(\frac{1}{7}\right) = \sqrt{50}\angle 8.1°$$

or

$$Z = Z_1 \cdot Z_2 = r_1 r_2 \angle(\varphi_1 + \varphi_2) = 5 \cdot \sqrt{2}\angle(53.1° + 45°) = \sqrt{50}\angle 8.1°$$

(d) $Z = \frac{Z_1}{Z_2} = \frac{r_1}{r_2}\angle(\varphi_1 - \varphi_2) = \frac{5}{\sqrt{2}}\angle(53.1° - (-45°)) = \frac{5\sqrt{2}}{2}\angle 98.1°$
(e) $Z = \frac{1}{Z_2} = \frac{1}{r_2}\angle - \varphi = \frac{1}{\sqrt{2}}\angle - (-45°) = \frac{\sqrt{2}}{2}\angle 45°$
(f) $\sqrt{Z_2} = \sqrt{r}\angle\frac{\varphi}{2} = \sqrt{\sqrt{2}}\angle - \left(\frac{45}{2}\right)^{\circ} = 2^{\frac{1}{4}}\angle - 22.5°$
(g) $Z_1^* = (1 - j)^* = 1 + j$

Problem 6.2.15 Transform $i(t) = -3\sin(314t + 60°)A$ into phasor.

Solution

$$-\sin x = \cos(x + 90°)$$
$$-\sin(\omega t + \emptyset) = \cos(\omega t + \varphi + 90°)$$
$$i(t) = -3\sin(314t + 60°) = 3\cos(314t + 60° + 90°) = 3\cos(314t + 150°)$$
$$\mathbf{I} = 3\angle 150°$$

Problem 6.2.16 Convert $I = (j2)e^{-j30°}$ into cartesian form and sinusoidal form.

Solution

$$j = 1\angle 90°,\ \mathbf{I} = (1\angle 90°)(2\angle -30°) = 2\angle(90 - 30)° = 2\angle 60°$$

Cartesian form: $\mathbf{I} = 2(\cos 60° + j\sin 60°) = 2\left(1/2 + j\sqrt{3}/2\right) = 1 + j\sqrt{3}$.
Sinusoidal form: $\mathbf{i}(t) = 2\cos(\omega t + 60°)$

Problem 6.2.17 Evaluate the following complex numbers (complex1.m):

(a) $(20\angle 30° + 10\angle - 60°)^{\frac{1}{2}}$

(b) $\dfrac{2\angle 60° + 1 - j}{(1 - j)^*}$.

Solution

(a) $(20\angle 30° + 10\angle - 60°)^{\frac{1}{2}} = \{20(\cos 30 + j\sin 30) + 10[\cos(-60) + j\sin(-60)]\}^{\frac{1}{2}}$
$\cos(-x) = \cos x, \sin(-x) = -\sin x$

$$(20\angle 30° + 10\angle - 60°)^{\frac{1}{2}} = \left[20\left(\frac{\sqrt{3}}{2} + j\frac{1}{2}\right) + 10\left(\frac{1}{2} - j\frac{\sqrt{3}}{2}\right)\right]^{\frac{1}{2}}$$
$$= \left[10\sqrt{3} + j10 + 5 - j5\sqrt{3}\right]^{\frac{1}{2}}$$
$$= \left[(10\sqrt{3} + 5) + j(10 - 5\sqrt{3}\right]^{\frac{1}{2}}$$

(b)

$$= \left[\sqrt{(10\sqrt{3} + 5)^2 + (10 - 5\sqrt{3})^2}\angle \tan^{-1}\left(\frac{10 - 5\sqrt{3}}{10\sqrt{3} + 5}\right)\right]^{\frac{1}{2}}$$
$$= \sqrt{\sqrt{(10\sqrt{3} + 5)^2 + (10 - 5\sqrt{3})^2}}\angle\frac{3.44°}{2}$$
$$= \sqrt{\sqrt{498.205 + 1.795}}\angle 1.72°$$
$$= \sqrt{\sqrt{500}}\angle 1.72° = \sqrt{22.361}\angle 1.72° = 4.729\angle 1.72°$$

(c)

$$\frac{2\angle 60^\circ + 1 - j}{(1-j)^*} = \frac{2(\cos 60^\circ + j\sin 60^\circ) + 1 - j}{1+j} = \frac{2\left(\frac{1}{2} + j\frac{\sqrt{3}}{2}\right) + 1 - j}{1+j}$$

$$= \frac{1 + j\sqrt{3} + 1 - j}{1+j} = \frac{2 + j(\sqrt{3}-1)}{1+j}$$

$$= \frac{\sqrt{(2)^2 + (\sqrt{3}-1)^2}\angle \tan^{-1}\left(\frac{\sqrt{3}-1}{2}\right)}{\sqrt{1^2+1^2}\angle \tan^{-1}\left(\frac{1}{1}\right)}$$

$$= \frac{\sqrt{4+3+1-2\sqrt{3}}\angle \tan^{-1}\left(\frac{\sqrt{3}-1}{2}\right)}{\sqrt{2}\angle 45^\circ}$$

$$= \frac{\sqrt{8-2\sqrt{3}}\angle \tan^{-1}\left(\frac{\sqrt{3}-1}{2}\right)}{\sqrt{2}\angle 45^\circ}$$

$$= \sqrt{\frac{8-2\sqrt{3}}{2}}\angle\left(\tan^{-1}\left(\frac{\sqrt{3}-1}{2}\right) - 45^\circ\right)$$

$$= \sqrt{4-\sqrt{3}}\angle - 24.89^\circ = 1.506\angle - 24.89^\circ$$

MATLAB script for the solution of part a:

```
c1=20*exp(i*pi/6); c2=10*exp(i*(-pi/3)); c=sqrt(c1+c2);
R=abs(c)
phir=angle(c);    %radian
phi=phir*180/pi %degrees

c=(2*exp(1i*pi/3)+1-1i)/(1+i);R=abs(c)
phir=angle(c);    %radian
phi=phir*180/pi %degrees
```

MATLAB script for the solution of part b:

```
c=(2*exp(1i*pi/3)+1-1i)/(1+i);R=abs(c)
phir=angle(c);      %radian
phi=phir*180/pi  %degrees
```

Problem 6.2.18 A current entering a junction is $I = 4 - j5$, while a current leaving the junction is $i_2(t) = 2.\cos(\omega t - 90°)$. Find the third current in sinusoidal form.

Solution
By KCL,

$$\begin{aligned}
&\boldsymbol{I_1} = \boldsymbol{I_2} + \boldsymbol{I_3}, \quad \boldsymbol{I_3} = \boldsymbol{I_1} - \boldsymbol{I_2} = (4 - j5) - (2\angle - 90°)\\
&\boldsymbol{I_3} = (4 - j5) - [2\cos(-90°) + j\,\sin(-90°)]\\
&\quad = (4 - j5) - [0 + j2(-1)] = (4 - j5) + j2 = 4 - j3\\
&\quad = \sqrt{3^2 + 4^2}\angle \tan^{-1}\left(\frac{-3}{4}\right) = 5\angle \tan^{-1}(-0.75) = 5\angle - 36.87°\\
&i_3(t) = 5\cos(\omega t - 36.87°)A
\end{aligned}$$

Problem 6.2.19 In an alternating current (AC) circuit, the current flowing through an impedance $Z = 5 + j5\,\Omega$ is $i(t) = -10\,\sin\omega t$. Find the voltage across the impedance in the time domain.

Solution

$$-10\sin\omega t = 10\cos(\omega t + 90°)$$

By Ohm's law, $\boldsymbol{V} = \boldsymbol{ZI} = (5 + j5)(10\angle 90°) = (5 + j5)(0 + j10) = -50 + j50$

$$\boldsymbol{V} = \sqrt{2500 + 2500}\angle \tan^{-1}\left(\frac{-50}{50}\right) = 70.7\angle \tan^{-1}(-1) = 70.7\angle - 45°$$

$$v(t) = 70.7.\cos(\omega t - 45°)$$

Problem 6.2.20 In an AC circuit the voltage across an impedance is $v(t) = 130 \cdot \cos(\omega t - 60°)V$.

If the current flowing through the impedance is $i(t) = 5\cos(\omega t + 40°)\,A$, determine the impedance in phasor form.

Solution

$$\boldsymbol{Z} = \frac{\boldsymbol{V}}{\boldsymbol{I}} = \frac{130\angle - 60°}{5\angle 40°} = 26\angle(-60° - 40°) = 26\angle - 100° = 26\angle 260°$$

Problem 6.2.21 Two impedances are series connected and voltage drop across one of these is

$$v_1(t) = 150 \cdot \cos\left(314t - \frac{\pi}{6}\right)V$$

If the voltage applied across these series-connected impedances is

$$v(t) = 250 \cdot \cos\left(314t + \frac{\pi}{3}\right) V$$

Determine the voltage drop across the second impedance.

Solution

$$\begin{aligned} \boldsymbol{V} &= \boldsymbol{V_1} + \boldsymbol{V_2} \\ 250\angle 60^\circ &= 150\angle -30^\circ + \boldsymbol{V_2} \end{aligned}$$

$$\begin{aligned} \boldsymbol{V_2} &= 250\angle 60^\circ - 150\angle -30^\circ = 250\left(\frac{1}{2} + j\frac{\sqrt{3}}{2}\right) - 150\left(\frac{\sqrt{3}}{2} - j\frac{1}{2}\right) \\ &= 125 + j125\sqrt{3} - 75\sqrt{3} + j75 = -4.90 + j291.51 \\ &= \sqrt{(4.9)^2 + (291.51)^2}\angle \tan^{-1}\left(\frac{-291.51}{4.9}\right) = \sqrt{85{,}000}\angle \tan^{-1}(-59.49) \\ &= 291.55\angle -89.04^\circ \\ v_2(t) &= 291.55 \cdot \cos(314t - 89.04) V \end{aligned}$$

Problem 6.2.22 Write down voltage–current relations for an inductor in time domain and frequency domain.

If the current through an inductor $L = 2H$ is given as $i_2(t) = 3\cos(314t + 30^\circ)$, what is the voltage across this inductor in time and frequency (phasor) domain?

Draw phasor diagram for the inductor. What is the phase angle between current and voltage? Which one is leading?

Solution

$$\begin{aligned} i_2(t) &= I_{\mathrm{m}}\cos(\omega t + \varphi), \\ i_{\mathrm{L}}(t) &= L\frac{\partial i}{\partial t} = -\omega L I_{\mathrm{m}}\sin(\omega t + \varphi) \end{aligned}$$

Since $-\sin(x) = \cos(x + 90^\circ)$,

$$\begin{aligned} v_L(t) &= \omega L I_{\mathrm{m}} \cos(\omega t + \varphi + 90^\circ) = 314 \times 2 \times 3 \cos(314t + 30^\circ + 90^\circ) \\ v_L(t) &= 1884 \cos(314t + 120^\circ) \\ \boldsymbol{V} &= \omega L I_{\mathrm{m}} e^{j(\varphi + 90^\circ)} = \omega L I_{\mathrm{m}} e^{j\varphi} e^{j90^\circ} = \omega L I_m \angle \varphi + 90^\circ \end{aligned}$$

The voltage has a magnitude of $\omega L I_{\mathrm{m}}$, the phase angle is $\varphi + 90^\circ$,

$$\boldsymbol{V} = (314 \times 2 \times 3)\angle(30^\circ + 90^\circ) = 1884\angle 120^\circ$$

Phase angle between I and V is 90°. Voltage phasor leads current phasor by 90°, Fig. 6.6.

Problem 6.2.23 Given that

$$e^{-j\frac{\pi}{2}} = \frac{1}{j}, \quad \ln(-10j) = ?$$

Solution

$$e^{-j\frac{\pi}{2}} = \frac{1}{j} = -j \quad \rightarrow \quad -10j = 10e^{-j\frac{\pi}{2}}$$

$$\ln(-10j) = \ln\left(-10\,e^{-j\frac{\pi}{2}}\right) = \ln 10 + \ln\left(e^{-j\frac{\pi}{2}}\right) = 2.3 - j\frac{\pi}{2} = 2.3 - j1.57$$

Problem 6.2.24 $V = 50\,e^{j53.1}, \quad I = 25\,e^{-j56.3}, \quad Z = ?$

Solution
By Ohm's law,

$$\boldsymbol{Z} = \frac{\boldsymbol{V}}{\boldsymbol{I}} = \frac{50e^{j53.1}}{25e^{-j56.3}} = 2e^{j109.4} = 2\angle 109.4^\circ$$

The angles are given in degrees with decimal fractions.

Problem 6.2.25

(a) Write down a list of voltage–current relationships in time and frequency domain.
(b) If $v(t) = 10\cos(314t - 60^\circ)$ is applied to a capacitor with $C = 100\,\mu\text{F}$, determine the current through capacitor in time and frequency domain.

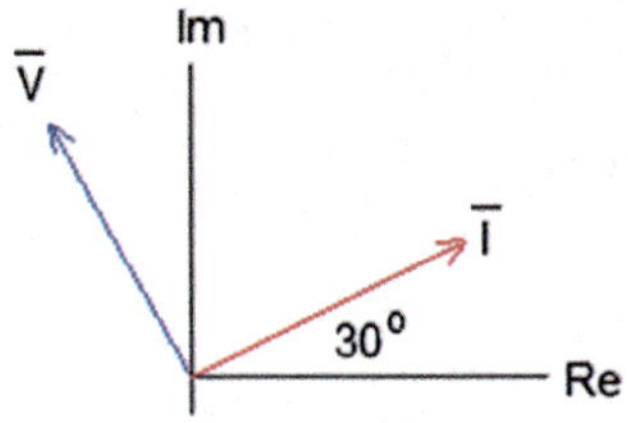

Fig. 6.6 The phasors of Problem 6.2.22

Solution

$$\text{(a)}\quad \begin{array}{ll} \text{Resistance} & v(t) = R \times i(t), \quad \boldsymbol{V} = R.\boldsymbol{I} \\ \text{Inductor} & v(t) = L \times \dfrac{\mathrm{d}i(t)}{\mathrm{d}t}, \quad \boldsymbol{V} = j\omega L.\boldsymbol{I} \\ \text{Capacitor} & i(t) = C \times \dfrac{\mathrm{d}V(t)}{\mathrm{d}t}, \quad \boldsymbol{V} = \dfrac{\boldsymbol{I}}{j\omega C} \end{array}$$

(b) In time domain,

$$\begin{aligned} v(t) &= A\cos(\omega t + \varphi) \\ i(t) &= C.\frac{\mathrm{d}V(t)}{\mathrm{d}t} = C \cdot \frac{\mathrm{d}[A\cos(\omega t + \varphi)]}{\mathrm{d}t} = -\omega CA\sin(\omega t + \varphi) \\ \sin x &= \cos(x + 90^\circ) \quad \rightarrow \quad i(t) = -\omega CA\cos(\omega t + \varphi + 90^\circ) \\ A &= 10, \quad \varphi = 60^\circ, \quad C = 100\,\mu\text{F}, \quad \omega = 314\,\text{rad/s}, \\ i(t) &= (314)\left(100 \times 10^{-6}\right)(10)\cos(314t - 60^\circ + 90^\circ)A \\ i(t) &= (314)\cos(314t + 30^\circ)\,\text{mA} \end{aligned}$$

In frequency domain,

$$\begin{aligned} \boldsymbol{V} &= \frac{\boldsymbol{I}}{j\omega C} \\ \boldsymbol{I} &= j\omega C\boldsymbol{V} = j(314)\left(100 \times 10^{-6}\right)(10\angle - 60^\circ)A = 314j\angle - 60^\circ\,\text{mA} \\ &= (314\angle 90^\circ) \times (1\angle - 60^\circ) = 314\angle(90^\circ - 60^\circ) = 314\angle 30^\circ\,\text{mA} \\ i(t) &= (314)\cos(314t + 30^\circ)\,\text{mA} \end{aligned}$$

Problem 6.2.26 The voltage across an inductor with $L = 4H$ is $V = 8\angle - 50\,\text{V}$. Find $i_L(t)$, the current through an inductor, if $\omega = 100$ rad/s.

Solution

$$\begin{aligned} \boldsymbol{I} &= \frac{\boldsymbol{V}}{j\omega L} = \frac{8\angle - 50}{j \cdot 100 \cdot 4} = \frac{8\angle - 50^\circ}{j \cdot 400} \\ (j &= 1\angle 90^\circ) \\ \boldsymbol{I} &= \frac{8\angle - 50^\circ}{400\angle 90^\circ} = 0.02\angle(-50^\circ - 90^\circ) = 0.02\angle - 140^\circ \\ i(t) &= 0.02\cos(100t - 140^\circ)A \end{aligned}$$

Problem 6.2.27 Determine admittance of the circuit of Fig. 6.7 (in mS), $R = 2\,\Omega, X_L = 2\,\Omega, X_C = 4\,\Omega$.

Fig. 6.7 The circuit for Problem 6.2.27

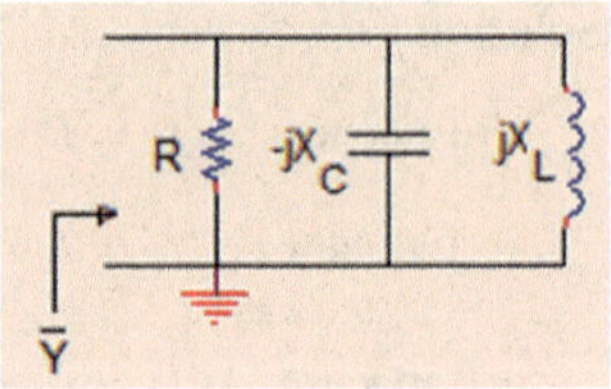

Solution

$$\begin{aligned}
Y &= \frac{1}{R} + \frac{1}{jX_L} + \frac{1}{-jX_C} = \frac{1}{2} + \frac{1}{j2} + \frac{1}{-j4} = 0.5 - j0.5 + j0.25 = 0.5 - j0.25\,\mathrm{S} \\
Y &= 500 - j250\,\mathrm{mS}
\end{aligned}$$

Problem 6.2.28 The current $i(t) = 5\cos(1000t + 5°)A$ and voltage $v(t) = 10\sin(1000t + 125°)V$ are observed across an impedance. Determine the associated impedance in cartesian form.

Solution

$$\begin{aligned}
v(t) &= 10\sin(1000t + 125°) = 10\cos(1000t + 125 - 90°) = 10\cos(1000t + 35°) \\
\mathbf{V} &= 10\angle 35° \\
\mathbf{Z} &= \frac{\mathbf{V}}{\mathbf{I}} = \frac{10\angle 35°}{5\angle 5°} = 2\angle(35° - 5°) = 2\angle 30° = 2(\cos 30° + j\sin 30°) \\
&= 2\left(\frac{\sqrt{3}}{2} + j\frac{1}{2}\right) = \sqrt{3} + j\,\Omega
\end{aligned}$$

Problem 6.2.29

(a) Calculate the voltage phase shift in the circuit of Fig. 6.8 with

$$R = 10\,\Omega,\ C = 10\,\mathrm{nF},\ f = 1\,\mathrm{MHz}.$$

(b) Find the frequency where the phase shift is—30° (RC1.cir).

Solution

(a) By voltage division rule,

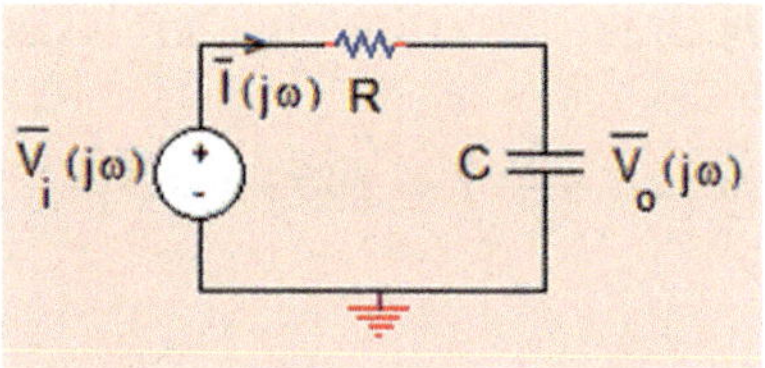

Fig. 6.8 The circuit for Problem 6.2.29

$$
\begin{aligned}
\boldsymbol{V} &= \frac{-jX_C}{R - jX_C}\\
X_C &= \frac{1}{2\pi \times 10^6 \times 10 \times 10^{-9}} = 15.9\,\Omega\\
\frac{\boldsymbol{V_o}}{\boldsymbol{V_i}} &= \frac{-j15.9}{10 - j15.9} = \frac{15.9\angle - 90^\circ}{\sqrt{10^2 + 15.9^2}\angle \arctan\left(\frac{-15.9}{10}\right)} = \frac{15.9\angle - 90^\circ}{18.78\angle - 57.8^\circ}\\
&= 0.847\angle - 32.17^\circ
\end{aligned}
$$

$v_o(t)$ is lagging $v_i(t)$ by 32.17° at 1 MHz. This is verified by SPICE (Fig. 6.9),

$$\Delta t = \frac{32.17}{180} \times 0.5\,\mu\text{s} \cong 0.09\,\mu\text{s})$$

SPICE netlist for RC1.cir

```
ac  voltage  signal
*SIN(V0  VA  FREQ  TD  THETA)
v1    1    0     sin(0    1    1meg    -0.25u    0)
R  1  2  10
C  2  0  10n
*.tran  1n  3u  1u  10n  uic
```

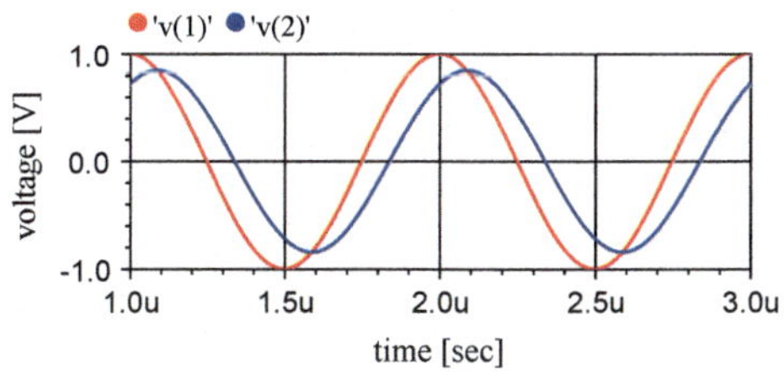

Fig. 6.9 Input output waveforms. $V(1) = V_i$, $V(2) = V_o$

(b) The frequency at which the phase shift is −30° is determined as follows:

$$\theta = -30 = -90 - \arctan\left(\frac{-X_C}{R}\right) = -90 + \arctan\left(\frac{X_C}{R}\right)$$

$$60^\circ = \arctan\left(\frac{X_C}{R}\right)$$

$$\tan(60^\circ) = \frac{X_C}{R} = \frac{1}{\omega CR} = 1.732 \quad \rightarrow \quad \omega = \frac{1}{(1.732)(10 \times 10^9)(10)}$$

$$\omega = 0.5774 \times 10^7 \,(\text{rad|sec}) \quad \rightarrow \quad f = \frac{0.5774 \times 10^7}{2\pi} = 0.092 \times 10^7 = 920\,\text{kHz}$$

Problem 6.2.30 An impedance consists of a 30 Ω resistive and 40 Ω of reactive components. Determine

(a) Squared magnitude of the impedance,
(b) Its admittance in mS

Solution

$$\mathbf{Z} = 30 + j40\,\Omega$$

(a) $|\mathbf{Z}| = \sqrt{30^2 + 40^2} = \sqrt{900 + 1600} = 50\,\Omega$

$$|\mathbf{Z}|^2 = 50^2 = 2500\,\Omega^2$$

(b) Admittance in mS,

$$\mathbf{Y} = \frac{\mathbf{Z}^*}{|\mathbf{Z}|^2} = \frac{30 - j40}{2500} = \frac{30}{2500} - \frac{j40}{2500}S = 0.012 - j0.016\,\text{S} = 12 - j16\,\text{mS}$$

Problem 6.2.31 In the circuit shown in Fig. 6.10, $v(t) = 10\cos(10t + 30^\circ)$, $L = 1\,\text{H}$, $R = 10\,\Omega$.

Determine the resistance, reactance, impedance, conductance, susceptance, and admittance of the circuit.

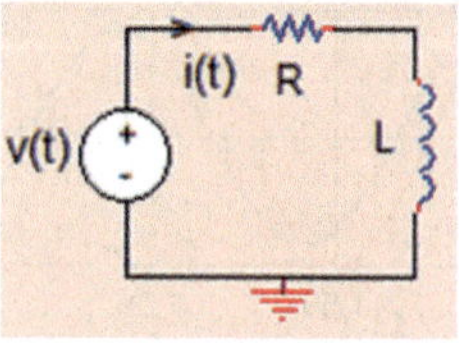

Fig. 6.10 The circuit for Problem 6.2.31

Solution

$$\boldsymbol{Z} = R + jX = |\boldsymbol{Z}|\angle\theta$$

$$|\boldsymbol{Z}| = \sqrt{R^2 + X^2}, \quad \theta = \arctan\left(\frac{X}{R}\right)$$

$$\boldsymbol{Y} = \frac{1}{\boldsymbol{Z}} = \frac{\boldsymbol{I}}{\boldsymbol{V}} = G + jB = \frac{1}{R + jX} = \frac{R}{R^2 + X^2} - j\frac{X}{R^2 + X^2}$$

Using given data, $L = 1\text{H}, \quad R = 10\,\Omega, \quad \omega = 10\,\text{rad/s}$

$$\text{Resistance} = R = 10\,\Omega$$

$$\text{Reactance} = X = \omega L = 10 \times 1 = 10\,\Omega$$

$$|\boldsymbol{Z}| = \sqrt{10^2 + 10^2} = \sqrt{200} = 10 \times \sqrt{2}\,\Omega\,(\text{Impedance magnitude})$$

$$\theta = \arctan\left(\frac{10}{10}\right) = 45^\circ$$

$$\boldsymbol{Y} = \frac{10}{100 + 100} - j\frac{10}{100 + 100} = 0.05 - j0.05\,S$$

$$\text{Conductance} = G = 0.05\,\text{S}, \quad \text{Susceptance} = B = 0.05\,S$$

$$\boldsymbol{Y} = \frac{1}{10\sqrt{2}\angle 45^\circ} = \frac{\sqrt{2}}{20}\angle - 45^\circ\ \text{S}\,(\text{Admittance})$$

Problem 6.2.32 Using phasors, determine the current flowing through a series-connected RL circuit, if the input voltage is given by $v(t) = 4\cos(10t - 45^\circ)$, $R = 5\,\Omega, \quad L = 2\,\text{H}$.

Solution

$$L\frac{\mathrm{d}i}{\mathrm{d}t} + R \cdot i(t) = v(t) \quad \rightarrow \quad j\omega L\boldsymbol{I} + R\boldsymbol{I} = \boldsymbol{V}$$

$$\boldsymbol{I}(j\omega L + R) = \boldsymbol{V} \quad \rightarrow \quad \boldsymbol{I} = \frac{\boldsymbol{V}}{j\omega L + R} = \frac{4\angle - 45^\circ}{j10 \times 2 + 5} = \frac{4\angle - 45^\circ}{20j + 5}$$

$$\boldsymbol{I} = \frac{4\angle - 45^\circ}{\sqrt{25 + 400}\angle\arctan\left(\frac{20}{5}\right)} = \frac{4\angle - 45^\circ}{\sqrt{425}\angle\arctan(4)} = \frac{4\angle - 45^\circ}{20.62\angle 75.96^\circ}$$

$$\boldsymbol{I} = 0.194\angle(-45^\circ - 75.96^\circ) = 0.194\angle(-120.96^\circ)$$

Problem 6.2.33 For a series RL circuit, shown in Fig. 6.10,

(a) If $v(t) = V \cdot \cos\omega t$, determine $i(t)$.
(b) If $V = 10 \times \sqrt{2}$ volts, $\quad R = 10\,\Omega, \quad L = 1\,\text{H}, \quad \omega = 10\,\text{rad/sec}, \quad i(t) = ?$

Solution

$$\boldsymbol{V} = \boldsymbol{I}\cdot(R + j\omega L) \quad \rightarrow \quad \boldsymbol{I} = \frac{V\angle 0^\circ}{R + j\omega L} = \frac{V\angle 0^\circ}{\sqrt{R^2 + \omega^2 L^2}\angle \tan^{-1}\left(\frac{\omega L}{R}\right)}$$

(a)
$$\boldsymbol{I} = \frac{V}{\sqrt{R^2 + \omega^2 L^2}}\angle - \tan^{-1}\left(\frac{\omega L}{R}\right)$$
$$i(t) = \frac{V}{\sqrt{R^2 + \omega^2 L^2}}\cos\left[\omega t - \tan^{-1}\left(\frac{\omega L}{R}\right)\right]$$

(b)
$$\boldsymbol{I} = \frac{10\sqrt{2}}{\sqrt{100 + 10^2 \times 1}}\angle - \tan^{-1}\left(\frac{10}{10}\right) = \frac{10\sqrt{2}}{\sqrt{200}}\angle - 45^\circ = \frac{10\sqrt{2}}{10\sqrt{2}}\angle - 45^\circ$$
$$i(t) = 1\cos(10\text{t} - 45^\circ)\text{A}$$

Problem 6.2.34 For a series RL circuit, shown in Fig. 6.10, $R = 1$ kΩ, $L = 10$ mH, $v_1(t) = 317\cos(314t)$. Plot $v_2(t)$ and (t) (phasor1.xlsx, RL2.cir).

Solution

$$\boldsymbol{V}_2 = \boldsymbol{V}_1\cdot\frac{j\omega L}{R + j\omega L} = \boldsymbol{V}_1\cdot\frac{j\omega L(R - j\omega L)}{R^2 + \omega^2 L^2} = \boldsymbol{V}_1\cdot\frac{j\omega LR + \omega^2 L^2}{R^2 + \omega^2 L^2}$$
$$= 317\frac{(314)^2(10^{-2})^2 + j314 \times 10^{-2} \times 10^3}{10^6 + (314)^2(10^{-2})^2}$$
$$= 317\frac{9.8596 + j3140}{1000009.8596} = 0.00313 + j0.995$$
$$\cong 0.995\angle 90^\circ$$
$$v_2(t) = 0.995\cos(314t + 90^\circ)$$

The current through the inductor,

$$\boldsymbol{I} = \frac{\boldsymbol{V}_1 - \boldsymbol{V}_2}{R} = \frac{317\angle 0^\circ - 0.995\angle 90^\circ}{1000} = 0.317 - 0.000995 \cong 0.317\angle 0^\circ$$
$$i(t) = 0.317\cos(314t)$$

A plot of $v_2(t)$ and $i(t)$, (i.e., current through and voltage across the inductor) is plotted in Fig. 6.11. Voltage across the inductor leads current through the inductor by 90.

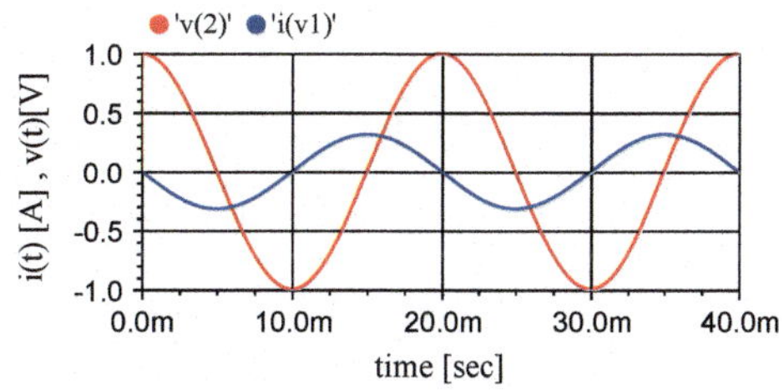

Fig. 6.11 Current and voltage waveforms on the inductor. Voltage leads current by 90°

SPICE netlist for RL2.cir:

```
voltage  and  current  over  L
v1    1  0    sin(0  317  50    0    0)
R1  1  2  1k
L1  2  0  10m
.tran(0.1m  40m  0  0.1m  uic)
```

Problem 6.2.35 Determine the time domain value of voltage across the capacitor in the circuit shown in Fig. 6.12a, if $v(t) = 2\cos 10t$, $R = 10k\,\Omega$, $C = 20\,\mu\text{F}$ (RC1.cir).

Solution
By voltage division,

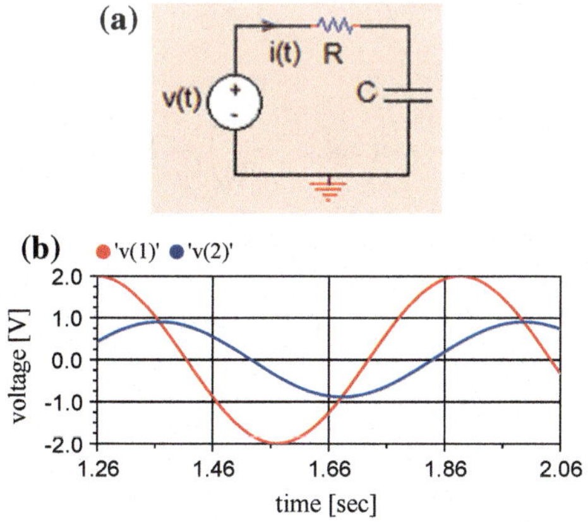

Fig. 6.12 **a** The circuit for Problem 6.2.35. **b** Voltage waveforms in the circuit

$$\begin{aligned}
\boldsymbol{V_c} &= \boldsymbol{V}\cdot\frac{\frac{1}{j\omega C}}{R+\frac{1}{j\omega C}} = \boldsymbol{V}\cdot\frac{1}{1+j\omega RC} = \boldsymbol{V}\cdot\frac{1-j\omega RC}{1+(\omega RC)^2} \\
&= 2\angle 0^\circ\,\frac{1-j10\times 10^4\times 20\times 10^{-6}}{1+(10\times 10^4\times 20\times 10^{-6})^2} \\
&= 2\frac{1-j200\times 10^{-2}}{1+(200\times 10^{-2})^2} = 2\frac{1-j2}{5} = 0.4-j0.8 \\
&= \sqrt{0.16+0.64}\angle\tan^{-1}\left(-\frac{0.8}{0.4}\right) = 0.894\angle-63.43^\circ \\
\boldsymbol{v_c}(\boldsymbol{t}) &= 0.894\cos(10t-63.43^\circ)V
\end{aligned}$$

Figure 6.12b shows the variation of the voltage across the capacitor, $v_c(t) = v(2)$, and input voltage to the circuit, $v(t) = v(1)$.

SPICE netlist (RC1.cir);

```
AC voltage  signal
v1   1   0    sin(0   2   1.59   -0.1572   0)
R  1  2  10k
C  2  0  20u
*.tran  1m  2  06  1.6  1  uic
```

Problem 6.2.36 In the circuit of Fig. 6.13, $i(t) = 4\cos(1000t)$, $L = 1\,\text{mH}$, $C = 10\,\mu\text{F}$, $v_C(0) = 0\,\text{V}$, $v(t) = ?$

Solution

$$\boldsymbol{V} = j\omega L\boldsymbol{I} + \frac{1}{j\omega C}\boldsymbol{I} = \boldsymbol{I}\cdot j\cdot\left(\omega L - \frac{1}{\omega C}\right) = \boldsymbol{I}\cdot j\left(\frac{\omega^2 LC-1}{\omega C}\right)$$

(a)
$$\begin{aligned}
&= (4\angle 0^\circ)(1\angle 90^\circ)\frac{10^6\times 10^{-3}\times 10^{-5}-1}{10^3\times 10^{-5}} = (4\angle 90^\circ)\frac{10^{-2}-1}{10^{-2}} \\
&= (4\angle 90^\circ)\left(\frac{-0.99}{0.01}\right) = (4\angle 90^\circ)(-99) = (396\angle 90^\circ)(-1) \\
&= (396\angle 90^\circ)(1\angle 180^\circ) = (396\angle 270^\circ)
\end{aligned}$$

$$v(t) = 396\cos(1000t+270^\circ) = 396\cos(1000t-90^\circ) = 396\sin(1000t)$$

Fig. 6.13 The circuit for Problem 6.2.36

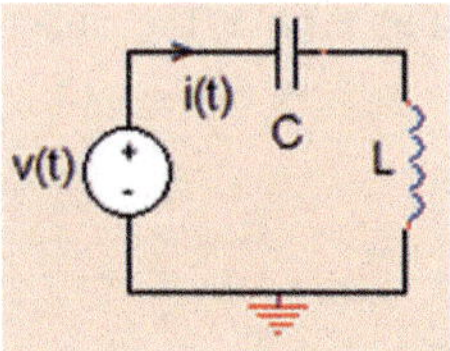

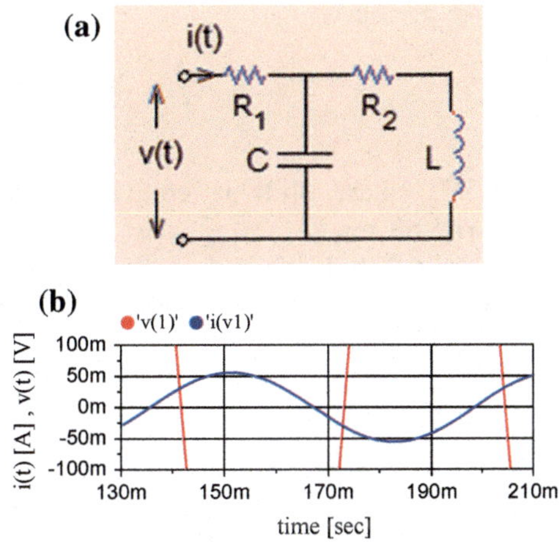

Fig. 6.14 **a** The circuit for Problem 6.2.37. **b** The input current and voltage waveforms for the circuit

(b) At resonance, voltage drop across series LC connection is null.

$$\omega^2 = \frac{1}{\mathrm{LC}} \quad \rightarrow \quad L = \frac{1}{\omega^2 C} = \frac{1}{10^6 10^{-5}} = 0.1\,\mathrm{H}$$

Problem 6.2.37 Determine the current $i(t)$, in the circuit shown in Fig. 6.14a, (mA) (ac2.cir).

$$v(t) = \cos(100t), \quad R_1 = 5\,\Omega, \quad R_2 = 10\,\Omega, \quad C = 1000\,\mu\mathrm{F}, \quad L = 100\,\mathrm{mH}.$$

Solution

$$L = 100\,\mathrm{mH} \quad \rightarrow \quad jX_L = j\omega L = j(100)\left(100 \times 10^{-3}\right) = j10\,\Omega$$

$$C = 1000\,\mu\mathrm{F} \quad \rightarrow \quad -jX_C = -j\frac{1}{\omega C} = -j\frac{1}{100 \times (1000 \times 10^{-6})} = \frac{-j}{10^{-1}} = -j10\,\Omega$$

$$\boldsymbol{Y_1} = \frac{1}{-j10} + \frac{1}{10 + j10} = j0.1 + \frac{10 - j10}{100 + 100} = j0.1 + \frac{10}{200} - j\frac{10}{200} = 0.05 + j0.05$$

$$\boldsymbol{Z_1} = \frac{1}{\boldsymbol{Y_1}} = \frac{1}{0.05 + j0.05} = \frac{0.05 - j0.05}{10^{-4}(25 + 25)} \cong \frac{0.05 - j0.05}{0,005} = 10 - j10$$

$$\boldsymbol{Z} = \boldsymbol{Z_1} + R_1 = 10 - j10 + 5 = 15 - j10\,\Omega = \sqrt{225 + 100}\angle\ \arctan\left(-\frac{10}{15}\right)$$

$$\boldsymbol{Z} = 18.03\angle - 33.7^\circ$$

$$\boldsymbol{I} = \frac{\boldsymbol{V}}{\boldsymbol{Z}} = \frac{1\angle 0^\circ}{18.03\angle - 33.7^\circ} = 0.055\angle(0^\circ + 33.7^\circ) = 0.055\angle(33.7^\circ)$$

$$i(t) = 55\cos(100t + 33.7^\circ)\,\mathrm{mA}$$

Since period is

$$T = \frac{2\pi}{\omega} = \frac{2\pi}{100} = 62.83\,\text{ms},$$

the 33.7° phase shift is equivalent to a time delay of (62.83 ms × 33.7°/ 360° =) 5.88 ms.

Figure 6.14b displays these current and input voltage waveforms.

Following is the SPICE netlist (ac2.cir):

```
ac voltage signal
ac voltage signal
*SIN(V0 VA FREQ TD THETA)
v1   1 0   sin(0  1  15.9  -15.7m  0)
R1 1 2 5
C   2 0 1000u
R2 2 3 10
L   3 0 100m
*.tran 1m 200m 100m 1m uic
```

Problem 6.2.38 In a series RLC ac circuit, $X_C = 1\,\Omega$, $X_L = 20\,\Omega$, $R = 10\,\Omega$.

(a) Determine the impedance of the circuit,

(b) If frequency is doubled, what is the new impedance (in phasor form)?

Solution

(a) The impedance,

$$\begin{aligned}\mathbf{Z} &= -j10 + j20 + 10 = j10 + 10 = \sqrt{200}\angle\tan^{-1}(1) = (\sqrt{2}\times 10)\angle 45^\circ \\ &= 14.1\angle 45^\circ\,\Omega\end{aligned}$$

(b) If ω is doubled,

$$X_c = \frac{1}{\omega C},\quad X_c \text{ is halved},\quad X_L = \frac{1}{\omega C},\quad X_L \text{ is doubled}$$

$$\mathbf{Z}_{\text{new}} = -j5 + j40 + 10 = j35 + 10 = \sqrt{35^2 + 100}\angle\tan^{-1}\left(\frac{10}{35}\right)$$

$$\mathbf{Z}_{\text{new}} = \sqrt{1325}\angle\tan^{-1}(0.2857) = 36.4\angle 15.95^\circ\,\Omega$$

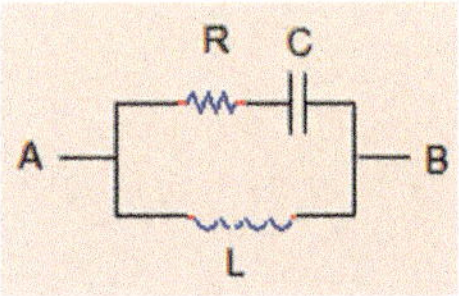

Fig. 6.15 The circuit for Problem 6.2.39

Problem 6.2.39 An inductor is shunted by a series-connected RC circuit, see Fig. 6.15.

(a) Determine the condition at which the circuit impedance is resistive at an angular frequency of ω.
(b) Calculate the value of resistance if $L = 1$ mH, $C = 100$ nF and frequency is 1000 Hz.

Solution

(a)

$$\boldsymbol{Y} = \frac{1}{j\omega L} + \frac{1}{R + \frac{1}{j\omega C}} = \frac{1}{j\omega L} + \frac{1}{\frac{\omega RC - j}{\omega C}}$$

$$= \frac{1}{j\omega L} + \frac{\omega C}{\omega RC - j} = \frac{\omega RC - j + j\omega L\omega C}{j\omega L(\omega RC - j)} = \frac{\omega RC + j(\omega^2 LC - 1)}{j\omega^2 RLC + \omega L}$$

$$\boldsymbol{Z} = \frac{1}{\boldsymbol{Y}} = \frac{j\omega^2 RLC + \omega L}{\omega RC + j(\omega^2 LC - 1)} = \frac{(j\omega^2 RLC + \omega L)[\omega RC + j(\omega^2 LC - 1)]}{\omega^2 R^2 C^2 + (\omega^2 LC - 1)^2}$$

$$= \frac{\omega^2 RLC - j\omega L(\omega^2 LC - 1) + j\omega^3 LR^2 C^2 + \omega^2 RLC(\omega^2 LC - 1)}{\omega^2 R^2 C^2 + \omega^4 L^2 C^2 + 1 - 2\omega^2 LC}$$

$$\mathrm{Im}(\boldsymbol{Z}) = \frac{-j\omega L(\omega^2 LC - 1) + j\omega^3 LR^2 C^2}{\omega^2 R^2 C^2 + \omega^4 L^2 C^2 + 1 - 2\omega^2 LC} = 0 = -\omega L(\omega^2 LC - 1) + \omega^3 LR^2 C^2$$

$$-L(\omega^2 LC - 1) + \omega^2 LR^2 C^2 = 0 \quad \rightarrow \quad -(\omega^2 LC - 1) + \omega^2 R^2 C^2 = 0$$

$$\omega^2 LC = \omega^2 R^2 C^2 + 1 \quad \rightarrow \quad L = \frac{\omega^2 R^2 C^2 + 1}{\omega^2 C}$$

$$L = R^2 C + \frac{1}{\omega^2 C}, \quad L \geq \frac{1}{\omega^2 C}$$

(b) The value of resistance is

$$R = \sqrt{\frac{L - \frac{1}{\omega^2 C}}{C}} = \sqrt{\frac{1 \times 10^{-3} - \frac{1}{6280^2 \times 100 \times 10^{-9}}}{100 \times 10^{-9}}} = 100\,\Omega$$

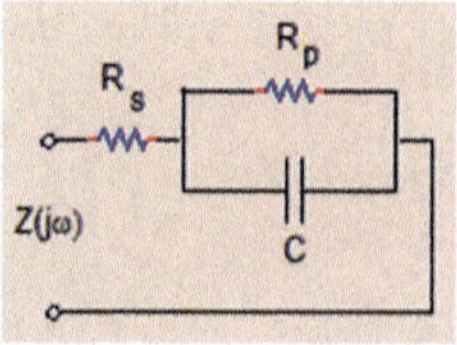

Fig. 6.16 The circuit for Problem 6.2.40

Problem 6.2.40 For the circuit of Fig. 6.16, determine the impedance function and plot its real and imaginary parts as well as its magnitude, and phase as a function of angular frequency.

Then, plot the modified Nyquist diagram of the impedance (Real_Imag_w.m).

$$R_s = 100\,\Omega, \quad R_p = 1000\,\Omega, \quad C = 10\,\text{nF}.$$

Solution

$$Z_1 = R_p \parallel C \quad \rightarrow \quad \frac{R_p \dfrac{1}{j\omega C}}{R_p + \dfrac{1}{j\omega C}} = \frac{R_p}{j\omega R_p C + 1}$$

$$Z = Z_1 + R_s = R_s + \frac{R_p}{1 + j\omega R_p C} = \frac{R_s + j\omega R_p R_s C + R_p}{1 + j\omega R_p C}$$

$$= \frac{(1 - j\omega R_p C)(j\omega R_p R_s C + R_p + R_s)}{1 + \omega^2 R_p^2 C^2}$$

$$= \frac{R_s + j\omega R_p R_s C + R_p - j\omega R_p^2 C - j\omega R_p R_s C + \omega^2 R_p^2 R_s C^2}{1 + \omega^2 R_p^2 C^2}$$

$$Z = \frac{R_s + R_p - j\omega R_p^2 C + \omega^2 R_p^2 R_s C^2}{1 + \omega^2 R_p^2 C^2} = \frac{R_s + R_p + \omega^2 R_p^2 R_s C^2}{1 + \omega^2 R_p^2 C^2} - j\frac{\omega R_p^2 C}{1 + \omega^2 R_p^2 C^2}$$

Figure 6.17a–d display real and imaginary parts of given impedance function, as well as its magnitude and phase plots as a function of angular frequency.

Figure 6.17e shows the modified Nyquist plot (Real part of complex impedance function versus negative imaginary part of impedance function). Modification of the Nyquist plot is the usual method employed in the field of impedance spectroscopy.

Part of the MATLAB script (Real_Imag_w.m) is given below.

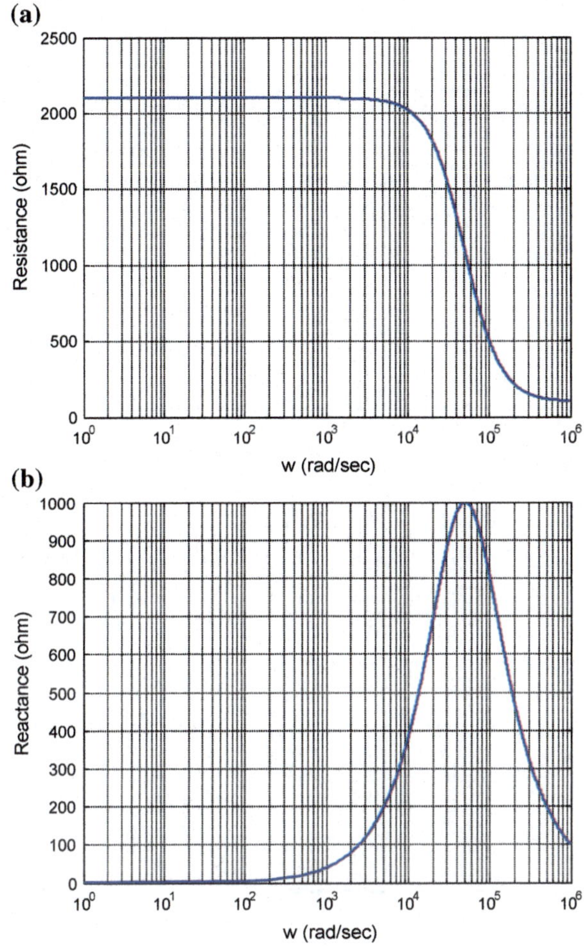

Fig. 6.17 **a**, **b**, **c**, **d** The real and imaginary parts of given impedance function, as well as its magnitude and phase plots as a function of angular frequency. **e** Nyquist diagram is plotted up to x = 10 Mrad/s in order to close the curve

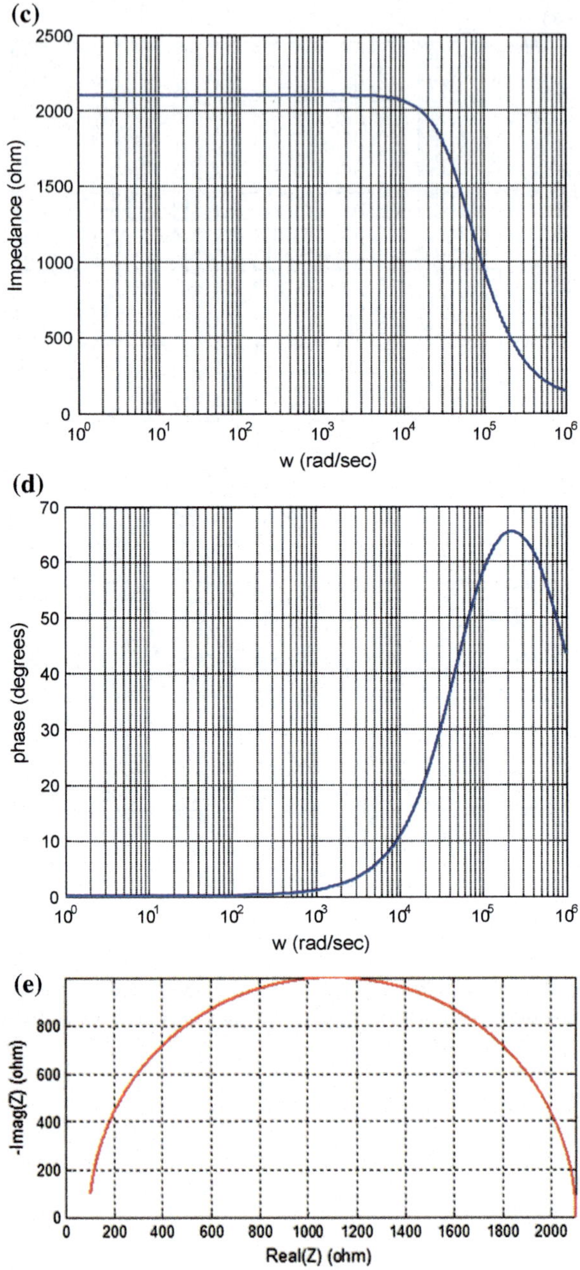

Fig. 6.17 (continued)

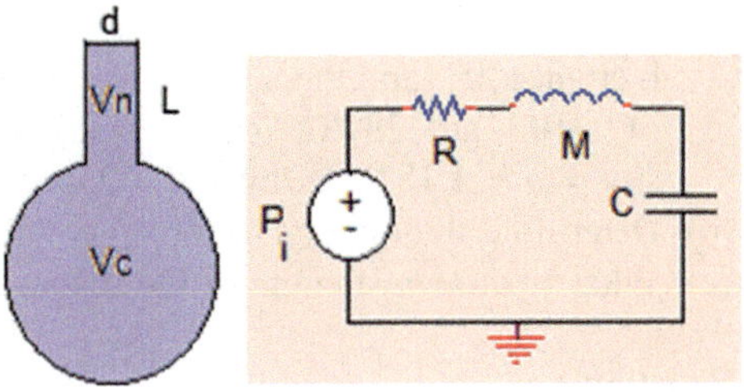

Fig. 6.18 Human vocal cord (*left*) and speech generation model (*right*)

```
%Plotting magnitude,phase,real and imaginary parts of Z(jw)
%and Nyquist plot of Z(jw)
Rs=100;Rp=2000;C=10e-9;w=1:100:1000000;
Ren=Rs+Rp+w.^2*Rp*Rp*Rs*C*C; Red=1+w.^2*Rp*Rp*C*C; Rea=Ren./Red;
Ima=Rp^2*C*w./Red; Magnitud=sqrt(Rea.^2+Ima.^2);
phas=atand(Ima./Rea);
semilogx(w,Rea,'linewidth',2);grid on;
xlabel('w (rad/sec)');ylabel('Resistance (ohm)');
semilogx(w,Ima,'linewidth',2);grid on;
xlabel('w (rad/sec)');ylabel('Reactance (ohm)');
semilogx(w,Magnitud,'linewidth',2);grid on;
xlabel('w (rad/sec)');ylabel('Impedance (ohm)');
semilogx(w,phas,'linewidth',2);grid on;
xlabel('w (rad/sec)');ylabel('phase (degrees)');
plot(Rea,Ima,'r','linewidth',2);axis('equal');grid on;
xlim([0 max(Rea)]);ylim([0 max(Ima)]);
xlabel('Real(Z) (ohm)');ylabel('-Imag(Z) (ohm)');
```

Problem 6.2.41 Human Vocal cord and speech generation can be modeled using a Helmholtz Resonator (HR), which consists of an orifice or a flanged neck connected to a larger volume, its dimensions being much smaller than the acoustic wavelength of interest. HR parameters are; d is the neck diameter, A is the neck area, D is the spherical cavity diameter, Vc the cavity volume, L the neck length. The effective length L_{eff} is bigger than the true length of the neck.

For the series electrical analogue circuit, the acoustic compliance C is analogous to electrical capacitance, the acoustic inertance M is analogous to electrical inductance and the acoustic resistance R is analogous to electrical resistance. P_1 is the incident acoustic pressure. See, Fig. 6.18.

$$M = \frac{\rho L_{\text{eff}}}{A},\quad C = \frac{V_c}{\rho c^2},\quad R = \frac{\rho c A}{2\pi V_c L_{\text{eff}}},\quad L_{\text{eff}} = L + 0.85d,\quad A = \frac{\pi d^2}{4},\quad V_c = \frac{\pi D^3}{6}$$

$$Q = 2\pi\sqrt{V_c\left(\frac{L_{\text{eff}}}{A}\right)^3}$$

(a) Determine the acoustic parameters (the acoustic compliance C, the acoustic inertance M , and the acoustic resistance R) and the tuned resonance frequency of HR at 28 °C (The speed of sound = c = 348.3 m/s; air density = ρ = 1.1839 kg/m^3), d = 2 cm, L = 0.5 cm, D = 6 cm.
(b) Determine the quality factor, Q of this HR.
(c) What is the wavelength at the resonant frequency? (helmholtz_resonator.m).

Solution

(a) Acoustic impedance Z of the HR (by analogy to electric resonator circuit) is

$$Z(j\omega) = R + j\left(\omega M - \frac{1}{\omega C}\right)$$

At resonance imaginary part vanishes,

$$\omega M - \frac{1}{\omega C} = 0$$

This yields the resonance frequency of the HR,

$$\omega = \frac{1}{\sqrt{MC}} = \frac{1}{\sqrt{\frac{\rho L_{\text{eff}}}{A} \cdot \frac{V_c}{\rho c^2}}} = c \cdot \sqrt{\frac{A}{V_c L_{\text{eff}}}}$$

Substituting given values,

$$L_{\text{eff}} = L + 0.85d = 0.5 + 0.85 \times 2 = 2.2\,\text{cm}, \quad A = \frac{\pi d^2}{4} = \frac{\pi \times 2^2}{4} = 3.1415\,\text{cm}^2,$$

$$V_c = \frac{\pi D^3}{6} = \frac{\pi \times 6^3}{6} = 113.1\,\text{cm}^3$$

$$\omega_0 = c \cdot \sqrt{\frac{A}{V_c L_{\text{eff}}}} = 348.3 \times \sqrt{\frac{3.1415 \times 10^{-4}}{1.131 \times 10^{-4} \times 2.2 \times 10^{-2}}} = 3913.7\,\text{rad/s}$$

$$f_0 = \frac{\omega}{2\pi} = \frac{3913.7}{2\pi} = 622.9\,\text{Hz}$$

$$M = \frac{\rho L_{\text{eff}}}{A} = \frac{1.1839 \times 2.2 \times 10^{-2}}{3.1415 \times 10^{-4}} = 82.91\,\text{kg/m}^4$$

$$C = \frac{V_c}{\rho c^2} = \frac{1.131 \times 10^{-4}}{1.1839 \times (348.3)^2} = 7.875 \times 10^{-10}\,\text{s}^2\text{m}^4/\text{kg}$$

$$R = \frac{\rho c A}{2\pi V_c L_{\text{eff}}} = \frac{1.1839 \times 348.3 \times 3.1415 \times 10^{-4}}{2\pi \times 1.131 \times 10^{-4} \times 2.2 \times 10^{-2}} = 8286\,\text{kg/s m}^4$$

(b) The quality factor of the resonator can be found by analogy to its electrical counterpart,

$$Q = \frac{\omega_0 M}{R} = \frac{3913.7 \times 82.91}{8286} = 39.2$$

(c) The wavelength at the resonant frequency is

$$\lambda = \frac{c}{f_0} = \frac{348.3}{622.9} = 0.559\,\text{m} = 55.9\,\text{cm}$$

Note that this wavelength is much larger than the geometric dimensions of the HR.

Part of a MATLAB script for computing these parameters is given below.

```
%helmholtz_resonator.m
ro=1.1839; c=348.3;
%d=neck diameter, L=neck length,D=cavity diameter
d=0.02; L=0.005; Leff=L+0.85*d; D=0.06; A=pi*d^2/4 % neck area
Vc=pi*D^3/6% cavity volume
M=ro*Leff/A; C = Vc / (ro*c^2);
R=ro*c*A /(2*pi*Vc*Leff); %acoustic resistance in the neck
disp (sprintf('Acoustic Inertance, M=  %2.3e ',M))
disp (sprintf('Acoustic Compliance,C=  %2.3e ',C))
disp (sprintf('Acoustic Resistance,R=  %2.3e ',R))
w0=c*sqrt (A/Vc/Leff); %rad/sec
disp (sprintf('acoustic resonance frequency = %2.1f rad/sec',w0))
f0=w0/2/pi; %Hz
disp (sprintf('acoustic resonance frequency=  %2.1f Hz',f0))
Q=w0*M/R; disp (sprintf ('Quality factor= %2.1f ', Q))
```

6.3 AC Analysis Methods, Resonance, and Matching

Problem 6.3.1 For the circuit shown in Fig. 6.19, determine V_1, V_2 (in complex form) and $i(t)$ using node voltages and Cramer's rule.

$$v(t) = 2\cos 2t, \quad k = 4, \quad R_1 = 1\,\Omega, \quad R_2 = 2\,\Omega, \quad C = 1F, L = 1H$$

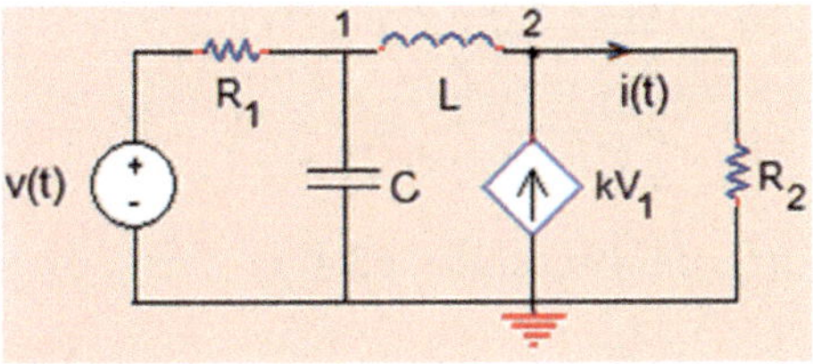

Fig. 6.19 The circuit for Problem 6.3.1

Solution
KCL at node 1:

$$\frac{2-V}{1}=\frac{V_1}{\frac{1}{j2\times 1}}+\frac{V_1-V_2}{j2\times 1}\quad\rightarrow\quad 2-V_1=\frac{V_1}{-j0.5}-j0.5(V_1-V_2)$$

$$2-V_1=j2V_1-j0.5V_1+j0.5V_2\quad\rightarrow\quad 2=V_1+j2V_1-j0.5V_1+j0.5V_2$$
$$2=V_1(1+j2-j0.5)+j0.5V_2$$

$$V_1\left(1+j\frac{3}{2}\right)+j\frac{1}{2}V_2=2\tag{6.1}$$

KCL at node 2:

$$4V_1+\frac{V_1-V_2}{j2\times 1}-\frac{V_2}{2}=0\quad\rightarrow\quad 4V_1-j\frac{V_1}{2}+j\frac{V_2}{2}-\frac{V_2}{2}=0$$
$$V_1\left(4-\frac{j}{2}\right)+V_2\left(-\frac{1}{2}+\frac{j}{2}\right)=0\tag{6.2}$$

From 1 and 2;

$$\begin{bmatrix}1+j\frac{3}{2} & j\frac{1}{2}\\ 4-j\frac{1}{2} & -\frac{1}{2}+\frac{j}{2}\end{bmatrix}\begin{bmatrix}V_1\\ V_2\end{bmatrix}=\begin{bmatrix}2\\ 0\end{bmatrix}$$

$$\Delta=\left(1+j\frac{3}{2}\right)\left(-\frac{1}{2}+\frac{j}{2}\right)-\left(4-j\frac{1}{2}\right)\left(j\frac{1}{2}\right)=-\frac{1}{2}+\frac{j}{2}-j\frac{3}{4}-\frac{3}{4}-\left[2j+\frac{1}{4}\right]$$
$$=-\frac{1}{2}-\frac{3}{4}-\frac{1}{4}+j\left(\frac{1}{2}-\frac{3}{4}-2\right)=-\frac{6}{4}-j\frac{9}{4}=-\frac{3}{4}(2+j3)$$

$$\Delta_2=\begin{vmatrix}1+j\frac{3}{2} & 2\\ 4-j\frac{1}{2} & 0\end{vmatrix}=0-(2)\left(4-j\frac{1}{2}\right)=-8+j$$

$$V_2=\frac{\Delta_2}{\Delta}=\frac{-8+j}{-\frac{3}{4}(2+j3)}=\frac{(-8+j)\times 4}{-3}\frac{(2-j3)}{4+9}=-\frac{4}{3}\frac{(-8+j)\times(2-j3)}{13}$$
$$=-\frac{4}{39}(-16+j24+j2+3)=-\frac{4}{39}(-13+j26)=-\frac{4\times 13}{39}(-1+j2)$$

$$V_2=\frac{4}{3}(1-j2)$$

$$I=\frac{V_2}{R_2}=\frac{4(1-j2)}{3\times 2}=\frac{2}{3}(1-j2)$$

$$i(t)=\frac{2}{3}\sqrt{1+4}\angle\tan^{-1}\left(-\frac{2}{1}\right)=\frac{2}{3}\sqrt{5}\angle\tan^{-1}(-2)=1.49\angle-63.4^\circ$$

$$i(t)=1.49\cos(2t-63.4^\circ)A$$

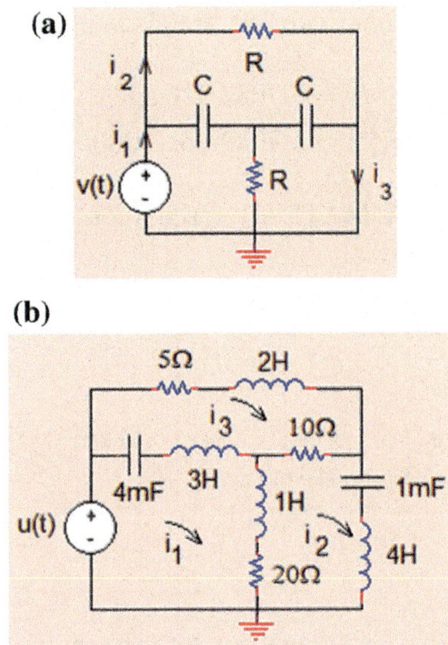

Fig. 6.20 **a** The circuit for Problem 6.3.2a. **b** The circuit for problem 6.3.2b

Problem 6.3.2

(a) In the circuit of Fig. 6.20a, determine mesh currents, if $v(t) = 10\cos(1000t)$, $R = 1\ \Omega$, $C = 1$ mF (complex_mesh1.m). Use Cramer's method for solving $i_3(t)$, then use MATLAB to determine all currents.

(b) In the circuit shown in Fig. 6.20b, determine mesh currents and express results in time domain, when $u(t) = 12\cos(6.28t)$ (complex_mesh2.m).

Solution

General equation of solving circuits for mesh currents in complex domain is

$$\boldsymbol{ZI} = \boldsymbol{V}$$

$$\boldsymbol{Z} = \begin{bmatrix} R - \frac{j}{\omega C} & +\frac{j}{\omega C} & 0 \\ \frac{j}{\omega C} & R - \frac{2j}{\omega C} & \frac{j}{\omega C} \\ 0 & \frac{j}{\omega C} & R - \frac{j}{\omega C} \end{bmatrix} = \begin{bmatrix} 1-j & j & 0 \\ j & 1-2j & j \\ 0 & j & 1-j \end{bmatrix}, \quad \boldsymbol{V} = \begin{bmatrix} 10 \\ 0 \\ 0 \end{bmatrix},$$

(a)

$$\boldsymbol{I}_3 = \frac{\Delta_3}{\Delta}, \quad \Delta_3 = \begin{vmatrix} 1-j & j & 10 \\ j & 1-2j & 0 \\ 0 & j & 0 \end{vmatrix} - 10(j)j = -10$$

$$\Delta = (1-j)^2(1-2j) - 2(j)(j)(1-j) = (1-1-2j)(1-2j) - 2(-1)(1-j)$$
$$= (-2j)(1-2j) + 2(1-j) = -2j - 4 + 2 - 2j = -2 - 4j$$

$$\boldsymbol{I}_3 = \frac{-10}{-2-4j} = \frac{10}{2+4j} = \frac{10(2-4j)}{4+16} = \frac{10(2-4j)}{20} = 1 - 2j$$

$$\boldsymbol{I}_3 = \sqrt{1+4}\angle \arctan(-2) = \sqrt{5}\angle -63.44^\circ = 2.2361\angle -63.44^\circ$$

$$i_3(t) = 2.2361\cos(1000t - 63.44^\circ)A$$

Other currents are also determined (using complex_mesh1.m), as follows:

$$\boldsymbol{I}_1 = 6.7082\angle 26.57^\circ, \quad \boldsymbol{I}_2 = 3.1623\angle 161.57^\circ$$
$$i_1(t) = 6.7082\cos(1000t + 26.57^\circ), \quad i_2(t) = 3.1623\cos(1000t + 161.57^\circ)$$

A MATLAB script (complex_mesh1.m), to solve equation $\boldsymbol{ZI} = \boldsymbol{V}$ is given below.

```
Z=[1-1j  -1j  0;
       -1j  1-2j  -1j;
       0  -1j  1-j]
V=[10  0  0]';  detZ=det(Z);  I=inv(Z)*V;  R=abs(I)
phir=angle(I);
phi=phir*180/pi
```

(b) Using analysis by inspection, and by letting $s = j\omega = j6.28$

$$\boldsymbol{V} = \begin{bmatrix} 12 \\ 0 \\ 0 \end{bmatrix}$$
$$= \begin{bmatrix} \frac{1}{sC} + sL_1 + sL_3 + R_2 & -(sL_3 + R_2) & -\left(\frac{1}{sC_1} + sL_1\right) \\ -(sL_3 + R_2) & R_2 + R_3 + L_3 + sL_4 + \frac{1}{sC_2} & -R_3 \\ -\left(\frac{1}{sC_1} + sL_1\right) & -R_3 & R_1 + \frac{1}{sC_1} + sL_1 + sL_2 + sL_3 \end{bmatrix} \begin{bmatrix} \boldsymbol{I}_1 \\ \boldsymbol{I}_2 \\ \boldsymbol{I}_3 \end{bmatrix}$$

Solving this equation for phasor currents yields

$$\boldsymbol{I}_1 = 0.1459 - j0.0289, \boldsymbol{I}_2 = 0.0418 + j0.0276, \boldsymbol{I}_3 = 0.1687 - j0.4849$$

Converting these currents from phasor domain to time domain,

$$i_1(t) = 0.1488\cos\left(t - 5.61^\circ\right), \quad i_2(t) = 0.0501\cos\left(t + 16.75^\circ\right),$$
$$i_3(t) = 0.5134\cos\left(t - 35.41^\circ\right),$$

MATLAB script (complex_mesh2.m) is given below.

```
w=6.28; s=i*w;
R1=5;  R2=20;  R3=10;
L1=3;  L2=2;  L3=1;  L4=4;
C1=4e-3;  C2=1e-3;
zC1=1/s/C1;  zC2=1/s/C2;  zL1=s*L1;  zL2=s*L2;  zL3=s*L3;  zL4=s*L4;
```

```
Z=[zC1+zL1+zL3+R2       -(zL3+R2)                      -(zC1+zL1);
     -(zL3+R2)      R2+R3+zL3+zL4+zC2                        -R3;
     -(zC1+zL1)             -R3              R1+zC1+zL1+zL2+zL3]
  V=[12;0;0]
I=Z\V
Magnitude=(abs(I))'
phase=phase(I)*(90/pi);phase=phase'
```

Problem 6.3.3

(a) Derive the impedance of a series RLC circuit in phasor form,
(b) Derive the expression for the resonance frequency, in Hz.

Solution

(a) For a series RLC circuit,

$$\boldsymbol{Z} = R + j\omega L + \frac{1}{j\omega C} = R + j\omega L - \frac{j}{\omega C} = R + j\left(\omega L - \frac{1}{\omega C}\right)$$
$$= \sqrt{R^2 + \left(\omega L - \frac{1}{\omega C}\right)^2}\angle\tan^{-1}\left(\frac{\omega L - \frac{1}{\omega C}}{R}\right)$$

(b) The frequency at which imaginary terms (reactances) cancel and the impedance becomes a pure resistance is named as the resonant frequency of a series RLC circuit,

$$\omega_0 L - \frac{1}{\omega_0 C} = 0 \quad \rightarrow \quad \omega_0 = \frac{1}{\sqrt{LC}} \quad \rightarrow \quad f_0 = \frac{1}{2\pi\sqrt{LC}}$$

Problem 6.3.4

(a) Derive the admittance of a parallel RLC circuit in phasor form,
(b) Derive the expression for the resonance frequency, in Hz.

Solution

(a) For a parallel RLC circuit,

$$\boldsymbol{Y} = G + j\omega C + \frac{1}{j\omega L} = G + j\omega C - \frac{j}{\omega L} = G + j\left(\omega C - \frac{1}{\omega L}\right)$$
$$= \sqrt{G^2 + \left(\omega C - \frac{1}{\omega L}\right)^2}\angle\tan^{-1}\left(\frac{\omega C - \frac{1}{\omega L}}{G}\right)$$

Fig. 6.21 Series RLC circuit

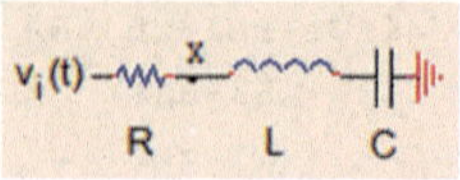

(b) The frequency at which imaginary parts (susceptances) cancel (the admittance becomes a pure conductance) is named as the resonant frequency of a parallel RLC circuit.

$$\omega_0 C - \frac{1}{\omega_0 L} = 0 \quad \rightarrow \quad \omega_0 = \frac{1}{\sqrt{LC}} \quad \rightarrow \quad f_0 = \frac{1}{2\pi\sqrt{LC}}$$

Problem 6.3.5 In the circuit of Fig. 6.21, what value of capacitance will cause the steady-state voltage V_X over the series RLC circuit to be zero if $v_i(t) = A.\cos\omega t$?

Solution

At resonance, $V_X = 0V$

$$\omega^2 = \frac{1}{LC} \quad \rightarrow \quad C = \frac{1}{\omega^2 L} F$$

Problem 6.3.6 In a series RC circuit excited by a sinusoidal voltage of U volts (rms), the voltage across the resistor is U/4 Volts (rms). Determine the voltage across the capacitor.

Solution

$$U^2 = V_R^2 + V_C^2 = \left(\frac{U}{4}\right)^2 + V_C^2$$

$$V_C^2 = U^2 - \left(\frac{U}{4}\right)^2 = U^2\left(1 - \frac{1}{16}\right) = \frac{15}{16}U^2$$

Problem 6.3.7 The same value of sinusoidal input voltage is separately applied to a series RL circuit (with $R = 40\ \Omega$, and reactance of 30 Ω) and a series LC circuit (with capacitive reactance of 20 Ω and inductive reactance of 30 Ω). Which one of these circuits has a larger current value?

Solution

$$\boldsymbol{Z}_1 = R + jX_L = \frac{\boldsymbol{V}}{\boldsymbol{I}_1}$$

$$\boldsymbol{I}_1 = \frac{\boldsymbol{V}}{40 + j30}$$

$$\boldsymbol{Z}_2 = jX_L - jX_C = j30 - j20 = j10$$

$$\boldsymbol{I}_2 = \frac{\boldsymbol{V}}{\boldsymbol{Z_2}} = \frac{\boldsymbol{V}}{j10}$$

Since the magnitude of impedance value of the first circuit will be larger, rms value of current flowing in the first (RL) circuit is less than the magnitude of current flowing in the second (LC) circuit.

Problem 6.3.8 In a series-connected RC circuit excited by a sinusoidal voltage, the voltage drop values across the resistor and capacitor are 40 and 30 V, respectively. Determine the phase angle between the current flowing through the circuit and the voltage applied to the circuit.

Solution
Voltage applied to the circuit, U,

$$U^2 = V_C^2 + V_R^2 = 30^2 + 40^2 = 50^2 \quad \rightarrow \quad U = 50\,\mathrm{V}$$

$$\theta = \arccos\left(\frac{V_R}{U}\right) = \arccos\left(\frac{40}{50}\right) = 36.87°$$

Problem 6.3.9 In a series RLC circuit operating in AC conditions, voltage drops across each element are the same (100 V). Determine the voltage applied to this circuit. Draw a conclusion.

Solution
Let U be the sinusoidal voltage applied to this series RLC circuit.

$$\boldsymbol{U} = V_R + jU_L - jU_C = V_R + j(U_L - U_C) = V_R = 100\,\mathrm{V}$$

It can be concluded that (in series RLC circuit operating in AC conditions, and all voltage drops are the same across each component) the voltage applied to the circuit equals the voltage drop over resistor. The voltages over inductor and capacitor cancel out due to 180° phase difference, although an AC voltmeter measures the same voltage drop across each element.

Problem 6.3.10 A series RLC circuit consists of equal valued impedance values of 100 Ω each, when a sinusoidal voltage of $U = 220$ V (rms) is applied to the circuit. Determine the effective values of voltages across the capacitor, inductor, and resistor.

Solution

$$\mathbf{Z} = R + jX_L - jX_C = R + j(X_L - X_C) = R = 100$$
$$\mathbf{U} = \mathbf{IZ}$$
$$220 = 100 \times I \quad \rightarrow \quad I = \frac{220}{100} = 2.2\,\text{A (rms)}$$
$$V_C = I \times X_C = 2.2 \times 100 = 220\,\text{V (rms)}$$
$$V_L = I \times X_L = 2.2 \times 100 = 220\,\text{V (rms)}$$
$$V_R = I \times R = 2.2 \times 100 = 220\,\text{V (rms)}$$

Problem 6.3.11 A series resonance circuit has the resonance frequency of 31.83 Hz with $L = 5$ mH and $R = 1$ Ω. Determine the bandwidth, quality factor, and the capacitance in this circuit.

Solution
Using the relationship between the bandwidth and inductance in a series resonance circuit,

$$BW = \frac{R}{2\pi L} = \frac{1}{2 \times \pi \times 5 \times 10^{-3}} = \frac{1000}{10 \times \pi} = 31.83\,\text{Hz}$$

The relationship between the bandwidth, quality factor, and the resonance frequency,

$$Q = \frac{f_0}{BW} = \frac{31.83}{31.83} = 1$$

The resonance frequency in terms of capacitance and inductance,

$$f_0 = \frac{1}{2\pi\sqrt{LC}} \quad \rightarrow \quad f_0^2 = \frac{1}{4\pi^2 LC} \quad \rightarrow \quad C = \frac{1}{4\pi^2 f_0^2 L}$$
$$C = \frac{1}{4 \times \pi^2 \times (31.83)^2 \times 5 \times 10^{-3}} = \frac{10^3}{199{,}987.57} \cong 5000\,\mu\text{F}$$

Alternatively, using relationship between the quality factor of the series resonance circuit and capacitance in series with the resistor,

$$Q = \frac{1}{\omega_0 RC} \quad \rightarrow \quad C = \frac{1}{\omega_0 RQ} = \frac{1}{2 \times \pi \times 31.83 \times 1 \times 1} \cong 5000\,\mu\text{F}$$

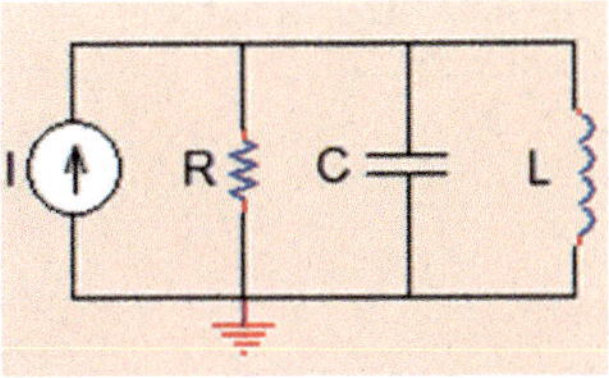

Fig. 6.22 Parallel resonance circuit

Problem 6.3.12 10 Inductors each with10 mH, 120 resistors each with 1 kΩ and 100 capacitors each of which has a capacitance of 1 nF are all connected in parallel.

(a) Find the inductance and capacitance of the system,
(b) Find the resonance frequency of the system,
(c) Find the impedance of the system at resonance frequency.

Solution

(a) $L_{eq} = L/10 = 1\,\text{mH}, \quad C_{eq} = 100 \cdot C = 100\,\text{nF}$

(b) $f_0 = \dfrac{1}{2\pi\sqrt{L_{eq}C_{eq}}} = \dfrac{0.159155}{[(10^{-3} \cdot 100 \cdot 10^{-9})]^{\frac{1}{2}}} = 0.159155 \cdot 10^5 \cong 15915.5\,\text{Hz}$

(c) $Z = R_{eq} = 120\,\text{k}\Omega$

Problem 6.3.13 Find the resonance frequency of a parallel RLC circuit (Fig. 6.22), if

$$R = 1\,\text{k}\Omega, \quad L = \left(\frac{1}{\pi}\right)^2 \text{mH}, \quad C = 0.25\,\text{nF}$$

Solution

$$f_0 = \frac{1}{2\pi\sqrt{LC}}, \quad \sqrt{LC} = \sqrt{\frac{1}{\pi^2} \cdot 10^{-3} \cdot 0,25 \cdot 10^{-9}} = \frac{1}{2\pi} \cdot 10^{-6}$$
$$f_0 = 1\,\text{MHz}$$

Problem 6.3.14 A mass–spring system is a mechanical analog for electrical coil and capacitor (LC) system:

m = Mass (kg), k = Spring constant (N/m), against L (in H) and capacitance (in F),

$$C = \frac{1}{k}$$

Determine the resonance frequency of a mass–spring system in Hz, if m = 1 kg, k = 100 (N/m), and write the equation for displacement $x(t)$, if the initial displacement is 1 mm.

What are the corresponding values of L and C?

Fig. 6.23 The circuit for Problem 6.3.15

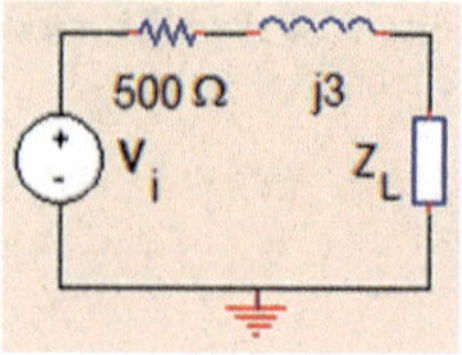

Solution

$$\omega_0 = \frac{1}{\sqrt{LC}} = \frac{1}{\sqrt{m \cdot \frac{1}{k}}} = \sqrt{\frac{k}{m}} = \sqrt{\frac{100\left(\frac{N}{m}\right)}{1(kg)}} = 10\,\text{rad/s}$$

$$f_0 = \frac{\omega_0}{2\pi} = \frac{10}{2\pi} = 1.59\,\text{Hz}$$

The equation of displacement is $x(t) = 0.001\cos(10t)$ (m). Similarly,

$$L = 1\,\text{H}, \quad C = \frac{1}{100} = 10\,\text{mF}$$

Problem 6.3.15 In the circuit shown in Fig. 6.23, if $v_i(t) = 3\cos(100t)V$, select components of the load for Maximum Power Transfer.

Solution

$$\bar{Z}_L = \bar{Z}_{\text{Th}} = 500 - j3 = 500\,\Omega\ \text{resistance} + (-\text{j}3)\Omega\,\text{capacitance};$$

Since $\omega = 100$ rad/s:

$$-j3 = \frac{1}{j\omega C} = \frac{1}{j \cdot 100 \cdot C} = -j\left(\frac{0.01}{C}\right) \quad \rightarrow \quad 3 = \frac{0.01}{C} \quad \rightarrow \quad C = \frac{0.01}{3} = 3.333\,\text{mF}$$

$$R = 500\,\Omega$$

Problem 6.3.16 In a series RLC circuit, variable capacitance values are C_L and C_H at half power points and C_O at resonance frequency. Express the quality factor of the circuit in terms of C_L, C_H, C_O, also specify relative capacitance values at respective frequencies.

Solution

$$\boldsymbol{Z} = R + j\left(\omega L - \frac{1}{\omega C}\right) = \frac{1}{\boldsymbol{Y}} \tag{6.3}$$

$$\text{At resonance} \quad \boldsymbol{Z_o} = R = \frac{1}{\boldsymbol{Y}} \tag{6.4}$$

Divide Eq. (6.4) by Eq. (6.3),

$$\frac{\boldsymbol{Z_o}}{\boldsymbol{Z}} = \frac{R}{R + j\left(\omega L - \frac{1}{\omega C}\right)} = \frac{1}{1 + j\left(\frac{\omega L}{R} - \frac{1}{\omega CR}\right)} \tag{6.5}$$

$$\frac{\boldsymbol{Z_o}}{\boldsymbol{Z}} = \frac{\boldsymbol{Y}}{\boldsymbol{Y_o}} = \frac{1}{1 + j\left(\frac{\omega_o L}{R} - \frac{1}{\omega_o CR} \cdot \frac{C_o}{C_o}\right)} = \frac{1}{1 + jQ\left(1 - \frac{C_o}{C}\right)}$$

At the half-power points,

$$\frac{Y}{Y_o} = \frac{1}{1 \mp j1}$$

$$1 - \frac{C_o}{C_L} = -\frac{1}{Q} \quad \text{and} \quad 1 - \frac{C_o}{C_H} = \frac{1}{Q}$$

A substraction operation yields,

$$\frac{C_o}{C_H} - \frac{C_o}{C_L} = \frac{2}{Q}$$

$$Q = \frac{2C_H C_L}{C_o(C_L - C_H)}, \quad f_L < f_0 < f_H, \quad C_H < C_0 < C_L$$

Problem 6.3.17 In the circuit shown in Fig. 6.24,

(a) Determine the load impedance for maximum power transfer condition.
(b) Find the general form of the current through the inductance if $Z_L = R$
(c) Find the current through $\overline{Z_L}$ at maximum power transfer condition.

Solution

(a) By Thévenin equivalent circuit concept,

$$\boldsymbol{Z_T} = \frac{R}{2} + j\omega L$$

$$\boldsymbol{Z_L} = \boldsymbol{Z_T^*} = \frac{R}{2} - j\omega L$$

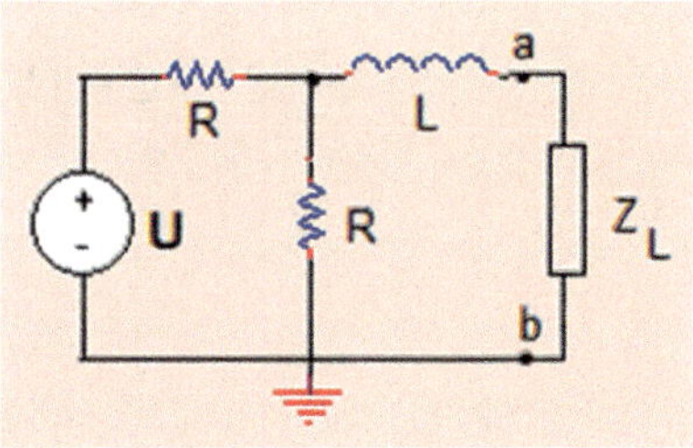

Fig. 6.24 The circuit for Problem 6.3.17

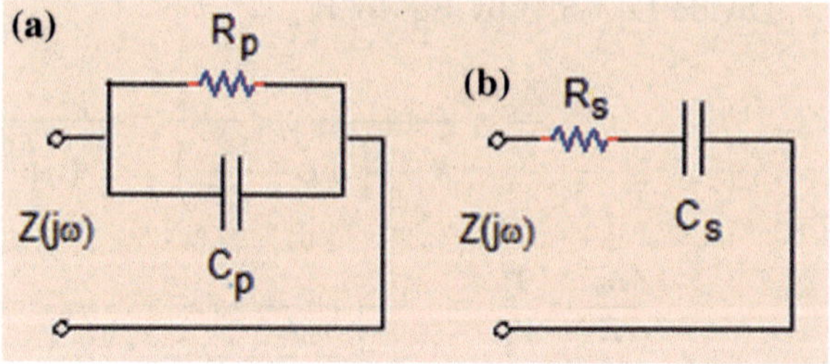

Fig. 6.25 The circuit for Problem 6.3.18

(b) Thévenin voltage,

$$\boldsymbol{V_T} = \frac{R}{2} \cdot \boldsymbol{U} = \boldsymbol{V_{oc}}$$

$$\boldsymbol{I} = \frac{\boldsymbol{V_T}}{\boldsymbol{Z_T} + R} = \frac{\frac{R}{2} \cdot \boldsymbol{U}}{\frac{R}{2} + j\omega L + R} = \frac{R \times \boldsymbol{U}}{3R + j2\omega L}$$

(c) $\boldsymbol{I_{mpt}} = \dfrac{\frac{R}{2} \times \boldsymbol{U}}{\frac{R}{2} + j\omega L + \frac{R}{2} - j\omega L} = \dfrac{\frac{R}{2} \times \boldsymbol{U}}{R} = \dfrac{\boldsymbol{U}}{2} A$

Problem 6.3.18 Determine the series–parallel RC circuit conversion parameters in terms of quality factor, Fig. 6.25.

Solution
Series circuit impedance,

$$Z_s = R_s + \frac{1}{j\omega C_s} = R_s - j\frac{1}{\omega C_s} \tag{6.6}$$

Parallel circuit impedance,

$$Z_p = \frac{R_p \frac{1}{j\omega C_p}}{R_p + \frac{1}{j\omega C_p}} = \frac{R_p}{j\omega C_p R_p + 1} = \frac{R_p (1 - j\omega C_p R_p)}{1 + (\omega C_p R_p)^2} \tag{6.7}$$

$$Q = Q_p = \frac{R_p}{\left(\frac{1}{\omega C_p}\right)} = \omega C_p R_p \tag{6.8}$$

Substituting (6.8) into (6.7)

$$Z_{\rm p} = \frac{R_{\rm p}(1 - jQ)}{1 + Q^2} \tag{6.9}$$

Equating real and imaginary parts of (6.6), (6.7) yields

$$R_{\rm s} = R_{\rm p}\frac{1}{1+Q^2}, \quad C_{\rm s} = C_{\rm p}\frac{1+Q^2}{Q^2} \tag{6.10}$$

$$R_{\rm p} = R_{\rm s}(1+Q^2), \quad C_{\rm p} = C_{\rm s}\frac{Q^2}{1+Q^2} \tag{6.11}$$

Problem 6.3.19 Determine the series–parallel RL circuit conversion parameters in terms of quality factor. See, Fig. 6.26.

Solution
Series circuit impedance,

$$Z_{\rm s} = R_{\rm s} + j\omega L_{\rm s}$$

Parallel circuit impedance,

$$Z_{\rm p} = \frac{j\omega L R_{\rm p}}{R_{\rm p} + j\omega L} = \frac{j\omega L R_{\rm p}(R_{\rm p} - j\omega L)}{R_{\rm p}^2 + (\omega L)^2} = \frac{(\omega L)^2 R_{\rm p} + j\omega L R_{\rm p}^2}{R_{\rm p}^2 + (\omega L)^2}$$

Equating real and imaginary parts of above equations,

$$R_{\rm s} = \frac{(\omega L)^2 R_{\rm p}}{R_{\rm p}^2 + (\omega L)^2} = \frac{R_{\rm p}}{1 + \left(\dfrac{R_{\rm p}}{\omega L}\right)^2} \tag{6.12}$$

$$\omega L_{\rm s} = \frac{\omega L_{\rm p} R_{\rm p}^2}{R_{\rm p}^2 + (\omega L)^2} \quad \rightarrow \quad L_s = \frac{R_{\rm p}}{1 + \left(\dfrac{\omega L_{\rm p}}{R_{\rm p}}\right)^2} \tag{6.13}$$

Fig. 6.26 **a** series RL circuit, **b** parallel RL circuit

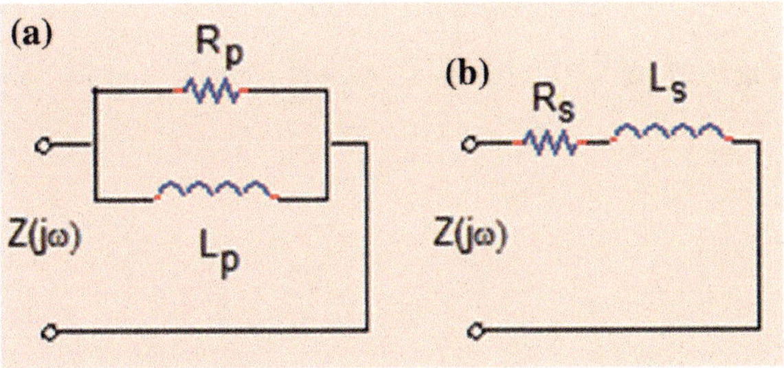

$$Q = Q_\mathrm{p} = \frac{R_\mathrm{p}}{X_\mathrm{p}} = \frac{R_\mathrm{p}}{\omega L_\mathrm{p}} \tag{6.14}$$

Substituting (6.14) into (6.13) and (6.12) yields

$$R_\mathrm{s} = R_\mathrm{p}\frac{1}{1+Q^2}, \quad L_\mathrm{s} = L_\mathrm{p}\frac{Q^2}{1+Q^2} \tag{6.15}$$

$$R_\mathrm{p} = R_\mathrm{s}(1+Q^2), \quad L_\mathrm{p} = L_\mathrm{s}\frac{1+Q^2}{Q^2} \tag{6.16}$$

Problem 6.3.20 Using an *L* matching circuit, design an interface between 50 Ω output of amplifier and a medical ultrasound probe with 250 Ω impedance (resistance) at 10 MHz (No DC transmission is allowed) (matching1.m).

Solution
Since no DC transmission is allowed and $R_L > R_g$, the *L* matching circuit of Fig. 6.27a, b can be used.

Using RL series–parallel transform relationship,

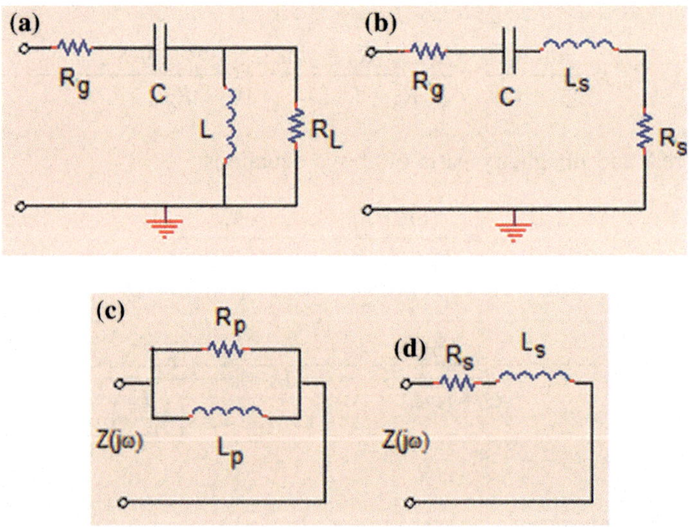

Fig. 6.27 **a**, **b** RL > Rg the L matching circuit. **c**, **d** RL series–parallel transform relationship

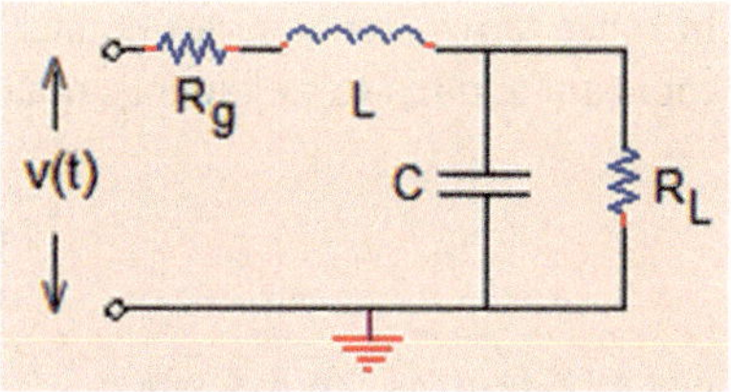

Fig. 6.28 "L" circuit (consisting of L, C elements) used for matching different source and load resistance values, particularly used at radio frequencies

$$R_g = R_s, \quad R_L = R_p, \quad L = L_p$$

$$L_s = L_p \frac{Q^2}{1+Q^2}, \quad R_s = R_p \frac{1}{1+Q^2}, \quad Q = \frac{R_p}{\omega L_p} = \frac{\omega L_s}{R_s}$$

$$Q = \sqrt{\frac{R_L}{R_g} - 1}, \quad R_L > R_g; \quad Q = \sqrt{\frac{250}{50} - 1} = \sqrt{5 - 1} = 2$$

$$Q = \frac{\omega L_s}{R_s} \quad \rightarrow \quad L_s = \frac{QR_s}{\omega} = \frac{2 \times 50}{2\pi \times 10^7} = \frac{100}{2\pi} \times 10^{-7} = 0.159 \times 10^{-5} H$$

$$L = L_p = L_s \frac{1+Q^2}{Q^2} = 0.159 \times 10^{-5} \times \frac{1+4}{4} = 0.159 \times 1.25 \times 10^{-5} H$$

$$L = 0.19875 \times 10^{-5} = 1.989\,\mu\text{H}$$

In maximum power transfer condition,

$$-j\frac{1}{\omega C} = -j\omega L_s \quad \rightarrow \quad \frac{1}{\omega C} = \omega L_s$$

$$C = \frac{1}{\omega^2 L_s} = \frac{1}{(2\pi \times 10^7)^2 \times 0.159 \times 10^{-5}} = 159.2\,\text{pF}$$

Problem 6.3.21 The circuit shown in Fig. 6.28 is an "*L*" circuit (consisting of *L*, *C* elements) used for matching different source and load resistance values, particularly used at radio frequencies. Determine component values (*L*, *C*) so that maximum power transfer occurs at $f_0 = 10\,\text{MHz}$, if the source and load resistances are 50 and 75 Ω respectively (matching1.m).

Solution

In the series RLC circuit, quality factor is defined by the following equation:

$$Q = \frac{\omega_0 L}{R} = \frac{1}{\omega_0 RL} \tag{6.17}$$

In other terms, for a series-connected inductor and resistor, and parallel-connected capacitor and resistor elements, quality factor in a matched circuit case can be given as

$$Q = \frac{X_L}{R_g} = \frac{R_p}{X_C} \tag{6.18}$$

where X term is the reactance.

In maximum power transfer condition, R_p, R_g match and reactances cancel with a conjugate match. Using standard equations of L, C at ω_0,

$$L = \frac{X_s}{\omega_0}, \quad C = \frac{1}{\omega_0 X_c} \tag{6.19}$$

or,

$$L = \frac{QR_g}{\omega_0}, \quad C = \frac{Q}{\omega_0 R_p} \tag{6.20}$$

On the other hand, series–parallel transformation of RC circuits yields

$$R_p = R_g(1 + Q^2), X_p = X_g\left(\frac{1 + Q^2}{Q^2}\right) \tag{6.21}$$

$$Q = \left(\frac{R_p}{R_g} - 1\right)^{1/2}, \quad R_p > R_g \tag{6.22}$$

Substituting R_p, R_g in (6.22) and then using (6.20)

$$L = \frac{\left(\frac{R_p}{R_g} - 1\right)^{1/2} R_g}{\omega_0} = \frac{\left(\frac{75}{50} - 1\right)^{1/2} 50}{2\pi \times 10^7} = 562.7\,\text{nH}$$

$$C = \frac{\left(\frac{R_p}{R_g} - 1\right)^{1/2}}{\omega_0 R_p} = \frac{\left(\frac{75}{50} - 1\right)^{1/2}}{2\pi \times 10^7 \times 75} = 150.1\,\text{pF}$$

Problem 6.3.22 Match the medical equipment transmitter with output impedance 50 Ω and operating at 433 MHz, to an antenna with 10 Ω impedance, using an L type matching circuit (i.e., determine the values of L and C, shown in Fig. 6.29) (matching1.m).

Solution

Since the output impedance of the equipment is the source resistance in this case, and it is larger than the impedance of the load (antenna input impedance); L circuit consisting of L, C elements can be used as shown in Fig. 6.29.

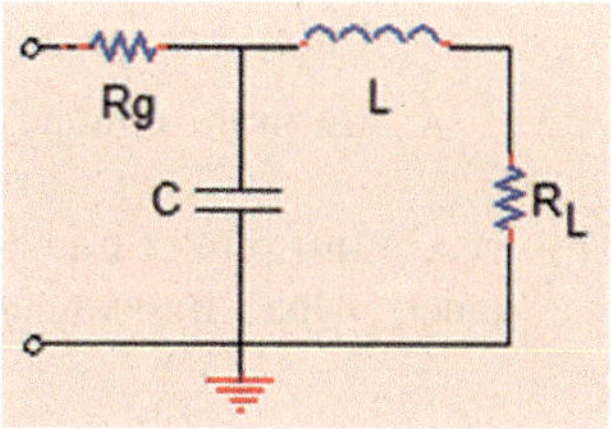

Fig. 6.29 The circuit for Problem 6.3.22

$$Q = \left(\frac{R_p}{R_g} - 1\right)^{1/2} = \left(\frac{50}{10} - 1\right)^{1/2} = 2$$

$$X_L = \omega_0 L = QR_L = 2 \times 10 = 20\,\Omega \quad \rightarrow \quad L = \frac{X_L}{\omega_0} = \frac{20}{2\pi \times 433 \times 10^6} = 7.4\,\text{nH}$$

$$X_c = \frac{1}{\omega_0 C} = \frac{R_g}{Q},$$

$$C = \frac{1}{\omega_0 X_c} = \frac{1}{\omega_0\left(\frac{R_g}{Q}\right)} = \frac{Q}{\omega_0 R_g} = \frac{2}{2\pi \times 433 \times 10^6 \times 50} = 14.7\,\text{pF}$$

Problem 6.3.23 In a magnetic resonance unit, the angular momentum vector processes about the external field axis at Larmor frequency, $f = \gamma B$, where γ = gyromagnetic ratio, and B is the applied magnetic field intensity. For a proton, $\gamma = 42.576\,\text{MHz} \cdot T^{-1}$.

(a) Determine Larmor frequency of a proton in $1.5T$ and $3T$ magnetic field strength.
(b) A manufacturer of MRI systems designs an LC matching circuit between an RF amplifier with 50 Ω output resistance and a coil with 500 Ω (see, Fig. 6.30). Determine the values of coupling circuit elements to provide maximum power transfer in a $3T$ MRI unit (without attempting to use Q-factor).

Solution

(a) $f = \gamma B$ since $\gamma = 42.576\,\text{MHz} \times T^{-1}$

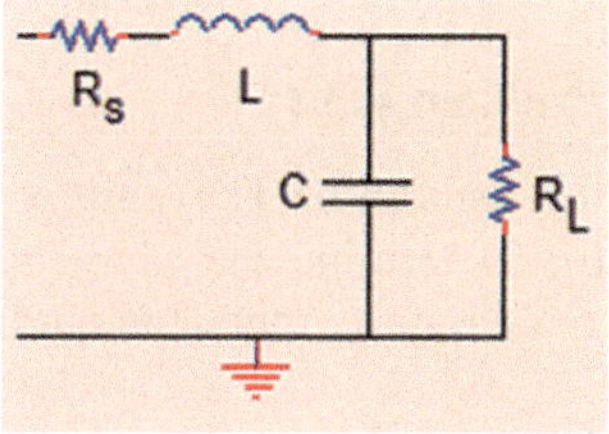

Fig. 6.30 The circuit for Problem 6.3.23

$$\text{Larmor frequency in a } 1.5\,T\,\text{unit} = 1.5 \times 42.576 = 63.864\,\text{MHz}$$
$$\text{Larmor frequency in a } 3\,T\,\text{unit} = 3 \times 42.576 = 127.728\,\text{MHz}$$

(b) Maximum power transfer condition, (complex conjugate of Thévenin's impedance = load impedance)

$$R_s - jX_L = R_L \parallel \left(\frac{1}{jX_C}\right) = \frac{R_L\left(\frac{1}{jX_C}\right)}{R_L + \left(\frac{1}{jX_C}\right)} = \frac{-jR_LX_C}{R_L - jX_C}$$
$$R_s - jX_L = \frac{-jR_LX_C(R_L + jX_C)}{R_L^2 + X_C^2} = \frac{R_LX_C^2}{R_L^2 + X_C^2} - \frac{R_L^2X_C}{R_L^2 + X_C^2} \tag{6.23}$$

Equating real parts,

$$R_s\left(R_L^2 + X_C^2\right) = X_C^2R_L \rightarrow R_sR_L^2 + R_sX_C^2 - X_C^2R_L = 0$$
$$X_C^2(R_s - R_L) = -R_sR_L^2 \rightarrow X_C = \left(\frac{-R_sR_L^2}{R_s - R_L}\right)^{1/2}$$

Substitute given values,

$$X_C = \left[\frac{50 \times (500)^2}{500 - 50}\right]^{1/2} = \frac{500}{3}\,\Omega$$
$$X_C = \frac{1}{\omega C} \quad \rightarrow \quad C = \frac{1}{\omega X_C} = \frac{1}{2\pi \times (127.728 \times 10^6)\left(\frac{500}{3}\right)} = 7.48\,\text{pF}$$

The value of inductance is found by equating the imaginary parts in Eq. (6.23),

$$X_L = \frac{R_L^2X_C}{R_L^2 + X_C^2} = \omega L$$
$$L = \frac{R_L^2X_C}{\omega(R_L^2 + X_C^2)} = \frac{(500)^2 \times (500/3)}{2\pi \times (127.728 \times 10^6)\left[(500)^2 \times (500/3)\right]}$$
$$L = \frac{3 \times 500^3}{2\pi \times (127.728 \times 10^6) \times (500)^2 \times 10} = 0.187 \times 10^{-6} H = 0.187\,\mu\text{H}$$

Problem 6.3.24

(a) Determine Thévenin equivalent of the circuit shown in Fig. 6.31a.
(b) Determine the phase angle expression for Thévenin impedance in terms of circuit components and angular frequency,

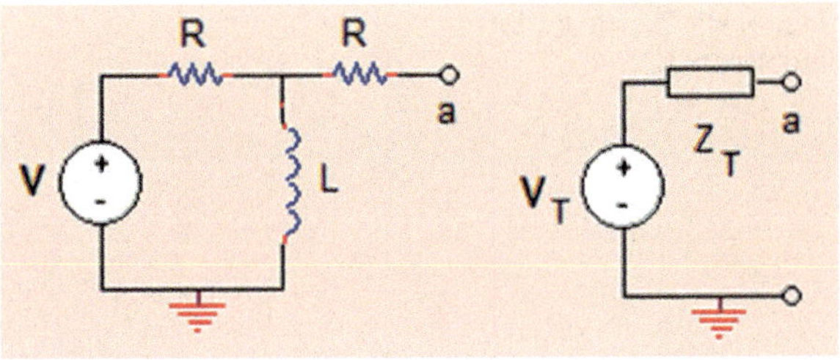

Fig. 6.31 **a** The circuit for Problem 6.3.24. **b** Thévenin circuit

(c) Determine the load impedance for maximum power transfer,
(d) Calculate the load impedance for maximum power transfer if $R = 1\,\Omega$, $L = 1\,H$, $\omega = 1$ rad/s.
(e) What is the magnitude of voltage at terminal a for $\omega = 0$ and $\omega = \infty$?

Solution

(a) In cartesian coordinates,

$$\boldsymbol{V_T} = \boldsymbol{V_a} = \frac{j\omega L}{R + j\omega L} \times \boldsymbol{V}$$

$$\boldsymbol{Z_T} = R + \frac{j\omega LR}{R + j\omega L} = \frac{2j\omega LR + R^2}{R + j\omega L} = \frac{(2j\omega LR + R^2)(R - j\omega L)}{R^2 + \omega^2 L^2}$$

$$= \frac{2j\omega LR^2 + R^3 + 2\omega^2 L^2 R - j\omega LR^2}{R^2 + \omega^2 L^2} = \frac{R^3 + 2\omega^2 L^2 R + j\omega LR^2}{R^2 + \omega^2 L^2}$$

$$\boldsymbol{Z_T} = \frac{R(R^2 + 2\omega^2 L^2)}{R^2 + \omega^2 L^2} + j\frac{\omega LR^2}{R^2 + \omega^2 L^2}$$

(b) $\varphi = \arctan\left[\dfrac{\omega LR^2}{R(R^2 + 2\omega^2 L^2)}\right] = \arctan\left(\dfrac{\omega LR}{R^2 + 2\omega^2 L^2}\right)$

(c) $\boldsymbol{Z_L} = \boldsymbol{Z_T^*} = \dfrac{R(R^2 + 2\omega^2 L^2)}{R^2 + \omega^2 L^2} - j\dfrac{\omega LR^2}{R^2 + \omega^2 L^2}$

(d) $R = 1\,\Omega$, $L = 1\,\text{H}$, $\omega = 1$ rad/sn

$$\boldsymbol{Z_L} = \frac{1(1 + 2 \times 1 \times 1)}{1 + 1 \times 1} - j\frac{1 \times 1 \times 1}{1 + 1 \times 1} = \frac{3}{2} - j\frac{1}{2}\Omega = \frac{1}{2}(3 - j)\Omega$$

(e) For DC conditions, inductance is a short circuit, therefore $V_a = 0$. For very high frequencies, ($\omega = \infty$), inductance behaves like an open circuit, therefore $V_a = |V|$

Problem 6.3.25 In the circuit of Fig. 6.32,

(a) If $\omega = 50$ rad/s, find input impedance, $\overline{Z_{ab}}$.

Fig. 6.32 The circuit for Problem 6.3.25

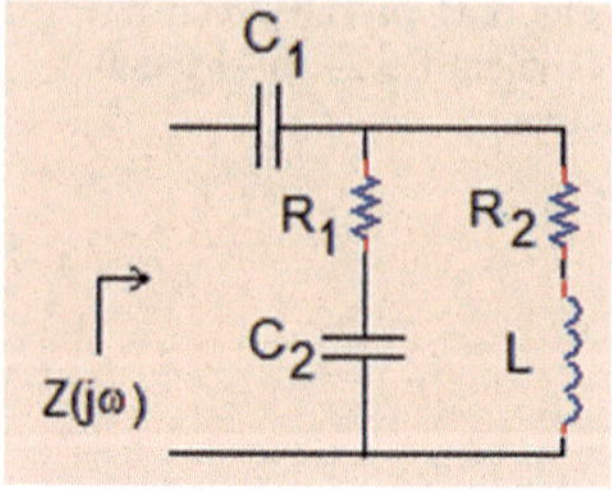

(b) At which frequency the parallel arm of the circuit is resonant?

$C_1 = 2$ mF, $C_2 = 10$ mF, $L = 0.2$ H, $R_1 = 8\ \Omega$, $R_2 = 3\ \Omega$.

Solution

(a)

$$\mathbf{Z}_1 = \frac{1}{j\omega C_1} = \frac{1}{j50 \times 2 \times 10^{-3}} = -j10\,\Omega,$$

$$\mathbf{Z}_2 = 3 + \frac{1}{j\omega C_2} = 3 + \frac{1}{j50 \times 10 \times 10^{-3}} = (3 - j2)\,\Omega$$

$$\mathbf{Z}_3 = 8 + j\omega L = 8 + j50 \times 0.2 = (8 + j10)\,\Omega$$

$$\mathbf{Z}_{ab} = \mathbf{Z}_1 + \mathbf{Z}_2 // \mathbf{Z}_3$$

$$\mathbf{Z}_{ab} = (-\mathrm{j}10) + \frac{(3 - \mathrm{j}2)(8 + \mathrm{j}10)}{11 + \mathrm{j}8} = -\mathrm{j}10 + \frac{(44 + \mathrm{j}14)(11 - \mathrm{j}8)}{11^2 + 8^2}$$

$$= -\mathrm{j}10 + 3.22 - \mathrm{j}1.07$$

$$\mathbf{Z}_{ab} = 3.22 - \mathrm{j}11.07\,\Omega$$

(b) The resonance frequency of the parallel arm,

$$f_0 = \frac{1}{2\pi\sqrt{LC}} = \frac{1}{2\pi\sqrt{10 \cdot 10^{-3} \cdot 0.2}} = \frac{0.159}{\sqrt{2 \times 10^{-3}}} = 3.55\,\mathrm{Hz}$$

Problem 6.3.26

(a) What is electrocauterization? What is an active and passive electrode in an electrosurgery unit (ESU)? Why the current density is a significant factor? What do monopolar and bipolar electrode mean? Are these terms correct in electrical sense?
(b) Describe the waveforms used for cutting and coagulation.
(c) Describe the operation of electrical circuit of a spark gap-based ESU. Comment on its advantages and disadvantages.
(d) Research and briefly discuss the present state of art on ESU technology.

Solution

(a) Electrocauterization is used in surgery to burn unwanted tissue, and to stop bleeding. ESU range spans from special dental units through systems for private clinical practices and outpatient departments to high-end units for the hospital operating rooms (OR). It is most suitable for neurosurgery, ophthalmic surgery, and plastic surgery.

During cutting, tissue is heated and the cells explode and evaporate. When there is a new tissue under contact, it tears apart, incision takes place.

In desiccation (coagulation) a needle (or ball) electrode is placed inside the tissue. When coagulating current flows through the cell, plasma evaporates slowly, the cell shrinks, causing coagulation.

An ESU is an alternating current (AC) source which is an indispensable surgical device for surgeons to cut tissue and ligate (cauterize) bleeding vessels. Most modern ESUs operate in a range of 200 kHz–3 MHz. For example, a brochure by Covidien/Medtronic Inc. for Force FX ESU describes an output for Monopolar Cut at pure sine wave of 390 kHz sinusoid, with maximum 2300 V peak-to-peak voltage at 300 ohms of rated load value and maximum of 300 W of power [2], while Maxium ESU (manufactured by Martin, Tuttlingen, Germany) has maximum cutting power of 360 W, at 300/400/600 kHz, [3] and Erbe Vio 300S (by Erbe Medizingeraete GmbH, Tübingen, Germany) provides 300 W at 500 Ω load, at an operating frequency of 350 kHz [4].

One of the two electrodes connected to generator is called the active electrode (pen). It is manipulated by the surgeon and has a few mm^2 cross-sectional area. On the other hand, the other electrode has a large area (100 cm^2 or higher) and called patient plate, return electrode, indifferent, neutral or dispersive electrode. New types of dispersive electrodes are disposable pads that are attached to the patient's thigh via conductive self-adhesive material. Patient is a part of the electrical circuit.

Although the current at both electrodes is the same, current densities in the vicinity of each electrode are quite different. It is much higher near the active electrode than it is for the passive electrode. The cutting and coagulation effects in tissue are due to the heating (power dissipation) in the tissue. For safe operation, passive (dispersive) electrode must have a large contact area and small contact resistance to skin.

Above description applies to monopolar electrodes. Bipolar scissors and forceps electrodes, on the other hand, do not require dispersive pads and look like tweezers. The RF current flows between the two tiny plates on the handpiece. Bipolar electrodes enable the precise, bloodless dissection of tissue. Since only the tissue that is located between the scissor's blades is integrated into the electric circuit, the energy consumed is reduced compared to the monopolar procedure. Therefore, the terms "monopolar" (unipolar) and "bipolar" electrodes are incorrect (in electrical circuits sense).

(b) The cut waveform is a continuous sinusoidal current, while coagulation waveform is damped sinusoidal current or chopped sine wave.

(c) Input to a spark gap ESU circuit is a 50–60 Hz sine wave at 2–3 kVrms value. This voltage ionizes the air in the space between spark gap electrodes. Arcing spark gap creates oscillations in a tank circuit which is transformer-coupled to the output to produce 30–400 W of power. See, Fig. 6.33 for a spark gap-based ESU circuit diagram.
Although spark gap-based ESUs are earlier designs, they are still used today (mostly in veterinary clinics), mainly due to their simple structures. It is also easy to grasp the principles of their electrical operation.
A significant disadvantage of such design is electromagnetic interference (harmonic signal) generation by spark gap action. This creates electromagnetic compatibility disturbances such as audible distortion on receivers tuned to frequencies even in VHF (very high frequency) region. The spark gap units have been largely superseded by solid-state devices. The solid-state units have been improved and are effective in all surgical procedures and decrease the risk of patient burns, they are safer to use than the spark gap ESUs, as solid-state units have integral return electrode contact-quality monitors that help to prevent patient burns, whereas most spark gap units do not have this monitoring capability. ECRI recommends surgeons to try the newer and safer technology [5, 6].
(d) Modern ESUs are equipped with embedded computer controlled circuitry (patient control system) and warning indication for missing or poor return electrode.

Recent technology concentrates on tissue response which uses a tissue feedback system that senses tissue impedance and adjusts the current and output voltage to maintain a consistent surgical effect. This system reduces the need to adjust power settings for different types of tissue. This technology provides surgeons with improved performance at lower power settings which minimize the risk of tissue damage and make a smoother cut through all tissue types possible. The ESU adjusts its output in response to tissue changes, maintaining power delivery, and minimizing drag. Capacitive coupling is also reduced when using this technology that is achieved by limiting the RMS voltage and the high-frequency harmonics. Lower voltage means less neuromuscular stimulation and more precise delivery of energy to reduce collateral damage [2].

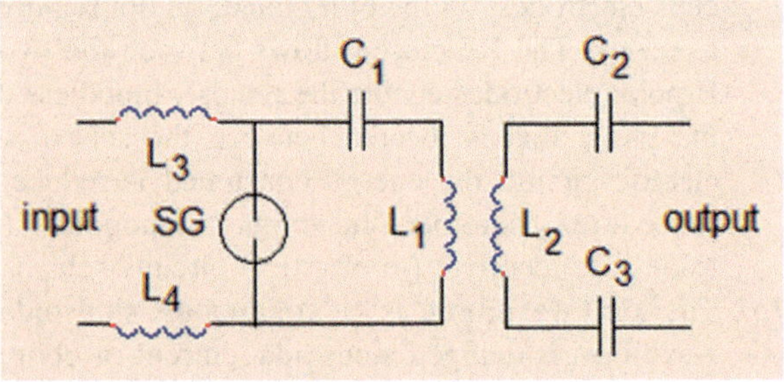

Fig. 6.33 Spark gap ESU basic circuit diagram

For example, a system so-called Instant Response technology (Covidien-Medtronic Inc.) features a feedback system that recognizes changes in tissue electrical resistance 200 times per second, and adjusts voltage and current accordingly to maintain appropriate power. This capability provides higher power efficiency rating (PER).

Multifunctional units:

Multifunctional units offer many regulated types of current for a variety of applications in various medical fields. Display screen lets the operator quickly recognize all of the settings made. As regards programming and program management, memory locations are available for user-defined settings (customized programs) that make such devices versatile. Using the medical discipline preselection function, the device is reduced to the current types and programs designed for the specific medical discipline [3].

Argon plasma surgery:

Argon is a chemically inert and nontoxic gas. Argon plasma technique is a monopolar electrosurgical procedure in which electrical energy is transferred to the target tissue using ionized and, thus, conductive argon gas (argon plasma), without the electrode coming into direct contact with the tissue. The argon plasma follows the path of least electrical resistance, regardless of whether the tissue lies directly in front of the electrode or lateral to it, and regardless of the direction of the flow of the argon gas [4].

When performing surgeries using an ESU of this mode, tissue carbonization is reduced and the wound healing process thus accelerated. Due to the low penetration depth, the risk of perforations is significantly reduced as well. In argon plasma surgery, argon gas is ionized with high-frequency current, producing conductive plasma. This results in a homogeneous and superficial coagulation. During application, the argon gas is always directed at tissue portions that are still bleeding and not yet coagulated. This process facilitates rapid hemostasis, which is important especially when dealing with diffuse bleeding. This operation significantly reduces blood loss and shortens the operating time. Maxium Beamer (Martin), APC (Erbe), and Force Argon II (Valleylab Medtronic Inc.) are examples for these types of equipment available in the medical device market (2017).

ESU with smoke evacuator:

Reliable protection from particles and viruses in the operating room can be accomplished in this mode. The ESU with smoke evacuator eliminates viruses and particles via its suction and filter technology. The suction process is initiated simultaneously with the activation of the electrosurgical unit. Typical systems are MarVAC (Martin), S-PILOT (Karl Storz, Tuttlingen, Germany).

Modern ESUs (in Europe, as of 2017) are Classified according to Medical Device Directive MDD II b, with Protection class I according to DIN EN 60-601-1 (Type of applied part CF; defibrillation-proof), and Approval/Mark of conformity with CE 0297 in compliance with 93/42 EEC.

Fig. 6.34 Equivalent circuit for the operation of an ESU

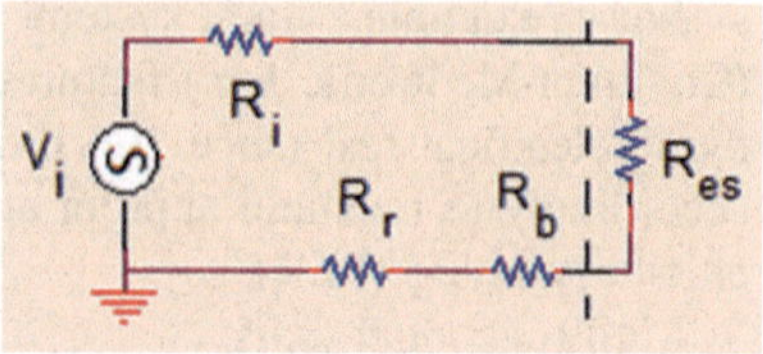

Problem 6.3.27 Equivalent circuit for the operation of an ESU is given in Fig. 6.34.

R_i = Internal resistance of ESU, R_b = Patient's body resistance,
R_r = Return electrode resistance, R_{es} = Electrode–skin interface resistance
V_i = Peak-to-peak sinusoidal output voltage of ESU

(a) Calculate the power absorbed in the electrode–skin interface resistance, P_{es}, as a function of R_{es}.
(b) Express the value of R_{es} at maximum power transfer.
(c) If $R_i = 530\ \Omega$, $R_b = 50\ \Omega$, $R_r = 20\ \Omega$, $V_i = 1200$ Vp−p, numerically calculate the power absorbed in the electrode–skin interface resistance, P_{es}.
(d) For the given data in part *c*, calculate the electrode–skin interface resistance, R_{es}.
(e) Find the maximum power transferred to electrode skin interface.
(f) For the given data in part c, calculate the electrode–skin interface resistance, R_{es}, when $P_{es} = 30$ W.
(g) Plot P_{es}, the power absorbed in the electrode–skin interface as a function of electrode–skin interface resistance, R_{es} is shown in Fig. 6.35. Then check the numerical results obtained.

Note that this equivalent circuit does not consider the capacitive effects of electrodes.

Solution

(a) $V_i = V_{\text{rms}} = \dfrac{\frac{V_{\text{pp}}}{2}}{\sqrt{2}} = \dfrac{V_{\text{pp}}}{2\sqrt{2}}$

$$P_{es} = i^2 R_{es} = \left(\frac{V_{\text{rms}}}{R_i + R_r + R_b + R_{es}}\right)^2 R_{es} = \left(\frac{\frac{V_{\text{pp}}}{2\sqrt{2}}}{R_i + R_r + R_b + R_{es}}\right)^2 R_{es}$$

$$P_{es} = \left[\frac{V_{\text{pp}}}{2\sqrt{2}(R_i + R_r + R_b + R_{es})}\right]^2 R_{es} = \frac{V_{\text{pp}}^2 R_{es}}{8(R_i + R_r + R_b + R_{es})^2}$$

(b) For maximum power transfer,

$$R_{es} = R_{Th} = R_i + R_r + R_b$$

(c) R_i = 530 Ω, R_b = 50 Ω, R_r = 20 Ω, V_i = 1200 Vp−p, the power absorbed in the electrode–skin interface resistance, P_{es}.

$$P_{es} = \frac{V_{pp}^2 R_{es}}{8(R_i + R_r + R_b + R_{es})^2} = \frac{1200^2 R_{es}}{8(530 + 20 + 50 + R_{es})^2} = \frac{1200^2 R_{es}}{8(600 + R_{es})^2}$$

(d) The electrode–skin interface resistance, for maximum power transfer is

$$R_{es} = R_{Th} = R_i + R_r + R_b = 530 + 20 + 50 = 600\,\Omega$$

(e) Maximum power transferred to electrode skin interface,

$$P_{es} = \frac{1200^2 R_{es}}{8(600 + R_{es})^2} = \frac{1200^2(600)}{8(600 + 600)^2} = 75\,\text{W}$$

(f) The electrode–skin interface resistance, R_{es}, when P_{es} = 30 W,

$$30 = \frac{1200^2 R_{es}}{8(600 + R_{es})^2}$$
$$1200^2 R_{es} = 1{,}440{,}000 R_{es} = 240(R_{es}^2 + 1200 R_{es} + 360{,}000)$$
$$6000 R_{es} = R_{es}^2 + 1200 R_{es} + 360{,}000$$
$$R_{es}^2 - 4800 R_{es} + 360{,}000 = 0$$
$$R_{es} = \frac{4800}{2} \mp \frac{\sqrt{23{,}040{,}000 - 4(360{,}000)}}{2} = 2400 \mp \frac{\sqrt{21{,}600{,}000}}{2} = 2400 \mp 2323.8$$
$$R_{es1} = 4723.8\,\Omega, \quad R_{es2} = 76.2\,\Omega$$

(g) A plot of P_{es}, the power absorbed in the electrode–skin interface as a function of electrode–skin interface resistance, R_{es} is shown in Fig. 6.35. Note that maximum power transferred to electrode skin interface resistance of 600 Ω is 75 W. The electrode–skin interface resistance, R_{es}, when P_{es} = 30 W occurs at 76.2 and 4723.8 Ω. MATLAB code for this plot is also given.

```
clc;
h=ezplot('1200.^2*x/(8*(600+x).^2)',[0 5000 ]);grid on;
ylabel('Pes [W]');xlabel('Res [ohm]');
set(h,'LineWidth',2);  %# Sets the line width to 2
```

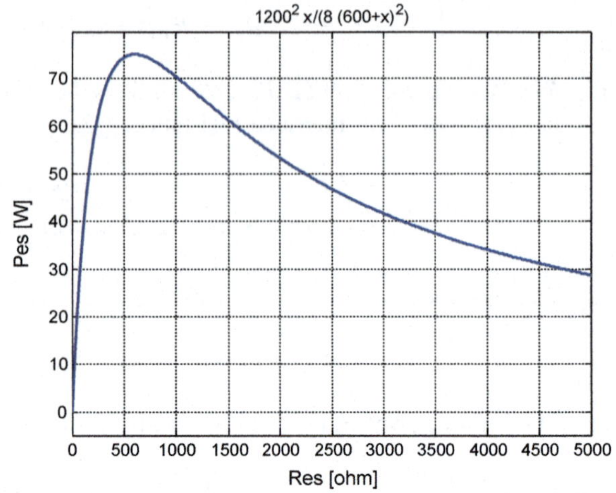

Fig. 6.35 Power change as a function of equivalent resistance

Problem 6.3.28 A return plate used in electrosurgery is 30 microns thick and has an area of 100 cm^2. What is the reactance of the plate at 400 kHz if the relative permittivity for the insulator is 3? What will be the reactance value at 800 kHz?

Solution

$$C = \frac{\varepsilon_o \varepsilon_r A}{d} = \frac{8.85 \times 10^{-12} \times 3 \times 100 \times 10^{-4}}{30 \times 10^{-6}} = 8.85 \times 10^{-9} F = 8.85\,\text{nF}$$

$$\mathbf{Z_c} = \frac{1}{j\omega C} = -jX_c$$

$$X_c = \frac{1}{\omega C} = \frac{1}{2\pi \times 400 \times 10^3 \times 8.85 \times 10^{-9}} = \frac{10^4}{8\pi \times 8.85} = 44.96\,\Omega$$

At 800 kHz, $f_2 = 2f_1$

$$X_{C2} = \frac{1}{2} \cdot X_c = 22.48\,\Omega$$

Problem 6.3.29

(a) What is the range of frequencies used in an electrosurgery unit (ESU)? Why?
(b) Assuming a homogenous and isotopic cubic material volume (V) of resistivity (ρ), determine the power dissipation if current density is i_d
(c) Calculate the power dissipation ratio near to the active and passive electrodes if current density around the active electrode is 0.5 A/m^2 while it is 10 mA/m^2 around the return plate. Assume a homogenous, isotropic material.
(d) Fig. 6.36 displays the basic principle of an electrosurgery unit. Calculate the power dissipated per cubic centimeter (in mW) of tissue that has resistivity of 1500 Ω m and current density of 400 mA/m^2. What is the total power dissipation (in W) if tissue volume is 0.1 m^3 at the frequency of operation?

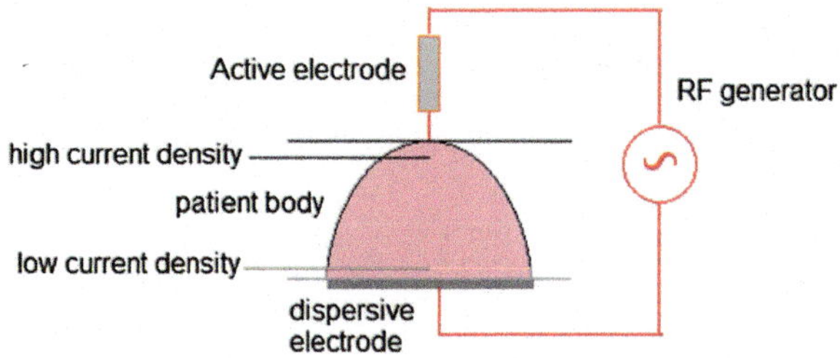

Fig. 6.36 Principle of (monopolar) electrosurgery unit

Solution

(a) Because muscular and neurological stimulation effects cease above 100 kHz, the range of frequencies used in an electrosurgery unit is chosen between 300 kHz and 3 MHz.

(b) $P = i^2R = (i_{\mathrm{d}}A)^2\left(\rho\dfrac{L}{A}\right) = \rho i_{\mathrm{d}}^2 AL = \rho V i_{\mathrm{d}}^2$

$P_a = \rho V i_{\mathrm{d}}^2, P_r = \rho V i_r^2$

(c) $\dfrac{P_a}{P_r} = \dfrac{i_{da}^2}{i_{dr}^2} = \left(\dfrac{i_{da}}{i_{dr}}\right)^2 = \left[\dfrac{0.5}{(10 \times 10^{-3})}\right]^2 = 500\,\mathrm{W/W}$

Current density near the active electrode is 500 times more than the current density around dispersive (= return = passive) electrode (PAD).

(d) $P = \rho V i_d^2 = 1500 \times \left(1 \times 10^{-2}\right)^3 \times \left(400 \times 10^{-3}\right)^2$

$P = 1500 \times 10^{-6} \times 16 \times 10^4 \times 10^{-6} = 2.4 \times 10^{-1} = 0.24\,\mathrm{mW}$

For a volume of $0.1\,\mathrm{m}^3$ tissue $V = 0.1\,\mathrm{m}^3 = 0.1 \times \left(10^2\,\mathrm{cm}\right)^3 = 10^5\,\mathrm{cm}^3$

$$P = 0.24 \times 10^5 = 24 \times 10^3\,\mathrm{mW} = 24\,\mathrm{W}$$

Problem 6.3.30 The principle of a "three parallel plates"-type capacitive sensor is illustrated in Fig. 6.37. The plates of equal areas are separated at equal distances of d units each. Two end plates are fixed but the mid plate can move to left or right by *d* under the influence of a force.

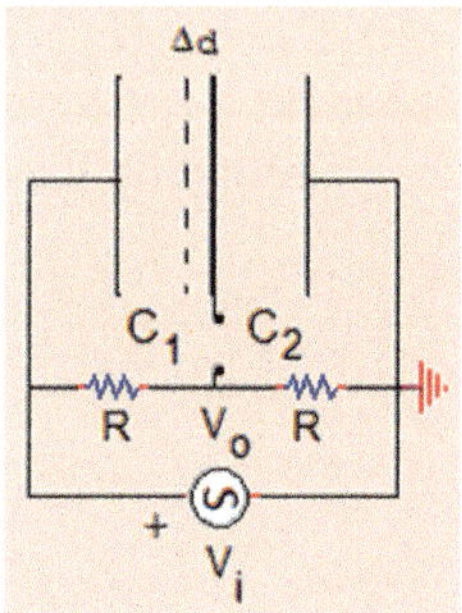

Fig. 6.37 Three parallel plates-type capacitive sensor

Fig. 6.38 Equivalent circuit diagram of the sensor construction

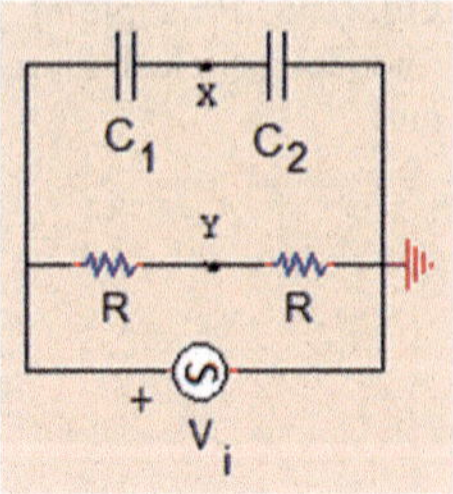

(a) Derive an expression for V_o in terms of V_i, d, and d.
(b) Is the measured voltage frequency dependent?
(c) Find V_o, if V_i = 1 V rms, d = 1 mm, and d = 0.2 mm

Solution

(a) Equivalent circuit diagram of the sensor construction is given in Fig. 6.38. This is an ac bridge circuit. Output voltage is obtained between X−Y,

$$V_o(j\omega) = V_X - V_Y$$

By voltage division,

$$V_o(j\omega) = \frac{X_{C2}}{X_{C1} + X_{C2}} V_i(j\omega) - \frac{R}{R+R} V_i(j\omega) = V_i(j\omega)\left(\frac{X_{C2}}{X_{C1} + X_{C2}} - \frac{1}{2}\right)$$

$$V_o(j\omega) = V_i(j\omega)\left(\frac{\frac{1}{j\omega C2}}{\frac{1}{j\omega C1} + \frac{1}{j\omega C2}} - \frac{1}{2}\right) = V_i(j\omega)\left(\frac{1}{\frac{C2}{C1} + 1} - \frac{1}{2}\right) \tag{6.24}$$

Noting that capacitance values for right and left parallel plate capacitors are given as

$$C_1 = \frac{\varepsilon A}{d - \Delta d}, \quad C_2 = \frac{\varepsilon A}{d + \Delta d}$$

Substituting these into Eq. (6.24),

$$\frac{C_2}{C_1} = \frac{\frac{\varepsilon A}{d + \Delta d}}{\frac{\varepsilon A}{d - \Delta d}} = \frac{d - \Delta d}{d + \Delta d} \tag{6.25}$$

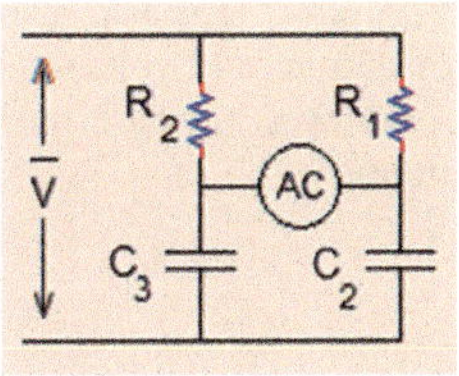

Fig. 6.39 A simple AC bridge circuit is De Sauty Bridge

$$V_o(j\omega) = V_i(j\omega)\left(\frac{1}{\dfrac{d-\Delta d}{d+\Delta d}+1} - \frac{1}{2}\right) = V_i(j\omega)\left(\frac{1}{\dfrac{d-\Delta d+d+\Delta d}{d+\Delta d}} - \frac{1}{2}\right)$$

$$V_o(j\omega) = V_i(j\omega)\left(\frac{1}{\dfrac{2d}{d+\Delta d}} - \frac{1}{2}\right) = V_i(j\omega)\left(\frac{d+\Delta d}{2d} - \frac{1}{2}\right) = \frac{V_i(j\omega)}{2}\left(\frac{d+\Delta d}{d} - 1\right)$$

$$V_o(j\omega) = \frac{V_i(j\omega)}{2}\left(\frac{d+\Delta d-d}{d}\right) = \left(\frac{\Delta d}{2d}\right)V_i(j\omega) \tag{6.26}$$

(b) The measured voltage is not frequency dependent.

(c) $|V_o(j\omega)| = \left(\frac{0.2}{2\times 1}\right)1 = 0.1$ Vrms

Problem 6.3.31 A simple AC bridge circuit is De Sauty Bridge, as shown in Fig. 6.39. Here C_2 is the unknown capacitance, and galvanometer is indicated by AC. Determine the value of C_2 for the balanced case.

Solution

For the balanced bridge, points B and D are at the same potential with respect to A or C,

$$I_1R_1 = I_2R_2 \tag{6.27}$$

$$I_1\left(\frac{-j}{\omega C_2}\right) = I_2\left(\frac{-j}{\omega C_3}\right) \quad \rightarrow \quad \frac{I_1}{\omega C_2} = \frac{I_2}{\omega C_3} \tag{6.28}$$

Dividing (6.27) by (6.28) yields

$$\omega R_1C_2 = \omega R_2C_3 \quad \rightarrow \quad R_1C_2 = R_2C_3 \quad \rightarrow \quad C_2 = \frac{R_2}{R_1}\cdot C_3 = k\cdot C_3$$

6.4 Power in AC Circuits

Problem 6.4.1 Rescaled display of current and voltage waveforms for an electric motor is shown in the graph of Fig. 6.40. Peak-to-peak voltage and current values

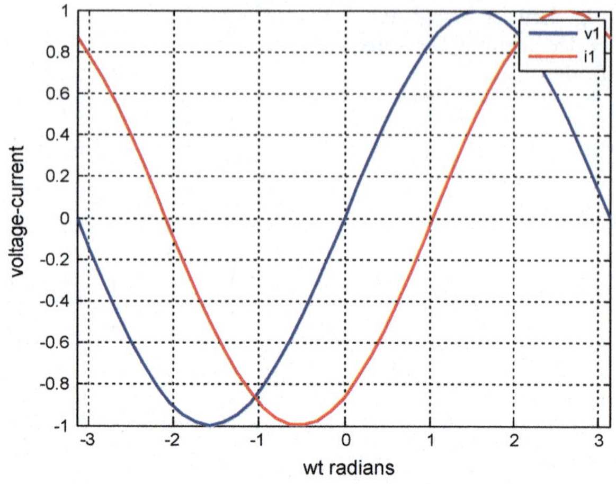

Fig. 6.40 Rescaled display of current and voltage waveforms for an electric motor

are measured as 650.6 V and 28.282 A, respectively. Determine the equation of current, and the voltage, apparent power value, the power factor, power consumed and the reactive power of the motor.

Solution
Inspection of the curves shows that the line frequency is 50 Hz, $\omega = 2\pi f = 314\,\text{rad/s}$.

$$V_{\text{pp}} = 650.6\,V \quad \rightarrow \quad V_p = 650.6/2 = 325.3V$$
$$v(t) = V_p \cos(314t) = 325.3\cos(314t)V$$
$$V_{\text{rms}} = \frac{V_p}{\sqrt{2}} = 325.3 \times 0.707 = 230V$$
$$I_{\text{pp}} = 28.282A \quad \rightarrow \quad I_{\text{p}} = 28.282/2 = 14.142A$$
$$I_{\text{rms}} = \frac{I_p}{\sqrt{2}} = 14.142 \times 0.707 = 10A$$

Apparent power of the load is $S = I_{\text{rms}} V_{\text{rms}} = 230 \times 10 = 2300\,\text{VA}$. From the graph,

$$\theta \cong \frac{\pi}{3}\text{rad} = 60°$$
$$i(t) = I_p \cos(314t - \theta) = 14.142\cos(314t - 60°)A$$
$$\text{power factor} = \cos\theta = \cos 60° = 0.5$$

Power consumed by the load is $P = S \cdot \cos\theta = 2300 \times \cos 60° = 2300 \times 0.5 = 1150\,\text{W}$

Reactive power of the load is $Q = S \cdot \sin\theta = 2300 \times \sin 60^\circ = 2300 \times 0.866 = 1992\,\text{VAR}$

Part of MATLAB script for plotting current and voltage waveforms:

```
%Plot_sine1.m
wt= -pi:pi/20:pi; v1=sin(wt); i1=sin(wt-pi/3);
plot(wt,v1,wt,i1,'r','linewidth',2);grid on; xlim([-pi,pi]);
```

Problem 6.4.2 One phase of a hospital electrical load operates at 1000 kVA, and 0.707 power factor.

Calculate the necessary reactive power in kVAR to correct this power factor to 0.95 lagging.

Solution

$$
\begin{aligned}
S &= \sqrt{P^2 + Q^2} = 1000\,\text{kVA} \quad (\text{Apparent power}) \\
\cos\theta &= 0.707, \quad \rightarrow \quad \theta = 45^\circ \\
P &= S \cdot \cos\theta = 1000 \times 0.707 = 707\,\text{kW} \quad (\text{Active power}) \\
Q &= S \cdot \sin\theta = 1000 \times 0.707 = 707\,\text{kW} \quad (\text{Reactive power})
\end{aligned}
$$

After correction (n index for "new"),

$$\cos\theta_n = 0.95, \quad \rightarrow \quad \theta_n = 18.2^\circ,$$

Active power consumption remains the same,

$$
\begin{aligned}
P &= S_n \cdot \cos\theta_n = S_n \times 0.95 = 707\,\text{kW} \quad \rightarrow \quad S_n = \frac{707}{0.95} = 744.2\,\text{kVA} \\
Q_n &= S_n \cdot \sin\theta = 1000 \times \sin(18.2^\circ) = 744.2 \times 0.312 = 232.4\,\text{kVAR}
\end{aligned}
$$

The necessary correction is

$$Q_C = Q_n - Q = 232.4 - 707 = -474.6\,\text{kVAR}$$

Figure 6.41 demonstrates the power triangle.

Problem 6.4.3 The coil of an AC contactor can be represented by an inductance L in series with resistance R. When 50 Hz current of 0.5 A is supplied to the coil at 230 V, a wattmeter indicates 10 W power delivered to the coil. Determine coil component values and the impedance of the coil.

Fig. 6.41 Power triangle

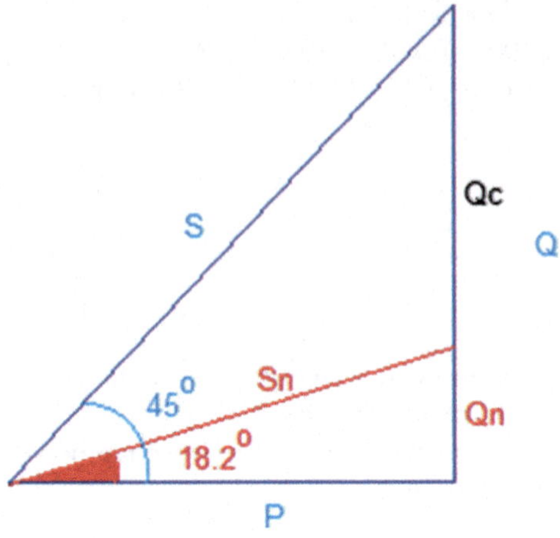

Solution

$$P = I^2R \quad \rightarrow \quad R = \frac{P}{I^2} = \frac{10}{0.5^2} = \frac{10}{0.25} = 40\,\Omega$$

$$P = IV\cos\theta \quad \rightarrow \quad \cos\theta = \frac{P}{IV} = \frac{10}{0.5 \times 230} = 0.087$$

$$Q = IV\sin\theta = 0.5 \times 230 \times \sin(85^\circ) = 115 \times 0.996 = 114.6VA$$

$$Q = I^2X_L \quad \rightarrow \quad X_L = \frac{Q}{I^2} = \frac{114.6}{0.5^2} = 458.4\,\Omega$$

$$X_L = \omega L \quad \rightarrow \quad L = \frac{X_L}{\omega} = \frac{X_L}{2\pi f} = \frac{458.4}{314} = 1.46\,\text{H}$$

$$Z = \frac{V}{I} = \frac{230}{0.5} = 460\,\Omega$$

$$\mathbf{Z} = Z\angle\theta = 460\angle 85^\circ\,\Omega$$

Problem 6.4.4 An induction furnace used in a (European) laboratory draws 10 A (rms) from the network. Nameplate reading indicates that the furnace power consumption is rated at 1800 W. Determine the resistance and inductance of the equivalent circuit of this furnace.

Solution

European network, $V_{\text{AC}} = 230\,\text{V}, f = 50\,\text{Hz}$

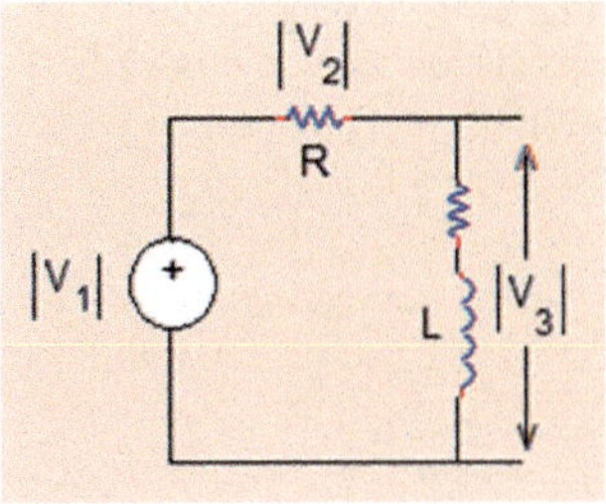

Fig. 6.42 Three voltmeters method

$$R = \frac{P}{I^2} = \frac{1800}{10} = 180\,\Omega$$

$$\theta = \cos^{-1}\left(\frac{P}{V \times I}\right) = \cos^{-1}\left(\frac{1800}{230 \times 10}\right) = \cos^{-1}(0.7826) = 38.5^{\circ}$$

$$Q = V \times I \times \sin\theta = 230 \times 10 \times \sin 38.5^{\circ} = 1431.78\ \mathrm{var}$$

$$X_L = \frac{Q}{I^2} = \frac{1431.78}{10^2} = 143.18\,\Omega,$$

$$L = \frac{X_L}{\omega} = \frac{X_L}{2\pi f} = \frac{143.18}{314} = 0.456\,H$$

Problem 6.4.5 Power consumed by a inductive circuit can be measured using three voltmeters method. As shown in Fig. 6.42, voltages across the input source, series resistor, and inductive load are measured and power is computed using these voltage values.

(a) What is the power factor in terms of measured voltages?
(b) Determine the power dissipated in the load in terms of measured voltages,
(c) Determine the load resistance and inductance.
(d) Calculate dissipated power, power factor of the circuit, load resistance, and load inductance for the values of $V_1 = 230\,\mathrm{V}$, $V_2 = 4.5\,\mathrm{V}$, $V_3 = 226\,\mathrm{V}$, $R_S = 9.1\,\Omega$, $f = 50\,\mathrm{Hz}$
(three_voltmeter1.m)

Solution

(a) By KVL, the sum of three rms voltages in the loop is zero.
Using the law of cosines, (removing absolute value signs for convenience) (Fig. 6.43),

$$V_1^2 = V_3^2 + V_2^2 - 2V_3V_2\cos(180^{\circ} - \varphi) \tag{6.29}$$

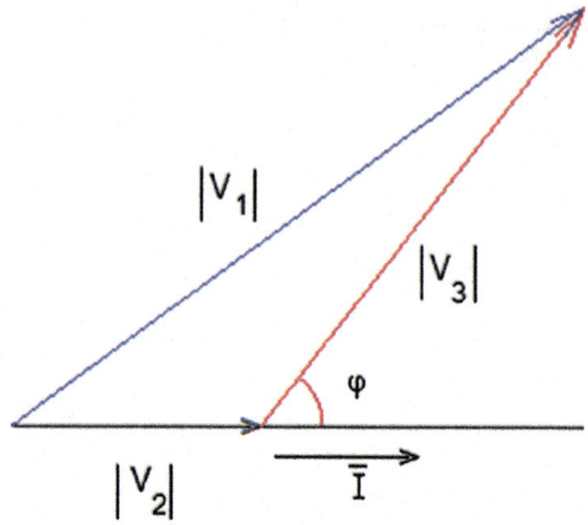

Fig. 6.43 Voltage calculation using the law of cosines

$$\cos(180^\circ - \varphi) = -\cos\varphi$$
$$V_1^2 = V_3^2 + V_2^2 + 2V_3V_2\cos\varphi$$

The power factor of the circuit

$$\cos\varphi = \frac{V_1^2 - V_2^2 - V_3^2}{2V_2V_3} \tag{6.30}$$

(b) The power dissipated in the load Z_L is

$$\begin{aligned} P &= IV_3\cos\varphi \\ V_2 &= IR_s \\ V_1^2 &= V_3^2 + V_2^2 + 2R_s(IV_3\cos\varphi) = V_3^2 + V_2^2 + 2R_s \cdot P \\ P &= \frac{V_1^2 - V_2^2 - V_3^2}{2R_s} \end{aligned} \tag{6.31}$$

(c) Since

$$\begin{aligned} P &= I^2R_L = \left(\frac{V_2}{R_s}\right)^2 R_L \\ R_L &= P\frac{R_s^2}{V_2^2} = \frac{V_1^2 - V_2^2 - V_3^2}{2R_s} \cdot \frac{R_s^2}{V_2^2} = \frac{V_1^2 - V_2^2 - V_3^2}{2V_2^2}R_s \end{aligned} \tag{6.32}$$

and

$$\begin{aligned}
\varphi &= \arccos\left(\frac{V_1^2 - V_2^2 - V_3^2}{2V_2V_3}\right) \\
Q &= I^2X_L = V_3 I \sin\varphi \\
IX_L &= V_3 \sin\varphi \\
\frac{V_2}{R_s}X_L &= V_3 \sin\varphi \\
\frac{V_2}{R_s}\omega L &= V_3 \sin\varphi \\
L &= \frac{V_3 \sin\varphi \cdot R_s}{\omega V_2}
\end{aligned} \tag{6.33}$$

(d) Using measured values of voltages and series resistance,

$$\begin{aligned}
\cos\varphi &= \frac{52{,}900 - 20.25 - 51{,}076}{2 \times 4.5 \times 226} = \frac{1803.75}{2034} = 0.8868 \\
P &= \frac{52{,}900 - 20.25 - 51076}{2 \times 9.1} = 99.1\,\mathrm{W} \\
R_L &= 99.1 \times \frac{9.1^2}{4.5^2} = 405.3\,\Omega \\
\varphi &= \arccos(0.8868) = 27.53^\circ \\
L &= \frac{226 \times \sin 27.53^\circ \times 9.1}{314 \times 4.5} = \frac{950.587}{314 \times 4.5} = 672.7\,\mathrm{mH}
\end{aligned}$$

This method can be implemented with op-amps, and power calculations with small errors are possible at frequencies below 5 kHz. However, the method requires the development of a resistive current shunt as well as voltage and current transformers whose phase defects are accurately known [7].

Problem 6.4.6 Determine the average power delivered by a generator of $v_o(t) = 220\cos(314t)$ with $R = 50\,\Omega$ output resistance to a load, as shown in Fig. 6.44.

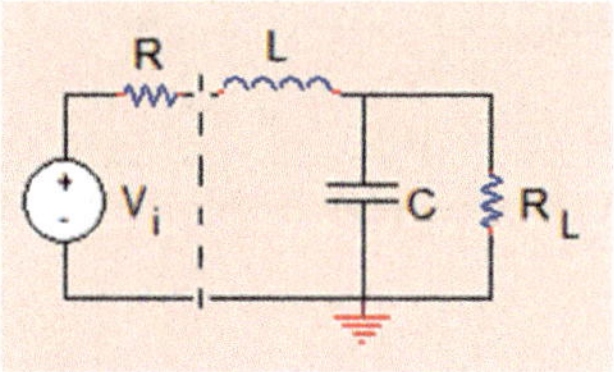

Fig. 6.44 The circuit for Problem 6.4.6

Solution
The load impedance is

$$Z = j100 + \frac{(-j50)(20)}{20 - j50} = j100 - \left[\frac{j1000(20 + j50)}{2900}\right] = j100 - \left(\frac{j20000 - 50000}{2900}\right)$$
$$Z = j100 - j6.897 + 17.24 = 17.24 + j93.1\,\Omega$$

The load current,

$$\boldsymbol{I} = \frac{220\angle 0^\circ}{50 + Z_L} = \frac{220\angle 0^\circ}{50 + 17.24 + j93.1} = \frac{220\angle 0^\circ}{67.24 + j93.1} = \frac{220\angle 0^\circ}{114.84\angle 54.16^\circ}$$
$$\boldsymbol{I} = 1.916\angle - 54.16^\circ\, A$$

The average power delivered to the load is

$$p = \frac{1}{2}R|I|^2 = \frac{1}{2}(20)(1.916)^2 = 36.71\,\text{W}$$

Problem 6.4.7 A signal generator with $u(t) = 10\cos(100t)V$ feeds a load resistor of $R_L = 100\ \Omega$ through an RC integrator circuit with $R = 100\ \Omega$ and $C = 1000\,\mu\text{F}$ (Fig. 6.45). Find the average power delivered to the load resistor (mW).

Solution
Thévenin equivalent parameters,

$$\boldsymbol{V}_{\text{Th}} = \frac{Z_C}{Z_C + R}\boldsymbol{U}$$
$$Z_C = \frac{-j}{\omega C} = \frac{-j}{100 \times 10^{-3}} = -j10\,\Omega$$
$$\boldsymbol{V}_{\text{Th}} = \frac{-j10}{-j10 + 100} = \frac{-j10(100 + j10)}{100 + 10000} = \frac{-j1000 + 100}{10100} = -0.099j + 0.0099$$
$$\boldsymbol{V}_{\text{Th}} = 0.099\angle - 84.29^\circ\ \text{V}$$

Short circuit output current,

$$\boldsymbol{I}_{\text{sc}} = \frac{10\angle 0^\circ}{100} = 0.1\angle 0^\circ\, A$$

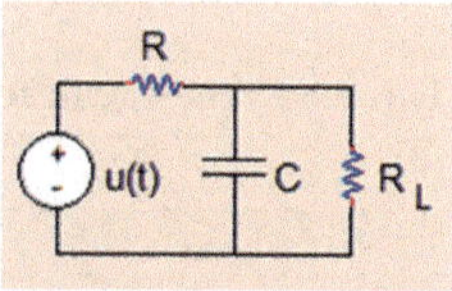

Fig. 6.45 The circuit for Problem 6.4.7

Thévenin impedance,

$$Z_{\mathrm{Th}} = \frac{\boldsymbol{V}_{\mathrm{Th}}}{\boldsymbol{I}_{\mathrm{sc}}} = \frac{0.099\angle -84.29^{\circ}}{0.1\angle 0^{\circ}} = 0.99\angle -84.29^{\circ} = 0.10 - j0.99\,\Omega$$

Load current,

$$\boldsymbol{I} = \frac{\boldsymbol{V}_{\mathrm{Th}}}{Z_{\mathrm{Th}} + R_L} = \frac{0.099\angle -84.29^{\circ}}{0.10 - j0.99 + 100} = \frac{0.099\angle -84.29^{\circ}}{100.1 - j0.99}$$

$$\boldsymbol{I} = \frac{0.099\angle -84.29^{\circ}}{100.105\angle -0.57} = 9.89 \times 10^{-4}\angle -83.72^{\circ}\ A$$

The average power delivered to load resistor,

$$P = \frac{1}{2}R_{\mathrm{L}}|\boldsymbol{I}|^2 = \frac{1}{2}(100)\left(9.89 \times 10^{-4}\right)^2 = 4890.605 \times 10^{-8}\,\mathrm{W} = 48.9\,\mathrm{mW}$$

Problem 6.4.8

(a) What is power superposition?
(b) Two current waveforms of different frequencies,

$$i_1(t) = 3\cos(314t - 45^{\circ})A, \quad i_2(t) = 4\cos(1570t)A,$$

flow through $R = 10\,\boldsymbol{\Omega}$ resistor. Determine the average power consumed on the resistor.

Solution

(a) The average power delivered to a circuit by multiple sinusoidal sources equals to the sum of the average power delivered to the circuit by each source acting alone, provided that all sources have *different* frequencies. This is called power superposition.

$$\boldsymbol{I} = \sum_{i=1}^{N} \boldsymbol{I}_i, \quad \rho = \frac{I_m^2 R}{2}, \quad |\boldsymbol{I}| = I_m$$

(b) Here, $\boldsymbol{I}_1 = 3\angle -45^{\circ}, \boldsymbol{I}_2 = 4\angle 0^{\circ}$ but they do not add as phasors. Although the average power could be calculated as

$$P = \frac{1}{T}\int_0^t i^2 R\mathrm{d}t = \frac{R}{T}\int_0^t [3\cos(314t - 45^{\circ}) + 4\cos(1570t)]^2 \mathrm{d}t$$

It can be calculated more easily using power superposition,

$$P = P_1 + P_2 = \frac{3^2 \times 10}{2} + \frac{4^2 \times 10}{2} = 125\,\text{W}$$

6.5 Inductive Coupling and Transformers

Problem 6.5.1

(a) What is mutual inductance?
(b) What is the unit and sign of mutual inductance?
(c) How is the polarity of mutually induced voltage determined?
(d) What is dot convention? Which factors determine the reference polarity of the mutual voltage?

Solution

(a) Mutual inductance is the ability of one inductor (coil) to induce a voltage in another inductor when they are in a close proximity to each other.
Assume that two coils with self-inductances L_1 and L_2 are in close proximity with each other and coil 1 has N_1 turns, while coil 2 has N_2 turns, and coil 1 carries a current i_1 and the second one carries no current. They are magnetically coupled. The magnetic flux generated by the first coil, φ_1, has one component linking only itself, φ_{11}, and another component linking both coils, φ_{12}, so that, $\varphi_1 = \varphi_{11} + \varphi_{12}$.
Since the fluxes are caused by the current i_1, the voltages induced in coil 1 and coil 2 are

$$v_1 = L_1 \frac{\mathrm{d}i_1}{\mathrm{d}t}, \quad v_2 = M_{21} \frac{\mathrm{d}i_1}{\mathrm{d}t}$$

M_{21} is the mutual inductance of coil 2 with respect to coil 1. Thus, the open-circuit induced voltage across coil 2 is v_2.
Now, let a current i_2 flow in coil 2, while coil 1 carries no current. The magnetic flux generated by the second coil, φ_2, has one component linking only itself, φ_{22}, and another component linking both coils, φ_{21}, so that, $\varphi_2 = \varphi_{22} + \varphi_{21}$.
Since the fluxes are caused by the current i_2, the voltages induced in coil 1 and coil 2 become

$$v_2 = L_2 \frac{\mathrm{d}i_2}{\mathrm{d}t}, \quad v_1 = M_{12} \frac{\mathrm{d}i_2}{\mathrm{d}t}$$

M_{12} is the mutual inductance of coil 1 with respect to coil 2. Thus, the open-circuit induced voltage across coil 1 is v_1. Here, $M_{12} = M_{21} = M$ is the mutual inductance between the two coils. In other words, mutual inductance results if a voltage is induced in one coil by a time-varying current in another coil.

(b) The unit of mutual inductance is Henry (H). It has always a positive value.

(c) The polarity for induced mutual voltage depends on the orientation or the way both coils are physically wound.

(d) A dot indicates the direction of the magnetic flux if a current enters that dotted coil terminal.[1] The dot convention can be stated as follows: If a current enters the dotted terminal of one coil, the reference polarity of the mutual voltage in the second coil is positive at the dotted terminal of the second coil. The reference polarity of the mutual voltage is a function of the reference direction of the inducing current and the dots on the coils.

Problem 6.5.2

(a) What is the meaning of sign convention in a transformer?

(b) What are the equations defining the current–voltage characteristics of the ideal transformer?

(c) If dots are placed on top of both coils in a transformer diagram, and primary current is directed toward dotted terminal of primary coil while secondary part has no load, determine the sign and value of the secondary voltage.

Solution

(a) Consider ideal transformers in Fig. 6.46.

One can write time domain ideal transformer voltage relationship as

$$\frac{v_2(t)}{v_1(t)} = \pm\frac{N_2}{N_1} = \pm n$$

where N_1 and N_2 are number of turns in primary and secondary winding, respectively, and n is the turns ratio. This ratio is always less than or equal to unity.

Since mutual inductance can be additive or subtractive, the sign depends on the direction of primary and secondary currents relative to the dots shown in the schematic of a transformer. Note that the dots indicate the relative winding orientations.

The sign in above given equation is positive ($n > 0$) when reference directions of both primary and secondary currents point toward or away from a dotted terminal. Otherwise, $n < 0$. Therefore, both of the ideal transformers in Fig. 6.46 have positive sign for voltage transfer ratio (= turns ratio).

---

[1]Here, it is assumed that the M and the placement of the dots are given parameters, like R, L, and C. For a given circuit, the dots are already placed beside the coils, therefore one does not have to bother about their placement.

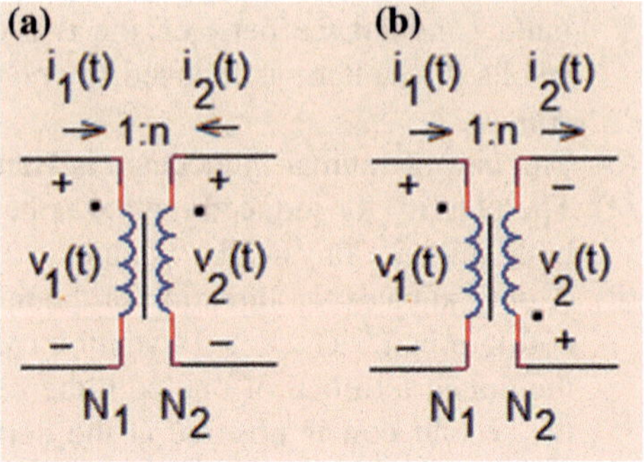

Fig. 6.46 Ideal transformer definition

(b) $v_2(t) = \pm n v_1(t), \quad i_2(t) = \mp \frac{1}{n} i_1(t)$

(The signs are opposite!)

(c) Let the primary current be $i_1(t)$. See, Fig. 6.46a, with $i_2(t) = 0$,

$$v_2 = M \frac{di_1}{dt}$$

The sign of the secondary voltage is positive.

Problem 6.5.3 A load $R_L = 120\ \Omega$ is connected to the secondary of an ideal transformer with a turns ratio of $n = 1/2$. What is the equivalent resistance seen on the primary side? Does the result depend on the location of the dot marks relative to current and voltage references?

Solution

$$R_{eq} = \frac{R_L}{n^2} = \frac{120}{0.5^2} = 480\ \Omega$$

The result is independent of the location of the dot marks relative to voltage and current reference directions.

Problem 6.5.4 An air core transformer has primary inductance of $L_1 = 16\,\text{mH}$, secondary winding inductance of $L_2 = 4\,\text{mH}$ and mutual inductance of $M = 4\,\text{mH}$. What is the coupling coefficient that indicates the degree of coupling between transformer windings?

Solution

$$k = \frac{M}{\sqrt{L_1 L_2}} = \frac{2}{\sqrt{16 \times 4}} = 0.25$$

Problem 6.5.5

(a) What is an ideal transformer? What is an isolation transformer? What is a linear transformer?
(b) What is the value of instantaneous power absorbed by the ideal transformer? What does this result imply?
(c) What are the values of internal resistance of primary and secondary windings of an ideal transformer? Inductances?

Solution

(a) An ideal transformer is a transformer with the coefficient of magnetic coupling equal to unity (Perfect coupling). An isolation transformer is a transformer whose turns ratio is unity.
A linear transformer linear is a transformer having a core made out of magnetically linear material so that its magnetic permeability is constant (such as air, wood, and plastics).
(b) $p(t) = v_1(t)i_1(t) + v_2(t)i_2(t) = v_1(t)[-ni_2(t)] + [nv_1(t)]i_2(t) = 0$

This result shows that the ideal transformer is lossless, since instantaneous absorbed power by ideal transformer is zero. It absorbs zero complex power, zero reactive power, and zero average power.

(c) $R_1 = R_2 = 0$ (The coils are lossless), $L_1, L_2 \to \infty$

Problem 6.5.6 If a voltage $v_i(t) = 10 + 0.2\sin(314t)$ is applied to the primary side of an ideal transformer with a turns ratio of 100, determine the magnitude of peak voltage, $V_{0\max}$, on its secondary winding.

Solution

Since only the AC signal is coupled through the ideal transformer from its primary to secondary side, and only the peak magnitude is of interest, $V_{0\max} = 100 \times 0.2 = 20\,\text{V}$

Problem 6.5.7

(a) Why and how impedance matching is realized using a transformer?
(b) If a load R_L is to be matched to a source with internal resistance R_s and $R_\text{s} > R_\text{L}$, what type of transformer should be used?
(c) A power amplifier with output resistance of 200 Ω is to be matched with internal resistance of 8 Ω. If a transformer with primary winding of 2000 turns is available, what should be its secondary winding turns number?
(d) Calculate the power delivered to loudspeaker in terms of amplifier's open-circuited output voltage if the amplifier is directly connected to the loudspeaker.

(e) Calculate the power delivered to loudspeaker by the power amplifiers in terms of amplifiers open-circuited output voltage if the matching transformer is in place.
(f) Compare the numerical results found in parts d and e, and comment.

Solution

(a) For maximum power transfer, the load resistance R_L is to be matched with the source resistance R_s. An iron core transformer matches the R_L to R_s, which is called transformer-based impedance matching.
Since an ideal transformer reflects its load to its primary with a scaling factor of n^2, setting source resistance equal to reflected resistance provides the matching condition,

$$R_s = \frac{R_L}{n^2}$$

(b) If $R_s > R_L$, and n = turns ratio, a step down transformer is required.
Here, n is the ratio of the number of turns in secondary winding to number of turns of primary winding of an ideal, iron core transformer.

(c)

$$R_{Th} = \frac{R_L}{n^2}, \quad n^2 = \frac{R_L}{R_{Th}} = \frac{R_L}{R_O} = \frac{8}{200} = \frac{1}{25} \quad \rightarrow \quad n = \frac{1}{5} = 0.2$$

$$n = \frac{N_{secondary}}{N_{primary}} = 0.2 \quad \rightarrow \quad N_{secondary} = 0.2 \times N_{primary} = 0.2 \times 2000 = 400 \text{ turns}$$

(d)

$$V_{oc} = V_{Th}, \quad R_s = R_{Th}$$

$$p_L = I^2 R_L = \left(\frac{V_{Th}}{R_{Th} + R_L}\right)^2 R_L = \left(\frac{1}{200+8}\right)^2 \times 8 \times V_{Th}^2 = 1.85 \times 10^{-4} \times V_{Th}^2 \text{ W}$$

(e) $R_s = R_{Th} = \dfrac{R_L}{n^2}$

Secondary current,

$$I_s = \frac{I_P}{n},$$

$$p_L = I_s^2 R_L = \left(\frac{I_P}{n}\right)^2 R_L = \left(\frac{\dfrac{V_{Th}}{R_{Th} + (R_L/n^2)}}{n}\right)^2 R_L = \left(\frac{n}{R_{Th} n^2 + R_L}\right)^2 R_L V_{Th}^2$$

$$= \left[\frac{0.2}{(200) \times \left(\dfrac{1}{25}\right) + 8}\right]^2 \times 8 \times V_{Th}^2 = 1.25 \times 10^{-3} \times V_{Th}^2 \text{ W}$$

(f) Without transformer, $p_L = 0.185 \times 10^{-3} \times V_{Th}^2$
with transformer in place, $p_L = 1.25 \times 10^{-3} \times V_{Th}^2$

Power transfer is increased by a factor of $1.25/0.185 = 6.76$.

Problem 6.5.8 Ideal transformer shown in Fig. 6.47 has a turns ratio of 10, and input voltage is

$$v_i(t) = 230 \sin 314t.$$

(a) Determine the secondary (load) voltage.
(b) Determine the primary current for a secondary load of 75 Ω.
(c) What type of ideal transformer is this?
(d) What is the input resistance seen at the primary side?

Solution

(a) Since the reference directions of both currents point toward a dotted terminal, the sign is positive, and $v_i(t) = v_1(t)$,

$$\frac{v_2(t)}{v_1(t)} = +n$$
$$v_2(t) = nv_1(t) = 10 \times 230\text{sin}314t = 2300\text{sin}314t$$

(b) Current–voltage characteristic equations of this ideal transformer are

$$v_2(t) = +nv_1(t) \quad \text{and} \quad i_2(t) = -\frac{1}{n} \times i_1(t)$$

Then, using Ohm's Law,

$$i_2(t) = \frac{-v_2(t)}{75} = \frac{-2300\text{sin}314t}{75} = -30.6667 \sin 314t\, A$$
$$i_1(t) = -ni_2(t) = -10 \times (-30.667 \sin 314t) = 306.667\text{sin}314t\, A$$

Fig. 6.47 Ideal transformer

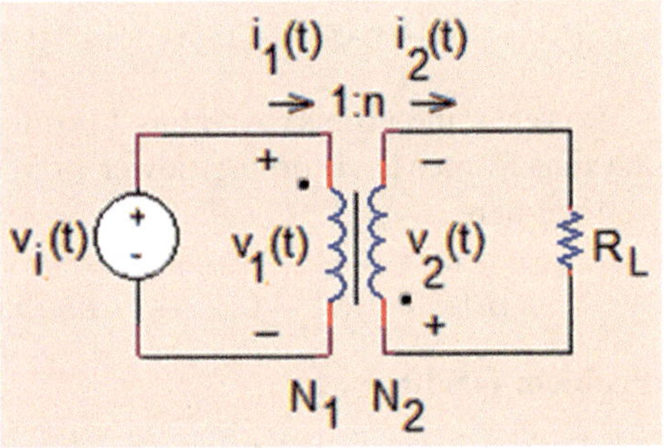

(c) This is a step up transformer, since the output (secondary) voltage is higher than the primary side.

(d) $R_{eq} = \frac{v_1(t)}{i_1(t)} = \frac{230\sin 314t}{306.667\sin 314t} = 0.75\,\Omega$

(Alternatively, $R_{eq} = R_L/n^2 = 75/10^2 = 0.75\,\Omega$)

Problem 6.5.9 The input to the primary side of and ideal transformer with a turns ratio of 5 is $v_1(t) = 230\ \sin(314t)$. The load is 100 Ω resistor, connected to secondary side. Transformer has additive coupling. Determine secondary and primary power in this element. What does a negative power mean here? Comment on the results.

Solution

Since additive coupling is present,

$$v_2(t) = +nv_1(t) = 5 \times 230\sin(314t) = 1150\sin 314t\ V$$

By Ohm's Law,

$$i_2(t) = -\frac{v_2(t)}{R_2} = -\frac{1150}{100}\sin(314t) = -11.5\sin(314t)A$$

The power in the secondary winding:

$$p_2(t) = i_2(t)v_2(t) = (-11.5) \times (1150) \times \sin^2(314t) = -13{,}225\sin^2(314t)[\mathrm{W}]$$

The negative sign shows that ideal transformer (secondary) delivers power to the load.

$$i_1(t) = -ni_2(t) = -5 \times [-11.5\sin(314t)] = 57.5\sin 314tA$$

since $v_1(t) = 230\sin(314t)$,

$$p_1(t) = i_1(t)v_1(t) = (57.5) \times (230) \times \sin^2(314t) = 13{,}225\sin^2(314t)[\mathrm{W}].$$

Power in the primary side has a positive sign, showing that primary winding of the transformer is absorbing power from the source. Zero power loss condition is verified here,

$$p_1(t) + p_2(t) = 0 \quad \rightarrow \quad 13{,}225\sin^2 314t - 13{,}225\sin^2 314t = 0$$

Problem 6.5.10

(a) Determine the resistance R_{xy} for the circuit of Fig. 6.48. Assume ideal transformers.

(b) Check the result found (in part a) using SPICE analysis (transformer2.cir).

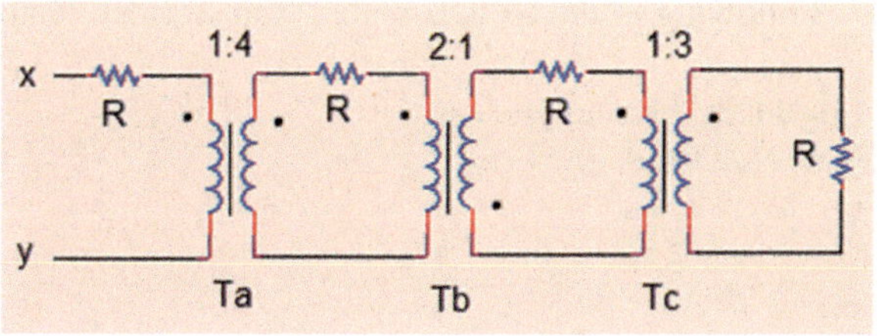

Fig. 6.48 Cascaded transformers

Solution

Reflect load R to the primary of T_C:

$$R_1 = \frac{R}{3^2} = \frac{R}{9}$$

Load at the secondary of T_B is $R_2 = R + R_1 = 10R/9$.

Reflect load R_2 to the primary of T_B:

$$R_3 = \frac{R_2}{\left(\frac{1}{2}\right)^2} = \frac{\frac{10R}{9}}{\frac{1}{4}} = \frac{40R}{9}$$

Load of the secondary of T_A is $R_4 = R + R_3 = 49R/9$.

Reflect R_4 to the primary of T_A:

$$R_5 = \frac{\frac{49R}{9}}{16} = \frac{49R}{144}$$

$$R_{xy} = R + R_5 = R + \frac{49R}{144} = \frac{193}{144}R = 1.34R$$

(b) In SPICE analysis of this cascade of transformers, $k = 1$ (due to perfect coupling condition) and $n_j = \sqrt{L_{Sj}/L_{Pj}}$, where L_{Sj}, L_{Pj} are secondary and primary inductance of *j*th ideal transformer, respectively. In order to measure the value of circuit input impedance, current source of unity value is input to the circuit and the resulting voltage at the input port is measured by performing an AC analysis. The voltage is measured (in a frequency region corresponding to flat response) as $V_{in} = 1.34$ V. This result is in agreement with the result obtained before, as this value divided by unity current value yields the input impedance, $Z_{in} = 1.34R$.

SPICE analysis also verifies that input impedance is not influenced by dot markings. Note that, changing current entry point (dotted terminal) for the secondary winding of transformer Tb (L4 of SPICE netlist) does not change the input impedance.

Following is the SPICE netlist (transformer2.cir):

```
cascaded transformers
*ac ANALYSIS
I1 0 1 1 AC
R1 1 2 1
R2 3 4 1
R3 5 6 1
R4 7 0 1
Ka L1 L2 1
Kb L3 L4 1
Kc L5 L6 1
L1 2 0 1
L3 4 0 1
L5 6 0 1
L2 3 0 16
L4 0 5 0.25
L6 7 0 9
*.ac dec 10 1 1000
```

Problem 6.5.11 An echo-cardiography device employs a power amplifier with 288 Ω output resistance which drives an 8 Ω, 10 W loudspeaker through an (ideal) transformer. Determine the turns ratio and magnitudes of voltages on the primary and secondary windings of this transformer.

Solution

$$\text{Since}\, p = \frac{V_s^2}{R_L}$$

Secondary voltage (and the load voltage) is

$$V_s = \sqrt{p \cdot R_L} = \sqrt{10 \times 8} = 8.9\,\text{V}$$

Maximum power transfer condition is realized if the output impedance of power amplifier is matched to the loudspeaker impedance.

$$Z_0 = \frac{Z_L}{n^2} \quad \rightarrow \quad 288 = \frac{8}{n^2} \quad \rightarrow \quad n = \frac{1}{6}$$

Primary voltage,

$$\frac{V_s}{V_p} = n \quad \rightarrow \quad V_p = \frac{V_s}{n} = \frac{8.9\,\text{V}}{\frac{1}{6}} = 53.4\,\text{V}$$

Problem 6.5.12 Calculate the currents $i_1(t)$ and $i_2(t)$ in the transformer circuit of Fig. 6.49.

$$v_1(t) = 10 \cos t, \quad L_1 = 2H, \quad L_2 = 3H, \quad R = 2\,\Omega, \quad M = 2H, \quad C = 0.5\,\mathrm{F}$$

where L_1, L_2 are primary and secondary winding inductance, respectively. Use Cramer's rule if required (phasor1.xlsx).

Solution
Since $\omega = 1$ rad/s

$$X_C = \frac{1}{\omega C} = \frac{1}{1 \times \frac{1}{2}} = 2\,\Omega,$$
$$X_{L1} = \omega L_1 = 2\,\Omega, \quad X_{L2} = \omega L_2 = 3\,\Omega, \quad X_M = \omega M = 2\,\Omega,$$

For loop 1, by KVL, $\boldsymbol{V} = (\boldsymbol{Z}_C + j\omega L_1)\boldsymbol{I_1} - j\omega M\boldsymbol{I_2}$
For loop 2, by KVL, $0 = -j\omega MI_1 + (R + j\omega L_2)I_2$
In matrix form,

$$\begin{bmatrix} \boldsymbol{Z}_C + j\omega L_1 & -j\omega M \\ -j\omega M & R + j\omega L_2 \end{bmatrix} \begin{bmatrix} \boldsymbol{I_1} \\ \boldsymbol{I_2} \end{bmatrix} = \begin{bmatrix} \boldsymbol{V} \\ 0 \end{bmatrix}$$
$$\begin{bmatrix} -j2 + j2 & -j2 \\ -j2 & 2 + j3 \end{bmatrix} \begin{bmatrix} \boldsymbol{I_1} \\ \boldsymbol{I_2} \end{bmatrix} = \begin{bmatrix} 10 \\ 0 \end{bmatrix} \quad \rightarrow \quad \begin{bmatrix} 0 & -j2 \\ -j2 & 2 + j3 \end{bmatrix} \begin{bmatrix} \boldsymbol{I_1} \\ \boldsymbol{I_2} \end{bmatrix} = \begin{bmatrix} 10 \\ 0 \end{bmatrix}$$
$$\Delta = (0 - (-1)4) = 4, \quad \Delta_1 = \begin{vmatrix} 10 & -j2 \\ 0 & 2 + j3 \end{vmatrix} = 20 + j30, \quad \Delta_2 = \begin{vmatrix} 0 & 10 \\ -j2 & 0 \end{vmatrix} = j20$$

Mesh currents are

$$\boldsymbol{I_1} = \frac{\Delta_1}{\Delta} = \frac{20 + j30}{4} = 5 + j7.5 = 9.014\angle 56.31^\circ\,A,$$
$$\boldsymbol{I_2} = \frac{\Delta_2}{\Delta} = \frac{j20}{4} = j5 = 5\angle 90^\circ\,A,$$

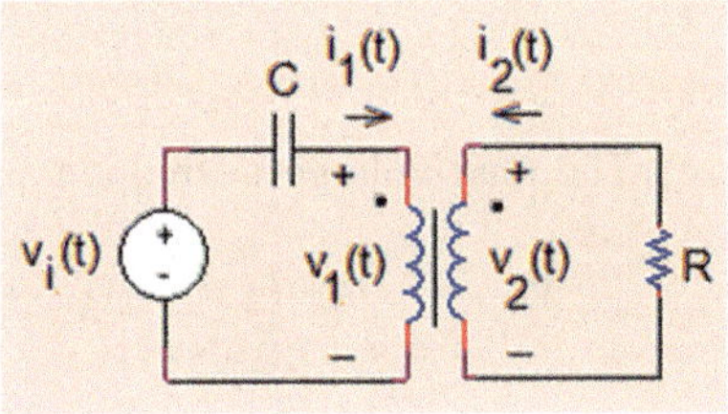

Fig. 6.49 The circuit for Problem 6.5.12

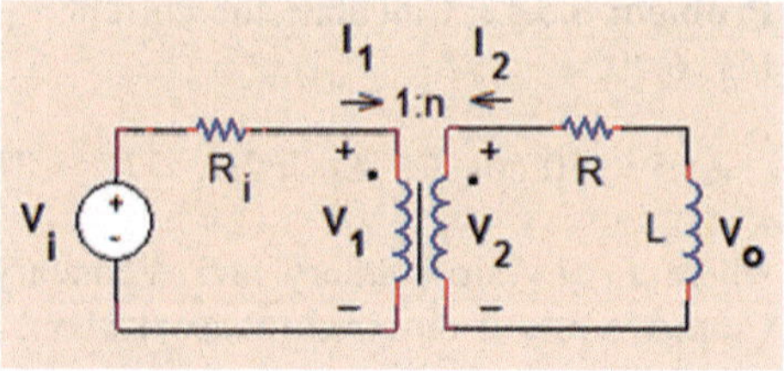

Fig. 6.50 Frequency domain representation

In time domain,

$$i_1(t) = 9.014\cos(t + 56.31°)A, \quad i_2(t) = 5\sin t A$$

Problem 6.5.13 An ideal transformer with a turns ratio of 2 connects a voltage source with 10 Ω internal resistance and Thévenin's voltage of $v_i(t) = 30\cos(314t + 45°)$ to an inductor having 10 Ω reactance. A resistor with 20 Ω resistance exists between the secondary winding of the transformer and the inductor. See Fig. 6.50.

(a) Determine primary current in phasor form.
(b) Determine secondary current in phasor form.
(c) Determine the inductor voltage (load voltage) in time domain (phasor1.xlsx).

Solution

(a) Impedance of secondary section is $\boldsymbol{Z}_s = 20 + j10\,\Omega$. Reflecting this impedance to the primary side gives equivalent reflected impedance,

$$\boldsymbol{Z}_{eq} = \frac{\boldsymbol{Z}_s}{n^2} = \frac{20 + j10}{2^2} = 5 + j2.5\,\Omega$$

$$\boldsymbol{I}_1 = \frac{\boldsymbol{V}_1}{10 + \boldsymbol{Z}_{eq}} = \frac{10\angle 45^{\circ}}{10 + 5 + j2.5} = \frac{30\angle 45^{\circ}}{15 + j2.5} = \frac{30\angle 45^{\circ}}{15.2\angle 9.46^{\circ}} = 1.974\angle 35.54^{\circ}$$

(b) The relationship between primary and secondary currents I_1 and I_2 is $I_1 = -nI_2$ Therefore,

$$I_2 = -\frac{I_1}{n} = -\frac{1.974\angle 35.54^{\circ}}{2} = -0.987\angle 35.54^{\circ} A$$

(c) The load voltage in frequency domain (using passive convention) is

$$\begin{aligned} \boldsymbol{V}_0 &= -(jX_L)\boldsymbol{I}_2 = -j10\left(-0.987\angle 35.54^{\circ}\right) = \left(10\angle 90^{\circ}\right)\left(0.987\angle 35.54^{\circ}\right) \\ &= 9.87\angle 125.54^{\circ} \end{aligned}$$

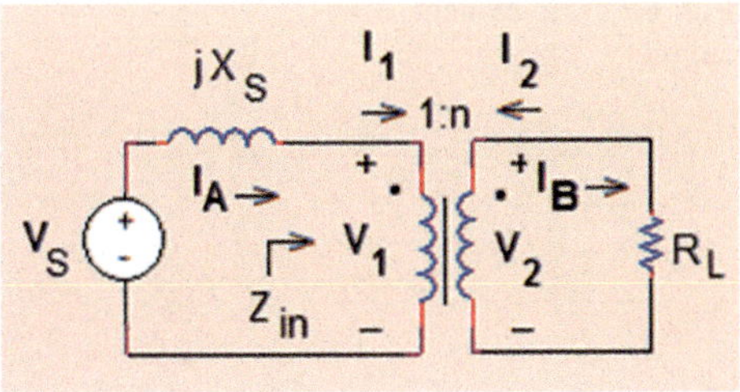

Fig. 6.51 The circuit for Problem 6.5.14

In time domain,

$$v_0(t) = 9.87\cos\left(314t + 125.54^{\circ}\right).$$

Problem 6.5.14 A source with Thévenin impedance of $Z_s = j10\,\Omega$ and $V_s(t) = 1200\cos(314t)$ feeds a nonideal transformer with $L_1 = 10.L_2 = 1H, M = 0.3H$, where L_1, L_2 and M are primary, secondary, and mutual inductances of transformer. A resistive load of 50 Ω is connected to secondary winding, see, Fig. 6.51.

(a) What is the value of coupling coefficient of transformer?
(b) Determine primary and secondary currents and voltages of transformer in phasor form.
(c) Calculate the input impedance of transformer seen as a load by the source circuit.
(d) Find the average power delivered by the source.
(e) Determine the equivalent *T*-circuit for transformer. Can this equivalent circuit be built out of ordinary inductors? (Sim_Lin_Eq_Solve.m, phasor1.xlsx).

Solution

(a) $k = \dfrac{M}{\sqrt{L_1 \times L_2}} = \dfrac{0.3}{\sqrt{1 \times 0.1}} = 0.949$

(b) Assigning CW current directions in both meshes, and noting that the relationship between defining currents of a transformer and mesh currents are $\mathrm{I}_1 = \mathrm{I}_A, \quad \mathrm{I}_2 = -\mathrm{I}_B,$
Reactances of each part of transformer are calculated as,

$$j\omega \mathrm{L}_1 = j314\,\Omega$$
$$j\omega \mathrm{L}_2 = j314 \times 0.1 = j31.4\,\Omega$$
$$j\omega M = j314 \times 0.3 = j94.2\,\Omega$$

Applying KVL for both meshes result in following matrix equation,

Fig. 6.52 Equivalent *T*-circuit

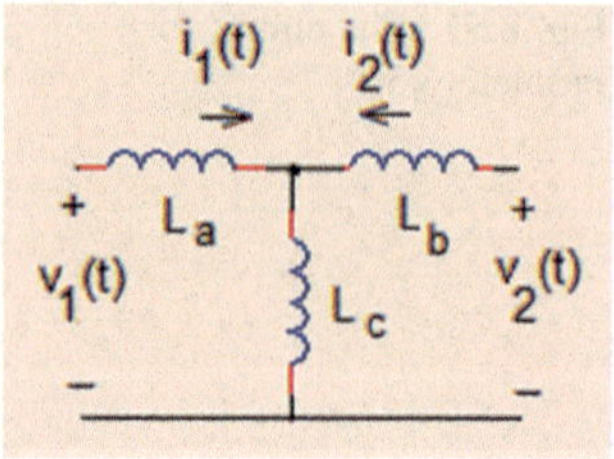

$$\begin{bmatrix} j10+j314 & -j94.2 \\ -j94.2 & 50+j31.4 \end{bmatrix} \begin{bmatrix} I_A \\ I_B \end{bmatrix} = \begin{bmatrix} 1200 \\ 0 \end{bmatrix}$$

$$\begin{bmatrix} j324 & -j94.2 \\ -j94.2 & 50+j31.4 \end{bmatrix} \begin{bmatrix} I_A \\ I_B \end{bmatrix} = \begin{bmatrix} 1200 \\ 0 \end{bmatrix}$$

Both mesh currents are computed by solving this matrix equation in MATLAB,

$$I_A = 2.0157 + j3.8655 = 4.3595\angle 62.46°A$$
$$I_B = 6.9331 + j0.5563 = 6.9554\angle 4.59°A$$
$$\boldsymbol{V_1} = \boldsymbol{V_S} - I_A Z_S = 1200 - (2.0157 + j3.8655)(j10) = 1200 - j20.157 + 38.655$$
$$\boldsymbol{V_1} = 1238.655 - j20.157 = 1238.819\angle 0.93°V$$
$$\boldsymbol{V_2} = \boldsymbol{R_L I_B} = 50 \times (6.9331 + j0.5563) = 346.655 + j27.815 = 347.769\angle 4.59°V$$

(c) The input impedance of transformer seen as a load by the source circuit is

$$Z_{in} = \frac{\boldsymbol{V_1}}{\boldsymbol{I_A}} = \frac{1238.819\angle 0.93°}{4.3595\angle 62.46°} = 284\angle -61.53°\,\Omega$$

$$R_{in} = 284 \times \cos(-61.53°) = 135.382\,\Omega$$

(d)
$$P = \frac{|\boldsymbol{I_A}|^2}{2} R_{in} = \frac{(4.3595)^2}{2} \times 135.382 = 1286.5\,\mathrm{W}$$

(e) Since mutual inductance is additive in this case,

$$L_a = L_1 - M, \quad L_b = L_2 - M, \quad L_c = M$$

Substituting given values of components yields (Fig. 6.52),

$$L_a = 1 - 0.3 = 0.7H, L_b = 0.1 - 0.3 = -0.2H, L_c = 0.3H.$$

This equivalent circuit cannot be built out of physical inductors, since a negative inductance is not a passive element.

Problem 6.5.15 What is Transcutaneous Energy Transfer (TET)? What is the main advantage of using TET? What are the present day medical applications of TET? What is resonant inductive coupling? Why is it employed in TET devices?

Solution

In Transcutaneous Energy Transfer (TET), energy is sent through an inductive coupling to a surgically embedded medical device in a patient's body, which uses that energy to charge batteries or run the device. Power Transmission via an electromagnetic field passing through the skin avoids the infection risks associated with wires penetrating the skin.

A variety of devices that require power can be surgically implanted within various portions of the body. These include a synthetic replacement heart, a circulatory blood pump or ventricular assist device (VAD), a cochlear implant, and a pacemaker. However, complications associated with repeated surgical entries and the risks of infection and/or dislodgment make TET systems attractive, as they are used to transfer energy from outside the body to inside the body in order to provide power to one or more implanted devices from an external power source.

Since TET systems operate without puncturing the skin, the possibility of infection is reduced while comfort and convenience for patients is increased.

TET systems typically use two inductive coils, where a primary coil is contained within a separate unit connected to an ac transmitter and a secondary coil is contained within the surgically implanted device and it may be connected to its own battery. When the two coils are in close proximity to each other, electrical energy is inductively transferred from the primary coil (the transmitter) to the secondary coil (in the implanted device).

TET and methods for inductively charging a rechargeable power supply has been a hot topic for many years and they are described in various articles and patents [8–19].

One problem with TET techniques is inefficient power transfer. This causes long charge times in the implant side. Increasing the inductive power transfer as a remedy tends to be more expensive, because it requires a larger transmitting device, and generates excessive heat and exacerbates electromagnetic compatibility (EMC) problems.

Greater distances between transmitter and receiver coils can be achieved when the TET system uses resonant inductive coupling. Series LC resonance method at the transmitter side increases the efficiency of power transfer between the primary and secondary coils. The inclusion of a capacitor connected in series with the primary inductive coil allows for the increased efficiency of inductive power transfer between the two coils due to resonance. A key characteristic of the LC resonance circuit is that the natural resonant frequency of the circuit selectively acts on the main harmonic frequency of the input voltage source.

The secondary coil is implanted inside the body of the patient. It can be connected to the implant device and/or charging circuit which controls the flow of power to the rechargeable battery (The details of the charging circuit and rechargeable power supply are application dependent).

The mutual inductance that is created due to the coupling of the coils (which varies with distance between the two coils, their orientation and secondary coil capacitance) causes variations on the resonance frequency. Therefore, LC circuit should be tuned by measuring secondary power across the load at different frequencies around the calculated natural resonant frequency for the specific application load.

An implantable medical device described in [12] utilizes about 1–100 kHz transcutaneous energy transfer (TET) for supplying power from an external module to an implantable medical device, avoiding power dissipation through eddy currents in a metallic case of an implant and/or in human tissue, thereby enabling smaller implants using a metallic case such as titanium and/or allowing TET signals of greater strength thereby allowing placement more deeply within a patient without excessive power transfer inefficiencies.

Due to increased complexity of medical implant device technology, it is also necessary to provide a wireless data communication between the implanted devices and an outside operator. These systems are components of the implanted system and use a separate RF antenna so that an external controller or programmer can communicate with internal sensors or control elements. Typically, the separate RF antenna is implanted in a patient away from the implanted secondary TET coil to avoid radio interference when the coil is in use [13].

Design tradeoffs in a TET system are governed by the conflicting demands of energy transfer efficiency, communication bandwidth, and insensitivity to changes in coil coupling variation, and concerns about patient quality of life. In high-energy transfer applications, selection of operating frequency involves tradeoffs between tissue thermal and electronic switching losses which tend to increase with frequency and losses in the tuned-circuits driving a coil which are inversely related to frequency [14].

Problem 6.5.16 A transformer with perfect coupling ($k = 1$) has a primary winding and secondary winding inductance values of $L_1 = 10\,\text{mH}$, $L_2 = 40\,\text{mH}$, respectively.

(a) Determine the turns ratio of the transformer and the mutual inductance.
(b) Plot the magnitude of voltage on the secondary winding of transformer as a function of frequency, under no load condition (neglecting core effects). Comment on this graph (transformer1.cir).

Solution

(a) $$n = \frac{N_2}{N_1} = \sqrt{\frac{L_1}{L_2}} = \sqrt{\frac{40}{10}} = 2$$
$$M = \sqrt{L_1 L_2} = \sqrt{10 \times 40}\,\text{mH} = 20\text{ mH}$$

(b) The magnitude of voltage on the secondary winding of the transformer as a function of input voltage frequency (under no load condition) is plotted as shown in Fig. 6.53 (frequency sweep at constant input voltage). Inspection of

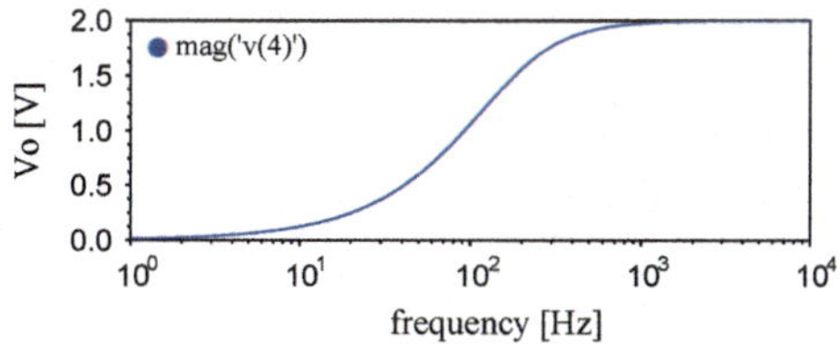

Fig. 6.53 The magnitude of voltage on the secondary winding of the transformer as a function of input voltage frequency sweep

this figure reveals that at lower frequencies transformer "step up" performance is not effective, but reduced due to inductive properties of transformer along with the resistance at the primary side (the transfer ratio is around n = 0.7 for the power line frequency, which is far below its calculated value). It should be noted also that the model used here does not include stray capacitances of any kind. Otherwise, one might expect low pass action at higher frequencies, resulting a frequency band pass structure as a whole.[2]

A SPICE netlist (transformer1.cir) is given below.

```
transformer1
*AC analysis
vi 1 2 1 AC
R1 1 3 10
L1 3 2 10M
L2 4 5 40M
K L1 L2 0.999
RA 2 0 1G
RB 5 0 1G
```

Problem 6.5.17 Direct connection of an oscilloscope probe to an electrosurgery unit (ESU) is not recommended during testing and calibrating the RF output of an ESU. Search for a method to perform scope-based pulse time and frequency measurements of an ESU, in order to check ESU pulse generation and its RF output frequency.

Solution
Using a dummy load resistor (which simulates a patient) and a ferrite toroidal (ring) transformer core, following setup shown in Fig. 6.54 can be used to test ESU output signal timing.

[2]Further SPICE examples on transformer applications are included in the SPICE netlists file. See, for example, blocking_oscillator.cir, which is a SPICE netlist for a blocking oscillator which uses a transformer in the feedback circuit. It is possible to test the influence of varying coupling coefficient and other circuit component values on the oscillator's output signal. Such circuits find various practical applications as they make (for example) the operation of higher voltage threshold LEDs possible, by using much lower supply voltages.

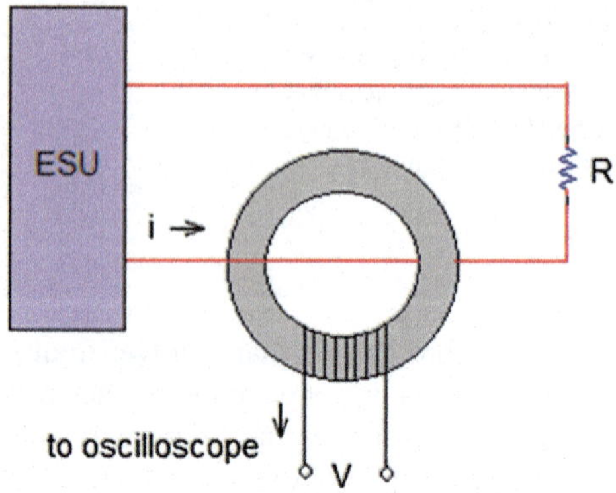

Fig. 6.54 Output pulse time and frequency measurements of an ESU

Here toroidal transformer is used to sample the RF waveform (*V*) and then the waveform is displayed on an oscilloscope. Otherwise, it can be dangerous to connect the oscilloscope to the load without using proper probes.

Single wire of ESU is passed through toroidal transformer core, and current passing in the cable induces some voltage across the winding on toroidal core. Here, it can be assumed that the cable passing through the ring constitutes the primary winding of transformer as a single turn. Therefore, the turns ratio is actually the number of turns of wire wound on toroidal core and this acts as the secondary winding of toroidal transformer. This makes pulse time and frequency measurements possible; however, amplitude measurements require more calibration.

References

1. Mallela VS, Ilankumaran V, Rao NS (2004) Trends in cardiac pacemaker batteries. Indian Pacing Electrophysiol J 4(4):201–212
2. Covidien Data Sheet (2007) LFORCEFX 945605267 Rev 2007/11
3. Maxium KlsMartin Group (2009) Document 02.15 90-288-02-09
4. ERBE Elektromedizin GmbH (2009) Document LIT/5204/01 05/09
5. Association for the Advancement of Medical Instrumentation (AAMI)/American National Standards Institute (ANSI) (1993) Standard for electrosurgical devices. ANSI/AAMI HF18
6. ECRI (1987) The Bovie CSV: still accepted? (consultants corner). Health Dev 6(9–10):340–341
7. Marzetta LA (1972) An evaluation of the three-voltmeter method for AC power measurement. IEEE Trans Instrum Measur IM-21(4):353–357
8. Knecht O et al (2014) Optimization of transcutaneous energy transfer coils for high power medical applications. In: IEEE 15th workshop on control and modeling for power electronics (COMPEL), pp 1–10
9. Elixmann I et al (2012) Transcutaneous energy transfer system incorporating a Datalink for a wearable autonomous implant. In: Ninth international conference on wearable and implantable body sensor networks, pp 1–5
10. Xue R-F et al (2011) High-efficiency transcutaneous wireless energy transfer for biomedical applications. In: Defense science research conference and expo (DSR), pp 1–4
11. Partovi A (2016) Chargers and methods for wireless power transfer. US Patent 9,356,659 B2

12. Hassler WL Jr, Dlugos DF Jr (2009) Low frequency transcutaneous energy transfer to implanted medical device. US Patent 7,599,743 B2
13. D'Ambrosio RL, Kortyka M (2013) Transcutaneous energy transfer coil with integrated radio frequency antenna. US Patent 8,620,447 B2
14. Gaumond RP (1994) Transcutaneous energy and information transfer for biomedical implants. In: Proceedings of 20th annual northeast bioengineering conference, p 42
15. Desai RH, Hassler WL Jr (2007) Spatially decoupled twin secondary coils for optimizing transcutaneous energy transfer (TET) power transfer characteristics. US Patent 7,191,007 B2
16. Wells BP (2005) Series resonant inductive charging circuit. US Patent 6,972,543 B1
17. Seelig A (1997) Method and arrangement for automatic contactless charging. US Patent 5,654,621
18. Klontz KW et al (1992) Contactless battery charging system. US Patent 5,157,319 A
19. Bolger JG, Ng LSB (1989) Inductive power coupling with constant voltage output. US Patent 4,800,328 A

Chapter 7
Laplace Transform Methods

7.1 LT, Inverse LT, Initial and Final Value Properties

Problem 7.1.1 Use definition of LT to determine LTs of given functions ($t > 0$).

$$\text{(a)}\, f(t) = 6, \quad \text{(b)}\, f(t) = \delta(t), \quad \text{(c)}\, f(t) = \mathrm{e}^{-2t}, \quad \text{(d)}\, f(t) = \sin(\omega t).$$

Solution

(a) $F(s) = \int\limits_{0^-}^{\infty} 6\mathrm{e}^{-st}\,\mathrm{d}t = -\frac{6}{s}\mathrm{e}^{-st}\Big|_0^{\infty} = -\frac{6}{s}(\mathrm{e}^{-\infty} - \mathrm{e}^0) = -\frac{6}{s}(0-1) = \frac{6}{s}$

(b) $F(s) = \int\limits_{0^-}^{\infty} \delta(t)\mathrm{e}^{-st}\,\mathrm{d}t = \mathrm{e}^{-0} = 1$

(c) $F(s) = \int\limits_{0^-}^{\infty} \mathrm{e}^{-2t}\mathrm{e}^{-st}\mathrm{d}t = \int\limits_{0^-}^{\infty} \mathrm{e}^{-(s+2)t}\,\mathrm{d}t, \quad \int \mathrm{e}^{ax}\,\mathrm{d}x = \frac{\mathrm{e}^{ax}}{a}$ with $a = -(s+2)$,

$F(s) = \frac{-1}{s+2}\mathrm{e}^{-(s+2)t}\Big|_0^{\infty} = \frac{-1}{s+2}(\mathrm{e}^{-\infty} - \mathrm{e}^0) = -\frac{1}{s+2}(0-1) = \frac{1}{s+2}$

(d)

$$F(s) = \int\limits_0^{\infty} \sin(\omega t)\mathrm{e}^{-st}\,\mathrm{d}t = \int\limits_0^{\infty} \left(\frac{\mathrm{e}^{j\omega t} - \mathrm{e}^{-j\omega t}}{2j}\right)\mathrm{e}^{-st}\,\mathrm{d}t$$

$$= \frac{1}{2j}\int\limits_0^{\infty} \left[\mathrm{e}^{-(s-j\omega)t} - \mathrm{e}^{-(s+j\omega)t}\right]\mathrm{d}t = \frac{1}{2j}\left(\frac{1}{s-j\omega} - \frac{1}{s+j\omega}\right)$$

$$F(s) = \frac{1}{2j}\left[\frac{s+j\omega - (s-j\omega)}{s^2+\omega^2}\right] = \frac{2j\omega}{2j(s^2+\omega^2)} = \frac{\omega}{s^2+\omega^2}$$

Problem 7.1.2

(a) Verify the periodicity property of Laplace transform.

(b) Find the Laplace transform of a periodic unit amplitude pulse waveform with a period of 3 s, and duty cycle (percent of the period in which the signal is positive) of 1/3.

A.Ü. Keskin, *Electrical Circuits in Biomedical Engineering*,
DOI 10.1007/978-3-319-55101-2_7

Solution

(a) $\mathcal{L}[y(t)] = \mathcal{L}\left[\sum_{0}^{\infty} x(t - kT)\right] = \sum_{0}^{\infty} x(s)\mathrm{e}^{-skT} = X(s)\sum_{k=0}^{\infty} \mathrm{e}^{-skT} = \frac{X(s)}{1-\mathrm{e}^{-sT}}$

(b) One period of $y(t)$ is $x(t) = u(t) - u(t - 1)$, $0 \le t \le 3$s.

$$X(s) = \frac{1}{s} - \frac{\mathrm{e}^{-s}}{s} \quad \rightarrow \quad Y(s) = \frac{X(s)}{1 - \mathrm{e}^{-sT}} = \frac{\frac{1}{s} - \frac{\mathrm{e}^{-s}}{s}}{1 - \mathrm{e}^{-3s}} = \frac{1 - \mathrm{e}^{-s}}{s(1 - \mathrm{e}^{-3s})}$$

Problem 7.1.3 Use the integral definition of LT and prove the transform of the time shifted function, $f(t - \tau)u(t - \tau) \leftrightarrow \mathrm{e}^{-st}F(s)$

Solution

$$\mathcal{L}[f(t - \tau)u(t - \tau)] = \int_0^{\infty} f(t - \tau)u(t - \tau)\mathrm{e}^{-st}\mathrm{d}t = \int_{\tau}^{\infty} f(t - \tau)\mathrm{e}^{-st}\mathrm{d}t$$

Substitute $u = t - \tau, \rightarrow t = u + \tau$

$$\mathcal{L}[f(t - \tau)u(t - \tau)] = \int_0^{\infty} f(u)\mathrm{e}^{-s(t+u)}\,du = \mathrm{e}^{-s\tau}\int_{\tau}^{\infty} f(u)\mathrm{e}^{-su}\,du = \mathrm{e}^{-s\tau}F(s)$$

Problem 7.1.4 Use table of LT pairs (see, Appendix) and calculate the LT of following functions $(t \geq 0)$

(a) $f(t) = \mathrm{e}^{-2t} + \sin 5t$, (b) $f(t) = t^2 \sin 3t$, (c) $f(t) = 5[u(t - 1) - u(t - 4)]$
(d) $f(t) = At$, $0 \leq t \leq 1$ s (Gated ramp function with max value of A)
(e) $f(t) = t\mathrm{e}^{-2t}\cos(4t)$, (f) $f(t) = 2t^3\mathrm{e}^{-t}$

Solution

(a) $F(s) = \dfrac{1}{s+2} + \dfrac{5}{s^2 + 2s} = \dfrac{s^2 + 5s + 35}{(s+2)(s^2 + 25)}$

(b) $G(s) = \mathcal{L}[\sin 3t] = \dfrac{3}{s^2 + 9}$ where $g(t) = \sin 3t$

By frequency differentiation property,

$$\mathcal{L}[tg(t)] = -\frac{\mathrm{d}}{\mathrm{d}s}G(s) - \frac{\mathrm{d}}{\mathrm{d}s}\left(\frac{3}{s^2 + 9}\right)$$

$$F(s) = \mathcal{L}\{t(tg(t))\} = -\frac{\mathrm{d}}{\mathrm{d}s}\left[-\frac{\mathrm{d}}{\mathrm{d}s}G(s)\right] = \frac{\mathrm{d}}{\mathrm{d}s}\left[\frac{(0 - 2s \times 3)}{(s^2 + 9)^2}\right]$$

$$= -\frac{\mathrm{d}}{\mathrm{d}s}\left[\frac{6s}{(s^2 + 9)^2}\right] = -\left[\frac{6(s^2 + 9)^2 - 2(s^2 + 9)^2 \times 2s \times 6s}{(s^2 + 9)^4}\right]$$

$$F(s) = -\frac{(s^2 + 9)[6(s^2 + 9) - 24s^2]}{(s^2 + 9)^4} = \frac{18s^2 - 54}{(s^2 + 9)^3} = 18\frac{s^2 - 3}{(s^2 + 9)^3}$$

(c) $\mathcal{L}[u(t)] = \frac{1}{s}$, apply time shift property,

$$F(s) = 5\frac{e^{-s}}{s} - 5\frac{e^{-4s}}{s} = \frac{5}{s}\left(e^{-s} - e^{-4s}\right)$$

(d) $f(t) = At \quad 0 \le t \le 1\,\mathrm{s}$

$$\begin{aligned} f(t) &= At[u(t) - u(t-1)] = At \times u(t) - At \times u(t-1) \\ &= At \times u(t) - A(t-1+1) \times u(t-1) \\ &= At \times u(t) - A(t-1) \times u(t-1) - A \times u(t-1) \\ F(s) = \mathcal{L}[f(t)] &= \frac{A}{s^2} - \frac{A}{s^2}e^{-s} - \frac{A}{s}e^{-s} = \frac{A}{s^2}(1 - e^{-s} - se^{-s}) \end{aligned}$$

(e) $F(s) = \mathcal{L}[t^n g(t)] = (-1)^n \frac{d}{ds^n}[G(s)], \quad g(t) = e^{-2t}\cos 4t, \quad n = 1$

$$\begin{aligned} \mathcal{L}\left[e^{-2t}\cos(4t)\right] &= \frac{s+2}{(s+2)^2 + 16} \\ \frac{d}{ds}\left[\frac{s+2}{(s+2)^2+16}\right] &= \frac{-s^2 - 4s + 12}{(s^2+4s+20)^2} \\ F(s) = (-1)^1 \frac{-s^2 - 4s + 12}{(s^2+4s+20)^2} &= \frac{s^2 + 4s - 12}{(s^2+4s+20)^2} \end{aligned}$$

(f) $\mathcal{L}[t^n g(t)] = (-1)^n \frac{d}{ds^n}[G(s)], \quad g(t) = e^{-t}, \quad n = 3$

$$\begin{aligned} F(s) &= 2\mathcal{L}\left[t^3 e^{-t}\right] = 2(-1)^3 \frac{d}{ds^3}\left(\frac{1}{s+1}\right) = -2\frac{d}{ds^3}\left(\frac{1}{s+1}\right) = -2\frac{d}{ds^2}\left(\frac{-1}{(s+1)^2}\right) \\ &= -2\frac{d}{ds}\left[\frac{-2(s+1)(-1)}{(s+1)^4}\right] = -2\frac{d}{ds}\left[\frac{2s+2}{(s+1)^4}\right] = -2\frac{d}{ds}\left[\frac{2}{(s+1)^3}\right] = -2\left[\frac{-3(s+1)^2 \times 2}{(s+1)^6}\right] \\ F(s) &= -2\left[\frac{-6(s+1)^2}{(s+1)^6}\right] = \frac{12}{(s+1)^4} \end{aligned}$$

Problem 7.1.5 Use MATLAB to find Laplace transforms of following functions (Laplace1.m):

(a) $f(t) = te^{-at}\sin(\omega t)$, (b) $f(t) = t^2\cos(\omega t)$, (c) $f(t) = t^2\cos\left(\omega t + \frac{\pi}{4}\right)$,
(d) $f(t) = \delta(t-2)$, (e) $f(t) = u(t-2)$, (f) $f(t) = u(t) - u(t-2)$,
(g) $f(t) = a[u(t) - u(t-2)]$, (h) $f(t) = t[u(t) - u(t-2)]$.

Solution

$$\text{(a) } F(s) = \frac{2\omega(s+a)}{[(s+a)^2+\omega^2]^2}, \quad \text{(b) } F(s) = \frac{s(2s^2-\omega^2)}{(s^2+\omega^2)^3},$$

$$\text{(c) } F(s) = \frac{\sqrt{2}(s^3 - 3s\omega^2 - 3\omega s^2 + \omega^3)}{(s^2+\omega^2)^3}, \quad \text{(d) } F(s) = \mathrm{e}^{-2s}, \quad \text{(e) } F(s) = \frac{\mathrm{e}^{-2s}}{s},$$

$$\text{(f) } F(s) = \frac{1-\mathrm{e}^{-2s}}{s}, \quad \text{(g) } F(s) = a\frac{1-\mathrm{e}^{-2s}}{s}, \quad \text{(h) } F(s) = \frac{1-\mathrm{e}^{-2s}-2s\mathrm{e}^{-2s}}{s^2}$$

Following is MATLAB script (Laplace1.m) to compute these Laplace transforms.

```
syms a w t s
Fa(s)=laplace(t*exp(-a*t)*sin(w*t))
Fb(s)=laplace(t^2*cos(w*t))
Fc(s)=laplace(t^2*cos(w*t+pi/4));simplify(Fc(s))
Fd(s)=laplace(dirac(t - 2)) %Dirac Delta Function
Fe(s)=laplace(heaviside(t - 2)) %Unit Step Function
Ff(s)=laplace((heaviside(t)-heaviside(t-2))); %Gate function
simplify(Ff(s))
fg = a*(heaviside(t)-heaviside(t-2))%Gated function
Fg(s)=laplace(fg); simplify(Fg(s))
fh = t*(heaviside(t)-heaviside(t-2))%Gated Ramp function
Fh(s)=laplace(fh); simplify(Fh(s))
```

Problem 7.1.6 Plot delayed step function $f(t) = Au(t-\tau)$ using alternative methods in MATLAB. Let $A = 2, \tau = 1$ s. (Hint: Use built in Heaviside function).

What is the Laplace transform of this waveform? (plot_delayed_step.m)

Solution

Part of MATLAB code (plot_delayed_step.m) for plotting this function is given below.

```
% delayed step function
tau=1; % delay time
A=2; % A = amplitude of step
%%Method 1: use fplot
%syms t
%fplot(heaviside(t-tau), [0, 2])
%Method 2:
t = 0:0.001:2; y =A*heaviside(t-tau);
plot (t,y,'r','linewidth',3);axis([0  2  0  1.5*A]);
```

Delayed step function $f(t) = 2u(t-1)$ is displayed in Fig. 7.1.

Its Laplace transform is calculated using time shifting property of LT,

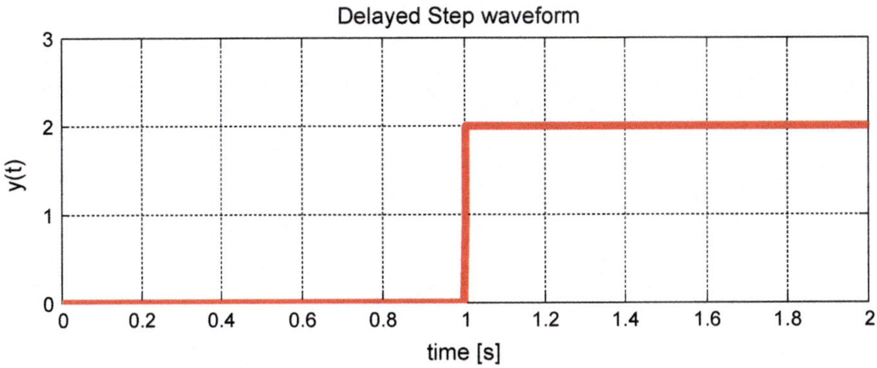

Fig. 7.1 Delayed step function $f(t) = 2u(t-1)$

$$F_1(s) = \mathcal{L}[2u(t)] = \frac{2}{s}, \quad F(s) = \mathcal{L}[2u(t-1)] = 2\frac{e^{-s}}{s}$$

Problem 7.1.7 Use step functions to determine time domain expressions and Laplace transforms of composite waveforms shown in Fig. 7.2 (plot_composite1_wf.m).

Solution

$$F(s) = \mathcal{L}[u(t)] = \frac{1}{s}, \quad G(s) = \mathcal{L}[tu(t)] = \frac{1}{s^2}, \quad A = 1, \quad \tau = 2,$$

(a) $G_1(s) = \mathcal{L}[(A/\tau)[tu(t) - (t-2)u(t-2)]] = \dfrac{1 - e^{-2s}}{2s^2}$

(b) $G_2(s) = \mathcal{L}\left[\left(\dfrac{A}{\tau}\right)[tu(t) - (t-2)u(t-2) - 2u(t-2)]\right]$

$$G_2(s) = \frac{1}{2}\left(\frac{1}{s^2} - \frac{e^{-2s}}{s^2} - 2\frac{e^{-2s}}{s}\right) = \frac{1 - e^{-2s} - 2se^{-2s}}{2s^2}$$

Following is part of MATLAB code (plot_composite1_wf.m) for the time domain solution.

```
%plotting  composite waveforms using step functions
ustep= @(x) 0<=x ;
A=1;tau=2;tmax=5;
%g1 = @(x) ( (A/tau)*( x.*ustep(x)-(x-tau).*ustep(x-tau)));
g2 = @(x) ( (A/tau)*(x.*ustep(x)- (x-tau).*ustep(x-tau)...
-2*ustep(x-tau)));
t = 0:0.001:tmax; plot(t,g2(t),'linewidth',3);
```

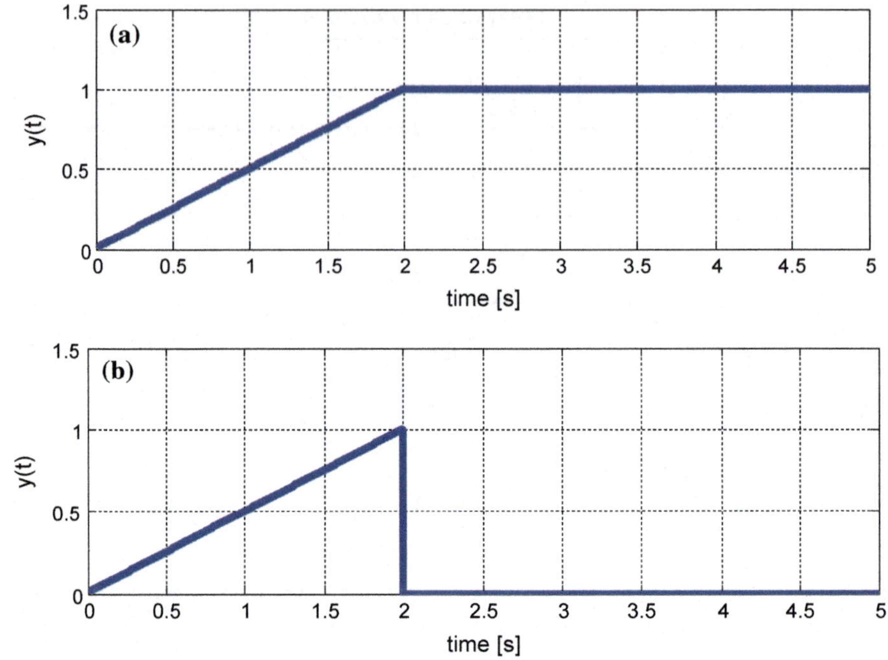

Fig. 7.2 **a** Composite waveform 1 **b** Composite waveform 2

Problem 7.1.8

(a) Use basic $u(t), r(t)$ (unit step and unit ramp) functions to find out expressions for each of the five composite waveforms shown in Fig. 7.3 (plot_composite_wf.m).
(b) Determine Laplace transform of each waveform.

Hint: In choosing basic function components of a composite function, try to use only addition operations. This procedure eliminates the products of functions, which is not so easy to compute. An example is decomposition of a ramp pulse function: It could be written as

$$\frac{A}{\tau}t[u(t) - u(t-\tau)] = \frac{A}{\tau}tu(t) - \frac{A}{\tau}tu(t-\tau)$$

but (although it yields the correct result) the last term on the right side of above equation is relatively harder to deal with.

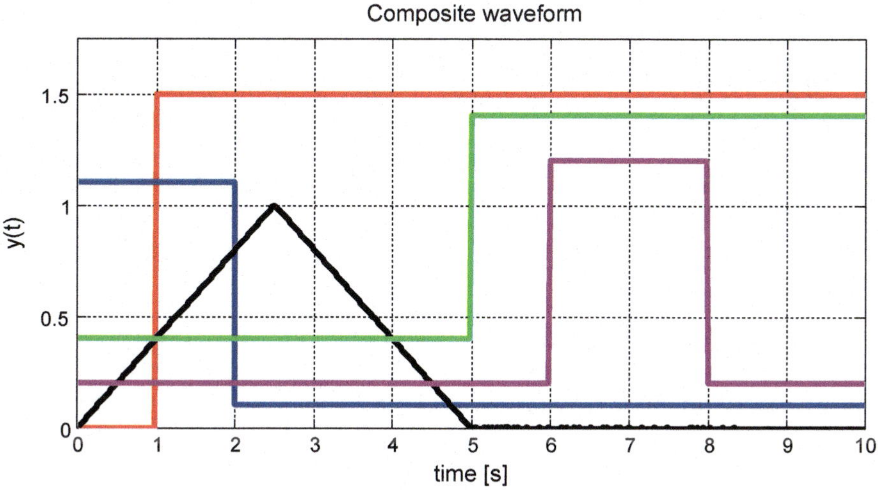

Fig. 7.3 Composite waveforms

Solution

(a) (i) Red colored waveform: $g_1(t) = 1.5u(t-1)$
(ii) Blue colored waveform: $g_2(t) = 1.1u(t) - u(t-2)$
(iii) Black colored waveform: $g_3(t) = 0.4r(t) - 0.8r(t-2.5) + 0.4r(t-5)$
(iv) Green colored waveform: $g_4(t) = 0.4u(t) - u(t-5)$
(v) Magenta colored waveform: $g_5(t) = 0.2u(t) + u(t-6) - u(t-8)$

Following is part of MATLAB code (plot_composite_wf.m):

```
ramp = @(x) max(0,x); ustep= @(x) 0 <= x ;
g1 = @(x) (1.5*ustep(x-1));
g2 = @(x) (1.1*ustep(x)- ustep(x-2));
g3 = @(x) (0.4*ramp(x)- .8*ramp(x-2.5)+0.4*ramp(x-5));
g4 = @(x) (0.4*ustep(x)+ustep(x-5));
g5 = @(x) (0.2*ustep(x)+ustep(x-6)-ustep(x-8));
t = 0:0.01:10;
plot(t,g1(t),'r',t,g2(t),'b',t,g3(t),'k',t,g4(t),'g',t,g5(t),'m',...
    'linewidth',3);axis([0  10  0  1.75]);
```

(b) Laplace transforms;

$$G(s) = \mathcal{L}[u(t)] = \frac{1}{s},$$

(i) $G_1(s) = \mathcal{L}[1.5u(t-1)] = 1.5\frac{\mathrm{e}^{-s}}{s}$

(ii) $G_2(s) = \mathcal{L}[1.1u(t) - u(t-2)] = \frac{1.1 - \mathrm{e}^{-2s}}{s}$

(iii) $G_3(s) = \mathcal{L}[0.4r(t) - 0.8r(t-2.5) + 0.4r(t-5)] = \frac{0.4 - 0.8\mathrm{e}^{-2.5s} + 0.4\mathrm{e}^{-5s}}{s^2}$

(iv) $G_4(s) = \mathcal{L}[0.4u(t) - u(t-5)] = \frac{0.4 - \mathrm{e}^{-5s}}{s}$

(v) $G_5(s) = \mathcal{L}[0.2u(t) + u(t-6) - u(t-8)] = \frac{0.2 + \mathrm{e}^{-6s} - \mathrm{e}^{-8s}}{s}$

Problem 7.1.9 Use a table of LT pairs (see, Appendix), and determine inverse LTs of the following s-domain functions.

(a) $F(s) = \frac{1}{s} - \frac{14}{s+1} + \frac{22}{(s+2)^2} + \frac{13}{s+2}$,

(b) $F(s) = \left[\frac{-2}{s+1} + \frac{2}{s+2} + \frac{2}{(s+1)^2}\right]$,

(c) $F(s) = \left[-\frac{4}{s+2} + 2s - 3\right]$,

(d) $F(s) = \left[-\frac{3}{(s+2)^2+9} + \frac{s+5}{(s+5)^2+16} + \frac{6}{(s+2)^4}\right]$,

(e) $F(s) = \left[\frac{10}{s^2+25} + \frac{s}{s^2+25} - 3\frac{\mathrm{e}^{-4t}}{s^2}\right]$,

(f) $F(s) = \left[\frac{6\mathrm{e}^{-2s}}{(s+5)^4} + \frac{24}{s^5}\right]$,

(g) $F(s) = \left[\frac{4s}{(s^2+4)^2} - \frac{1}{s-3}\right]$.

Solution

(a) $v(t) = \mathcal{L}^{-1}\left[\frac{1}{s} - \frac{14}{s+1} + \frac{22}{(s+2)^2} + \frac{13}{s+2}\right] = 1 - 14\mathrm{e}^{-t} + 22t\mathrm{e}^{-2t} + 13\mathrm{e}^{-2t}$

(b) $f(t) = \mathcal{L}^{-1}\left[\frac{-2}{s+1} + \frac{2}{s+2} + \frac{2}{(s+1)^2}\right] = \left(-2\mathrm{e}^{-t} + 2\mathrm{e}^{-2t} + 2t\mathrm{e}^{-t}\right)u(t)$

(c) $f(t) = \mathcal{L}^{-1}\left[-\frac{4}{s+2} + 2s - 3\right] = \left[-4\mathrm{e}^{-2t} + 2\frac{\mathrm{d}}{\mathrm{d}t}\delta(t) - 3\delta(t)\right]u(t)$

(d) $f(t) = \mathcal{L}^{-1}\left[-\frac{3}{(s+2)^2+9} + \frac{s+5}{(s+5)^2+16} + \frac{6}{(s+2)^4}\right]$

$= \left[-\mathrm{e}^{-2t}\sin(3t) + \mathrm{e}^{-5t}\cos(4t) + \mathrm{e}^{-2t}t^3\right]u(t)$

$$\text{(e) } f(t) = \mathcal{L}^{-1}\left[\frac{10}{s^2+25} + \frac{s}{s^2+25} - 3\frac{e^{-4s}}{s^2}\right] = [2\sin(5t) + \cos(5t) - 3(t-4)]u(t)$$

$$\text{(f) } f(t) = \mathcal{L}^{-1}\left[\frac{6e^{-2s}}{(s+5)^4} + \frac{24}{s^5}\right] = \left[e^{-5(t-2)}(t-2)^3 + t^4\right]u(t)$$

$$\text{(g) } f(t) = \mathcal{L}^{-1}\left[\frac{4s}{(s^2+4)^2} - \frac{1}{s-3}\right] = \left[t\sin(2t) - e^{3t}\right]u(t)$$

Problem 7.1.10 Expand $F(s)$ into its partial fractions:

$$F(s) = \frac{2(s+5)(s+12)}{s(s+1)(s+6)}$$

Solution

$$F(s) = \frac{2(s+5)(s+12)}{s(s+1)(s+6)} = \frac{N(s)}{D(s)} = \frac{a_n s^n + a_{n-1}s^{n-1} + \cdots + a_1 s + a_0}{a_m s^m + a_{m-1}s^{m-1} + \cdots + b_1 s + b_0}$$

This is a case for distinct real roots of $D(s)$.
$N(s)/D(s)$ = proper rational function if $m > n$, (otherwise improper);

$$F(s) = \frac{K_1}{s} + \frac{K_2}{s+1} + \frac{K_3}{s+6}$$

$$K_1 = \frac{2\cdot(0+5)\cdot(0+12)}{(1)\cdot(6)} = \frac{120}{6} = 20, \quad K_2 = \frac{2\cdot(-1+5)\cdot(-1+12)}{(-1)\cdot(-1+6)} = \frac{88}{5}$$

$$K_3 = \frac{2\cdot(-6+5)\cdot(-6+12)}{(-6)\cdot(-6+1)} = -\frac{2}{5}$$

$$F(s) = \frac{20}{s} - \frac{88/5}{s+1} + \frac{-2/5}{s+6}$$

$$f(t) = \mathcal{L}^{-1}[F(s)] = \left[20 + \frac{88}{5}e^{-t} - \frac{2}{5}e^{-6t}\right]\cdot u(t)$$

Problem 7.1.11 Determine Inverse Laplace transform,

$$x(t) = \mathcal{L}^{-1}\left[\frac{s+5}{(s+1)^3}\right]$$

using partial fractions expansion for repeated linear factors in dominator.

Hint: If

$$H(s) = \frac{N(s)}{D(s)} = \frac{N(s)}{(s-p)^n} = \frac{A_n}{(s-p)^n} + \frac{A_{n-1}}{(s-p)^{n-1}} \cdots + \frac{A_1}{s-p}$$

the partial fractions expansion of the n fractions (repeated linear factors) are found by defining

$$\varphi_p = (s-p)^n \cdot H(s)$$
$$A_{n-k} = \frac{1}{k!}\frac{\mathrm{d}^k}{\mathrm{d}s^k}\varphi_p\Big|_{s=p}, \quad k = 0, 1, \ldots, n-1$$

where n is the order of polynomial and p is the pole, $p \leq 0$.

Solution

$$X(s) = \frac{s+5}{(s+1)^3} = \frac{A_3}{(s+1)^3} + \frac{A_2}{(s+1)^2} + \frac{A_1}{s+1}$$
$$\varphi_{-1} = (s+1)^3 \cdot X(s) = s+5$$
$$A_3 = \frac{1}{0!}\varphi_{-1}\big|_{s=-1} = s+5\big|_{s=-1} = 4$$
$$A_2 = \frac{1}{1!}\frac{\mathrm{d}}{\mathrm{d}s}\varphi_{-1}\big|_{s=-1} = \frac{\mathrm{d}}{\mathrm{d}s}(s+5)\big|_{s=-1} = 1$$
$$A_1 = \frac{1}{2!}\frac{\mathrm{d}^2}{\mathrm{d}s^2}\varphi_{-1}\big|_{s=-1} = \frac{\mathrm{d}^2}{\mathrm{d}s^2}(s+5)\big|_{s=-1} = 0$$
$$X(s) = \frac{4}{(s+1)^3} + \frac{1}{(s+1)^2}$$

Since

$$\mathcal{L}^{-1}\left[\frac{1}{(s+a)^m}\right] = \frac{t^{m-1}\cdot \mathrm{e}^{-at}}{(m-1)!}$$
$$x(t) = \mathcal{L}^{-1}[X(s)] = \left[4\cdot\frac{t^2-\mathrm{e}^{-t}}{(2)!} + \frac{t\cdot \mathrm{e}^{-t}}{(1)!}\right]\cdot u(t) = \left[2t^2\mathrm{e}^{-t} + t\mathrm{e}^{-t}\right]\cdot u(t)$$
$$x(t) = \mathrm{e}^{-t}\left(2t^2 + t\right)u(t)$$

where $u(t)$ is unit step operator.

Problem 7.1.12 Use MATLAB to determine the inverse LT of the following s-domain function (invlaplace1.m).

$$F(s) = \left[\frac{5(-1+j1)}{2[s-(-3+j)]} + \frac{5(-1-j1)}{2[s-(-3-j)]}\right]$$

Solution

$$f(t) = \frac{5}{2}\mathcal{L}^{-1}\left[\frac{-1+j1}{s-(-3+j)} + \frac{5(-1-j1)}{s-(-3-j)}\right] = \frac{5}{2}\left[-2(\cos t + \sin t\mathrm{e}^{-3t}\right]u(t)$$
$$= -5(\cos t + \sin t)\,\mathrm{e}^{-3t}u(t)$$

Problem 7.1.13

(a) $x(t) = \mathcal{L}^{-1}[X(s)] = \mathcal{L}^{-1}\left[\dfrac{s^2+4s+3}{s^4+10s^3+36s^2+56s+32}\right] = ?$

(b) Compute the partial fractions expansion of $X(s)$ and verify the result found in part (a) using a MATLAB script (pfe2.m).

Solution

$$X(s) = \frac{s^2+4s+3}{s^4+10s^3+36s^2+56s+32} = \frac{(s+1)(s+3)}{(s+2)^3(s+4)}$$

Partial fractions expansion;

$$X(s) = \frac{B}{s+4} + \frac{A_3}{(s+2)^3} + \frac{A_2}{(s+2)^2} + \frac{A_1}{s+2}$$

$$B = \left.\frac{(s+1)(s+3)}{(s+2)^3}\right|_{s=-4} = -\frac{3}{8}$$

$$A_3 = \left.\frac{(s+1)(s+3)}{(s+4)}\right|_{s=-2} = -\frac{1}{2}$$

$$A_2 = \frac{1}{1!}\frac{\mathrm{d}}{\mathrm{d}s}\varphi_{-2}\big|_{s=-2} = \left.\frac{\mathrm{d}}{\mathrm{d}s}\frac{(s^2+4s+3)}{(s+4)}\right|_{s=-2}$$

$$= \left.\frac{(2s+4)(s+4)-1\cdot(s^2+4s+3)}{(s+4)^2}\right|_{s=-2} = \left.\frac{(s^2+8s+13)}{(s+4)^2}\right|_{s=-2} = \frac{1}{4}$$

$$A_1 = \frac{1}{2!}\frac{\mathrm{d}^2}{\mathrm{d}s^2}\varphi_{-2}\big|_{s=-2} = \left.\frac{1}{2}\frac{\mathrm{d}}{\mathrm{d}s}\frac{(s^2+8s+13)}{(s+4)^2}\right|_{s=-2}$$

$$= \left.\frac{(2s+8)(s+4)^2 - 2\cdot(s^2+8s+13)(s+4)}{2(s+4)^2}\right|_{s=-2} = \frac{3}{8}$$

$$X(s) = \frac{-\frac{3}{8}}{s+4} + \frac{-\frac{1}{2}}{(s+2)^3} + \frac{\frac{1}{4}}{(s+2)^2} + \frac{\frac{3}{8}}{s+2}$$

Using the Laplace transform pair,

$$\mathcal{L}^{-1}\left[\frac{m!}{(s+a)^{m+1}}\right] = t^m\cdot \mathrm{e}^{-at},$$

$$x(t) = \mathcal{L}^{-1}[X(s)] = \left[-\frac{3}{8}\mathrm{e}^{-4t} - \frac{1}{2}\cdot\frac{1}{2!}t^2\cdot\mathrm{e}^{-2t} + \frac{1}{4}\cdot t\cdot\mathrm{e}^{-2t} + \frac{3}{8}\mathrm{e}^{-2t}\right]\cdot u(t)$$

$$x(t) = \left[-\frac{3}{8}\mathrm{e}^{-4t} - \frac{1}{4}t^2\cdot\mathrm{e}^{-2t} + \frac{1}{4}\cdot t\cdot\mathrm{e}^{-2t} + \frac{3}{8}\mathrm{e}^{-2t}\right]\cdot u(t)$$

MATLAB script for computing partial fraction expansion for $X(s)$ and then $x(t)$ is given below (pfe2.m).

```
%N is numerator, D is denominator polynomial
syms s t
N =[ 1  4  3];u1=[1 2]; u2=conv(u1,u1); u3=conv(u2,u1); v=[1 4];
D=conv(u3,v);
H = tf(N,D)%Transfer function (for command window check purpose)
%  partial fraction expansion
[r,p,k] = residue(N,D);
disp(sprintf('   residue       pole'))
[r  p]
%Alternative solution is to use directly ILT,(ilaplace)
G= (s^2 + 4*s + 3)/ ((s+2)^3*(s+4)); ilaplace(G)
```

Problem 7.1.14 Blood plasma concentration of an analyte following a bolus injection is described by

$$P(s) = \frac{(s+a)(s+b)}{(s+1)(s+a)(s+b) - cd(s+b) - ef(s+a)}$$

(a) What is the order of this system?
(b) Given that $a = 2,\ b = 1.5,\ c = d = 1,\ e = 5,\ f = 0.1,$ determine the system function in s-domain.
(c) Find out clearance of the analyte from plasma, i.e., time response of the system to analyte injection for $t \geq 0$ (bolus1.m).

Solution

(a) The system is a third order one, since degree of denominator polynomial is 3.
(b) Substituting given data into the given function yields

$$P(s) = \frac{s^2+3.5s+3}{s^3+4.5s^2+5s+0.5} = \frac{A}{s+2.7446} + \frac{B}{s+1.6446} + \frac{C}{s+0.1108}$$

$$A = 0.3199, \quad B = 0.05305, \quad C = 0.6497$$

$$p(t) = \mathcal{L}^{-1}[P(s)] = 0.3199\mathrm{e}^{-2.7446t} + 0.0305\mathrm{e}^{-0.1108t} \quad (t \geq 0)$$

(c) $$p(t) = \mathcal{L}^{-1}[P(s)] = 0.3199\mathrm{e}^{-2.7446t} + 0.0305\mathrm{e}^{-0.1108t} \quad (t \geq 0)$$

Problem 7.1.15

(a) State inital and final value properties of Laplace transforms.
(b) What do these properties mean in time and frequency domains?
(c) What restrictions apply on the initial and final value properties?
(d) Explain reasons why final value property can not be applied to following functions.

$$\text{(i) } f(t) = 3\cos(628t), \quad \text{(ii) } F(s) = \frac{(s+1)(s+2)}{s(s^2+9)},$$

$$\text{(iii) } F(s) = \frac{s^2+2s+1}{s(s^2-4)}, \quad \text{(iv) } F(s) = \frac{s^2+s+1}{s^2(s+16)}$$

Solution

(a) Initial value property: $\lim_{t\to 0^+} f(t) = \lim_{s\to\infty} sF(s)$
Final value property: $\lim_{t\to\infty} f(t) = \lim_{s\to 0} sF(s)$

(b) Initial value in the time domain is the same as the value of $sF(s)$ at infinity in the s-plane, while $f(\infty)$ is the same as the value of $sF(s)$ at the origin in the s-plane.

(c) The initial value property is valid when $F(s)$ is a proper rational function (i.e., a rational function numerator degree of which is less than the degree of its denominator), or alternatively, initial value property is applicable when $f(t)$ and its first derivative are transformable.

The final value property is valid when $F(s)$ has all poles in the left hand plane (LHP), but there can be a simple pole at $s = 0$.

(d)

(i) $$\lim_{t\to\infty} 3\cos(628t) = \lim_{s\to 0}\left[s \cdot \frac{3s}{s^2+(628)^2}\right] = 0$$

This result is inconsistent with the function properties in time domain, since the time function oscillates between ∓ 3. Note also that the poles of $sF(s)$ are on the j-axis at $s = \mp j628$. Thus, final value property can not be applied here.

(ii) Since $F(s)$ has j-axis poles at $s = \mp j3$, final value property does not apply.

(iii) A right hand pole exists at $s = 2$.

(iv) Multiple poles exist at the origin of the s-plane.

Problem 7.1.16 Find

(a) initial value,

(b) final value of $f(t)$ whose Laplace transform, $F(s)$ is given as

$$F(s) = \frac{s^2+5s+2}{s(s+1)(s+2)}$$

(c) Determine $f(t) = \mathcal{L}^{-1}[F(s)]$. Verify results found in part a.

Solution

Since given function in s-domain is a proper rational form and has two poles in left hand of s-plane and a single pole at the origin, both initial and final value properties hold.

(a) Initial value of $f(t)$ is calculated as follows,

$$f(0)=\lim_{s\to\infty} s\cdot F(s)=\lim_{s\to\infty}\left(s\cdot\frac{s^2+5s+2}{s(s+1)(s+2)}\right)=\lim_{s\to\infty}\left(\frac{s^2+5s+2}{(s+1)(s+2)}\right)$$
$$f(0)=\lim_{s\to\infty}\left(\frac{s^2+5s+2}{s^2+3s+2}\right)=\lim_{s\to\infty}\frac{s^2(1+5/s+2/s^2)}{s^2(1+3/s+2/s^2)}=1$$

(b) Final value of $f(t)$ is calculated as follows,

$$f(\infty)=\lim_{s\to 0} s\cdot F(s)=\lim_{s\to 0}\left(s\cdot\frac{s^2+5s+2}{s(s+1)(s+2)}\right)=\lim_{s\to 0}\left(\frac{s^2+5s+2}{(s+1)(s+2)}\right)$$
$$f(\infty)=\lim_{s\to 0}\left(\frac{s^2+5s+2}{s^2+3s+2}\right)=1$$

(c) Inverse Laplace transform of time function yields $f(t)$,

$$F(s)=\frac{s^2+5s+2}{s(s+1)(s+2)}=\frac{A}{s}+\frac{B}{s+1}+\frac{C}{s+2}$$
$$\text{for } s=0,\quad A=\frac{2}{1\times 2}=1, \qquad \text{for } s=-1,\quad B=\frac{1-5+2}{(-1)\times 1}=2$$
$$\text{for } s=-2,\quad C=\frac{4-10+2}{(-2)\times(-1)}=-2$$
$$F(s)=\frac{1}{s}+\frac{2}{s+1}-\frac{2}{s+2}$$
$$f(t)=\mathcal{L}^{-1}F(s)=1+2\mathrm{e}^{-t}-2\mathrm{e}^{-2t}, \qquad t\geq 0$$

Note that $f(0)=f(\infty)=1$, as determined before.

Problem 7.1.17 Find the initial and final values of $f(t)$ if

$$F(s)=\frac{4s^2+7s+1}{s\cdot(s+1)^2}$$

Solution

Since $F(s)$ is a proper rational function and has two poles in left hand s-plane and only one pole at the origin of the s-plane, initial and final value properties hold.

Initial Value:

$$\lim_{s\to\infty} s\cdot F(s)=\lim_{s\to\infty}\frac{4s^3+7s^2+s}{s\cdot(s^2+2s+1)}=\lim_{s\to\infty}\frac{4s^3+7s^2+s}{(s^3+2s^2+s)}$$
$$=\lim_{s\to\infty}\left(\frac{4+7/s+1/s^2}{1+2/s+1/s^2}\right)=4$$

Final Value:

$$\lim_{s\to 0} s\cdot F(s) = \frac{4s^2+7s^2+1}{(s+1)^2} = \frac{0+0+1}{(0+1)^2} = \frac{1}{1} = 1$$

Problem 7.1.18 Determine initial and final values of $h(t) = \mathcal{L}^{-1}[H(s)]$ if,

$$H(s) = \frac{s^2+5s+14}{s^3+8s^2+12s}$$

Solution

$$H(s) = \frac{s^2+5s+14}{s^3+8s^2+12s} = \frac{s^2+5s+14}{s(s^2+8s+12)}$$

Since $H(s)$ is a proper rational function and has two poles in left hand s-plane and one pole at origin, initial and final value properties hold. Inital value:

$$\lim_{s\to\infty} sH(s) = \lim_{s\to\infty}\frac{s(s^2+5s+14)}{s(s^2+8s+12)} = \lim_{s\to\infty}\frac{1+\frac{5}{s}+\frac{14}{s^2}}{1+\frac{8}{s}+\frac{12}{s^2}} = \frac{1}{1} = 1$$

Final value:

$$\lim_{s\to 0} sH(s) = \lim_{s\to 0}\frac{s^2+5s+14}{s^2+8s+12} = \frac{0+0+14}{0+0+12} = \frac{7}{6}$$

Problem 7.1.19 Compute the initial and final values of $F(s)$, if

$$F(s) = \frac{20}{(s+3)(s^2+8s+25)}$$

Solution

Since $F(s)$ is a proper rational function and has all poles in left hand s-plane, initial and final value properties hold. Inital value:

(a) $\lim_{s\to\infty} s\cdot F(s) = \lim_{s\to\infty}\frac{20s}{(s+3)(s^2+8s+25)} = \lim_{s\to\infty}\frac{20s}{s^3+8s^2+25s+3s^2+24s+75}$

$\lim_{s\to\infty}\frac{20s}{s^3+11s^2+49s+75} = \frac{\infty}{\infty}(\text{L'Hospital}) = \lim_{s\to\infty}\frac{20}{3s^2+22s+49} = \frac{20}{\infty} = 0$

(b) $\lim_{s\to 0} s\cdot F(s) = \lim_{s\to 0}\frac{20s}{(s+3)(s^2+8s+25)} = \frac{0}{75} = 0$

Problem 7.1.20

$$I(s) = \frac{6s^4 + 5s^3 + 4s^2 + 3s + 2}{s^5 + 4s^4 + 3s^2 + 2s + 1}$$

Find,

(a) Initial value,
(b) Final value of $i(t) = \mathcal{L}^{-1}[I(s)]$ (root_poly.m).

Solution

(a) Since $I(s)$ is a proper rational function, initial value property holds,

$$i(0) = \lim_{s \to \infty} sI(s) = \frac{6s^5 + 5s^4 + 4s^3 + 3s^2 + 2s}{s^5 + 4s^4 + 3s^2 + 2s + 1}$$

Divide numerator and denominator by s^5,

$$\lim_{s \to \infty} sI(s) = \frac{6 + \frac{5}{s} + \frac{4}{s^2} + \frac{3}{s^3} + \frac{2}{s^4}}{1 + \frac{4}{s} + \frac{3}{s^3} + \frac{2}{s^4} + \frac{1}{s^5}} = i(0) = 6$$

(b) One must check if $I(s)$ has all of its poles in left hand s-plane. This is not so easy by manual calculation, therefore computer (e.g., MATLAB) methods can be applied. A Simple solution is to find the roots of the denominator polynomial, and check if all the real parts of these roots have negative values (one pole at the origin is also acceptable). Following is a short MATLAB script for this method:

```
c=[1 4 0 3 2 1]% polynomial coefficients
roots(c)
```

The five roots are computed as (-4.1496), $(0.4171 \pm 0.8717j)$, $(-0.3423 \pm 0.3754j)$.

It is seen that two roots out of five have non-negative real parts (=0.4171). Therefore, final value property is not applicable for $I(s)$.

A blind attempt to find out the final value $i(\infty)$ would be,

$$i(\infty) = \lim_{s \to 0} sI(s) = \frac{6s^5 + 5s^4 + 4s^3 + 3s^2 + 2s}{s^5 + 4s^4 + 3s^2 + 2s + 1} = \frac{0}{1} = 0$$

Problem 7.1.21

$$I(s) = \frac{2s^4 + 3s^3 + 4s^2 + 5s + 6}{s^5 + 2s^4 + 3s^2 + 4s + 5}$$

Find,

(a) Initial value,
(b) Final value of $i(t) = \mathcal{L}^{-1}[I(s)]$ (root_poly.m).

Solution

(a) Since given function is in proper rational form, initial value property holds,

$$i(0) = \lim_{s\to\infty} sI(s) = \lim_{s\to\infty} \frac{2s^5 + 3s^4 + 4s^3 + 5s^2 + 6s}{s^5 + 2s^4 + 3s^2 + 4s + 5}$$

Divide numerator and denominator by s^5,

$$\lim_{s\to\infty} \frac{2 + \frac{3}{s} + \frac{4}{s^2} + \frac{5}{s^3} + \frac{6}{s^4}}{1 + \frac{2}{s} + \frac{3}{s^3} + \frac{4}{s^4} + \frac{5}{s^5}} = \frac{2}{1} = 2$$

(b) The five roots of denominator polynomial are computed as (−2.3866), (0.8314 ± 1.1415*j*) and (−0.6381 ± 0.8021*j*). Two roots out of five have non-negative real parts (=0.8314). Therefore, final value property is not applicable for $I(s)$.

A blind attempt to find out the value of $i(\infty)$ would be,

$$i(\infty) = \lim_{s\to 0} \to sI(s) = \lim_{s\to 0} \frac{2s^5 + 3s^4 + 4s^3 + 5s^2 + 6s}{s^5 + 2s^4 + 3s^2 + 4s + 5} = \frac{0}{5} = 0$$

Problem 7.1.22 Let $I(s) \leftrightarrow i(t)$, (that is, they form a Laplace transform pair), and

$$I(s) = \frac{3s^3 + 4s^2 + 3s + 1}{As^4 + s^3 + 2s^2 + s + 5}$$

Find the value of A if $i(0) = 0.3$.

Solution

$$i(0) = \lim_{s\to\infty} s \cdot I(s) = \lim_{s\to\infty} \frac{3s^3 + 4s^2 + 3s + 1}{As^4 + s^3 + 2s^2 + s + 5}$$

Divide both numerator and denominator by the highest power of s,

$$i(0) = \lim_{s\to\infty} \frac{3 + \frac{4}{s} + \frac{3}{s^2} + \frac{1}{s^3}}{A + \frac{1}{s} + \frac{2}{s^2} + \frac{1}{s^3} + \frac{5}{s^4}} = \frac{3}{A} = 0.3 \quad \to \quad A = \frac{3}{0.3} = 10$$

Problem 7.1.23

$$f(t) = \mathcal{L}^{-1}\left[\frac{3}{(s+1)(s+2)^2}\right] = ?$$

Solution

$F(s) = \dfrac{3}{(s+1)(s+2)^2}$ and $\dfrac{3}{(s+1)(s+2)^2} = \dfrac{A}{(s+1)} + \dfrac{B}{(s+2)^2} + \dfrac{C}{(s+2)}$

$3 = A(s+2)^2 + B(s+1) + C(s+2)(s+1)$ and for $s = -2$: $-B = 3$ and $B = -3$

For $s = -1$: $A = 3$ and with using A and B for $s = 0$: $9 + 2C = 3$ and $C = -3$

$$\mathcal{L}^{-1}[F(s)] = \mathcal{L}^{-1}\left[\frac{3}{s+1}\right] + \mathcal{L}^{-1}\left[\frac{-3}{(s+2)^2}\right] + \mathcal{L}^{-1}\left[\frac{-3}{s+2}\right]$$

$\mathcal{L}^{-1}\left[\dfrac{3}{s+1}\right] = 3\cdot \mathrm{e}^{-t}$ and $\mathcal{L}^{-1}\left[\dfrac{-3}{(s+2)^2}\right] = -3t\mathrm{e}^{-2t}$ and $\mathcal{L}^{-1}\left[\dfrac{-3}{s+2}\right] = -3\mathrm{e}^{-2t}$

$\mathcal{L}^{-1}[F(s)] = f(t) = 3\cdot \mathrm{e}^{-t} - 3t\mathrm{e}^{-2t} - 3\mathrm{e}^{-2t}$

Problem 7.1.24

(a) $v_i(t) = (1-\mathrm{e}^{-t})\cdot u(t), \quad i(t) = \mathrm{e}^{-t}\cdot u(t), \quad Z(s) = \dfrac{V(s)}{I(s)} = ?$

(b) If $i(t) = \delta(t)$, $t \geq 0$, $V_i(t) = ?$

Solution

(a) $Z(s) = \dfrac{V_i(s)}{I(s)} = \dfrac{\dfrac{1}{s} - \dfrac{1}{s+1}}{\dfrac{1}{s+1}} = \dfrac{1}{s}$

(b) $I(s) = \mathcal{L}[i(t)] = \mathcal{L}[\delta(t)] = 1$, $V_i(s) = I(s)\cdot Z(s) = \dfrac{1}{s}$

$v_i(t) = \mathcal{L}^{-1}[V_i(s)] = \mathcal{L}^{-1}\left[\dfrac{1}{s}\right] = 1\cdot u(t)V$

7.2 Circuits in s-Domain

Problem 7.2.1 In the circuit shown in Fig. 7.4, $v_C\ (0^-) = -2$ V, $C = 1/2$ F, $R = 3\ \Omega$.

(a) Find $v_C(t)$,
(b) Find t when $v_C(t) = 0$ V.

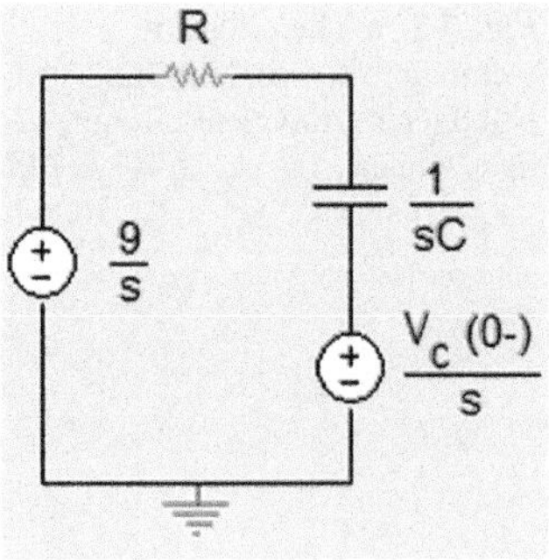

Fig. 7.4 The circuit of Problem 7.2.1

Solution

(a) $$\frac{9}{s}+\frac{2}{s}=\left(3+\frac{2}{s}\right)\cdot I(s) \quad\rightarrow\quad \frac{11}{s}=\left(\frac{3s+2}{s}\right)\cdot I(s) \quad\rightarrow\quad I(s)=\frac{11}{3s+2}$$

$$V_c(s)=I(s)\cdot\left(\frac{2}{s}\right)-\frac{2}{s}=\left(\frac{11}{3s+2}\cdot\frac{2}{s}\right)-\frac{2}{s}=\left(\frac{22}{s(3s+2)}\right)-\frac{2}{s}=\frac{-6s+18}{s(3s+2)}$$

$$V_c(s)=\frac{-6(s-3)}{3s\left(s+\frac{2}{3}\right)}=\frac{-2(s-3)}{s\left(s+\frac{2}{3}\right)}=\frac{A}{s}+\frac{B}{\left(s+\frac{2}{3}\right)}$$

$$\text{For } s=0,\ A=\frac{-2(s-3)}{\left(s+\frac{2}{3}\right)}=\frac{6}{\frac{2}{3}}=9;$$

$$\text{For } s=-\frac{2}{3},\quad B=\frac{-2(s-3)}{s}=-11$$

$$V_c(s)=\frac{9}{s}-\frac{11}{\left(s+\frac{2}{3}\right)}$$

$$v_c(t)=\left(9-11e^{-\frac{2t}{3}}\right)u(t)V$$

(b) $t=0;\quad v_c(0)=-2\,\text{V};\quad t=\infty;\quad v_c(\infty)=9\,\text{V}$

Problem 7.2.2 In the circuit shown in Fig. 7.5a,

$$v_i(t)=3e^{-8t}u(t), i(0^-)=1A;\ L=2\,\text{H},\ R=1\,\Omega, v_L(t)=?$$

Solution

Inductor equivalent circuit in s-domain is shown in Fig. 7.5b,

$3/(s+8)+2=(1+2s)\cdot I(s)$

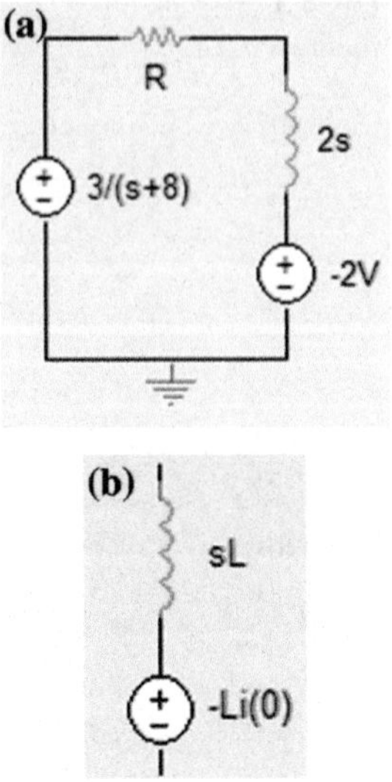

Fig. 7.5 **a** The circuit in s-domain, for Problem 7.2.2 **b** Inductor equivalent circuit in s-domain

$$I(s) = \left(\frac{2s+19}{(2s+1)\cdot(s+8)}\right) = \left(\frac{s+9.5}{(s+8)\cdot(s+0.5)}\right)$$

$$V_L(s) = 2s \cdot I(s) - 2 = 2s \cdot \left(\frac{s+9.5}{(s+8)\cdot(s+0.5)}\right) - 2 = 2s^2 + 19s - 2(s^2 + 0.5s + 8s + 4)$$

$$V_L(s) = \frac{2s-8}{(s+8)\cdot(s+0.5)} = \frac{A}{s+8} + \frac{B}{s+0.5}$$

$$A = \left.\frac{2s-8}{(s+0.5)}\right|_{s=-8} = \frac{-16-8}{-8+0.5} = 3.2; \quad B = \left.\frac{2s-8}{(s+8)}\right|_{s=-0.5} = \frac{-1-8}{-0.5+8} = -1.2$$

$$V_L(s) = \frac{3.2}{s+8} - \frac{1.2}{(s+0.5)}$$

$$v_L(t) = \mathcal{L}^{-1}[V_L(s)] = \left[3.2\mathrm{e}^{-8t} - 1.2\mathrm{e}^{-0.5t}\right]u(t)\ \mathrm{V}$$

Problem 7.2.3 If $v_i(t) = 4 \cdot \mathrm{e}^{-6t} \cdot u(t)$, $i(0) = -1/2$ A, $R = 2\ \Omega$, $L = 2$ H, in the circuit shown in Fig. 7.6a, find, $I(s)$, $v_L(t)$, $V_R(s)$.

Solution

Using equivalent circuit in s-domain, Fig. 7.6b,

$$v(0^-) = -L \cdot i(0^-) = -2\left(\frac{1}{2}\right) = -1\,\mathrm{V}$$

$$\text{(a)}\ I(s) = \frac{V(s)}{R+sL} = \frac{\frac{4}{(s+6)}+1}{(s+2)} = \frac{(s+10)}{(s+2)(s+6)}$$

$$\text{(b)}\ V_L(s) = 1s \cdot \frac{(s+10)}{(s+2)(s+6)} - 1 = \frac{2s-12}{(s+2)(s+6)} = \frac{6}{s+6} - \frac{4}{s+2}$$

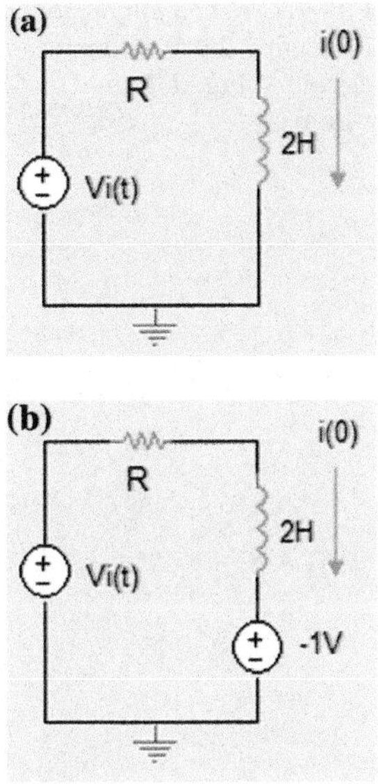

Fig. 7.6 **a** The circuit for Problem 7.2.3 **b** Equivalent circuit in s-domain

$$v_L(t) = \mathcal{L}^{-1}[V_L(s)] = \mathcal{L}^{-1}\left[\frac{6}{s+6} - \frac{4}{s+2}\right] = \left[6 \cdot e^{-6t} - 4 \cdot e^{-2t}\right] \cdot u(t)$$

(c)

$$I_R(s) = I_L(s) = \frac{V(s)}{sL} = \frac{s+10}{s(s+6)} \quad \rightarrow \quad V_R(s)$$

$$= I_R(s) \cdot R = \frac{s+10}{s(s+6)} \cdot 2 = \frac{2s+20}{s(s+6)}$$

Problem 7.2.4 In the RC circuit shown in Fig. 7.7a, if $U(t) = 4 \cdot u(t)$, $R = 6\,\Omega, C = 0.25\,\mathrm{F}$,, Find the voltage at the junction of the resistor and capacitor with respect to ground. $v_c(0) = -1$ V.

Solution

Re-drawing the circuit in s-domain (Fig. 7.7b), and calculating the current first,

$$I(s) = \frac{U(s)}{R + \frac{1}{Cs}} = \frac{\frac{4}{s} + \frac{1}{s}}{6 + \frac{4}{s}} = \frac{5}{6s+4}$$

The voltage at the junction is calculated as

$$V_x(s) = \left(I(s) \cdot \frac{4}{s}\right) - \frac{1}{s} = \left(\frac{5}{6s+4} \cdot \frac{4}{s}\right) - \frac{1}{s} = \frac{16 - 6s}{s(6s+4)}$$

Fig. 7.7 **a** The circuit for Problem 7.2.4 **b** Redrawn circuit of Fig. 7.7a in s-domain

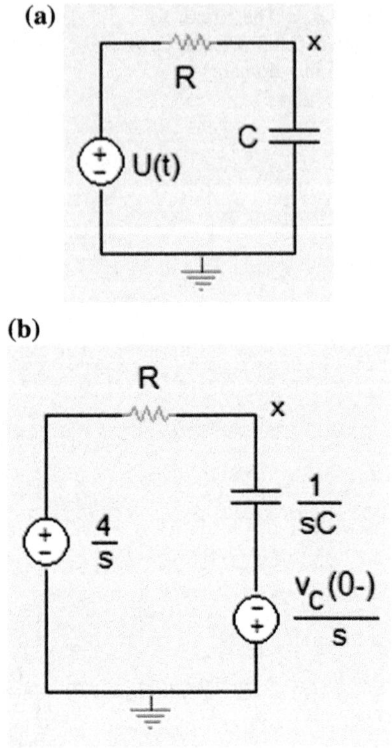

Inverse Laplace transform using partial fraction expansion,

$$v_x(t) = \mathcal{L}^{-1}[V_C(s)] = \mathcal{L}^{-1}\left[\frac{16 - 6s}{s(6s + 4)}\right] = \mathcal{L}^{-1}\left(\frac{4}{s} - \frac{5}{s + \frac{2}{3}}\right)$$

$$v_x(t) = \left[4 - 5\mathrm{e}^{(-2/3)t}\right] \cdot u(t)$$

Problem 7.2.5 Determine Thévenin equivalent to the left of *x*-*y* terminals of the circuit shown in Fig. 7.8, assuming no initially stored energy in the circuit.

Solution

In Laplace domain,

Fig. 7.8 The circuit for Problem 7.2.5

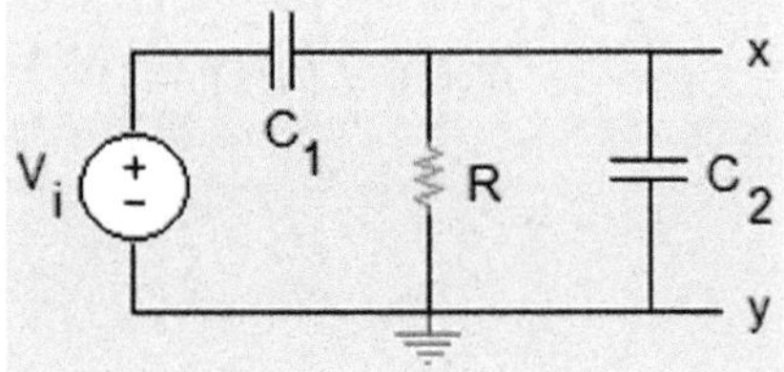

$$Z_{\text{Th}}(s) = \frac{V_{oc}(s)}{I_{sc}(s)}$$

Drop the s operator for convenience, and use voltage division rule,

$$V_{\text{Th}} = V_{oc} = V_i \frac{Z_1}{Z_1 + \frac{1}{sC_2}}, \quad Z_1 = \frac{1}{sC_1} \Big\| R = \frac{R}{sRC_1 + 1}$$

$$V_{\text{Th}} = V_{oc} = V_i \frac{\frac{R}{sRC_1+1}}{\frac{R}{sRC_1+1} + \frac{1}{sC_2}} = \frac{sRC_2 V_i}{sR(C_1+C_2)+1} = \left(\frac{C_2}{C_1+C_2}\right) \frac{s}{s + \frac{1}{R(C_1+C_2)}} V_i \tag{7.1}$$

In order to determine Thévenin equivalent impedance, short circuit current at the output of the circuit is needed,

$$I_{sc} = \frac{V_i}{\frac{1}{sC_2}} = sC_2 V_i$$

$$Z_{\text{Th}} = \frac{V_{oc}}{I_{sc}} = \frac{V_i \left[\frac{sRC_2}{sR(C_1+C_2)+1}\right]}{V_i sC_2} = \frac{sRC_2}{sC_2[sR(C_1+C_2)+1]}$$

$$Z_{\text{Th}} = \frac{R}{sR(C_1+C_2)+1} = \left(\frac{1}{C_1+C_2}\right) \frac{1}{s + \frac{1}{R(C_1+C_2)}} \tag{7.2}$$

Equations (7.1) and (7.2) constitute the Thévenin equivalent circuit in s-domain.

Problem 7.2.6

(a) If there is no initially stored energy in the circuit of Fig. 7.9, find the voltage expression at the output of the circuit, $v_o(t) = v_{C2}(t)$, after the switch is closed, if $v_i(t) = U, t \geq 0$
(b) If $C_2 = 10C_1 = 10$ F, $R_1 = 2R_2 = 2\ \Omega$, $U = 100$ V, $v_o(t) = ?\ t \geq 0$

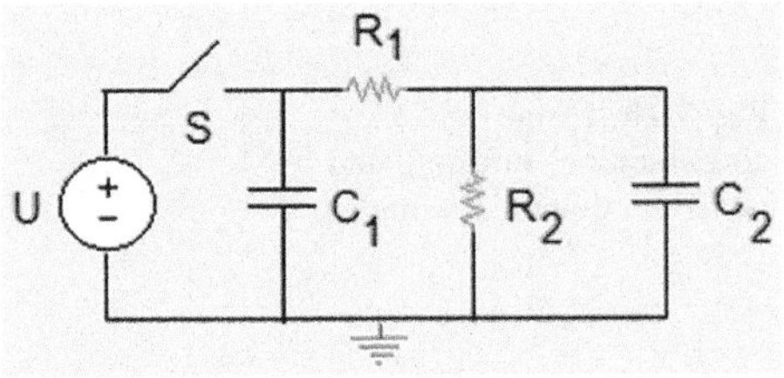

Fig. 7.9 The circuit for Problem 7.2.6

Solution

(a) Node equation at the output terminal (drop s operator for convenience),

$$\frac{V_i - V_0}{R_1} - \frac{V_0}{R_2} - sC_2V_0 = \frac{V_i}{R_1} - \frac{V_0}{R_1} - \frac{V_0}{R_2} - sC_2V_0 = 0 \quad \rightarrow \quad \frac{V_i}{R_1} = V_0\left(\frac{1}{R_1} + \frac{1}{R_2} + sC_2\right)$$

$$V_0 = \frac{V_i}{R_1\left(\frac{1}{R_1} + \frac{1}{R_2} + sC_2\right)} = \frac{V_i}{\left(1 + \frac{R_1}{R_2} + sR_1C_2\right)} V_0 = \frac{1}{R_1C_2}\frac{V_i}{\left(\frac{1+\frac{R_1}{R_2}}{R_1C_2} + s\right)}, \quad V_i = \frac{U}{s},$$

$$V_0 = \frac{U}{R_1C_2s\left(\frac{1+\frac{R_1}{R_2}}{R_1C_2} + s\right)}$$

By partial fractions expansion,

$$v_0(t) = \mathcal{L}^{-1}[V_0(s)] = \mathcal{L}^{-1}\left[\frac{A}{s} + \frac{B}{s+a}\right], \quad a = \frac{1+\frac{R_1}{R_2}}{R_1C_2}$$

$$A = \frac{U}{\left(\frac{1+\frac{R_1}{R_2}}{R_1C_2}\right)R_1C_2} = \frac{U}{1+\frac{R_1}{R_2}}, \quad B = \frac{U}{R_1C_2\left(-\frac{1+\frac{R_1}{R_2}}{R_1C_2}\right)} = -\frac{U}{1+\frac{R_1}{R_2}}$$

$$v_0(t) = \mathcal{L}^{-1}\left[\frac{U}{1+\frac{R_1}{R_2}}\left(\frac{1}{s} - \frac{1}{s+a}\right)\right] = A(1 - e^{-at})u(t), \quad A = \frac{U}{1+\frac{R_1}{R_2}}, \quad a = \frac{1+\frac{R_1}{R_2}}{R_1C_2}$$

(b) Inserting numerical values of components into the last voltage expression of part a yields

$$v_0(t) = 33.334\left(1 - e^{-1.667t}\right)u(t)$$

Problem 7.2.7 The circuit in Fig. 7.10 is used as an electrical analog for the studies of drug concentration, infusion and perfusion dynamics in the organs and circulatory system. Here; v_i, v_0 can be drug dose at a specific site of the body and the

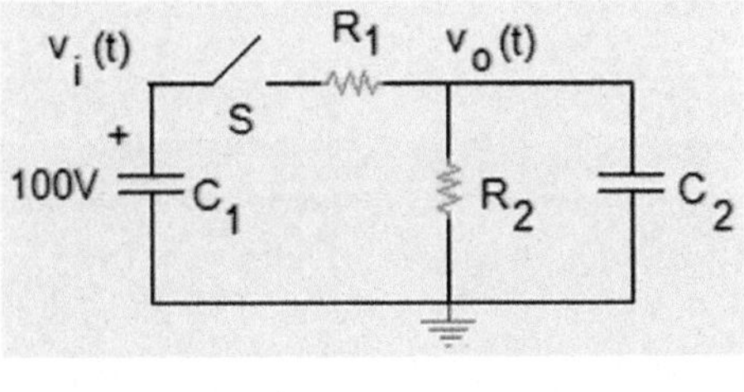

Fig. 7.10 Drug concentration, infusion, and perfusion dynamics simulator

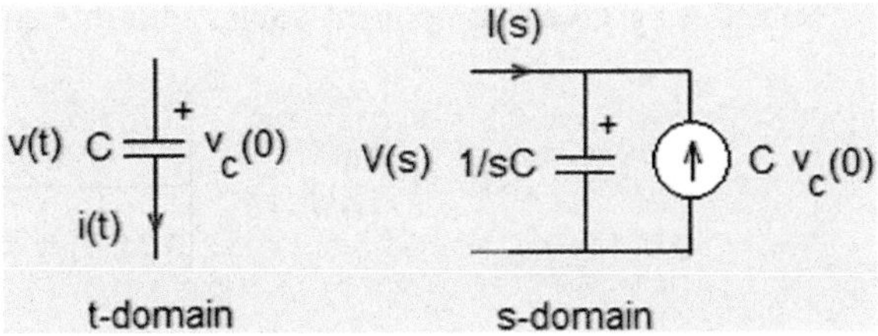

Fig. 7.11 s-domain equivalent circuit of capacitor

drug concentration in the blood, respectively. C_1, C_2 represent drug volume and blood volume in a region of interest, respectively, and R_1 is the resistance to drug passage into blood stream, while R_2 is the excretion resistance.

If $C_1 = 1$ F, $C_2 = 10$ F, $R_1 = 2\ \Omega$, $R_2 = 1\ \Omega$, $v_{C1}(0) = 100$ V, determine the characteristic equation in s-domain, its poles and $v_0(t), t \geq 0$ (invlaplace.m, root_-poly1.m, plot_t_function.m, RLC1.cir).

Solution

After the switch closes for $t \geq 0$, the capacitor begins to discharge.

Using s-domain equivalent circuit of capacitor with a current source (see, Fig. 7.11), application of analysis by inspection procedure is possible in this circuit, because it does not contain any dependent sources. Therefore, one can write a nodal equation at the output node of the circuit in matrix form as

$$\begin{bmatrix} sC_1 + \dfrac{1}{R_1} & -\dfrac{1}{R_1} \\ -\dfrac{1}{R_1} & \dfrac{1}{R_1} + \dfrac{1}{R_2} + sC_2 \end{bmatrix} \begin{bmatrix} V_1 \\ V_2 \end{bmatrix} = \begin{bmatrix} C_1 \cdot V_{i0} \\ 0 \end{bmatrix}$$

where $V_1 = V_i(s)$, $V_2 = V_o(s)$, $V_{i0} = v_{C1}(0) = v_i(0) = 100$ V.

Using Cramer's rule,

$$V_2(s) = \frac{\Delta_2}{\Delta},$$

$$\Delta = \begin{vmatrix} \dfrac{sC_1R_1 + 1}{R_1} & -\dfrac{1}{R_1} \\ -\dfrac{1}{R_1} & \dfrac{R_1 + R_2 + sC_2R_1R_2}{R_1R_2} \end{vmatrix} = \frac{s^2C_1C_2R_1R_2 + s(C_1R_1 + C_1R_2 + C_2R_2) + 1}{R_1R_2}$$

$$\Lambda_2 = \begin{vmatrix} sC_1 + \dfrac{1}{R_1} & C_1 \cdot V_{i0} \\ -\dfrac{1}{R_1} & 0 \end{vmatrix} = \frac{C_1 \cdot V_{i0}}{R_1}$$

$$V_2(s) = \frac{\Delta_2}{\Delta} = \frac{C_1R_2V_{i0}}{s^2C_1C_2R_1R_2 + s(C_1R_1 + C_1R_2 + C_2R_2) + 1}$$

$$V_2(s) = \left(\frac{V_{i0}}{C_2R_1}\right) \frac{1}{s^2 + s\left(\dfrac{1}{C_2R_2} + \dfrac{1}{C_2R_1} + \dfrac{1}{C_1R_1}\right) + \dfrac{1}{C_1C_2R_1R_2}}$$

$$v_2(t) = \mathcal{L}^{-1}[V_2(s)] = \left(\frac{V_{i0}}{C_2R_1}\right) \cdot \mathcal{L}^{-1}\left[\frac{1}{s^2 + s\left(\dfrac{1}{C_2R_2} + \dfrac{1}{C_2R_1} + \dfrac{1}{C_1R_1}\right) + \dfrac{1}{C_1C_2R_1R_2}}\right]$$

Substituting given component values into this equation,

$$v_O(t) = v_2(t) = \left(\frac{100}{20}\right) \cdot \mathcal{L}^{-1}\left[\frac{1}{s^2 + s\left(\frac{1}{2} + \frac{1}{10} + \frac{1}{20}\right) + \frac{1}{20}}\right]$$

$$= 5 \cdot \mathcal{L}^{-1}\left[\frac{1}{s^2 + 0.65s + 0.05}\right]$$

The roots of denominator polynomial (poles) are $s_1 = -0.5608$, $s_2 = -0.0892$.

$$v_o(t) = \mathcal{L}^{-1}\left[\frac{5}{(s + 0.5608)(s + 0.0892)}\right] = \mathcal{L}^{-1}\left[\frac{A}{(s + 0.5608)} + \frac{B}{(s + 0.0892)}\right]$$

Using partial fractions expansion, $A = -B = -10.6$
Taking inverse LT yields,

$$v_O(t) = 10.6(\mathrm{e}^{-0.0892t} - \mathrm{e}^{-0.5608t})u(t)$$

Figure 7.12 displays the resulting waveform. SPICE netlist (RLC1.cir) of this solution is given below.

```
drug concentration simulator
R1 1 2 2
R2 2 0 1
C1 1 0 1 IC=100
C2 2 0 10
*.tran .01 30 0 .1 uic
```

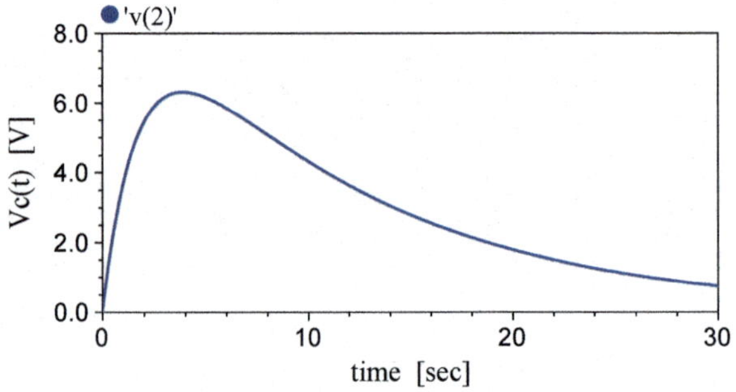

Fig. 7.12 Simulated drug concentration as a function of time

7.3 Transfer Functions, Poles and Zeros, Frequency Response

Problem 7.3.1 In a series RLC circuit with no initial conditions (i.e., all voltages and currents have zero values), $V_i(t) = \mathrm{e}^{-t}u(t)$, $R = 1\,\Omega$, $L = 1\,\mathrm{H}$, $C = 1\,\mathrm{F}$. $I(s) = ?$

Solution

$$\frac{V(s)}{I(s)} = Z(s) = R + sL + \frac{1}{sC} = 1 + s + \frac{1}{s} = \frac{s^2+s+1}{s}$$

$$I(s) = \frac{V(s)}{Z(s)} = \frac{\left(\frac{1}{s+1}\right)}{\left(\frac{s^2+s+1}{s}\right)} = \frac{s}{(s+1)(s^2+s+1)}$$

Problem 7.3.2 In a circuit with no (non-zero) initial conditions, Voltage Transfer Function is;

$$H_v(s) = \frac{1}{s^2+3s+2}$$
$$\text{For } v_i(t) = \delta(t),\ v_O(t) = ?$$

Solution

$$\frac{V_O(s)}{V_i(s)} = H(s) = \frac{1}{s^2+3s+2} = \frac{1}{(s+1)(s+2)}$$
$$V_O(s) = V_i(s)\frac{1}{(s+1)(s+2)} = 1 \times \frac{1}{(s+1)(s+2)} = \frac{1}{(s+1)(s+2)}$$
$$v_O(t) = \mathcal{L}^{-1}[H(s)] = \frac{A}{(s+1)} + \frac{B}{(s+2)}$$

Using partial fraction expansions,

$$A = \left.\frac{1}{(s+2)}\right|_{s=-1} = 1, B = \left.\frac{1}{(s+1)}\right|_{s=-2} = -1,$$
$$V_O(t) = \left[\mathrm{e}^{-t} - \mathrm{e}^{-2t}\right]u(t)$$

Problem 7.3.3 In a series RLC circuit shown in Fig. 7.13 (no initial conditions), if $i(t) = \mathcal{L}^{-1}[I(s)] = 1$ A, $v_{ab}(t) = ?$ $(t \geq 0)$

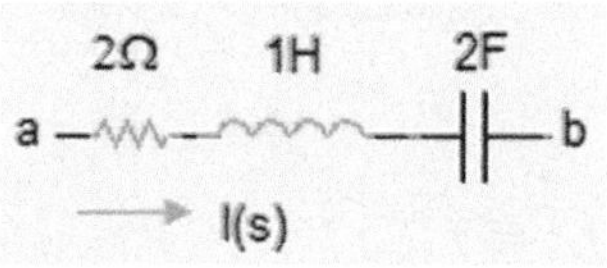

Fig. 7.13 Series RLC circuit

Solution

$$I(s) = \mathcal{L}[i(t)] = \mathcal{L}[1] = \frac{1}{s}$$

$$Z(s) = 2 + s + \frac{1}{2s} = \frac{2s^2 + 4s + 1}{2s}$$

$$V(s) = I(s) \cdot Z(s) = \frac{1}{s} \cdot \frac{2s^2 + 4s + 1}{2s} = \frac{2s^2 + 4s + 1}{2s^2}$$

$$V(s) = 1 + \frac{4s}{2s^2} + \frac{1}{2s^2} = 1 + \frac{4}{2s} + \frac{1}{2s^2} = 1 + \frac{2}{s} + \frac{1}{2s^2}$$

$$v(t) = \mathcal{L}^{-}[V(s)] = \delta(t) + 2u(t) + \frac{1}{2}t$$

Problem 7.3.4 In the circuit shown in Fig. 7.14,

(a) $Z_{ab}(s) = \dfrac{V(s)}{I(s)} = ?\quad (R_1 = 1\,\Omega, C_1 = 10\,\text{F})$.
(b) *If* $i(t) = 1\,A, v(t) = ?\quad t \geq 0,\ v(0) = 0\,V$.

Solution

(a) $$Z(s) = \frac{V(s)}{I(s)} = \frac{R_1}{sR_1C_1 + 1} = \frac{\frac{1}{C_1}}{s + \frac{1}{R_1C_1}} = \frac{0.1}{s + 0.1}$$

(b) If the current through the circuit is 1 A, $t \geq 0$,

$$V(s) = \mathcal{L}[v(t)] = Z(s)I(s) = \frac{0.1}{s + 0.1} \cdot \frac{1}{s} = \frac{A}{s} + \frac{B}{s + 0.1} = \frac{1}{s} - \frac{1}{s + 0.1}$$

$$v(t) = \mathcal{L}^{-1}[V(s)] = \left(1 - e^{-0.1t}\right)u(t)$$

Problem 7.3.5 In the circuit shown in Fig. 7.15, all initial conditions are zero.

(a) Determine *I*(*s*) as a function of *R, L* and *s*.
(b) If $R = 1/2\ \Omega$ and $L = 1$ H, determine the poles and zeros of *I*(*s*).
(c) Use final value property of LT and find the value of $i(\infty)$ where $i(t) \leftrightarrow I(s)$.

Fig. 7.14 The circuit for Problem 7.3.4

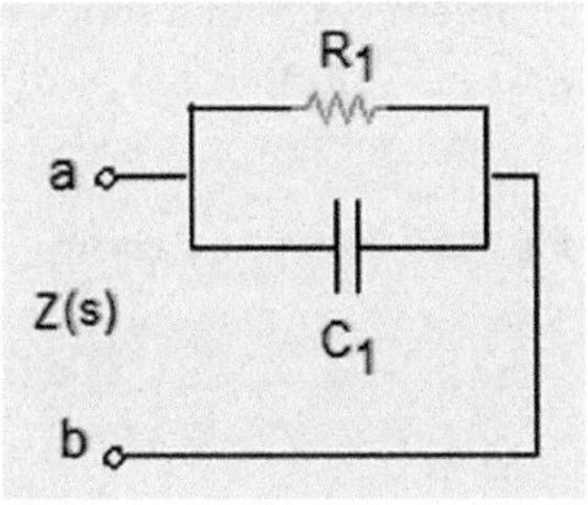

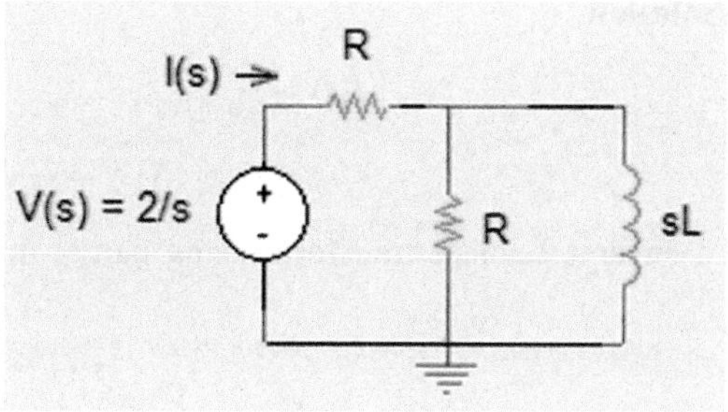

Fig. 7.15 The circuit for Problem 7.3.5

Solution

(a) $Z(s) = V(s)/I(s)$

$$I(s) = V(s)/Z(s) = \frac{\frac{2}{s}}{R + (sL||R)} = \frac{\frac{2}{s}}{R + \frac{sLR}{R+sL}} = \frac{\frac{2}{s}}{\frac{2sLR+R^2}{R+sL}} = \frac{2(sL+R)}{s(2RLs+R^2)}$$

$$= \frac{2L\left(s+\frac{R}{L}\right)}{2LRs\left(s+\frac{R}{2L}\right)} = \frac{1}{R}\cdot\frac{\left(s+\frac{R}{L}\right)}{s\left(s+\frac{R}{2L}\right)}$$

(b) $I(s) = \dfrac{2\left(s+\frac{1}{2}\right)}{s\left(s+\frac{1}{4}\right)}$, Poles $= \left(0, -\frac{1}{4}\right)$, zero $= -\frac{1}{2}$

(c) $i(\infty) = \lim\limits_{s\to 0}[s \cdot I(s)] = \left.\dfrac{2\left(s+\frac{1}{2}\right)}{\left(s+\frac{1}{4}\right)}\right|_{s=0} = 4\,\text{A}$

Problem 7.3.6 What is the differential equation of a circuit with the transfer function of

$$H(s) = \frac{4}{s^2+3s+6}$$

Solution

$$H(s) = \frac{Y(s)}{X(s)} = \frac{4}{s^2+3s+6}$$

By cross-multiplication, one obtains $s^2Y(s) + 3s(Ys) + 6Y(s) = 4X(s)$. Since $H(s)$ is defined at resting system, all initial conditions are zero. Taking inverse Laplace transform of each term yields,

$$\frac{d^2y(t)}{dt^2} + 3\frac{dy(t)}{dt} + 6y(t) = 4x(t)$$

Problem 7.3.7 Unit Impulse response of a circuit is $h(t) = 3e^{-t} \cdot u(t)$. Find its unit ramp response.

Solution

$$H(s) = \frac{3}{s+1},$$
$$Y(s) = H(s)\cdot X(s) = \frac{3}{s+1}\cdot\frac{1}{s^2} = \frac{3}{s^2(s+1)}$$

Inverse Laplace transform using partial fractions expansion,

$$y(t) = \mathcal{L}^{-1}[Y(s)] = \mathcal{L}^{-1}\left[\frac{3}{s^2(s+1)}\right]$$
$$= \mathcal{L}^{-1}\left[\frac{3}{(s+1)} + \frac{3}{s^2} - \frac{3}{s}\right] = 3\cdot\left[e^{-2t} + t - 1\right]\cdot u(t)$$

Problem 7.3.8 In the circuit of Fig. 7.16, find $v_X(t)$, $t \geq 0$, using Laplace transform analysis, if all initial conditions are zero.

Solution

Parallel RC impedance is

$$\frac{\frac{2}{s}}{2+\left(\frac{1}{s}\right)} = \frac{2}{1+2s}$$

By voltage division,

$$\frac{V_x(s)}{V_i(s)} = \frac{\frac{2}{1+2s}}{\frac{2}{1+2s}+1} = \frac{\frac{2}{1+2s}}{\frac{2+1+2s}{1+2s}} = \frac{2}{3+2s} = \frac{2}{2\left(\frac{3}{2}+s\right)} = \frac{1}{s+\frac{3}{2}}$$

since $V_i(s) = \frac{1}{s}$,

$$V_x(s) = \frac{A}{s} + \frac{B}{s+\frac{3}{2}}$$

$$A = \left.\frac{1}{s+\frac{3}{2}}\right|_{s=0} = \frac{2}{3},\quad B = \left.\frac{1}{s}\right|_{s=-\frac{3}{2}} = -\frac{2}{3},\quad V_x(s) = \frac{\frac{2}{3}}{s} - \frac{2}{3}\cdot\frac{1}{s+\frac{3}{2}}$$

$$v_x(t) = \mathcal{L}^{-1}[V_x(s)] = \left[\frac{2}{3} - \frac{2}{3}\cdot e^{-\frac{3}{2}t}\right]u(t) = \frac{2}{3}\left(1 - e^{-\frac{3}{2}t}\right)u(t)$$

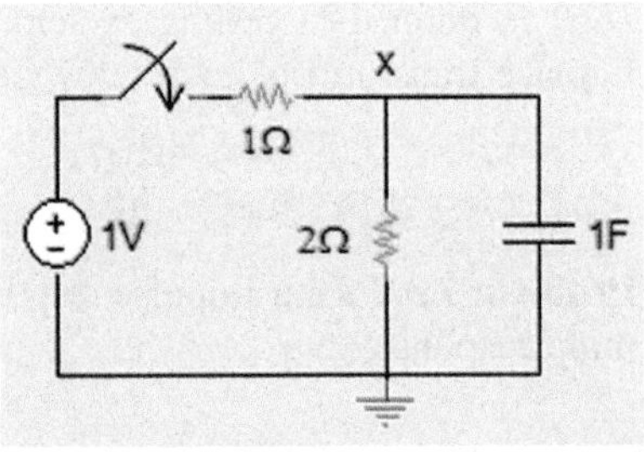

Fig. 7.16 The circuit for Problem 7.3.8

Problem 7.3.9 Determine the voltage over impedance $Z(s)$, if a unit step current is applied to the circuit and

$$Z(s) = \frac{s^3+9}{(s+1)(s+3)}$$

Solution

$$V(s) = Z(s)I(s) = \frac{s^3+9}{(s+1)(s+3)} \cdot \frac{1}{s} = \frac{A}{s} + \frac{B}{s+1} + \frac{C}{s+3}$$

$$A = \left.\frac{s^3+9}{(s+1)(s+3)}\right|_{s=0} = \frac{9}{1\cdot 3} \cdot 3, \quad B = \left.\frac{s^3+9}{s(s+3)}\right|_{s=-1} = \frac{-1+9}{(-1)\cdot(2)} = -4$$

$$C = \left.\frac{s^3+9}{s(s+1)}\right|_{s=-3} = \frac{-18}{6} = -3$$

$$v(t) = \mathcal{L}^{-1}[V(s)] = \left[3 - 4e^{-t} - 3e^{-3t}\right]u(t)$$

Problem 7.3.10 Voltage Transfer Function of a circuit is

$$H(s) = \frac{V_1(s)}{V_{\text{in}}(s)} = \frac{8s^3+18s^2+8s+8}{10s^3+34s^2+18s+12}$$

If input voltage $v_{\text{in}}(t)$ is unit step function, and $V_1(s) = \mathcal{L}[v_1(t)]$.

$$v_1(0+) = ?, \quad v_1(\infty) = ?$$

Solution

$$v_1(0^+) = \lim_{s\to\infty} s\cdot V_1(s) = \lim_{s\to\infty} s\cdot H(s)\cdot V_{\text{in}}(s) = \lim_{s\to\infty} s\cdot H(s)\cdot\frac{1}{s} = \lim_{s\to\infty} H(s)$$

$$v_1(0^+) = \lim_{s\to\infty} \frac{8+\frac{18}{s}+\frac{8}{s^2}+\frac{8}{s^3}}{10+\frac{34}{s}+\frac{18}{s^2}+\frac{12}{s^3}} = \frac{8}{10} = 0.8\,\text{V}$$

Similarly, steady-state voltage output in time domain is

$$v_1(\infty) = \lim_{s\to 0} s\cdot V_1(s) = \lim_{s\to 0} s\cdot H(s)\cdot V_{\text{in}}(s) = \lim_{s\to 0} s\cdot H(s)\cdot\frac{1}{s} = \lim_{s\to 0} H(s)$$

$$v_1(\infty) = \lim_{s\to 0}\frac{0+0+0+8}{0+0+0+12} = \frac{8}{12} = 0.667\,\text{V}$$

Problem 7.3.11 Voltage transfer function of a circuit is

$$H(s) = \frac{V_o(s)}{V_i(s)} = \frac{As^3+18s^2+8s+8}{10s^3+34s^2+18s+B}$$

Find coefficients A and B, if $V_o(0^+) = 0.8$, $V_o(\infty) = 0.6667$ V and $v_i(t) = u(t)$ (unit step input), $V_i(s) = \mathcal{L}[v_i(t)]$.

Solution

$$V_o(s) = H(s) \cdot V_i(s) = H(s) \cdot \frac{1}{s} = \frac{H(s)}{s}$$

$$V_o(0^+) = 0.8V = \lim_{s\to\infty} s \cdot \frac{H(s)}{s} = \lim_{s\to\infty} H(s) = \lim_{s\to\infty} \frac{A + \frac{18}{s} + \frac{8}{s^2} + \frac{8}{s^3}}{10 + \frac{34}{s} + \frac{18}{s^2} + \frac{B}{s^3}} = \frac{A}{10}$$

$$A = (0.8) \times 10 = 8,$$

$$V_o(\infty) = 0.6667 = \frac{2}{3} = \lim_{s\to 0} s \cdot \frac{H(s)}{s} = \lim_{s\to 0} H(s) = \lim_{s\to 0} \frac{0+0+0+8}{0+0+0+B} = \frac{8}{B}$$

$$B = \frac{8}{0.6667} = 12.$$

Problem 7.3.12 Poles of a transfer function $H(s)$ are at $s = 0, (-2 + j), -2 - j$ and zero is at (-1).

$H(s) = N(s)/D(s) = ?$

Solution

$$H(s) = \frac{(s+1)}{s(s+2+j)(s+2-j)} = \frac{s+1}{s^3 + 4s^2 + 5s}$$

Problem 7.3.13 Plot the pole-zero diagram of a function $Z(s)$, if

$$Z(s) = \frac{s^4 + 10s^2 + 9}{s^5 + 20s^3 + 64s}$$

Solution

$$Z(s) = \frac{s^4 + 10s^2 + 9}{s^5 + 20s^3 + 64s} = \frac{(s^2+1)(s^2+9)}{s(s^2+4)(s^2+16)}$$

This is a function with purely imaginary poles and zeros, i.e., an LC circuit. Figure 7.17 displays the resulting pole-zero diagram of a function $Z(s)$.

Problem 7.3.14 The poles of a circuit transfer function are at $p1 = -5$, $p2 = -2$ while the zeros are at

$z1 = -1$, $z2 = -4$.

(a) Determine circuit rational transfer function, $H(s) = Y(s)/X(s)$
(b) Determine its unit step response, $y(t)$.

Solution

(a) $H(s) = \dfrac{(s+1)(s+4)}{(s+2)(s+5)}$

(b) $Y(s) = X(s) \times H(s) = \dfrac{1}{s}\dfrac{(s+1)(s+4)}{(s+2)(s+5)} = \dfrac{A}{s} + \dfrac{B}{s+2} + \dfrac{C}{s+5}$

Fig. 7.17 The pole-zero diagram of a function Z(s) for Problem 7.3.13

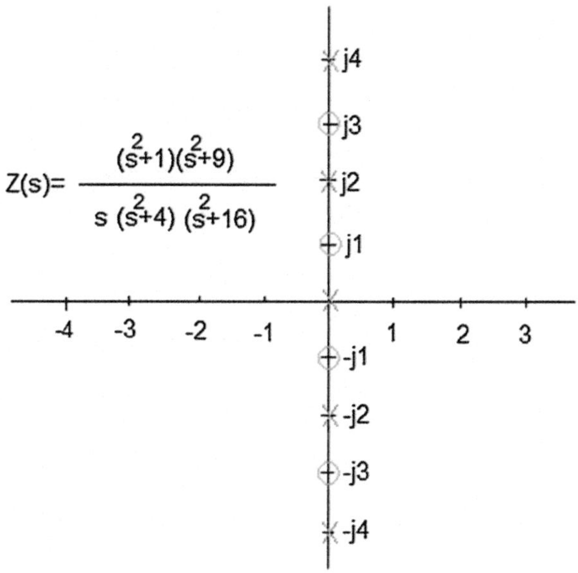

$$A = \frac{1 \times 4}{2 \times 5} = 0.4, \quad B = \frac{(-1) \times 2}{(-2) \times 3} = \frac{1}{3}, \quad C = \frac{(-1) \times (-4)}{(-3) \times (-5)} = \frac{4}{15}$$

$$y(t) = \mathcal{L}^{-1}\left[\frac{0.4}{s} + \frac{0.333}{s+2} + \frac{4/15}{s+5}\right] = \left(0.4 + 0.333\mathrm{e}^{-2t} + \frac{4}{15}\mathrm{e}^{-5t}\right)u(t)$$

Problem 7.3.15 Poles and zeros of a voltage transfer function are at $p_1 = -6$, $p_2 = -3$, $z_1 = -2$, $z_2 = -5$ rad/s.

If the input to this circuit is a unit impulse, determine the output signal $y(t)$.

Solution

$$H(s) = \frac{(s+2)(s+5)}{(s+3)(s+6)}$$

$$Y(s) = H(s) \cdot X(s) = H(s) \cdot 1 = H(s)$$

$$y(t) = \mathcal{L}^{-1}[Y(s)] = \mathcal{L}^{-1}\left[\frac{(s+2)(s+5)}{(s+3)(s+6)}\right] = \frac{A}{s+3} + \frac{B}{s+6}$$

$$A = \frac{(-6+2)(-6+5)}{(-6+3)} = \frac{(-4)(-1)}{(-3)} = -\frac{4}{3}$$

$$B = \frac{(-3+2)(-3+5)}{(-3+6)} = \frac{(-1)(2)}{(3)} = -\frac{2}{3}$$

$$Y(s) = -\frac{4}{3} \times \frac{1}{s+3} - \frac{2}{3} \times \frac{B}{s+6}$$

$$y(t) = \mathcal{L}^{-1}[Y(s)] = \left(-\frac{4}{3}\mathrm{e}^{-3t} - \frac{2}{3}\mathrm{e}^{-6t}\right)u(t) = -\frac{2}{3}\left(2\mathrm{e}^{-3t} + \mathrm{e}^{-6t}\right)u(t)$$

Problem 7.3.16 A circuit with a current transfer function of

$$T(s) = \frac{I_0(s)}{I_i(s)} = \frac{s+4}{(s+1)^2(s+2)}$$

has an input $i_{in}(t) = 2e^{-t} \cdot u(t)$. Determine the output current $i_0(t)u(t)$, and its initial value. Check the result by using initial value property in s-domain.

Solution

$$I_0(s) = T(s) \cdot I_{in}(s) = \frac{s+4}{(s+1)^2(s+2)} \cdot \frac{2}{(s+1)} = \frac{2(s+4)}{(s+1)^3(s+2)}$$

$$I_0(s) = \frac{A}{(s+1)} + \frac{B}{(s+1)^2} + \frac{C}{(s+1)^3} + \frac{D}{(s+2)}$$

Partial fractions expansion with multiple poles,

$$D = (s+2) \cdot I_0(s)|_{s=-2} = \left.\frac{2(s+4)}{(s+1)^3}\right|_{s=-2} = \frac{2(-2+4)}{(-2+1)^3} = \frac{2 \times 2}{-1} = -4$$

$$C = (s+1)^3 \cdot I_0|_{s=-1} = \left.\frac{2(s+4)}{s+2}\right|_{s=-1} = 6$$

$$B = \frac{d}{ds}\left[(s+1)^3 \cdot I_0(s)\right]\Big|_{s=-1} = \frac{d}{ds}\left(\frac{2s+8}{s+2}\right)\Big|_{s=-1} = \left.\frac{2(s+2)-(2s+8)}{(s+2)^2}\right|_{s=-1} = -4$$

$$2A = \frac{d^2}{ds^2}\left[(s+1)^3 \cdot I_0(s)\right]\Big|_{s=-1} = \frac{d^2}{ds^2}\left(\frac{2s+8}{s+2}\right)\Big|_{s=-1} = \frac{d}{ds}\left[\frac{-4}{(s+2)^2}\right]\Bigg|_{s=-1}$$

$$2A = \left.\frac{-2(s+2)(-4)}{(s+2)^4}\right|_{s=-1} = \left.\frac{8}{(s+2)^3}\right|_{s=-1} = 8$$

$$A = 4$$

$$I_0(s) = \frac{4}{(s+1)} - \frac{4}{(s+1)^2} + \frac{6}{(s+1)^3} - \frac{4}{(s+2)}$$

$$i_0(t) = \mathcal{L}^{-1}[I_0(s)] = \left[4e^{-t} - 4te^{-t} + 6t^2e^{-t} - 4e^{-2t}\right]u(t)$$

$$i_0(0) = (4 - 0 + 0 - 4) = 0\,\text{A}$$

Check using initial value property in s-domain:

$$f(0) = \lim_{s\to\infty} s \cdot I(s) = \frac{2s(s+4)}{(s+1)^3(s+2)} = \lim_{s\to\infty}\left(\frac{2s^2+8s}{s^4+\cdots}\right) = \lim_{s\to\infty}\frac{2s^2/s^4+8s/s^4}{s^4/s^4+\cdots} = 0$$

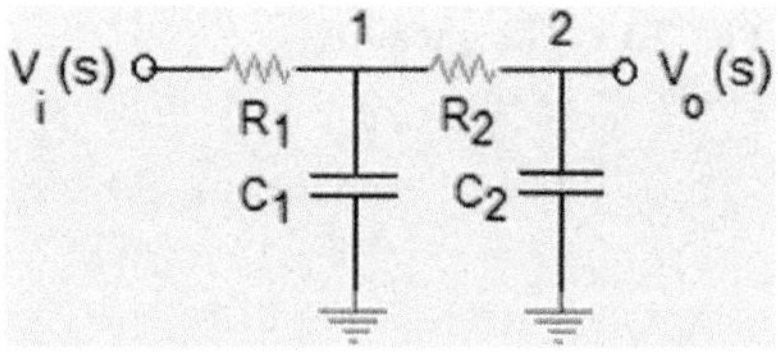

Fig. 7.18 Second-order passive RC low-pass filter circuit

Problem 7.3.17 Figure 7.18 shows a second-order passive RC low-pass filter. Determine its Voltage Transfer Function using nodal analysis.

$$H(s) = \frac{V_o(s)}{V_i(s)}$$

Solution

Applying Kirchhoff's current law at the output,

$$\frac{V_1(s) - V_o(s)}{R_2} - sC_2 \cdot V_o(s) = 0 \tag{7.3}$$

Kirchhoff's current law on the middle node 1,

$$\frac{V_i(s) - V_1(s)}{R_1} - sC_1 \cdot V_1(s) - \frac{V_1(s) - V_o(s)}{R_2} = 0 \tag{7.4}$$

$$\frac{V_i(s) - V_1(s)}{R_1} - sC_1 \cdot V_1(s) - sC_2 \cdot V_o(s) = 0$$

$$V_1(s)(1 + sC_1R_1) + sC_2R_1 \cdot V_o(s) = V_i(s) \tag{7.5}$$

From (7.3),

$$V_1(s) = V_o(s)(1 + sC_2R_2) \tag{7.6}$$

Substituting (7.6) into (7.5),

$$V_o(s)(1 + sC_2R_2)(1 + sC_1R_1) + sC_2R_1 \cdot V_o(s) = V_i(s) \tag{7.7}$$

Dividing all terms by $V_i(s)$ yields

$$\frac{V_o(s)}{V_i(s)} = \frac{1}{(1 + sC_2R_2)(1 + sC_1R_1) + sC_2R_1} \tag{7.8}$$

$$\frac{V_o(s)}{V_i(s)} = \frac{1}{s^2C_1R_1C_2R_2 + s[C_1R_1 + C_2(R_1 + R_2)] + 1} \tag{7.9}$$

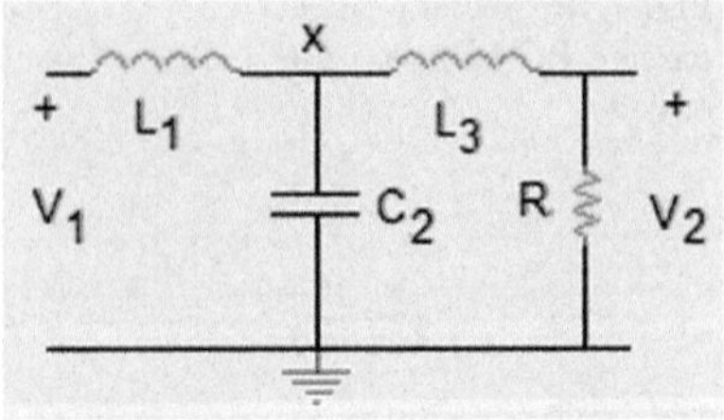

Fig. 7.19 The circuit for Problem 7.3.18

Problem 7.3.18 Determine the Voltage Transfer Function (VTF), $H(s) = V_2(s)/V_1(s)$, for the circuit shown in Fig. 7.19. $R = 1\ \Omega$, $L_1 = 1$ H, $C_2 = 1/2$ F, $L_2 = 2$ H.

Solution

$$(R + sL_3) \parallel \frac{1}{sC_2} = \frac{\left((R + sL_3) \times \frac{1}{sC_2}\right)}{\left((R + sL_3) + \frac{1}{sC_2}\right)} = \frac{(R + sL_3)}{s^2L_3C_2 + sRC_2 + 1}$$

By voltage division,

$$V_x = V_1 \frac{\frac{R + sL_3}{s^2L_3C_2 + sRC_2 + 1}}{sL_1 + \frac{R + sL_3}{s^2L_3C_2 + sRC_2 + 1}} = V_1 \frac{R + sL_3}{s^3L_1L_3C_2 + s^2RL_1C_2 + sL_1 + sL_3 + R}$$

$$V_2 = V_x \frac{R}{R + sL_3}$$

$$= V_1 \frac{R + sL_3}{s^3L_1L_3C_2 + s^2RL_1C_2 + s(L_1 + L_3) + R} \times \frac{R}{R + sL_3}$$

$$\frac{V_2}{V_1} = \frac{R}{s^3L_1L_3C_2 + s^2RL_1C_2 + s(L_1 + L_3) + R}$$

Substituting given component values, $R = 1\ \Omega$, $L_1 = 1$ H, $C_2 = 1/2$ F, $L_3 = 2$ H,

$$\frac{V_2(s)}{V_1(s)} = \frac{1}{s^3 \times 1 \times 2 \times \frac{1}{2} + s^2 \times 1 \times 1 \times \frac{1}{2} + 3s + 1} = \frac{2}{2s^3 + 1s^2 + 6s + 2}$$

Problem 7.3.19 Find the voltage transfer function in s-domain for the circuit shown in Fig. 7.20, using mesh current equations. Assume no initial conditions exist.

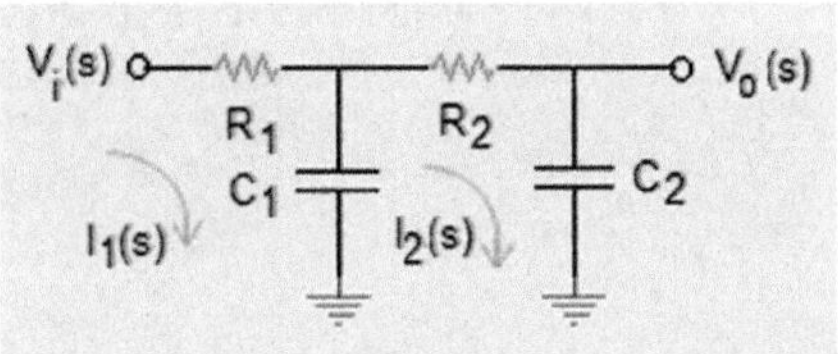

Fig. 7.20 The circuit for Problem 7.3.19

Solution

$$\begin{bmatrix} R_1 + 1/sC_1 & -1/sC_1 \\ -1/sC_1 & R_2 + 1/sC_2 + 1/sC_1 \end{bmatrix} \begin{bmatrix} I_1 \\ I_2 \end{bmatrix} = \begin{bmatrix} V_i \\ 0 \end{bmatrix}$$

$$V_o = I_2 \cdot \frac{1}{sC_2}$$

$$I_2 = \frac{\Delta_2}{\Delta}$$

$$\Delta_2 = \begin{bmatrix} R_1 + 1/sC_1 & V_i \\ -1/sC_1 & 0 \end{bmatrix} = 0 - \left(-\frac{V_i}{sC_1}\right) = \frac{V_i}{sC_1}$$

$$\begin{aligned}
\Delta &= \frac{sR_1C_1 + 1}{sC_1} \cdot \frac{sR_2C_2C_1 + C_1 + C_2}{sC_1C_2} - 1/(s^2C_1^2) \\
&= \frac{(sR_1C_1 + 1)(sR_2C_2C_1 + C_1 + C_2)}{s^2C_1^2C_2} - \frac{1}{s^2C_1^2} \\
&= \frac{s^2R_1C_1^2C_2R_2 + sR_1C_1^2 + sR_1C_1C_2 + sR_2C_2C_1 + C_1 + C_2 - C_2}{s^2C_1^2C_2} \\
&= \frac{C_1(s^2R_1R_2C_1C_2 + sR_1C_1 + sR_1C_2 + sR_2C_2 + 1)}{s^2C_1^2C_2} \\
&= \frac{s^2R_1R_2C_1C_2 + s(R_1C_1 + R_1C_2 + R_2C_2) + 1}{s^2C_1C_2}
\end{aligned}$$

$$I_2 = \frac{\Delta_2}{\Delta} = \frac{\dfrac{V_i}{sC_1}}{\dfrac{(s^2C_1R_1R_2C_2 + s(R_1C_1 + R_1C_2 + R_2C_2) + 1)}{s^2C_1C_2}}$$

$$I_2 = \frac{sC_2V_i}{s^2R_1R_2C_1C_2 + s(R_1C_1 + R_1C_2 + R_2C_2) + 1}$$

$$V_o = I_2 \cdot \frac{1}{sC_2} = \frac{sC_2V_i}{s^2R_1R_2C_1C_2 + s(R_1C_1 + R_1C_2 + R_2C_2) + 1} \cdot \frac{1}{sC_2}$$

$$\frac{V_o(s)}{V_i(s)} = H(s) = \frac{1}{s^2R_1R_2C_1C_2 + s(R_1C_1 + R_1C_2 + R_2C_2) + 1}$$

$$\frac{V_o(s)}{V_i(s)} = \frac{1}{s^2C_1R_1C_2R_2 + s[C_1R_1 + C_2(R_1 + R_2)] + 1}$$

Problem 7.3.20 An input impedance function $Z(s)$ is given as

$$Z(s)=\frac{3s^4+60s^3+420s^2+1200s+1152}{8s^5+200s^4+1840s^3+7600s^2+13512s+7560}$$

Write this function as the ratio of two polynomials in factored form (Gain_H0.m).

Solution

Determining its poles and zeros and then finding constant term (DC gain constant) will yield the result as

$$Z(s)=K\frac{(s+2)(s+4)(s+6)(s+8)}{(s+1)(s+3)(s+5)(s+7)(s+9)}$$

The constant, K, is calculated as

$$Z(0)=\frac{1152}{7560}=0.1524=K\frac{(0+2)(0+4)(0+6)(0+8)}{(0+1)(0+3)(0+5)(0+7)(0+9)}=\frac{384}{945}=0.4064$$
$$K=\frac{0.4064}{0.1524}=0.375$$

Following is the m-file script for computing the poles, zeros, and (dc gain) constant K.

```
N=[3 60 420 1200 1152]
D=[8 200 1840 7600 13512 7560]
H=tf(N,D)
H0=polyval(N,0)/polyval(D,0);%dc gain
%format rational
format short
p=abs(pole(H))
z=abs(zero(H))
HZ=prod(z);  HP=prod(p); K=H0/(HZ/HP)
```

Problem 7.3.21 Determine component values of a circuit which has the voltage transfer function $H(s)$,

$$H(s)=\frac{s+1}{s+2}$$

Solution

The voltage transfer function can be set up using voltage division rule,

$$H(s)=\frac{V_0(s)}{V_i(s)}=\frac{Z_0(s)}{Z_0(s)+Z_1(s)}=\frac{Z_0(s)}{Z(s)}$$

Since,

$Z_0(s) = s + 1, \quad \rightarrow \quad Z(s) = s + 2 = Z_0(s) + Z_1(s) = (s + 1) + Z_1(s)$ If $Z_1(s) = R_1,$

$$Z(s) = s + (1 + R_1) = s + 2,$$
$$Z_1(s) = R_1 = 1\,\Omega$$

Numerator of $H(s)$ is a series RL circuit,

$$Z_0(s) = sL + R_2,$$
$$H(s) = \frac{Z_0(s)}{Z(s)} = \frac{sL + R_2}{(sL + R_2) + R_1} = \frac{sL + R_2}{sL + (R_1 + R_2)}$$

Therefore,

$$L = 1\,\text{H}, R_2 = 1\,\Omega.$$

Problem 7.3.22

(a) Determine input admittance $Y(s)$ of the circuit shown in Fig. 7.21.
(b) Find out the component values if

$$Y(s) = \frac{s + 1}{3s + 2}$$

Solution

(a) $Z(s) = R_2 + \dfrac{sL \cdot R_1}{sL + R_1} = \dfrac{R_1R_2 + sR_2L + sR_1L}{R_1 + sL} = \dfrac{sL(R_1 + R_2) + R_1R_2}{sL + R_1}$

$$Y(s) = \frac{1}{Z(s)} = \frac{sL + R_1}{sL(R_1 + R_2) + R_1R_2}$$

(b) Since

$$Y(s) = \frac{s + 1}{3s + 2}, \quad R_1R_2 = 2\,\Omega, \quad L = 1\,\text{H}, \quad R_1 = 1\,\Omega$$

Then, this admittance function results in $R_2 = 2\,\Omega$.

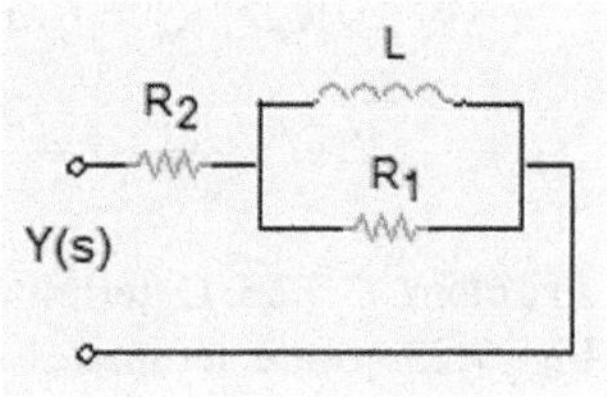

Fig. 7.21 The circuit for Problem 7.3.22

Problem 7.3.23 Determine the number of decades and octaves between the following frequencies:

100 Hz, 1 kHz
5 Hz, 200 Hz
2 rad/s, 200 rad/s

Solution

The number of decades between two frequencies is given as

$$\Delta f_{10} = \log_{10}\left(\frac{f_2}{f_1}\right), \quad f_2 > f_1$$

The number of octaves between two frequencies is

$$\Delta f_2 = \log_2\left(\frac{f_2}{f_1}\right), \quad f_2 > f_1$$

(a) $\log_{10}\left(\frac{1000}{100}\right) = \log_{10}(10) = 1\,\text{decade}$

$$\log_2\left(\frac{f_2}{f_1}\right) = \frac{\log_{10}\left(\frac{f_2}{f_1}\right)}{\log_{10}(2)} = 3.322\log_{10}\left(\frac{f_2}{f_1}\right) = 3.322 \times \log_{10}(10) = 3.322\,\text{octave}$$

(b) $\log_{10}\left(\frac{200}{5}\right) = \log_{10}(40) = 1.6\,\text{decade}$

$$\log_2\left(\frac{200}{5}\right) = 3.322 \times \log_{10}\left(\frac{200}{5}\right) = 5.322\,\text{octave}$$

(c) $\log_{10}\left(\frac{200}{2}\right) = \log_{10}(100) = 2\,\text{decades}$

$$\log_{10}\left(\frac{200}{2}\right) = 3.322\ \log_{10}\left(\frac{200}{2}\right) = 3.322 \times \log_{10}(100) = 6.644\,\text{octave}$$

Problem 7.3.24 Determine *H*(*s*) from given Bode magnitude plot of Fig. 7.22 (Bode_template1.m).

Solution

Poles: 1, 10; zeros: 2, constant term, *K*:

$$20\log_{10}|K| = 6\,\text{dB} \quad \rightarrow \quad \log_{10}|K| = \frac{6}{20} = 0.3 \quad \rightarrow \quad K = 2$$

$$H(s) = \frac{1}{s+1}\cdot\frac{10}{s+10}\cdot\frac{s+2}{2}\cdot 2 = \frac{10(s+2)}{(s+1)(s+10)}$$

Problem 7.3.25 Determine *H*(*s*) from given Bode magnitude plot shown in Fig. 7.23 (Bode_template1.m).

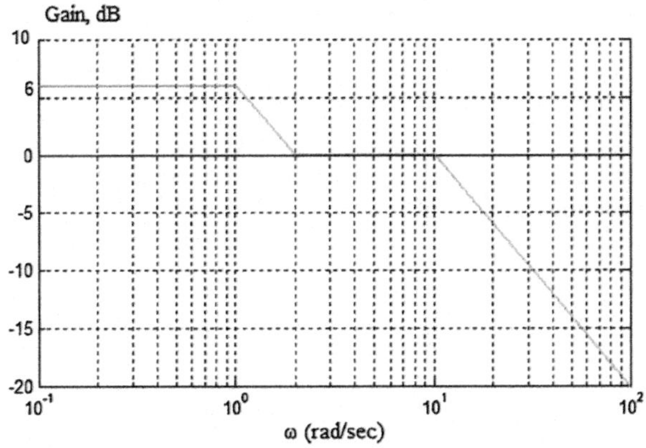

Fig. 7.22 Bode magnitude plot for Problem 7.3.24

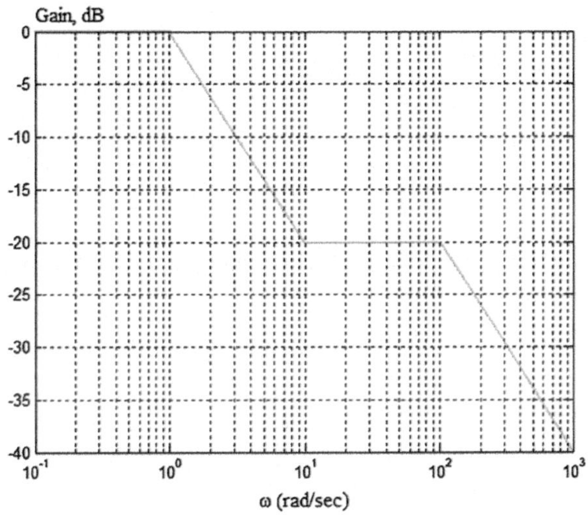

Fig. 7.23 Bode magnitude plot for Problem 7.3.25

Use MATLAB and plot *H*(*s*) magnitude and phase to prove the result obtained in part a.

Solution

Poles: 1, 100, zero: 10

$$20\log_{10}|K| = 0\,\text{dB} \quad \rightarrow \quad \log_{10}|K| = 0 \quad \rightarrow \quad K = 1$$

$$H(s) = \frac{1}{s+1}\cdot\frac{100}{s+100}\cdot\frac{s+10}{10}\cdot 1 = \frac{10(s+10)}{(s+1)(s+100)}$$

Problem 7.3.26 Determine the straight-line approximation to the gain of the transfer function

$$H(s) = \frac{250(s+2.5)}{(s+25)(s+250)}$$

Solution

The form of a $T(s)$ with real poles and zeros is approximated by $T(j\omega)$ using straight-lines as

$$T(s) = \frac{ks(s+a)}{(s+b)(s+c)} \quad \rightarrow \quad T(j\omega) = K_0 \frac{j\omega\left(1+\frac{j\omega}{a}\right)}{\left(1+\frac{j\omega}{b}\right)\left(1+\frac{j\omega}{c}\right)}$$

here $a<b<c$ and

$$K_0 = \frac{Ka}{bc}$$

Therefore,

$$H(j\omega) = \left(\frac{2500 \times 2.5}{25 \times 250}\right) \frac{\left(1+\frac{j\omega}{2.5}\right)}{\left(1+\frac{j\omega}{25}\right)\left(1+\frac{j\omega}{250}\right)} = \frac{\left(1+\frac{j\omega}{2.5}\right)}{\left(1+\frac{j\omega}{25}\right)\left(1+\frac{j\omega}{250}\right)}$$

The scale factor is $K_0 = 1$. For $\omega < 2.5$ rad/s, $|H(j\omega)|_{SL} = K_0 = 1$.
At $\omega < 2.5$ rad/s, there is a zero.

$$|H(j\omega)|_{SL} = K_0\left(\frac{\omega}{2.5}\right) = \frac{\omega}{2.5}$$

This applies until it passes $\omega = 25$ rad/s. For $\omega \geq 25$ rad/s the gain is $|H(j\omega)|_{SL} = 1(\omega/25) = 10$. It applies until it passes $\omega = 250$ rad/s. Beyond this angular frequency the gain is

$$|H(j\omega)|_{SL} = \frac{1 \times \left(\frac{\omega}{2.5}\right)}{\frac{\omega}{25} \times \frac{\omega}{250}} = \frac{\frac{\omega}{2.5}}{\frac{\omega^2}{6250}} = \frac{2500}{\omega}$$

The straight-line approximation to the gain is summarized below.

$$|H(j\omega)|_{SL} = \begin{cases} 1 & \text{if } 0<\omega\leq 2.5 \\ \frac{\omega}{2.5} & \text{if } 2.5<\omega\leq 25 \\ 10 & \text{if } 25<\omega\leq 250 \\ \frac{2500}{\omega} & \text{if } 250<\omega \end{cases}$$

Problem 7.3.27 Determine the straight-line approximation of the gain of the transfer function

$$H(s) = \frac{100s^2}{(s+10)(s+500)}$$

Solution

Since the order of critical angular frequencies are $\omega \to 0$, 10, 50 rad/s, and

$$K_0 = \frac{K}{ab} = \frac{100}{10 \times 500} = \frac{1}{50}$$

$$H(j\omega) = \frac{1}{50}\frac{(j\omega)^2}{\left(1+\frac{j\omega}{10}\right)\left(1+\frac{j\omega}{500}\right)}$$

For $\omega \leq 100\,\text{rad/s}$, $|H(j\omega)|_{SL} = \omega^2/50$

For $10 < \omega \leq 500\,\text{rad/s}$,

$$|H(j\omega)|_{SL} = \frac{\frac{\omega^2}{50}}{\frac{\omega}{10}} = \frac{\omega}{5}$$

For $\omega > 500\,\text{rad/s}$,

$$|H(j\omega)|_{SL} = \frac{\frac{\omega^2}{50}}{\left(\frac{\omega}{10}\right)\left(\frac{\omega}{500}\right)} = \frac{\frac{\omega^2}{50}}{\frac{\omega^2}{5000}} = 100$$

The straight-line approximation to the gain response is summarized below.

$$|H(j\omega)|_{SL} = \begin{cases} \frac{\omega^2}{50} & \text{if } \omega \leq 10 \\ \frac{\omega}{5} & \text{if } 10 < \omega < 500 \\ 100 & \text{if } 500 < \omega \end{cases}$$

Problem 7.3.28 An amplifier which is used in a wireless sensor network has a voltage transfer function with zeros at $\omega = 0$, $\omega = \infty$, and poles at $\omega = 100$ and $\omega = 100{,}000$ rad/s.

Use straight-line method and determine approximate values of the voltage transfer function magnitude (gain) at $\omega = 10^4$ and $\omega = 10^7$ rad/s if $|T(j10)| = 40\,\text{dB}$

Solution

Since there is a zero at the origin, an s term exists in numerator of the gain function. Along with two poles, voltage transfer function $T(s)$ has the form

$$T(s) = K\frac{s}{\left(1+\frac{s}{100}\right)\left(1+\frac{s}{100,000}\right)}$$

Since by definition of dB gain,

$$|T(j\omega)|_{\text{dB}} = 20\log|T(j\omega)|,$$

$|T(j10)| = 40\,\text{dB} = 100\,(\text{V/V})$, which implies that $T(j1) = 20\,\text{dB} = 10\,(\text{V/V})$, or the scale factor is $K = 10$. Therefore,

$$T(s) = 10\frac{s}{\left(1+\frac{s}{100}\right)\left(1+\frac{s}{100,000}\right)}$$

The gain at $\omega = 10^4$ rad/s is in the flat region of $|T(j\omega)|$. This region is between $100 \leq \omega < 100{,}000$ rad/s, where

$$|T(j100)| = |T(j100,000)| = 60\,\text{dB} = 1000\,(\text{V/V})$$

The gain drops −20 dB/decade after $\omega = 10^5$ rad/s. Therefore, the gain at two decades after $\omega = 10^5$ is $|T(j10^7)| = |T(j1)| = 60 + 2\times(-20) = 20\,\text{dB}$ (V/V).

Problem 7.3.29

(a) Determine the voltage transfer function (VTF) of a series resonant RLC notch filter, shown in Fig. 7.24.
(b) Design a series resonant RLC circuit with a notch at 50 Hz, $R = 100\ \Omega$.
(c) Find the coefficients of VTF, in part b.
(d) Determine the poles and zeros of the VTF, in part b.
(e) Plot the magnitude of VTF versus frequency (log–log), in part b.
(f) Repeat parts b–e for $R = 1\ \Omega$ and compare results with those obtained before (RLCnotch1.m).

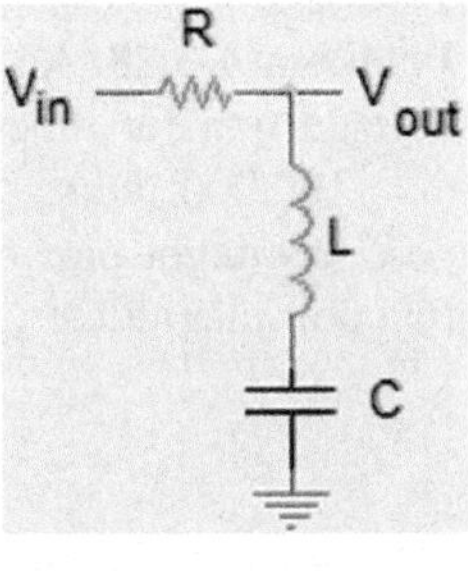

Fig. 7.24 A series resonant RLC notch filter

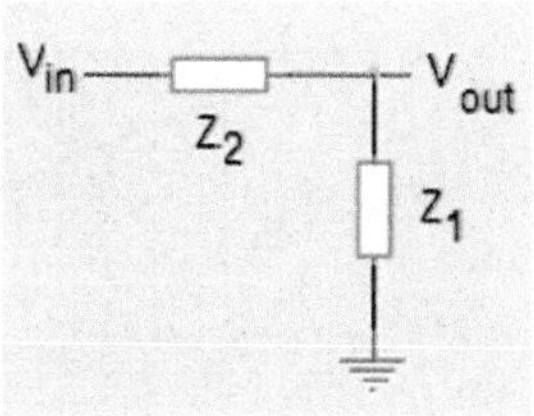

Fig. 7.25 A series resonant RLC notch filter topology as a voltage divider

Solution

If the circuit topology is visualized as a voltage divider (see, Fig. 7.25),

$$V_{out} = \frac{Z_1}{Z_1 + Z_2} \cdot V_{in}, \quad Z_1 = sL + \frac{1}{sC} = \frac{s^2LC + 1}{sC}, \quad Z_2 = R$$

$$V_{out} = V_{in} \cdot \frac{s^2LC + 1}{s^2LC + sRC + 1} = \frac{LC\left(s^2 + \frac{1}{LC}\right)}{LC\left(s^2 + \frac{RC}{LC}s + \frac{1}{LC}\right)} \cdot V_{in}$$

$$\frac{V_{out}(s)}{V_{in}(s)} = H(s) = \frac{s^2 + \frac{1}{LC}}{s^2 + \frac{R}{L}s + \frac{1}{LC}}$$

$$\text{Let}\, \omega_0^2 = \frac{1}{LC}, \quad L = \frac{1}{\omega_0^2 C} \quad (\omega_0 = \text{notch frequency, rad/s})$$

$$H(s) = \frac{s^2 + \omega_0^2}{s^2 + \frac{R}{\frac{1}{\omega_0^2 C}}s + \omega_0^2} = \frac{s^2 + \omega_0^2}{s^2 + \omega_0^2 RCs + \omega_0^2}$$

A series resonant RLC circuit with a notch at 50 Hz, and $R = 100\ \Omega$,

$$\omega_0 = 2\pi f_0, \quad f_0 = 50\,\text{Hz}, \quad \omega_0 = \frac{1}{\sqrt{LC}}, \ \sqrt{LC} = \frac{1}{2\pi f_0}, \ LC = \left(\frac{1}{4\pi^2}\right) \cdot \frac{1}{f_0^2} = \frac{k}{f_0^2}$$

$$k = \frac{1}{4\pi^2} = 0.025330296$$

$$C = \frac{k}{L \cdot f_0^2}$$

In the last equation let $L = 1$ mH, then

$$C = \frac{0.25330296}{10^{-3} \cdot 50^2} = 10{,}132\,\mu\text{F}.$$

$$H(s) = \frac{V_{out}(s)}{V_{in}(s)} = \frac{s^2 + 98{,}700}{s^2 + 100000s + 98{,}700}$$

Poles = −10,000 rad/s, −1 rad/s (roots of denominator polynomial)
Zeros = ±314.16j (roots of numerator polynomial)

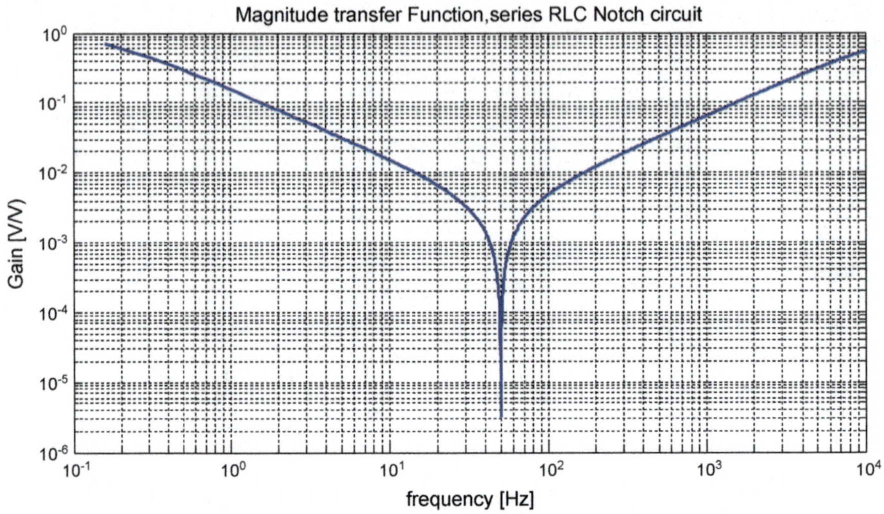

Fig. 7.26 Graph of the magnitude of VTF versus frequency (R = 100 Ω, f_0 = 50 Hz)

The frequency response behavior of the circuit is shown in Fig. 7.26.
When $R = 1\ \Omega$, $f_0 = 50$ Hz, $C = 10{,}132\ \mu$F, $L = 1$ mH.
Voltage Transfer function for $R = 1\ \Omega$,

$$H(s) = \frac{s^2 + 98{,}700}{s^2 + 1000s + 98{,}700}$$

Poles: −888.98 rad/s, −111.02 rad/s
Zeros: $\pm 314.16j$ rad/s

The frequency response behavior of the circuit for the magnitude of the VTF when $R = 1\ \Omega$, $f_0 = 50$ Hz, $C = 10{,}132\ \mu$F, $L = 1$ mH is displayed in Fig. 7.27.

It is noted that as R becomes smaller in value, the bandwidth gets smaller and suppression at 50 Hz becomes more effective as compared to VTF at neighboring reference frequencies.

Problem 7.3.30

(a) Determine the voltage transfer function (VTF) of a parallel resonant RLC notch filter shown in Fig. 7.28.
(b) Design a parallel resonant RLC notch circuit with a notch at 50 Hz, $R = 1\ \Omega$.
(c) Find the coefficients of VTF, in part b.
(d) Determine the poles and zeros of the VTF, in part b.
(e) Plot the magnitude of VTF versus frequency (log–log), in part b (RLCnotch2.m).

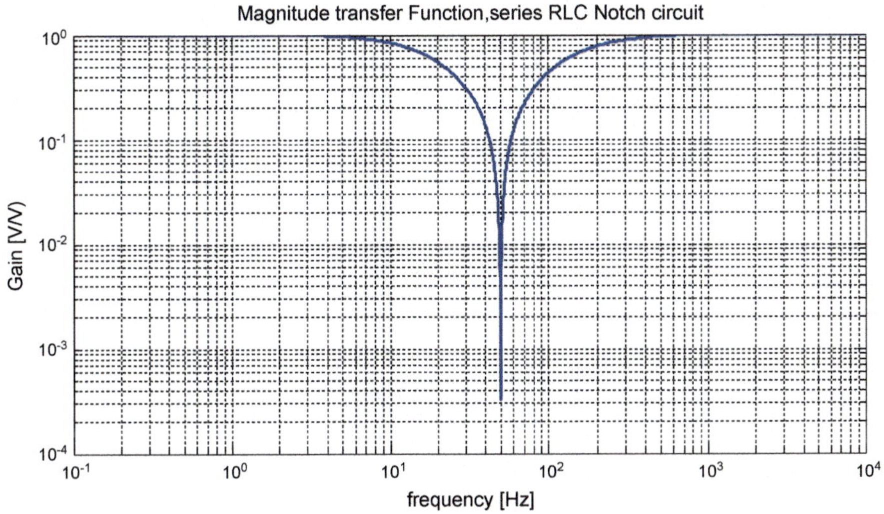

Fig. 7.27 Graph of the magnitude of VTF versus frequency (R = 1 Ω, C = 10, 132 μF, L = 1 mH)

Fig. 7.28 A parallel resonant RLC notch filter

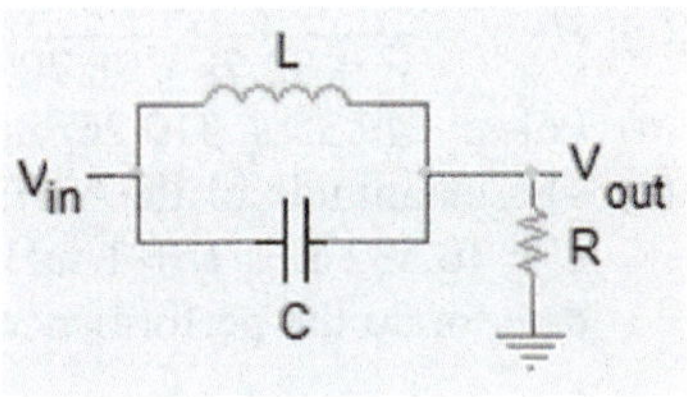

Solution

(a) Using the voltage divider circuit topology of Fig. 7.29,

$$V_{\text{out}} = \frac{Z_1}{Z_1 + Z_2} \cdot V_{\text{in}},\ Z_1 = R,\ Z_2 = \frac{1}{Y_2} = \frac{1}{sC + \dfrac{1}{sL}} = \frac{sL}{s^2LC + 1}$$

$$\frac{V_{\text{out}}}{V_{\text{in}}} = \frac{R}{R + \dfrac{sL}{s^2LC + 1}} = \frac{R(s^2LC + 1)}{s^2RLC + sL + R} = \frac{s^2RLC + R}{s^2RLC + sL + R}$$

$$\frac{V_{\text{out}}}{V_{\text{in}}} = \frac{RLC\left(s^2 + \dfrac{R}{RLC}\right)}{RLC\left(s^2 + \dfrac{sL}{RLC} + \dfrac{R}{RLC}\right)} = \frac{s^2 + \dfrac{1}{LC}}{s^2 + \dfrac{1}{RC}s + \dfrac{1}{LC}}$$

Fig. 7.29 Voltage divider circuit topology for parallel resonant RLC notch filter

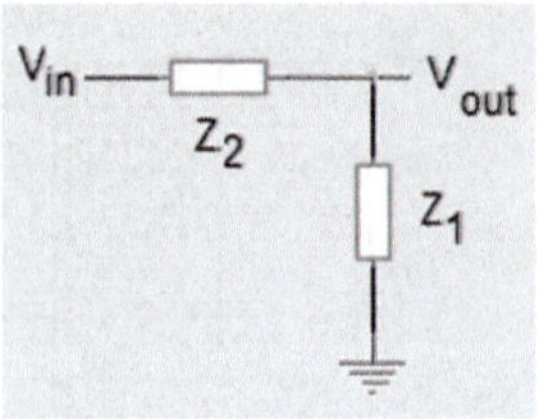

(b) $\omega_0 = 2\pi f_0$, $f_0 = 50$ Hz, $\omega_0 = \frac{1}{\sqrt{LC}}$, $\sqrt{LC} = \frac{1}{2\pi f_0}$, $LC = \left(\frac{1}{4\pi^2}\right) \cdot \frac{1}{f_0^2} = \frac{k}{f_0^2}$

$$k = \frac{1}{4\pi^2} = 0.025330296$$
$$C = \frac{k}{L \cdot f_0^2}$$
$$\text{Let} \quad L = 1\,\text{mH},$$
$$C = \frac{0.25330296}{10^{-3} \cdot 50^2} = 10{,}132\,\mu\text{F}$$

(c) $H(s) = \frac{s^2 + 98{,}700}{s^2 + 98.7s + 98{,}700}$

(d) Poles: $-49.35 + 310.26j$ rad/s, $-49.35 - 310.26j$ rad/s, Zeros: $\pm 314.16j$

(e) The magnitude of the VTF versus frequency plot when $R = 1\ \Omega$, $f_0 = 50$ Hz, $C = 10{,}132\ \mu$F, $L = 1$ mH is displayed in Fig. 7.30. Larger resistor values deteriorate the performance of the circuit.

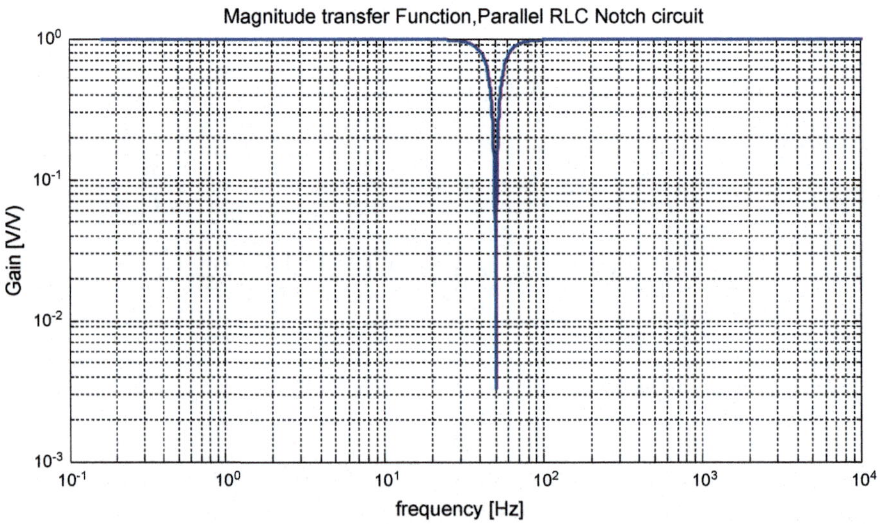

Fig. 7.30 Magnitude transfer Function of parallel RLC notch circuit

Problem 7.3.31 Determine the poles and zeros of the following voltage transfer functions. Then, plot their respective pole-zero diagrams. Comment on their frequency behavior (polezero2.m).

(a) $H(s) = \dfrac{V_o(s)}{V_{\text{in}}(s)} = \dfrac{s^3 + 2s^2 + 2s + 1}{s^3 + 5s^2 + 5s + 1}$

(b) $H(s) = \dfrac{V_o(s)}{V_{\text{in}}(s)} = \dfrac{s^3 + s^2 + s + 1}{s^3 + 5s^2 + 5s + 1}$

Solution

(a) $H(s) = \dfrac{V_o(s)}{V_{\text{in}}(s)} = \dfrac{s^3 + 2s^2 + 2s + 1}{s^3 + 5s^2 + 5s + 1}$

Zeros are the roots of numerator polynomial of the transfer function,

$$z_1 = -1, \quad z_2 = -\frac{1 + j\sqrt{3}}{2}, \quad z_3 = -\frac{1 - j\sqrt{3}}{2}$$

Poles are the roots of denominator polynomial of the transfer function (Fig. 7.31),

$$p_1 = -3.7321, \quad p_2 = -1, \quad p_3 = -0.2679$$

$$H(s) = \frac{s^3 + 2s^2 + 2s + 1}{s^3 + 5s^2 + 5s + 1} = \frac{(s+1)(s^2 + s + 1)}{(s + 3.7321)(s + 1)(s + 0.2679)}$$

$$H(s) = \frac{s^2 + s + 1}{(s + 3.7321)(s + 0.2679)} = \frac{s^2 + s + 1}{s^2 + 4s + 1}$$

(b) $H(s) = \dfrac{V_o(s)}{V_{\text{in}}(s)} = \dfrac{s^3 + s^2 + s + 1}{s^3 + 5s^2 + 5s + 1}$

Fig. 7.31 Pole zero diagram for part a of Problem 7.3.31

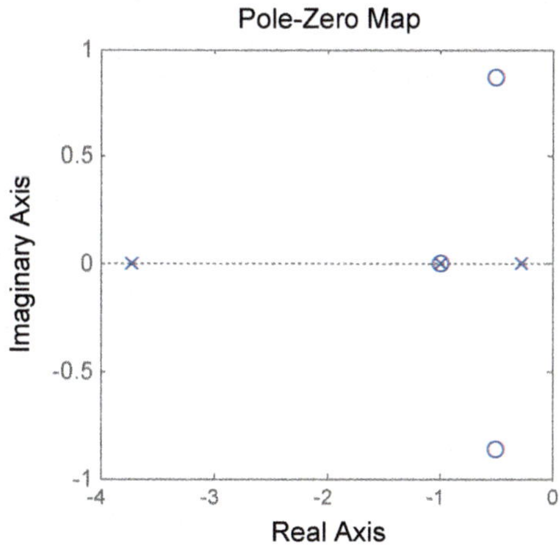

zeros are the roots of numerator polynomial of the transfer function,

$$z_1 = -1, \quad z_2 = j, \quad z_3 = -j$$

Poles are the roots of denominator polynomial of the transfer function,

$$p_1 = -3.7321, \quad p_2 = -1, \quad p_3 = -0.2679$$
$$H(s) = \frac{s^3+s^2+s+1}{s^3+5s^2+5s+1} = \frac{(s+1)(s^2+1)}{(s+3.7321)(s+1)(s+0.2679)} = \frac{s^2+1}{(s+3.7321)(s+0.2679)}$$
$$H(s) = \frac{s^2+1}{(s+4s+1)}$$

Both of these VTFs have the same poles, but different zeros. The second VTF has purely complex conjugate zeros. The overlapping of poles and zeros for both VTFs on the real axis at $\sigma = -1$ indicates that the pole and the zero at $\sigma = -1$ cancel out each other. For both VTFs, the critical frequency is at $\omega = 1$ rad/s. At this frequency both of the VTFs are null, this means that a circuit possessing these transfer functions blocks the signal at $\omega = 1$ rad/s (=0.1591 Hz). However, both VTFs pass the signals at dc $(s = \infty)$ and high frequencies $(s = \infty)$.

An ideal twin-T notch filter with a stop angular frequency $\omega = a = 1/(RC)$ has the following general form that shows that a third-order response reduces into a second-order response (Fig. 7.32),

$$H(s) = \frac{V_o(s)}{V_{\text{in}}(s)} = \frac{s^3+as^2+a^2s+a^3}{s^3+5as^2+5a^2s+a^3} = \frac{(s+a)(s^2+a^2)}{(s+a)(s+4as+a^2)} = \frac{s^2+a^2}{s^2+4as+a^2}$$

Problem 7.3.32

(a) Determine the voltage transfer function for twin-T notch filter topology of Fig. 7.33, in terms of its component values (in s-domain), using nodal analysis.

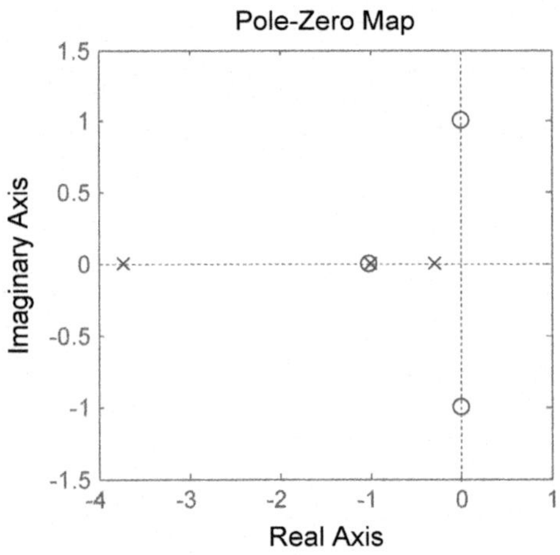

Fig. 7.32 Pole zero diagram for part b of Problem 7.3.31

Fig. 7.33 Twin-T notch filter

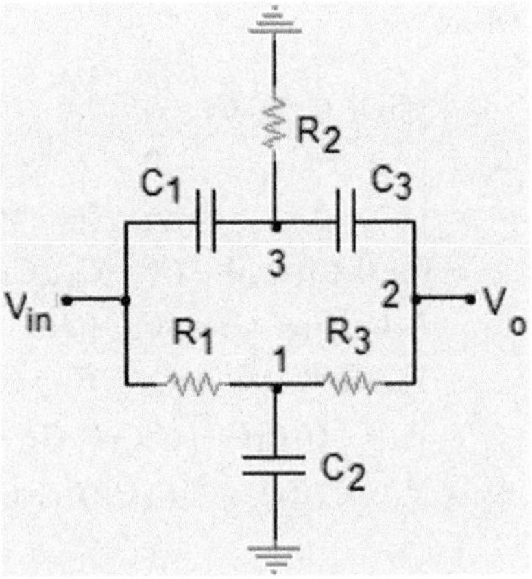

(b) Evaluate the VTF for $C_2 = 2C_1 = 2C_3 = 2C$, $R_2 = 0.5R_1 = 0.5R_3 = 0.5R$
(c) Numerically compute the coefficients of VTF in part b, for $R = 1\ \Omega$, $C = 1$ F.

Solution

(a) Transform the input voltage source into current source. This operation results in *two* current sources to account for R_1 and C_1, (see, Fig. 7.34)

$$I_1 = \frac{V_{\text{in}}}{R_1} = G_1 \cdot V_{\text{in}}, \quad I_2 = V_{\text{in}} \cdot sC_1$$

Apply KCL for the nodes,

$$\begin{bmatrix} G_1 + G_3 + sC_2 & -G_3 & 0 \\ -G_3 & G_3 + sC_3 & -sC_3 \\ 0 & sC_3 & G_2 + sC_1 + sC_3 \end{bmatrix} \begin{bmatrix} V_1 \\ V_2 \\ V_3 \end{bmatrix} = \begin{bmatrix} I_1 \\ 0 \\ I_2 \end{bmatrix}$$

Use Cramer's rule to find out the output voltage,

$$V_o = V_2 = \frac{\Delta_2}{\Delta} = \frac{V_{\text{in}} \cdot \Delta_2'}{\Delta}$$

This yields the general voltage transfer function relationship,

$$\frac{V_o}{V_{\text{in}}} = H(s) = \frac{\Delta_2'}{\Delta} \tag{7.10}$$

Fig. 7.34 Source transformation results in two current sources in twin-T notch filter

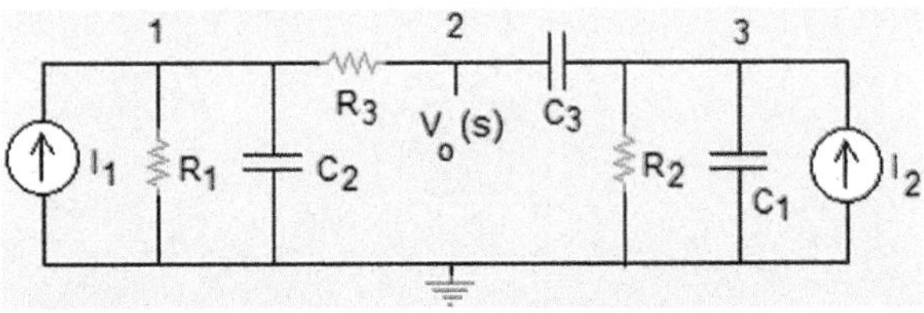

where

$$\Delta_2' = \begin{vmatrix} G_1+G_3+sC_2 & I_1 & 0 \\ -G_3 & 0 & -sC_3 \\ 0 & I_2 & G_2+sC_1+sC_3 \end{vmatrix}$$

$$= 0+0+0-[0+I_2(-sC_3)(G_1+G_3+sC_2)+I_1(G_2+sC_1+sC_3)(-G_3)]$$

$$= I_2 sC_3(G_1+G_3+sC_2)+I_1G_3(G_2+sC_1+sC_3)$$

$$= V_{\text{in}}sC_1.sC_3(G_1+G_3+sC_2)+G_1V_{in}.G_3(G_2+sC_1+sC_3)$$

$$= V_{\text{in}}\left[\left(s^2G_1C_1C_3+s^2C_1C_3G_3+s^3C_1C_3C_2\right)+G_1G_2G_3+sC_1G_1G_3+sC_3G_1G_3\right]$$

$$\Delta_2' = s^3(C_1C_2C_3)+s^2(C_1C_3G_1+C_1C_3G_3)+s(C_1G_1G_3+C_3G_1G_3)+G_1G_2G_3 \tag{7.11}$$

Determinant of the admittance matrix is

$$\Delta = \begin{vmatrix} G_1+G_3+sC_2 & -G_3 & 0 \\ -G_3 & G_3+sC_3 & -sC_3 \\ 0 & sC_3 & G_2+sC_1+sC_3 \end{vmatrix}$$

$$= (G_1+G_3+sC_2)[(G_3+sC_3)(G_2+sC_1+sC_3)+(sC_3\cdot sC_3)]$$
$$- G_3^2[(G_2+sC_1+sC_3)]$$

$$= (G_1+G_3+sC_2)\left[G_2G_3+sC_1G_3+sC_3G_3+sC_3G_2+s^2C_1C_3+s^2C_3^2-s^2C_3^2\right]$$
$$- G_2G_3^2-sC_1G_3^2-sC_3G_3^2$$

$$= G_1G_2G_3+sC_1G_1G_3+sC_3G_1G_3+sC_3G_1G_2+s^2C_1C_3G_1+G_2G_3^2+sC_1G_3^2$$
$$+sC_3G_3^2+sC_3G_2G_3+s^2C_1C_3G_3+sC_2G_2G_3+s^2C_1C_2G_3+s^2C_2C_3G_3$$
$$+s^2C_2C_3G_2+s^3C_1C_2C_3-G_2G_3^2-sC_1G_3^2-sC_3G_3^2$$

Factoring terms into the powers of s,

$$\Delta = s^3(C_1C_2C_3)+s^2(C_1C_3G_1+C_1C_3G_3+C_1C_2G_3+C_2C_3G_3+C_2G_3G_2)$$
$$+s(C_1G_1G_3+C_3G_1G_3+C_3G_1G_2+C_2G_2G_3+C_3G_2G_3)+(G_1G_2G_3) \tag{7.12}$$

Using Eqs. (7.10)–(7.12),

$$H(s) = \frac{V_o(s)}{V_{\text{in}}(s)} = \frac{\Delta_2'}{\Delta} = \frac{As^3+Bs^2+Cs+D}{Es^3+Fs^2+Gs+H} \tag{7.13}$$

$$A = C_1C_2C_3$$
$$B = C_1C_3G_1 + C_1C_3G_3$$
$$C = C_1G_1G_3 + C_3G_1G_3$$
$$D = G_1G_2G_3$$
$$E = A = C_1C_2C_3$$
$$F = C_1C_3G_1 + C_1C_3G_3 + C_1C_2G_3 + C_2C_3G_3 + C_2C_3G_2$$
$$G = C_1G_1G_3 + C_3G_1G_3 + C_3G_1G_2 + C_2G_2G_3 + C_3G_2G_3$$
$$H = D = G_1G_2G_3$$

(The term G should not be confused with conductivity symbol)

Alternative form of VTF expression is obtained by dividing both numerator and denominator polynomials of $H(s)$ by the coefficient $a = e$,

$$H(s) = \frac{V_o(s)}{V_{\text{in}}(s)} = \frac{s^3 + b_2s^2 + b_1s + b_0}{Es^3 + a_2s^2 + a_1s + a_0} \tag{7.14}$$

$$b_2 = \frac{1}{R_1C_2} + \frac{1}{R_3C_2}$$
$$b_1 = \frac{1}{R_1R_3C_2C_3} + \frac{1}{R_1R_3C_1C_2}$$
$$b_0 = a_0 = \frac{1}{R_1R_2R_3C_1C_2C_3}$$
$$a_2 = \frac{1}{R_1C_2} + \frac{1}{R_3C_2} + \frac{1}{R_3C_3} + \frac{1}{R_3C_1} + \frac{1}{R_2C_1}$$
$$a_1 = \frac{1}{R_1R_3C_2C_3} + \frac{1}{R_1R_3C_1C_2} + \frac{1}{R_1R_2C_1C_2} + \frac{1}{R_2R_3C_1C_3} + \frac{1}{R_2R_3C_1C_2}$$

(b) $C_2 = 2C_1 = 2C_3 = 2C$, $R_2 = 0.5R_1 = 0.5R_3 = 0.5R$

$$H(s) = \frac{V_o(s)}{V_i(s)} = \frac{s^3 + \frac{1}{CR}s^2 + \frac{1}{R^2C^2}s + \frac{1}{R^3C^3}}{s^3 + \frac{5}{CR}s^2 + \frac{5}{R^3C^3}s + \frac{1}{R^3C^3}}$$

In this last equation, let

$$\omega = a = \frac{1}{RC},$$
$$H(s) = \frac{V_o(s)}{V_{\text{in}}(s)} = \frac{s^3 + as^2 + a^2s + a^3}{s^3 + 5as^2 + 5a^2s + a^3} = \frac{(s+a)(s^2+a^2)}{(s+a)(s+4as+a^2)} = \frac{s^2+a^2}{s^2+4as+a^2}$$

This shows that a third-order response reduces into a second-order response.

(c) Let $R = 1\ \Omega$, $C = 1$ F,

$$H(s) = \frac{V_o(s)}{V_i(s)} = \frac{s^3 + s^2 + s + 1}{s^3 + 5s^2 + 5s + 1} = \frac{(s+1)(s^2+1)}{(s+3.7321)(s+1)(s+0.2679)}$$
$$= \frac{s^2+1}{(s+3.7321)(s+0.2679)} = \frac{s^2+1}{(s+4s+1)}$$

$\omega_0 = b_0 = a_0 = 1$ rad/s.

Problem 7.3.33 Derive the general VTF of a twin-T notch filter of Fig. 7.33 (in s-domain) using the concept of wye to delta transformation.

Solution

Wye to delta ($Y \rightarrow \Delta$) transformation rule (Fig. 7.35);

$$Y_A = \frac{Y_1 \cdot Y_3}{Y}, \quad Y_B = \frac{Y_1 \cdot Y_2}{Y}, \quad Y_C = \frac{Y_2 \cdot Y_3}{Y}, \quad Y = Y_1 + Y_2 + Y_3$$

Note that $Y \rightarrow \Delta$ transformation is done using admittances, while $\Delta \rightarrow Y$ transformation is done using impedances

$$Z_1 = \frac{Z_A \cdot Z_B}{Z}, \quad Z_2 = \frac{Z_B \cdot Z_C}{Z}, \quad Z_3 = \frac{Z_A \cdot Z_C}{Z}, \quad Z = Z_1 + Z_2 + Z_3$$

If the node at (c) is assumed to be the reference voltage (ground) (Figs. 7.36, 7.37 and 7.38),

$$Y'_A = \frac{Y'_1 \cdot Y'_3}{Y'} = \frac{(sC_2) \cdot G_3}{G_1 + G_3 + sC_2}$$
$$Y'_B = \frac{Y'_1 \cdot Y'_2}{Y'} = \frac{(sC_2) \cdot G_1}{G_1 + G_3 + sC_2}$$

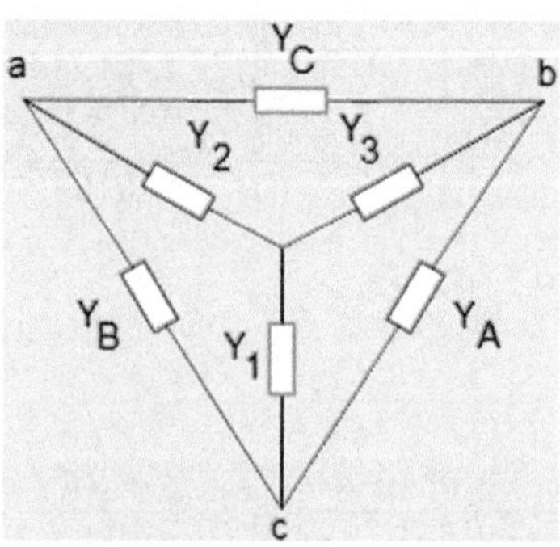

Fig. 7.35 Wye to delta ($Y \rightarrow \Delta$) transformation

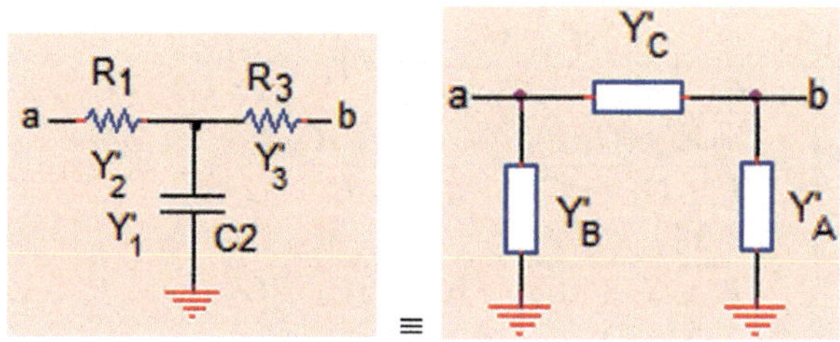

Fig. 7.36 First step of star – delta transformation

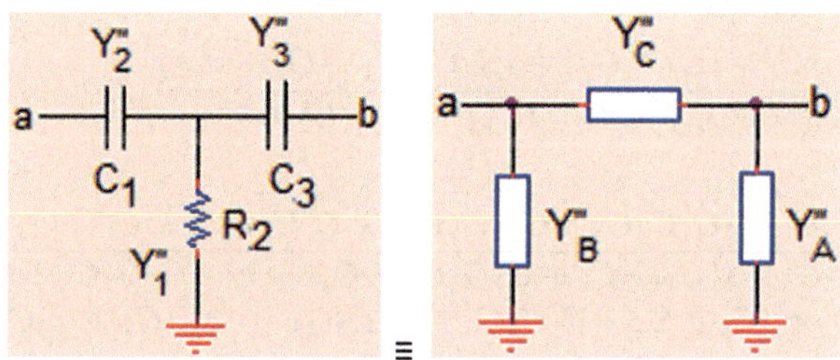

Fig. 7.37 Second step of star – delta transformation

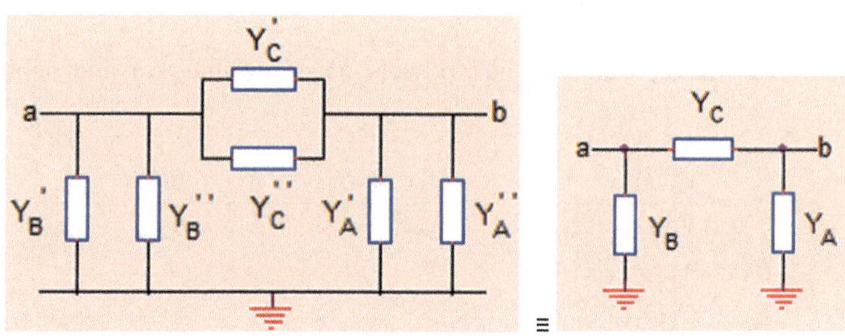

Fig. 7.38 Third step of star – delta transformation

$$Y_c' = \frac{Y_2' \cdot Y_3'}{Y'} = \frac{G_1 \cdot G_3}{G_1 + G_3 + sC_2}$$

$$Y_A'' = \frac{Y_1'' \cdot Y_3''}{Y''} = \frac{G_2 \cdot sC_3}{sC_1 + sC_3 + G_2}$$

$$Y_B'' = \frac{Y_1'' \cdot Y_2''}{Y''} = \frac{G_2 \cdot sC_1}{sC_1 + sC_3 + G_2}$$

$$Y_C'' = \frac{Y_2'' \cdot Y_3''}{Y''} = \frac{(sC_1)(sC_3)}{sC_1 + sC_3 + G_2}$$

$$Y_C = \frac{G_1 G_3}{G_1 + G_3 + sC_2} + \frac{sC_1 \cdot sC_3}{sC_1 + sC_3 + G_2} = Y_C' + Y_C''$$

$$Y_B = Y_B' + Y_B'' = \frac{sC_2 \cdot G_1}{G_1 + G_3 + sC_2} + \frac{G_2 \cdot sC_1}{sC_1 + sC_3 + G_2}$$

$$Y_A = Y_A' + Y_A'' = \frac{sC_2 \cdot G_3}{G_1 + G_3 + sC_2} + \frac{G_2 \cdot sC_3}{sC_1 + sC_3 + G_2}$$

$$V_{\text{out}} = V_b = V_{\text{in}} \cdot \left(\frac{Y_C}{Y_A + Y_C}\right) \quad \text{(Laplace operator } (s) \text{ is dropped for convenience)}$$

$$Y_C = \frac{G_1 G_3 (sC_1 + sC_3 + G_2) + s^2 C_1 C_3 (G_1 + G_3 + sC_2)}{(G_1 + G_3 + sC_2)(sC_1 + sC_3 + G_2)}$$

$$Y_A = \frac{sC_2 G_3 (sC_1 + sC_3 + G_2) + sG_2 C_3 (G_1 + G_3 + sC_2)}{(G_1 + G_3 + sC_2)(sC_1 + sC_3 + G_2)}$$

$$Y_C = \frac{sG_1G_3C_1 + sG_1G_3C_3 + G_1G_2G_3 + s^2C_1C_3G_1 + s^2C_1C_3G_3 + s^3C_1C_2C_3}{\begin{array}{c} G_1G_2G_3 + s(G_1G_3C_1 + G_1G_2C_3 + C_2G_2G_3 + G_1G_2C_3 + G_2G_3C_3) + \\ s^2(C_1C_3G_1 + C_1C_3G_3 + C_1C_2G_3 + C_2C_3G_3 + C_2C_3G_2) \end{array}}$$

$$Y_A = \frac{s^2C_1C_2G_3 + s^2C_2C_3G_3 + sC_2G_2G_3 + sG_1G_2C_3 + sG_2G_3C_3 + s^2C_2C_3G_2}{\begin{array}{c} G_1G_2G_3 + s(G_1G_3C_1 + G_1G_2C_3 + C_2G_2G_3 + G_1G_2C_3 + G_2G_3C_3) + \\ s^2(C_1C_3G_1 + C_1C_3G_3 + C_1C_2G_3 + C_2C_3G_3 + C_2C_3G_2) \end{array}}$$

Collecting terms according to like powers of s in numerator and denominator yields

$$H(s) = \frac{V_o(s)}{V_i(s)} = \frac{Y_C}{Y_A + Y_C} = \frac{B_3 s^3 + B_2 s^2 + B_1 s + B_0}{A_3 s^3 + A_2 s^2 + A_1 s + A_0}$$

$$B_3 = A_3 = C_1 C_2 C_3$$

$$B_2 = C_1 C_3 G_1 + C_1 C_3 G_3$$

$$B_1 = G_1 G_3 C_1 + G_1 G_3 C_3$$

$$B_0 = A_0 = G_1 G_2 G_3$$

$$A_2 = C_1C_3G_1 + C_1C_3G_3 + C_1C_2G_3 + C_2C_3G_3 + C_2C_3G_2$$

$$A_1 = C_1G_1G_3 + C_3G_1G_3 + C_2G_2G_3 + C_3G_1G_2 + C_3G_2G_3$$

Further dividing each term by $C_1C_2C_3$,

$$H(s) = \frac{V_o(s)}{V_i(s)} = \frac{s^3 + b_2 s^2 + b_1 s + b_0}{s^3 + a_2 s^2 + a_1 s + a_0}$$

where,

$$b_2 = \frac{1}{C_2R_1} + \frac{1}{C_2R_3}$$
$$b_1 = \frac{1}{R_1R_3C_2C_3} + \frac{1}{R_1R_3C_1C_2}$$
$$b_0 = a_0 = \frac{1}{C_1C_2C_3R_1R_2R_3}$$
$$a_2 = \frac{1}{C_2R_1} + \frac{1}{C_2R_3} + \frac{1}{C_3R_3} + \frac{1}{C_1R_3} + \frac{1}{C_1R_2}$$
$$a_1 = \frac{1}{R_1R_3C_2C_3} + \frac{1}{R_1R_3C_1C_2} + \frac{1}{R_1R_2C_1C_2} + \frac{1}{R_2R_3C_1C_3} + \frac{1}{R_2R_3C_1C_2}$$

Problem 7.3.34 Derive the voltage transfer function (VTF) in s-domain for a twin-T notch filter circuit shown in Fig. 7.39, using mesh analysis (symbolic_TT_mesh.m).

Solution

(a) Equivalent circuit of twin-T notch filter for mesh analysis is shown in Fig. 7.3.34.
KVL around four meshes,

$$[Z][I] = [V]$$

$$\begin{bmatrix} R_2 + 1/sC_1 & 0 & 0 & -R_2 \\ 0 & R_1 + 1/sC_1 & -1/sC_2 & 0 \\ 0 & -1/sC_2 & R_3 + R_L + 1/sC_2 & -R_L \\ -R_2 & 0 & -R_L & R_2 + R_L + 1/sC_3 \end{bmatrix} \begin{bmatrix} I_1 \\ I_2 \\ I_3 \\ I_4 \end{bmatrix} = \begin{bmatrix} -V_{in} \\ V_{in} \\ 0 \\ 0 \end{bmatrix}$$

$$V_o = (I_3 - I_4) \cdot R_4 \tag{7.15}$$

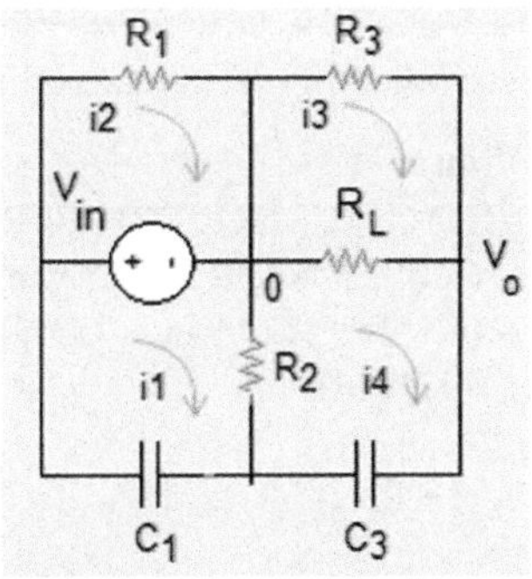

Fig. 7.39 Equivalent circuit of twin-T notch filter for mesh analysis

Using Cramer's rule,

$$i_3 = \frac{\Delta_3}{\Delta}, \quad i_4 = \frac{\Delta_4}{\Delta}$$
$$V_o = \frac{R_L}{\Delta} \cdot (\Delta_3 - \Delta_4) \tag{7.16}$$

$$\Delta_3 = \begin{vmatrix} R_2 + 1/sC_1 & 0 & V_{\text{in}} & -R_2 \\ 0 & R_1 + 1/sC_1 & -V_{\text{in}} & 0 \\ 0 & -1/sC_2 & 0 & -R_L \\ -R_2 & 0 & 0 & R_2 + R_L + 1/sC_3 \end{vmatrix}$$

$$\Delta_4 = \begin{vmatrix} R_2 + 1/sC_1 & 0 & 0 & V_{\text{in}} \\ 0 & R_1 + 1/sC_1 & -1/sC_2 & -V_{\text{in}} \\ 0 & -1/sC_2 & R_3 + R_L + 1/sC_2 & 0 \\ -R_2 & 0 & -R_L & 0 \end{vmatrix}$$

Determinant,

$$\Delta = \begin{vmatrix} R_2 + 1/sC_1 & 0 & 0 & -R_2 \\ 0 & R_1 + 1/sC_1 & -1/sC_2 & 0 \\ 0 & -1/sC_2 & R_3 + R_L + 1/sC_2 & -R_L \\ -R_2 & 0 & -R_L & R_2 + R_L + 1/sC_3 \end{vmatrix}$$

An easier way to find out this transfer function using these values in Eq. (7.16) is to employ MATLAB symbolic toolbox. Following are the results in doing so:

```
Vout =
(C1*C2*C3*RL*s^3*(((C1*C2*R1*R2*R3*Vin + C1*C2*R1*R2*RL*Vin)
*s^2 + (C1*R1*R2*Vin + C1*R2*R3*Vin)*s - RL*Vin)/(C1*C2*s^2) + (- C1*C2*
C3*R1*R2*RL*Vin*s^3 + (C1*R2*Vin + C3*R2*Vin + C3*RL*Vin)*s + Vin)/
(C1*C2*C3*s^3)))
/(C1*C2*C3*R1*R2*R3*RL*s^3 + (C1*C2*R1*R2*R3 + C2*C3*R1*R2*R3 +
C1*C2*R1*R2*RL + C1*C3*R1*R2*RL + C2*C3*R1*R2*RL + C1*C3*R2*
R3*RL + C2*C3*R1*R3*RL)*s^2 + (C1*R1*R2 + C1*R2*R3 + C2*R1*R3 +
C3*R1*R2 + C3*R2*R3 + C1*R2*RL + C2*R1*RL + C3*R1*RL + C3*R2*RL
+ C3*R3*RL)*s + R1 + R3 + RL)
```

Collecting terms according to powers of s

```
Vout =
((C1*C2*C3*R1*R2*R3*RL*Vin)*s^3
+ (C1*C3*R1*R2*RL*Vin + C1*C3*R2*R3*RL*Vin)*s^2
+ (C1*R2*RL*Vin + C3*R2*RL*Vin)*s
+ RL*Vin)
/
```

((C1*C2*C3*R1*R2*R3*RL)*s^3
+ (C1*C2*R1*R2*R3 + C2*C3*R1*R2*R3 + C1*C2*R1*R2*RL + C1*C3*R1*R2*RL + C2*C3*R1*R2*RL + C1*C3*R2*R3*RL + C2*C3*R1*R3*RL)*s^2
+ (C1*R1*R2 + C1*R2*R3 + C2*R1*R3 + C3*R1*R2 + C3*R2*R3 + C1*R2*RL
+ C2*R1*RL + C3*R1*RL + C3*R2*RL + C3*R3*RL)*s
+ R1 + R3 + RL)

Factoring out input voltage term in denominator,

$$H(s) = \frac{V_{\text{out}}(s)}{V_{\text{in}}(s)} = \frac{N(s)}{D(s)}$$

= ((C1*C2*C3*R1*R2*R3*RL)*s^3
+ (C1*C3*R1*R2*RL + C1*C3*R2*R3*RL)*s^2
+ (C1*R2*RL + C3*R2*RL)*s
+ RL)
/
((C1*C2*C3*R1*R2*R3*RL)*s^3
+ (C1*C2*R1*R2*R3 + C2*C3*R1*R2*R3 + C1*C2*R1*R2*RL + C1*C3*R1*R2*RL + C2*C3*R1*R2*RL + C1*C3*R2*R3*RL + C2*C3*R1*R3*RL)*s^2
+ (C1*R1*R2 + C1*R2*R3 + C2*R1*R3 + C3*R1*R2 + C3*R2*R3 + C1*R2*RL + C2*R1*RL + C3*R1*RL + C3*R2*RL + C3*R3*RL)*s + R1 + R3
+ RL)

R_L factorization both numerator and denominator polynomials finally result in canceling of R_L coefficients in numerator and denominator,

$$\begin{aligned}
N(s) &= C_1C_2C_3R_1R_2R_3s^3 + (C_1C_3R_1R_2 + C_1C_3R_2R_3)s^2 + (C_1R_2 + C_3R_2)s + 1 \\
D(s) &= C_1C_2C_3R_1R_2R_3s^3 \\
&\quad + \left[\frac{C_1C_2R_1R_2R_3 + C_2C_3R_1R_2R_3}{R_L} + C_1C_2R_1R_2 + C_1C_3R_1R_2\right. \\
&\quad \left. + C_2C_3R_1R_2 + C_1C_3R_2R_3 + C_2C_3R_1R_3\right]s^2 \\
&\quad + \left(C_1R_2 + C_2R_1 + C_3R_1 + C_3R_2 + C_3R_3\right. \\
&\quad \left. + \frac{C_1R_1R_2 + C_1R_2R_3 + C_2R_1R_3 + C_3R_1R_2 + C_3R_2R_3}{R_L}\right)s \\
&\quad + 1 + \frac{R_1 + R_3}{R_L}
\end{aligned}$$

Let the load resistance be very large,

$$H(s) = \frac{V_{\text{out}}(s)}{V_{\text{in}}(s)} = \frac{N(s)}{D(s)}$$

$$N(s) = C_1C_2C_3R_1R_2R_3s^3 + (C_1C_3R_1R_2 + C_1C_3R_2R_3)s^2 + (C_1R_2 + C_3R_2)s + 1$$

$$\begin{aligned} D(s) = {} & C_1C_2C_3R_1R_2R_3s^3 \\ & + [C_1C_2R_1R_2 + C_1C_3R_1R_2 + C_2C_3R_1R_2 + C_1C_3R_2R_3 + C_2C_3R_1R_3]s^2 \\ & + (C_1R_2 + C_2R_1 + C_3R_1 + C_3R_2 + C_3R_3)s + 1 \end{aligned}$$

Dividing each term by $(C_1C_2C_3R_1R_2R_3)$,

$$H(s) = \frac{V_o(s)}{V_i(s)} = \frac{s^3 + b_2s^2 + b_1s + b_0}{s^3 + a_2s^2 + a_1s + a_0}$$

where,

$$b_2 = \frac{1}{C_2R_1} + \frac{1}{C_2R_3}$$

$$b_1 = \frac{1}{R_1R_3C_2C_3} + \frac{1}{R_1R_3C_1C_2}$$

$$b_0 = a_0 = \frac{1}{C_1C_2C_3R_1R_2R_3}$$

$$a_2 = \frac{1}{C_2R_1} + \frac{1}{C_2R_3} + \frac{1}{C_3R_3} + \frac{1}{C_1R_3} + \frac{1}{C_1R_2}$$

$$a_1 = \frac{1}{R_1R_3C_2C_3} + \frac{1}{R_1R_3C_1C_2} + \frac{1}{R_1R_2C_1C_2} + \frac{1}{R_2R_3C_1C_3} + \frac{1}{R_2R_3C_1C_2}$$

Problem 7.3.35 Investigate the effect of load resistance, R_L, connected at the output of a twin-T notch filter of Fig. 7.33 on its voltage transfer function ($C_2 = 2C_1 = 2C_3$, $R_2 = 0.5R_1 = 0.5R_3 = 0.5R$).

Solution

Consider the General VTF of a twin-T notch filter including the load resistance R_L connected to the output terminal (one should refer to the derivation of VTF for a twin-T notch filter, for instance, using mesh analysis);

$$H(s) = \frac{V_{\text{out}}(s)}{V_{\text{in}}(s)} = \frac{N(s)}{D(s)}$$

Load resistance term does not appear in numerator, but in denominator of the VTF,

$$N(s) = C_1C_2C_3R_1R_2R_3s^3 + (C_1C_3R_1R_2 + C_1C_3R_2R_3)s^2 + (C_1R_2 + C_3R_2)s + 1$$

Table 7.1 Denominator polynomials for various values of load resistors

R_L (Ω)	Denominator polynomial
1	$s^3 + 7s^2 + 9s + 3$
10	$s^3 + 5.2s^2 + 5.4s + 1.2$
100	$s^3 + 5.02s^2 + 5.04s + 1.02$
1000	$s^3 + 5.002s^2 + 5.004s + 1.002$
10,000	$s^3 + 5.0002s^2 + 5.0004s + 1.0002$
100,000	$s^3 + 5s^2 + 5s + 1$

$$\begin{aligned}
D(s) = {} & C_1C_2C_3R_1R_2R_3s^3 \\
& + \left[\frac{C_1C_2R_1R_2R_3 + C_2C_3R_1R_2R_3}{R_L} + C_1C_2R_1R_2 + C_1C_3R_1R_2\right. \\
& \left. + C_2C_3R_1R_2 + C_1C_3R_2R_3 + C_2C_3R_1R_3\right]s^2 \\
& + \left(C_1R_2 + C_2R_1 + C_3R_1 + C_3R_2 + C_3R_3\right. \\
& \left. + \frac{C_1R_1R_2 + C_1R_2R_3 + C_2R_1R_3 + C_3R_1R_2 + C_3R_2R_3}{R_L}\right)s \\
& + 1 + \frac{R_1 + R_3}{R_L}
\end{aligned}$$

Numerical evaluation of denominator polynomial for various values of load resistors is given in Table 7.1.

Therefore, good rejection performance is exhibited by the filter (even under variable load conditions), as long as $R_L \gg R$.

Problem 7.3.36 A twin-T notch filter shown in Fig. 7.33 has the following voltage transfer function;

$$H(s) = \frac{V_o(s)}{V_i(s)} = \frac{b_3s^3 + b_2s^2 + b_1s + b_0}{a_3s^3 + a_2s^2 + a_1s + a_0}$$

$$\begin{aligned}
b_3 &= a_3 = C_1C_2C_3 \\
b_2 &= C_1C_3G_1 + C_1C_3G_3 \\
b_1 &= G_1G_3C_1 + G_1G_3C_3 \\
b_0 &= a_0 = G_1G_2G_3 \\
a_2 &= C_1C_3G_1 + C_1C_3G_3 + C_1C_2G_3 + C_2C_3G_3 + C_2C_3G_2 \\
a_1 &= C_1G_1G_3 + C_3G_1G_3 + C_2G_2G_3 + C_3G_1G_2 + C_3G_2G_3
\end{aligned}$$

Plot the voltage transfer function (VTF) magnitude (i.e., $|H(s)|$) in s-domain on the same log–log graph using the following component values:

$C_1 = C_2 = C_3 = 1$ F, $R_1 = R_2 = R_3 = 1$ Ω
$C_1 = C_2 = C_3 = 1$ F, $R_1 = R_2 = 1$ Ω, $R_3 = 0.5$ Ω
$C_1 = C_2 = C_3 = 1$ F, $R_2 = R_3 = 1$ Ω, $R_1 = 0.5$ Ω

$C_1 = C_2 = C_3 = 1$ F, $R_1 = R_2 = 1\ \Omega$, $R_3 = 0.25\ \Omega$
$C_1 = C_3 = 1$ F, $R_1 = R_3 = 1\ \Omega$, $C_2 = 2$ F, $R_2 = 0.5\ \Omega$

Draw possible conclusions out of these magnitude versus frequency plots (plot_transfer_function1.m).

Solution
Following transfer functions are obtained using the general form of voltage transfer function for a twin-T notch filter

(a) $$H(s) = \frac{s^3 + 2s^2 + 2s + 1}{s^3 + 5s^2 + 5s + 1} = \frac{(s+1)(s^2+s+1)}{(s+3.7321)(s+1)(s+0.2679)} = \frac{s^2+s+1}{(s+3.7321)(s+0.2679)}$$

$$H(s) = \frac{s^2+s+1}{s^2+4s+1}$$

(b) $$H(s) = \frac{s^3+3s^2+4s+2}{s^3+8s^2+9s+2} = \frac{(s+1)(s^2+2s+2)}{(s+1)(s^2+7s+2)} = \frac{s^2+2s+2}{s^2+7s+2}$$

(c) $$H(s) = \frac{s^3+3s^2+4s+2}{s^3+6s^2+8s+2} = \frac{(s+1)(s^2+2s+2)}{s^3+6s^2+8s+2}$$

Denominator polynomial does not factor out with (s + 1).

(d) $$H(s) = \frac{s^3+5s^2+8s+4}{s^3+14s^2+17s+4} = \frac{(s+1)(s^2+4s+4)}{(s+1)(s^2+13s+4)} = \frac{s^2+4s+4}{s^2+13s+4}$$

(e) $$H(s) = \frac{s^3+s^2+s+1}{s^3+5s^2+5s+1} = \frac{(s+1)(s^2+1)}{(s+3.7321)(s+1)(s+0.2679)} = \frac{s^2+1}{(s+3.7321)(s+0.2679)}$$

$$H(s) = \frac{s^2+1}{(s+4s+1)}$$

These are plotted in Fig. 7.3.35 (using MATLAB, plot_transfer_function1.m file).

It is seen here that a twin-T filter topology is practically efficient only if proper component values are used to realize such a filter function. Therefore, VTF of part (e) is the best option which satisfies the need for a good rejection at the notch frequency while passing other signals at lower or higher frequencies than the notch frequency.

If $R_1 = R_3 = R$, $R_2 = R/2$, $C_1 = C_3 = C$, $C_2 = 2C$ the notch frequency is $f_0 = 1/(2\pi RC)$

Using $C = C_1 = C_3 = 1$ F, $R = R_1 = R_3 = 1\ \Omega$, $C_2 = 2$ F, $R_2 = 0.5\ \Omega$, $f_0 = 1/(2\pi.1.1) = 0.159\,\text{Hz}$ which is validated by the plot of this VTF. Because of the fact that the condition stated here is not met for $H(s)$ given in (a), a sharp notch is not obtained for this VTF, despite the notch frequencies are the same for both of the filter transfer functions in (a) and (e) (Fig. 7.40).

Problem 7.3.37 The VTF of a bridged-T notch filter circuit shown in Fig. 7.41 is

$$H(s) = \frac{V_o}{V_{\text{in}}} = \frac{C_1C_2R_1R_2s^2 + (C_2R_2 + C_2R_1)s + 1}{C_1C_2R_1R_2s^2 + (R_1C_2 + R_2C_2 + R_1C_1)s + 1}$$

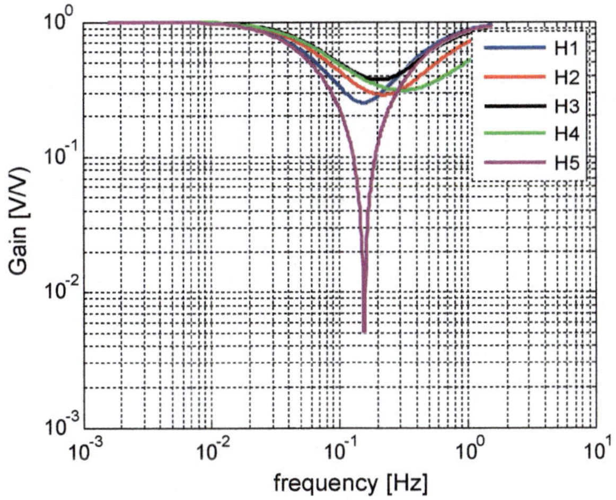

Fig. 7.40 Twin-T notch filter gain functions for different parameters

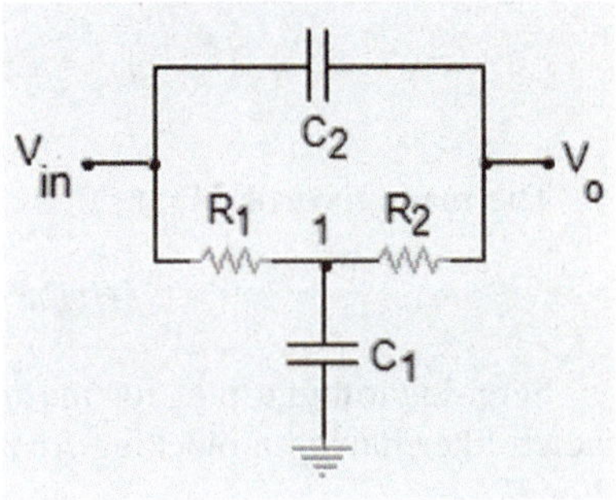

Fig. 7.41 Bridged-T notch filter

The critical (rejection) frequency is

$$\omega_c = \frac{1}{\sqrt{R_1 R_2 C_1 C_2}}$$

Design a filter to reject 50 Hz (hum) with $C_1/C_2 = 10$, $C_2 = 1$ nF, $R_1 = R_2$, calculate (numerically) $H(s)$, and Plot TF magnitude versus frequency on a semi-log graph.

What are the poles and zeros of this VTF? Plot the pole-zero diagram.

Plot separate Bode magnitude and phase diagrams for this VTF (bridgedT_TF1.m).

Solution

$$\omega_c = 2\pi f_c = \frac{1}{\sqrt{R_1 R_2 C_1 C_2}} = \frac{1}{\sqrt{R_1 \cdot R_2 \cdot 10 \cdot C_2 \cdot C_2}}$$

Let $R_1 = R_2$,

$$f_c = \frac{1}{2\pi \cdot R_2 \cdot C_2 \cdot \sqrt{10}} = 50\,[\text{Hz}]$$

$$R_2 = \frac{1}{2\pi \cdot \sqrt{10} \cdot 50 \cdot C_2} = \frac{1}{(993.4588)(10^{-9})} = 1.0066\,\text{M}\Omega$$

$$\text{Thus,} \quad R_2 = R_1 \cong 1.0066\,\text{M}\Omega,\ C_2 = 1\,\text{nF}, \quad C_1 = 10\,\text{nF}$$

Notch frequency = 50 Hz = 314.154 rad/s
Notch frequency squared = 98,693 rad/s
The voltage transfer function of the bridged-T circuit having these component values is

$$H(s) = \frac{0.0000101324s^2 + 0.002013s + 1}{0.0000101324s^2 + 0.01208s + 1}$$

or alternatively,

$$H(s) = \frac{s^2 + 198.7\text{s} + 98{,}693}{s^2 + 1192\text{s} + 98{,}693}$$

The magnitude of TF at 50 Hz is

$$|H(j2\pi \cdot 50)| = 0.1667\,\text{V/V}$$

Semi-logarithmic plot for the magnitude response of the VTF for a bridged T notch filter having a blocking frequency at f = 50 Hz (ω = 314 rad/s) is shown in Fig. 7.42.

The poles are $p_1 = -1102.6$, $p_2 = -89.5$ (rad/s)

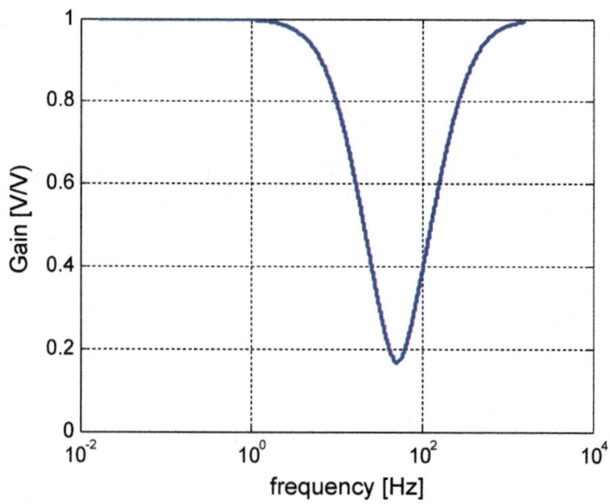

Fig. 7.42 Semi-log plot of magnitude response for a bridged T notch filter

The zeros are $z_1 = -99.34 + 298.03i$, $z_2 = -99.34 - 298.03i$ (rad/s), see Fig. 7.43.

In addition to semi-log plot of magnitude function, Bode magnitude and phase plots for the VTF of a bridged T notch filter having a blocking frequency at $f = 50$ Hz ($\omega = 314$ rad/s) is shown in Fig. 7.44.

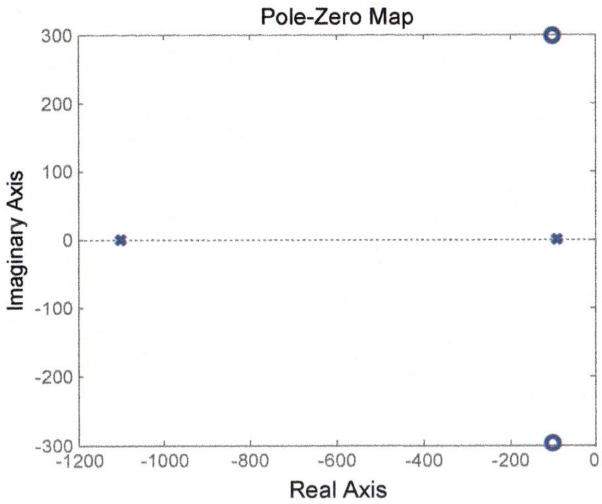

Fig. 7.43 Pole-zero plot for a bridged T notch filter

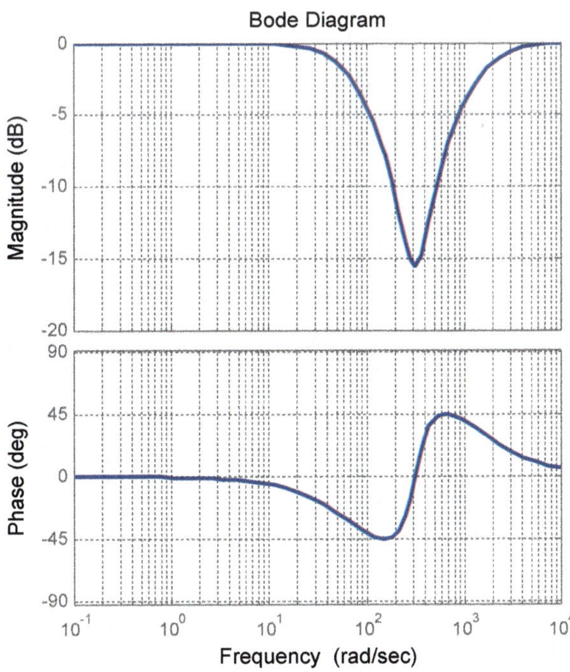

Fig. 7.44 Bode magnitude and phase plots for the VTF of a bridged T notch filter With the blocking frequency at 50 Hz (=314 rad/s)

7.4 Impedance and Admittance

Problem 7.4.1 Determine $Y(j\omega)$, the admittance of the circuit shown in Fig. 7.45, and determine the real and imaginary parts of $Y(j\omega)$. Plot them as functions of angular frequency.

(R_0 = 20 kΩ; C_1 = 1 nF; R_1 = 1 MΩ, at 101 measurement points) (GB1.m, RRC.m).

Solution

$$Z(j\omega) = R_0 + \frac{R_1}{1 + j\omega R_1 C_1} = \frac{R_0 + R_1 + j\omega R_1 C_1 R_0}{1 + j\omega R_1 C_1}$$

$$Y(j\omega) = \frac{1}{Z(j\omega)} = \frac{1 + j\omega R_1 C_1}{(R_0 + R_1) + j\omega R_0 C_1 R_1} = \frac{(1 + j\omega R_1 C_1)(R_0 + R_1 - j\omega R_1 C_1 R_0)}{(R_0 + R_1)^2 + \omega^2 R_0^2 R_1^2 C_1^2}$$

$$Y(j\omega) = \frac{R_0 + R_1 + \omega^2 R_0 R_1^2 C_1^2}{(R_0 + R_1)^2 + \omega^2 R_0^2 R_1^2 C_1^2} + j\omega \frac{C_1 R_1^2 + c_1 R_0 R_1 - R_0 R_1 C_1}{(R_0 + R_1)^2 + \omega^2 R_0^2 R_1^2 C_1^2} = Y'(\omega) + jY''(\omega)$$

Figure 7.46 displays conductance and susceptance as functions of angular frequency.

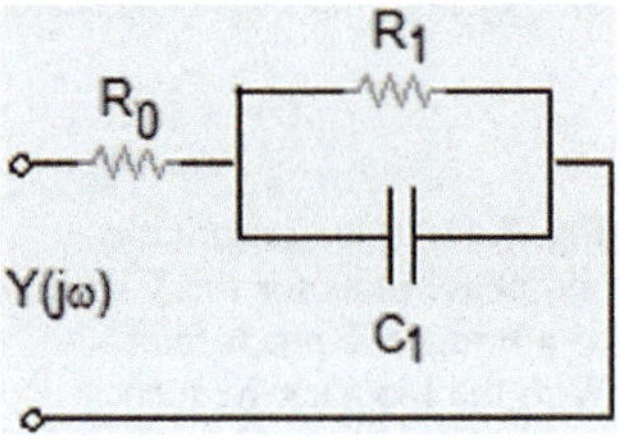

Fig. 7.45 The circuit for Problem 7.4.1

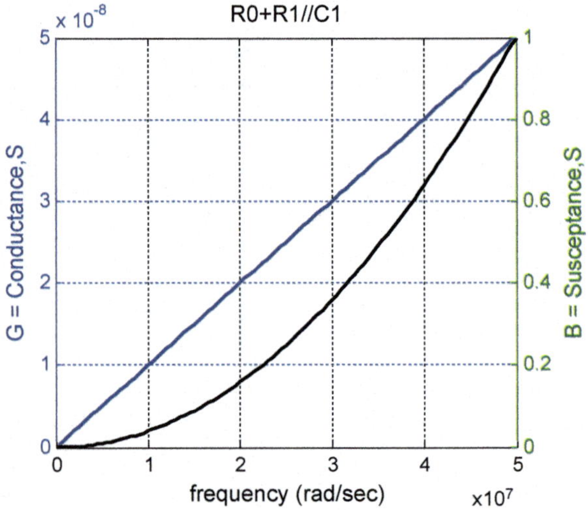

Fig. 7.46 Conductance (*blue*) and Susceptance (*black*) (R_0 = 20 kΩ; C_1 = 1 nF; R_1 = 1 MΩ)

(R_0 = 20 kΩ; C_1 = 1 nF; R_1 = 1 MΩ, at 101 measurement points).
A MATLAB script (GB1.m) is given below.

```
%GB1.m
%Real and imaginary parts of admittance
%n = number of intervals, h the distance between n+1 measurements
% program requires separate function call: RRC
R0=2e4; C1=1e-9; R1=1e6; w0=1/ (R1*C1) % R0+ (R1//C1) circuit
b=50000*w0; n=100;x=0:b/n:b;m=length(x);
[G,B]=RRC(x,R0,R1,C1,m);% R0+(R1//C1)circuit
[AX,H1,H2]=plotyy(x,B,x,G,'plot');grid on;
set(get(AX(1),'Ylabel'),'String','G = Conductance,S')
set(get(AX(2),'Ylabel'),'String','B = Susceptance,S')
xlabel('frequency (rad/sec)')
title('R0+R1//C1')
function [G,B] = RRC(x,R0,R1,C1,m)
%Susceptance and conductance terms of admittance model
x = x';
numG = R0+R1 + x.^2*(R0*R1*C1)^2;numB=x*C1*R1^2;
den =(R0+R1)^2 + x.^2*(R0*R1*C1)^2;
B = numB/den; G=numG/den; G=G(:,m); B=B(:,m);
```

Problem 7.4.2 Determine the admittance of a series RLC circuit, $Y(j\omega)$. Show the conductance and susceptance terms, and plot their variations as functions of angular frequency for C = 1 µF; R = 40 Ω; L = 1 mH, at 11 measurement points (GB2.m, BRLC.m).

Solution

$$Z(j\omega) = R + j\omega L + \frac{1}{j\omega C} = \frac{j\omega RC - \omega^2 LC + 1}{j\omega C}$$

$$Y(j\omega) = \frac{j\omega C}{1 - \omega^2 LC + j\omega RC} = \frac{j\omega C - j\omega^3 LC^2 + \omega^2 RC^2}{(1 - \omega^2 LC)^2 + \omega^2 R^2 C^2} = G(\omega) + jB(\omega)$$

The conductance G, and susceptance B, are

$$G(\omega) = \frac{\omega^2 RC^2}{(1 - \omega^2 LC)^2 + \omega^2 R^2 C^2}, \quad B(\omega) = \frac{\omega C - \omega^3 LC^2}{(1 - \omega^2 LC)^2 + \omega^2 R^2 C^2}$$

The results for a series RLC circuit with C = 1 µF; R = 40 W; L = 1 mH, at 11 measurement points is displayed in Fig. 7.47.

MATLAB script (GB2.m and function BRLC.m):

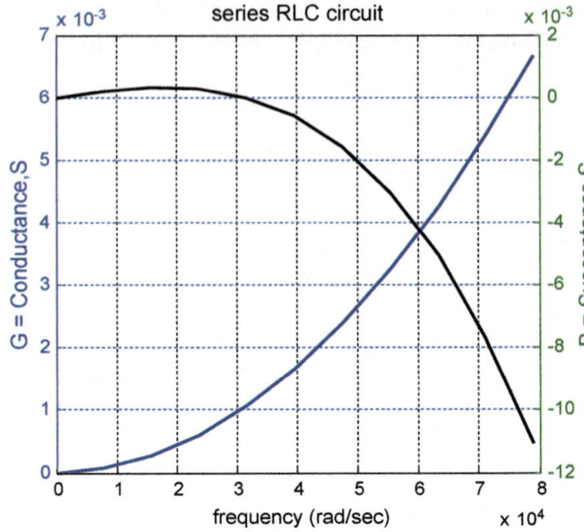

Fig. 7.47 Conductance(*blue*) and susceptance (*black*) for a series RLC circuit

```
%Real and imaginary parts of admittance
%n = number of intervals, h the distance between n+1 measurements
% program requires separate function call: BRLC
clc; clear all;
C=1e-6; R=40; L=1e-3; w0=1/sqrt(L*C) %series RLC circuit
b=2.5*w0; n=10;x=0:b/n:b;m=length(x);
[G,B]=BRLC(x,R,L,C,m) %series RLC circuit
[AX,H1,H2]=plotyy(x,G,x,B,'plot');grid on;
set(get(AX(1),'Ylabel'),'String','G = Conductance,S');
set(get(AX(2),'Ylabel'),'String','B = Susceptance,S') ;
xlabel('frequency (rad/sec)'); title('series RLC circuit');
function [G,B]=BRLC(x,R,L,C,m)
%Susceptance and conductance terms of admittance model
x=x';
numB = x*C-x.^3*L*C^2;den=(1-x.^2*L*C).^2+(x*R*C).^2;
numG = x.^2*R*C^2; B=numB/den; G=numG/den; G=G(:,m); B=B(:,m);
```

Problem 7.4.3 For the given impedance $Z(s)$, determine $Z(j\omega) = Z'(\omega) + jZ''(\omega)$ and its phase angle as a function of ω (Bode6.m).

$$Z(s) = \frac{s+2}{(s+1)(s+5)}$$

Solution

$$Z(j\omega) = \left.\frac{s+2}{s^2+6s+5}\right|_{s=j\omega} = \frac{j\omega+2}{(j\omega)^2+6j\omega+5} = \frac{j\omega+2}{-\omega^2+6j\omega+5}$$

$$= \frac{j\omega+2}{(5-\omega^2)+6j\omega} = \frac{(2+j\omega)[(5-\omega^2)+6j\omega]}{(5-\omega^2)^2+(6j\omega)^2} = \frac{(2+j\omega)(5-\omega^2)-(2+j\omega)6j\omega}{(5-\omega^2)^2+36\omega^2}$$

$$= \frac{(10-2\omega^2+j5\omega-j\omega^3)-12j\omega+6\omega^2}{25+\omega^4-10\omega^2+36\omega^2} = \frac{10-2\omega^2+6\omega^2+j(5\omega-\omega^3-12\omega)}{\omega^4+26\omega^2+25}$$

$$Z(j\omega) = \frac{4\omega^2+10}{\omega^4+26\omega^2+25} - j\frac{\omega^3+7\omega}{\omega^4+26\omega^2+25}$$

The phase angle is (Fig. 7.48),

$$\varphi(\omega) = \tan^{-1}\left[\frac{\mathrm{Im}[Z(j\omega)]}{\mathrm{Re}[Z(j\omega)]}\right] = \tan^{-1}\left(\frac{-\omega^3-7\omega}{4\omega^2+10}\right) = -\tan^{-1}\left(\frac{\omega^3+7\omega}{4\omega^2+10}\right)$$

Following is the MATLAB script (Bode6.m) for plotting the phase angle:

```
Num = [0 1 2]; den= [1 6 5]; Z=tf (Num, den); bode (Z)
```

Problem 7.4.4 Determine the phase angle (as a function of angular frequency) for the impedance function given below, and plot it (display the grid lines) (Bode6.m).

$$Z(s) = \frac{200s}{s^2+12s+20}$$

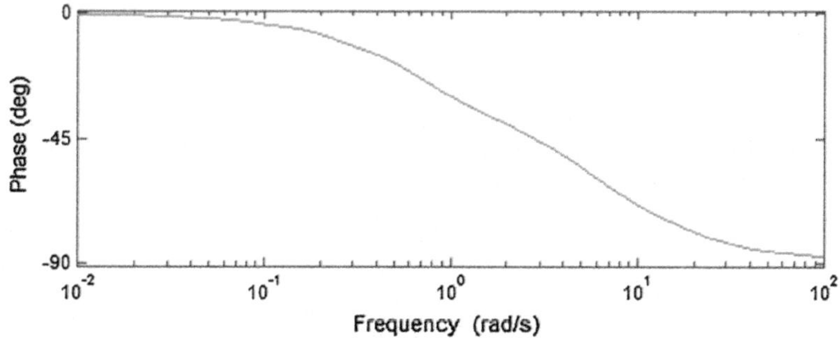

Fig. 7.48 The graph of phase angle versus frequency for *Z*(*s*) of Problem 7.4.3

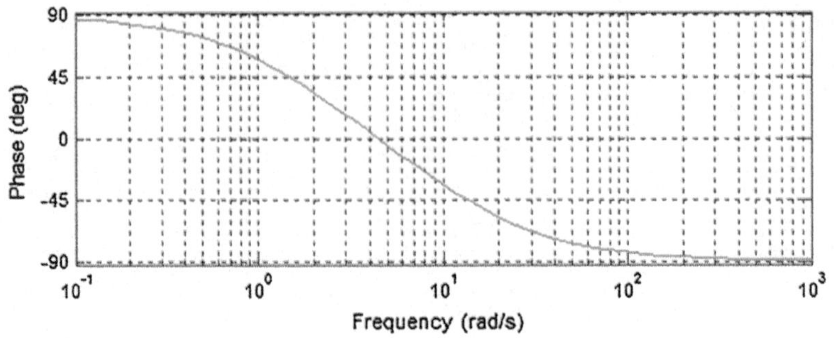

Fig. 7.49 Phase angle versus frequency for $Z(s) = 200s/(s^2 + 12s + 20)$

Solution

$$Z(j\omega) = \frac{200j\omega}{-\omega^2 + 12j\omega + 20} = \frac{200j\omega[(20 - \omega^2) - 12j\omega]}{[(20 - \omega^2) - 12j\omega][(20 - \omega^2) + 12j\omega]}$$

$$= \frac{4000j\omega - 200j\omega^3 + 2400\omega^2}{(20 - \omega^2)^2 - 144\omega^2} = \frac{2400\omega^2}{(20 - \omega^2)^2 - 144\omega^2} + j\frac{4000\omega - 200\omega^3}{(20 - \omega^2)^2 - 144\omega^2}$$

$$\varphi(\omega) = \tan^{-1}\left\{\frac{\mathrm{Im}[Z(j\omega)]}{\mathrm{Re}[Z(j\omega)]}\right\} = \tan^{-1}\left(\frac{-200\omega^3 + 4000\omega}{2400\omega^2}\right) = -\tan^{-1}\left(\frac{200\omega^3 - 4000\omega}{2400\omega^2}\right)$$

MATLAB script (Bode6.m) for Fig. 7.49:

```
num=[ 0 200 0];den=[ 1 12 20];Z=tf(num,den);bode(Z);grid on
```

Problem 7.4.5 Determine the phase angle expression for the following impedance function (Bode6.m):

$$Z(s) = \frac{11s + 2}{10s^2 + 11s + 1}$$

Solution

$$s = j\omega,$$

$$Z(j\omega) = \frac{11(j\omega) + 2}{10(j\omega)^2 + 11(j\omega) + 1} = \frac{2 + j11\omega}{1 - 10\omega^2 + j11\omega}$$

$$= \frac{(2 + j11\omega)[(1 - 10\omega^2) - j11\omega]}{(1 - 10\omega^2)^2 + 121\omega^2}$$

$$= \frac{2 - 20\omega^2 - 22\omega j + 11\omega j - 110j\omega^3 + 121\omega^2}{1 + 100\omega^4 - 20\omega^2 + 121\omega^2}$$

$$= \frac{101\omega^2 + 2 + j(-110\omega^3 - 11\omega)}{100\omega^4 + 101\omega^2 + 1}$$

$$\varphi(\omega) = \frac{\mathrm{Re}[Z(j\omega)]}{\mathrm{Im}[Z(j\omega)]} = -\tan^{-1}\left(\frac{110\omega^3 + 11\omega}{101\omega^2 + 2}\right)$$

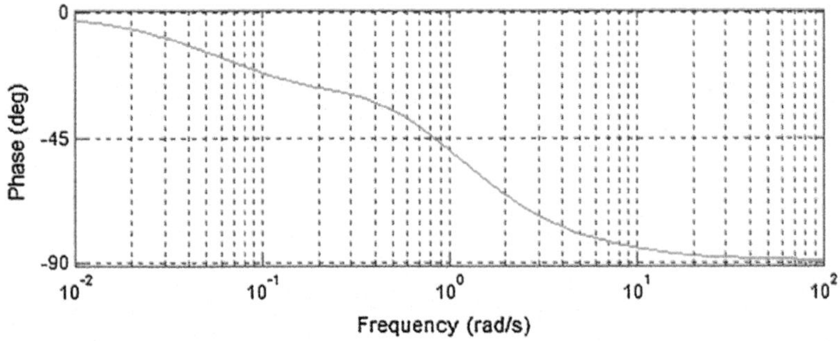

Fig. 7.50 Phase angle versus frequency for $Z(s) = (11s+2)/(10s^2+11s+1)$

Figure 7.50 displays phase function as a function of angular frequency.

Problem 7.4.6

(a) Determine the phase function from corresponding impedance functions given below.

$$\text{(i) } Z(s) = \frac{s^2+5s+4}{s^3+7s^2+10s} \quad \text{(ii) } Z(s) = \frac{s+1}{s^2+2s} \quad \text{(iii) } Z(s) = \frac{s^2+5s+4}{s^2+2s}$$

$$\text{(iv) } Z(s) = \frac{1}{s+1} \quad \text{(v) } Z(s) = \frac{s^3+12s^2+35s+24}{s^3+6.5s^2+11s+4} \quad \text{(vi) } Z(s) = \frac{11s+2}{10s^2+11s+1}$$

$$\text{(vii) } Z(s) = \frac{(s+3)(s+10)}{(s+1)(s+4)} \quad \text{(viii) } Z(s) = \frac{(s+3)(s+10)}{s(s+4)}$$

$$\text{(ix) } Z(s) = \frac{(s+1)(s+4)}{(s+3)(s+10)} \quad \text{(x) } Z(s) = \frac{s(s+4)}{(s+3)(s+10)}$$

$$\text{(xi) } Z(s) = \frac{s+2}{s^2+6s+5} \quad \text{(xii) } Z(s) = \frac{(s+1)(s+3)}{(s+2)(s+4)}$$

(b) Which forms of $Z(s)$ yield o phase function with an odd numerator polynomial? (Conversely, when dominator of a phase function is an even polynomial?)
(c) Which forms of $Z(s)$ yield o phase function with a degree of its numerator polynomial greater than the degree of its denominator?
(d) Which forms of $Z(s)$ yield o phase function with a degree of its denominator polynomial greater than the degree of its nominator? (Z_to_phase1.m) (Z_to_phase2.m).

Solution

(a)

$$\text{(i)} \quad Z(s) = \frac{s^2+5s+4}{s^3+7s^2+10s}, \quad \varphi(\omega) = -\tan^{-1}\left(\frac{\omega^4+21\omega^2+40}{2\omega^3+22\omega}\right)$$

$$\text{(ii)} \quad Z(s) = \frac{s+1}{s^2+2s} = \frac{s+1}{s(s+2)}, \quad \varphi(\omega) = -\tan^{-1}\left(\frac{\omega^2+2}{\omega}\right)$$

(iii) $Z(s) = \dfrac{s^2+5s+4}{s^2+2s} = \dfrac{(s+1)(s+4)}{s(s+2)}, \quad \varphi(\omega) = -\tan^{-1}\left(\dfrac{3\omega^2+8}{\omega^3+6\omega}\right)$

(iv) $Z(s) = \dfrac{1}{s+0.1}, \quad \varphi(\omega) = -\tan^{-1}(10\omega)$

(v)

$$Z(s) = \frac{s^3+12s^2+35s+24}{s^3+6.5s^2+11s+4}, \quad \varphi(\omega) = -\tan^{-1}\left(\frac{11\omega^5+151\omega^3+298\omega}{2\omega^6+64\omega^4+362\omega^2+192}\right)$$

(vi) $Z(s) = \dfrac{11s+2}{10s^2+11s+1}, \quad \varphi(\omega) = -\tan^{-1}\left(\dfrac{110\omega^3+11\omega}{101\omega^2+2}\right)$

(vii) $Z(s) = \dfrac{(s+3)(s+10)}{(s+1)(s+4)}, \quad \varphi(\omega) = -\tan^{-1}\left(\dfrac{8\omega^3+98\omega}{\omega^4+31\omega^2+120}\right)$

(viii) $Z(s) = \dfrac{(s+3)(s+10)}{s(s+4)}, \quad \varphi(\omega) = -\tan^{-1}\left(\dfrac{9\omega^2+120}{\omega^3+22\omega}\right)$

(ix) $Z(s) = \dfrac{(s+1)(s+4)}{(s+3)(s+10)}, \quad \varphi(\omega) = +\tan^{-1}\left(\dfrac{8\omega^3+98\omega}{\omega^4+31\omega^2+120}\right)$

(x) $Z(s) = \dfrac{s(s+4)}{(s+3)(s+10)}, \quad \varphi(\omega) = +\tan^{-1}\left(\dfrac{9\omega^3+120\omega}{\omega^4+22\omega^2}\right)$

(xi) $Z(s) = \dfrac{s+2}{s^2+6s+5} = \dfrac{s+2}{(s+1)(s+5)}, \quad \varphi(\omega) = -\tan^{-1}\left(\dfrac{\omega^3+7\omega}{4\omega^2+10}\right)$

(xii) $Z(s) = \dfrac{(s+1)(s+3)}{(s+2)(s+4)}, \quad \varphi(\omega) = +\tan^{-1}\left(\dfrac{2\omega^3+14\omega}{\omega^4+12\omega^2+24}\right)$

(b) Numerator of a phase function is an odd polynomial when there is no pole of denominator located at $s = 0$. This is the case in parts (iv), (v), (vi), (vii), (ix), (x), (xi), (xii). Conversely, numerator of a phase function is an even polynomial if there is a pole of denominator located at $s = 0$. This is the case in parts (i), (ii), (iii), and (viii).

(c) The order of the numerator polynomial of a phase function is greater than the degree of its numerator polynomial if the order of denominator polynomial of corresponding impedance function is greater than the order of its numerator polynomial. In other words, let

$$Z(s) = \frac{N(s)}{D(s)} = \frac{\sum_{j=0}^{m} b_j s^j}{\sum_{i=0}^{n} a_i s^i}$$

and corresponding phase function be

$$\varphi(\omega) = \tan^{-1}\left[\frac{P(\omega)}{Q(\omega)}\right]$$

If $n > m$, then, $[\text{order of } P(\omega)] > [\text{order of } Q(\omega)]$. This is the case in parts (i), (ii), (iv), (vi), (xi).

(d) The order of the numerator polynomial of a phase function is less than the degree of its denominator polynomial if the order of denominator polynomial of

corresponding impedance function is equal to the order of its numerator polynomial; in other words,
if $n = m$, then, $[\text{order of } P(\omega)] < [\text{order of } Q(\omega)]$. This is the case in parts (iii), (v), (vii), (viii), (ix), (x), (xii).
Note that, the conditions stated here are valid for the given functions, and for RC and RL impedances.

Problem 7.4.7 Given a driving point impedance function

$$Z(s) = \frac{V(s)}{I(s)} = \frac{s+2}{s+1}$$

Compute magnitude and phase angle values of $Z(j\omega)$ at 10 different angular frequencies between $0.01 \leq \omega \leq 100$ rad/s, and plot magnitude versus angular frequency. Print m-file script (freq_resp0.m).

Solution

Table 7.2 is a list of computed magnitude and phase angle values of $Z(j\omega)$ at 10 different angular frequencies between $0.01 \leq \omega \leq 100$ rad/s (phase angles are in degrees).

Figure 7.51 displays magnitude versus angular frequency for $Z(s) = (s+2)/(s+1)$

A MATLAB script (freq_resp0.m) is given below.

```
%N is numerator, D is denominator polynomial
%Generate an impedance transfer function in s domain
N=[1 2];D=[1 1];w=logspace(-2,2,10);H = tf(N,D) %Transfer function
[mag,phase] = bode(H,w); mag = squeeze(mag);phase = squeeze(phase);
disp(sprintf('w(rad/sec)    Mag    Phase (deg)'));
Z=[w'  mag   phase]
semilogx(w,mag,'r*-');grid on;
xlabel('w, rad/sec');ylabel('Magnitude, ohm');
```

Table 7.2 Magnitude and phase angle values

ω (rad/s)	Mag	Phase
0.01	1.9999	−0.2865
0.0278	1.9994	−0.7968
0.0774	1.9955	−2.2104
0.2154	1.9665	−6.0099
0.5995	1.7908	−14.2563
1.6681	1.3391	−19.2282
4.6416	1.0645	−11.1524
12.9155	1.0089	−4.3751
35.9381	1.0012	−1.5914
100.0000	1.0000	−0.5728

Fig. 7.51 Magnitude versus angular frequency for $Z(s) = (s+2)/(s+1)$

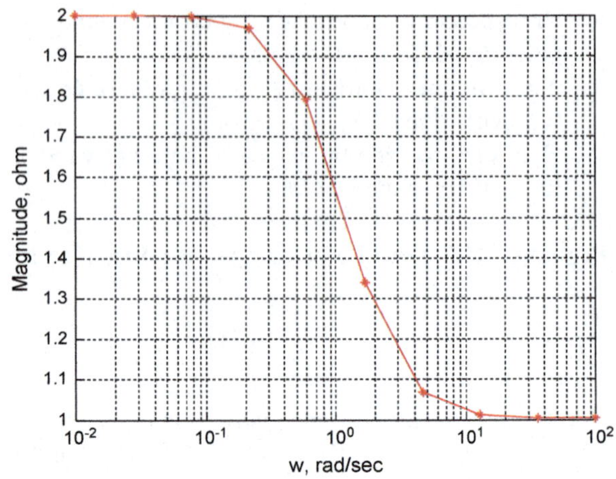

Problem 7.4.8 Let an DPI impedance function *Z*(*s*) be

$$Z(s) = \frac{s+3}{s+1}$$

(a) Determine the real and imaginary parts of $Z(j\omega)$.
(b) Check the results found above using a MATLAB symbolic m-file, and plot Re $[Z(j\omega)]$ and Im$[Z(j\omega)]$ on the same graph, $0.01 < \omega < 100$ rad/s (Real_Imag_TF_symb.m).
(c) Determine the magnitude and phase functions of $Z(j\omega)$.
(d) Plot magnitude and phase of $Z(j\omega)$ on the same graph, $0.01 < \omega < 100$ rad/s (Mag_phase_from_TF1.m).
(e) Use MATLAB and prepare an m-file script to sample $s = j\omega$ at arbitrary points between $0.01 < \omega < 100$ rad/s, then compute magnitude and phase angle of $Z(j\omega)$ at these angular frequencies, and print in a tabulated form.
(f) Plot (on the same graph), the magnitude and phase angle of $Z(j\omega)$ at the frequencies calculated above (Mag_phase_from_TF.m).

Solution

(a) The general form of $Z(j\omega)$;

$$Z(j\omega) = \frac{j\omega+3}{j\omega+1} = \frac{(j\omega+3)(1-j\omega)}{(1+j\omega)(1-j\omega)} = \frac{3+\omega^2-j2\omega}{\omega^2+1} = \frac{\omega^2+3}{\omega^2+1} - j\frac{2\omega}{\omega^2+1}$$

$$\text{Re}[Z(j\omega)] = \frac{\omega^2+3}{\omega^2+1}, \quad \text{Im}[Z(j\omega)] = \frac{-2\omega}{\omega^2+1}$$

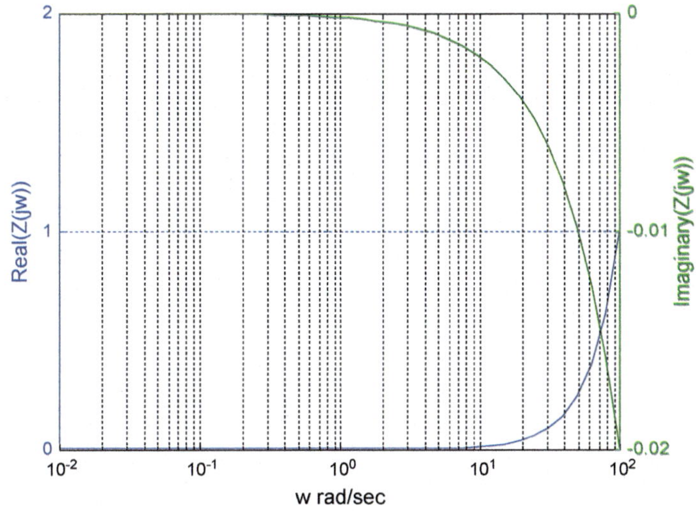

Fig. 7.52 Real and imaginary parts of Z(jω)

(b) A MATLAB script (Mag_phase_from_TF1.m) is given below. Real and imaginary parts of $Z(j\omega)$ are shown in Fig. 7.52.

```
%determine real and Imaginary parts of a TF
w = sym('w','real'); s=w*1i; H=(3+s)/(1+s)
R=factor(real(H))
I=factor(imag(H))
w=logspace(-2,2,40);w=w';
Re=(w.^2 + 3)/(w.^2 + 1);Re=Re(:,40);Im=-(2*w)/(w.^2 + 1);
Im=Im(:,40); [ax]=plotyy(w,Re,w,Im,@semilogx);grid on;
ylabel(ax(1),'Real(Z(jw))'); ylabel(ax(2),'Imaginary(Z(jw))');
xlabel(ax(2),'w rad/sec');
```

(c) Magnitude and phase functions of $Z(j\omega)$;

$$|Z(j\omega)| = \sqrt{\mathrm{Re}[Z(j\omega)]^2 + \mathrm{Im}[Z(j\omega)]^2} = \sqrt{\left(\frac{\omega^2+3}{\omega^2+1}\right)^2 + \left(\frac{-2\omega}{\omega^2+1}\right)^2} = \frac{\sqrt{\omega^4+10\omega^2+9}}{\omega^2+1}$$

$$\varphi[Z(j\omega)] = \arctan\left(\frac{\mathrm{Im}[Z(j\omega)]}{\mathrm{Re}[Z(j\omega)]}\right) = \arctan\left(\frac{\frac{-2\omega}{\omega^2+1}}{\frac{\omega^2+3}{\omega^2+1}}\right) = \arctan\left(\frac{-2\omega}{\omega^2+3}\right)$$

(d) Fig. 7.53 displays plots of magnitude and phase of $Z(j\omega)$ on the same graph, $0.01 < \omega < 100$ rad/s.

Table 7.3 Magnitude and phase angle of Z(jω)

ω (rad/s)	Mag	Phase
0.01	3.00	−0.38
0.03	3.00	−1.15
0.10	2.99	−3.80
0.30	2.89	−10.99
1.00	2.24	−26.57
2.00	1.61	−29.74
3.00	1.34	−26.57
10.00	1.04	−10.99
30.00	1.00	−3.80
100.00	1.00	−1.15

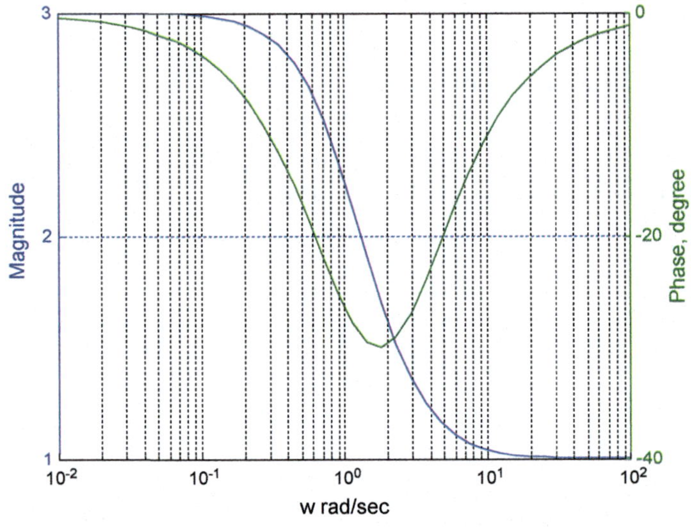

Fig. 7.53 Magnitude and phase graph of $Z(j\omega)$, $0.01 < \omega < 100$ rad/s

(e) Magnitude and phase angle of $Z(j\omega)$ at different angular frequencies are listed in Table 7.3 (phase angles are in degrees).

(f) Following is the MATLAB script to generate Table 7.3, as well as Fig. 7.53 displaying magnitude and phase of $Z(s)$ as a function of angular frequency while Fig. 7.54 displays data of Table 7.3.

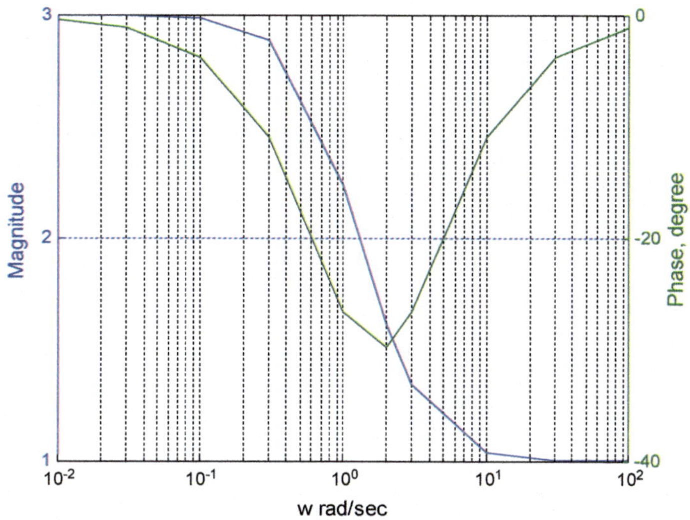

Fig. 7.54 Magnitude and phase graph of Z(jω), 0.01 < ω < 100 rad/s

```
% Computes mag and phase from a given tf H(s)=N(s)/D(s)
b=[1 1]; d=[1 3]; N=d; D=b;
H = tf(N,D);    %Transfer function
bode(H);
%nyquist(H)
w=[0.01 .03 .1 .3  1 2 3 10 30 100];
[mag,phase]=bode(H,w);
mag=squeeze(mag(1,1,:)); phase=squeeze(phase(1,1,:));
w=w';
disp(sprintf('w(rad/sec)    Mag   Phase (deg)'));
format bank
[w mag phase]
%plot mag and phase on the same graph (Double axis graph)
[ax]=plotyy(w,mag,w,phase,@semilogx);grid on;
ylabel(ax(1),'Magnitude');ylabel(ax(2),'Phase,  degree')
xlabel(ax(2),'w rad/sec')
```

Problem 7.4.9 Given a driving point impedance function

$$Z(s) = \frac{V(s)}{I(s)} = \frac{(s+1)(s+100)}{(s+0.1)(s+10)}$$

Determine Z(jω) and magnitude and phase angle values at 17 different angular frequencies between 0.01 $\leq$ ω $\leq$ 100 rad/s, and plot magnitude and phase angles versus angular frequency (freq_resp1.m).

Solution

$$Z(s) = \frac{V(s)}{I(s)} = \frac{(s+1)(s+100)}{(s+0.1)(s+10)}$$

Poles are at 0.1 and 10, zeros are at 1 and 100 rad/s.

Table 7.4 lists and Fig. 7.55 displays plots of magnitude and phase angles versus angular frequency for Z(jω).

Following is the MATLAB script (freq_resp1.m) used for the solution:

```
%driving point impedance (1-port)
%Z(s)=V(s)/I(s)=N(s)/D(s)
%Generate an impedance transfer function in s domain
a=[1 0.1];b=[1 1];c=[1 10];d=[1 100];
N=conv(b,d);D=conv(a,c);H = tf(N,D)    %Transfer function (for checking)
w = logspace(-2,2,17);  % 17 points on w
Z = freqs(N,D,w); mag = abs(Z); phase = angle(Z);
%       convert to hertz, degrees, and decibels,
%f = w/(2*pi);
%mag = 20*log10(mag);
phase = phase*180/pi;
%format long;
format short;
[w' Z'  mag'  phase']
subplot(2,1,1), semilogx(w,mag);grid on;
xlabel('w, rad/sec');ylabel('Magnitude, ohm');
subplot(2,1,2), semilogx(w,phase);grid on;xlabel('w, rad/sec');
ylabel('Phase, degree')
```

Table 7.4 Magnitude and phase angles versus angular frequency for Z(jω)

ω (rad/s)	Z(jω)	|Z(jω)| (Ω)	Phase (rad)
0.01	0.9910 + 0.0900i	0.9951	−0.0519
0.02	0.9722 + 0.1567i	0.9847	−0.0916
0.03	0.9174 + 0.2613i	0.9539	−0.1590
0.06	0.7818 + 0.3885i	0.8730	−0.2642
0.10	0.5459 + 0.4549i	0.7106	−0.3981
0.18	0.3100 + 0.3895i	0.4978	−0.5148
0.32	0.1743 + 0.2637i	0.3161	−0.5653
0.56	0.1194 + 0.1611i	0.2006	−0.5346
1.00	0.1000 + 0.0980i	0.1400	−0.4443
1.78	0.0922 + 0.0650i	0.1128	−0.3520
3.16	0.0853 + 0.0522i	0.1000	−0.3147
5.62	0.0724 + 0.0511i	0.0887	−0.3520
10.00	0.0510 + 0.0500i	0.0714	−0.4443
17.78	0.0297 + 0.0401i	0.0499	−0.5346
31.62	0.0174 + 0.0264i	0.0316	−0.5653
56.23	0.0125 + 0.0157i	0.0201	−0.5148
100.00	0.0108 + 0.0090i	0.0141	−0.3981

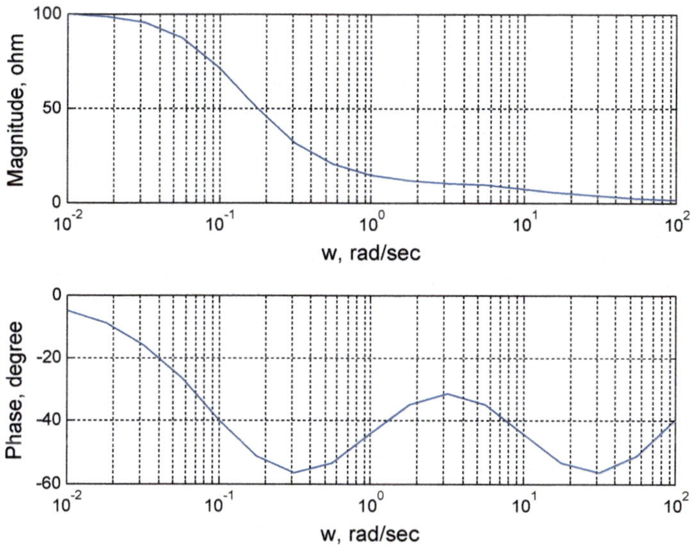

Fig. 7.55 Bode magnitude and phase graph of Z(jω), 0.01 < ω < 100 rad/s

Problem 7.4.10

(a) Draw a simplified electrical equivalent circuit for the electrical properties of human body. Indicate corresponding elements.
(b) Describe its behavior at low and high frequencies. What is the general form of its impedance in Laplace domain? What is the critical frequency?
(c) What is the definition of total body fluid volume $V_{TBF}(V_{TBW})$?
(d) Determine the ratio of high frequency impedance to the DC resistance of tissue, i.e, R_∞/R_E.
(e) Determine the ratio of total body fluid volume to the extracellular fluid volume, i.e., $V_{\mathrm{TBF}}/V_{\mathrm{ECF}}$. Note that,

$$R_E = K\rho_{\mathrm{ECF}}\left(\frac{V}{V_{\mathrm{ECF}}}\right)^{3/2}\frac{H^2}{V}, \quad V = \frac{W}{d}$$

where $K, \rho_{\mathrm{ECF}}, V, H, W, d$ are geometry factor (De Lorenzo's factor = 4.3), extracellular fluid resistivity, body volume, body height, body weight, and body density, respectively.
(f) Determine total body fluid volume, V_{TBF} in terms of V_{ECF}, ρ_{TBF}, ρ_{ECF}, R_E, and R_∞
(g) Rewrite equation for V_{TBF} in terms of V_{ECF}, ρ_{TBF}, ρ_{ECF}, R_E, and R_{I}.
(h) It has been shown that [1],

$$\frac{V_{\text{ECF}}}{V_{\text{TBF}}} = \left(\frac{\rho_{\text{ICF}} - \rho_{\text{TBF}}}{\rho_{\text{ICF}} - \rho_{\text{ECF}}}\right)\left(\frac{\rho_{\text{ECF}}}{\rho_{\text{TBF}}}\right)^{2/3}$$

Determine total body fluid resistivity, ρ_{TBF}, in terms of ρ_{ECF}, ρ_{ICF}, R_∞, and R_E.

(i) Express V_{TBF} in terms of ρ_{ECF}, ρ_{ICF}, R_I ,and R_E.

(j) If $V_{\text{TBF}} = 13.166$liters, $k_\rho = \rho_{\text{ICF}}/\rho_{\text{ECF}} = 6.76$, $R_E = 800\,\Omega$, $R_\infty = 450\,\Omega$, determine V_{TBF} and V_{ICF}.

(k) Calculate intracellular fluid resistance, R_I, using data given in (j).

(l) Determine percent fat volume of this person with W = 60 kg, using fat-free mass (FFM) equation

$$\text{FFM} = d_{\text{ECF}} \cdot V_{\text{ECF}} + d_{\text{ICF}} \cdot V_{\text{ICF}}$$

where, d_{ECF} is density of extracellular fluid (=1.106 g cm^{-3}), d_{ICF} is density of intracellular fluid (=1.521 g cm^{-3}) [2].

Solution

(a) Figure 7.56 shows an equivalent circuit for the electrical behavior of a biological tissue. Here, R_E, R_I, C_m denotes the resistance of extracellular fluid, resistance of intracellular fluid, and the cell membrane capacitance, respectively.

(b) At zero frequency (DC), electric current runs around the cell, therefore equivalent impedance consists of only R_E. At very high frequencies $(f = \infty)$, C_m acts as a short circuit and equivalent electrical impedance of tissue consists of parallel connected resistance, i.e, $R_I \| R_E$.

The general form of electrical impedance in s-domain is

$$Z = R_E \parallel \left(R_I + \frac{1}{sC_m}\right) = \frac{R_E + sR_IR_EC_m}{s(R_IC_m + R_EC_m) + 1}$$

With the critical frequency of

$$f_c = \frac{1}{2\pi C_m(R_E + R_I)}$$

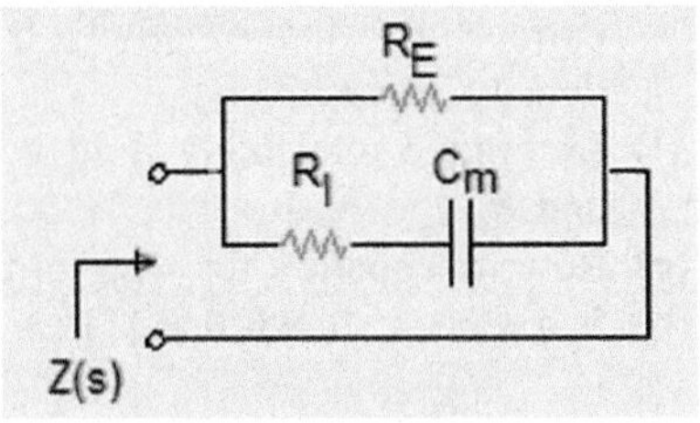

Fig. 7.56 Simplified electrical equivalent circuit for the electrical properties of human body

(c) TBF volume is the sum of the volumes of fluids contained within the cells and outside these cells (intra/extracellular fluids).

(d) $\frac{R_\infty}{R_E} = \frac{\frac{R_E R_I}{R_E + R_I}}{R_E} = \frac{R_I}{R_E + R_I}$

(e) Since

$$R_E = K\rho_{\text{ECF}} \left(\frac{V}{V_{\text{ECF}}}\right)^{3/2} \frac{H^2}{V}, \quad V = \frac{W}{d} \quad \rightarrow$$
$$V_{\text{ECF}} = \left(\frac{K\rho_{\text{ECF}}}{\sqrt{d}}\right)^{2/3} \left(\frac{H^2\sqrt{W}}{R_E}\right)^{2/3}$$

Because TBF is the conducting material at high frequencies one may replace R_E with R_∞, ECF with TBF in these equations,

$$R_\infty = K\rho_{\text{TBF}} \left(\frac{V}{V_{\text{TBF}}}\right)^{\frac{3}{2}} \frac{H^2}{V} \quad \rightarrow \quad V_{\text{TBF}} = \left(\frac{K\rho_{\text{TBF}}}{\sqrt{d}}\right)^{\frac{2}{3}} \left(\frac{H^2\sqrt{W}}{R_\infty}\right)^{2/3}$$
$$\frac{V_{\text{TBF}}}{V_{\text{ECF}}} = \left(\frac{\frac{K\rho_{\text{TBF}}}{\sqrt{d}}}{\frac{K\rho_{\text{ECF}}}{\sqrt{d}}}\right)^{\frac{2}{3}} \left(\frac{\frac{H^2\sqrt{W}}{R_\infty}}{\frac{H^2\sqrt{W}}{R_E}}\right)^{2/3} = \left(\frac{\rho_{\text{TBF}}}{\rho_{\text{ECF}}}\right)^{\frac{2}{3}} \left(\frac{R_E}{R_\infty}\right)^{2/3}$$

(f) Using the result found above, volume of total body fluid is

$$V_{\text{TBF}} = V_{\text{ECF}} \left(\frac{\rho_{\text{TBF}}}{\rho_{\text{ECF}}}\right)^{\frac{2}{3}} \left(\frac{R_E}{R_\infty}\right)^{2/3}$$

(g) Since

$$\frac{R_E}{R_I} = \frac{R_E + R_I}{R_I},$$

Substituting this into the resulting equation found in (f) yields,

$$V_{\text{TBF}} = V_{\text{ECF}} \left(\frac{\rho_{\text{TBF}}}{\rho_{\text{ECF}}}\right)^{\frac{2}{3}} \left(\frac{R_E + R_I}{R_I}\right)^{2/3}$$

(h) $\frac{V_{\text{ECF}}}{V_{\text{TBF}}} = \left(\frac{\rho_{\text{ICF}} - \rho_{\text{TBF}}}{\rho_{\text{ICF}} - \rho_{\text{ECF}}}\right) \left(\frac{\rho_{\text{ECF}}}{\rho_{\text{TBF}}}\right)^{\frac{2}{3}}$

On the other hand, the resulting equation of part (e) is

$$\frac{V_{\mathrm{TBF}}}{V_{\mathrm{ECF}}}=\left(\frac{\rho_{\mathrm{TBF}}}{\rho_{\mathrm{ECF}}}\frac{R_E}{R_\infty}\right)^{\frac{2}{3}}$$

Note that these two equations are reciprocal of each other. Therefore,

$$\left(\frac{\rho_{\mathrm{ICF}}-\rho_{\mathrm{TBF}}}{\rho_{\mathrm{ICF}}-\rho_{\mathrm{ECF}}}\right)\left(\frac{\rho_{\mathrm{ECF}}}{\rho_{\mathrm{TBF}}}\right)^{\frac{2}{3}}=\left(\frac{\rho_{\mathrm{ECF}}}{\rho_{\mathrm{TBF}}}\frac{R_E}{R_\infty}\right)^{\frac{2}{3}}$$

or,

$$\frac{\rho_{\mathrm{ICF}}-\rho_{\mathrm{TBF}}}{\rho_{\mathrm{ICF}}-\rho_{\mathrm{ECF}}}=\left(\frac{R_\infty}{R_E}\right)^{\frac{2}{3}}$$

Solving this equation for ρ_{TBF}:

$$\rho_{\mathrm{ICF}}-\rho_{\mathrm{TBF}}=(\rho_{\mathrm{ICF}}-\rho_{\mathrm{ECF}})\left(\frac{R_\infty}{R_E}\right)^{\frac{2}{3}}$$

$$\rho_{\mathrm{TBF}}=(\rho_{\mathrm{ECF}}-\rho_{\mathrm{ICF}})\left(\frac{R_\infty}{R_E}\right)^{\frac{2}{3}}+\rho_{\mathrm{ICF}}$$

(i) Using results found in part (h) and in part (d),

$$\frac{R_\infty}{R_E}=\frac{R_I}{R_E+R_I}$$

$$\rho_{\mathrm{TBF}}=(\rho_{\mathrm{ECF}}-\rho_{\mathrm{ICF}})\left(\frac{R_I}{R_E+R_I}\right)^{\frac{2}{3}}+\rho_{\mathrm{ICF}}$$

(j) Since total body fluid volume is given by the equation found in part (f) as

$$V_{\mathrm{TBF}}=V_{\mathrm{ECF}}\left(\frac{\rho_{\mathrm{TBF}}}{\rho_{\mathrm{ECF}}}\right)^{\frac{2}{3}}\left(\frac{R_E}{R_\infty}\right)^{\frac{2}{3}}$$

and total body resistivity is

$$\rho_{\mathrm{TBF}}=(\rho_{\mathrm{ECF}}-\rho_{\mathrm{ICF}})\left(\frac{R_\infty}{R_E}\right)^{\frac{2}{3}}+\rho_{\mathrm{ICF}}$$

Because the value of intracellular fluid resistivity is not given, one may divide both sides of this equation by ρ_{ECF},

$$\frac{\rho_{\text{TBF}}}{\rho_{\text{ECF}}} = \left(1 - \frac{\rho_{\text{ICF}}}{\rho_{\text{ECF}}}\right)\left(\frac{R_\infty}{R_E}\right)^{\frac{2}{3}} + \left(\frac{\rho_{\text{ICF}}}{\rho_{\text{ECF}}}\right)$$

Let

$$k_p = \frac{\rho_{\text{ICF}}}{\rho_{\text{ECF}}}$$

$$\frac{\rho_{\text{TBF}}}{\rho_{\text{ECF}}} = (1 - k_p)\left(\frac{R_\infty}{R_E}\right)^{\frac{2}{3}} + k_p$$

Since $k_p = 6.76 > 1$,

$$\frac{\rho_{\text{TBF}}}{\rho_{\text{ECF}}} = k_p - (k_p - 1)\left(\frac{R_\infty}{R_E}\right)^{\frac{2}{3}} = 6.76 - 5.76\left(\frac{450}{800}\right)^{\frac{2}{3}} = 6.76 - 3.924 = 2.836$$

Substituting measured data in total body fluid volume equation yields,

$$V_{\text{TBF}} = 13{,}166 \times (2.836)^{\frac{2}{3}} \times \left(\frac{800}{450}\right)^{\frac{2}{3}} = 38{,}732.64\,\text{cm}^3$$

$$V_{\text{TBF}} = 38.733\,\text{l}$$

Intracellular fluid volume is calculated as

$$V_{\text{ICF}} = V_{\text{TBF}} - V_{\text{ECF}} = 38.733 - 13.166 = 25.567\,\text{l}$$

(k) The ratio of DC to high frequency body resistance values is

$$\frac{R_E}{R_\infty} = \frac{R_E + R_I}{R_I},$$

Solving this equation for R_I and substituting known numerical values yields

$$R_I = \frac{R_E}{\frac{R_E}{R_\infty} - 1} = \frac{800}{\frac{800}{450} - 1} = 1028.6\,\Omega$$

(l) Substituting calculated volume values into given equation for FFM,

$$\begin{aligned}\text{FFM} &= 1.106 \cdot V_{\text{ECF}} + 1.521 \cdot V_{\text{ICF}} = 1.106 \cdot 13{,}166 + 1.521 \cdot 25{,}567 \\ &= 53{,}449.0\,\text{g}\end{aligned}$$

Fat = W − FFM = 60,000 − 53,449 = 6551 g

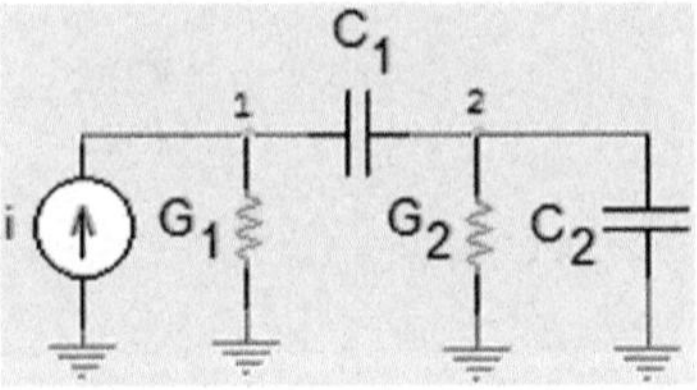

Fig. 7.57 The circuit for Problem 7.4.11

$$\%\text{Fat} = \frac{6551}{60{,}000} \times 100\% = 10.92\%$$

Problem 7.4.11

(a) Find the input impedance of the circuit shown in Fig. 7.57.
(b) Find the dual of this circuit.
(c) Find the input admittance of the dual circuit.
(d) Compare the results in parts (a) and (c).

Solution

$$Y_2 = G_2 \,//\, sC_2 = G_2 + sC_2$$

$$Y_{12} = \frac{sC_1 + Y_2}{Y_2 + sC_1} = \frac{sC_1(G_2 + sC_2)}{G_2 + SC_2 + sC_1}$$

(a)
$$Y_1 = G_1 + Y_{12} = G_1 + \frac{sC_1(G_2 + sC_2)}{G_2 + SC_2 + sC_1}$$

$$= \frac{G_1G_2 + sG_1C_1 + sG_1C_2 + sG_2C_1 + s^2C_1C_2}{G_2 + s(C_1 + C_2)}$$

$$Z_1 = \frac{V_1}{I} = \frac{G_2 + s(C_1 + C_2)}{s^2C_1C_2 + s(G_1C_1 + G_1C_2 + G_2C_1) + G_1G_2}$$

(b) Duality principle implies that,

current source → voltage source; i // G_1 branch → V − R_1 branch;
G_2 // C_2 branch → R_2 − L_2 branch; Series C_1 → shunt L_1;
Node voltages → Mesh currents.
Application of these conversions results in the circuit of Fig. 7.58.

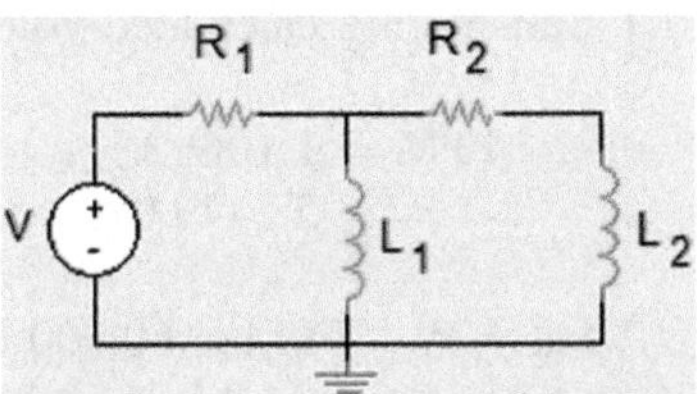

Fig. 7.58 Dual circuit

(c) $Z_1 = R_2 + sL_2, \quad Z_2 = Z_1 \parallel sL_1 = \dfrac{(R_2 + sL_2)(sL_1)}{(R_2 + sL_1) + sL_1}$

$$Z = R_1 + Z_2 = R_1 + \frac{sR_2L_1 + s^2L_1L_2}{R_2 + s(L_1 + L_2)} = \frac{R_1R_2 + s(L_1 + L_2)R_1 + sR_2L_1 + s^2L_1L_2}{R_2 + s(L_1 + L_2)}$$

$$Y = \frac{1}{Z} = \frac{R_2 + s(L_1 + L_2)}{s^2L_1L_2 + s(L_1R_1 + L_2R_1 + R_2L_1) + R_1R_2}$$

(d) Input impedance of the original circuit has the same form of the input admittance of the dual circuit, with (R → G, L → C, C → L).

7.5 Frequency and Component Scaling

Problem 7.5.1 Show that magnitude (impedance) scaling of resistance and inductance requires multiplication, while for capacitance division by the same factor.

Solution

Let individual R, L, C impedances in an AC circuit be $Z_R = R$, $Z_L = j\omega L$, $Z_C = 1/(j\omega C)$

Then, multiply R, L and divide C by some scalar factor "a,"

$$Z'_R = aR,\ Z_L = j\omega(aL),\ Z_C = \frac{1}{j\omega\left(\frac{C}{a}\right)} = \frac{a}{j\omega C}$$

Comparing these two sets of impedances shows that increasing the impedance of each element by same factor is equivalent to an operation of increasing R and L, and decreasing C by the same factor, frequency being unchanged.

Problem 7.5.2 Determine magnitude scale factors for L = 2 mH and C = 10 μF to L′ = 1 H, C = 1 F

Solution

$$L' = a_L \cdot L$$

$$1 = a_L \times 2 \times 10^{-3} \quad \rightarrow \quad a_L = \frac{1}{2 \times 10^{-3}} = 500$$

$$C' = \frac{C}{a_C} \quad \rightarrow \quad a_C = \frac{C}{C'} = \frac{10 \times 10^{-6}}{1} = 10^{-5}$$

Problem 7.5.3 In an RC low-pass filter R = 1 Ω, C = 1 F. Its cut-off frequency is to be shifted to

$\omega'_c = 10\,\text{rad}\,\text{s}^{-1}$ using a 1 kΩ resistor. What is the scaled value of the capacitor?

Fig. 7.59 RLC circuit for Problem 7.5.4

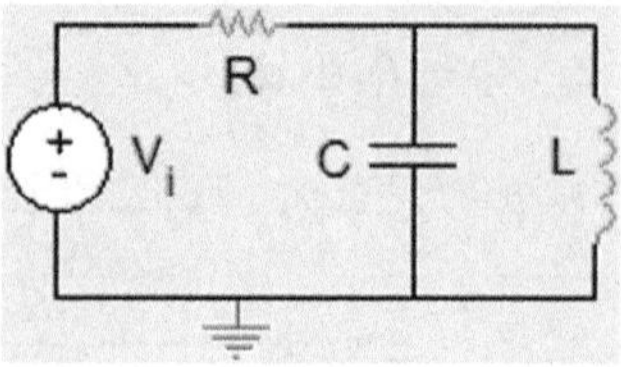

Solution

$$
\begin{aligned}
H(s) &= \frac{\frac{1}{sC}}{R+\frac{1}{sC}} = \frac{1}{sRC+1} = \frac{\frac{1}{RC}}{s+\frac{1}{RC}} \\
\omega_c &= \frac{1}{RC} = \frac{1}{1.1} = 1\,\text{rad/s} \\
\omega_c' &= 10\,\text{rad/s} \\
k_f &= \frac{\omega_c'}{\omega_c} = 10 \\
k_m &= \frac{R'}{R} = \frac{1000\,\Omega}{1\,\Omega} = 1000
\end{aligned}
$$

Applying frequency and magnitude scaling to capacitor yields

$$
C' = \frac{C}{k_m \times k_f} = \frac{1}{1000 \times 10} = 10^{-4} = 100\,\mu\text{F}
$$

Problem 7.5.4 An RLC circuit shown in Fig. 7.59 is designed such that $\omega_0 = 1\,\text{rad/s}$.

(a) Scale this circuit for a resonance frequency of 500 Hz using 1 kΩ resistor (R = 1 Ω, L = 1 H, C = 1 F).
(b) Prove that scaled resonance frequency is indeed 500 Hz.

Solution

Frequency scale factor = b,

(a) $b = \frac{\omega_0'}{\omega_0} = \frac{2\pi \cdot 500}{1} = 1000\pi$

Magnitude scale factor = a,

$$
\begin{aligned}
a &= \frac{R'}{R} = \frac{1000}{1} = 1000 \\
L' &= \frac{a}{b} \cdot L = \frac{1000}{1000\pi} \cdot 1 = \frac{1}{\pi}\,\text{H} \\
C' &= \frac{C}{a \cdot b} = \frac{1}{(1000)(1000\pi)} = \frac{10^{-6}}{\pi}\,\text{F}
\end{aligned}
$$

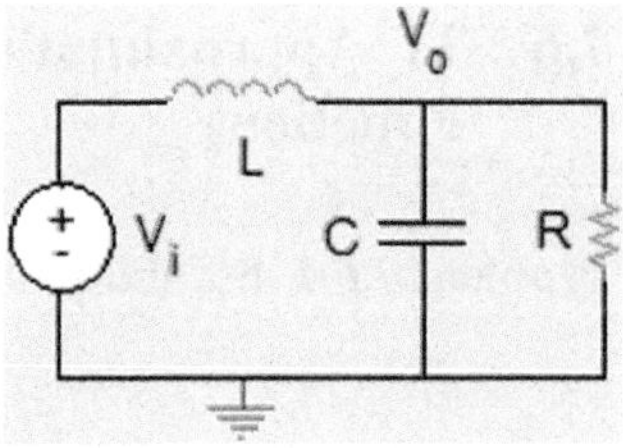

Fig. 7.60 The circuit for Problem 7.5.5

(b) $\omega_0' = \dfrac{1}{\sqrt{L'C'}} = \sqrt{\dfrac{1}{(1/\pi)\cdot 10^{-6}/\pi}} = (\pi^2\cdot 10^6)^{\frac{1}{2}} = 10^3\pi = 2\pi f = 10^3\pi$

$$f = \frac{10^3\pi}{2\pi} = 500\,\text{Hz}$$

Problem 7.5.5

(a) Determine the voltage transfer function of the circuit shown in Fig. 7.60, with L = 1 H, C = 1 F, R = 1 Ω.
(b) Determine the voltage transfer function, if L = 2 H, R = 2 Ω, C = 0.5 F.
(c) Determine the voltage transfer function, if L = 4 H, R = 4 Ω, C = 0.25 F.

Solution

(a) By voltage division ,

$$Z_1(s) = \frac{R\cdot\dfrac{1}{sC}}{R+\dfrac{1}{sC}} = \frac{R}{sRC+1}$$

$$V_0(s) = V_i(s)\frac{Z_1(s)}{Z_1(s)+sL} = \frac{\dfrac{R}{sRC+1}}{\dfrac{R}{sRC+1}+sL} = \frac{R}{s^2RLC+sL+R}$$

(b) L′ = 2L, R′ = 2R, C′ = 0.5C,

$$H(s) = \frac{V_0(s)}{V_i(s)} = \frac{R}{s^2RLC+sL+R}$$

$$H'(s) = \frac{V_0'(s)}{V_i'(s)} = \frac{2R}{s^2(2R)(2L)(C/2)+s(2L)+2R} = \frac{2R}{s^2RLC+2sL+2R}$$

$$H'(s) = \frac{2R}{2(s^2RLC+sL+R)} = \frac{R}{s^2RLC+sL+R}$$

(c) $H''(s) = \dfrac{4R}{s^2(4R)(4L)(L/4)} = \dfrac{R}{s^2RLC+sL+R}$

In conclusion, if L′ = aL, C′ = C/a, R′ = aR, voltage transfer function (VTF) does not change. This is due to impedance (magnitude) scaling.

7.6 TF Approximation (Reducing the Order of a Transfer Function)

Problem 7.6.1 For the following transfer function

$$H(s) = \frac{10}{(s+1)(s+10)} = \frac{Y(s)}{X(s)}$$

(a) Find y(t) for a unit step input.
(b) Determine dominant pole of H(s) for unit step input.
(c) Determine the output of system y′(t) using an approximate transfer function H′(s).

Solution

(a) $Y(s) = H(s) \cdot X(s) = \dfrac{10}{s(s+1)(s+10)} = \dfrac{A}{s} + \dfrac{B}{s+1} + \dfrac{C}{s+10}$

by partial fractions expansion,

$$A = 1, \quad B = \frac{10}{9}, \quad C = \frac{1}{9};$$

$$y(t) = \mathcal{L}^{-1}[Y(s)] = \left[1 - \frac{10}{9}\mathrm{e}^{-t} + \frac{1}{9} \cdot \mathrm{e}^{-10t}\right] \cdot u(t)$$

(b) Since the pole at $s_1(= -1) \gg s_2(= -10)$, dominant pole is at $s_1 = -1$

(c) $H'(s) = \dfrac{K}{s+1} \quad \rightarrow \quad K = \dfrac{10}{10} = 1 \quad \rightarrow \quad H'(s) = \dfrac{1}{s+1} = \dfrac{Y'(s)}{X'(s)}$

$Y'(s) = \dfrac{1}{s(s+1)} = \dfrac{A'}{s} + \dfrac{B'}{s+1}$ by partial fractions expansion,

$$A' = 1, \quad B' = -1,$$

$$Y'(s) = \frac{1}{s} - \frac{1}{s+1}$$

$$y'(t) = \mathcal{L}^{-1}[Y'(s)] = [1 - \mathrm{e}^{-t}] \cdot u(t)$$

Problem 7.6.2

(a) Find a simplified transfer function of H(s), (root_poly.m)

$$H(s) = \frac{1000}{s^3 + 61s^2 + 560s + 500}$$

(b) Plot step responses of both transfer functions (dominant_pole0.m). Print the script of m-file for this plot.

Solution

$$H(s) = \frac{1000}{s^3 + 61s^2 + 560s + 500} = \frac{1000}{(s+1)(s+10)(s+50)}$$

First step is to find dominant pole of TF. The pole closest to the imaginary axis in the s-plane is selected as dominant pole. In this case, the pole at s = −1 is dominant pole, as it dominates the step response of H(s).

Second step is to determine dc gain of Z(s). The rule is to ignore the s-dependence of the insignificant poles, but keep the constant part to maintain the correct steady-state response.

Therefore, approximate transfer function of H(s) is

$$H(s)' = \frac{1000}{(s+1)(0+10)(0+50)} = \frac{1000}{(s+1) \times 10 \times 50}$$

$$H(s)' = \frac{2}{(s+1)}$$

In summary, when $p_d \leq 10p_2$, (where p_2 is the next largest pole of t.f. while p_d is dominant pole), transfer function exhibits dominant first-order behavior. The approximate lower order transfer function is

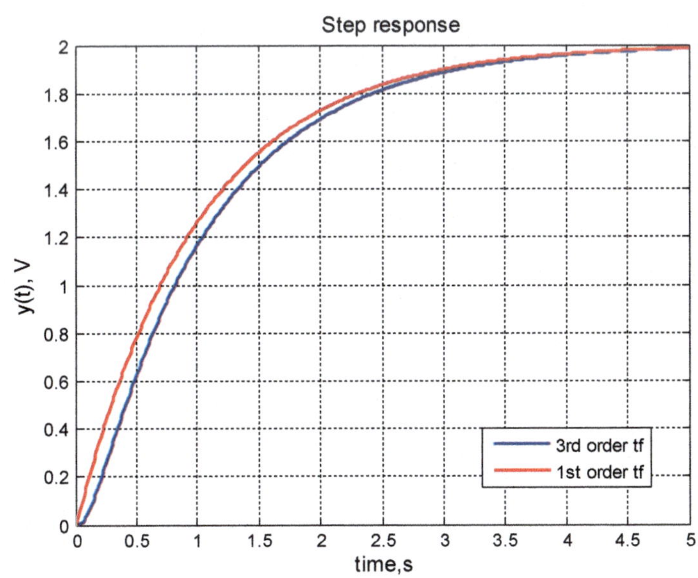

Fig. 7.61 Step responses of first and third-order transfer functions for Problem 7.6.2

$$H(s)' = \frac{\frac{K}{p_2 p_3}}{(s+p)}$$

The script of m-file for the plots of step responses of both transfer functions as shown in Fig. 7.61 is given below

```
%real dominant pole in a third order tf
%third order tf is approximated by a first order tf
K=1000;
num1=[0 K]; den1=[1 61 560 500]; H1=tf(num1,den1)
p2=10; p3=50; num2=[0 K/p2/p3]; den2=[1 1]; H2=tf(num2,den2)
t=0:.01:5;
%STEP RESPONSE
ys1=step(H1,t); ys2=step(H2,t);
plot(t,ys1,t,ys2,'r','linewidth',1.5);grid on;
xlabel('time,s');ylabel('y(t), V');
title('Step response');legend('3rd order tf','1st order tf')
```

Problem 7.6.3 For the third-order transfer function,

$$H(s) = \frac{400}{s^3 + 25s^2 + 424s + 400}$$

(a) Determine its approximate transfer function (root_poly.m)
(b) Plot step responses of both transfer functions (dominant_pole1.m). Print the script of m-file for this plot.

Solution

$$H(s) = \frac{400}{s^3 + 25s^2 + 424s + 400} = \frac{400}{(s+1)(s^2 + 24s + 400)}$$

First step is to find dominant pole of TF. The pole closest to the imaginary axis in the s-plane is selected as dominant pole.

Since given transfer function is of the form

$$H(s) = \frac{K}{(s+p)(s^2 + 2\zeta\omega_n s + \omega_n^2)},$$

with quadratic term,

$$s^2 + 24s + 400 = s^2 + 2\zeta\omega_n s + \omega_n^2,$$
$$2\zeta\omega_n = 24, \quad \zeta\omega_n = 12, \quad p = 1$$
$$\zeta\omega_n = 12p$$

This means that $\zeta\omega_n \gg p$, in this case, the pole at s = −1 is the dominant pole. Second step is to determine dc gain of H(s). The rule is to drop the s-dependence of the insignificant pole, but keep the constant part to maintain the correct steady-state response.

Therefore, approximate transfer function of H(s) is

$$H(s)' = \frac{400}{(s+1)(0+0+400)} = \frac{1}{(s+1)}$$

In summary, when $p \leq 10\zeta\omega_n$, transfer function exhibits dominant first-order behavior. The approximate lower order transfer function is

$$H(s)' = \frac{\frac{K}{\omega_n^2}}{(s+p)}$$

The script of m-file for the plots of step responses of both transfer functions shown in Fig. 7.62 is given below

```
%real dominant pole in a third order tf
%third order tf is approximated by a first order tf
K=400; wn2=400; % natural freq.squared
num1=[0 K]; den1=conv([1 1],[1 24 400]); H1=tf(num1,den1)
num2=[0 K/wn2]; den2=[1 1]; H2=tf(num2,den2)
t=0:.01:5;
%STEP RESPONSES
ys1=step(H1,t); ys2=step(H2,t);
plot(t,ys1,t,ys2,'r','linewidth',1.5);grid on;
xlabel('time,s');ylabel('y(t), V');
title('Step response');legend('3rd order tf','1st order tf')
```

Problem 7.6.4 For the third-order transfer function,

$$H(s) = \frac{400}{s^3 + 144s^2 + 3280s + 48{,}000},$$

(a) Determine its approximate (reduced order) transfer function (root_poly.m)
(b) Plot step responses of both transfer functions (dominant_pole2.m).

Print the script of m-file for this plot.

Solution

$$H(s) = \frac{400}{s^3 + 144s^2 + 3280s + 48{,}000} = \frac{400}{(s+120)(s^2 + 24s + 400)}$$

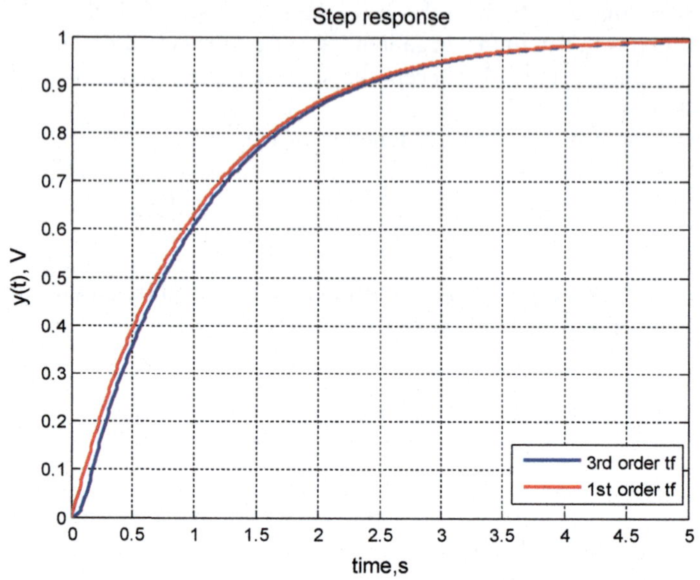

Fig. 7.62 Step responses of first- and third-order transfer functions of Problem 7.6.3

First step is to find dominant pole of transfer function. Since transfer function is of the form

$$H(s) = \frac{K}{(s+p)(s^2 + 2\zeta\omega_n s + \omega_n^2)},$$

with quadratic term,

$$s^2 + 24s + 400 = s^2 + 2\zeta\omega_n s + \omega_n^2, \quad \rightarrow \quad 2\zeta\omega_n = 24, \ \zeta\omega_n = 12, \ p = 120$$
$$p = 10\zeta\omega_n$$

This means that $\zeta\omega_n \ll p$. In this case, the poles at s = $(-12 \pm j16)$ are dominant poles, as they are much closer to imaginary axis of s-plane than the real pole.

Second step is to determine dc gain of H(s). The rule is to drop the s-dependence of the insignificant pole, but keep the constant part to maintain the correct steady-state response.

Therefore, approximate transfer function of H(s) is

$$H(s)' = \frac{400}{120(s^2 + 24s + 400)}$$

In summary, when $p \geq 10\zeta\omega_n$, transfer function exhibits dominant second-order behavior. The approximate lower order transfer function is

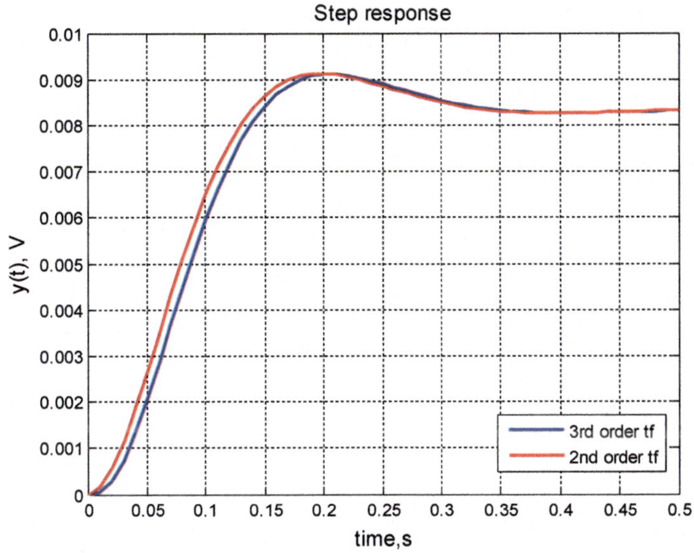

Fig. 7.63 Step responses of second- and third-order transfer functions

$$H(s)' = \frac{\frac{K}{p}}{s^2 + 2\zeta\omega_n s + \omega_n^2}$$

The script of m-file for the plots of step responses of both transfer functions as shown in Fig. 7.63 is given below

```
%complex conjugate dominant poles in a third order tf
%third order tf is approximated by a second order tf
K=400; p=120;% non-dominant real pole (absolute value)
num1=[0 K];den1=conv([1 120],[1 24 400]);H1=tf(num1,den1)
num2=[0 K/p];den2=[1 24 400];H2=tf(num2,den2)
t=0:.01:0.5;
%STEP RESPONSE of H(s)
ys1=step(H1,t);ys2=step(H2,t);
plot(t,ys1,t,ys2,'r','linewidth',1.5);grid on;
xlabel('time,s');ylabel('y(t), V');
title('Step response');legend('3rd order tf','2nd order tf')
```

Problem 7.6.5 The open-circuit (or zero-value) time constant (OCTC) method is used to determine the approximate pole (3 dB frequency) of relatively large circuits that contain many poles in their transfer functions [3, 4].

The dominant pole frequency of the original circuit is determined by the linear frequency term in the denominator of its transfer function by summing the

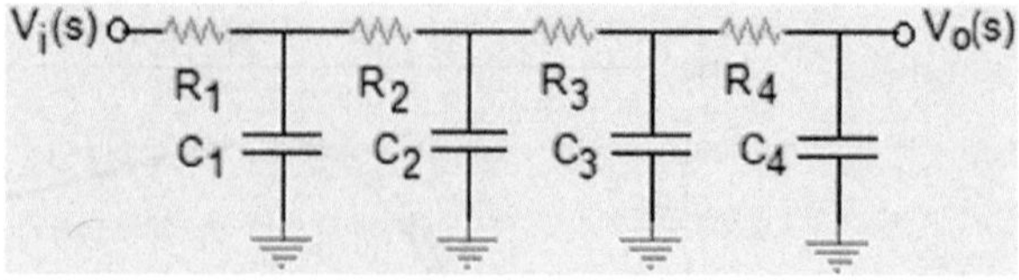

Fig. 7.64 A fourth-order filter circuit

RC-products for each capacitor in the circuit, while the resistor R for a selected capacitor is the resistance found by inserting a voltage source in place of that capacitor, and removing all other capacitors from the circuit.

A fourth-order filter circuit displayed in Fig. 7.64 has voltage transfer function (VTF)

$$\frac{V_o}{V_i} = \frac{1}{s^4 + 7s^3 + 15s^2 + 10s + 1}$$

with the component values of $R_1 = R_2 = R_3 = R_4 = 1\ \Omega$, $C_1 = C_2 = C_3 = C_4 = 1$ F

(a) Determine its approximate VTF.
(b) Show that the filter is approximated by a first-order circuit with a time constant equal to the sum of time constants associated with each pole of the filter.
(c) Compare this result with the one obtained by performing exact analysis (polezero2.m), (Mag2_plot.m).

Solution

(a) Approximate VTF is determined using OCTC rule.
Equivalent resistance values for each "time constant" are calculated as follows: Set the input signal source to zero.
Select C_1, replace it by V_x, remove all other capacitors from the circuit. The resistance seen by the test voltage V_x is calculated as

$$R_a = \frac{V_x}{I_x} = R_1$$

Repeating this procedure for each capacitor in the circuit gives following results:

$$R_b = R_1 + R_2, \quad R_c = R_1 + R_2 + R_3, \quad R_d = R_1 + R_2 + R_3 + R_4$$

each time constant becomes,

$$\tau_1 = R_a C_1, \quad \tau_2 = R_b C_2, \quad \tau_3 = R_c C_3, \quad \tau_4 = R_d C_4$$
$$\tau = \tau_1 + \tau_2 + \tau_3 + \tau_4$$

Approximate pole (3 dB frequency) of the circuit is

$$p = \frac{1}{\tau} = \frac{1}{\tau_1 + \tau_2 + \tau_3 + \tau_4}$$

The value of approximate pole frequency is obtained by substituting given component values into these equations,

$$\tau_1 = 1\,\Omega \times 1\,\text{F} = 1\,\text{s}, \quad \tau_2 = 2\,\Omega \times 1F = 2\,\text{s}, \quad \tau_3 = 3\,\Omega \times 1\,\text{F} = 3\,\text{s}$$
$$\tau_4 = 4\,\Omega \times 1\,\text{F} = 4\,\text{s}$$
$$p = \frac{1}{\tau} = \frac{1}{1+2+3+4} = 0.1\,\text{rad}\,\text{s}^{-1}$$

Approximate VTF of the circuit is

$$\frac{V_o}{V_i} \cong \frac{K}{s+0.1}$$

DC gain constant is calculated as

$$1 = \frac{K}{0+0.1} \quad \rightarrow \quad K = 0.1$$

(b) VTF of the circuit can be written as

$$\frac{V_o}{V_i} = \frac{1}{s^4 + 7s^3 + 15s^2 + 10s + 1} = \frac{1}{(s+3.5321)(s+2.3473)(s+1)(s+0.1206)}$$
$$= \frac{1}{3.5321 \times 2.3473 \times 0.1206 \times \left(1 + \frac{s}{3.5321}\right)\left(1 + \frac{s}{2.3473}\right)(1+s)\left(1 + \frac{s}{0.1206}\right)}$$
$$\frac{V_o}{V_i} = \frac{1}{(1+0.2831s)(1+0.426s)(1+s)(1+8.2919s)}$$
$$\tau = 0.2831 + 0.426 + 1 + 8.2919 = 10\,\text{s}$$

On the other hand, approximation is

$$H(s)_{\text{app}} = \frac{0.1}{s+0.1} = \frac{1}{1+10s}$$

This shows that

$$H(s)_{\text{app}} = \frac{0.1}{s+0.1} = \frac{1}{1+\tau s}$$
$$\frac{V_o}{V_i} = \frac{1}{s^4 + 7s^3 + 15s^2 + 10s + 1} \cong \frac{1}{10s+1}$$

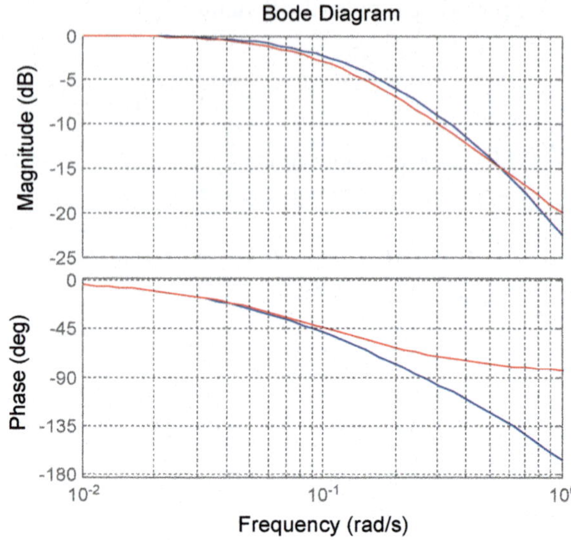

Fig. 7.65 Bode plots display how good OCTS approximation is made within the frequency range around the pole frequency of 0.1 rad/s. *Red line* shows the first-order transfer function as the approximation to fourth-order filter (*blue line*)

(c) Since

$$\frac{V_o}{V_i} = \frac{1}{s^4 + 7s^3 + 15s^2 + 10s + 1} \cong \frac{0.1}{s + 0.1}$$

A capacitor with C = 1 F and a resistor with R = 10 Ω can be used to realize such a filter. This is also equivalent to eliminating first three capacitors in the original circuit and leaving only the fourth one, plus adding a 6 Ω resistor in series to R_1.

Bode magnitude and phase plots in Fig. 7.65 display how good this approximation is made within the frequency range around the pole frequency of 0.1 rad/s. Red line shows the first-order transfer function as the approximation to fourth-order filter (blue line). It has been noted that the approximation largely deteriorates beyond this frequency range.

MATLAB script for this analysis (Mag2_plot.m);

```
% Plotting magnitude and phase of two TF on the same graph
N1=[1]; D1=[1 7 15 10 1]; N2=[.1];D2=[1 .1]; H1 = tf(N1,D1)
H2 = tf(N2,D2)
bode(H1,H2,'r',{0.01,1});grid on;
```

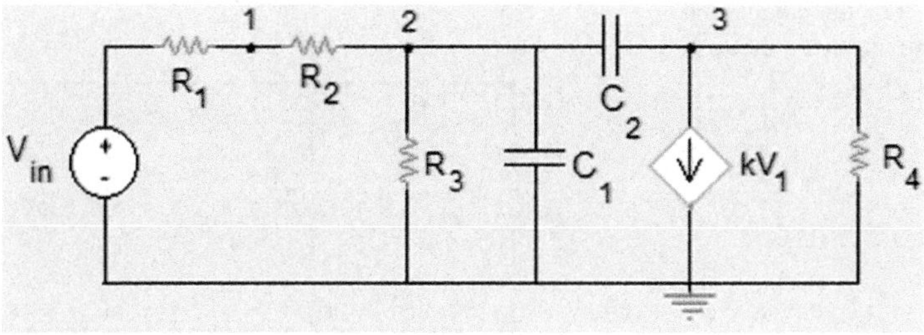

Fig. 7.66 The circuit for Problem 7.6.6

Problem 7.6.6

(a) Determine approximate pole of the circuit shown in Fig. 7.66, using the OCTC method.
(b) For the values of components in the circuit,

$$R_1 = R_2 = 1\,\Omega, \quad R_3 = 2\,\Omega, \quad R_4 = \frac{1}{2}\Omega, \quad C_1 = 10\,\text{F}, \quad C_2 = 1\,\text{F},$$
$$k = 0.5\,\text{S}$$

Calculate approximate 3-dB frequency of the circuit (OCTC1.m).
(c) For the given component values, exact voltage transfer function of the circuit is

$$H(s) = -0.225\frac{s + 0.1667}{(s + 2.2355)(s + 0.0895)}$$

Show that the filter is approximated by a first-order circuit with a time constant equal to the sum of time constants associated with each pole of this filter.

Solution

(a) Set the input signal source to zero.

Select C_1, replace it by V_x, remove all other capacitors from the circuit. See, Fig. 7.67. The resistance seen by the test voltage V_x is calculated as

Fig. 7.67 For the calculation of resistance seen by the test voltage in place of $C1$

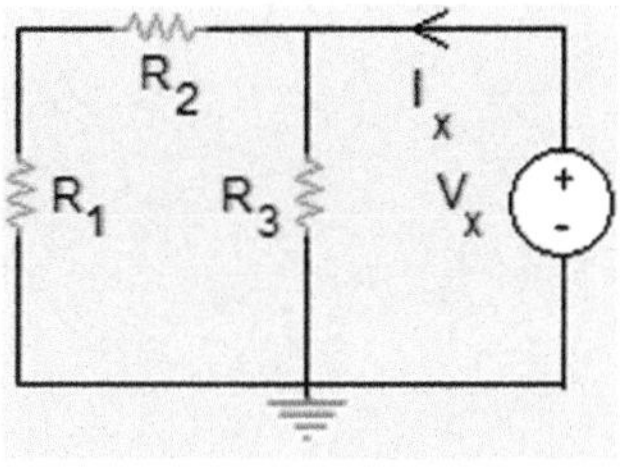

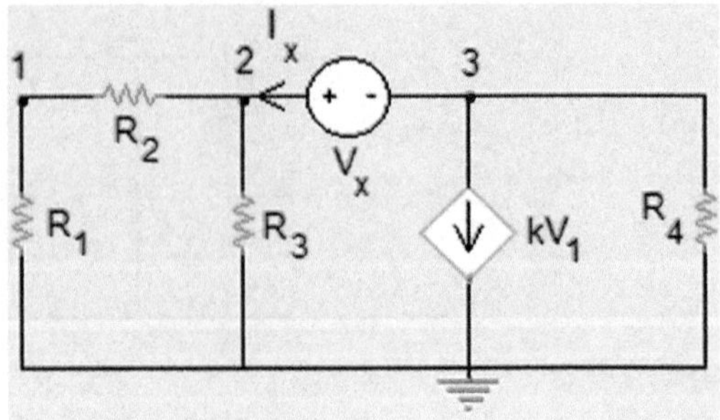

Fig. 7.68 For the calculation of resistance seen by the test voltage in place of *C*2

$$R_a = \frac{V_x}{I_x} = \frac{(R_1 + R_2) \cdot R_3}{R_1 + R_2 + R_3}$$

Then, first time constant is $\tau_a = R_a C_1$

Next, select C_2, replace it by V_x test voltage, and remove all other capacitors from the circuit. See, Fig. 7.68. The resistance seen by the test voltage is calculated as

$$V_2 = I_x \cdot \left(\frac{(R_1 + R_2) \cdot R_3}{R_1 + R_2 + R_3} \right) \tag{7.17}$$

$$V_x = V_2 - V_3 \tag{7.18}$$

KCL at node 3

$$-I_x - kV_1 - \frac{V_3}{R_4} = 0 \tag{7.19}$$

$$V_3 = -R_4 \cdot I_x - kV_1 R_4 \tag{7.20}$$

$$V_1 = V_2 \cdot \frac{R_1}{R_1 + R_2} \tag{7.21}$$

$$V_3 = -R_4 \cdot I_x - k \cdot V_2 \cdot \frac{R_1}{R_1 + R_2} R_4 \tag{7.22}$$

(6) → (2);

$$V_x = V_2 + R_4 \cdot I_x + k \cdot V_2 \cdot \frac{R_1 R_4}{R_1 + R_2} \tag{7.23}$$

$$= V_2 \left(1 + k \frac{R_1 R_4}{R_1 + R_2} \right) + R_4 \cdot I_x \tag{7.24}$$

divide all terms of (7.24) by I_x,

$$\frac{V_x}{I_x} = R_x = \frac{V_2}{I_x}\left(1 + k\frac{R_1 R_4}{R_1 + R_2}\right) + R_4 \tag{7.25}$$

Replace $\frac{V_2}{I_x}$ by (7.17),

$$R_b = R_x = \frac{(R_1 + R_2)\cdot R_3}{R_1 + R_2 + R_3}\left(1 + \frac{kR_1 R_4}{R_1 + R_2}\right) + R_4 \tag{7.26}$$

Second time constant is $\tau_b = R_b C_2$.
Finally,

$$\tau = \tau_a + \tau_b = R_a C_1 + R_b C_2$$

Approximate pole (3 dB frequency)

$$\omega_c = \frac{1}{\tau} = \frac{1}{R_a C_1 + R_b C_2}$$

(b) Using component values, $R_1 = R_2 = 1\ \Omega$, $R_3 = 2\ \Omega$, $R_4 = \frac{1}{2}\Omega$, $C_1 = 10$ F,

$C_2 = 1\,\text{F}, k = 0.5\,\text{S}$, approximate 3-dB frequency of the circuit is numerically computed as,

$$R_a = \frac{(R_1 + R_2)\cdot R_3}{R_1 + R_2 + R_3} = 1\,\Omega, \quad R_b = \frac{(R_1 + R_2)\cdot R_3}{R_1 + R_2 + R_3}\left(1 + \frac{kR_1 R_4}{R_1 + R_2}\right) + R_4 = 1.625\,\Omega,$$

$$\tau_a = R_a C_1 = 10\,\text{s}, \quad \tau_b = R_b C_2 = 1.625\,\text{s}, \tau_{\text{app}} = \tau_a + \tau_b = 11.625\,\text{s}$$

$$\omega_c = \frac{1}{\tau} = \frac{1}{5.625} = 0.086\,\text{rad s}^{-1}$$

(c) For the given component values, exact voltage transfer function of the circuit is

$$H(s) = -0.225\frac{s + 0.1667}{(s + 2.2355)(s + 0.0895)}$$

$$\tau = \frac{1}{2.2355} + \frac{1}{0.0895} = 0.4473 + 11.1732 = 11.621\,\text{s}$$

This shows that time constant of the approximate circuit is equal to the sum of time constants associated with each pole of given filter circuit (Small difference in values may be due to truncation and round off errors).

References

1. Mathie JR (2005) Second generation mixture theory equation for estimating intra-cellular water using bioimpedance spectroscopy. J Appl Physiol 99:780–781
2. Jodal L (2010) Lecture notes on electrical theory behind the measurement of body fluids with BIS, with applications to the measurement device 4200 Hydra BIS analyzer From Xitron Technologies
3. Sedra AS, Smith KC (2004) Microelectronic circuits, 5th edn. New York, Oxford University Press, pp 575–278
4. Salvatori S, Conte G (2009) On the SCTC-OCTC method for the analysis and design of circuits. IEEE Trans Educ 52(3):318–326

Chapter 8
Network Synthesis

8.1 Positive Real Functions, Hurwitz Polynomials

Problem 8.1.1 A Hurwitz polynomial (HP) is a polynomial whose coefficients are positive real numbers and whose roots (zeros) are located in the left half (LH)-complex plane or on the $j\omega$ axis. In other words, (1) the real part of every root of HP is zero or negative. (2) The coefficients of a HP are all real and positive. (3) Imaginary axis roots are simple (have a multiplicity of one). (4) A strictly Hurwitz polynomial (SHP) has no missing coefficients of s, while a weak Hurwitz polynomial (WHP) can have missing coefficients of s.

A HP can be either SHP or WHP. The term "Strictly Hurwitz polynomial, SHP" is used for polynomials whose roots have real parts that are strictly negative, excluding the $j\omega$ axis (i.e., no roots on imaginary axis).

The term "Weak or Modified Hurwitz polynomial, WHP" is used for polynomials that may have roots with negative real parts and simple roots on the $j\omega$ axis.

HPs represent the characteristic equations of stable linear systems. Denominator polynomial of a positive real-rational transfer function is a HP.

Whether a polynomial is HP can be determined by solving the equation to find the roots.

Note that all second-degree polynomials with real, positive, and no missing coefficients are SHP. A product of two HPs is HP.

(Before the advent of computers, so-called "Routh–Hurwitz stability criterion" is used to test if a polynomial is a HP, using the coefficients of a given polynomial without solving the equation for its roots.)

Find out if the following polynomials are Weak (modified) Hurwitz, strictly Hurwitz or neither (hurwitz1.m).

(a) $s^2 + 5s - 4$,
(b) $(s^2 + 1)^2(s + 1)$,
(c) $s^3 + 3s^2 + 4$,

A.Ü. Keskin, *Electrical Circuits in Biomedical Engineering*,
DOI 10.1007/978-3-319-55101-2_8

(d) $s^3 + js + 7$
(e) $s^4 + s^2 + 1$,
(f) $s^2 + s + 1$,
(g) $s^6 + s^5 + 3s^4 + 2s^3 + 3s^2 + s + 1$
(h) $(s^2 + s + 1)(s + 1)$,
(i) $s^2 + 1$,
(j) $s^3 + s$,
(k) $s^4 + 3s^3 + 4s^2 + 2s$.

Solution

(a) Not Hurwitz, negative coefficient.
(b) Not Hurwitz, multiple roots on imaginary axis $(2 \times \mp j)$.
(c) Not Hurwitz, missing coefficient, roots in RH complex plane.
(d) Not Hurwitz, nonreal coefficient.
(e) Not HP, missing coefficient, roots in RH complex plane.
(f) Strictly Hurwitz (SHP), roots in left-hand complex plane.
(g) Not Hurwitz, multiple roots on imaginary axis $(2 \times \mp j)$.
(h) HP, product of two SHP.
(i) WHP, missing term, simple roots on imaginary axis.
(j) WHP, missing term, simple roots on imaginary axis.
(k) WHP, missing term, a simple root on imaginary axis ($s = 0$).

Problem 8.1.2 Let $F(s) = (a_2 s^2 + a_1 s + a_0)^2$, and $a_i \geq 0$. Is $F(s)$ always a HP?

Solution

$[F(s)]^{1/2} = a_2 s^2 + a_1 s + a_0$ is at least a weak (modified) HP (WHP). However, there is no guarantee that multiplication of two WHPs generates another WHP. For example, let

$$F(s) = (s^2 + 4)^2, \quad a_2 = 1, \quad a_1 = 0, \quad a_0 = 4.$$

Then, $F(s) = (s^2 + 4)(s^2 + 4) = s^4 + 8s^2 + 16$.

This polynomial has roots on the imaginary axis with multiplicity of more than 1, namely,

$$F(s) = (s \pm j2)^2.$$

Nonexistence of simple roots on $j\omega$ axis precludes the polynomial from being a HP.

Problem 8.1.3 For the polynomial $P(s) = s^4 + 5s^2 + k$, k is a positive integer, write a MATLAB script and determine the range of values for k so that $P(s)$ is hurwitz (hurwitz1.m).

Solution

Using MATLAB script (hurwitz1.m), it is shown that $P(s)$ has simple roots on imaginary axis when $0 < k \leq 6$ and RH roots (not Hurwitz) if $k \geq 7$.

Problem 8.1.4 Let $Z(s) = N(s)/D(s)$ be a lossless input impedance function (containing inductors and capacitors, only), then $p(s) = N(s) + D(s)$ is a HP. If a lossless input impedance function is

$$Z(s) = \frac{s^2 + 2}{(s^2 + 1)(s^2 + 3)}, \quad p(s) = ?$$

Solution

$$p(s) = s^2 + 2 + \left(s^2 + 1\right)\left(s^2 + 3\right) = s^2 + 2 + s^4 + 3s^2 + s^2 + 3 = s^4 + 5s^2 + 5$$

which is a HP.

Problem 8.1.5 Let $p(s) = M(s) + N(s)$ be a HP where $M(s)$ and $N(s)$ are, respectively, the even and odd parts of $p(s)$. Then,

$$Z(s) = \frac{M(s)}{N(s)}$$

can be realized as a lossless (one port) input impedance function (containing inductors and capacitors, only).

If, $p(s) = 8s^2 + 4s + 1$, determine $Z(s)$ and realize it.

Solution

Given polynomial is a HP, with $M(s) = 8s^2 + 1,\ N(s) = 4s$,

$$Z(s) = \frac{M(s)}{N(s)} = \frac{8s^2 + 1}{4s} = 2s + \frac{1}{4s} = Ls + \frac{1}{Cs}.$$

This is a (one port) series impedance with $L = 2$ H, and $C = 4$ F.

Problem 8.1.6 Determine if following rational functions are positive real (PR) (PosRealTest1.m).

(a) $G(s) = \dfrac{1}{s + a}$

(b) $G(s) = \dfrac{s^2}{s^2 + s + 1}$

(c) $G(s) = \dfrac{1}{s}$

(d) $G(s) = \dfrac{3s^2 + s + 12}{6s^3 + 32s^2 + 39s + 99}$

(e) $G(s) = \dfrac{(s+2)(s+4)(s+5)}{(s+3)^3}$

(f) $G(s) = \dfrac{s(s^2+4)}{s^3}$

(g) $G(s) = \dfrac{s^4+s^3+s^2+s+1}{s^4+s^2+1}$

(h) $G(s) = \dfrac{s^4+8s^3+23s^2+28s+12}{s^4+2s^3+3s^2+2s+1}$

(i) $G(s) = \dfrac{1}{3s^2+s+12}$

(j) $G(s) = \dfrac{s^3+4s^2+3s+12}{2.7s^3+3.1s^2+9.6s+1.1}$

(k) $G(s) = \dfrac{s+4}{s^2+4s+0.25}$

(l) $G(s) = \dfrac{s(s^2+4)}{(s^2+16)(s^2+25)}$

(m) $G(s) = \dfrac{s^2+s-12}{32s^2+3s+9}$

(n) $G(s) = \dfrac{s(s^2+4)}{s^3+4s^2+3s+12}$

(o) $G(s) = \dfrac{1}{s^2+s+1}$.

Solution

(a) $G(s) = \dfrac{1}{s+a}$

(i) Denominator polynomial is Hurwitz (HP),
(ii) Absolute difference between the orders of numerator and denominator polynomials is 1,
(iii) $\mathrm{Re}[G(j\omega)] = \mathrm{Re}\left[\dfrac{1}{j\omega+a}\right] = \mathrm{Re}\left[\dfrac{a-j\omega}{\omega^2+a^2}\right] = \dfrac{a}{\omega^2+a^2} > 0.$

(For all angular frequencies between 0 and ∞)
Therefore, $G(s)$ is PR.

(b) $G(s) = \dfrac{s^2}{s^2+s+1}$
Because of multiple zeros on the imaginary axis, $G(s)$ is not a PRF.

(c) $G(s) = \dfrac{1}{s}$

(i) Denominator polynomial is weak (modified) hurwitz,
(ii) Absolute difference between the orders of numerator and denominator polynomials is 1,

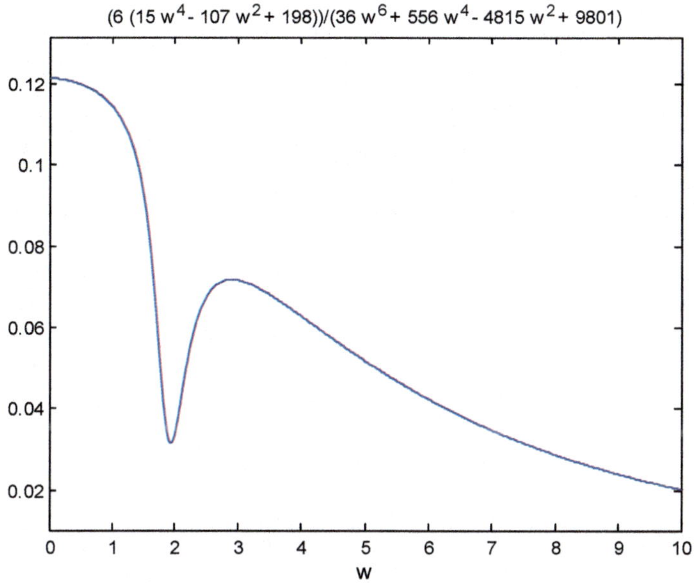

Fig. 8.1 The graph of the real part of the function in Problem 8.1.6d

(iii) $\mathrm{Re}[G(j\omega)] = \mathrm{Re}[1/(j\omega)] = 0,\ \forall\omega[0,\infty)$, therefore, $G(s)$ is a positive real function.

(d) $G(s) = \dfrac{3s^2 + s + 12}{6s^3 + 32s^2 + 39s + 99}$

(i) Both numerator and denominator polynomials are Hurwitz,
(ii) Absolute difference between the orders of numerator and denominator polynomials is 1,
(iii) $\mathrm{Re}[G(j\omega)] = 6\dfrac{15\omega^4 - 107\omega^2 + 198}{36\omega^6 + 556\omega^4 - 4815 + \omega^2 + 9801} > 0,\ \forall\omega[0,\infty)$.

Therefore, $G(s)$ is a PRF.
Figure 8.1 shows the graph of $\mathrm{Re}[G(j\omega)]$, $0 < \omega < 10\,\mathrm{rad/s}$.

(e) $G(s) = \dfrac{(s+2)(s+4)(s+5)}{(s+3)^3}$

(i) Both numerator and denominator polynomials are HP,
(ii) Absolute difference between the orders of numerator and denominator polynomials is 0,
(iii) $\mathrm{Re}[G(j\omega)] = \dfrac{\omega^6 + 34\omega^4 + 369\omega^2 + 1080}{(\omega^2 + 9)^3} > 0,\ \forall\omega[0,\infty)$.

Therefore, $G(s)$ is a PRF.

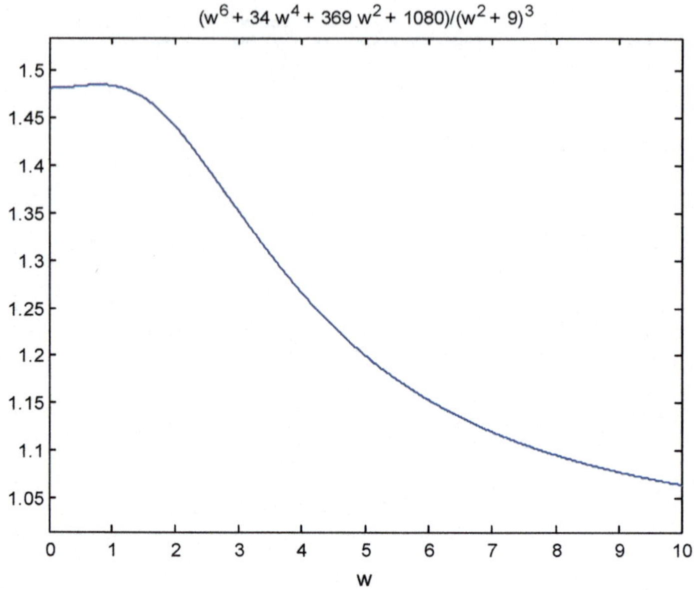

Fig. 8.2 The graph of the real part of the function in Problem 8.1.6e

Figure 8.2 shows the graph of $\text{Re}[G(j\omega)],\ 0<\omega<10\,\text{rad/s}$.

(f) $G(s) = \dfrac{s(s^2+4)}{s^3}$

Because of multiple poles on the imaginary axis, $G(s)$ is not a PRF.

(g) $G(s) = \dfrac{s^4+s^3+s^2+s+1}{s^4+s^2+1}$

Both numerator and denominator polynomials of $G(s)$ are not a HP (right-hand roots).

Therefore, $G(s)$ is not a PRF.

(h) $G(s) = \dfrac{s^4+8s^3+23s^2+28s+12}{s^4+2s^3+3s^2+2s+1}$

(i) Both numerator and denominator polynomials are HP,

(ii) Absolute difference between the orders of numerator and denominator polynomials is 0,

(iii) $\text{Re}[G(j\omega)] = \dfrac{\omega^8 - 10\omega^6 + 10\omega^4 - 3\omega^2 + 12}{(\omega^4 - \omega^2 + 1)^2}$.

However, $\text{Re}[G(j\omega)] \geq 0$ condition is not satisfied at all angular frequencies. Figure 8.3 shows the graph of $\text{Re}[G(j\omega)],\ 0<\omega<10\,\text{rad/s}$, $G(s)$ is not a PRF.

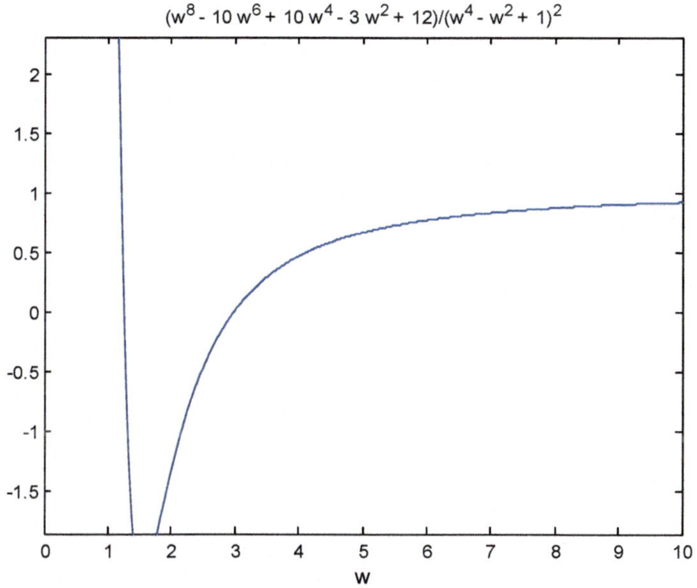

Fig. 8.3 The graph of the real part of the function in Problem 8.1.6h

(i) $G(s) = \dfrac{1}{3s^2 + s + 12}$

- (i) Denominator polynomial is HP,
- (ii) Absolute difference between the orders of numerator and denominator polynomials is 2. $G(s)$ is <u>not</u> a PRF.

(j) $G(s) = \dfrac{s^3 + 4s^2 + 3s + 12}{2.7s^3 + 3.1s^2 + 9.6s + 1.1}$

- (i) Both numerator and denominator polynomials are HP,
- (ii) Absolute difference between the orders of numerator and denominator polynomials is 0,
- (iii) $\mathrm{Re}[G(j\omega)] = \dfrac{10(\omega^2 - 3)(27\omega^4 + 28\omega^2 - 44)}{729\omega^6 - 4223\omega^4 + 8534\omega^2 + 121}$.

However, $\mathrm{Re}[G(j\omega)] \geq 0$ condition is <u>not</u> satisfied at all angular frequencies. Figure 8.4 shows the graph of $\mathrm{Re}[G(j\omega)]$, $0 < \omega < 10\,\mathrm{rad/s}$, $G(s)$ is <u>not</u> a PRF.

(k) $G(s) = \dfrac{s + 4}{s^2 + 4s + 0.25}$

- (i) Both numerator and denominator polynomials are HP,
- (ii) Absolute difference between the orders of numerator and denominator polynomials is 1,

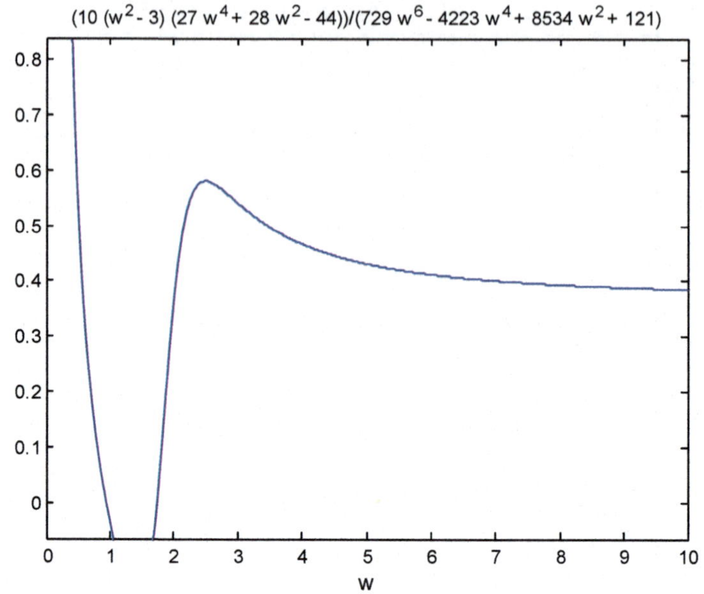

Fig. 8.4 The graph of the real part of the function in Problem 8.1.6j

(iii) $\mathrm{Re}[G(j\omega)] = \dfrac{16}{16\omega^4 + 248\omega^2 + 1} > 0,\ \forall\omega[0, \infty).$

Therefore, $G(s)$ is a PRF.

(l) $G(s) = \dfrac{s(s^2+4)}{(s^2+16)(s^2+25)}$

(i) Both numerator and denominator polynomials are HP,
(ii) Absolute difference between the orders of numerator and denominator polynomials is 1,
(iii) $\mathrm{Re}[G(j\omega)] = 0,\ \forall\omega[0, \infty)$, Therefore, $G(s)$ is a PRF.

(m) $G(s) = \dfrac{s^2+s-12}{32s^2+3s+9}$
Since there is a negative coefficient in numerator polynomial, G(s) is not PRF.

(n) $G(s) = \dfrac{s(s^2+4)}{s^3+4s^2+3s+12}$

(i) Both numerator and denominator polynomials are HP,
(ii) Absolute difference between the orders of numerator and denominator polynomials is 0,
(iii) $\mathrm{Re}[G(j\omega)] = \dfrac{\omega^2(\omega^2-4)}{(\omega^2+16)(\omega^2-3)}.$

$\mathrm{Re}[G(j\omega)] \leq 0$ for $\sqrt{3} \leq \omega \leq 2$, Therefore, $G(s)$ is not a PRF.

(o) $G(s) = \dfrac{1}{s^2 + s + 1}$

(i) Dominator is HP,
(ii) Absolute difference between the orders of numerator and denominator polynomials is 2, therefore, $G(j\omega)$ is not PR.

Problem 8.1.7 A complex function $F(s)$ with positive and real coefficients,

$$F(s) = \frac{as^2 + bs + c}{ds^3 + es^2 + fs}$$

is a PRF, if and only if $bf - ce \geq 0,\ ae - bd \geq 0,\ af - cd \geq 0$.

Let

$$F(s) = \frac{s^2 + 2s + 5}{s^3 + 4s^2 + fs}.$$

Determine the value of f if $F(s)$ is to be a PRF (PosRealTest1.m).

Solution

$$2f - 5 \times 4 = 2f - 20 \geq 0 \tag{8.1}$$

$$1 \times 4 - 2 \times 1 = 4 - 2 \geq 0 \text{ (checks)} \tag{8.2}$$

$$1 \times f - 5 \times 1 = f - 5 \geq 0. \tag{8.3}$$

From (8.1) to (8.3),

$$f \geq 10 \text{ and } f \geq 5.$$

Select $f \geq 10$, then

$$F(s) = \frac{s^2 + 2s + 5}{s^3 + 4s^2 + 10s}, \quad \mathrm{Re}[F(j\omega)] = \frac{2\omega^2}{\omega^4 - 4\omega^2 + 100}.$$

Note that $\mathrm{Re}[F(j\omega)] > 0,\ \forall\omega[0, \infty)$.

Figure 8.5 shows the graph of $\mathrm{Re}[G(j\omega)]$, $0 < \omega < 10\,\mathrm{rad/s}$ which demonstrates that it is a regular PRF.

Problem 8.1.8 Pole-zero diagram of a function $F(s)$ is shown in Fig. 8.6. If $F(0) = -1$,

(a) Find $F(s)$.
(b) Can $F(s)$ be a realizable DPI function, consisting of real passive elements?

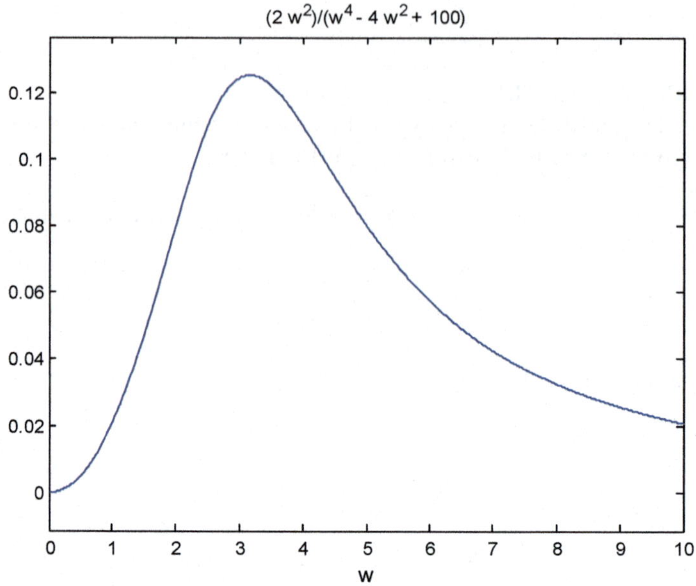

Fig. 8.5 The graph of the real part of the function in Problem 8.1.7

Fig. 8.6 Pole-zero diagram of a function F(s) in Problem 8.1.8

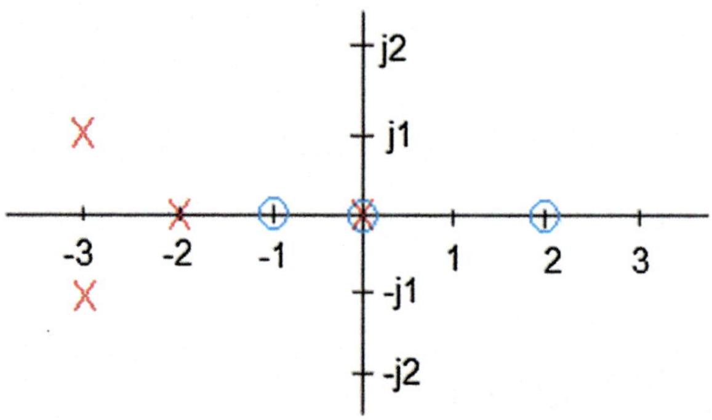

Solution

(a) Poles: 0, (−2), (−3 + $j1$), (−3 − $j1$), zeros: 0, −1, +2

$$F(s) = k\frac{s(s+1)(s-2)}{s(s+2)(s+3+j1)(s+3-j1)} = k\frac{s(s+1)(s-2)}{s(s+2)(s^2+6s+10)}$$

$$F(s) = k\frac{s^2-s-2}{s^3+8s^2+22s+20}$$

$$F(0) = -1 = k\frac{(-2)}{20} \quad \rightarrow \quad k = 10$$

$$F(s) = 10\frac{s^2-s-2}{s^3+8s^2+22s+20}.$$

(b) $F(s)$ cannot be a driving point impedance function (consisting of real passive elements), since it contains a zero on the right half of s-plane.

Problem 8.1.9 Given a second-order real-rational function where all coefficients are non-negative (and at least one denominator polynomial coefficient is positive),

$$Z(s) = \frac{as^2 + bs + c}{ds^2 + es + f}.$$

$Z(s)$ is a PRF if and only if

$$be \geq \left(\sqrt{af} - \sqrt{cd}\right)^2. \tag{8.4}$$

(a) Check if the following functions are PRFs:

(i) $\frac{s^2 + 3s + 1}{s^2 + 1}$,
(ii) $\frac{s^2 + 4}{s^2 + s + 1}$,
(iii) $\frac{s^2 + 5s + 4}{s^2 + 2s + 2}$.

(b) Determine possible integer values of coefficient b, if

(i) $Z(s) = \frac{s^2 + bs + 4}{s^2 + 2s + 2}$,
(ii) $Z(s) = \frac{4s^2 + bs + 4}{s^2 + s + 16}$.

are to be a PRF.

Solution

(a) Substituting given coefficient values into (8.4) yields the following results:

(i) $0 = 0$, $Z(s)$ is PRF,
(ii) $be = 0$, $\left(\sqrt{af} - \sqrt{cd}\right)^2 = 1$, $be < \left(\sqrt{af} - \sqrt{cd}\right)^2$, $Z(s)$ is not PRF,
(iii) $10 \geq (6 - 4\sqrt{2})$, $Z(s)$ is PRF.

(b) Using given coefficient values in (8.4) and solving it for b yields

(i) $2b \geq 2(3 - 2\sqrt{2}) \quad \rightarrow \quad b \geq 0.172 \quad \rightarrow \quad b \geq 1$,
(ii) $1 \times b \geq \left(\sqrt{4 \times 16} - \sqrt{4 \times 1}\right)^2 \quad \rightarrow \quad b \geq 36$.

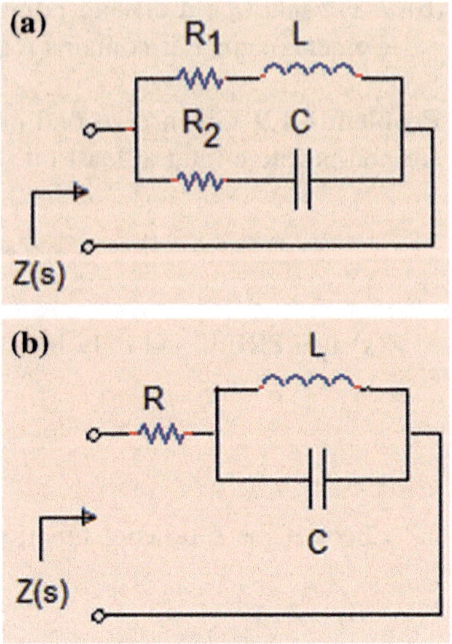

Fig. 8.7 **a** circuit having a regular input impedance function in Problem 8.1.10 **b** Reduced circuit in Problem 8.1.10

Problem 8.1.10 A PRF function $F(s)$ is said to be regular if $\min\{\mathrm{Re}[F(j\omega)]\}$ occurs at $\omega = 0$ or at $\omega = \infty$. Design a second-order circuit having a regular input impedance function. Determine its impedance function (Is it unique?).

Solution

Impedance is to be minimum at $\omega = 0$; A simple circuit that realizes this condition is a series RL circuit. Impedance is to be minimum at very high frequencies (that is, at $\omega = \infty$) means that a series RC circuit can be the impedance under consideration. A parallel connection of these two circuits is a possible solution for this problem, as shown in Fig. 8.7a.

The impedance is calculated as

$$Z(s) = \frac{LCR_2s^2 + (R_1R_2C + L)s + R_1}{CLs^2 + C(R_1 + R_2)S + 1}.$$

Note that this circuit consists of four components with two energy-storing elements. A reduction of component count can be established using a resistor in series with an LC tank circuit (Fig. 8.7b).

This new circuit has the input impedance function of

$$Z(s) = \frac{LCRs^2 + Ls + R}{CLs^2 + 1}.$$

If $\min\{\mathrm{Re}[Z(j\omega)]\} = 0$, then $R = 0\,\Omega$, and minimum component count circuit to fulfill the PRF regularity requirement becomes a simple LC tank circuit.

Problem 8.1.11 Let a function *F*(*s*) be

$$F(s) = \left(\frac{s+a}{s+b}\right)^n. \tag{8.5}$$

F(*s*) is a positive real function, if and only if

$$\tan^{-1}\left(\frac{|a-b|}{2\sqrt{ab}}\right) \le \frac{\pi}{2n}. \tag{8.6}$$

(a) Prove the condition given by Eq. (8.6).
(b) Determine n if $a = 1$, and $b = 2$ (n integer).
(c) Determine n if $a = 9$, and $b = 10$ (n integer).

Solution

(a) The conditions under which the phase function of $F(j\omega)$ satisfies

$$|\varphi(\omega)| \le \frac{\pi}{2} \tag{8.7}$$

(for positive angular frequencies),

$$\varphi(\omega) = n\tan^{-1}\left(\frac{\omega}{a}\right) - n\tan^{-1}\left(\frac{\omega}{b}\right) \tag{8.8}$$

which is an odd function. Its maximum values are

$$\varphi_{\max}(\omega) = n \cdot \tan^{-1}\left(\frac{|a-b|}{2\sqrt{ab}}\right) \tag{8.9}$$

and

$$|\varphi(\omega)| \le \varphi_{\max}(\omega) \tag{8.10}$$

or, by using (8.7), one obtains Eq. (8.6):

(b) $F(s) = \left(\frac{s+1}{s+2}\right)^n$

$$\tan^{-1}\left(\frac{|2-1|}{2\sqrt{1\times 2}}\right) \le \frac{\pi}{2n} \quad \rightarrow \quad \tan^{-1}\left(\frac{1}{2\sqrt{2}}\right) \le \frac{\pi}{2n} \quad \rightarrow \quad n \le \frac{\pi}{2\tan^{-1}\left(\frac{\sqrt{2}}{4}\right)}$$

$$n \le \frac{\pi}{2 \times 0.339837} = 4.6222.$$

For integer values of n, $n \le 4$ yields a positive real *F*(*s*).

(c) $F(s) = \left(\frac{s+9}{s+10}\right)^n$

$$\tan^{-1}\left(\frac{1}{2\sqrt{90}}\right) \le \frac{\pi}{2n} \quad \rightarrow \quad n \le \frac{\pi}{2\tan^{-1}\left(\frac{1}{2\sqrt{90}}\right)}$$

$$n \le \frac{\pi}{0.052656} = 29.8038.$$

For integer values of n, $n \le 29$ yields positive real $F(s)$.

Problem 8.1.12 An admittance function

$$Y(s) = k\frac{a_0s^2 + a_1s + 1}{s(b_0s^2 + b_1s + 1)} = \frac{p(s)}{q(s)} \tag{8.11}$$

is PR function if and only if $(a_0b_1 - a_1b_0) \ge 0$, $a_0 \ge b_0$, and $a_1 \ge b_1$.

The resultant of $p(s)$ and $q(s)$ is defined as

$$R_k = \begin{vmatrix} a_0 & a_1 & 1 & 0 \\ 0 & a_0 & a_1 & 1 \\ b_0 & b_1 & 1 & 0 \\ 0 & b_0 & b_1 & 1 \end{vmatrix} = (a_0 - b_0)^2 - (a_0b_1 - a_1b_0)(a_1 - b_1). \tag{8.12}$$

If $R_k = 0$, common factors exist between $p(s)$ and $q(s)$.

Canonic forms of realization for (8.11) are possible with five components.

The circuit in Fig. 8.8A is valid for $R_k > 0$. The circuit in Fig. 8.8B is valid for $R_k < 0$.

However, two circuits realizing $Y(s)$ using four elements are given in Fig. 8.9 [1].

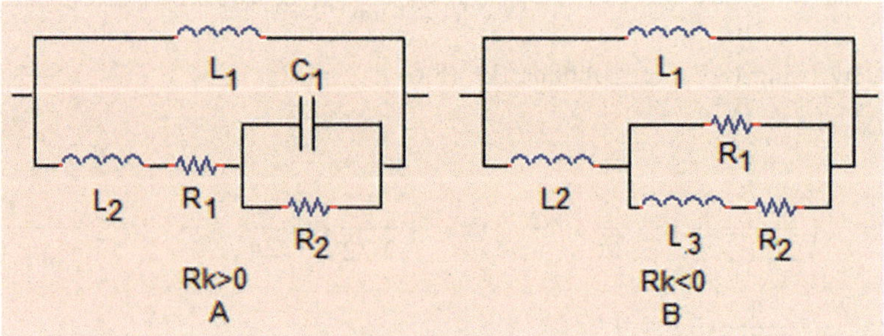

Fig. 8.8 **A, B** Canonic forms of realization in Problem 8.1.12

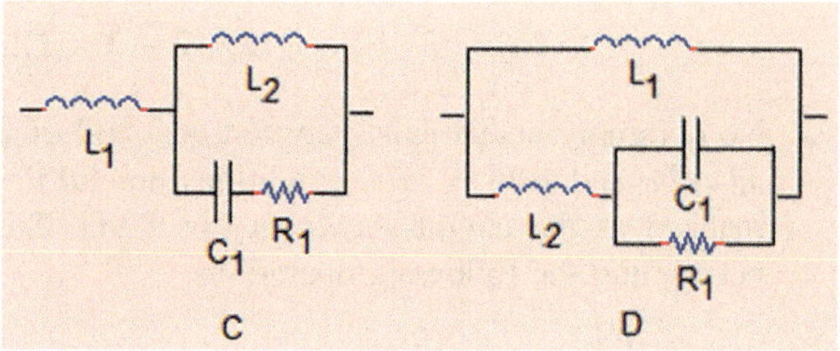

Fig. 8.9 **C, D** Alternative forms of realization in Problem 8.1.12

Realizability condition for the circuit shown in Fig. 8.9C is $a_0 > b_0$ and $a_1 > b_1$. The component values for this circuit shown in Fig. 8.9C are given as

$$R_1 = \frac{a_1(a_0 - b_0)}{ka_0^2}, \quad L_1 = \frac{b_0}{ka_0}, \quad L_2 = \frac{a_0 - b_0}{ka_0}, \quad C_1 = \frac{ka_0^2}{a_0 - b_0}.$$

The realizability condition for the second circuit is given as $a_0 > b_0$ and $a_0b_1 = a_1b_0$.

The circuits shown in Fig. 8.9C, D are frequency-inverse duals (i.e., dual circuits with corresponding elements are of the same kind, but have the inverse values).

For the following admittance functions, check their realizability and construct them if they are realizable.

(a) $Y(s) = \dfrac{3s^2 + s + 1}{s(2s^2 + s + 1)}$,

(b) $Y(s) = \dfrac{3s^2 + 3s + 1}{s(2s^2 + 2s + 1)}$.

Solution

(a) $a_0 = 3,\ a_1 = 1,\ b_0 = 2,\ b_1 = 1$, $Y(s)$ is PRF.

$$R_k = (3 - 2)^2 - (3 \times 1 - 1 \times 2)(1 - 1) = 1 > 0.$$

No common factors exist between $p(s)$ and $q(s)$.
Since $a_0 > b_0$ and $a_1 = b_1$, $Y(s)$ can be realized using the configuration shown in Fig. 8.9C,

$$R_1 = \frac{(1)(3-2)}{(1)(3)^2} = \frac{1}{9}\,\Omega, \quad L_1 = \frac{2}{(3)(1)} = \frac{2}{3}\,\text{H}, \quad L_2 = \frac{3-2}{(3)(1)} = \frac{1}{3}\,\text{H},$$

$$C_1 = \frac{1(3)^2}{3-2} = 9\,\text{F}.$$

(b) $a_0 = 3,\ a_1 = 3,\ b_0 = 2,\ b_1 = 2$, $Y(s)$ is PRF

$$R_k = (3-2)^2 - (3\times 2 - 3\times 2)(3-2) = 1 > 0.$$

No common factors exist between $p(s)$ and $q(s)$.
$a_0 > b_0$ and $a_0 b_1 = a_1 b_0$ conditions are fulfilled, and therefore $Y(s)$ can be realized as the circuit shown in Fig. 8.9D. Component values are found by performing the following operations,

$$a_0 \to \frac{1}{b_0}, \quad a_1 \to \frac{b_1}{b_0}, \quad b_0 \to \frac{1}{a_0}, \quad b_1 \to \frac{a_1}{a_0}$$

$$R_1 = \frac{\left(\frac{b_1}{b_0}\right)\left(\frac{1}{b_0} - \frac{1}{a_0}\right)}{k\left(\frac{1}{b_0}\right)^2} = \frac{b_1(a_0 - b_0)}{k a_0} = \frac{2(3-2)}{(3)(1)} = \frac{2}{3}\ \Omega,$$

$$L_1 = \frac{b_0}{k a_0} = \frac{2}{(1)(3)} = \frac{2}{3}\ \mathrm{H}, \quad L_2 = \frac{a_0 - b_0}{k a_0} = \frac{1}{3}\ \mathrm{H},$$

$$C_1 = \frac{k\left(\frac{1}{b_0}\right)^2}{\left(\frac{1}{b_0} - \frac{1}{a_0}\right)} = \frac{k a_0}{b_0(a_0 - b_0)} = \frac{(1)(3)}{(2)(3-2)} = \frac{3}{2}\ \mathrm{F}.$$

8.2 LC Driving Point Impedance (DPI)

Problem 8.2.1 Given the lossless impedance function

$$Z(s) = \frac{(s^2+4)(s^2+16)}{s(s^2+9)(s^2+25)}.$$

Check if $Z(s)$ is a LC DPI function (pfe1.m).

Solution

1. $Z(s)$ is an odd rational function,
2. $Z(s)$ has a pole at $s = 0$, a zero at $s = \infty$.
3. All poles and zeros are simple and alternate on $j\omega$ axis.
4. To find the residues, let $p = s^2$,

$$\frac{Z(s)}{s} = \frac{(s^2+4)(s^2+16)}{s^2(s^2+9)(s^2+25)} = \frac{(p+4)(p+16)}{p(p+9)(p+25)} = \frac{A}{p} + \frac{B}{(p+9)} + \frac{C}{(p+25)}$$

$$A = 0.2844, \quad B = 0.2431, \quad C = 0.4725.$$

Therefore,

$$Z(s) = \frac{0.2844}{s} + \frac{0.2431s}{(s^2+9)} + \frac{0.4725s}{(s^2+25)}.$$

All residues are real and positive.
Hence, $Z(s)$ is a LC DP function.
A MATLAB script pfe1.m for the solution of this problem is given as follows:

```
%pfe1
x1=[1 4]; x2=[1 16]; y1=[1 0]; y2=[1 9]; y3=[1 25];
N=conv(x1,x2); ya=conv(y1,y2); D=conv(ya,y3);
[r,p,k] = residue(N,D)
Z = tf(N,D)   %Impedance function
```

Problem 8.2.2 Realize LC DPI using Foster 1 form (pfe1.m).

$$Z(s) = \frac{s^4+20s^2+64}{s^5+34s^3+225s}.$$

Solution

$$Z(s) = \frac{s^4+20s^2+64}{s(s^4+34s^2+225)} = \frac{s^4+20s^2+64}{s(s^2+9)(s^2+25)}$$

$$\frac{Z(s)}{s} = \frac{A}{p} + \frac{B}{(p+9)} + \frac{C}{(p+25)}$$

$A = 0.2844, \quad B = 0.2431, \quad C = 0.4725$ (partial fraction expansion.)

$$\begin{aligned} Z(s) &= \frac{0.2844}{s} + \frac{0.2431}{(s^2+9)} + \frac{0.4725}{(s^2+25)} \\ &= \frac{1}{\frac{1}{0.2844}s} + \frac{1}{\frac{s}{0.2431} + \frac{1}{\left(\frac{0.2431}{9}\right)s}} + \frac{1}{\frac{s}{0.4725} + \frac{1}{\left(\frac{0.4725}{2s}\right)s}} \end{aligned}$$

$$C_0 = \frac{1}{k_0} = \frac{1}{0.2844} = 3.5162\,\text{F}, \quad C_2 = \frac{1}{k_2} = \frac{1}{0.2431} = 4.1135\,\text{F}$$

$$C_4 = \frac{1}{k_4} = \frac{1}{0.4725} = 2.1164\,\text{F},$$

$$L_2 = \frac{k_2}{\omega_2^2} = \frac{0.2431}{9} = 0.0270\,\text{H}, \quad L_4 = \frac{k_4}{\omega_4^2} = \frac{0.4725}{25} = 0.0189\,\text{H}.$$

Synthesized LC DPI is shown in Fig. 8.10.

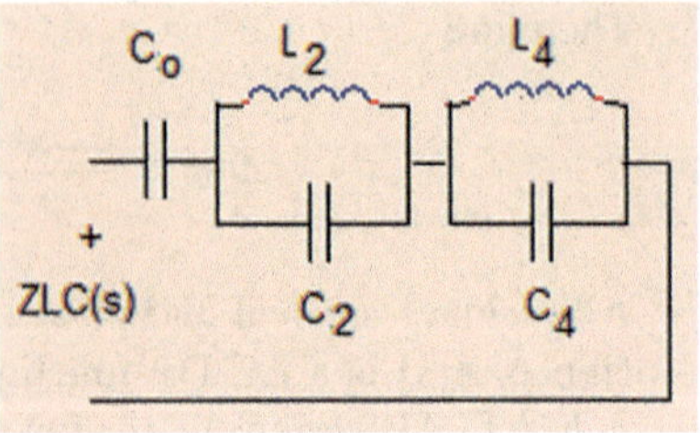

Fig. 8.10 Synthesized LC DPI

Problem 8.2.3

(a) Perform manual continuous fraction expansion (CFE) on $F(s)$,

$$F(s) = \frac{s^5 + 20s^3 + 64s}{s^4 + 10s^2 + 9}.$$

(b) Write a MATLAB script for CFE on $F(s)$ (cfe.m).

Solution

(a) Manual continuous fraction expansion on $F(s)$,

$$\begin{array}{ll}
s^5 + 20s^3 + 64s & /s^4 + 10s^2 + 9 \\
\underline{-s^5 \mp 20s^3 \mp 64s} & s \\
10s^3 + 55s & \\
s^4 + 10s^2 + 9 & /10s^3 + 55s \\
\underline{-s^4 \mp \frac{11}{2}s^2} & \frac{1}{10}s \\
\frac{9}{2}s^2 + 9 & \\
10s^3 + 55s & /\frac{9}{2}s^2 + 9 \\
\underline{-10s^3 \mp 20s} & \frac{20}{9}s \\
35s & \\
\frac{9}{2}s^2 + 9 & /35s \\
\underline{-\frac{9}{2}s^2} & \frac{9}{70}s \\
9 & \\
35s & /9 \\
\underline{-35s} & \frac{35}{9}s \\
0 &
\end{array}$$

Collecting quotients together in fractional form gives $F(s)$,

$$F(s) = s + \cfrac{1}{\cfrac{1}{10}s + \cfrac{1}{\cfrac{20}{9}s + \cfrac{1}{\cfrac{9}{70}s + \cfrac{1}{\cfrac{35}{9}s}}}}.$$

(b) A MATLAB m file for continuous fraction expansion of a rational function:

```
%cfe.m
%Continuous fraction expansion (division method)
clear; clc;  format rational;
%F=B/A is  given rational function
%[q,r]=deconv(B,A)  is main equation of this script
%Warning:First coefficient of A must be non-zero.
B=[1 0 20 0 64 0]; A=[1 0 10 0 9 ];
F=tf(B,A)% Given rational function
x{1}=B;
x{2}=A;
for i=1:size(B,2)-1
[tmp1,tmp2]=deconv(x{i},x{i+1});
loc=find(tmp2~=0);
q{i}=tmp1(1);%quotients
x{i+2}=tmp2(min(loc):end);
end
q=cell2mat(q)'
```

Problem 8.2.4

(a) Use Cauer's first form to realize LC DPI impedance,

$$Z(s) = \frac{s^4 + 20s^2 + 64}{s^5 + 34s^3 + 225s}.$$

(b) Plot $|Z(j\omega)|, \varphi(j\omega)$ (Cauer1_LC1.m).

Solution

Since the order of denominator polynomial of $Z(s)$ is higher than the order of numerator polynomial, procedure starts by inverting $Z(s)$,

$$Y(s) = \frac{1}{Z(s)} = \frac{s^5 + 34s^3 + 225s}{s^4 + 20s^2 + 64}.$$

Applying continued fraction expansion on the corresponding admittance function,

$$Y(s) = 1s + \cfrac{1}{\frac{1}{14}s + \cfrac{1}{\frac{28}{17}s + \cfrac{1}{\frac{289}{1890}s + \cfrac{1}{\frac{945}{1088}s}}}}.$$

Since procedure starts by inverting $Z(s)$, first element is a shunt capacitor. Quotient terms yield the component values as

$$C_1 = 1\,\text{F}, \quad L_2 = \frac{1}{14}\,\text{H}, \quad C_3 = \frac{28}{17}\,\text{F}, \quad L_4 = \frac{289}{1890}\,\text{H}, \quad C_5 = \frac{945}{1088}\,\text{F}.$$

Synthesized LC DPI circuit and its magnitude and phase response are shown in Figs. 8.11 and 8.12.

A MATLAB script used for the solution of this problem is (cauer1_LC1.m):

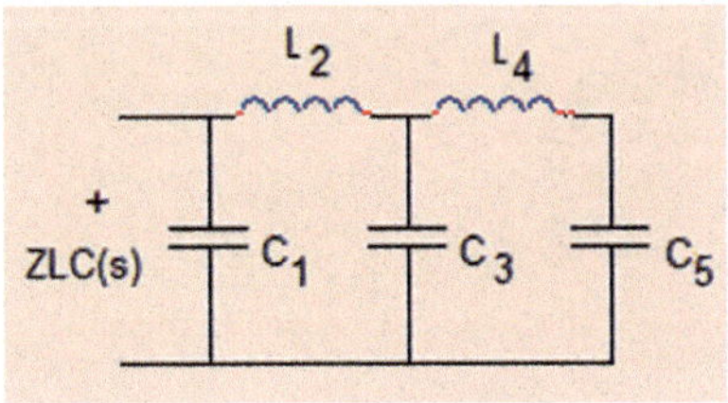

Fig. 8.11 Synthesized LC DPI circuit

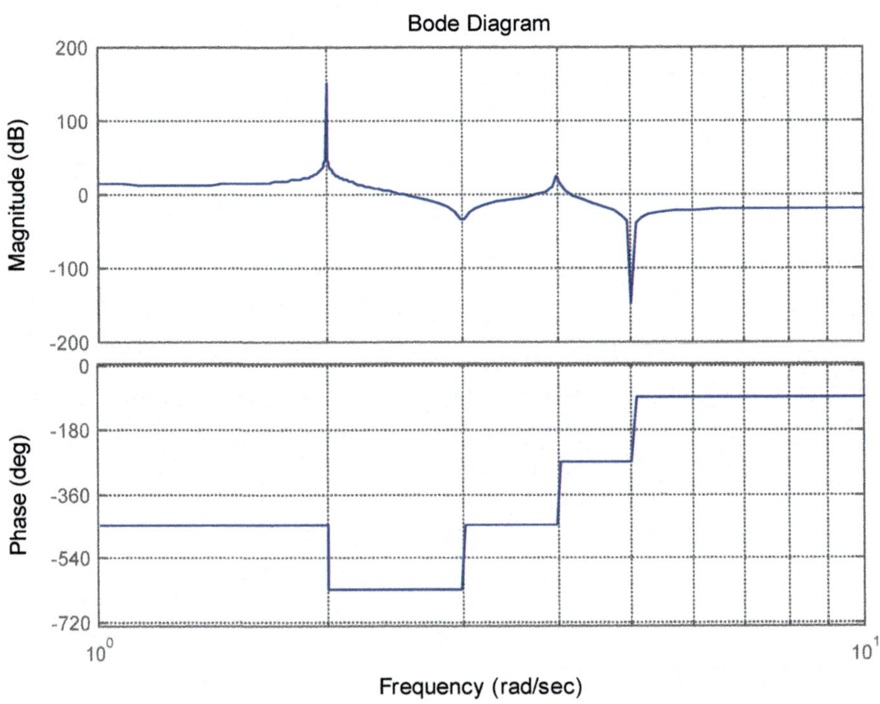

Fig. 8.12 Bode magnitude and phase diagrams

```
%cauer1_LC1.m
%Cauer 1 LC DP function realization
%Given: Z(s)=p(s)/q(s). If the order of p(s)<q(s), then N=q(s),D=p(s)
%In this case first element is a shunt Capacitor
%If the  order of p(s)>q(s), then N=p(s),D=q(s)
%In this case, first element is a series inductor
%Enter polynomials without typing their zero coefficients
%Last enty must be 0
%Example: s4+3s2+12   [1 3 12 0]  (not [1 0 3 0 12])
clc;clear all;%format short
format rat
p=[1 0 34 0 225 ];    q=[1 0 20 0 64 0];
N=[ 1  34  225  0 ]; D=[ 1  20  64  0 ];
Z = tf(p,q) %original Z
A = [N;D]
m=size(A,2);
% Continuous Fraction Expansion (matrix method)
for i = 3:m+3
for j = 1:m-1
B = A(i-2,j+1)*A(i-1,1);
C = A(i-2,1)  *A(i-1,j+1);
A(i,j) = (B-C)/A(i-1,1);
end
end
n=size(A,1);
for i = 1:n-1
Q(1,i) = A(i,1)/A(i+1,1);
end
Q=Q';
if(Q(end)>1e12)
    Q(end)=[];
end
Q
bode(Z,{1,10});grid on;
```

Problem 8.2.5 In the circuit shown in Fig. 8.13a,

(a) Determine $Z(s) = V(s)/I(s)$.
if $L_1 = 1\,\text{H}$, $L_2 = 2\,\text{H}$, $C_3 = 1\,\text{F}$, $C_4 = 2\,\text{F}$, $C_5 = 1/2\,\text{F}$, $L_6 = 1\,\text{H}$ (Z_symbolic1.m).
(b) Determine Cauer's first form LC equivalent of this circuit (cauer1_LC1.m).
(c) If C_3 is short circuited, what will be the new $Z(s)$? Determine its Cauer's first form equivalent LC circuit.

Solution

(a) The general circuit topology can be drawn as shown in Fig. 8.13b.
Converting Y circuit of Z_1, Z_2, Z_3 into Δ circuit,

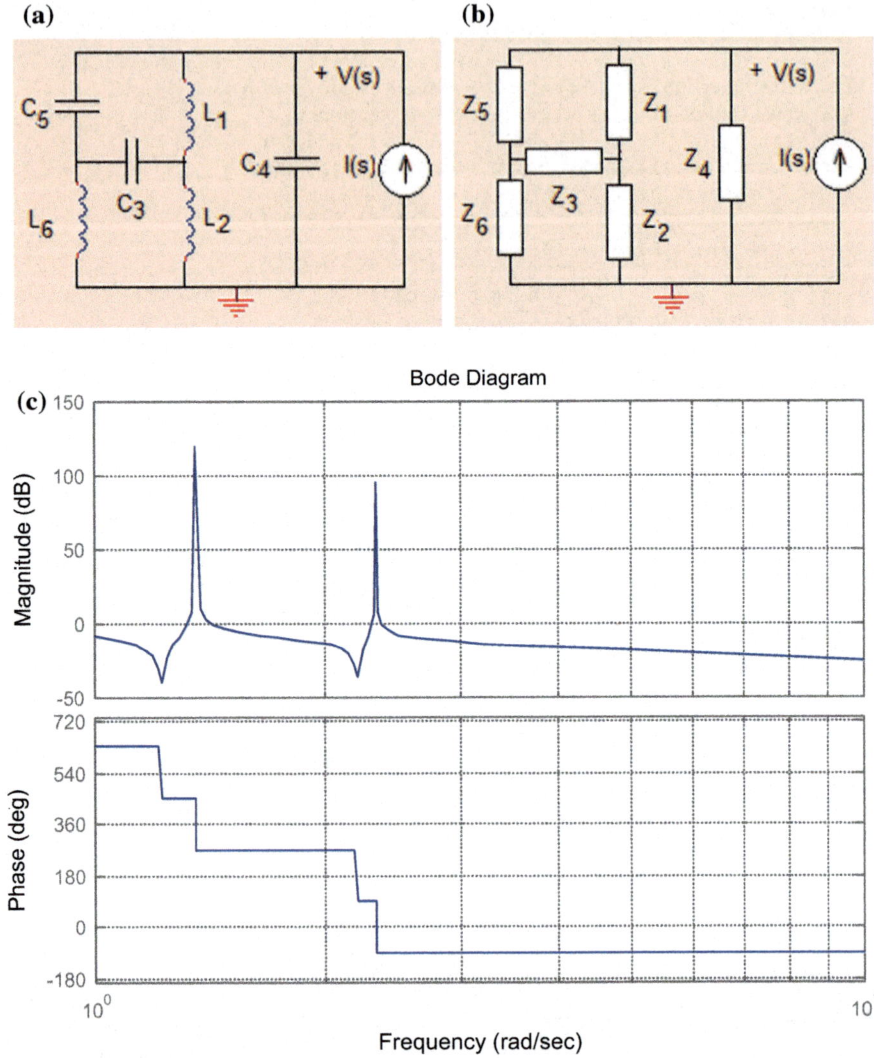

Fig. 8.13 **a**. **b** General topology. **c** Bode magnitude and phase diagrams

$$Z_a = \frac{Z_1Z_2 + Z_2Z_3 + Z_3Z_1}{Z_1}$$
$$Z_b = \frac{Z_1Z_2 + Z_2Z_3 + Z_3Z_1}{Z_2}$$
$$Z_c = \frac{Z_1Z_2 + Z_2Z_3 + Z_3Z_1}{Z_3}$$
$$Z_d = Z_4 \,||\, Z_c = \frac{Z_4 \times Z_c}{Z_4 + Z_c}$$
$$Z_e = Z_5 \,||\, Z_b = \frac{Z_5 \times Z_b}{Z_5 + Z_b}$$
$$Z_f = Z_6 \,||\, Z_a = \frac{Z_6 \times Z_a}{Z_6 + Z_a}, \quad Z = (Z_e + Z_f) \,||\, Z_d = \frac{(Z_e + Z_f) \times Z_d}{Z_e + Z_f + Z_d}.$$

(In these expressions, s-operator is dropped for convenience.)
Using given component values in (Z_symbolic1.m) script, one obtains input impedance function as

$$Z(s) = \frac{4s^6 + 28s^4 + 45s^2 + 18}{8s^7 + 64s^5 + 122s^3 + 57s}.$$

MATLAB script for the solution of part (a) of this problem is given below.

```
%Z_symbolic1.m
clc;clear all;
syms s
L1=1; L2=2; L6=1; C4=2;  C5=1/2;%C3=1;
C3=inf;
Z1=s*L1; Z2=s*L2; Z3=1/(s*C3); Z6=s*L6; Z5=1/(s*C5); Z4=1/(s*C4);
Za=(Z1*Z2+Z2*Z3+Z3*Z1)/Z1; Zb=(Z1*Z2+Z2*Z3+Z3*Z1)/Z1;
Zc=(Z1*Z2+Z2*Z3+Z3*Z1)/Z1;
Zd=Z4*Zc/(Z4+Zc); Ze=Z5*Zb/(Z5+Zb); Zf=Z6*Za/(Z6+Za);
Z=(Ze+Zf)*Zd/(Ze+Zf+Zd); Z=collect(Z,s)
```

(b) Applying continued fraction expansion on $Z(s)$, and using (cfe.m) MATLAB script (or cauer1_LC.m);

$$Y(s) = 2s + \cfrac{1}{\frac{1}{2}s + \cfrac{1}{\frac{2}{3}s + \cfrac{1}{\frac{4}{3}s + \cfrac{1}{\frac{2}{5}s + \cfrac{1}{\frac{25}{2}s + \cfrac{1}{\frac{1}{10}s}}}}}}.$$

Since denominator of $Z(s)$ has higher degree than its numerator, first element is a shunt capacitor. Cauer's equivalent circuit is given in Fig. 8.14, with component values of

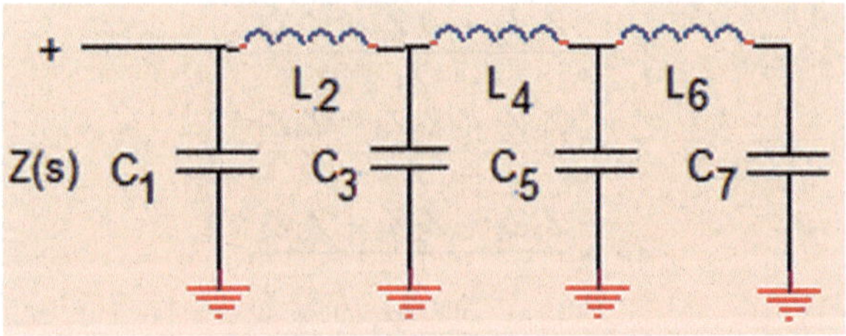

Fig. 8.14 Cauer's equivalent circuit

$$C_1 = 2\,\text{F},\ L_2 = \frac{1}{2}\,\text{H},\ C_3 = \frac{2}{3}\,\text{F},\ L_4 = \frac{4}{3}\,\text{H},\ C_5 = \frac{2}{5}\,\text{F},\ L_6 = \frac{25}{2}\,\text{H},$$
$$C_7 = \frac{1}{10}\,\text{F}.$$

Note that these component values are different than those given in the statement of problem.

Cauer's equivalent circuit contains seven components, while original circuit has six reactive components.

(c) If the capacitor $C_3 = \infty, Z_3(s) = 0$, entering $C_3 = \text{inf}$ in the MATLAB script Z_symbolic1.m, results in new form of $Z(s)$ as

$$Z(s) = \frac{2s^3 + 8s}{4s^2 + 20s^2 + 7}.$$

Applying continuous fraction expansion on new $Z(s)$ yields

$$Y(s) = 2s + \cfrac{1}{\cfrac{1}{2}s + \cfrac{1}{\cfrac{8}{9}s + \cfrac{1}{\cfrac{9}{14}s}}}.$$

Since denominator polynomial of $Z(s)$ has higher order than its numerator, first element is a shunt capacitor:

$$C_1 = 2\,\text{F},\quad L_2 = \frac{1}{2}\,\text{H},\quad C_3 = \frac{8}{9}\,\text{F},\quad L_4 = \frac{9}{14}\,\text{H}.$$

Cauer's first form of $Z(s)$ is shown in Fig. 8.15. It is noted that the values of first two reactances of Cauer's first form equivalent circuits remain the same.

Problem 8.2.6 For the lossless bridge configuration shown in Fig. 8.16,

(a) Determine the number of possible combinations of L's and C's.
(b) Find out the largest possible numerator to denominator orders of input impedance, $Z(s) = P^m(s)/Q^n(s)$.

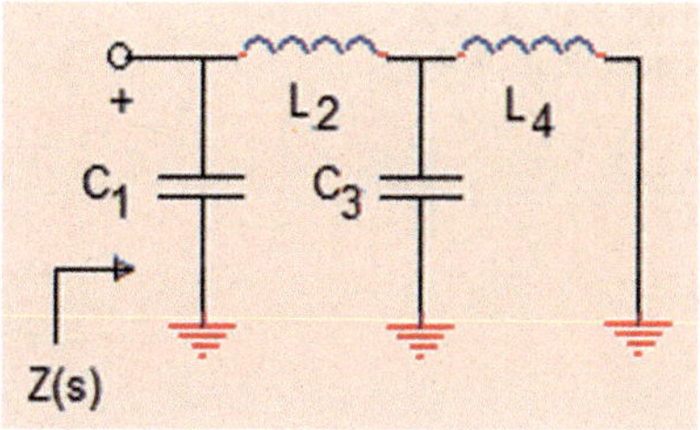

Fig. 8.15 Cauer's first form of $Z(s)$

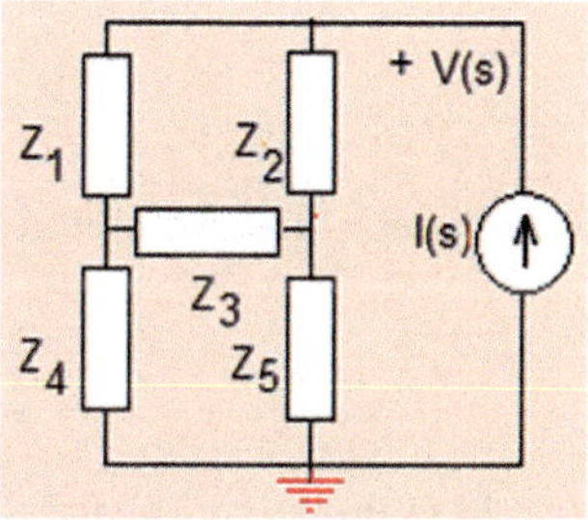

Fig. 8.16 Lossless bridge configuration

(c) Select a configuration with a ratio of the largest denominator order of $Z(s)$, and determine its continuous fraction expansion.
(d) Draw Cauer's first form equivalent circuit of $Z(s)$.
(e) Plot magnitude and phase response of lossless input impedance (Z_symbolic3.m).
(f) If lossless circuit is source terminated by $Rs = 1\ \Omega$, determine the new input impedance of the network, $Zs(s)$, and poles and zeros of this function.
(g) Plot magnitude and phase response of $Zs(s)$ as obtained (in part (f)), above.

Solution

(a) The number of combination of lossless bridge circuits is $2^5 = 32$.
(b) A MATLAB study (using Z_symbolic3.m) shows that largest possible orders for $Z(s)$ are $P^4(s)/Q^5(s)$ and $P^5(s)/Q^4(s)$.
(c) Let the selection of capacitors be in place of $Z_1(s), Z_3(s), Z_5(s)$ and inductors for $Z_2(s)$ and $Z_4(s)$ ($C_1 = 1\,\text{F},\ L_2 = 2\,\text{H},\ C_3 = 3\,\text{F},\ L_4 = 4\,\text{H},\ C_5 = 5\,\text{F}$)

$$Z(s) = \frac{184s^4 + 32s^2 + 1}{120s^5 + 96s^3 + 9s}.$$

Continuous fraction expansion yield

$$Y(s) = \frac{1}{Z(s)} = \frac{15}{23}s + \cfrac{1}{\frac{529}{216}s + \cfrac{1}{\frac{1944}{299}s + \cfrac{1}{\frac{169}{27}s + \cfrac{1}{\frac{24}{13}s}}}}.$$

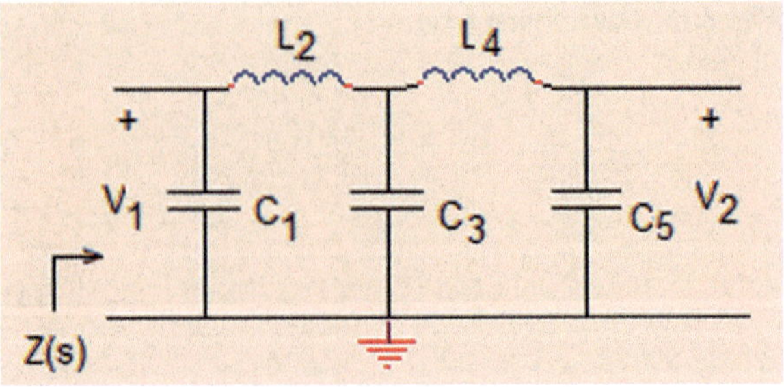

Fig. 8.17 Cauer's first form equivalent of $Z(s)$

(d) Cauer's first form equivalent of $Z(s)$ is shown in Fig. 8.17.

$$C_1 = \frac{15}{23}\,\text{F},\quad L_2 = \frac{529}{216}\,\text{H},\quad C_3 = \frac{1944}{299}\,\text{F},\quad L_4 = \frac{169}{27}\,\text{H},\quad C_5 = \frac{24}{13}\,\text{F}.$$

(e) Magnitude and phase response of lossless $Z(s)$ is shown in Fig. 8.18.

(f) If lossless circuit is source terminated by $Rs = 1\ \Omega$, the new input impedance of the network is $Zs(s) = Z(s) + Rs$, or

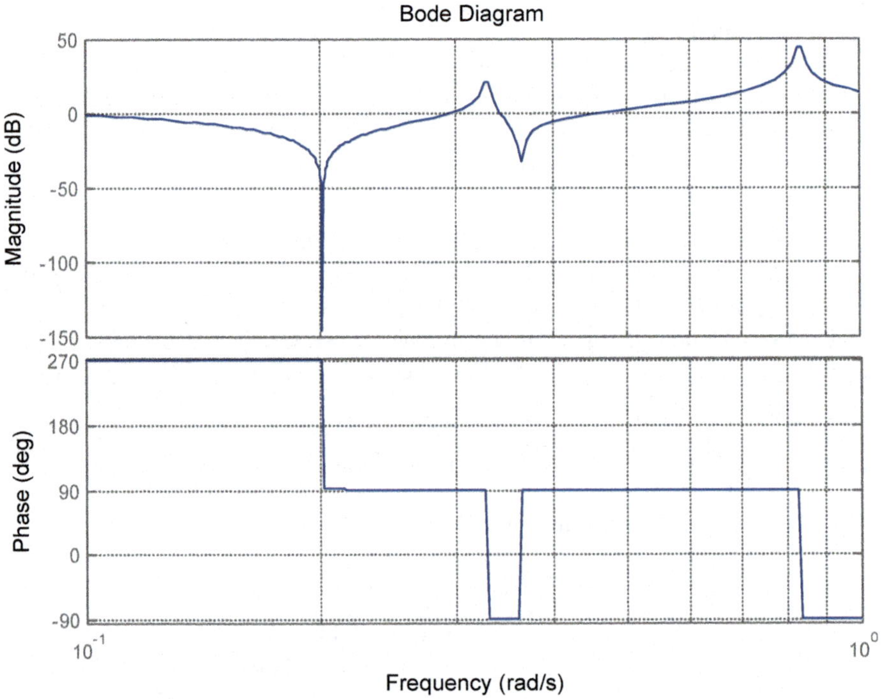

Fig. 8.18 Magnitude and phase response of lossless $Z(s)$

$$
\begin{aligned}
Zs(s) &= \frac{120s^5 + 184s^4 + 96s^3 + 32s^2 + 9s + 1}{120s^5 + 96s^3 + 9s} \\
\text{zeros}: &\quad -0.8623, \quad -0.4467, \quad -0.1889, \quad -0.0177 \mp 0.3379i \\
\text{poles}: &\quad 0, \quad \mp 0.8316i, \quad \mp 0.3293i.
\end{aligned}
$$

(g) Magnitude and phase responses of $Zs(s) = Z(s) + Rs$ are shown in Fig. 8.19.

```
%Z_symbolic3.m
%Equivalent resistance of an unbalanced bridge
clc;clear all;
syms s
L1=1; L2=2; L3=3; L4=4; L5=5; C1=1; C2=2; C3=3; C4=4; C5=5;
Z1=1/(s*C1);Z3=1/(s*C3);Z2=s*L2;Z5=1/(s*C5);Z4=s*L4;
Za=(Z3*Z5+Z2*Z3+Z2*Z5)/Z5; Zb=(Z3*Z5+Z2*Z3+Z2*Z5)/Z2;
Zc=(Z3*Z5+Z2*Z3+Z2*Z5)/Z3;
Zd=Z1*Za/(Z1+Za); Ze=Z4*Zb/(Z4+Zb); Zf=Zd+Ze; ZZ=Zf*Zc/(Zc+Zf);
factor(ZZ)
N=sym2poly(184*s^4 + 32*s^2 + 1);
D=sym2poly(3*s*(40*s^4 + 32*s^2 + 3));
%Continuous fraction expansion (division method)
format rational;
B=D; A=N;%order of B>order of A for cfe
F=tf(B,A); x{1}=B; x{2}=A;
for i=1:size(B,2)-1
[tmp1,tmp2]=deconv(x{i},x{i+1});
loc=find(tmp2~=0);
q{i}=tmp1(1);%quotients
x{i+2}=tmp2(min(loc):end);
end
q=cell2mat(q)'
Z=tf(N,D)
bode(Z,{.1,1});grid on;figure; Zs=Rs+Z
bode(Zs,{.1,1});grid on;format short
zero(Zs)
pole(Zs)
pzmap(Zs);
```

Problem 8.2.7 Realize $Z_{LC}(s)$ using Cauer's first form (cfe.m).

$$
Z_{LC}(s) = \frac{s^9 + 20s^7 + 140s^5 + 400s^3 + 384s}{s^8 + 16s^6 + 86s^4 + 176s^2 + 105}.
$$

Solution

By continuous fraction expansion,

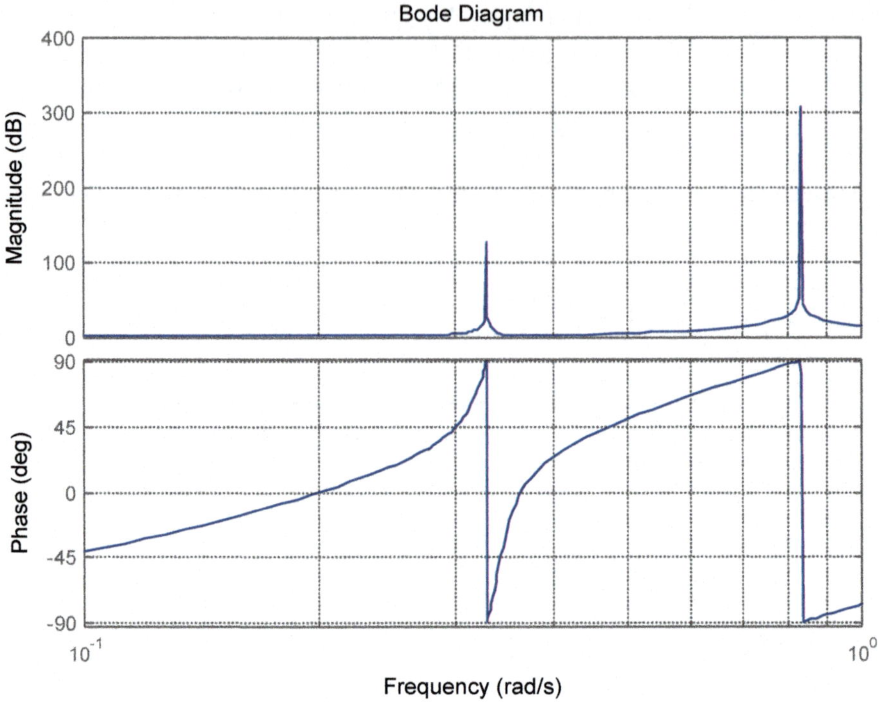

Fig. 8.19 Magnitude and phase responses of $Zs(s) = Z(s) + Rs$

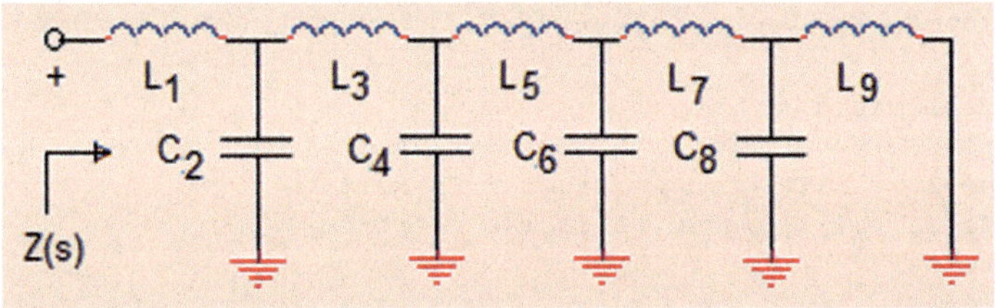

Fig. 8.20 Cauer's first form ladder circuit

$$Z_{LC}(s) = s + \cfrac{1}{\cfrac{1}{4}s + \cfrac{1}{\cfrac{8}{5}s + \cfrac{1}{\cfrac{5}{12}s + \cfrac{1}{\cfrac{4}{5}s + \cfrac{1}{\cfrac{5}{4}s + \cfrac{1}{\cfrac{8}{35}s + \cfrac{1}{\cfrac{34}{4}s + \cfrac{1}{\cfrac{1}{35}s}}}}}}}}.$$

Since numerator of $Z_{LC}(s)$ has a higher degree of than its denominator, first element of Cauer's first form ladder circuit is an inductor.

Synthesized circuit is shown in Fig. 8.20, with

$$L_1 = 1\,\text{H},\quad C_2 = \frac{1}{4}\,\text{F},\quad L_3 = \frac{8}{5}\,\text{H},\quad C_4 = \frac{5}{12}\,\text{F},\quad L_5 = \frac{4}{5}\,\text{H},$$
$$C_6 = \frac{5}{4}\,\text{F},\quad L_7 = \frac{8}{35}\,\text{H},\quad C_8 = \frac{34}{4}\,\text{F},\quad L_9 = \frac{1}{35}\,\text{H}.$$

8.3 RC Driving Point Impedance

Problem 8.3.1 What is the form of equivalent canonical realizations of a general nth degree RC driving point impedance function? How many forms of realizations are possible? Show two of these circuit realizations. Describe the procedures of obtaining these structures. Which of these forms achieve minimum total resistance? Include references.

Solution

Equivalent canonical realizations of a general nth degree RC driving point impedance function $Z(s)$ is specified as

$$Z(s) = \prod_{i=1}^{n} \frac{(s + \sigma_{2i})}{(s + \sigma_{2i-1})} \qquad \sigma_{2i-1} < \sigma_{2i} \text{ for each } i$$

are given for two types of circuits in Fig. 8.21.

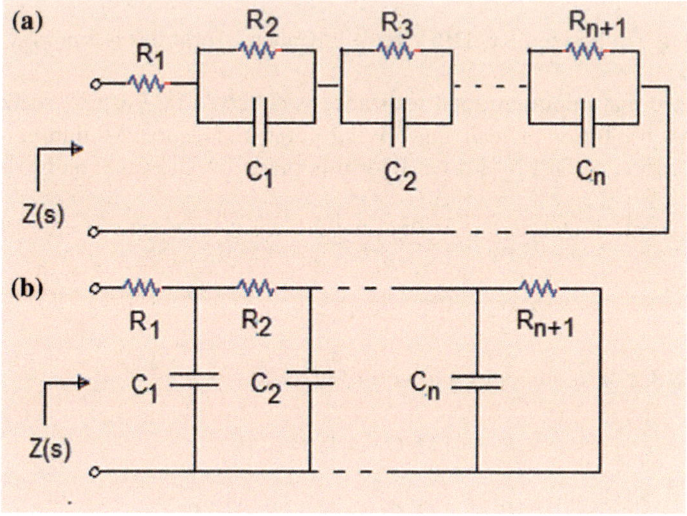

Fig. 8.21 **a** Foster-1 form; **b** Cauer-1 form

Note that there are other realizations, namely Foster 2 and Cauer 2 circuits [2, 3]; however, they are not included in this figure.

Foster form realizations of a general nth degree RC driving point impedance function $Z(s)$ involves partial fractions expansion of $Z(s)$ rational function. Cauer's synthesis is performed by continued fractions of the same function. Transformations between these equivalent forms are also possible [4–6].

Foster 1 form for the synthesis of RC driving point impedance functions (DPIFs) can be described as follows:

If $Z(s) = N(s)/D(s)$ is an RC DPIF, then

$$Z_{RC}(s) = \frac{a_n s^n + a_{n-1}s^{n-1} + \cdots + a_1 s + a_0}{b_n s^n + b_{n-1}s^{n-1} + \cdots + b_1 s + b_0} = k_R + \frac{k_R}{s} + \sum_i^n \frac{k_i}{s + \sigma_i},$$

where

$$k_C = \frac{1}{C}, \quad k_R = R, \quad k_i = \frac{1}{C_i}, \quad R_i = \frac{k_i}{\sigma_i}.$$

Resulting synthesized RC DPI circuit of Foster-1 form has a topology shown in Fig. 8.21a.

Cauer's form of an nth order DPIF can be described as

$$Z_{RC}(s) = \frac{a_n s^n + a_{n-1}s^{n-1} + \cdots + a_1 s + a_0}{b_n s^n + b_{n-1}s^{n-1} + \cdots + b_1 s + b_0} = Y_1 + \cfrac{1}{Z_2 + \cfrac{1}{Y_3 + \cfrac{1}{Z_4 + \cdots \cfrac{1}{Y_n}}}}.$$

Resulting synthesized RC DPI circuit in Cauer 1 form has a topology shown in Fig. 8.21b.

It is noted that minimum total resistance is achieved by Foster-1 realization out of four possible forms of realizing driving point impedance. Minimum total resistance realization of DPI by Foster 1 form is given by Hagopian and Frisch [7],

$$R_{T\min} = \frac{a_1 b_1 - a_0 b_2}{b_1^2}, \quad Z_{RC}(s) = \infty$$

$$R_{T\min} = \frac{a_0}{b_0}, \quad Z_{RC}(s) < \infty.$$

Problem 8.3.2 In the circuits shown in Fig. 8.22,

(a) $Z_1(s) = \dfrac{V(s)}{I(s)} = ? \quad (R_1 = 1\,\Omega, C_1 = 10\,\text{F})$

(b) $Z_2(s) = \dfrac{V(s)}{I(s)} = ? \quad (R_1 = 1\,\Omega, C_1 = 10\,\text{F}, R_2 = 1\,\Omega, C_2 = 1\,\text{F}).$

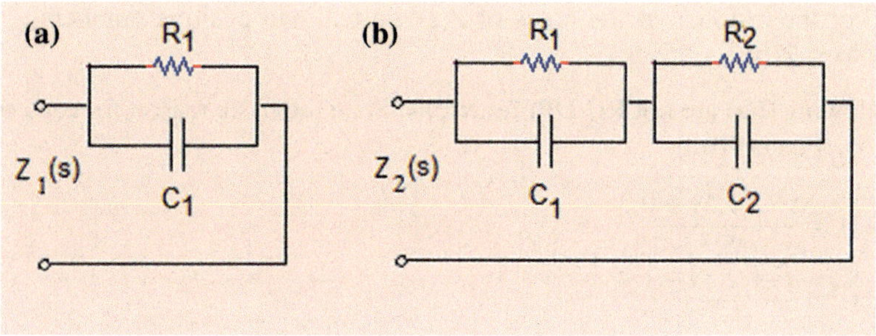

Fig. 8.22 **a, b** The circuits for Problem 8.3.2

Solution

(a) $$Z_1(s) = \frac{V(s)}{I(s)} = \frac{R_1}{sR_1C_1+1} = \frac{R_1}{R_1C_1\left(s+\frac{1}{R_1C_1}\right)} = \frac{\frac{1}{C_1}}{s+\frac{1}{R_1C_1}} = \frac{0.1}{s+0.1}$$

(b) $$Z_2(s) = \frac{V(s)}{I(s)} = \frac{R_1}{sR_1C_1+1} + \frac{R_2}{sR_2C_2+1} = \frac{R_1R_2C_2s+R_1+R_1R_2C_2s+R_2}{(sR_1C_1+1)(sR_2C_2+1)}$$

$$Z_2(s) = \frac{sR_1R_2(C_1+C_2)+(R_1+R_2)}{R_1C_1\left(s+\frac{1}{R_1C_1}\right)\cdot R_2C_2\left(s+\frac{1}{R_2C_2}\right)} = \frac{R_1R_2\left[s(C_1+C_2)+\frac{R_1+R_2}{R_1R_2}\right]}{R_1R_2C_1C_2\left(s+\frac{1}{R_1C_1}\right)\left(s+\frac{1}{R_2C_2}\right)}$$

$$= \frac{s(C_1+C_2)+\frac{R_1+R_2}{R_1R_2}}{C_1C_2\left(s+\frac{1}{R_1C_1}\right)\left(s+\frac{1}{R_2C_2}\right)} = \frac{(C_1+C_2)\left[s+\frac{R_1+R_2}{R_1R_2(C_1+C_2)}\right]}{C_1C_2\left(s+\frac{1}{R_1C_1}\right)\left(s+\frac{1}{R_2C_2}\right)}$$

$$Z_2(s) = \frac{s(C_1R_1R_2+C_2R_1R_2)+(R_1+R_2)}{s^2(C_1C_2R_1R_2)+s(R_1C_1+R_2C_2)+1}, \quad C_1 = 10\,\text{F},\ C_2 = 1\,\text{F},\ R_1 = R_2 = 1\,\Omega$$

$$Z_2(s) = \frac{1.1s+0.2}{s^2+1.1s+0.1} = 1.1\left(\frac{s+0.1818}{s^2+1.1s+0.1}\right) = 1.1\frac{s+0.1818}{(s+0.1)(s+1)}.$$

Problem 8.3.3 Properties of RC driving point (input) impedance functions (DPIF):

If $Z(s) = N(s)/D(s)$ is a driving point impedance function, $N(s)$ and $D(s)$ are polynomials, then

1. Poles and zeros of $Z(s)$ are all real numbers,
2. All poles and zeros of $Z(s)$ are located on left half of s-plane,
3. Poles and zeros are interlaced,
4. Lowest critical frequency is a pole,
5. Highest critical frequency is a zero,

6. All of the residues of the poles of $Z(s)$ are real and positive numbers,
7. $Z(\infty) < Z(0)$.

Following $Z(s)$ are not RC DPI functions. State the basic reason for each transfer function (Foster11.m).

(a) $Z(s) = \dfrac{s^2 + 1.5s + 0.5}{s^3 + 8s^2 + 15s}$,

(b) $Z(s) = \dfrac{s^2 + 2.5s + 1}{s^2 + 1}$,

(c) $Z(s) = \dfrac{s^2 + 3s + 2}{s + 0.5}$,

(d) $Z(s) = \dfrac{s^2 + 5s + 4}{s^2 - 6s}$,

(e) $Z(s) = \dfrac{s^3 + 11s^2 + 26s + 16}{s^3 + 7.5s^2 + 15.5s + 6}$,

(f) $Z(s) = \dfrac{s^2 + 3s + 2}{s^3 + s}$,

(g) $Z(s) = \dfrac{s + 3}{s^3 + 8s^2 + 16s}$,

(h) $Z(s) = \dfrac{s^2 + 4s + 3}{s^4 + 11s^3 + 38s^2 + 40s}$.

Solution

(a) $Z(s) = \dfrac{s^2 + 1.5s + 0.5}{s^3 + 8s^2 + 15s} = \dfrac{(s + 0.5)(s + 1)}{s(s + 3)(s + 5)}$, noninterlaced roots

(b) $Z(s) = \dfrac{s^2 + 2.5s + 1}{s^2 + 1} = \dfrac{(s + 0.5)(s + 2)}{(s + i)(s - i)}$, complex roots in $D(s)$

(c) $Z(s) = \dfrac{s^2 + 3s + 2}{s + 0.5} = \dfrac{(s + 2)(s + 1)}{s + 0.5}$, Num. order > Denom. order

(d) $Z(s) = \dfrac{s^2 + 5s + 4}{s^2 - 6s} = \dfrac{(s + 1)(s + 4)}{s(s - 6)}$, root in right hand s-plane

(e) $Z(s) = \dfrac{s^3 + 11s^2 + 26s + 16}{s^3 + 7.5s^2 + 15.5s + 6}$

$Z(s) = \dfrac{(s + 1)(s + 2)(s + 8)}{(s + 0.5)(s + 3)(s + 4)}$, noninterlaced roots

(f) $Z(s) = \dfrac{s^2 + 3s + 2}{s^3 + s} = \dfrac{(s + 2)(s + 1)}{s(s^2 + 1)}$, complex roots in $D(s)$

(g) $Z(s) = \dfrac{s + 3}{s^3 + 8s^2 + 16s} = \dfrac{s + 3}{s(s + 4)^2}$, noninterlaced multiple roots

(h) $Z(s) = \dfrac{s^2 + 4s + 3}{s^4 + 11s^3 + 38s^2 + 40s} = \dfrac{(s + 1)(s + 3)}{s(s + 2)(s + 4)(s + 5)}$, noninterlaced roots.

Problem 8.3.4 Given driving point (input) impedance *Z*(*s*),

$$Z(s) = \frac{(s+2)(s+4)}{s(s+3)(s+5)}$$

(a) $|Z(j\omega = \infty)| = ?$
(b) Is this an RC type of driving point impedance function?
(c) Synthesize *Z*(*s*) (Foster11.m).

Solution

(a) $|Z(j\omega = \infty)| = 0$, because the (order of num) < (order of denum).
(b) Since the poles and zeros of *Z*(*s*) are interlaced in alternating ($p \to z \to p \to z$) order, and all poles and zeros are real numbers located in left half of *s*-plane, this is an RC driving point impedance circuit.
(c) Using PFE,

$$Z(s) = \frac{(s+2)(s+4)}{s(s+3)(s+5)} = \frac{A}{s} + \frac{B}{s+3} + \frac{C}{s+5}$$

$$s = 0: \quad A = \frac{(s+2)(s+4)}{(s+3)(s+5)} = \frac{(2)(4)}{(3)(5)} = \frac{8}{15}$$

$$s = -3: \quad B = \frac{(s+2)(s+4)}{s(s+5)} = \frac{(-1)(1)}{(-3)(2)} = \frac{1}{6}$$

$$s = -5: \quad C = \frac{(s+2)(s+4)}{s(s+3)} = \frac{(-3)(-1)}{(-5)(-2)} = \frac{3}{10}$$

$$Z(s) = \frac{\frac{8}{15}}{s} + \frac{\frac{1}{6}}{s+3} + \frac{\frac{3}{10}}{s+5}. \tag{8.13}$$

It is further verified that *Z*(*s*) is an RC transfer function, since all residues are real and positive.

First term in the right side of (8.13) is

$$\frac{1}{sC} = \frac{1}{s\left(\frac{15}{8}\right)} \quad \rightarrow \quad C = \frac{15}{8}\ \text{F.}$$

Second term of Eq. (8.13)

$$Z_1 = \frac{R_1\left(\frac{1}{sC_1}\right)}{R_1 + \frac{1}{sC_1}} = \frac{R_1}{sR_1C_1 + 1} = \frac{\frac{1}{C_1}}{s + \frac{1}{R_1C_1}} = \frac{\frac{1}{6}}{s+3}$$

$$\frac{1}{C_1} = \frac{1}{6} \quad \rightarrow \quad C_1 = 6\,\mathrm{F}$$

$$\frac{1}{R_1C_1} = 3 \quad \rightarrow \quad R_1C_1 = \frac{1}{3} \quad \rightarrow \quad R_1 = \frac{1}{3C_1} = \frac{1}{3\cdot 6} = \frac{1}{18}\,\Omega.$$

Similarly, the third term in Eq. (8.13) is a ($R_2 // C_2$) circuit,

$$Z_2 = \frac{\frac{1}{C_2}}{s + \frac{1}{R_2C_2}} = \frac{\frac{3}{10}}{s+5}$$

$$\frac{1}{C_2} = \frac{3}{10} \quad \rightarrow \quad C_2 = \frac{10}{3}\,\mathrm{F}$$

$$\frac{1}{R_2C_2} = 5 \quad \rightarrow \quad R_2 = \frac{1}{5\cdot C_2} = \frac{1}{5\cdot\frac{10}{3}} = \frac{3}{50}\,\Omega.$$

Synthesized circuit is shown in Fig. 8.23.

Problem 8.3.5 Foster 1 form for the synthesis of RC driving point impedance functions (DPIFs) can be described as follows: If $Z(s) = N(s)/D(s)$ is an RC DPIF, then

$$Z(s) = k_R + \frac{k_R}{s} + \sum_i^n \frac{k_i}{s+\sigma_i}$$

where

$$k_C = \frac{1}{C}, \quad k_R = R, \quad k_i = \frac{1}{C_i}, \quad R_i = \frac{k_i}{\sigma_i}.$$

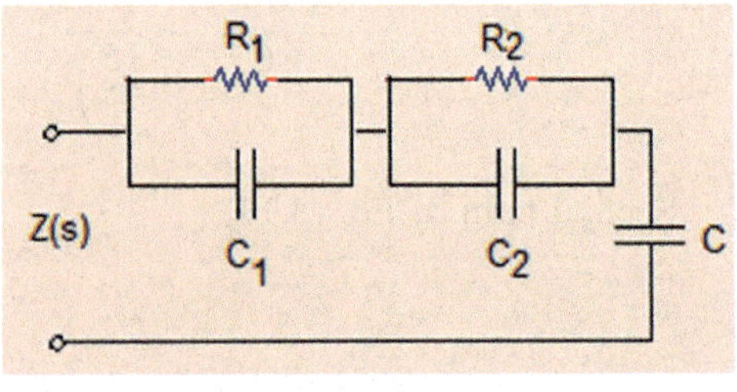

Fig. 8.23 Synthesized circuit of Problem 8.3.4

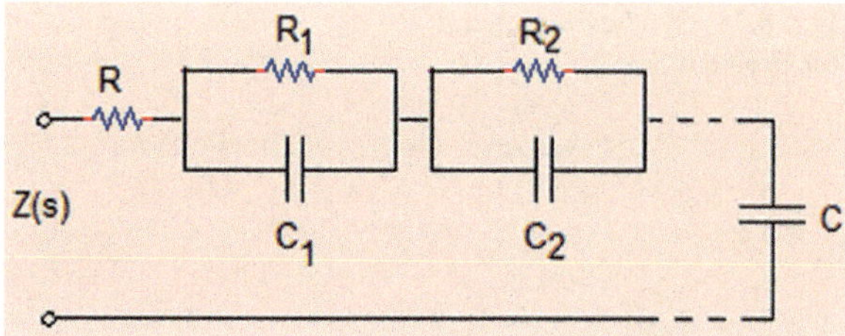

Fig. 8.24 Foster I type RC DPI circuit topology

Resulting synthesized RC DPI circuit has a topology shown in Fig. 8.24.

Synthesize following RC DPI circuits using Foster 1 form and prepare and print a script file in MATLAB for such a procedure (Foster11.m).

(a) $Z(s) = \dfrac{s+2}{s+1}$,

(b) $Z(s) = \dfrac{s+1}{s^2+2s}$,

(c) $Z(s) = \dfrac{s^2+5s+4}{s^2+2s}$

(d) $Z(s) = \dfrac{s^2+5s+4}{s^3+7s^2+10s}$,

(e) $Z(s) = \dfrac{s^3+12s^2+35s+24}{s^3+6.5s^2+11s+4}$.

Solution

(a) $Z(s) = \dfrac{s+2}{s+1} = 1 + \dfrac{1}{s+1} = R + \dfrac{k_1}{s+\sigma_i}$ (Fig. 8.25)

$$C_1 = \frac{1}{k_1} = 1\,\mathrm{F}, \quad R = k_R = 1\,\Omega, \quad R_1 = \frac{k_1}{\sigma_1} = \frac{1}{1} = 1\,\Omega.$$

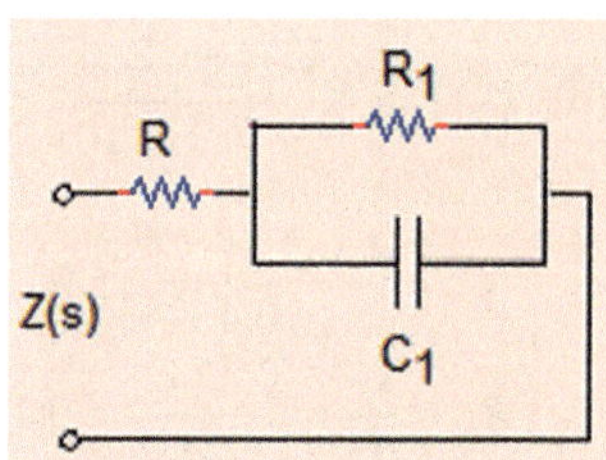

Fig. 8.25 Synthesized circuit for Problem 8.3.5a

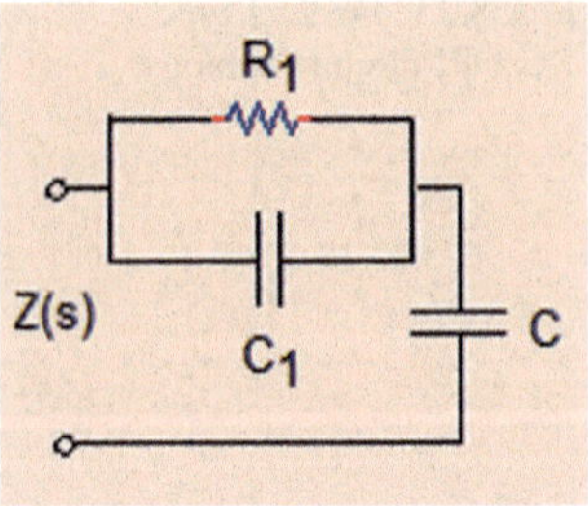

Fig. 8.26 Synthesized circuit for Problem 8.3.5b

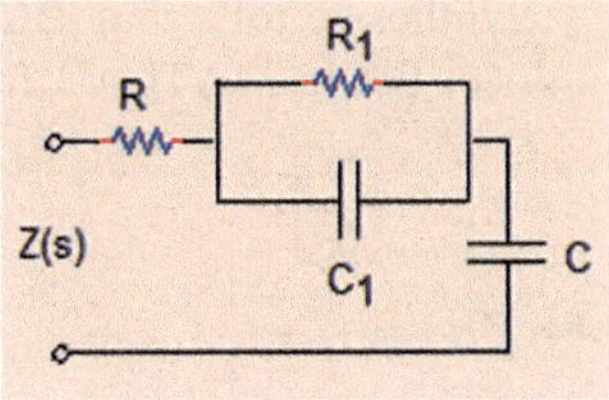

Fig. 8.27 Synthesized circuit for Problem 8.3.5c

(b) $Z(s) = \dfrac{s+1}{s^2+2s} = \dfrac{s+1}{s(s+2)} = \dfrac{0.5}{s} + \dfrac{0.5}{s+2}$ (Fig. 8.26)

$$C = \frac{1}{k_C} = \frac{1}{0.5} = 2\,\mathrm{F}, \quad C_1 = \frac{1}{k_1} = \frac{1}{0.5} = 2\,\mathrm{F}, \quad R_1 = \frac{k_1}{\sigma_1} = \frac{0.5}{2} = 0.25\,\Omega.$$

(c) $Z(s) = \dfrac{s^2+5s+4}{s^2+2s} = \dfrac{(s+1)(s+4)}{s(s+2)} = 1 + \dfrac{2}{s} + \dfrac{1}{s+2}$ (Fig. 8.27)

$$R = 1\,\Omega, \quad C = \frac{1}{k_C} = \frac{1}{2} = 0.5\,\mathrm{F}, \quad C_1 = \frac{1}{k_1} = \frac{1}{1} = 1\,\mathrm{F},$$
$$R_1 = \frac{k_1}{\sigma_1} = \frac{1}{2} = 0.5\,\Omega.$$

(d) $Z(s) = \dfrac{s^2+5s+4}{s^3+7s^2+10s} = \dfrac{(s+1)(s+4)}{s(s+2)(s+5)} = \dfrac{0.4}{s} + \dfrac{0.3333}{s+2} + \dfrac{0.2667}{s+5}$ (Fig. 8.28)

$$C = \frac{1}{k_C} = \frac{1}{0.4} = 2.5\,\mathrm{F}, \quad C_1 = \frac{1}{k_1} = \frac{1}{0.3333} = 3\,\mathrm{F}, \quad C_2 = \frac{1}{k_2} = \frac{1}{0.2667} = 3.75\,\mathrm{F},$$
$$R_1 = \frac{k_1}{\sigma_1} = \frac{0.3333}{2} = 0.1667\,\Omega, \quad R_2 = \frac{k_2}{\sigma_2} = \frac{0.2667}{5} = 0.0533\,\Omega.$$

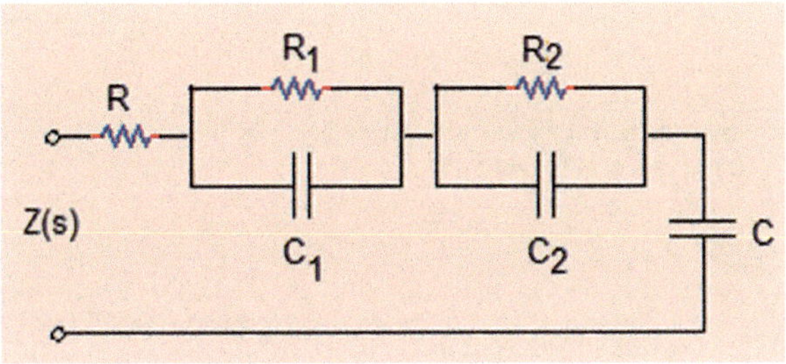

Fig. 8.28 Synthesized circuit for Problem 8.3.5d

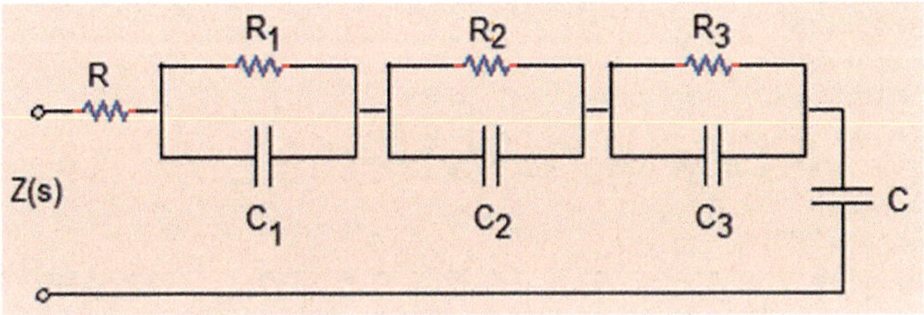

Fig. 8.29 Synthesized circuit for Problem 8.3.5e

(e) $Z(s) = \dfrac{s^3 + 12s^2 + 35s + 24}{s^3 + 6.5s^2 + 11s + 4} = \dfrac{(s+1)(s+3)(s+8)}{(s+0.5)(s+2)(s+4)}$ (Fig. 8.29)

$$Z(s) = 1 + \frac{1.7857}{s+0.5} + \frac{2}{s+2} + \frac{1.7143}{s+4}$$

$$R = 1\,\Omega, \quad C_1 = \frac{1}{k_1} = \frac{1}{1.7857} = 0.56\,\text{F}, \quad C_2 = \frac{1}{k_2} = \frac{1}{2} = 0.5\,\text{F},$$

$$C_3 = \frac{1}{k_3} = \frac{1}{1.7143} = 0.5833\,\text{F},$$

$$R_1 = \frac{k_1}{\sigma_1} = \frac{1.7857}{0.5} = 3.5714\,\Omega, \quad R_2 = \frac{k_2}{\sigma_2} = \frac{2}{2} = 1\,\Omega, \quad R_3 = \frac{k_3}{\sigma_3} = \frac{1.7143}{4} = 0.4286\,\Omega.$$

Following is a MATLAB script file for checking the RC DPI function realizability conditions for a given *Z*(*s*). Then, it is synthesized if *Z*(*s*) is an RC DPI function (Foster11.m).

```
%Foster 1 RC DPI synthesis
%N is numerator, D is denominator polynomial
clc;clear;
%Generate an impedance transfer function in s domain
b=[1 1];c=[1 2]; e=[1 4];h=[1 0];
N=conv(b,e);
D=conv(h,c);
% N =[ 1 5 4];
% D =[ 1 2 0];
H = tf(N,D)    %Transfer function
%Test if numerator order <= denominator order
if(length(N)>length(D))
    warning('Num order > Denum order')
end
%pzmap(H)       %plots poles and zeros
Nroots=roots(N)'
Droots=roots(D)'
q=length(Nroots);%Numerator poly order
p=length(Droots);%denominator poly order
%Test for realizability of the circuit
for i=1:p
if (Droots(i)>0)
warning('positive roots in RH s-plane observed, TF not realizable')
end
end
for i=1:q
if (Nroots(i)>0)
warning('positive roots observed, TF not realizable')
end
end
%test if all roots are real
for i=1:p
if(imag(Droots(i))~=0)
warning('Z(s) is not an RC circuit, complex root in numerator')
end
end
for i=1:q
if(imag(Nroots(i))~=0)
warning('Z(s) is not an RC circuit, complex root in denominator')
end
end
```

```
%Test if poles and zeros are interlaced: (abs(Den->Num->Den->Num))
DR=sort(abs(Droots));
NR=sort(abs(Nroots));
if((p-q)>1)
warning('Z(s) is not an RC circuit, non-interlaced roots')
end
for i=1:min(length(DR),length(NR))
if(DR(i)>NR(i))
warning('Z(s) is not an RC circuit, non-interlaced roots')
end
end
% partial fraction expansion
[r,p,k] = residue(N,D);
if(numel(k)>0)
R0=k(1)
else
R0=0
end
r=r';
p=p';
i=1:length(p);
C(i)=1./r(i);
R(i)=-1./(p(i).*C(i));
R=[R(i)]
C=[C(i)]
Sample output:
Transfer function:
s^2 + 5 s + 4
-------------
  s^2 + 2 s
Nroots =     -4     -1
Droots =      0     -2

R0 =       1
R =      0.5000       -Inf
C =      1.0000     0.5000
```

Problem 8.3.6 Determine the sum of exponential rising voltages from the given sum of exponential decaying voltages of the form, $v(t) = -0.5\mathrm{e}^{-t} - 0.1\mathrm{e}^{-5t} + 0.6$.

Solution

$$v(t) = -0.5\mathrm{e}^{-t} - 0.1\mathrm{e}^{-5t} + 0.6 = -0.5\mathrm{e}^{-t} - 0.1\mathrm{e}^{-5t} + (0.5 + 0.1)$$
$$v(t) = 0.5 - 0.5\mathrm{e}^{-t} + 0.1 - 0.1\mathrm{e}^{-5t} = 0.5(1 - \mathrm{e}^{-t}) + 0.1(1 - \mathrm{e}^{-5t}).$$

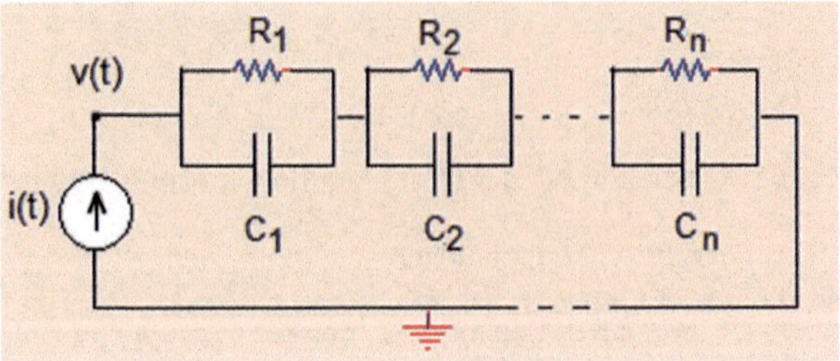

Fig. 8.30 The circuit for Problem 8.3.7

Problem 8.3.7 Determine $v(t)$ for a unit step current response of a driving point impedance with n parallel RC sections. Then, determine the equation of "transient impedance", z(t) (Fig. 8.30).

Solution

For a single section,

$$V(s) = I(s) \cdot Z(s)$$

$$V(s) = I(s) \cdot \frac{\frac{1}{C}}{s + \frac{1}{RC}} = \frac{1}{s} \cdot \frac{\frac{1}{C}}{s + \frac{1}{RC}} = \frac{A}{s} + \frac{B}{s + \frac{1}{RC}}$$

$$A = \left. \frac{\frac{1}{C}}{s + \frac{1}{RC}} \right|_{s=0} = \frac{\frac{1}{C}}{\frac{1}{RC}} = R, \quad B = \left. \frac{\frac{1}{C}}{s} \right|_{s=-\frac{1}{RC}} = -R$$

$$V(s) = R\left(\frac{1}{s} - \frac{1}{s + \frac{1}{RC}}\right) = R\left(\frac{1}{s} - \frac{1}{s + \frac{1}{\tau}}\right), \quad \tau = RC$$

$$V(t) = \mathcal{L}^{-1}[V(s)] = R(1 - \mathrm{e}^{-t/\tau}).$$

For two series-connected RC sections, (Fig. 8.31b), $V(s) = V_1(s) + V_2(s)$

$$V(s) = R_1\left(\frac{1}{s} - \frac{1}{s + \frac{1}{\tau_1}}\right) + R_2\left(\frac{1}{s} - \frac{1}{s + \frac{1}{\tau_2}}\right).$$

Extending the analysis for n sections,

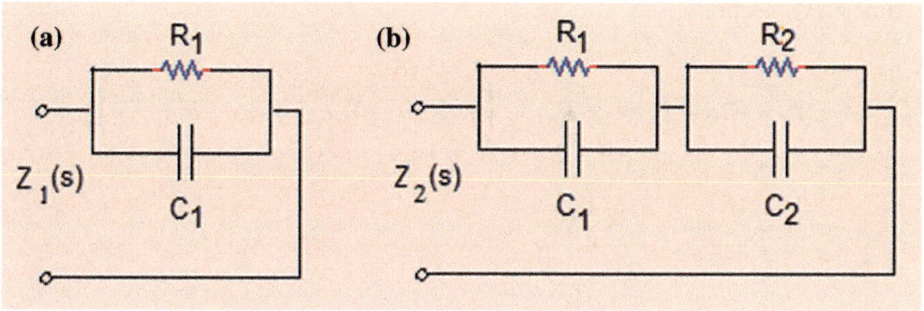

Fig. 8.31 **a** Single **b** double circuit subsections for Problem 8.3.7

$$V(s) = V_1(s) + V_2(s) + \ldots V_n(s)$$
$$V(s) = R_1\left(\frac{1}{s} - \frac{1}{s + \frac{1}{\tau_1}}\right) + R_2\left(\frac{1}{s} - \frac{1}{s + \frac{1}{\tau_2}}\right) + \cdots R_n\left(\frac{1}{s} - \frac{1}{s + \frac{1}{\tau_n}}\right)$$

$$v(t) = \mathcal{L}^{-1}[V(s)] = R_1\left(1 - \mathrm{e}^{-\frac{t}{\tau_1}}\right) + R_2\left(1 - \mathrm{e}^{-\frac{t}{\tau_2}}\right) + \cdots R_n\left(1 - \mathrm{e}^{-\frac{t}{\tau_n}}\right)$$

$$v(t) = \sum_{i=1}^{n} R_i\left(1 - \mathrm{e}^{-\frac{t}{\tau_i}}\right) \text{ (V)}.$$

The so-called "transient impedance" in time domain becomes

$$z(t) = \sum_{i=1}^{n} R_i\left(1 - \mathrm{e}^{-\frac{t}{\tau_i}}\right) \ (\Omega).$$

For two RC sections, alternative expression is

$$z(t) = R_1 - R_1\mathrm{e}^{-\frac{t}{\tau_1}} + R_2 - R_2\mathrm{e}^{-\frac{t}{\tau_2}}$$
$$z(t) = (R_1 + R_2) - \left(R_1\mathrm{e}^{-\frac{t}{\tau_1}} + R_2\mathrm{e}^{-\frac{t}{\tau_2}}\right).$$

For n RC sections,

$$z(t) = (R_1 + R_2 + \cdots R_n) - \left(R_1 e^{-\frac{t}{\tau_1}} + R_2 e^{-\frac{t}{\tau_2}} + \cdots R_n e^{-\frac{t}{\tau_n}} \right)$$

$$z(t) = \sum_{i=1}^{n} (R_i) - \sum_{i=1}^{n} R_i e^{-\frac{t}{\tau_i}}$$

$$z(t) = \sum_{i=1}^{n} R_i (1 - e^{-\frac{t}{\tau_i}})$$

Problem 8.3.8 Plot magnitude and phase angle as functions of angular frequency (on the same graph) for the following RC driving point impedances, and then synthesize them using Foster 1 form.

(a) $Z(s) = \dfrac{1}{s + 0.1}$,

(b) $Z(s) = \dfrac{s + 1.1}{s + 0.1}$,

(c) $Z(s) = \dfrac{2(s + 0.05)}{s(s + 0.1)}$

(d) $Z(s) = \dfrac{(s + 2.05)(s + 0.05)}{s(s + 0.1)}$.

Solution

Using MATLAB script files (Foster11.m), (Mag_phase_fom_TF1.m);

(a) $R_1 = 10\ \Omega$, $C_1 = 1$ F (Figs. 8.32 and 8.33)
(b) $R = 1\ \Omega$, $R_1 = 10\ \Omega$, $C_1 = 1$ F (Figs. 8.34 and 8.35)
(c) $C = 1$ F, $R_1 = 10\ \Omega$, $C_1 = 1$ F (Figs. 8.36 and 8.37)
(d) $R = 1\ \Omega$, $R_1 = 10\ \Omega$, $C_1 = 1$ F, $C = 1$ F (Figs. 8.38 and 8.39).

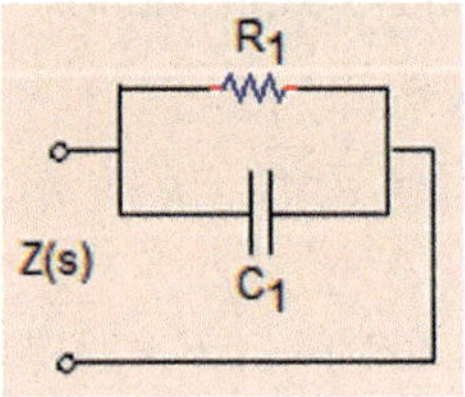

Fig. 8.32 Foster 1 form of $Z(s)$ (part a)

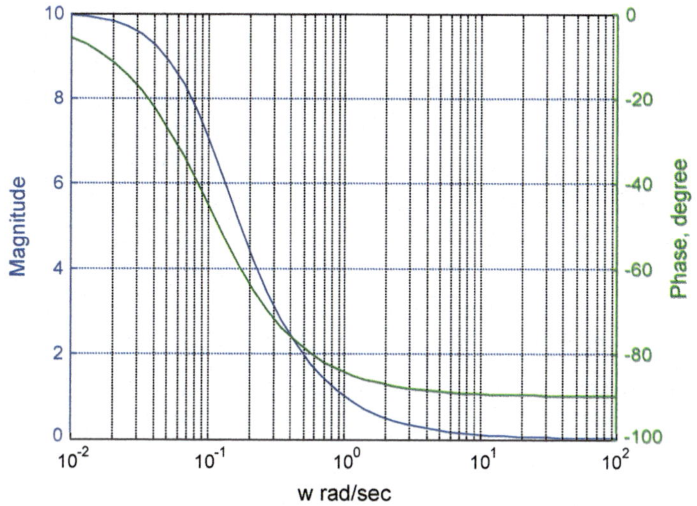

Fig. 8.33 Magnitude and phase functions

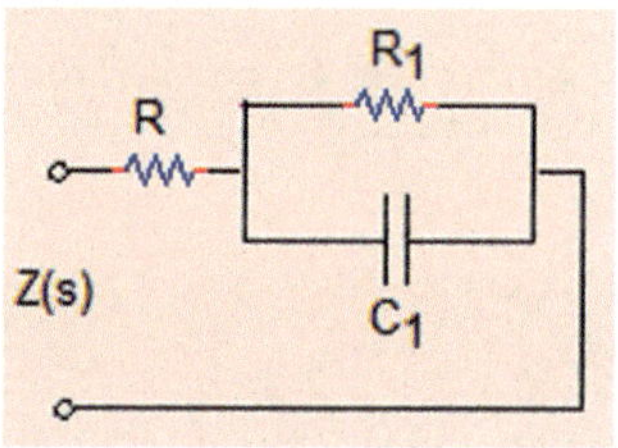

Fig. 8.34 Foster 1 form of *Z*(*s*) (part b)

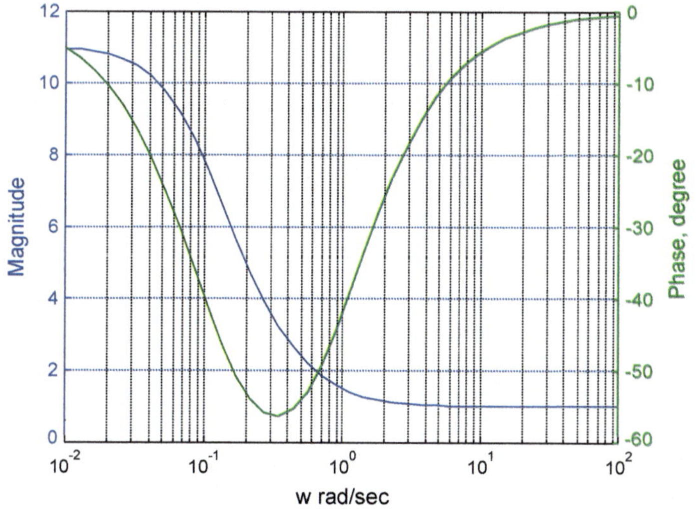

Fig. 8.35 Magnitude and phase response plots for Problem 8.3.8b

Fig. 8.36 Foster 1 form of *Z*(*s*) (part c)

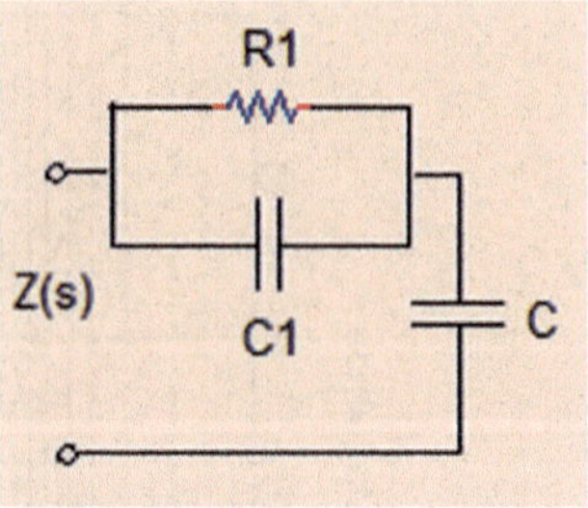

Following is the MATLAB script for m.file for (Mag_phase_fom_TF1.m):

```
Following is the MATLAB script for m.file for (Mag_phase_fom_TF1.m):
%Computing Mag and phase of a given transfer function H(s)=N(s)/D(s)
clc;clear;
%Generate an impedance transfer function in s domain
a=[1 .1];b=[1 1.1];
%N=b;
%D=a;
N =[ 1 2.1 0.1];
D =[ 1 0.1 0];
w = logspace(-2,2,40);%generates 40 points between 10^-2 and 10^2
H = tf(N,D)    %Transfer function
[mag,phase]=bode(H,w);
mag=squeeze(mag(1,1,:));
phase=squeeze(phase(1,1,:));
%magdb = 20*log10(mag);
w=w';
%plot mag and phase on the same figure (Double axis graph)
[ax]=plotyy(w,mag,w,phase,@semilogx);grid on;
ylabel(ax(1),'Magnitude')
ylabel(ax(2),'Phase,  degree')
xlabel(ax(2),'w rad/sec')
```

Problem 8.3.9 For the driving point (input) impedance function

$$Z(s) = \frac{s^3 + 6s^2 + 10s + 4}{s^4 + 7s^3 + 15s^2 + 10s + 1}.$$

(a) Write $Z(s)$ in factored form.
(b) Plot magnitude and phase functions as functions of angular frequency.
(c) What is the resistance seen at input?
(d) Nyquist plot of $Z(s)$? At which angular frequency $\mathrm{Re}[Z(j\omega)] = 0.553\ \Omega$, and $-\mathrm{Im}[Z(j\omega)] = 0.966\ \Omega$? (Bode1.m)
(e) Synthesize the circuit of given input impedance function in ladder form. Print MATLAB script for this m file. (CauerRC1.m).

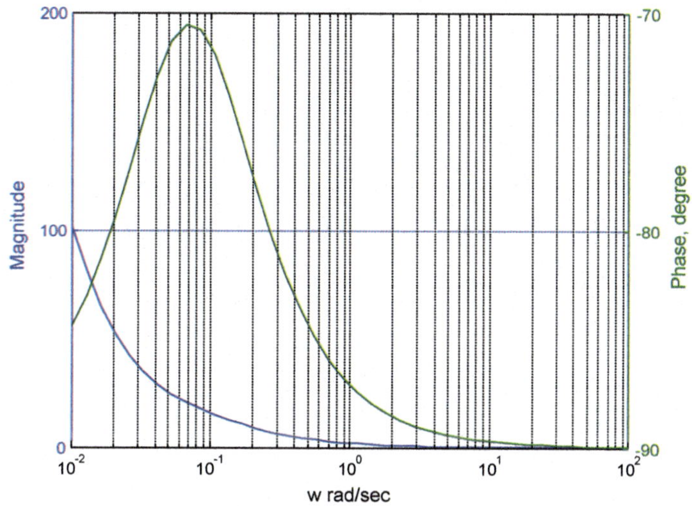

Fig. 8.37 Magnitude and phase response plots for Problem 8.3.8c

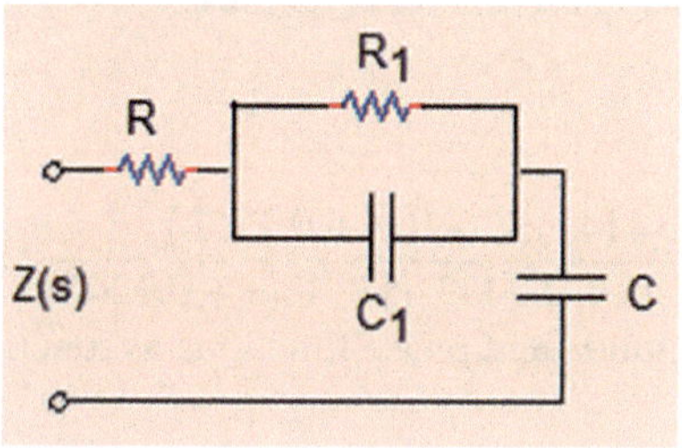

Fig. 8.38 Foster 1 form of $Z(s)$ (part d)

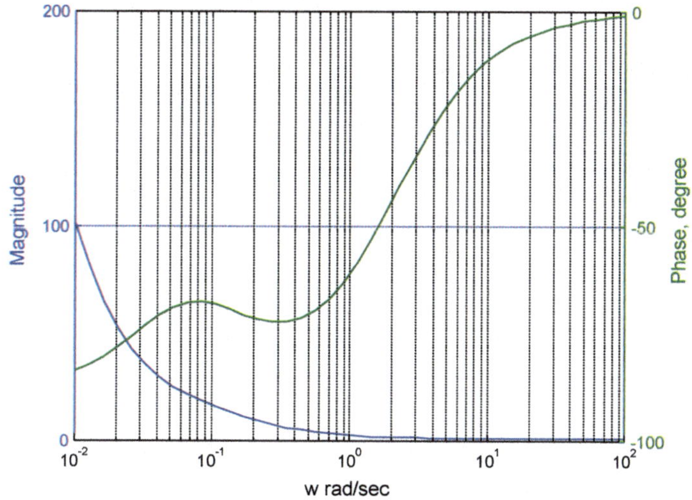

Fig. 8.39 Magnitude and phase response plots for Problem 8.3.8c

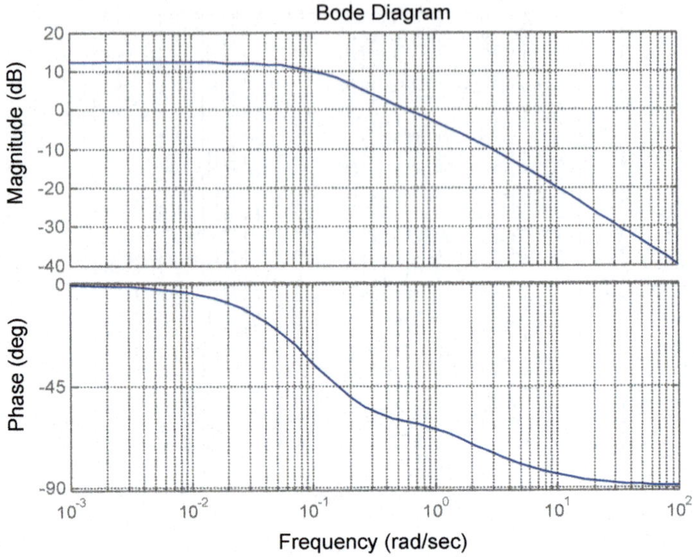

Fig. 8.40 Bode plots for magnitude and phase functions

Solution

(a) $Z(s) = \dfrac{(s+3.4142)(s+2)(s+0.5858)}{(s+3.5321)(s+2.3473)(s+1)(s+0.1206)}$.

(b) Bode plots for magnitude and phase functions as functions of angular frequency (Fig. 8.40);

(c) The resistance seen at input is the dc gain of $Z(0)$,

$$Z(0) = \frac{4}{1} = 4\,\Omega.$$

(d) Nyquist plot of $Z(s)$ is shown in Fig. 8.41.
From the Nyquist plot it is seen that at the angular frequency of $\omega = 0.5$ rad/s., $\mathrm{Re}[Z(j\omega)] = 0.553\ \Omega$, and $-\mathrm{Im}[Z(j\omega)] = 0.966\ \Omega$.
MATLAB m.file script:

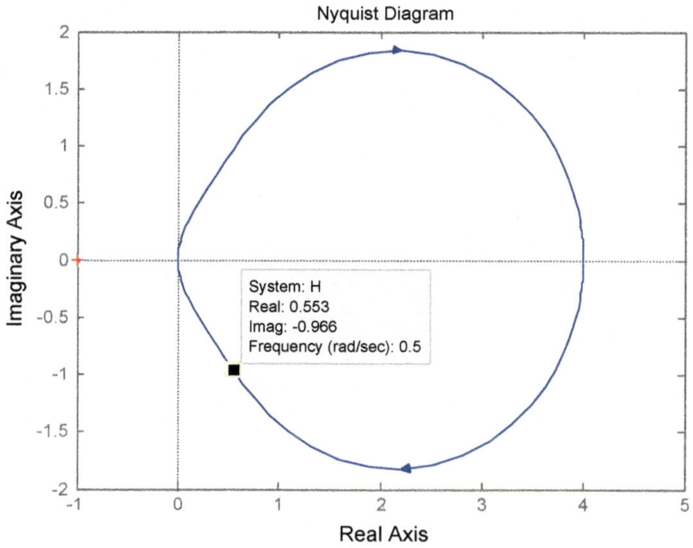

Fig. 8.41 Nyquist plot

```
%Plotting pole-zero diagram, mag and phase,
%Nyquist  from a given transfer function H(s)=N(s)/D(s)
clc;clear;
N3=[1 6 10 4];
D4=[1 7 15 10 1];
H = tf(N3,D4)    %Transfer function
zero(H)
pole(H)
pzmap(H);figure;
bode(H);grid on;figure;
nyquist(H);
```

(e) Synthesized RC input impedance circuit is shown in Fig. 8.42. Computed component values are

$$R_1 = 0,\ R_2 = R_3 = R_4 = R_5 = 1\ \Omega,\ \ C_1 = C_2 = C_3 = C_4 = 1\ \text{F}.$$

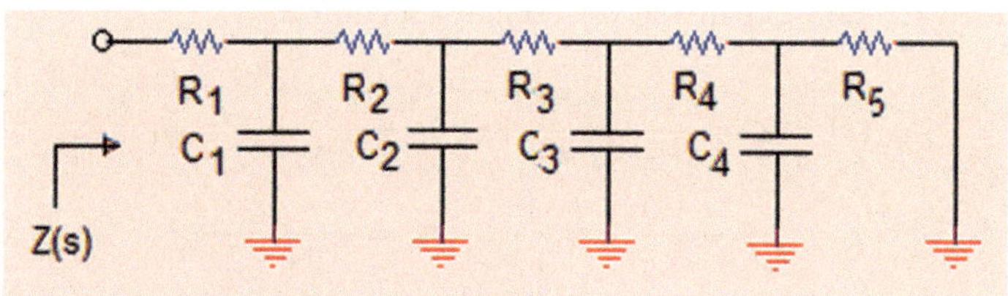

Fig. 8.42 Synthesized RC input impedance circuit

MATLAB script for this m file is (CauerRC1.m):

```
%Cauer's first form RC synthesis, Z(s)=N(s)/D(s)
%Order of N(s) must be less than or equal to the order of D(s).
%m=order of D(s)
%Result is in the following order:Z=[R1 C1 R2 C2 ...Rm Cm R(m+1)]
%R(m+1)=Inf  means it is open circuit, Ex.: m=2; Z=[R1 C1 R2 C2 R3]
N=[0 1 6 10 4];    %numerator poly.
D=[1 7 15 10 1 ]; %denominator poly.
C = [N; D]
% Continuous Fraction Expansion
for i = 3:size(C,2)+length(D)
for j = 1:size(C,2)-1
A = C(i-2,j+1)*C(i-1,1); B = C(i-2,1)*C(i-1,j+1);
C(i,j) = (A-B)/C(i-1,1);
end
end
n=size(C,1);
for i = 1:n-1
Z(1,i) = C(i,1)/C(i+1,1);
end
%format rational
format short
Z
```

8.4 RC Driving Point Impedance Synthesis

Problem 8.4.1 Show that following DPI function is obtained by its continued fraction expansion:

$$Z_{RC}(s) = \frac{s^2+6s+8}{s^2+4s+3} = 1 + \cfrac{1}{\frac{1}{2}s + \cfrac{1}{\frac{4}{3} + \cfrac{1}{\frac{3}{2}s + \frac{1}{3}}}}.$$

Solution

By continued fraction expansion of rational function Z_{RC},

$$\begin{aligned} Z_{RC}(s) = Z = \frac{s^2+6s+8}{s^2+4s+3} &= 1 + \frac{2s+5}{s^2+4s+3} = q_1 + Z_1 \\ Z_1 = \frac{2s+5}{s^2+4s+3} = \frac{1}{Y_1}, \quad Y_1 &= \frac{s^2+4s+3}{2s+5} \end{aligned} \tag{8.14}$$

$$\begin{aligned} Y_1 &= \frac{s^2+4s+3}{2s+5} = \frac{1}{2}s + \frac{\frac{3}{2}s+3}{2s+5} = \frac{1}{2}s + Y_2 = q_2 + Y_2 \\ Y_2 &= \frac{\frac{3}{2}s+3}{2s+5} = \frac{1}{Z_2}, \quad Z_2 = \frac{2s+5}{\frac{3}{2}s+3} \end{aligned} \tag{8.15}$$

$$Z_2 = \frac{2s+5}{\frac{3}{2}s+3} = \frac{4}{3} + \frac{1}{\frac{3}{2}s+3} = \frac{4}{3} + Z_3 = q_3 + Z_3 \tag{8.16}$$

$$Z_3 = \frac{1}{\frac{3}{2}s+3} = \frac{1}{Y_3}, \quad Y_3 = \frac{3}{2}s + 3 = \frac{3}{2}s + Y_4 \tag{8.17}$$

$$Y_4 = 3 = \frac{1}{Z_4}, \quad Z_4 = \frac{1}{3} = q_5 \tag{8.18}$$

$$Z_{RC}(s) = \frac{s^2+6s+8}{s^2+4s+3} = q_1 + \cfrac{1}{q_2 + \cfrac{1}{q_3 + \cfrac{1}{q_4 + \cfrac{1}{q_5}}}}. \tag{8.19}$$

Substituting quotient terms found in (8.14)–(8.18) into (8.19) yield the expected result.

Problem 8.4.2 Synthesize the following RC driving point impedance function, $Z(s)$,

$$Z(s) = 1 + \cfrac{1}{\frac{1}{2}s + \cfrac{1}{\frac{2}{3} + \cfrac{1}{\frac{9}{4}s + \frac{15}{2}}}}.$$

Solution

This function can be synthesized using Cauer I form, since it is a continued fraction. Final ratio is seen to be

$$\frac{15}{2} = \cfrac{1}{\frac{2}{15}}$$

$$R_1 = 1\,\Omega, \quad C_1 = \frac{1}{2}\,\text{F}, \quad R_2 = \frac{2}{3}\,\Omega, \quad C_2 = \frac{9}{4}\,\text{F}, \quad R_3 = \frac{2}{15}\,\Omega.$$

Figure 8.43 displays the circuit realization.

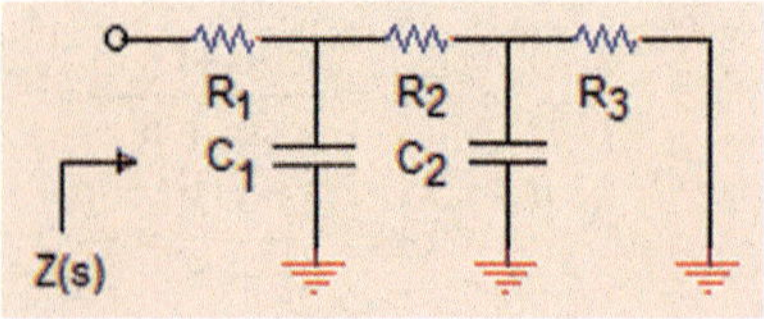

Fig. 8.43 Cauer I form circuit realization of RC DPI function

Problem 8.4.3 Synthesize the following input impedance functions using Cauer 1 form (CauerRC.m).

(a) $Z(s) = \dfrac{(s+1)(s+4)}{s(s+2)}$,

(b) $Z(s) = \dfrac{(s+5)}{(s+1)(s+10)}$,

(c) $Z(s) = \dfrac{(s+1)(s+3)}{s(s+2)}$,

(d) $Z(s) = \dfrac{(s+2)(s+4)}{(s+1)(s+3)}$,

(e) $Z(s) = \dfrac{(s+2)(s+3)}{(s+1)\left(s+\frac{5}{2}\right)}$.

Solution

(a) $Z(s) = \dfrac{(s+1)(s+4)}{s(s+2)} = \dfrac{s^2+5s+4}{s^2+2s} = 1 + \dfrac{1}{\frac{1}{3}s + \dfrac{1}{\frac{9}{2} + \dfrac{1}{\frac{1}{6}s}}}$ (Fig. 8.44)

$$R_1 = 1\,\Omega,\quad C_1 = \frac{1}{3}\,\text{F},\quad R_2 = \frac{9}{2}\,\Omega,\quad C_2 = \frac{1}{6}\,\text{F},\quad R_3 = \infty$$

(b) $Z(s) = \dfrac{(s+5)}{(s+1)(s+10)} = \dfrac{(s+5)}{s^2+11s+10}$ (Fig. 8.45)

$$R_1 = 0,\quad C_1 = 1\,\text{F},\quad R_2 = \frac{1}{6}\,\Omega,\quad C_2 = \frac{9}{5}\,\text{F},\quad R_3 = \frac{1}{3}\,\Omega$$

(c) $Z(s) = \dfrac{(s+1)(s+3)}{s(s+2)} = \dfrac{s^2+4s+3}{s^2+2s}$

$$R_1 = 1\,\Omega,\quad C_1 = \frac{1}{2}\,\text{F},\quad R_2 = 4\,\Omega,\quad C_2 = \frac{1}{6}\,\text{F},\quad R_3 = \infty$$

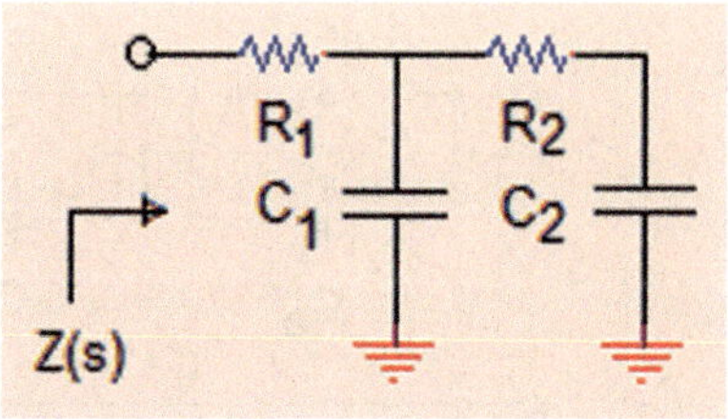

Fig. 8.44 Synthesized circuit for Problem 8.4.3a

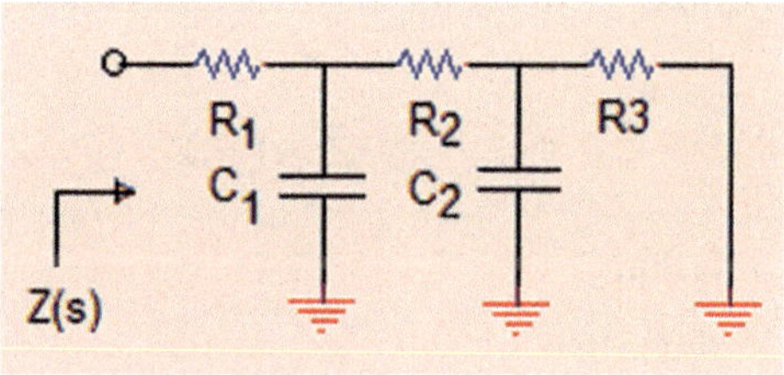

Fig. 8.45 Synthesized circuit for Problem 8.4.3b

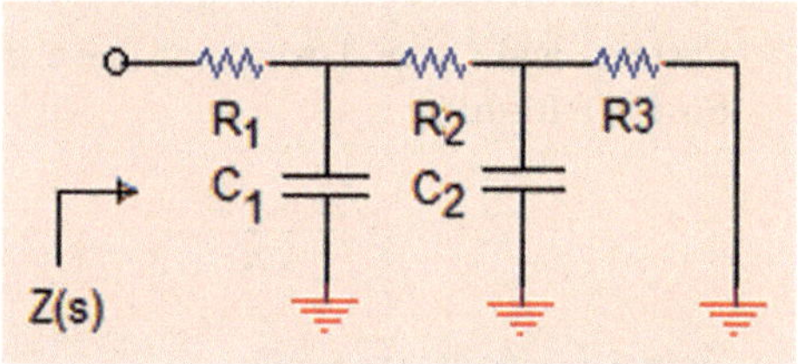

Fig. 8.46 Synthesized circuit for Problem 8.4.3c

Circuit schematic is similar to the one in part (a), and redrawn in Fig. 8.46.

(d) $Z(s) = \dfrac{(s+2)(s+4)}{(s+1)(s+3)} = \dfrac{s^2+6s+8}{s^2+4s+3}$

$$R_1 = 1\,\Omega, \quad C_1 = \frac{1}{2}\,\text{F}, \quad R_2 = \frac{4}{3}\,\Omega, \quad C_2 = \frac{3}{2}\,\text{F}, \quad R_3 = \frac{1}{3}\,\Omega$$

(e) $Z(s) = \dfrac{(s+2)(s+3)}{(s+1)\left(s+\frac{5}{2}\right)} = \dfrac{s^2+5s+6}{s^2+\frac{7}{2}s+\frac{5}{2}}$

$$R_1 = 1\,\Omega, \quad C_1 = \frac{2}{3}\,\text{F}, \quad R_2 = 1.2857\,\Omega, \quad C_2 = 4.0833\,\text{F}, \quad R_3 = 0.1143\,\Omega$$

Circuit schematic is similar to the one in part (d).

Problem 8.4.4 Determine the values of components in the following driving point impedances so that both transfer functions are numerically identical (Fig. 8.47),

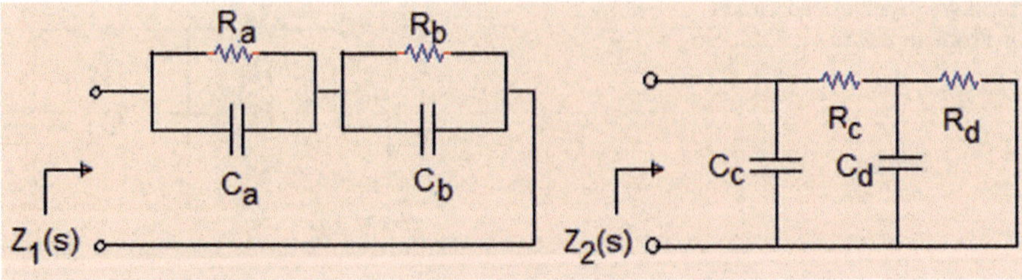

Fig. 8.47 Two circuits with identical DPI functions for Problem 8.4.4

$$Z_1(s) = Z_2(s) = Z(s) = \frac{s+2}{s^2+4s+3}.$$

Solution

In order that

$$Z_1(s) = Z_2(s) = Z(s),$$

Foster 1 and Cauer 1 forms of these impedances should be realized using $Z(s)$. Foster 1 form:

$$Z_1(s) = \frac{s+2}{s^2+4s+3} = \frac{s+2}{(s+1)(s+3)} = \frac{k_1}{s+1} + \frac{k_2}{s+3}$$

by partial fractions expansion,

$$k_1 = \frac{-1+2}{-1+3} = \frac{1}{2}, \quad k_2 = \frac{-3+2}{-3+1} = \frac{1}{2}$$

$$Z_1(s) = \frac{1/2}{s+1} + \frac{1/2}{s+3}$$

$$R_a = \frac{k_1}{1} = \frac{1/2}{1} = \frac{1}{2}\,\Omega, \quad R_b = \frac{k_1}{3} = \frac{1/2}{3} = \frac{1}{6}\,\Omega, \quad C_a = \frac{1}{k_1} = 2\,\mathrm{F},$$
$$C_b = \frac{1}{k_2} = 2\,\mathrm{F}.$$

Cauer 1 form:

$$Z_2(s) = Z_1(s) = Z(s) = \frac{s+2}{s^2+4s+3}.$$

Using continuous fractions expansion, since order of denominator is greater than the order of numerator, synthesis procedure begins by Y (admittance) removal, and then continues by alternating Z and Y removals:

$$\frac{s^2+4s+3}{s+2} = s + \frac{2s+3}{s+2}$$

$$\frac{s+2}{2s+3} = \frac{1}{2} + \frac{\frac{1}{2}}{2s+3}$$

$$\frac{2s+3}{\frac{1}{2}} = 4s+6$$

$$\frac{s^2+4s+3}{s+2} = Y_1 + \frac{1}{Z_2 + \frac{1}{Y_3+Y_4}} = 1s + \frac{1}{\frac{1}{2} + \frac{1}{4s + \frac{1}{\frac{1}{6}}}}$$

$$C_c = \frac{Y_1}{s} = 1\,\mathrm{F}, \quad R_c = Z_2 = \frac{1}{2}\,\Omega, \quad C_d = \frac{Y_3}{s} = 4\,\mathrm{F}, \quad R_d = Z_4 = 1/6\,\Omega.$$

Problem 8.4.5 Synthesize $Z(s)$ input impedance function,

$$Z(s) = k\frac{(s+2)(s+6)}{(s+1)(s+4)}$$

(a) for $k = 1$,
(b) for $k = 10$,
(c) Draw a conclusion from these results.

Solution

(a) for $k = 1$,

$$Z(s) = 1 \cdot \frac{(s+2)(s+6)}{(s+1)(s+4)} = \frac{s^2+8s+12}{s^2+5s+4}$$

$R_1 = 1\,\Omega, \quad C_1 = 0.3333\,\mathrm{F}, \quad R_2 = 1.2857\,\Omega, \quad C_2 = 0.8167\,\mathrm{F},$
$R_3 = 0.7143\,\Omega.$

(b) for k = 10,

$$Z(s) = 10 \cdot \frac{s^2+8s+12}{s^2+5s+4}$$

$R_1 = 10\,\Omega, \quad C_1 = 0.03333\,\mathrm{F}, \quad R_2 = 12.8571\,\Omega, \quad C_2 = 0.0817\,\mathrm{F},$
$R_3 = 7.1429\,\Omega.$

(c) It can be concluded that $Z(0)$ ($=k$) term causes R's to increase, C's to decrease by a factor of k.

Problem 8.4.6 Determine the driving point (input) impedance $Z(s)$ for each circuit that has been realized in Foster 1 form, given in Figs. 8.48, 8.49, 8.50 and 8.51, using a MATLAB script (Foster 1RC_to_ TF.m).

(a) $R_1 = 3\ \Omega$, $R_2 = 4\ \Omega$, $C_1 = C_2 = 1$ F
(b) $R = 1\ \Omega$, $R_1 = 3\ \Omega$, $R_2 = 4\ \Omega$, $C_1 = C_2 = 1$ F
(c) $R_1 = 3\ \Omega$, $R_2 = 4\ \Omega$, $C_1 = C_2 = C = 1$ F
(d) $R = 1\ \Omega$, $R_1 = 3\ \Omega$, $R_2 = 4\ \Omega$, $C_1 = C_2 = C = 1$ F.

Solution
General form of Foster 1 realization of a driving point impedance function is

$$Z(s) = R + \frac{1}{sC} + \sum_{i=1}^{m} \frac{R_i}{1 + sR_iC_i}.$$

A series component inclusion can be thought as a parallel subcircuit topology, as shown in Fig. 8.52.

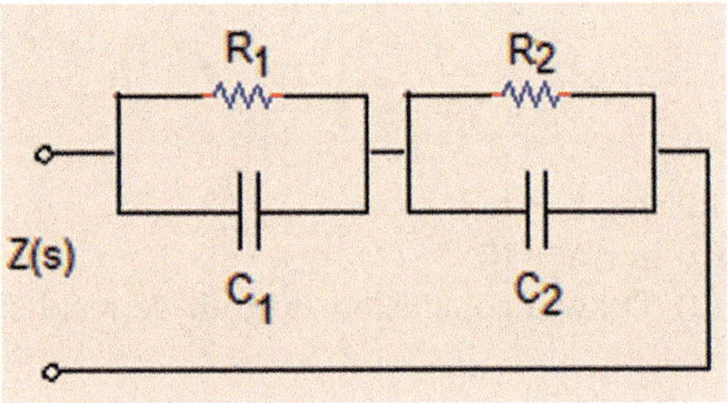

Fig. 8.48 The circuit for Problem 8.4.6a

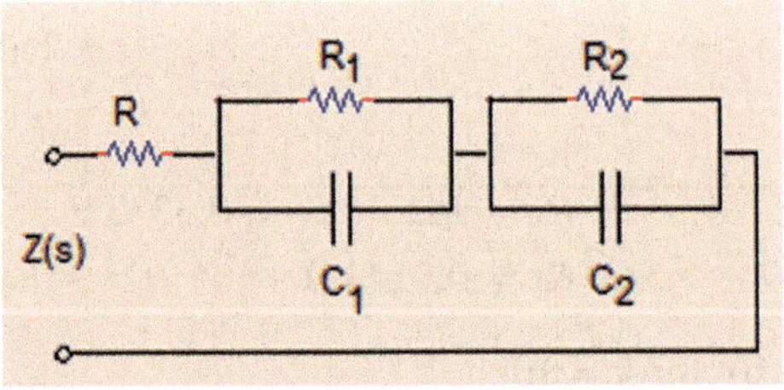

Fig. 8.49 The circuit for Problem 8.4.6b

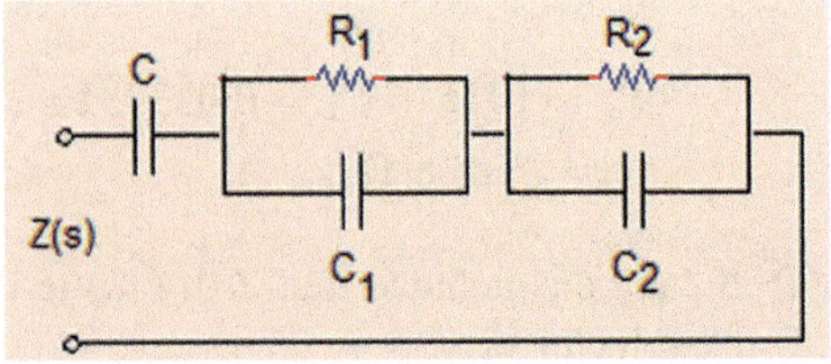

Fig. 8.50 The circuit for Problem 8.4.6c

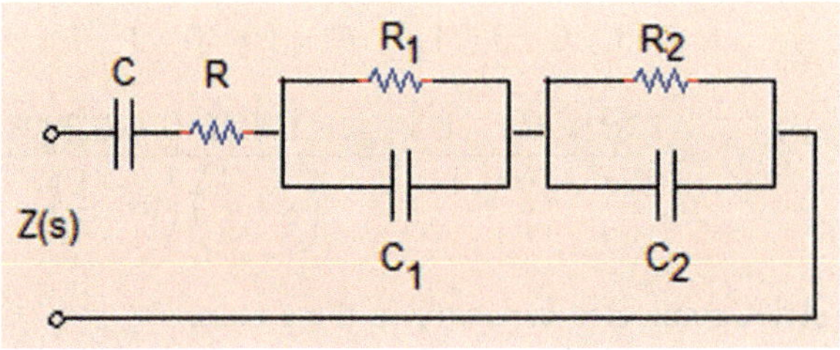

Fig. 8.51 The circuit for Problem 8.4.6d

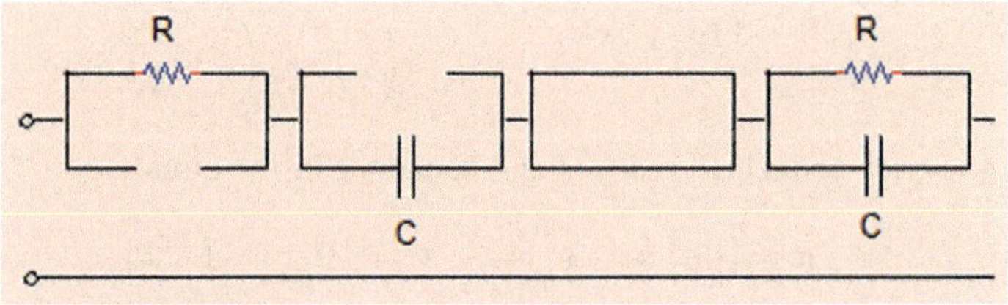

Fig. 8.52 Model development

A series resistor $\rightarrow \quad R_i = R,\ C_i = 0$; A series capacitor $\rightarrow \quad R_i = \infty,\ C_i = C$; A short $\rightarrow \quad R_i = C_i = 0$

Therefore,

$$Z(s) = \sum_{i=1}^{m+2} \frac{R_i}{1 + sR_iC_i} = \sum_{i=1}^{m+2} \frac{\frac{1}{C_i}}{s + \frac{1}{R_iC_i}}.$$

This means that first two summations denote series-connected elements (i = 1, 2); the rest are parallel-connected R, C elements.

(a) Case 1: No series-connected component,

$$R = [0 \quad 0 \quad 3 \quad 4], \quad C = [0 \quad 0 \quad 1 \quad 1],$$

$$Z(s) = \frac{24s + 7}{12s^2 + 7s + 1} = 2\frac{\left(s + \frac{7}{24}\right)}{\left(s + \frac{1}{3}\right)\left(s + \frac{1}{4}\right)}.$$

(b) Case 2: A series-connected resistor in the circuit,

$$R = [1 \quad 0 \quad 3 \quad 4], \quad C = [0 \quad 0 \quad 1 \quad 1],$$

$$Z(s) = \frac{12s^2 + 31s + 8}{12s^2 + 7s + 1} = \frac{(s + 2.2925)(s + 0.2908)}{\left(s + \frac{1}{3}\right)\left(s + \frac{1}{4}\right)}.$$

(c) Case 3: A series-connected capacitor in the circuit,

$$R = [0 \quad \infty \quad 3 \quad 4], \quad C = [0 \quad 1 \quad 1 \quad 1],$$

$$Z(s) = \frac{36s^2 + 14s + 1}{12s^3 + 7s^2 + s} = \frac{(s + 0.2946)(s + 0.0943)}{s\left(s + \frac{1}{3}\right)\left(s + \frac{1}{4}\right)}.$$

(d) Case 4: Series-connected resistor and capacitor in the circuit,

$$R = [0 \quad \infty \quad 3 \quad 4], \quad C = [0 \quad 1 \quad 1 \quad 1],$$

$$Z(s) = \frac{12s^3 + 43s^2 + 15s + 1}{12s^3 + 7s^2 + s} = \frac{(s + 3.2010)(s + 0.2937)(s + 0.0886)}{s\left(s + \frac{1}{3}\right)\left(s + \frac{1}{4}\right)}.$$

Problem 8.4.7 A third-order driving point impedance is given as

$$Z(s) = \frac{N_1 s^2 + N_2 s + N_3}{D_1 s^3 + D_2 s^2 + D_3 s + 1}$$

$$\begin{aligned}
N_1 &= C_2 C_3 R_1 R_2 R_3 \\
N_2 &= C_2 R_1 (R_2 + R_3) + C_2 R_3 (R_1 + R_2) \\
N_3 &= R_1 + R_2 + R_3 \\
D_1 &= R_1 R_2 R_3 C_1 C_2 C_3 \\
D_2 &= C_1 C_2 R_1 (R_2 + R_3) + C_1 C_3 R_3 (R_1 + R_2) + C_2 C_3 R_2 R_3 \\
D_3 &= C_1 (R_1 + R_2 + R_3) + C_2 (R_2 + R_3) + C_3 R_3.
\end{aligned}$$

If $Z(s) = \dfrac{s^2 + 40s + 300}{s^3 + 50s^2 + 600s + 1000}$,

(a) Determine component values of $Z(s)$ using Cauer 1 form. Draw the circuit.
(b) Prove that components values found in part (a) satisfy polynomial coefficients of $Z(s)$.
(c) Synthesize $Z(s)$ in Foster 1 form, and draw the circuit.
(d) Prove that the values of found in part (c) satisfy polynomial coefficients of $Z(s)$, if

Fig. 8.53 Synthesized DPI circuit using Cauer 1 form

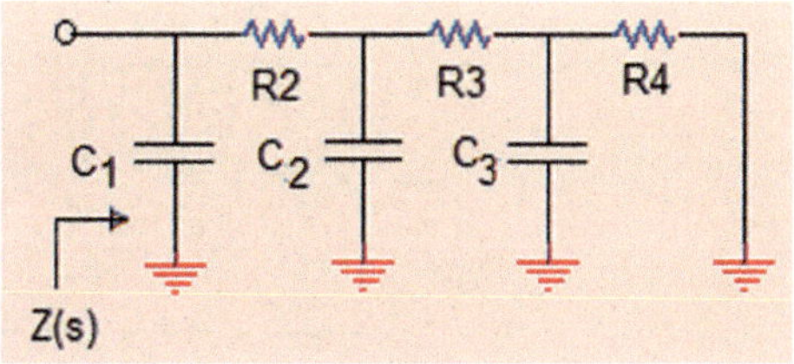

$$
\begin{aligned}
N_1 &= R_1R_2R_3(C_1C_2 + C_1C_3 + C_2C_3) \\
N_2 &= C_1R_1(R_2 + R_3) + C_2R_2(R_1 + R_3) + C_2R_3(R_1 + R_2) \\
N_3 &= R_1 + R_2 + R_3 \\
D_1 &= R_1R_2R_3C_1C_2C_3 \\
D_2 &= C_1C_2R_1R_2 + C_1C_3R_3R_1 + C_2C_3R_2R_3 \\
D_3 &= C_1R_1 + C_2R_2 + C_3R_3
\end{aligned}
$$

(Foster_Cauer1.m, Foster_proof1.m).

Solution

(a) Cauer 1 form of $Z(s)$ yields $R_1 = 0$, $C_1 = C_2 = C_3 = 1\,\text{F}$, $R_2 = R_3 = R_4 = 0.1\,\Omega$ (Fig. 8.53).

(b) Cauer 1 polynomial coefficients are calculated as

$$
\begin{aligned}
N_1 &= 10^{-3}, \quad N_2 = 4 \times 10^{-2}, \quad N_3 = 3 \times 10^{-1}, \\
D_1 &= 10^{-3}, \quad D_2 = 5 \times 10^{-2}, \quad D_3 = 6 \times 10^{-1}
\end{aligned}
$$

$$
Z(s) = \frac{10^{-3}s^2 + (4 \times 10^{-2})s + (3 \times 10^{-1})}{10^{-3}s^3 + (5 \times 10^{-2})s^2 + (6 \times 10^{-1})s + 1} = \frac{s^2 + 40s + 300}{s^3 + 50s^2 + 600s + 1000}
$$

(c) Foster 1 form of realization of $Z(s)$ yields

$$
\begin{aligned}
&R_{f0} = 0, \quad R_{f1} = 0.0033\,\Omega, \quad R_{f2} = 0.0225\,\Omega, \quad R_{f3} = 0.2742\,\Omega \\
&C_{f1} = 9.2959\,\text{F}, \quad C_{f2} = 2.8629\,\text{F}, \quad C_{f3} = 1.8412\,\text{F}
\end{aligned}
$$

in the circuit shown in Fig. 8.54,

$$
R_{f1} = R_1, \; R_{f2} = R_2, \; R_{f3} = R_3, \; C_{f1} = C_1, \; C_{f2} = C_2, \; C_{f3} = C_3.
$$

(d) Substituting numerical values of components into polynomial coefficient equations yields

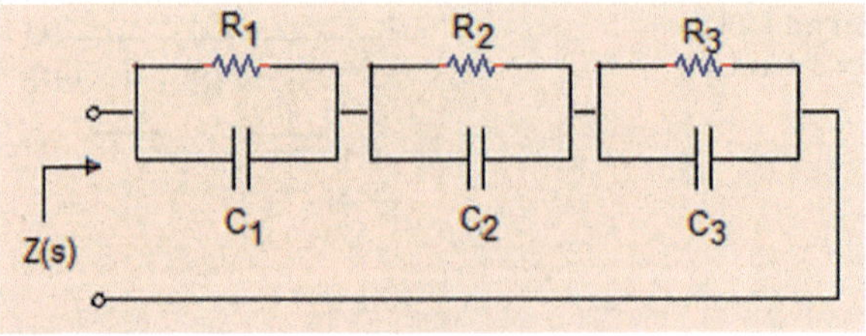

Fig. 8.54 Synthesized DPI circuit using Foster 1 form

$$Z(s) = \frac{s^2 + 40.1s + 300.7}{s^3 + 50.1s^2 + 601.4s + 1002}.$$

Differences in the numerical values of polynomial coefficients are due to truncation and rounding off errors in computations.

Following is a MATLAB script for proving transfer function synthesis used in this problem.

```
%Foster_proof1
clc;
R1=0.0033;R2=0.0225;R3=0.2742;C1=9.2959;C2=2.8629;C3=1.8412;
N1=C2*C1*R1*R2*R3 + C3*C1*R1*R2*R3 + C3*C2*R1*R2*R3;
N2=C1*R1*R3 + C1*R1*R2 + C2*R2*R3 + C2*R1*R2 + C3*R2*R3 + C3*R1*R3;
N3=R1+R2+R3;
D1=C3*C2*C1*R1*R2*R3;
D2= C2*C1*R1*R2 + C3*C1*R1*R3 + C3*C2*R2*R3;
D3=C1*R1 + C2*R2 + C3*R3;
N=[N1 N2 N3]/D1;
D=[D1 D2 D3 1]/D1;
H=tf(N,D)
```

Problem 8.4.8 Use MATLAB and synthesize RC input impedance

$$Z(s) = \frac{s^3 + 6s^2 + 10s + 4}{s^4 + 7s^3 + 15s^2 + 10s + 1}.$$

(a) In Foster 1 form,
(b) In Cauer 1 form (Foster_Cauer1.m),
(c) Factorize polynomial and plot magnitude (in ohms) and phase (in deg) of $Z(j\omega)$ (Bode2.m).

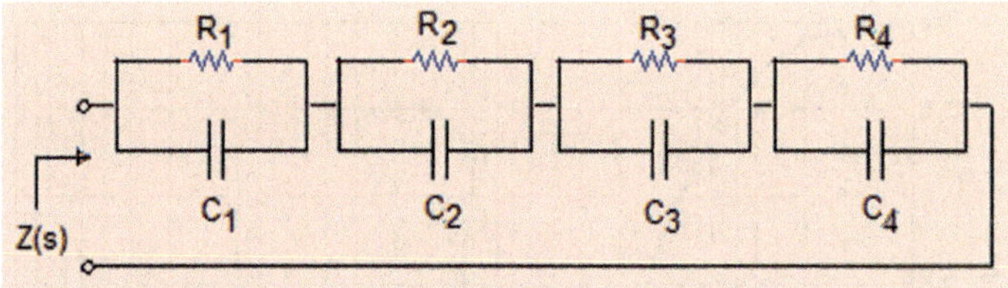

Fig. 8.55 Foster I form

Fig. 8.56 Cauer I form

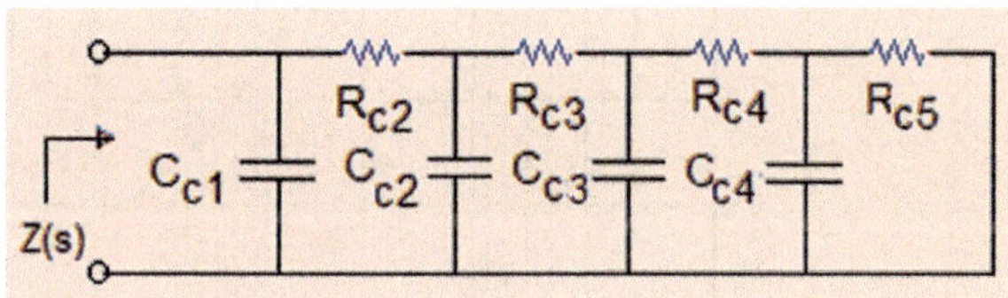

Solution

(a) $R_0 = 0, \quad R_{f1} = 0.0147\,\Omega,\ R_{f2} = 0.0782\,\Omega,\ R_{f3} = 0.3333\,\Omega,\ R_{f4} = 3.5737\,\Omega,$ (Fig. 8.55)

$$C_{f1} = 19.2344\,\text{F}, \quad C_{f2} = 5.4456\,\text{F}, \quad C_{f3} = 3\,\text{F}, \quad C_{f4} = 2.32\,\text{F}$$

(b) $R_{C1} = 0,\ R_{C2} = R_{C3} = R_{C4} = R_{C5} = 1\,\Omega,\ C_{C1} = C_{C2} = C_{C3} = C_{C4} = 1\,\text{F}$ (Fig. 8.56)

(c) $Z(s) = \dfrac{(s+3.4142)(s+2)(s+0.5858)}{(s+3.5321)(s+2.3473)(s+1)(s+0.1206)}.$

Figures 8.57 and 8.58 display the magnitude and phase response of given impedance function.

Problem 8.4.9 Determine driving point impedance of the ladder circuit shown in Fig. 8.59, in terms of its symbolic components.

Determine $Z(s)$ if $C_1 = C_2 = C_3 = 1$ F, $R_1 = R_2 = R_2 = 1\ \Omega$.

Determine poles and zeros of $Z(s)$.

Plot $|Z(j\omega)|$ and $\varphi(j\omega)$ using given component values above.

Plot Nyquist diagram using given component values. Find the angular frequency at which (Im[$Z(j\omega)$]) is a maximum. Find Re[$Z(j\omega)$] at that frequency (Bode2.m).

Solutions

The circuit can be viewed as $Z(s) = (1/sC_1)//[R_1 + [(1/sC_2)//[R_2 + [(1/sC_3)//R_3]]]]$

Using given component values, this yields,

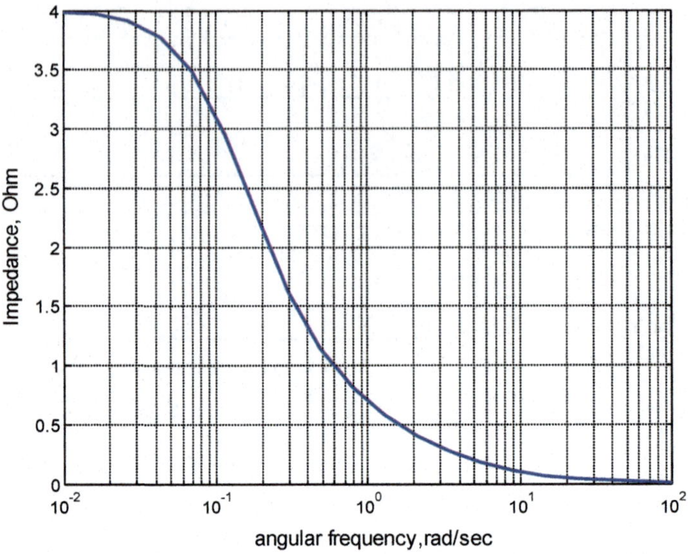

Fig. 8.57 Impedance magnitude plot

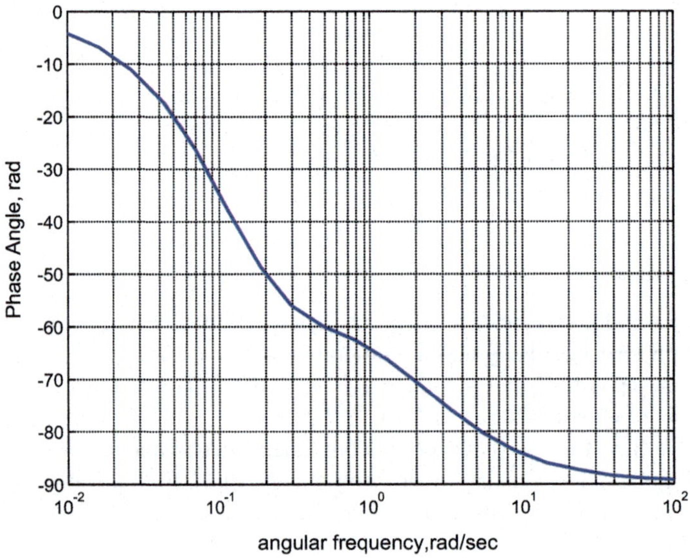

Fig. 8.58 Phase diagram

Fig. 8.59 Ladder circuit

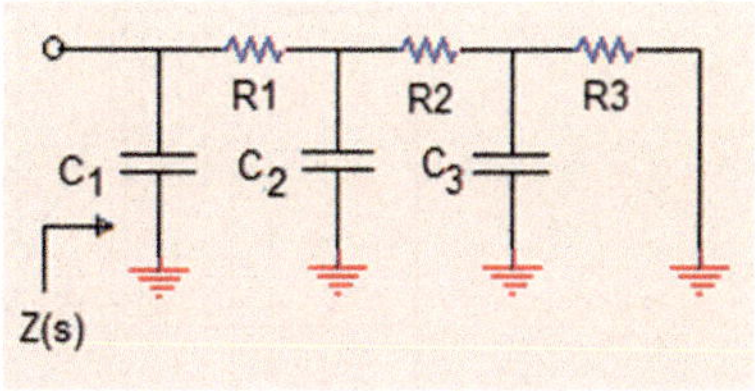

$$Z(s) = \frac{s^2 + 4s + 3}{s^3 + 5s^2 + 6s + 1} = \frac{(s + 1)(s + 3)}{(s + 3.247)(s + 1.555)(s + 0.1981)}.$$

The zeros are at (−1, −3) rad/s, and the poles are at (−3.247, −1.555, 0.1981) rad/s.

Plottings of magnitude and phase angle of $Z(j\omega)$ as functions of angular frequency are shown in Figs. 8.60, 8.61 and 8.62. First two of these graphs are obtained using a standard MATLAB command "bode (sys)". Additional couple of complementary figures is given to provide further insight to magnitude plots having y-axis in terms of (linear scaled) impedance unit, rather than decibels.

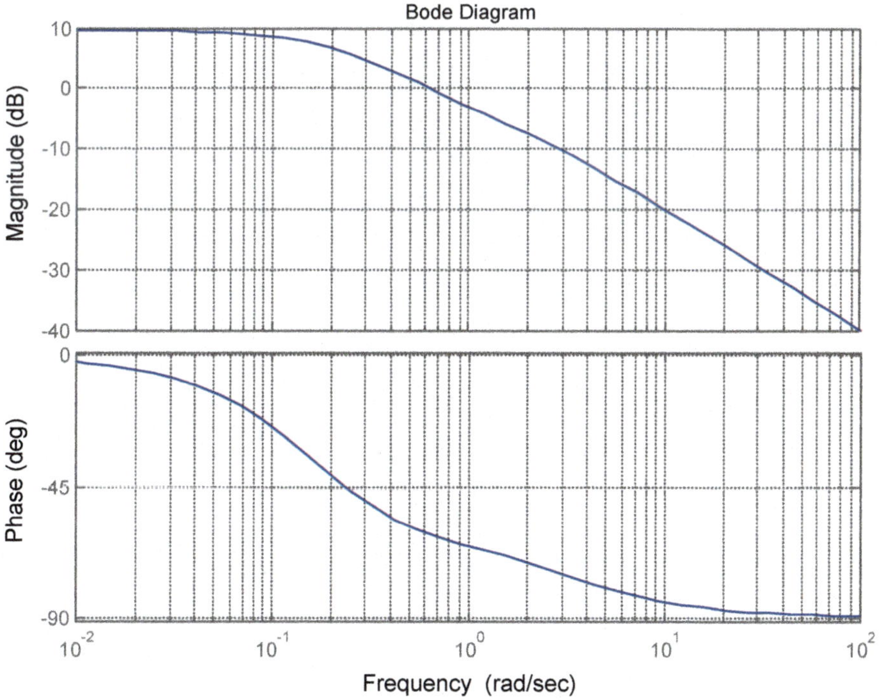

Fig. 8.60 Bode magnitude and phase plots

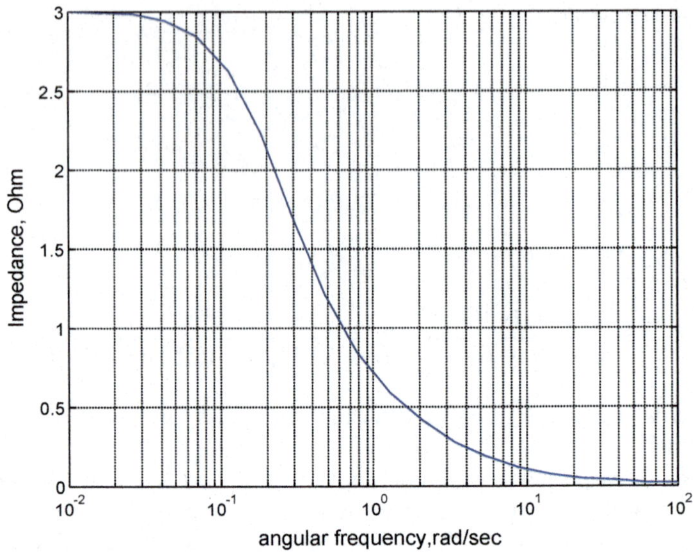

Fig. 8.61 The graph of impedance versus frequency

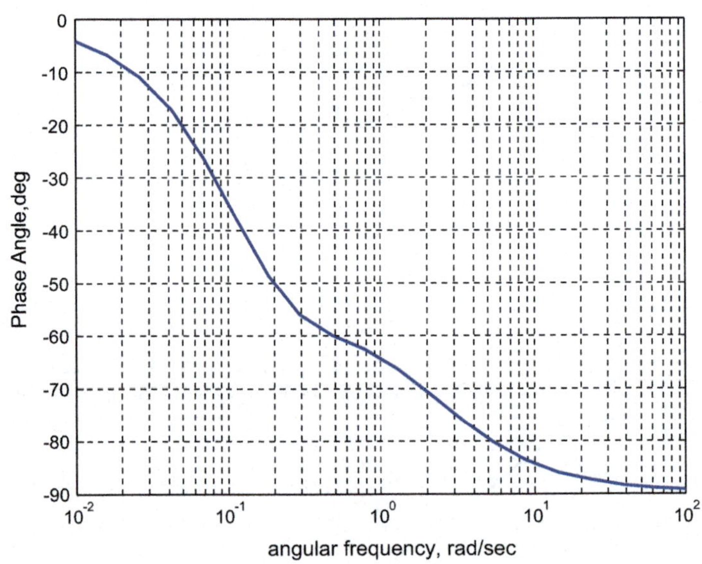

Fig. 8.62 Alternative plot for the phase angle against angular frequency

Nyquist diagram for driving point impedance having component data of part (b) is given in Fig. 8.63.

Note that dc gain (=resistance) of $Z(0)$ is 3 **Ω**. This is seen from the magnitude and Nyquist plots. However, it is also obvious from given $Z(s)$ that $Z(0) = 3$ **Ω**.

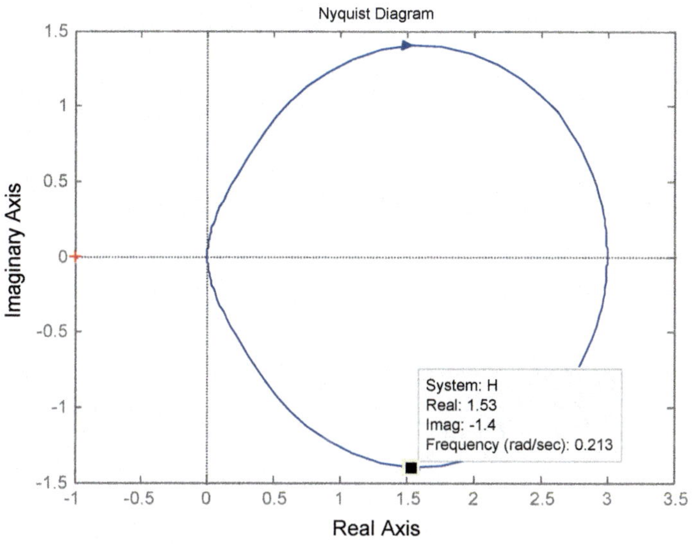

Fig. 8.63 Nyquist diagram

The angular frequency at which (Im[$Z(j\omega)$]) is a maximum can be obtained from Nyquist plot. The Re[$Z(j\omega)$] at that frequency is also measurable in Nyquist plot of $Z(j\omega)$ (see Fig. 8.63).

MATLAB script for Bode2.m is given below.

```
%Bode2
% Plotting pole-zero diagram, mag and phase
%Nyquist plot of a given transfer function H(s) = N(s)/D(s)
clc;clear;
N=[1 4 3];
D=[1 5 6 1];
 H = tf(N,D)    %Transfer function
 zero(H)
 pole(H)
%pzmap(H);figure; bode(H);grid on;figure; nyquist(H);
 w = logspace(-2,2,20);
 h = freqs(N,D,w);
 mag = abs(h);
 phase = angle(h);
%   To convert to hertz, degrees, and decibels, use
%   f = w/(2*pi);  mag = 20*log10(mag);
phase = phase*180/pi; semilogx(w,mag);grid on;
xlabel('angular frequency,rad/sec');ylabel('Impedance, Ohm');
figure;
semilogx(w,phase);grid on;
xlabel('angular frequency,rad/sec');ylabel('Phase Angle, deg');
```

Problem 8.4.10 Current unit step response of an RC driving point impedance is measured to be

$$v(t) = 0.6 - 0.5e^{-t} - 0.1e^{-5t}.$$

Synthesize this driving point impedance $Z(s)$, in two different forms.
Check the results using MATLAB m file (Foster_Cauer1.m).

Solution

$$\mathcal{L}[v(t)] = V(s) = \frac{0.6}{s} - \frac{0.5}{s+1} - \frac{0.1}{s+5}$$

$$V(s) = I(s)Z(s) = \frac{1}{s}\cdot Z(s) = \frac{0.6}{s} - \frac{0.5}{s+1} - \frac{0.1}{s+5}$$

$$sV(s) = Z(s) = 0.6 - \frac{0.5s}{s+1} - \frac{0.1s}{s+5}.$$

Inverse Laplace transform of the last expression denotes differential operation in time domain,

$$\frac{dv(t)}{dt} = 0.6\delta(t) - \frac{d}{dt}(0.5e^{-t}) - \frac{d}{dt}(0.1e^{-5t}) = 0.6\delta(t) + 0.5e^{-t} + 0.5e^{-5t}.$$

Dropping the first (impulse) term, and taking the Laplace transform of remaining terms in right side of above equation yields

$$Z(s) = \frac{\frac{1}{2}}{s+1} + \frac{\frac{1}{2}}{s+5} \tag{8.20}$$

$$Z(s) = \frac{s+3}{(s+1)(s+5)}. \tag{8.21}$$

Foster 1 type synthesis can be performed using (8.20), i.e., partial fractions expanded form of $Z(s)$ (Fig. 8.64),

$$Z(s) = \frac{s+b}{(s+a)(s+c)} = \frac{k_1}{s+a} + \frac{k_2}{s+c} = \frac{\frac{1}{2}}{s+1} + \frac{\frac{1}{2}}{s+5}$$

$$C_{f1} = \frac{1}{k_1} = 2\,\mathrm{F},\ C_{f2} = \frac{1}{k_2} = 2\,\mathrm{F},\ R_{f1} = \frac{k_1}{a} = \frac{\frac{1}{2}}{1} = \frac{1}{2}\,\Omega,\ R_{f2} = \frac{k_2}{c} = \frac{\frac{1}{2}}{5} = \frac{1}{10}\,\Omega.$$

Cauer 1 type synthesis can be performed using (8.21), and by continued fractions expansion of $Z(s)$,

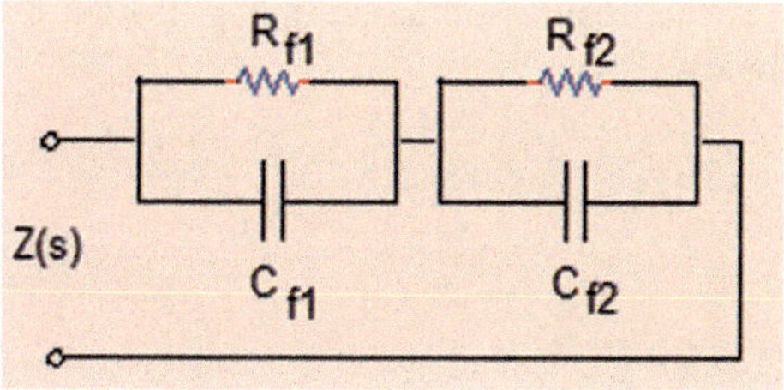

Fig. 8.64 Foster 1 type synthesis

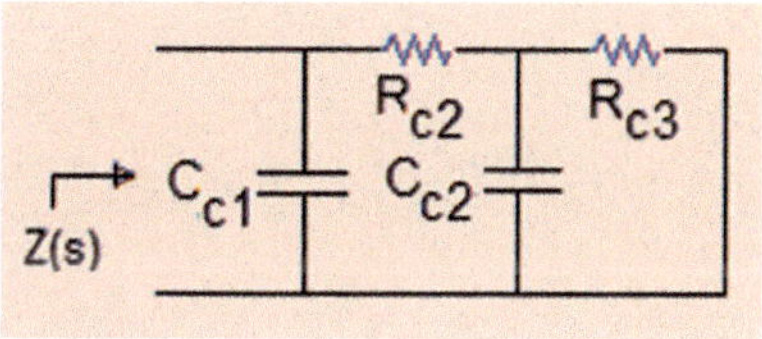

Fig. 8.65 Cauer 1 type synthesis

$$Z(s) = \frac{s+3}{(s+1)(s+5)} = \frac{s+3}{s^2+6s+5}.$$

Since order of denominator is greater than the order of numerator, synthesis procedure begins by Y (admittance) removal, and then continues by alternating Z and Y removals (Fig. 8.65),

$$\frac{s^2+6s+5}{s+3} = s + \frac{3s+5}{s+3}$$
$$\frac{s+3}{3s+5} = \frac{1}{3} + \frac{\frac{4}{3}}{3s+5}$$
$$\frac{3s+5}{\frac{4}{3}} = \frac{9}{4}s + \frac{15}{4}$$
$$\frac{s^2+6s+5}{s+3} = Y_1 + \frac{1}{Z_2 + \frac{1}{Y_3+Y_4}} = 1s + \frac{1}{\frac{1}{3} + \frac{1}{\frac{9}{4}s + \frac{1}{\frac{4}{15}}}}$$

$$C_{c1} = \frac{Y_1}{s} = 1\,\text{F}, \quad R_{c2} = Z_2 = \frac{1}{3}\,\Omega, \quad C_{c2} = \frac{Y_3}{s} = \frac{9}{4}\,\text{F}, \quad R_{c3} = Z_4 = \frac{4}{15}\,\Omega.$$

The results obtained using Foster_Cauer.m file in MATLAB are displayed below:

```
Transfer function:
   s + 3
-------------
s^2 + 6 s + 5
Foster 1 circuit



R0 = 0
R = 1/10   1/2
C = 2     2



Cauer 1 circuit
R1  C1   R2   C2   R3



0   1   1/3  9/4   4/15
```

Problem 8.4.11 Current unit step response of an RC driving point impedance is measured to be

$$v(t) = 0.6667 - 0.5e^{-t} - 0.1667e^{-3t}$$

Synthesize this driving point impedance $Z(s)$, in two different forms.
Check the results using MATLAB m file (Foster_Cauer1.m).

Solution

$$\mathcal{L}[v(t)] = V(s) = \frac{0.6667}{s} - \frac{0.5}{s+1} - \frac{0.1667}{s+3}$$
$$V(s) = I(s)Z(s) = \frac{1}{s} \cdot Z(s) = \frac{0.6667}{s} - \frac{0.5}{s+1} - \frac{0.1667}{s+3}$$
$$sV(s) = Z(s) = 0.6667 - \frac{0.5s}{s+1} - \frac{0.1667s}{s+3}.$$

Inverse Laplace transform of the last expression denotes differential operation in time domain,

$$\begin{aligned}\frac{\mathrm{d}v(t)}{\mathrm{d}t} &= 0.6667\delta(t) - \frac{\mathrm{d}}{\mathrm{d}t}(0.5\mathrm{e}^{-t}) - \frac{\mathrm{d}}{\mathrm{d}t}(0.1667\mathrm{e}^{-3t}) \\ &= 0.6667\delta(t) + 0.5\mathrm{e}^{-t} + 0.5\mathrm{e}^{-3t}.\end{aligned}$$

Concentrating on the right side of this equation, dropping the first (impulse) term, and taking the Laplace Transform of remaining terms yields

$$Z(s) = \frac{\frac{1}{2}}{s+1} + \frac{\frac{1}{2}}{s+3} \tag{8.22}$$

$$Z(s) = \frac{s+2}{(s+1)(s+3)}. \tag{8.23}$$

Foster 1 type synthesis can be performed using (8.22), i.e., partial fractions expanded form of $Z(s)$ (Fig. 8.66),

$$Z(s) = \frac{s+b}{(s+a)(s+c)} = \frac{k_1}{s+a} + \frac{k_2}{s+c} = \frac{\frac{1}{2}}{s+1} + \frac{\frac{1}{2}}{s+3}$$

$$C_{f1} = \frac{1}{k_1} = 2\,\mathrm{F},\quad C_{f2} = \frac{1}{k_2} = 2\,\mathrm{F},\quad R_{f1} = \frac{k_1}{a} = \frac{\frac{1}{2}}{1} = \frac{1}{2}\,\Omega,\quad R_{f2} = \frac{k_2}{c} = \frac{\frac{1}{2}}{3} = \frac{1}{6}\,\Omega$$

Cauer 1 type synthesis can be performed using (8.23), and by continued fractions expansion of $Z(s)$,

$$Z(s) = \frac{s+2}{(s+1)(s+3)} = \frac{s+2}{s^2+4s+3}.$$

Since order of denominator is greater than the order of numerator, synthesis procedure begins by Y (admittance) removal, and then continues by alternating Z and Y removals (Fig. 8.67),

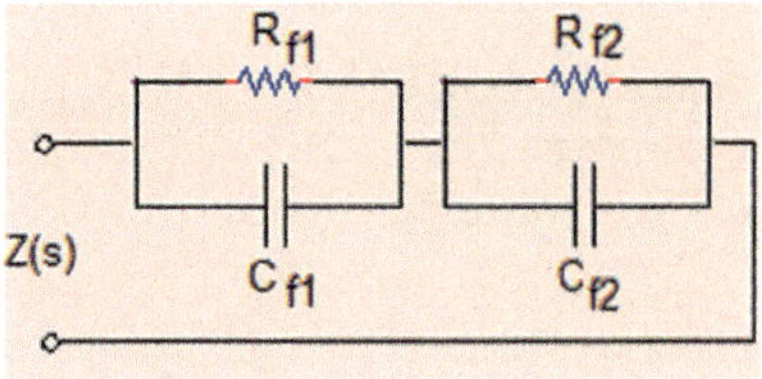

Fig. 8.66 Foster 1 type synthesis

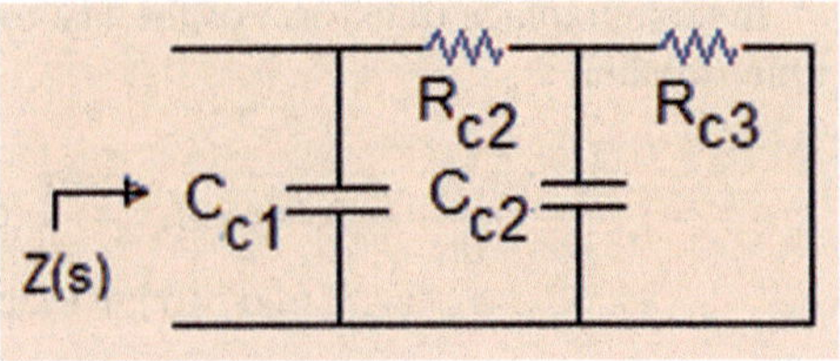

Fig. 8.67 Cauer 1 type synthesis

$$\frac{s^2+4s+3}{s+2} = s + \frac{2s+3}{s+2}$$

$$\frac{s+2}{2s+3} = \frac{1}{2} + \frac{\frac{1}{2}}{2s+3}$$

$$\frac{2s+3}{\frac{1}{2}} = 4s+6$$

$$\frac{s^2+4s+3}{s+2} = Y_1 + \frac{1}{Z_2 + \frac{1}{Y_3+Y_4}} = 1s + \frac{1}{\frac{1}{2} + \frac{1}{4s + \frac{1}{\frac{1}{6}}}}$$

$$C_{c1} = \frac{Y_1}{s} = 1\,\text{F}, \quad R_{c2} = Z_2 = \frac{1}{2}\,\Omega, \quad C_{c2} = \frac{Y_3}{s} = 4\,\text{F}, \quad R_{c3} = Z_4 = 1/6\,\Omega.$$

The results obtained using Foster_Cauer.m file in MATLAB are displayed below:

```
Transfer function:
  s + 2
-------------
s^2 + 4 s + 3
Foster 1 circuit


R0 = 0
R =  1/6   1/2
C =  2     2


Cauer 1 circuit
R1  C1  R2  C2   R3
0   1   1/2  4    1/6
```

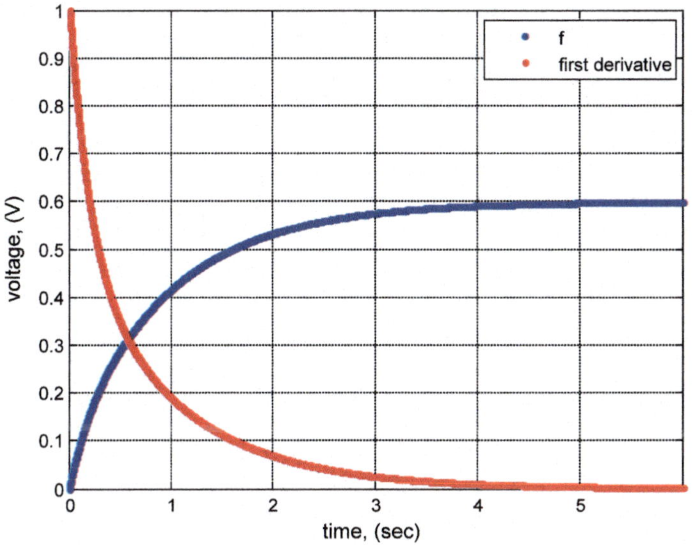

Fig. 8.68 Graph of voltage at the input (response) and its derivative. (The lines appear thick due to printing of dot symbols on plot function.)

Problem 8.4.12 Current unit step response (voltage) of a second-order RC driving point impedance is sampled at equal time intervals (h) and recorded (blue curve in Fig. 8.68).

Determine this driving point impedance $Z(s)$ (diff_fit1.m).

Solution

Voltage at the input (response) and its derivative are plotted as shown in Fig. 8.68. Here, blue-colored "f" in legend indicates the voltage response of the driving point impedance to a unit step current input, while the "first derivative" is indicated in red color (Lines appear thick due to printing of dot symbols on the plot function).

Using curve fitting application toolbox in MATLAB, and implementing built-in library function "exp2", on the derivative,

General model: $z(t) = a\mathrm{e}^{-bt} + c\mathrm{e}^{-dt}$.

Coefficients: $a = 0.5$, $b = -5$, $c = 0.5$, $d = -1$

$$\frac{\mathrm{d}v(t)}{\mathrm{d}t} = 0.5\mathrm{e}^{-t} + 0.5\mathrm{e}^{-5t}.$$

Taking the Laplace transform of the right side of above equation yields

$$Z(s) = \frac{\frac{1}{2}}{s+1} + \frac{\frac{1}{2}}{s+5} = \frac{s+3}{(s+1)(s+5)}.$$

This function can be synthesized in Foster/Cauer forms.

Response voltage data (due to unit step current excitation) is generated using the following MATLAB (diff_fit1.m) script. The same script takes the numerical derivative simulated "sampled data," and fits the double decaying exponential model into this derivative vector.

It must be pointed out here that decreasing step size (h) from $h = 0.1$ to $h = 0.00001$ improves the fit, mainly the two coefficients of exponents (but not decay constants). In order to provide more accuracy, this (derivative fitting and transformation) approach necessitates relatively larger number of samples to be input to computer (which does not create a problem for real-time data import, but may be quite cumbersome—if not impossible—for manual data input).

```
h = 0.00001;      % step size
x = (0:h:6);     % domain
f = 0.5*(1-exp(-x))+0.1*(1-exp(-5*x)); % range generating
Y = diff(f)/h ;   % first derivative
plot(x,f,'.',x(:,1:length(Y)),Y,'r.');xlim([0 6-h]);
legend('f','first derivative');grid on;
xlabel('time, (sec)');ylabel('voltage, (V)');
x(end)=[]; f(end)=[];
A=[x(1) f(1) Y(1)] % checking for approximation
x=x';Y=Y';
% Curve fitting(sum of two decaying exponentials)
f = fit(x,Y,'exp2')
```

Problem 8.4.13 Current unit step response of a second-order RC driving point impedance is sampled at equal time intervals and recorded, see Table 8.1 (time in seconds, voltage in volts).

Table 8.1 Sampled data values

Time	Voltage
0.0	0.0000
0.5	0.2885
1.0	0.4154
1.5	0.4884
2.0	0.5323
2.5	0.5590
3.0	0.5751
3.5	0.5849
4.0	0.5908
4.5	0.5944
5.0	0.5966
5.5	0.5980
6.0	0.5988

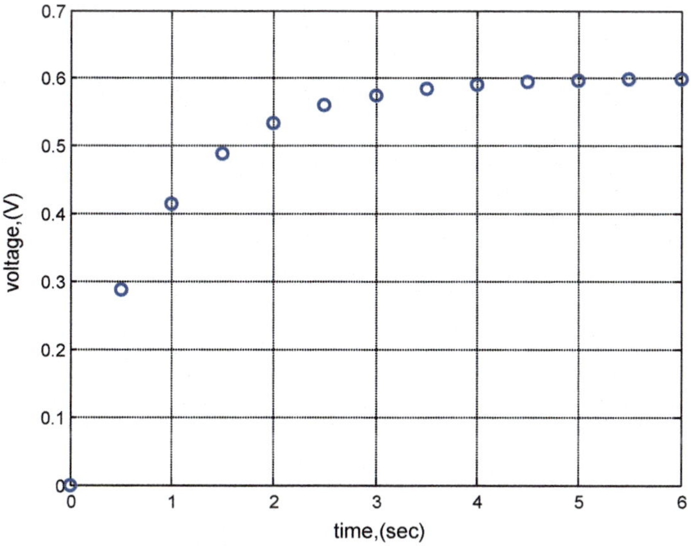

Fig. 8.69 Voltage data sample points against time

Synthesize this driving point impedance $Z(s)$, in two different forms.
Check the results using MATLAB (Foster_Cauer1.m), (fit_exp.m).

Solution
Data is plotted as shown in Fig. 8.69.

Using curve fitting application toolbox in MATLAB, and implementing custom equation,

General model: $f(x) = ae^{-bx} + ce^{-dx} + e$.

Coefficients: $a = -0.1$, $b = 5$, $c = -0.5$, $d = 1$, $e = 0.6$.

Goodness of fit:

SSE: 1.523e−18, R-square: 1, Adjusted R-square: 1, RMSE: 4.363e–10.

(Fit options: Levenberg–Marquardt algorithm is used as fit algorithm, start points for $d = 1$, $e = 1$). Figure 8.70 displays the fitting result.

$$v(t) = -0.5e^{-t} - 0.1e^{-5t} + 0.6$$

$$\mathcal{L}[v(t)] = V(s) = \frac{0.6}{s} - \frac{0.5}{s+1} - \frac{0.1}{s+5}$$

$$V(s) = I(s)Z(s) = \frac{1}{s} \cdot Z(s) = \frac{0.6}{s} - \frac{0.5}{s+1} - \frac{0.1}{s+5}$$

$$sV(s) = Z(s) = 0.6 - \frac{0.5s}{s+1} - \frac{0.1s}{s+5}.$$

Inverse Laplace Transform of the last expression denotes differential operation in time domain,

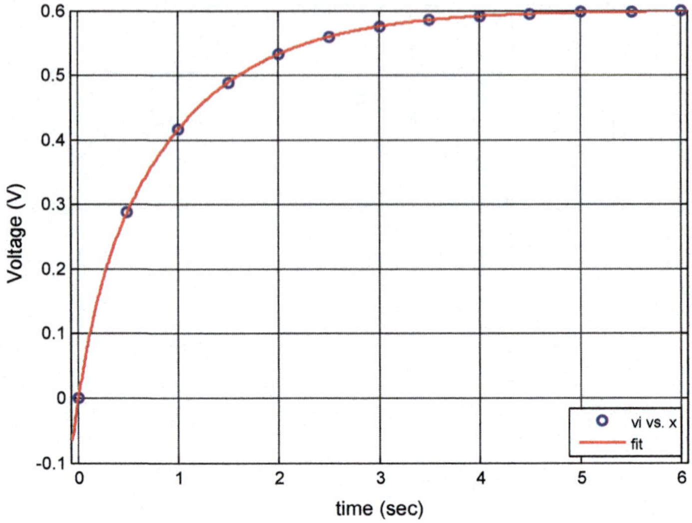

Fig. 8.70 Curve fitting result

$$\frac{\mathrm{d}v(t)}{\mathrm{d}t} = 0.6\delta(t) - \frac{\mathrm{d}}{\mathrm{d}t}(0.5\mathrm{e}^{-t}) - \frac{\mathrm{d}}{\mathrm{d}t}(0.1\mathrm{e}^{-5t}) = 0.6\delta(t) + 0.5\mathrm{e}^{-t} + 0.5\mathrm{e}^{-5t}.$$

Dropping the first (impulse) term, and taking the Laplace transform of remaining terms in right side of above equation yields

$$Z(s) = \frac{\frac{1}{2}}{s+1} + \frac{\frac{1}{2}}{s+5} \tag{8.24}$$

$$Z(s) = \frac{s+3}{(s+1)(s+5)}. \tag{8.25}$$

Foster 1 type synthesis can be performed using (8.24), i.e., partial fractions expanded form of $Z(s)$,

$$Z(s) = \frac{s+b}{(s+a)(s+c)} = \frac{k_1}{s+a} + \frac{k_2}{s+c} = \frac{\frac{1}{2}}{s+1} + \frac{\frac{1}{2}}{s+5}$$

$$C_{f1} = \frac{1}{k_1} = 2\,\mathrm{F},\quad C_{f2} = \frac{1}{k_2} = 2\,\mathrm{F},\quad R_{f1} = \frac{k_1}{a} = \frac{\frac{1}{2}}{1} = \frac{1}{2}\,\Omega,\quad R_{f2} = \frac{k_2}{c} = \frac{\frac{1}{2}}{5} = \frac{1}{10}\,\Omega.$$

Figure 8.71 shows the synthesized Foster I circuit.

Cauer 1 type of synthesis can be performed using (8.25), and by continued fractions expansion of $Z(s)$,

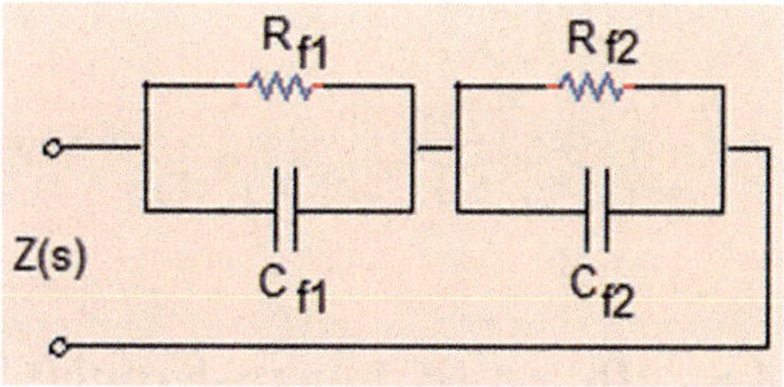

Fig. 8.71 Synthesized Foster I circuit

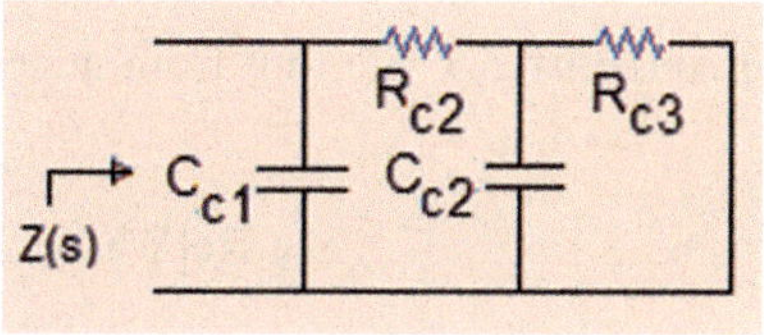

Fig. 8.72 Synthesized Cauer I circuit

$$Z(s) = \frac{s+3}{(s+1)(s+5)} = \frac{s+3}{s^2+6s+5}.$$

Since order of denominator is greater than the order of numerator, synthesis procedure begins by Y (admittance) removal, and then continues by alternating Z and Y removals,

$$\frac{s^2+6s+5}{s+3} = s + \frac{3s+5}{s+3}, \quad \frac{s+3}{3s+5} = \frac{1}{3} + \frac{\frac{4}{3}}{3s+5}, \quad \frac{3s+5}{\frac{4}{3}} = \frac{9}{4}s + \frac{15}{4}$$

$$\frac{s^2+6s+5}{s+3} = Y_1 + \frac{1}{Z_2 + \frac{1}{Y_3+Y_4}} = 1s + \frac{1}{\frac{1}{3} + \frac{1}{\frac{9}{4}s + \frac{1}{\frac{4}{15}}}}$$

$$C_{c1} = \frac{Y_1}{s} = 1\,\text{F}, \quad R_{c2} = Z_2 = \frac{1}{3}\,\Omega, \quad C_{c2} = \frac{Y_3}{s} = \frac{9}{4}\,\text{F}, \quad R_{c3} = Z_4 = \frac{4}{15}\,\Omega.$$

Figure 8.72 shows the synthesized Cauer I circuit.

The preceding results obtained are checked by running Foster_Cauer.m file in MATLAB:

```
Foster 1 circuit
R0 = 0
R =  1/10    1/2
C =  2       2
Cauer 1 circuit
R1   C1     R2     C2     R3
0    1      1/3    9/4    4/15
```

Data is generated using the following MATLAB script:

```
clc;clear all;
% Curve fitting
x = (0:0.5:6)';
vi = (-1/2)*exp(-1*x)+ (-1/10)*exp(-5*x)+6/10;
plot(x,vi,'o');grid on; xlabel('time,(sec)');ylabel('voltage,(V)');
```

8.5 Hilbert (Kramers–Kronig) Transforms

Problem 8.5.1 Starting from general form of the relations, show that

$$\mathrm{Re}[X(j\omega)] = \frac{-2}{\pi}\mathcal{P}\int_0^\infty \frac{x\mathrm{Im}[X(jx)]}{x^2-\omega^2}\mathrm{d}x$$

$$\mathrm{Im}[X(j\omega)] = \frac{2\omega}{\pi}\mathcal{P}\int_0^\infty \frac{\mathrm{Re}[X(jx)]}{x^2-\omega^2}\mathrm{d}x,$$

where $\mathcal{P}$ is cauchy principle value of the integral. Comment on these forms.

Solution

$$X(j\omega) = \mathrm{Re}[X(j\omega)] + j\mathrm{Im}[X(j\omega)]$$
$$\mathrm{Re}[X(-\omega)] + j\mathrm{Im}[X(-\omega)] = \mathrm{Re}[X(\omega)] - j\mathrm{Im}[X(\omega)].$$

In other words, $\mathrm{Re}[X(\omega)]$ is even, while $\mathrm{Im}[X(j\omega)]$ is an odd function.

$$\begin{aligned}\mathrm{Re}[X(j\omega)] &= \frac{-1}{\pi}\mathcal{P}\int_{-\infty}^{\infty} \frac{\mathrm{Im}[X(jx)]}{x-\omega}\mathrm{d}x = \frac{-1}{\pi}\mathcal{P}\int_{-\infty}^{0} \frac{\mathrm{Im}[X(jx)]}{x-\omega}\mathrm{d}x - \frac{1}{\pi}\mathcal{P}\int_0^{\infty} \frac{\mathrm{Im}[X(jx)]}{x-\omega}\mathrm{d}x \\
&= \frac{-1}{\pi}\mathcal{P}\int_{-\infty}^{0} \frac{\mathrm{Im}[X(-jx)]}{-x-\omega}\mathrm{d}x - \frac{1}{\pi}\mathcal{P}\int_0^{\infty} \frac{\mathrm{Im}[X(jx)]}{x-\omega}\mathrm{d}x \\
&= \frac{-1}{\pi}\mathcal{P}\int_0^{\infty} \frac{\mathrm{Im}[X(jx)]}{x+\omega}\mathrm{d}x - \frac{1}{\pi}\mathcal{P}\int_0^{\infty} \frac{\mathrm{Im}[X(jx)]}{x-\omega}\mathrm{d}x \\
&= \frac{-1}{\pi}\mathcal{P}\int_0^{\infty} \left\{\frac{\mathrm{Im}[X(jx)]}{x+\omega} + \frac{\mathrm{Im}[X(jx)]}{x-\omega}\right\}\mathrm{d}x = -\frac{\mathcal{P}}{\pi}\int_0^{\infty} \frac{(x-\omega+x+\omega)\mathrm{Im}[X(jx)]}{x^2-\omega^2}\mathrm{d}x \\
&= -\frac{2\mathcal{P}}{\pi}\int_0^{\infty} \frac{x\mathrm{Im}[X(jx)]}{x^2-\omega^2}\mathrm{d}x.\end{aligned}$$

Similarly,

$$\begin{aligned}\mathrm{Im}[X(jx)] &= \frac{1}{\pi}\mathcal{P}\int_{-\infty}^{\infty}\frac{\mathrm{Re}[X(jx)]}{x-\omega}\mathrm{d}x = \frac{1}{\pi}\mathcal{P}\int_{-\infty}^{0}\frac{\mathrm{Re}[X(-jx)]}{-x-\omega}\mathrm{d}x + \frac{1}{\pi}\mathcal{P}\int_{0}^{\infty}\frac{\mathrm{Re}[X(jx)]}{x-\omega}\mathrm{d}x\\
&= \frac{1}{\pi}\mathcal{P}\int_{0}^{\infty}\frac{-\mathrm{Re}[X(jx)]}{x+\omega}\mathrm{d}x + \frac{1}{\pi}\mathcal{P}\int_{0}^{\infty}\frac{\mathrm{Re}[X(jx)]}{x-\omega}\mathrm{d}x\\
&= \frac{1}{\pi}\mathcal{P}\int_{0}^{\infty}\left\{\frac{\mathrm{Re}[X(jx)]}{x-\omega} + \frac{\mathrm{Re}[X(jx)]}{x+\omega}\right\}\mathrm{d}x = \frac{\mathcal{P}}{\pi}\int_{0}^{\infty}\frac{(x-\omega+x+\omega)\mathrm{Im}[X(jx)]}{x^2-\omega^2}\mathrm{d}x\\
&= \frac{2\omega\mathcal{P}}{\pi}\int_{0}^{\infty}\frac{\mathrm{Re}[X(jx)]}{x^2-\omega^2}\mathrm{d}x.\end{aligned}$$

Because integral limits start at 0, rather than $-\infty$, these relations can be more helpful in numerical computations, in a real life situation, as they consist of only positive frequencies.

Problem 8.5.2 Derive a form of Hilbert (Kramers–Kronig) transform that eliminates the need for computing Cauchy principal value of integral.

Solution

$$\begin{aligned}\mathcal{P}\int_{0}^{\infty}\frac{1}{x^2-\omega^2}\mathrm{d}x &= \frac{\mathcal{P}}{2\omega}\int_{0}^{\infty}\frac{1}{x-\omega}-\frac{1}{x+\omega}\mathrm{d}x\\
&= \frac{1}{2\omega}\lim_{h\to 0}\left[\int_{0}^{\omega-h}\left(\frac{1}{x-\omega}-\frac{1}{x+\omega}\right)\mathrm{d}x + \int_{\omega+h}^{\infty}\left(\frac{1}{x-\omega}-\frac{1}{x+\omega}\right)\mathrm{d}x\right]\\
&= \frac{1}{2\omega}\lim_{h\to 0}\left[(\ln|x-\omega|-\ln|x+\omega|)|_{0}^{\omega-h} + (\ln|x-\omega|-\ln|x+\omega|)|_{\omega+h}^{\infty}\right]\\
&= \frac{1}{2\omega}\lim_{h\to 0}\left[(\ln|x-\omega|-\ln|x+\omega|)|_{\omega-h}^{\omega+h}\right] + (\ln|x-\omega|-\ln|x+\omega|)|_{0}^{\infty}\\
&= \frac{1}{2\omega}(\ln|x-\omega|-\ln|x+\omega|)|_{0}^{\infty} = \frac{1}{2\omega}\left(\lim_{x\to\infty}\ln\left|\frac{x-\omega}{x+\omega}\right| - (\ln|-\omega|-\ln|\omega|)\right)\\
&= 0-0=0.\end{aligned}$$

Thus,

$$-\frac{2\mathcal{P}}{\pi}\int_{0}^{\infty}\frac{-\omega\mathrm{Im}[X(j\omega)]}{x^2-\omega^2}\mathrm{d}x = 0,\quad \frac{2\mathcal{P}}{\pi}\int_{0}^{\infty}\frac{-\mathrm{Re}[X(j\omega)]}{x^2-\omega^2}\mathrm{d}x = 0.$$

Adding the equation to general form of integral relations (with positive frequencies only), one obtains the following results:

$$\text{Re}[X(j\omega)] = -\frac{2}{\pi}\int_0^{\infty} \frac{x\text{Im}[X(jx)] - \omega\text{Im}[X(j\omega)]}{x^2 - \omega^2}\text{d}x$$

$$\text{Im}[X(j\omega)] = \frac{2\omega}{\pi}\int_0^{\infty} \frac{\text{Re}[X(jx)] - \text{Re}[X(j\omega)]}{x^2 - \omega^2}\text{d}x.$$

These equations can be helpful in numerical computations (using different methods of computing integrals such as Simpson's rule, trapezoidal method, or interpolative methods).

Problem 8.5.3 Given the real part of a causal and stable circuit function $Z(s)$,

$$R(\omega) = \frac{2}{4+\omega^2}.$$

(a) Determine the imaginary part of $Z(s)$;
(b) Determine the $Z(s)$;
(c) Realize $Z(s)$.

Hint: Use Hilbert transform pair:

$$\frac{a^2}{a^2+\omega^2} \quad \leftrightarrow \quad \frac{-\omega}{a^2+\omega^2}.$$

Solution

(a) $Z(j\omega) = R(\omega) + jX(\omega)$.
Using Hilbert transform,

$$X(\omega) = -\frac{1}{\pi}\int_{-\infty}^{\infty} \frac{R(x)}{\omega - x}\text{d}x = -\frac{1}{\pi}\int_{-\infty}^{\infty} \frac{2}{(4+\omega^2)(\omega - x)}\text{d}x = -\frac{\omega}{4+\omega^2}.$$

(b) $$Z(j\omega) = \frac{2}{4+\omega^2} + j\left(-\frac{\omega}{4+\omega^2}\right) = \frac{2-j\omega}{4+\omega^2} = \frac{2-j\omega}{4-(j\omega)^2} = \frac{2-j\omega}{(2-j\omega)(2+j\omega)} = \frac{1}{2+j\omega}.$$

Since

$$s = j\omega \rightarrow \omega = \frac{s}{j}, \quad \text{(analytic continuation)}, \quad Z(s) = \frac{1}{2+j\left(\frac{s}{j}\right)} = \frac{1}{s+2}.$$

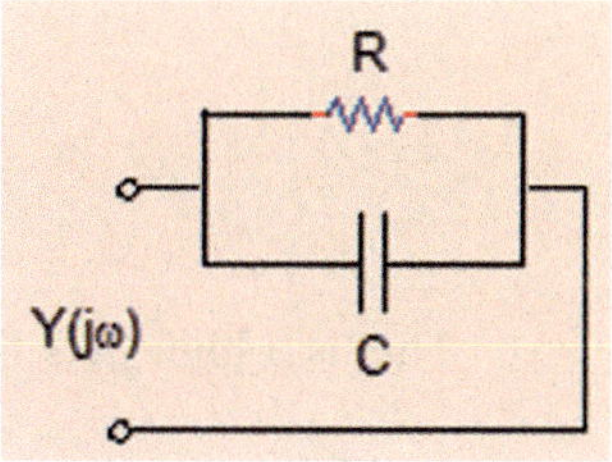

Fig. 8.73 Circuit realization via Hilbert transform

(c) Using admittance,

$$Y(s) = \frac{1}{Z(s)} = s + 2 = sC + G, C = 1\,\text{F}, \quad G = 2\,\text{S} \;\; or \;\; C = 1\,\text{F}, \quad R = 0.5\,\Omega.$$

Figure 8.73 shows the circuit realization.

Problem 8.5.4 If the real part of a complex causal function $H(j\omega)$ is

$$R(\omega) = \frac{\omega^2}{\omega^2 + a^2}.$$

(a) Determine the imaginary part of the function, $I(\omega)$.
(b) Determine $H(s)$.

Solution

(a) Using Hilbert transform and $x \to \omega$,

$$I(\omega) = -\frac{1}{\pi}\int_{-\infty}^{\infty} \frac{R(x)}{\omega - x}\mathrm{d}x \tag{8.26}$$

$$\left[\frac{x^2}{(a^2 + x^2)}\right]\left[\frac{1}{\omega - x}\right] = \left[\frac{1}{a^2 + \omega^2}\right]\left[\frac{-a^2(x - \omega)}{a^2 + x^2} + \frac{\omega^2}{\omega - x}\right]. \tag{8.27}$$

Using (8.27) in (8.26),

$$I(\omega) = -\frac{1}{\pi}\left[\frac{1}{a^2 + \omega^2}\right]\int_{-\infty}^{\infty}\left[\frac{-a^2(x - \omega)}{a^2 + x^2} + \frac{\omega^2}{\omega - x}\right]\mathrm{d}x = \frac{a\omega}{a^2 + \omega^2}.$$

(b) $H(j\omega) = R(\omega) + jI(\omega) = \dfrac{\omega^2}{a^2 + \omega^2} + j\dfrac{a\omega}{a^2 + \omega^2} = \dfrac{j\omega(a - j\omega)}{(a + j\omega)(a - j\omega)} = \dfrac{j\omega}{a + j\omega},$

since $\omega = s/j$,

$$H(s) = \frac{j\left(\frac{s}{j}\right)}{a + j\left(\frac{s}{j}\right)} = \frac{s}{s + a}.$$

This is a high-pass filter with cutoff frequency at a (rad/s).

Problem 8.5.5

(a) Determine the Hilbert transform of the real part of a causal function

$$A(\omega) = \frac{a^2}{\omega^2 + a^2}.$$

(b) Determine the complex function $H(s)|_{s=j\omega} = A(\omega) + jB(\omega)$ in s-domain.

Solution

(a) $\mathcal{H}\left[\frac{a^2}{\omega^2 + a^2}\right] = -\frac{1}{\pi}\int_{-\infty}^{\infty} \frac{A(x)}{\omega - x}\mathrm{d}x$

$$\frac{a^2}{(a^2 + x^2)(\omega - x)} = \frac{a^2}{\omega^2 + a^2}\left[\frac{\omega + x}{a^2 + x^2} + \frac{1}{\omega - x}\right]$$

$$B(\omega) = \frac{a^2}{\omega^2 + a^2}\int_{-\infty}^{\infty}\left(\frac{\omega + x}{a^2 + x^2} - \frac{1}{\omega - x}\right)\mathrm{d}x = -\frac{a\omega}{\omega^2 + a^2}$$

(b) $H(j\omega) = \frac{a^2}{\omega^2 + a^2} - j\frac{a\omega}{\omega^2 + a^2} = \frac{a(a - j\omega)}{(a - j\omega)(a + j\omega)} = \frac{a}{(a + j\omega)}$

$$s = j\omega \quad \rightarrow \quad \omega = \frac{s}{j},$$

$$H(s) = \frac{a}{a + j\left(\frac{s}{j}\right)} = \frac{a}{s + a}.$$

This is a low-pas filter (voltage transfer function) with a cutoff frequency at $\omega = a$ rad/sn.

Problem 8.5.6 In the circuit shown in Fig. 8.74, $Z(j\omega) = Z'(\omega) + jZ''(\omega)$ where

$$Z'(\omega) = \mathrm{Re}[Z(j\omega)], \; Z''(\omega) = \mathrm{Im}[Z(j\omega)]$$

.

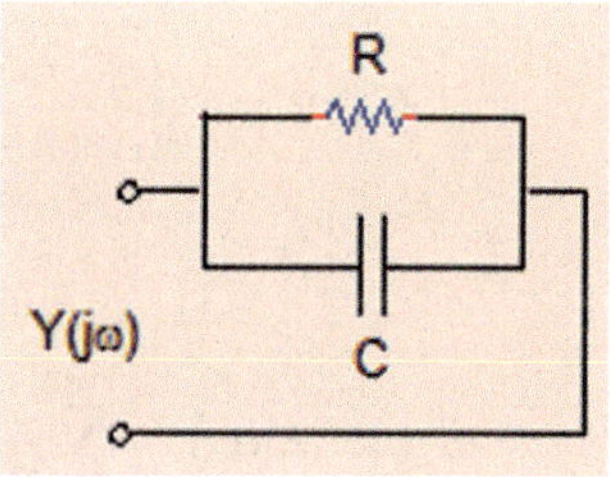

Fig. 8.74 The circuit for Problem 8.5.6

(a) Determine $Z'(\omega)$ and $Z''(\omega)$.
(b) Show that $Z'(\omega)$ and $Z''(\omega)$ are Hilbert transforms of each other.
(c) Generalize the expressions for $Z'(\omega)$ and $Z''(\omega)$ for a driving point impedance consisting of n pieces of RC tank sections.

Solution

(a) For this single-tank circuit (Voigt circuit or Foster 1 form),

$$Z(s) = \frac{1}{sC}|| = \frac{R}{sRC+1}, \quad s = j\omega,$$

$$Z(j\omega) = \frac{R(1 - j\omega RC)}{1+(\omega RC)^2} = \frac{R}{1+(\omega RC)^2} - j\frac{\omega R^2 C}{1+(\omega RC)^2}$$

$$Z'(\omega) = \frac{R}{1+(\omega RC)^2}, \tag{8.28}$$

$$Z''(\omega) = -\frac{\omega R^2 C}{1+(\omega RC)^2} \tag{8.29}$$

(b)
$$Z'(\omega) = \frac{R}{1+(\omega RC)^2} = \frac{R/(RC)^2}{1/(RC)^2 + \omega^2} = R\frac{a^2}{a^2+\omega^2}, \quad a = \frac{1}{RC} \tag{8.30}$$

$$Z''(\omega) = \mathcal{H}\left[R\frac{a^2}{a^2+\omega^2}\right] = -R\frac{a\omega}{a^2+\omega^2}. \tag{8.31}$$

Substituting the value assigned for $a = 1/(RC)$ in (8.31) yields Eq. (8.29):

$$Z''(\omega) = -R\frac{\frac{1}{RC}\omega}{\frac{1}{(RC)^2}+\omega^2} = -\frac{\frac{1}{C}\omega}{\frac{1}{(RC)^2}+\omega^2} = -\frac{\omega R^2 C}{1+(\omega RC)^2}$$

(c) $$Z'(\omega) = \frac{R_1}{1+(\omega R_1 C_1)^2} + \frac{R_2}{1+(\omega R_2 C_2)^2} + \cdots + \frac{R_n}{1+(\omega R_n C_n)^2}$$

$$Z''(\omega) = -\left(\frac{\omega R_1^2 C_1}{1 + (\omega R_1 C_1)^2} + \frac{\omega R_2^2 C_2}{1 + (\omega R_2 C_2)^2} + \cdots + \frac{\omega R_n^2 C_n}{1 + (\omega R_n C_n)^2}\right)$$

or,

$$Z'(\omega) = \sum_{i=1}^{n} \frac{R_i}{1 + (\omega R_i C_i)^2}, \quad Z''(\omega) = -\sum_{i=1}^{n} \frac{\omega R_i^2 C_i}{1 + (\omega R_i C_i)^2}.$$

Problem 8.5.7 What are the practical biomedical engineering applications of Hilbert transforms?

Answer
The Hilbert transform finds applications when complex signal (quadrature) processing is necessary. Speech processing and medical imaging are among these fields of applications.

For example, in single sideband (SSB) modulation, there are two approaches to eliminate one of the sidebands; (a) the filter method and (b) the phasing method. Filtering can be difficult due to tough filtering requirements, while phasing method uses a Hilbert transform to implement SSB modulation which requires the shifting of the message signal to another center frequency without creating lower and upper pairs of frequency components.

In discrete Hilbert transform, frequencies are phase-delayed/advanced by 90°. Shifting the transformed results and then adding it to the original signal creates a complex and analytic signal which is made up of the original real-valued signal plus the Hilbert transform of that real signal.

Digital beamforming techniques are used for image acquisition in an ultrasound system. Digital beamforming is the process of phase alignment and summation of signals that are generated by a source, and received at different times by a multi-element ultrasound transducer. Beamformer has the functions of directivity of the transducer and focusing the source within the body. An accurate way of obtaining the quadrature components is to pass the echo signal through a Hilbert transformer [8, 10].

Note: Applications in the field of impedance spectroscopy are discussed separately.

8.6 Constructing a Network Function from Its Real Part

Problem 8.6.1 For an impedance function $Z(s)$,
$\mathrm{Re}[Z(j\omega)] = [Z(s) + Z(-s)]/2 = F(s) \cdot F(-s)$

$$\text{If } Z(s) = \frac{s+4}{s+1}, \quad F(s) = ?$$

Solution

$$\begin{aligned}
\mathrm{Re}[Z(j\omega)] &= \frac{1}{2}\left[\frac{s+4}{s+1} + \frac{-s+4}{-s+1}\right] = \frac{1}{2}\left[\frac{(s+4)(1-s)+(4-s)(s+1)}{(s+1)(1-s)}\right] \\
&= \frac{1}{2}\left[\frac{-2s^2+4}{(s+1)(1-s)}\right] \\
&= \frac{1}{2} \times 2 \times \frac{4-s^2}{(1+s)(1-s)} = \frac{(2-s)(s+2)}{(1+s)(1-s)} \\
&= \frac{s+2}{s+1} \cdot \left(\frac{-(s-2)}{-(1-s)}\right) = \frac{s+2}{s+1} \cdot \frac{2-s}{1-s}.
\end{aligned}$$

It follows that

$$F(s) = \frac{s+2}{s+1}, \quad F(-s) = \frac{2-s}{1-s}.$$

Problem 8.6.2 Given a parallel RC circuit,

(a) Determine $Y(j\omega)$, the admittance of this circuit.
(b) If a shunt inductor is connected to the circuit, determine the new admittance of the circuit $Y_1(j\omega)$.
(c) Based on the results obtained in this problem, determine if a circuit function is uniquely specified when its real part is given.

Solution

(a) $Y(j\omega) = \frac{1}{R} + j\omega C_1$

(b) $Y_1(j\omega) = \frac{1}{j\omega L} + Y(j\omega) = \frac{1}{R} + j\omega C_1 - j\frac{1}{\omega L} = \frac{1}{R} + j\left(\omega C_1 - \frac{1}{\omega L}\right)$

(c) $\mathrm{Re}[Y_1(j\omega)] = \mathrm{Re}[Y(j\omega)] + \mathrm{Re}\left[\frac{1}{j\omega L}\right] = \mathrm{Re}[Y(j\omega)].$

The real parts of both admittances are the same, but the admittances themselves are different. Therefore, if the real part is given, it is not possible to know $Y_1(s)$ or $Y(s)$ which corresponds to given real part. There can be an infinite number of functions that differ from $Y(s)$.

It should be noted here that, when an impedance function is given, its real part is determined uniquely. However, converse may not be true always. For example, the real (even) part of $Z(s)$ is unaltered by the addition of an arbitrary rational reactance function. Therefore, $Z(s)$ should not have a pole on the imaginary axis.

Problem 8.6.3 Make a literature survey if there are simple methods other than Hilbert transforms (Kramers–Kronig relations) to construct a minimum phase network function if its real part is available. Indicate related references.

Solution

The integral calculations of Hilbert transforms are not easy. In fact, the strength of the Hilbert transform is found in what it says rather than what it does. It states that if the real or the even part (conversely, if the imaginary or odd part) of a circuit is given, then the circuit is specified completely.

Alternative methods to construct a minimum phase function from its real part are due to Bode [3, 10], Brune-Gewertz [3, 10, 11], Miyata [3, 11], and Ho [12].

Problem 8.6.4 Describe Bode's procedure to construct a minimum phase network function if its real part is available.

Solution

Bode's procedure to construct a minimum phase network function from its real part:

Let an immitance function be $Z(j\omega)$,

$$Z(j\omega) = \frac{N(j\omega)}{D(j\omega)} = \mathrm{Re}[Z(j\omega)] + j\mathrm{Im}[Z(j\omega)]$$

$$\mathrm{Re}[Z(j\omega)] = \frac{1}{2}[Z(j\omega) + Z(-j\omega)]$$

$$\omega = -js,$$

Even part of $Z(s)$,

$$\begin{aligned} Ev[Z(s)] &= \mathrm{Re}[Z(j\omega)]|_{\omega=-js} = \frac{1}{2}[Z(s) + Z(-s)] \\ &= \frac{N(s)}{[(s-p_1)(s-p_1^*)\ldots][(s+p_1)(s+p_1^*)\ldots]}. \end{aligned}$$

Assume that all poles in $Ev[Z(s)]$ are simple, by partial fraction expansion of $Ev[Z(s)]$,

$$\begin{aligned} Ev[Z(s)] = &\left(\frac{k_1}{s-p_1} + \frac{k_1^*}{s-p_1^*} + \frac{k_2}{s-p_2} + \frac{k_2^*}{s-p_2^*} + \ldots + \frac{R_\infty}{2}\right) \\ &+ \left(\frac{k_{-1}}{s+p_1} + \frac{k_{-1}^*}{s+p_1^*} + \frac{k_{-2}}{s+p_2} + \frac{k_{-2}^*}{s+p_2^*} + \ldots + \frac{R_\infty}{2}\right) \end{aligned}$$

$$R_\infty = \lim_{s\to\infty} Ev[Z(s)] = \lim_{s\to\infty}[Z(s) + Z(-s)]$$

$$Z(s) = 2Z(s)|_{s\le 0} = 2\left(\frac{k_1}{s-p_1} + \frac{k_1^*}{s-p_1^*} + \frac{k_2}{s-p_2} + \frac{k_2^*}{s-p_2^*}\right) + R_\infty.$$

Note that Bode's method is conceptually simple and it is based on expanding the even part of transfer function into its partial fractions.

Problem 8.6.5 Write a MATLAB script for Bode's procedure to construct a minimum phase network function if its real part is available. Use this m.file (R_to_Z1.m) to compute transfer functions $Z(s)$ from their real parts $R(\omega)$ given below:

$$Z(j\omega) = R(\omega) + jX(\omega)$$

(i) $\dfrac{1}{\omega^2 + 1}$

(ii) $\dfrac{1}{\omega^6 + 1}$

(iii) $\dfrac{\omega^2}{\omega^6 + 1}$

(iv) $\dfrac{\omega^4}{\omega^6 + 1}$

(v) $\dfrac{1}{2\omega^6 + 1}$

(vi) $\dfrac{3\omega^4 - 6\omega^2 + 6}{\omega^6 + 1}$

(vii) $\dfrac{2\omega^2 - 1}{\omega^6 + 1}$

(viii) $\dfrac{\omega^2 + 1}{\omega^4 - \omega^2 + 1}$

(ix) $\dfrac{\omega^2 - 1}{\omega^4 - \omega^2 + 1}$

(x) $\dfrac{\omega^2}{\omega^4 - \omega^2 + 1}$

(xi) $\dfrac{-\omega^2 + 16}{\omega^4 - 16\omega^2 + 16}$

(xii) $\dfrac{\omega^4 - 16\omega^2 + 16}{\omega^6 + 1}$

(xiii) $\dfrac{\omega^4 - 16}{\omega^6 + \omega^4 + \omega^2 + 1}$

(xiv) $\dfrac{-\omega^8 - 16}{\omega^8 + \omega^6 + \omega^4 + \omega^2 + 1}$

(xv) $\dfrac{-6\omega^2 + 6}{\omega^8 + \omega^6 + \omega^2 + 1}$

(xvi) $\dfrac{-\omega^4 + \omega^2 + 2}{\omega^4 + 16}$

(xvii) $\dfrac{\omega^4 + 128}{\omega^6 + 64}.$

Solution

(i) $R(\omega) = \dfrac{1}{\omega^2 + 1}, \quad Z(s) = \dfrac{1}{s+1}$

(ii) $R(\omega) = \dfrac{1}{\omega^6 + 1}, \quad Z(s) = \dfrac{1}{3} \cdot \dfrac{2s^2 + 4s + 3}{s^3 + 2s^2 + 2s + 1}$

(iii) $R(\omega) = \dfrac{\omega^2}{\omega^6 + 1}, \quad Z(s) = \dfrac{1}{3} \cdot \dfrac{s^2 + 2s}{s^3 + 2s^2 + 2s + 1}$

(iv) $R(\omega) = \dfrac{\omega^4}{\omega^6 + 1}, \quad Z(s) = \dfrac{1}{3} \cdot \dfrac{2s^2 + s}{s^3 + 2s^2 + 2s + 1}$

(v) $R(\omega) = \dfrac{1}{2\omega^6 + 1}, \quad Z(s) = \dfrac{0.4454s^2 + 0.2646s + 0.1179}{s^3 + 1.782s^2 + 1.587s + 0.7071}$

(vi) $R(\omega) = \dfrac{3\omega^4 - 6\omega^2 + 6}{\omega^6 + 1} \quad Z(s) = \dfrac{4s^2 + 5s + 6}{s^3 + 2s^2 + 2s + 1}$

(vii) $R(\omega) = \dfrac{2\omega^2 - 1}{\omega^6 + 1} \quad Z(s) = \dfrac{-1}{s^3 + 2s^2 + 2s + 1}.$

Note that this result can also be obtained by adding the results of (ii) and (iii).

(viii) $R(\omega) = \dfrac{\omega^2 + 1}{\omega^4 - \omega^2 + 1} \quad Z(s) = \dfrac{2s + 1}{s^2 + s + 1}$

(ix) $R(\omega) = \dfrac{\omega^2 - 1}{\omega^4 - \omega^2 + 1} \quad Z(s) = \dfrac{-1}{s^2 + s + 1}$

(x) $R(\omega) = \dfrac{\omega^2}{\omega^4 - \omega^2 + 1} \quad Z(s) = \dfrac{s}{s^2 + s + 1}.$

This result can also be obtained by adding the results of (viii) and (ix), and dividing by 2.

(xi) $R(\omega) = \dfrac{-\omega^2 + 16}{\omega^4 - 16\omega^2 + 16} \quad Z(s) = \dfrac{2s^2 + 32}{s^4 + 16s^2 + 16}$

(xii) $R(\omega) = \dfrac{\omega^4 - 16\omega^2 + 16}{\omega^6 + 1} \quad Z(s) = \dfrac{6s^2 + 11s + 16}{s^3 + 2s^2 + 2s + 1}$

(xiii) $R(\omega) = \dfrac{\omega^4 - 16}{\omega^6 + \omega^4 + \omega^2 + 1} \quad Z(s) = 4.25 \dfrac{(s+1)}{s^3 + 2.414s^2 + 2.414s + 1}$

(xiv) $R(\omega) = \dfrac{-\omega^8 + 16}{\omega^8 + \omega^6 + \omega^4 + \omega^2 + 1} \quad Z(s) = \dfrac{10.61s^3 + 24.77s^2 + 25.07s + 12.3}{s^4 + 3.078s^3 + 4.236s^2 + 3.078s + 1}$

(xv) $R(\omega) = \dfrac{-6\omega^2 + 6}{\omega^8 + \omega^6 + \omega^2 + 1} \quad Z(s) = \dfrac{4s^3 + 10s^2 + 10s + 6}{s^4 + 3s^3 + 4s^2 + 3s + 1}$

(xvi) $R(\omega) = \dfrac{-\omega^4 + \omega^2 + 2}{\omega^4 + 16} \quad Z(s) = \dfrac{1.945s + 4.5}{s^2 + 2.828s + 4}$

(xvii) $R(\omega) = \dfrac{\omega^4 + 128}{\omega^6 + 64} \quad Z(s) = \dfrac{3s^2 + 11s + 16}{s^3 + 4s^2 + 8s + 8}.$

8.7 Constructing a Network Function from Its Phase

Problem 8.7.1

(a) Determine the admittance of a series RL circuit, $Y(j\omega)$.
(b) Show its real and imaginary parts, $Y'(j\omega)$ and $Y''(j\omega)$.
(c) Find the phase function of this admittance.

Solution

$$Z(j\omega) = R + j\omega L$$

(a) $$Y(j\omega) = \frac{1}{Z(j\omega)} = \frac{1}{R + j\omega L} = \frac{1(R - j\omega L)}{(R + j\omega L)(R - j\omega L)} = \frac{R - j\omega L}{R^2 + (\omega L)^2}$$

(b) $$Y(j\omega) = \frac{R}{R^2 + (\omega L)^2} - j\frac{\omega L}{R^2 + (\omega L)^2} = Y'(\omega) + jY''(\omega)$$

(c) $$\varphi(\omega) = a\tan\left(-\frac{\omega L}{\mathrm{R}}\right).$$

Problem 8.7.2 Determine the phase function of a parallel RC impedance circuit.

Solution

$$Z(s) = R||\frac{1}{sC} = \frac{R}{1 + sRC}$$

$$Z(j\omega) = \frac{R}{1 + j\omega RC} = \frac{R(1 - j\omega RC)}{1 + (\omega RC)^2} = \frac{R}{1 + (\omega RC)^2} - j\frac{\omega R^2 C}{1 + (\omega RC)^2}$$

$$\varphi(\omega) = \tan^{-1}\left(-\frac{\omega R^2 C}{R}\right) = \tan^{-1}(-\omega RC).$$

Problem 8.7.3

(a) Determine the phase function of $Z(s)$ shown in Fig. 8.75.
(b) Calculate the frequency of minimum phase angle.
(c) Plot the phase angle change if $R_1 = 1\ \Omega$, $C_1 = 1$ F and the value of resistor R_2 is varied 3 decades between 0.01 and 10 Ω (phase_analysis1.m).

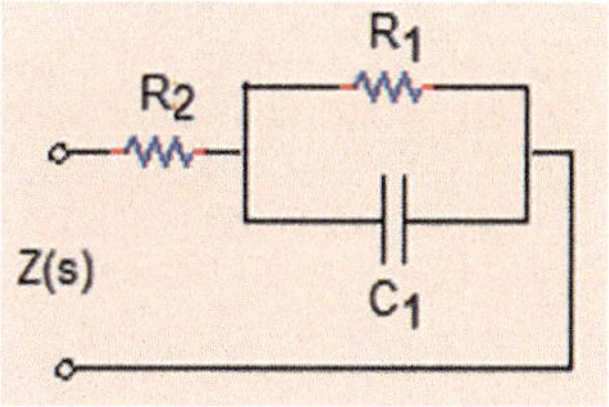

Fig. 8.75 The circuit for Problem 8.7.3

Solution

(a) $$Z(s) = R_2 + \frac{R_1\left(\frac{1}{sC_1}\right)}{R_1 + \frac{1}{sC_1}} = R_2 + \frac{R_1}{sC_1R_1 + 1} = \frac{sR_1R_2C_1 + R_2 + R_1}{sR_1C_1 + 1}$$

$$\begin{aligned}
Z(j\omega) &= \frac{j\omega R_1R_2C_1 + (R_2 + R_1)}{1 + j\omega R_1C_1} \times \frac{1 - j\omega R_1C_1}{1 - j\omega R_1C_1} \\
&= \frac{j\omega R_1R_2C_1 + R_2 + R_1 + \omega^2R_1^2C_1^2R_2 - j\omega R_1C_1(R_2 + R_1)}{1 + (\omega R_1C_1)^2} \\
&= \frac{(R_1 + R_2) + \omega^2R_1^2C_1^2R_2}{1 + (\omega R_1C_1)^2} + j\frac{\omega R_1C_1(R_2 - R_1 - R_2)}{1 + (\omega R_1C_1)^2} \\
&= \frac{(R_1 + R_2) + (\omega R_1C_1)^2R_2}{1 + (\omega R_1C_1)^2} - j\frac{\omega R_1^2C_1}{1 + (\omega R_1C_1)^2}
\end{aligned}$$

$$\varphi(\omega) = \tan^{-1}\left[-\frac{\omega R_1^2C_1}{(R_1 + R_2) + (\omega R_1C_1)^2R_2}\right].$$

(b) This phase angle equation is of the form

$$\varphi(x) = \tan^{-1}\left[\frac{-ax}{cx^2 + b}\right]$$

$$\frac{d\varphi(x)}{dx} = \frac{d}{dx}\left[\tan^{-1}\left(\frac{-ax}{cx^2 + b}\right)\right] = \frac{a(cx^2 - b)}{c^2x^4 + (a^2 + 2bc)x^2 + b^2} = 0$$

$$cx^2 - b = 0 \quad \rightarrow \quad x = \sqrt{\frac{b}{c}}.$$

Therefore,

$$\omega_x = \left(\frac{R_1 + R_2}{R_1^2C_1^2R_2}\right)^{1/2} = \frac{1}{R_1C_1}\left(\frac{R_1 + R_2}{R_2}\right)^{1/2}.$$

(c) Phase functions as functions of R_2 and frequency are displayed in Fig. 8.76. Minimum phase extrema are marked on each phase plot. Impedance functions for different values of R_2 are given as follows:

$R_2 = 0.01\,\Omega$; $Z(s) = \frac{0.01s + 1.01}{s + 1}$ (blue), $R_2 = 0.1\,\Omega$; $Z(s) = \frac{0.1s + 1.1}{s + 1}$ (green)

$R_2 = 1\,\Omega$; $Z(s) = \frac{s + 2}{s + 1}$ (red), $R_2 = 10\,\Omega$; $Z(s) = \frac{10s + 11}{s + 1}$ (cyan).

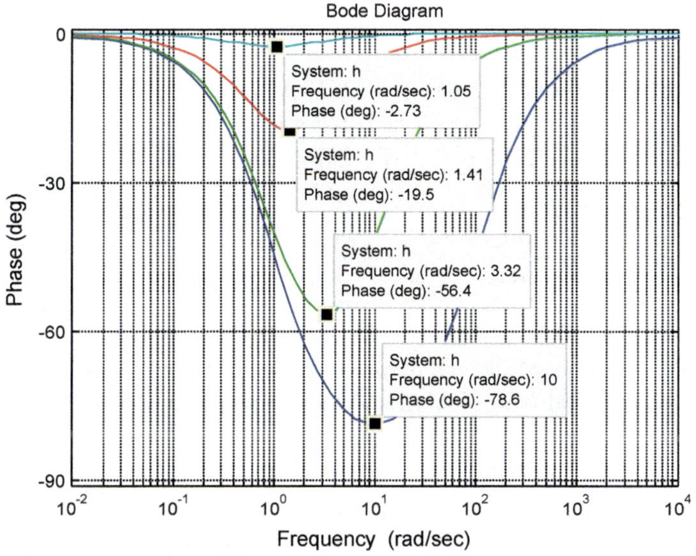

Fig. 8.76 Phase functions as functions of R_2 and frequency

```
%phase_analysis1.m
%phase response analysis of  Z = R2 + (R1//C1)
R1=1; C1=1; n=2 % R2 is evaluated at n+2 decades
for i=-n:1
R2=10.^(i)
z=(R1+R2)/(R1*R2*C1); p=1/(R1*C1); N=R2*[1 z];D=[1 p];
h=tf(N,D)
P = bodeoptions; % Set magnitude visiblity to off
P.MagVisible = 'off';
h = bodeplot(h,P);grid on;
hold on
wx=sqrt((R1+R2)/(R1*R1*C1*C1*R2))
phix=atan(-wx*R1*R1*C1/((R1+R2)+wx*wx*R1*R1*R2*C1*C1));
phix = radtodeg(phix)
end
```

Problem 8.7.4 Determine the $Z(s)$ from its phase function $\varphi(\omega) = -\tan^{-1}(-\omega RC)$.

Solution

$$\varphi(\omega) = -\tan^{-1}(-\omega RC)$$
$$C(j\omega) = \frac{1+j(-\omega RC)}{1-j(-\omega RC)} = \frac{1-j\omega RC}{1+j\omega RC}$$
$$C(s) = \frac{1-sRC}{1+sRC} = \frac{RC\left(s-\frac{1}{RC}\right)}{RC\left(s+\frac{1}{RC}\right)} = \frac{s-\frac{1}{RC}}{s+\frac{1}{RC}} = \frac{\frac{1}{C}\left(sC-\frac{1}{R}\right)}{s+\frac{1}{RC}}.$$

The pole is in LH s-plane, and the zero is in RH s-plane. There is no unique way to assign the zero and pole of $Z(s)$, and the following may be suitable for $Z(s)$:

$$\frac{\left(sC-\frac{1}{R}\right)}{s+\frac{1}{RC}}, \quad \frac{\frac{1}{C}}{s+\frac{1}{RC}}.$$

The second one is a minimum phase function and it can be selected as the answer

$$Z(s) = \frac{\frac{1}{C}}{s+\frac{1}{RC}} = \frac{R}{sRC+1}.$$

Problem 8.7.5 A general expression for the angle of a transfer function $F(j\omega)$ is

$$\varphi(\omega) = \frac{1}{j2}\cdot \ln\frac{F(j\omega)}{F(-j\omega)} = \frac{1}{j2}\cdot C(j\omega). \tag{8.32}$$

Show that

(a)
$$C(j\omega) = \frac{1+\tan\varphi(\omega)}{1-\tan\varphi(\omega)} \tag{8.33}$$

(b) If $C(s) = \dfrac{(s^2-s+1)(1-s)}{(s^2+s+1)(s+1)}$.

Is there a unique way to assign the zeros and poles of $F(s)$? Comment.

Solution

(a) $F(j\omega) = R(\omega) + jX(\omega)$

$$\tan\varphi(\omega) = \frac{X(\omega)}{R(\omega)}.$$

From (8.32),

$$C(j\omega) = \frac{F(j\omega)}{F(-j\omega)} = e^{j2\varphi(\omega)} \tag{8.34}$$

$$j\tan\varphi(\omega) = \frac{e^{j\varphi(\omega)} - e^{-j\varphi(\omega)}}{e^{j\varphi(\omega)} + e^{-j\varphi(\omega)}}. \tag{8.35}$$

Dividing numerator and denominator terms by $e^{-j\varphi(\omega)}$ and using (8.34),

$$j\tan\varphi(\omega) = \frac{e^{j2\varphi(\omega)} - 1}{e^{j2\varphi(\omega)} + 1} = \frac{C(j\omega) - 1}{C(j\omega) + 1} \tag{8.36}$$

$$\begin{aligned}
&[C(j\omega) + 1]\cdot j\tan\varphi(\omega) = C(j\omega) - 1\\
&jC(j\omega)j\tan\varphi(\omega) + j\tan\varphi(\omega) = C(j\omega) - 1\\
&-C(j\omega) + jC(j\omega)\tan\varphi(\omega) = -1 - j\tan\varphi(\omega)\\
&C(j\omega)[-1 + j\tan\varphi(\omega)] = -1 - j\tan\varphi(\omega)\\
&C(j\omega) = \frac{-[1 + j\tan\varphi(\omega)]}{-1 + j\tan\varphi(\omega)} = \frac{1 + j\tan\varphi(\omega)}{1 - j\tan\varphi(\omega)}.
\end{aligned}$$

(b) $$C(s) = \frac{(s^2 - s + 1)(1 - s)}{(s^2 + s + 1)(s + 1)} = \frac{-(s-1)\left[s - \left(\frac{1}{2} + \frac{\sqrt{3}}{2}i\right)\right]\left[s - \left(\frac{1}{2} - \frac{\sqrt{3}}{2}i\right)\right]}{(s+1)\left[s - \left(-\frac{1}{2} + \frac{\sqrt{3}}{2}i\right)\right]\left[s - \left(-\frac{1}{2} - \frac{\sqrt{3}}{2}i\right)\right]}.$$

All the poles of $C(s)$ are in LH s-plane, whereas all the zeros are located in the RH s-plane. Therefore, there is no unique way of assigning zeros and poles of transfer function $F(s)$. The following functions are suitable,

$$F_1(s) = \frac{1 - s}{s^2 + s + 1}, \quad F_2(s) = \frac{1}{(s^2 + s + 1)(s + 1)}, \quad F_3(s) = \frac{s^2 - s + 1}{s + 1}.$$

Each of them has the same phase angle for all values of angular frequency but different magnitudes.

The only condition that provides unique transfer function is the minimumphase requirement imposed upon $F(s)$. Then, the unique transfer function will be $F_2(s)$.

Problem 8.7.6 Determine the impedance function that has the phase response of (phase_to_Z1.m)

$$\varphi(\omega) = -\tan^{-1}\left(\frac{200\omega^3 - 4000\omega}{2400\omega^2}\right).$$

Solution

If $Z(s)$ is a minimum phase function (i.e., no zero or pole of $Z(s)$ is located in the right half s-plane), it can be constructed uniquely (without resorting to the Hilbert transform) from its phase or the magnitude function:

$$Z(s) = \frac{\sum_{i=0}^{m} a_i s^i}{\sum_{i=0}^{n} b_i s^i} = \frac{M_1(s) + N_1(s)}{M_2(s) + N_2(s)} = \frac{A(s)}{B(s)}.$$

$M_1(s)$ and $N_1(s)$ are even and odd parts of numerator polynomial, while $M_2(s), N_2(s)$ are the even and odd parts of denominator polynomial:

$$\varphi(\omega) = -\tan^{-1}\frac{\mathrm{Im}[Z(j\omega)]}{\mathrm{Re}[Z(j\omega)]} = -\tan^{-1}\frac{\varphi_o(s)}{\varphi_e(s)}\Big|_{s=j\omega},$$

where $\varphi_o(s)$ and $\varphi_e(s)$ are odd and even polynomials, with $\varphi_o(s) + \varphi_e(s) = A(s) \times B(-s)$.

Procedure:

1. Let $p(s) = \varphi_o(s) + \varphi_e(s)$
2. Locate the roots of $p(s)$
3. Assign factors in left half of s-plane roots of $p(s)$ to $A(s)$ and right half s-plane roots of $p(s)$ to $B(-s)$
4. Find $B(s)$ (i.e., replace s with its negative in step 3)
5. Set up

$$Z(s) = \frac{A(s)}{B(s)}.$$

Using this information, let $\omega = s/j$,

$$\varphi(\omega) = -\tan^{-1}\left[\frac{200\left(\frac{s}{j}\right)^3 - 4000\left(\frac{s}{j}\right)}{2400\left(\frac{s}{j}\right)^2}\right] = -\tan^{-1}\left[\frac{200js^3 + 4000js}{-2400s^2}\right]$$

$$= -\tan^{-1}\left(j\frac{200s^3 + 4000s}{-2400s^2}\right) = -\tan^{-1}\left(j\frac{s^3 + 20s}{-12s^2}\right)$$

$$p(s) = s^3 + 20s - 12s^2 = s^3 - 12s^2 + 20s.$$

Roots of $p(s)$: $r1 = 0,\ r2 = 2,\ r3 = 10,$

$$p(s) = s(s-2)(s-10).$$

Assign $A(s) \to s,\ B(-s) = (s-2)(s-10)$

$$B(s) = (-s-2)(-s-10) = -(s+2)[-(s+10)] = (s+2)(s+10) = s^2+12s+20$$

$$Z(s) = \frac{A(s)}{B(s)} = \frac{s}{s^2+12s+20}.$$

Replacing common divisor of 200,

$$Z(s) = \frac{200s}{s^2+12s+20}.$$

Problem 8.7.7 Determine the impedance function that has the following phase response (phase_to_Z1.m)

$$\varphi(\omega) = -\tan^{-1}\left(\frac{\omega^3+7\omega}{4\omega^2+10}\right).$$

Solution

$$\omega = \frac{s}{j},$$

$$\varphi(\omega) = -\tan^{-1}\left[\frac{\left(\frac{s}{j}\right)^3 + 7\left(\frac{s}{j}\right)}{4\left(\frac{s}{j}\right)^2 + 10}\right] = -\tan^{-1}\left[\frac{\frac{s^3}{-j} - j7s}{-4s^2+10}\right]$$

$$= -\tan^{-1}\left(\frac{js^3 - j7s}{-4s^2+10}\right) = -\tan^{-1}\left(j\frac{s^3-7s}{-4s^2+10}\right)$$

$$p(s) = s^3 - 7s + \left(-4s^2+10\right) = s^3 - 4s^2 - 7s + 10 = (s-1)(s+2)(s-5).$$

Left-half plane root is at $s = -2$,

$$A(s) = s - (-2) = s + 2.$$

Right half plane roots are at $s = 1,\ s = 5$

$$B(-s) = (s-1)(s-5)$$

$$B(s) = (-s-1)(-s-5) = -(s+1)[-(s+5)] = (s+1)(s+5)$$

$$Z(s) = \frac{A(s)}{B(s)} = \frac{s+2}{(s+1)(s+5)} = \frac{s+2}{s^2+6s+5}.$$

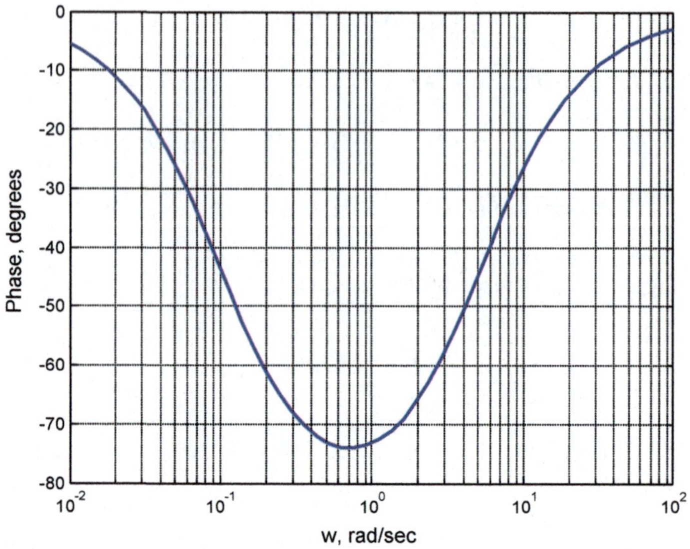

Fig. 8.77 Phase response obtained during an implant corrosion study

Problem 8.7.8 Data from a phase response measurement that was obtained during a metal oxide implant corrosion study is displayed in Fig. 8.77. It is known that unknown impedance $Z(s)$ of this system has the equivalent circuit as given in Fig. 8.75. Determine the component values of $Z(s)$ and its phase function (phi. synthesis1.m, phi.m).

Solution
Impedance function of this circuit is

$$Z(s) = R_2 + \frac{R_1\left(\frac{1}{sC_1}\right)}{R_1 + \frac{1}{sC_1}} = R_2 + \frac{R_1}{sC_1R_1 + 1} = \frac{R_2\left(s + \frac{R_1 + R_2}{R_1R_2C_1}\right)}{s + \frac{1}{R_1C_1}} = \frac{k(s+z)}{s+p}.$$

Its phase function is calculated as

$$\varphi(\omega) = \tan^{-1}\left[-\frac{\omega R_1^2C_1}{(R_1 + R_2) + (\omega R_1C_1)^2R_2}\right].$$

These functions are used in equations of predictions in the following computations.

This is a nonlinear optimization problem that searches for the suitable values of resistor and capacitor values to satisfy given phase and impedance functions. In other terms, objective is to minimize the error between experimental (available) phase response and predicted phase response function in least squares sense.

The results are obtained using Nelder–Mead algorithm,

$$R_1 = 5.3742\,\Omega, \quad R_2 = 0.1075\,\Omega, \quad C_1 = 1.8607\,\mathrm{F}.$$

The phase function

$$\varphi(\omega) = \tan^{-1}\left[-\frac{53.7423\omega}{5.4817 + 10.7485\omega^2}\right].$$

Impedance function

$$Z(s) = \frac{0.1075(s + 5.0995)}{s + 0.1}.$$

MATLAB scripts that are used in the solution of the problem have also been included here.

```
%phi_synthesis1.m
%phase response synthesis of Z = R2+(R1//C1)
clc;clear all;
global w f ff pnorm i E
R1=5; R2=0.1; C1=2; Z0=[R1 R2 C1]'% Actual data
z=(R1+R2)/(R1*R2*C1); %ZERO
p=1/(R1*C1);            %POLE
N=R2*[1 z]; D=[1 p]; h=tf(N,D)
i=50; % number of  samples
w = logspace(-2,2,i);[mag,phase]=bode(h,w); %phase in degrees
j=1:i;
f(j)=phase(:,:,j); %phase vector,degrees
%-------------------------optimization---------------------
%Make a guess for initial estimate x0, then invoke FMINSEARCH.
u=2.7;
x0=[u u u]';% initial guess parameters [R1 R2 C1]
pnorm=2;% norm value  can be  set to 2,4,8,16,32
options = optimset('TolX',0.1,'MaxFunEvals',200,'MaxIter',100);
[Z,fval,pass] = fminsearch(@(x)phi(x),x0,options)
semilogx(w,f,w,ff,'r','linewidth',2);ylabel('Phase,degrees');
xlabel('w, rad/sec');
title('Actual Phase (blue), Predicted Phase (red)');figure;
plot(E);
title('Prediction phase Error  = Actual phase - Predicted phase');
ylabel('Phase Error, degrees');xlabel('w, rad/sec');
%----------resulting functions-------------------
z=(Z(1)+Z(2))/(Z(1)*Z(2)*Z(3)); %ZERO
p=1/(Z(1)*Z(3));            %POLE
Num=Z(2)*[1 z]; Den=[1 p];
H=tf(Num,Den)
%phi=atan(-aw/(b+cw^2))
```

```
a=Z(1)^2*Z(3)
b=Z(1)+Z(2)
c=Z(1)^2*Z(2)*Z(3)^2

function err = phi(x)
global i w f pnorm ff  E
R1=x(1); C1=x(3); R2=x(2);z=(R1+R2)/(R1*R2*C1); p=1/(R1*C1);
NN=R2*[1 z]; DD=[1 p];hh=tf(NN,DD);
[mag,phase]=bode(hh,w); %phase in degrees
for j=1:i;
ff(j)=phase(:,:,j); %Predicted phase vector, degrees
end
E=f-ff;
%norm(E,pnorm) = sum(abs(E).^p)^(1/p) %definition of p norm
err=norm(E,pnorm);
```

Notes:

Initial estimate vector $x_0 = [R_1 \quad R_2 \quad C_1] = [2.7 \quad 2.7 \quad 2.7]$ provides a good approximation to phase response (at 2 norm with maximum function evaluation number = 200, with 100 iterations). Phase response is evaluated at i = 50 points between 0.01 and 100 rad/s.

Playing with these numbers presents different phase response graphics and provides extensive details about the optimization procedure. The reader is urged to run these m files and demonstrate various results of changing parameters described in this program.

The scripts used also provide an error curve which is not displayed here for space saving reasons.

Problem 8.7.9 Use circuit in Fig. 8.75 to demonstrate numerically that an impedance function $Z(s)$ with the same poles and zeros can be realized using different sets of component values (phi.synthesis1.m, phi.m).

Solution
Impedance function of this circuit is

$$Z(s) = R_2 + \frac{R_1\left(\frac{1}{sC_1}\right)}{R_1 + \frac{1}{sC_1}} = \frac{R_2\left(s + \frac{R_1 + R_2}{R_1 R_2 C_1}\right)}{s + \frac{1}{R_1 C_1}} = \frac{k(s+z)}{s+p}$$

Let, $R_1 = 5\,\Omega,\ R_2 = 0.1\,\Omega,\ C_1 = 2\,\text{F}$, substituting these values in $Z(s)$ equation yields

$$Z(s) = \frac{0.1(s+5.1)}{s+0.1}.$$

Let $R_1 = 5.3742\,\Omega$, $R_2 = 0.1075\,\Omega$, $C_1 = 1.8607\,\text{F}$, these values yield

$$Z(s) = \frac{0.1075(s+5.1)}{s+0.1}.$$

Let $R_1 = 11.9304\,\Omega$, $R_2 = 0.2386\,\Omega$, $C_1 = 0.8382\,\text{F}$, these values yield

$$Z(s) = \frac{0.2386(s+5.1)}{s+0.1}.$$

It is seen that the same form of impedance function with identical pole and zero values are obtained although different sets of component values are used, only difference being the coefficient value of k, which equals to the value of resistor R_2.

The solution of this problem is made easier using MATLAB scripts phi.synthesis1.m and phi.m, and varying content of input vector (x_0), and noting at the end of each nonlinear optimization procedure that the approximation of phase function remains within acceptable limits.

The reader is urged to run these MATLAB m files individually, and observe the result of changing input vector (x_0) described in this program.

8.8 Impedance Spectroscopy

Problem 8.8.1 What are the basic equations defining an impedance and admittance?

What do microstructural models describe? Describe layer models and effective medium models.

Solution
Basic equations defining impedance (based on impedance diagram of Fig. 8.78),

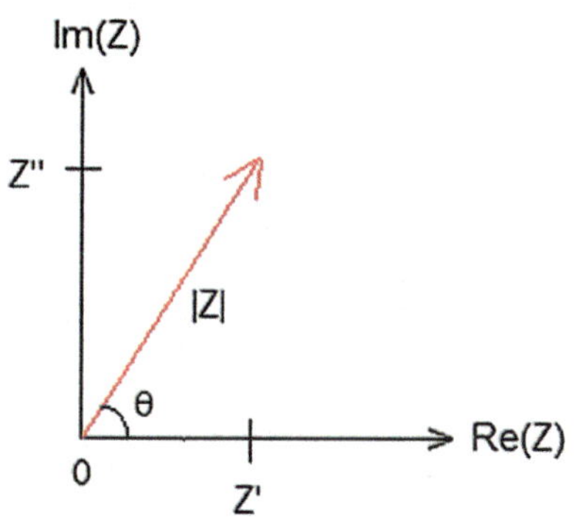

Fig. 8.78 Impedance diagram

$$j = \sqrt{-1} = e^{j\frac{\pi}{2}}$$
$$Z(\omega) = Z' + jZ'', \quad \mathrm{Re}(Z) = Z' = |Z|\cos\theta, \quad \mathrm{Im}(Z) = Z'' = |Z|\sin\theta$$
$$\theta = \tan^{-1}\frac{Z''}{Z'}, \quad |Z| = [(Z')^2 + (Z'')^2]^{1/2}$$
$$Z(\omega) = |Z|e^{j\theta},$$
$$e^{j\theta} = \cos\theta + j\sin\theta.$$

(Euler's identity)

Admittance: $Y = Z^{-1} = Y' + jY''$

Impedance spectra of polycrystalline solids are related to microstructure of these materials. Microstructural models are used to describe grains, grain boundaries, suspensions and porous materials which fall into two types: layer models and effective medium models.

It should be noted here that the dc parameter models are not the subject of Impedance Spectroscopy , alone.

Complex conductivity in a microstructural model is expressed for phase i as $\Psi_i = \sigma_i + j\omega \in_i$

where σ_i = dc conductivity, $\in_i$ = permittivity.

An example to macroscopic property is the resistivity of a sample as a function of its grain boundaries (measurable), while an example for microscopic property is the resistivity of grain boundary (phase) (estimated using a model).

A model provides a hypothetical microstructure. Overall complex conductivity is calculated for such a structure, which can be later converted to complex resistivity. These are represented by equivalent electrical circuits whose parameters are found by IS.

As an example, a Voigt (Foster 1) circuit for each phase is an equivalent circuit to represent the complex conductivity.

Layer Models

(a) **Series layer model**: The phases are assumed to be stacked in layers parallel to the measurement electrodes, see, Fig. 8.79.

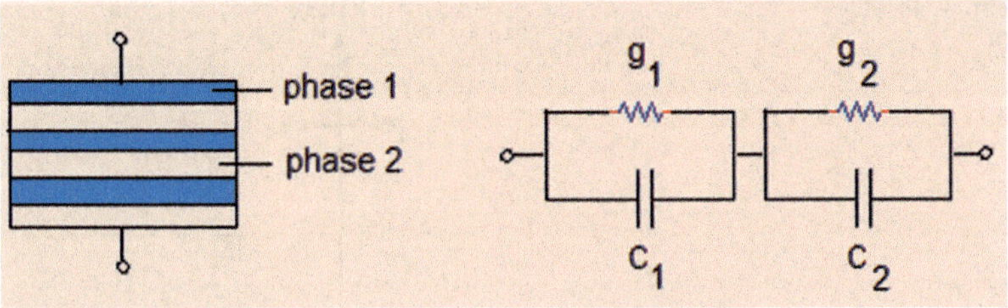

Fig. 8.79 Series layer model

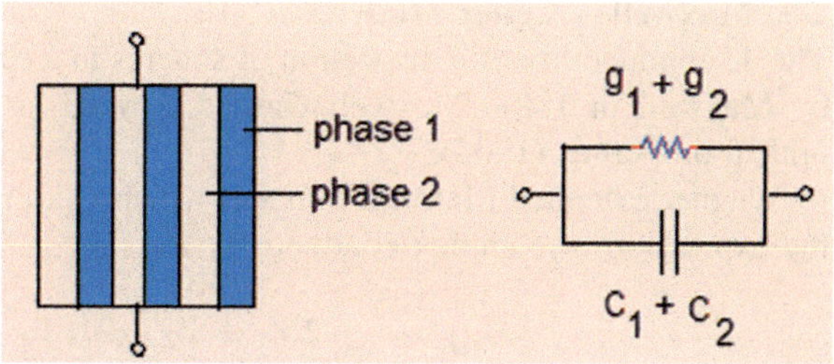

Fig. 8.80 Parallel layer model

For a two-phase model, using linear mixing rule ($x_1 + x_2 = 1$) complex conductivity is

$$\Psi^{-1} = x_1 \Psi_1^{-1} + x_2 \Psi_2^{-1} \tag{8.37}$$

complex resistivity,

$$\rho = x_1 \rho_1 + x_2 \rho_2 \tag{8.38}$$

with

$$\sigma_1 = x_1 g_1, \quad \sigma_2 = x_2 g_2 \tag{8.39}$$

$$\in_1 = x_1 C_1, \quad \in_2 = x_2 C_2 \tag{8.40}$$

$$\tau_1 = C_1/g_1, \quad \tau_2 = C_2/g_2. \tag{8.41}$$

(b) **Parallel layer model**: The phases are assumed to be stacked across the measurement electrodes, see, Fig. 8.80.

$$\Psi = x_1 \Psi_1 + x_2 \Psi_2 \tag{8.42}$$

$$\sigma_1 = g_1/x_1, \quad \sigma_2 = g_2/x_2 \tag{8.43}$$

$$\in_1 = C_1/x_1, \quad \in_2 = C_2/x_2. \tag{8.44}$$

This model shows only a single relaxation. Individual relaxations cannot be resolved. It fails to describe the behavior of grain boundaries in ceramics.

Effective Medium Models

Layer models are derived under unrealistic assumptions about current distribution. Effective medium technique assumes an effective (continuous) medium of conductivity Ψ_{eff} and when a part of it is removed and then replaced by some heterogeneous multiphase material, the current distribution does not change (Overall conductivity of the multiphase material equals to that of effective medium, i.e., $\Psi_t = \Psi_{\mathrm{eff}}$).

The Maxwell–Wagner Model

The dc conductivity of a dispersion of spheres in a continuous medium was derived by Maxwell in 1881. Maxwell Garnett derived an expression for dielectric and optical properties (1904).

Wagner extended Maxwell's model which is known as M–W model, and gives the expression of complex conductivity as

$$\Psi_t = \Psi_1 \frac{2\Psi_1 + \Psi_2 - 2x(\Psi_1 - \Psi_2)}{2\Psi_1 + \Psi_2 + x(\Psi_1 - \Psi_2)}. \tag{8.45}$$

For two-phase heterogeneous media, the maximum volume fraction of dispersed phase is $x < 0.30$. It is shown that M–W model for two-phase heterogeneous media can be represented by two banks RC Voigt (Foster 1) or 4-element ladder (Cauer 1) circuits [13].

The Fricke Model (Two-Phase Dispersions)

This model relates to dispersion of ellipsoidal particles of random orientation in a continuous medium:

$$\Psi_t = \Psi_2 + \frac{(\Psi_1 - \Psi_2)(1 - x)}{1 + \left(\frac{x}{3}\right) \sum_{i=1}^{3} (\Psi_1 - \Psi_2)/(\varphi_i \Psi_1 + \Psi_2)} \tag{8.46}$$

Ψ_t complex conductivity of the dispersion
Ψ_1 complex conductivity of the medium (matrix)
Ψ_2 complex conductivity of the dispersed phase
x volume fraction of the dispersed phase
φ_i form factors (depend on the axial ratios of the ellipsoids, a, b, c; $a \geq b \geq c$).

This model can be represented by a Voigt (Foster 1) circuit, whose component values are estimated by nonlinear least squares fit to the spectrum.

The double-layer capacitance coupled with a faradic reaction resistance and the series resistance of the electrolyte in contact with the test electrode is represented by a three-component electrical equivalent circuit, see Fig. 8.81.

A way of representing the impedance behavior of materials is to implement complex impedance diagrams [14]. For a simple parallel RC circuit, such a diagram

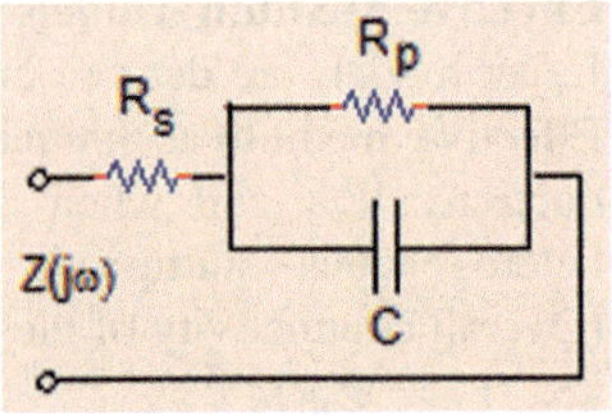

Fig. 8.81 Two-phase dispersions model

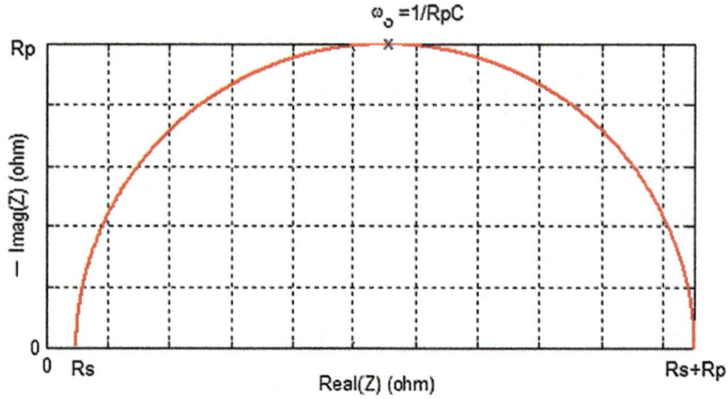

Fig. 8.82 Complex impedance diagram for a simple RC circuit

(with negated part of imaginary axis and omitted positive imaginary values) consists of a semicircle (Fig. 8.82).

Problem 8.8.2 Describe briefly some applications of IS in medical field.

Solution
In the medical field, electrical impedance spectroscopy has applicability in neoplastic conditions to differentiate between normal, precancerous, and cancerous tissues (neoplasias). In these applications, EIS exploits the different electrical resistivities of each specific tissue type based on its cellular structure. It can be applied for the diagnosis of various cancers and precancerous conditions (e.g., within the cervical and gastro-esophageal cancer diagnostic pathway).

Electrical spectroscopic properties are known to change during the development of neoplasia, with more dysplastic tissue exhibiting a reduced resistance to the flow of current as a result of physical changes in its structure. If one measures the electrical impedance at different frequencies, then it is possible to create a characteristic spectrum according to the structure of the tissue. Each reading is given a nominal value which is compared to a reference, providing a semi-quantitative, reproducible assessment of the tissue such as cervical tissue.

In one research, [15] a pencil probe (diameter 5 mm) was used to measure electrical impedance spectra from eight points on the cervix in 124 women with abnormal cervical smears. Variables that should be sensitive to the expected tissue changes were calculated. These were compared with the colposcopic results. The measured electrical impedance changes were those predicted on the basis of the expected tissue structures. Measurements made on normal squamous tissues were well separated from those made on precancerous tissues. characteristics of the electrical impedance spectra of tissues can be explained by changes in cell arrangements (layering) and in the size of the nuclei. This relation opens the way to deriving tissue structure from electrical impedance spectral measurements. EIS approach can be used to give good separation of normal and precancerous cervical tissues.

Another study [16] recruited women referred to colposcopy with an abnormal Papanicolaou smear. A pencil probe incorporating four gold electrodes was used to measure electrical impedance spectra from cervical epithelium. Colposcopy examinations, including probe positioning, were video recorded to allow for correlation between results obtained from colposcopic impression, histopathologic examination of colposcopic punch biopsies, and impedance measurements. Cervical impedance-derived parameters were assessed to see if significant difference in values obtained in CIN and normal epithelium existed. The performance of the probe in identifying women with CIN was also assessed. 176 women were recruited and 1168 points analyzed. It was concluded that cervical impedance spectrometry provides a potentially promising real-time screening tool for CIN with similar sensitivity and specificity to currently used screening tests.

Balasubramani et al. [17] reported their EIS studies in which a pencil type probe was used to record impedance spectra from 12 points on the cervix before and after the application of 5% acetic acid (AA). Spectra were also recorded from tissue boundaries (165 women, either with a clinical indication or abnormal cervical cytology, were recruited into the study). The results showed no significant difference, indicating that application of AA does not produce a large change in spectra. The probe could distinguish tissue boundaries from homogeneous tissue points.

Tidy et al. [18] presented a prospective, comparative, multi-center (two in England and one in Ireland) clinical study recruiting 474 women (referred with abnormal cytology). They concluded that if EIS is used as an adjunct to colposcopy, it improves colposcopic performance, and the addition of EIS could lead to more appropriate patient management with lower intervention rates.

References [19–28] are other examples of continuing research related to this field.

Problem 8.8.3 In impedance spectroscopy, the circuits shown in Fig. 8.83 are named as Voigt and Maxwell models, respectively. In electrical circuits (synthesis) these are known as Foster 1 and Foster 2 forms of an RC driving point impedance.

Driving point impedances of both circuits are the same at all frequencies if

$$R_a = R_1 + R_2 \tag{8.47}$$

$$C_a = \frac{C_1 C_2}{C_1 + C_2} \tag{8.48}$$

$$R_b = \frac{R_1 R_2 (R_1 + R_2)(C_1 + C_2)^2}{(R_1 C_1 - R_2 C_2)^2} \tag{8.49}$$

$$C_b = \frac{(R_1 C_1 - R_2 C_2)^2}{(C_1 + C_2)(R_1 + R_2)^2}. \tag{8.50}$$

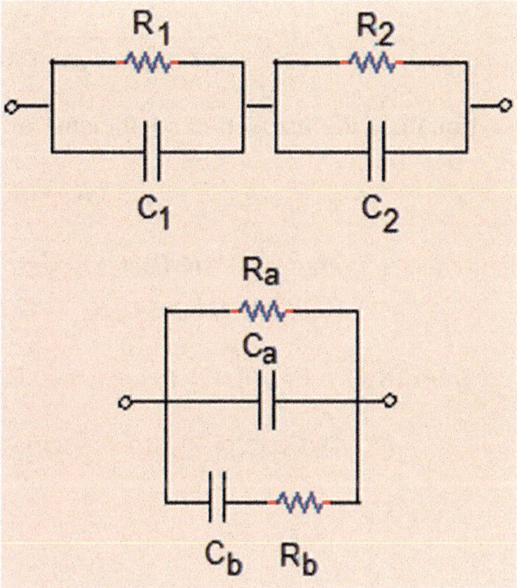

Fig. 8.83 Voigt and Maxwell models (Foster 1 and Foster 2 forms of an RC driving point impedance)

Prove the validity of Eqs. (8.47) and (8.48).

Solution

Driving point impedance for the first circuit (above) is

$$Z_f(s) = \frac{R_1R_2(C_1+C_2)s + (R_1+R_2)}{R_1C_1R_2C_2s^2 + (R_1C_1+R_2C_2)s+1}. \tag{8.51}$$

Driving point impedance for the second (lower) circuit of Fig. 8.83 is calculated as follows:

$$Z_a(s) = C_a \parallel R_a = \frac{\dfrac{R_a}{sC_a}}{R_a + \dfrac{1}{sC_a}} = \frac{R_a}{1+sR_aC_a}$$

$$Z_b(s) = \frac{1}{sC_b} + R_b = \frac{1+sR_bC_b}{sC_b}$$

$$Z_m(s) = Z_a(s) \parallel Z_b(s) = \frac{\dfrac{R_a}{1+sR_aC_a} \cdot \dfrac{1+sR_bC_b}{sC_b}}{\dfrac{R_a}{1+sR_aC_a} + \dfrac{1+sR_bC_b}{sC_b}}$$

$$Z_m(s) = \frac{R_a(1+R_bC_bs)}{sC_bR_a + (1+sC_bR_b)(1+sC_aR_a)}$$

$$= \frac{sR_aR_bC_b + R_a}{sC_bR_a + 1 + sC_aR_a + sC_bR_b + s^2C_aC_bR_aR_b}$$

$$Z_m(s) = \frac{sR_aR_bC_b + R_a}{s^2C_aC_bR_aR_b + s(C_aR_a + C_bR_b + R_aC_b) + 1}. \tag{8.52}$$

Equating the respective coefficients of s in polynomials in (8.51) and (8.52),

$$R_a = R_1 + R_2 \tag{8.53}$$

$$R_aR_bC_b = R_1R_2(C_1 + C_2), \quad C_1C_2R_1R_2 = C_aC_bR_aR_b$$
$$C_aR_a + C_bR_a + C_bR_b = C_1R_1 + C_2R_2.$$

From (8.53), Eq. (8.47) is apparent. From (8.48) and (8.49),

$$C_a \cdot R_1R_2(C_1 + C_2) = C_1C_2R_1R_2 \quad \rightarrow \quad C_a(C_1 + C_2) = C_1C_2$$
$$C_a = \frac{C_1C_2}{C_1 + C_2}.$$

This is the relationship in (8.48).

Problem 8.8.4 Determine Maxwell (Foster 2) circuit elements in terms of ladder (Cauer 1) circuit elements so that both driving point impedances (DPI) are equal at all frequencies, see Fig. 8.84.

Solution
DPIs for both circuits are

$$Z_c(s) = \frac{C_2R_1R_2s + (R_1 + R_2)}{C_1C_2R_1R_2s^2 + (R_1C_1 + R_2C_2 + C_1R_2)s + 1} \tag{8.54}$$

$$Z_m(s) = \frac{C_bR_aR_bs + R_a}{C_aC_bR_aR_bs^2 + (C_aR_a + C_bR_b + C_bR_a) + 1}. \tag{8.55}$$

Equating respective numerator and denominator polynomial coefficients yields

$$R_a = R_1 + R_2 \tag{8.56}$$

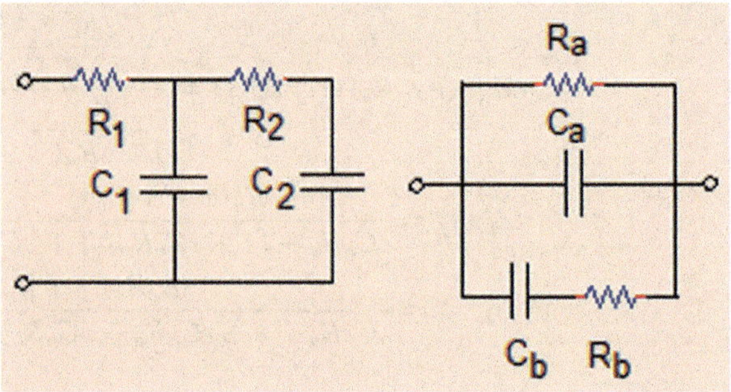

Fig. 8.84 Foster and Cauer circuits' equivalence

$$C_bR_aR_b = C_2R_1R_2 \tag{8.57}$$

$$C_aC_bR_aR_b = C_1C_2R_1R_2 \tag{8.58}$$

$$C_aR_a + C_bR_b + C_bR_a = R_1C_1 + R_2C_2 + C_1R_2. \tag{8.59}$$

From (8.57), (8.58) ,

$$C_a(C_2R_1R_2) = C_1C_2R_1R_2$$

$$C_a = C_1. \tag{8.60}$$

Form (8.56), (8.57) ,

$$C_bR_b = \frac{C_2R_1R_2}{R_1 + R_2}. \tag{8.61}$$

From (8.59), (8.60), (8.61),

$$C_1(R_1 + R_2) + \frac{C_2R_1R_2}{R_1 + R_2} + C_b(R_1 + R_2) = R_1C_1 + R_2C_2 + C_1R_2$$

$$C_b(R_1 + R_2) = R_1C_1 + R_2C_2 + C_1R_2 - R_1C_1 - C_1R_2 - \frac{C_2R_1R_2}{R_1 + R_2}$$

$$C_b = \frac{R_2C_2\left(1 - \frac{R_1}{R_1 + R_2}\right)}{R_1 + R_2} = C_2\left(\frac{R_2}{R_1 + R_2}\right)^2. \tag{8.62}$$

From (8.61), (8.62),

$$C_2\left(\frac{R_2}{R_1 + R_2}\right)^2 \cdot R_b = \frac{C_2R_1R_2}{R_1 + R_2}$$

$$\frac{R_2}{R_1 + R_2} R_b = R_1$$

$$R_b = \frac{R_1}{R_2}(R_1 + R_2). \tag{8.63}$$

In summary,

$$R_a = R_1 + R_2, \quad C_a = C_1$$

$$R_b = \frac{R_1}{R_2}(R_1 + R_2), \quad C_b = C_2\left(\frac{R_2}{R_1 + R_2}\right)^2.$$

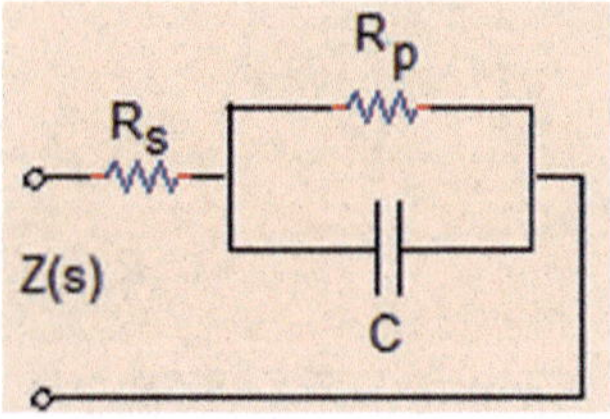

Fig. 8.85 Double-layer model

Problem 8.8.5 An electrode/electrolyte interface (double layer) can be modeled using the circuit shown in Fig. 8.85.

R_s Resistance of electrolyte solution
R_p Resistance of double layer
C Capacitance of double layer

Show that analytical equation of this system in a Nyquist plot consists of a half circle.

Solution
The impedance is

$$Z(j\omega) = \frac{V(j\omega)}{I(j\omega)} = R_s + \frac{R_p}{j\omega R_p C + 1} = R_s + \frac{R_p}{(\omega R_p C)^2 + 1} - j\frac{(\omega R_p C)^2}{1 + (\omega R_p C)^2}$$

$$Z' = \mathrm{Re}[Z(j\omega)] = R_s + \frac{R_p}{(\omega R_p C)^2 + 1} \tag{8.64}$$

$$-Z'' = -\mathrm{Im}[Z(j\omega)] = \frac{(\omega R_p C)^2}{1 + (\omega R_p C)^2}. \tag{8.65}$$

Eliminating ω from (8.64) and (8.65) yields half circle equation,

$$\left[Z' - (R_s + \frac{R_p}{2})\right]^2 + (Z'')^2 = \left(\frac{R_p}{2}\right)^2.$$

A modified plot of this equation is shown in Fig. 8.86. Note that the sign of ordinate axis is minus. As a rule, this sign convention is used for drawing spectrograms in IS.

Problem 8.8.6 An electrical circuit model used to simulate impedance spectroscopic data of a composite material is shown in Fig. 8.87 [29]. These kinds of circuits represent time constants as a measure of accessibility of different parts of an internal layer surface. The circuit includes an electrolytic resistance in the outermost layer (=R_1) and subsequent resistances related to conductivity in the inner layers.

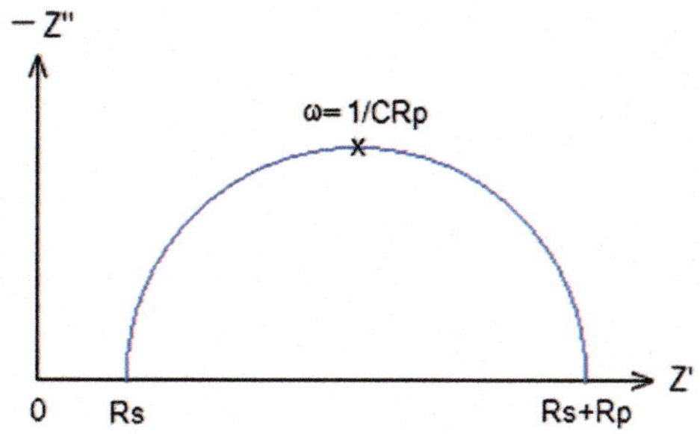

Fig. 8.86 Nyquist diagram

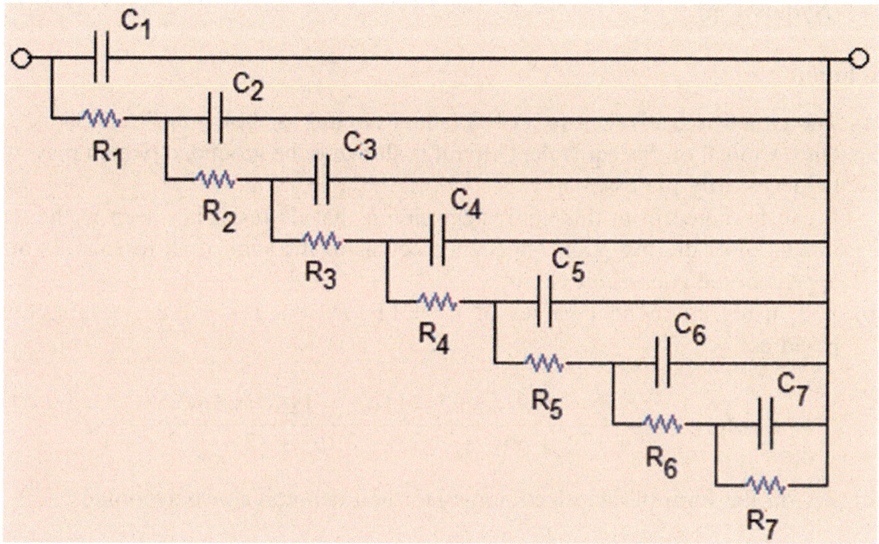

Fig. 8.87 Model for impedance spectroscopic data of a composite material

Number of resistor elements is a function of morphology of filler matrix, particle size, and porosity. Composite surface capacitance is represented by a network of shunt capacitors.

(a) Show that the circuit is equivalent to an RC ladder circuit.
(b) Determine its driving point impedance rational function, $Z(s)$, using magnitude and frequency scaled component values of $C_i = 1\,\text{F},\ R_i = 1\,\Omega,\ i = 1, 2, \ldots, 7$ (Impedance_calc1.m).
(c) Determine poles and zeros of $Z(s)$ and write this rational function in the form of factorized numerator and denominator polynomials (Gain_H0.m).
(d) Convert the equivalent ladder circuit into Foster 1 form (Foster11.m).

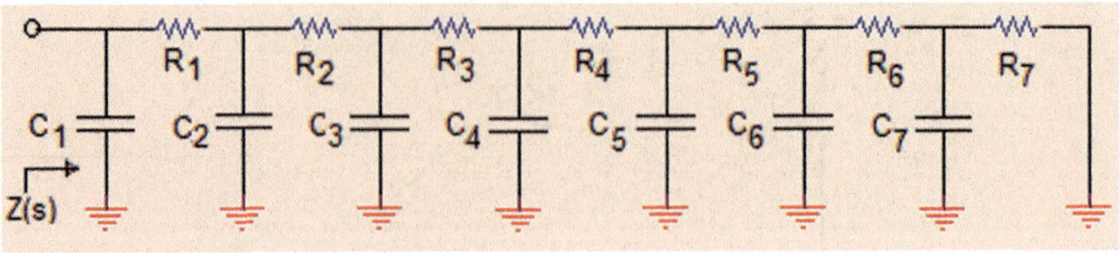

Fig. 8.88 Cauer I (ladder) equivalent circuit

(e) Plot magnitude and phase functions of $Z(j\omega)$ for data set given in part (b).
(f) Plot $\text{Im}[Z(j\omega)]$ versus $\text{Re}[Z(j\omega)]$ for the same data set given in part (b). List the values of real and imaginary parts of impedance function at pole frequencies (Nyquist1.m).

Solution

(a) The circuit is equivalent to an RC ladder circuit, as shown in Fig. 8.88.
One terminal of this equivalent circuit is shown to be grounded (which may not be necessarily the case).
It can be noted from this equivalent circuit that dc resistance seen at the terminals of of driving point impedance equals to the sum of all resistances that appear in the equivalent circuit.
(b) $Z(s)$, using component values of $C_i = 1\,\text{F}$, $R_i = 1\,\Omega$, $i = 1, 2, \ldots, 7$ is computed as

$$Z(s) = \frac{s^6 + 12s^5 + 55s^4 + 120s^3 + 126s^2 + 56s + 7}{s^7 + 13s^6 + 66s^5 + 165s^4 + 210s^3 + 126s^2 + 28s + 1}.$$

(c) $Z(s)$ in the form of factorized numerator and denominator polynomials,

$$Z(s) = \frac{(s+3.8019)(s+3.247)(s+2.445)(s+1.555)(s+0.753)(s+0.1981)}{(s+3.8271)(s+3.3383)(s+2.618)(s+1.7909)(s+1)(s+0.382)(s+0.0437)}.$$

(d) Foster 1 equivalent form circuit component values of Fig. 8.89 are:

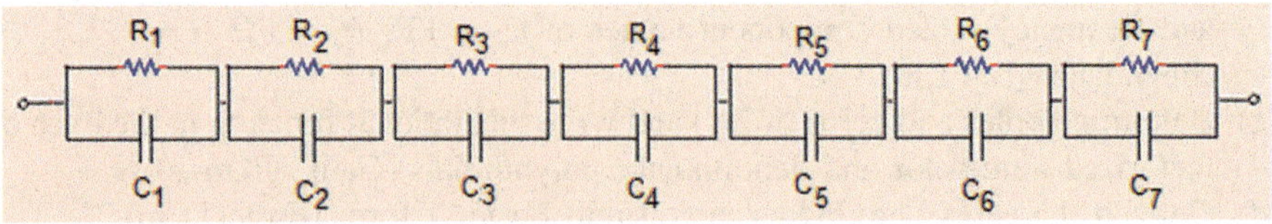

Fig. 8.89 Foster I equivalent circuit

$$R = [R_1 \quad R_2 \quad \ldots \quad R_7], \quad C = [C_1 \quad C_2 \quad \ldots \quad C_7];$$
$$R = [0.0030 \quad 0.0132 \quad 0.0352 \quad 0.0822 \quad 0.2000 \quad 0.6315 \quad 6.0349]\,(\Omega)$$
$$C = [86.7508 \quad 22.6676 \quad 10.8541 \quad 6.7902 \quad 5.0000 \quad 4.1459 \quad 3.7914]\,(\mathrm{F}).$$

MATLAB script for ladder circuit transfer function calculation from component values is given below (Impedance_calc1.m).

```
syms s
R=1;C=1;
z1=R; y1=1/z1; y2=s*C+y1; z2=1/y2;
z3=R+z2; y3=1/z3; y4=s*C+y3; z4=1/y4;
z5=R+z4; y5=1/z5; y6=s*C+y5; z6=1/y6;
z7=R+z6; y7=1/z7; y8=s*C+y7; z8=1/y8;
z9=R+z8; y9=1/z9; y10=s*C+y9; z10=1/y10;
z11=R+z10; y11=1/z11; y12=s*C+y11; z12=1/y12;
z13=R+z12; y13=1/z13; y14=s*C+y13; z14=1/y14;
%N=numerator, D=Denominator
[N,D] = numden(z14)
```

The plot of Im[$Z(j\omega)$] versus Re[$Z(j\omega)$] for data set given in part (b) is shown in Fig. 8.90. The graph consists of 50 data points and it is drawn between angular frequencies of 0.001–1000 rad/s.

The values of real and imaginary parts of impedance function at pole frequencies are listed as follows:

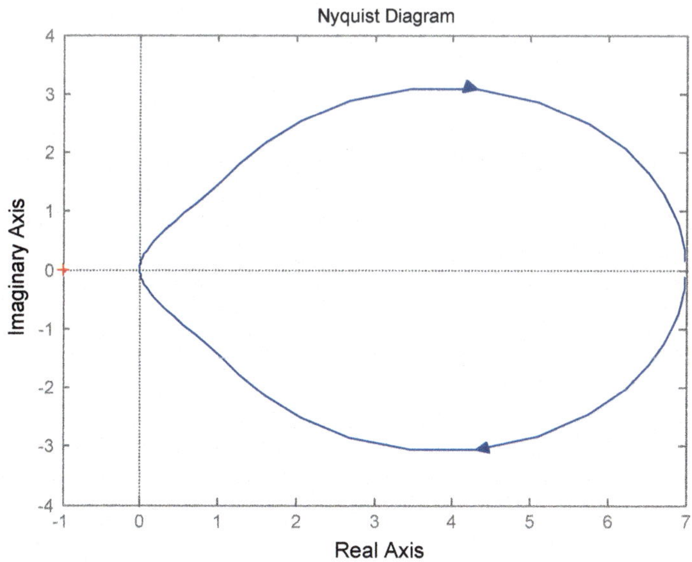

Fig. 8.90 Nyquist diagram

ω	Re($j\omega$)	Im($j\omega$)
3.8271	0.0530	0.2362
3.3383	0.0658	0.2647
2.6180	0.0944	0.3213
1.7909	0.1564	0.4253
1.0000	0.3002	0.6248
0.3820	0.6974	1.0875
0.0437	3.9743	3.1003

```
%N is numerator, D is denominator polynomial of H
%Generate an impedance transfer function in s domain
N=[1 12 55 120 126 56 7]; D=[1 13 66 165 210 126 28 1];
w=logspace(-3,3,50);
H = tf(N,D)    %Transfer function
p=pole(H)
z=zero(H)
%w=[3.8271  3.3383  2.6180  1.7909   1  0.3820  0.0437];
[RE,IM] = nyquist(H,w); nyquist(H,w); %Plot
RE = squeeze(RE);IM = abs(squeeze(IM)); U=[w'  RE  IM]
```

Problem 8.8.7

(a) Discuss the use of Kramers–Kronig (K–K) relations (Hilbert transform) for impedance spectrum validation.
(b) What are the limitations of K–K method for impedance spectrum (EIS) consistency verification?

Solution

(a) Electrical impedance spectroscopy (EIS) data analysis requires the fulfillment of causality, linearity, finiteness, and stability conditions. Noncompliance with any of these conditions may cause wrong interpretations. The Kramers–Kronig relations are helpful tools in EIS consistency verification.
Let $Z(j\omega) = Z'(\omega) + jZ''(\omega)$ be an analytic complex function. The Kramers–Kronig relations are used to calculate the imaginary part from real part of the complex physical (stable) functions (or vice versa). The real and imaginary parts of such a function are dependent, and the full function can be reconstructed given just one of its parts. $Z'(\omega)$ is an even and $Z''(\omega)$ is an odd function of frequency.
The Kramers–Kronig relations (in a physically realistic form) are

$$Z'(\omega) - Z'(\infty) = \frac{2}{\pi} \int_0^\infty \frac{xZ''(x) - \omega Z''(\omega)}{x^2 - \omega^2} \mathrm{d}x$$

$$Z''(\omega) = -\frac{2\omega}{\pi} \int_0^\infty \frac{Z'(x) - Z'(\omega)}{x^2 - \omega^2} \mathrm{d}x.$$

Evaluating if the experimental data satisfies K–K relationships is a consistency test [30–33].

The Kramers–Kronig relations give a condition that is both necessary and sufficient. Therefore, even before applying an inverse Fourier integral, one can determine whether a given frequency response will yield a causal or a non-causal impulse response. If the real and imaginary parts are Hilbert transforms of each other, the impulse response is causal, and not otherwise. This fact is very useful because it allows us to test whether or not a frequency response is causal without ever having to leave the frequency domain [34].

In performing K–K transformation, one may set an equivalent circuit model and then fit it to EIS data. The circuit model needs to match the data near the frequency limits. It may be possible to match the model and data at high and low frequencies by adjusting the R and C values.

(b) These equations require the integration from zero to infinite frequency. Since experimental data is not available over an infinite frequency range, application of the K–K relations requires an approximation to be used at frequencies outside the range of measured frequencies. The K–K method fails if the wrong approximation is used. Limited frequency range is a major problem of the K–K validation.

Other fundamental limitations of K–K transformations are the following:

(1) Many data artifacts may cause a "false positive" result.
(2) Note also that an ideal capacitor cannot be transformed since a capacitor has infinite reactance at zero frequency, while the K–K relations require that the impedance be finite at all frequencies.

Problem 8.8.8 A microporous isotropic layer of Titanium Oxide has been produced on a Titanium substrate. Four points in-line probe measurements (with equal interelectrode spacing of 1 cm) yield data as shown in Table 8.2.

(a) Compute and plot bulk electrical resistivity of the sample as a function of frequency.
(b) If thickness of the oxide layer is 1 μm, compute and plot sheet resistance as function of frequency (Rho Tio2 1.xlsx).

Table 8.2 Impedance values obtained at different frequencies

f (Hz)	Z (Ω)
1000	88.50
500	116.43
250	156.44
100	241.57
50	353.02
25	531.85
10	970.47
5	1564.59
2	2991.16
1	4982.88

Solution

(a) $\rho(\omega) = 2\pi d|Z(j\omega)| = 2 \times 3.141592 \times 0.01 \times |Z(j\omega)|\ \Omega\text{m}$
Resistivity values computed from given data are plotted as shown in Fig. 8.91.

(b) If TiO_2 film thickness is $t = 1\ \mu\text{m}$, $t \ll d$,

$$R_{sh}(\omega) = \frac{\pi}{\ln(2)} \times |Z(j\omega)| = \frac{3.141592}{0.69315} \times |Z(j\omega)| = 4.5324 \times |Z(j\omega)|\ \Omega\text{m}.$$

Figure 8.92 displays sheet resistivity values as function of frequency.

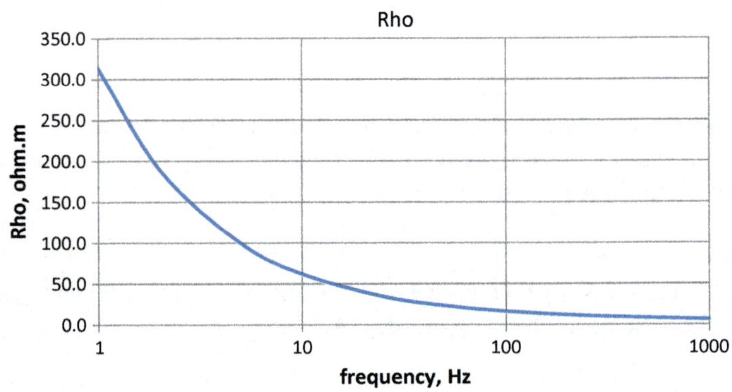

Fig. 8.91 Resistivity as a function of frequency

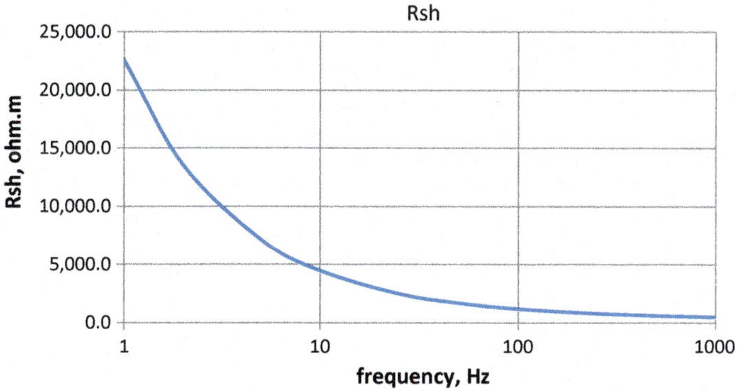

Fig. 8.92 Sheet resistivity values as a function of frequency

References

1. Chen MZQ, Wang K, Shu Z, Li C (2013) Realizations of a special class of admittances with strictly lower complexity than canonical forms. IEEE Trans Circuits Syst I Regul Pap 60 (9):2465–2473
2. Lamm HYF (1979) Analog and digital filters: Design and realization. Prentice Hall
3. Balabanian N, Bickart TA, Seshu S (1969) Electrical network theory. Wiley, New Jersey
4. Tavsanoglu V (1973) Application of Howitt transformation to two-element-kind networks. Electron Lett 9:206–208
5. Tavsanoglu V (1976) Transformation of Cauer forms. In: Proceedings of IEEE international symposium on circuits and systems, pp 175–178
6. Murthy KVV, Bedford RE (1978) Transformation between Foster and Cauer equivalent networks. IEEE Trans Circuits Syst CAS-25:238–239
7. Hagopian JD and Frisch IT (1970) Capacitance and resistance minimization in one-port RC networks. IEEE Trans Circuits Theor CT-17:386–392
8. Ali M, Magee D, Dasgupta U (2008) Signal processing overview of ultrasound systems for medical imaging. SPRAB12, Texas Instruments
9. Zhu Y-S, Chen W-K (2000) Computer-aided design of communication networks (Chapter 9). World Scientific, pp 507–514
10. Yarman BS (2010) Gewertz procedure to generate a minimum function from its even part: generation of minimum function in rational form (Chapter 8). In: Design of ultra-wideband power transfer networks. Wiley, Chichester, UK
11. Miyata FA (1952) New system of two-terminal synthesis. J Inst Electr Eng (Japan) 35:211–218
12. Ho CF (1983) Determination of a driving-point function from its specified real part. Int J Electr Eng Educ 20:53–57
13. Bonanos N, Lilley E (1981) Conductivity relaxations in single crystals of sodium chloride containing suzuki phase precipitates. J Phys Chem Solids 42:943–952
14. Barsoukov E, Macdonald JR (2005) Impedance spectroscopy, theory, experiment and applications. Wiley Interscience
15. Brown BH, Tidy J, Boston K, Blackett AD, Smallwood RH, F Sharp F (2000) The relationship between tissue structure and imposed electrical current flow in cervical neoplasia. Lancet 355:892–895
16. Abdul S, Brown BH, Milnes P, Tidy JA (2006) The use of electrical impedance spectroscopy in the detection of cervical intraepithelial neoplasia. Int J Gynaecol Cancer 16:1823–1832

17. Balasubramani L, Brown BH, Healey J, Tidy JA (2009) The detection of cervical intraepithelial neoplasia by electrical impedance spectroscopy: the effects of acetic acid and tissue homogeneity. Gynecol Oncol 115(2):267–271
18. Tidy JA, Brown BH, Healey TJ, Daayana S, Martin M, Prendiville W, Kitchener HC (2013) Accuracy of detection of high-grade cervical intraepithelial neoplasia using electrical impedance spectroscopy with colposcopy. Br J Obstet Gynaecol 120:400–411
19. Gonzalez-Correa CA, Brown BH, Smallwood RH, Kalia N, Stoddard CJ, Stephenson TJ, Haggie SJ, Slater DN, Bardhan KD (1999) Virtual biopsies in Barret's esophagus using an impedance probe. Ann N Y Acad Sci 873:313–321
20. Nebuya S, Brown BH, Smallwood RH, Milnes P, Waterworth AR, Noshiro M (1999) Measurement of high frequency electrical transfer impedances from biological tissues. Electron Lett 35(23):1985–1987
21. Walker DC, Brown BH, Hose DR, Smallwood RH (2000) Modelling the electrical impedivity of normal and premalignant cervical tissue. Electron Lett 1603–1604
22. Brown BH, Wilson AJ, Bertemes-Filho P (2000) Bipolar and tetrapolar transfer impedance measurements from a volume conductor. Electron Lett 36(25):2060–2062
23. Jones DM, Smallwood RH, Hose DR, Brown BH (2001) Constraints on tetrapolar tissue impedance measurements. Electron Lett 37(25):1515–1517
24. Brown BH (2001) Measurement of the electrical properties of tissue—new developments in impedance imaging and spectroscopy. IEICE Trans Inf Syst E85-D:2–5
25. Gonzalez-Correa CA, Brown BH, Smallwood RH, Stephenson TJ, Stoddard CJ, Bardhan KD (2003) Low frequency electrical bioimpedance for the detection of inflammation and dysplasia in Barrett's oesophagus. Physiol Meas 24(2):291–296
26. Gandhi SV, Walker D, Milnes P, Mukherjee S, Brown BH, Anumba DOC (2006) Electrical impedance spectroscopy of the cervix in non-pregnant and pregnant women. Eur J Obstet Gynaecol 129:145–149
27. Lundin P, Karpefors M, Carllson K, Hansen MB, Ruth M (2011) Bioimpedance spectroscopy: a new tool to assess early esophageal changes linked to gastroesophageal reflux disease? Dis Esophagus 24:462–469
28. Tidy JA (2015) Utilization of electrical impedance spectroscopy (EIS) in the detection of dysplasia. Oncology News March–April issue
29. Pavlovic MM et al (2012) Electrical conductivity of lignocellulose composites loaded with electrodeposited copper powders. Part III. Influence of particle morphology on appearance of electrical conductive layers. Int J Electrochem Sci 7:8894–8904
30. Macdonald DD, Macdonald MU (1986) Application of Kramers-Kronig transformations in the analysis of electrochemical impedance data. Part 1. J Electrochem Soc 132(10):2317
31. Macdonald DD, Macdonald MU (1986) Application of Kramers-Kronig transformations in the analysis of electrochemical impedance data. Part 2. J Electrochem Soc 133(10):2023
32. Esteban JM, Orazem ME (1991) On the application of the Kramers-Kronig relations to evaluate the consistency of electrochemical impedance data. J Electrochem Soc 138
33. Giner-Sanz JJ, Ortega EM, Pérez-Heranz V (2015) Montecarlo based quantitative Kramers-Kronig test for PEMFC impedance spectrum validation. Int J Hydrogen Energy 40:11279–11293
34. Warwick C (2010) Understanding the Kramers-Kronig relation using a pictorial proof, White paper. Agilent Technologies, Inc. 5990-5266EN

Chapter 9
Two-Port Networks

Problem 9.1 True or False (T, F)?

(a) Driving-point impedances are z_{11} and $z_{22.}$
(b) Transfer impedance parameters are z_{21} and $z_{12.}$
(c) A two-port network is symmetrical if $z_{11} \neq z_{22.}$
(d) When the two-port network is linear and no dependent sources, the transfer impedances are not equal.
(e) If $z_{12} = z_{21}$, the two-port is said to be reciprocal.
(f) In a reciprocal circuit, if the points of excitation and response are interchanged, the transfer impedances remains the same.
(g) Any two-port consisting of only passive *R*, *L*, *C* components is not reciprocal.
(h) An ideal transformer has all nonzero z-parameters.
(i) An ideal transformer has all nonzero *h*-parameters.
(j) A circuit with $z_{11} = 4\ \Omega$, $z_{22} = 4\ \Omega$, $z_{12} = j2\ \Omega$, $z_{21} = j3\ \Omega$ is reciprocal.

Answers

(a) T, (b) T, (c) F, ($z_{11} = z_{22}$), (d) F, ($z_{12} = z_{21}$), (e) T, (f) T, (g) F, reciprocal, (h) F, (no *z*-parameters), (i) T, (j) F, ($z_{12} \neq z_{21}$).

Problem 9.2 Determine *z*-parameters of the two-port network shown in Fig. 9.1 ($R_1 = 10\ \Omega$, $R_2 = 20\ \Omega$), and draw its equivalent circuit.

A.Ü. Keskin, *Electrical Circuits in Biomedical Engineering*,
DOI 10.1007/978-3-319-55101-2_9

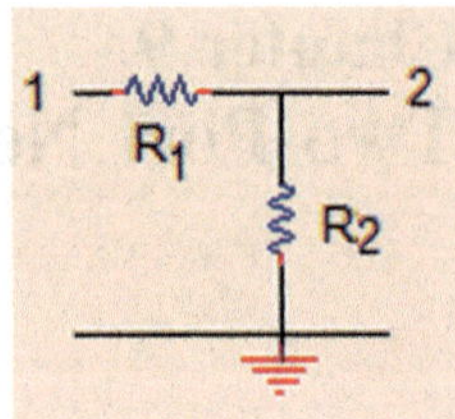

Fig. 9.1 Two-port network

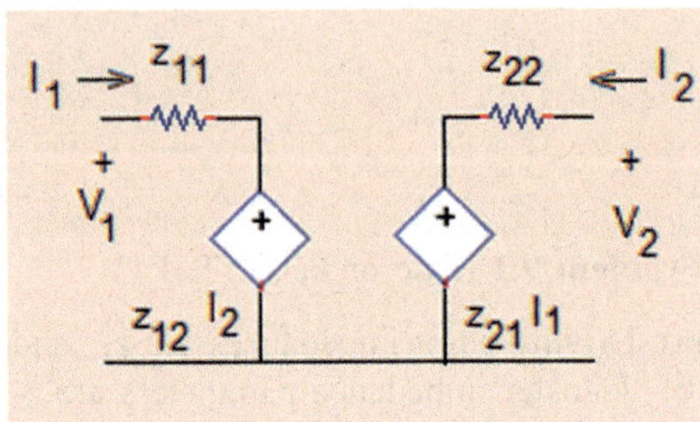

Fig. 9.2 Open-circuit transfer impedence

Solution

$$z_{11} = \left.\frac{V_1}{I_1}\right|_{I_2=0} = \text{open circuit input impedance} = R_1 + R_2 = 30\,\Omega$$

$$z_{22} = \left.\frac{V_2}{I_1}\right|_{I_1=0} = \text{open circuit output impedance} = R_2 = 20\,\Omega$$

$$z_{12} = \left.\frac{V_1}{I_2}\right|_{I_1=0} = \text{open circuit transfer impedance from port 1 to port 2} = R_2 = 20\,\Omega$$

$$z_{21} = \left.\frac{V_2}{I_1}\right|_{I_2=0} = \text{open circuit transfer impedance from port 2 to port 1} = R_2 = 20\,\Omega$$

$$[z] = \begin{bmatrix} 30 & 20 \\ 20 & 20 \end{bmatrix}\Omega$$

(Fig. 9.2).

Problem 9.3 Determine the z-parameters for the circuit shown in Fig. 9.3.

$$(R_1 = 50\,\Omega, R_2 = 60\,\Omega, R_3 = 70\,\Omega).$$

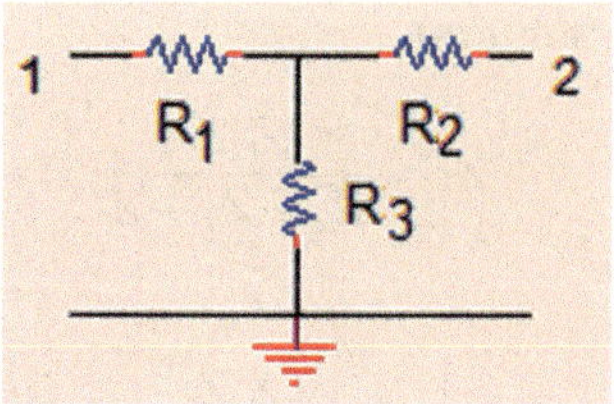

Fig. 9.3 Circuit of Problem 9.3 for the calculation of z-parameters

Solution

$$z_{11} = \text{input impedance} = R_1 + R_3 = 50 + 70 = 120\,\Omega$$
$$z_{22} = \text{input impedance} = R_2 + R_3 = 60 + 70 = 130\,\Omega$$
$$z_{12} = z_{21} = \text{open circuit transfer impedance} = R_3 = 70\,\Omega$$
$$z = \begin{bmatrix} 120 & 70 \\ 70 & 130 \end{bmatrix}\Omega.$$

Problem 9.4 Calculate z-parameters of the circuit as shown in Fig. 9.4, using pi to $T\ (\Delta - Y)$ network transformation, with

$$Z_a = R,\ Z_b = sL,\ Z_c = \frac{1}{sC}$$

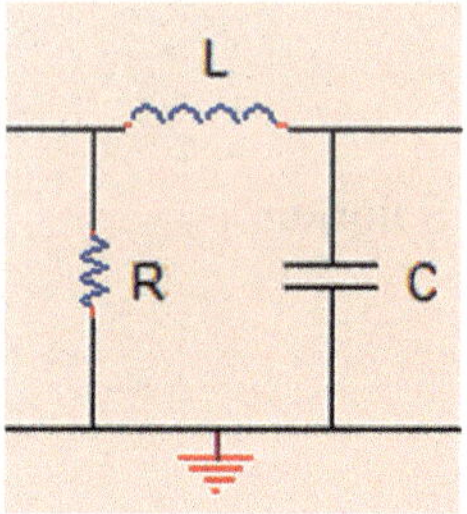

Fig. 9.4 pi to $T(\Delta - Y)$ network transformation

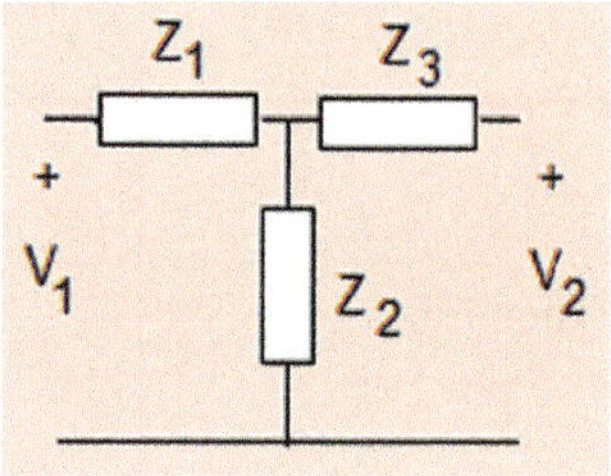

Fig. 9.5 T circuit

Solution

$$Z_1 = \frac{Z_a Z_b}{Z_a + Z_b + Z_c} = \frac{R \times sL}{R + sL + \frac{1}{sC}} = \frac{s^2 RLC}{sRC + s^2 LC + 1}$$

$$Z_2 = \frac{Z_a Z_c}{Z_a + Z_b + Z_c} = \frac{R \times \frac{1}{sC}}{R + sL + \frac{1}{sC}} = \frac{R}{sRC + s^2 LC + 1}$$

$$Z_3 = \frac{Z_b Z_c}{Z_a + Z_b + Z_c} = \frac{sL \times \frac{1}{sC}}{R + sL + \frac{1}{sC}} = \frac{\frac{\frac{L}{C}}{sRC + s^2 LC + 1}}{sC} = \frac{sL}{sRC + s^2 LC + 1}$$

$$z_{11} = Z_1 + Z_2 = \frac{s^2 RLC + R}{s^2 LC + sRC + 1} = \frac{R(s^2 LC + 1)}{s^2 LC + sRC + 1}$$

$$z_{22} = Z_2 + Z_3 = \frac{R + sL}{s^2 LC + sRC + 1}$$

$$z_{12} = z_{21} = Z_2 = \frac{R}{s^2 LC + sRC + 1}.$$

(Fig. 9.5)

Problem 9.5 Determine z-parameters of a circuit, the y-parameters of which are given as

$$\boldsymbol{y} = \begin{bmatrix} s & -1/2 \\ -1/2 & 2 \end{bmatrix}.$$

Solution

$$z_{11} = \frac{y_{22}}{\Delta y} = \frac{2}{\left(2s - \frac{1}{4}\right)} = \frac{8}{8s - 1}$$

$$z_{12} = z_{21} = \frac{-y_{12}}{\Delta y} = \frac{1/2}{\left(2s - \frac{1}{4}\right)} = \frac{2}{8s - 1}$$

$$z_{22} = \frac{y_{11}}{\Delta y} = \frac{s}{\left(2s - \frac{1}{4}\right)} = \frac{4s}{8s - 1}.$$

Therefore,

$$[z] = \begin{bmatrix} \frac{8}{8s - 1} & \frac{2}{8s - 1} \\ \frac{2}{8s - 1} & \frac{4s}{8s - 1} \end{bmatrix}.$$

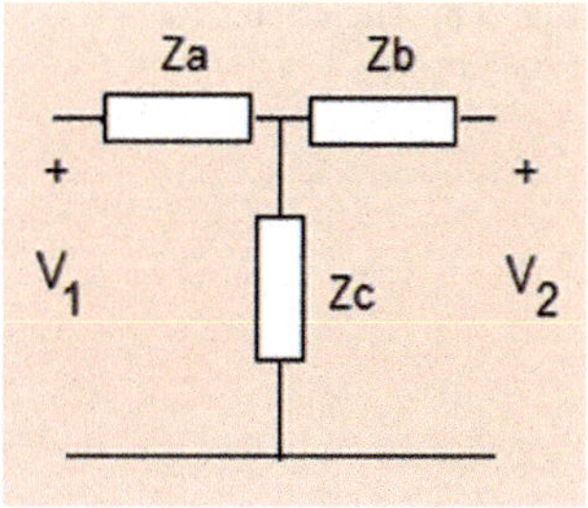

Fig. 9.6 *T*(*Y*) equivalent circuit

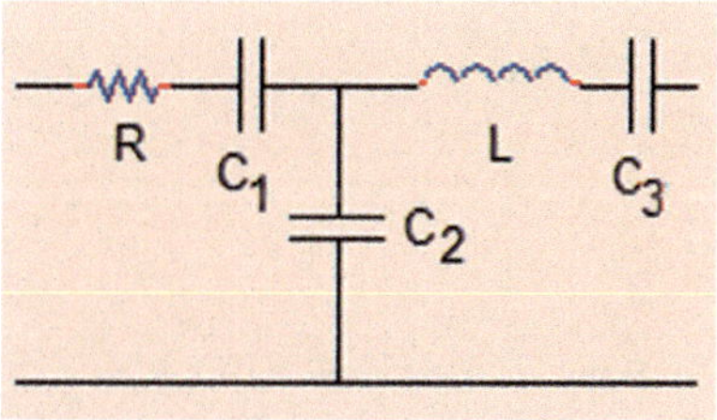

Fig. 9.7 The circuit for Problem 9.6

Problem 9.6 Synthesize a two-port circuit that realizes following z-parameters.

$$[z] = \begin{bmatrix} 1+\frac{6}{s} & \frac{2}{s} \\ \frac{2}{s} & s+\frac{3}{s} \end{bmatrix}.$$

Solution

Since z-matrix is reciprocal $(z_{12} = z_{21})$, *T*(*Y*) equivalent circuit can be constructed (Fig. 9.6).

$$Z_a = z_{11} - z_{12} = 1 + \frac{6}{s} - \frac{2}{s} = 1 + \frac{4}{s}$$
$$Z_b = z_{22} - z_{12} = s + \frac{3}{s} - \frac{2}{s} = s + \frac{1}{s}$$
$$Z_c = z_{12} = z_{21} = \frac{2}{s}.$$

The circuit of Fig. 9.7 can be constructed using following component values:

$$R_1 = 1\,\Omega,\ C_1 = \frac{1}{4}\,\mathrm{F},\ C_2 = \frac{1}{2}\,\mathrm{F},\ C_3 = 1\,\mathrm{F},\ L = 1\,\mathrm{H}.$$

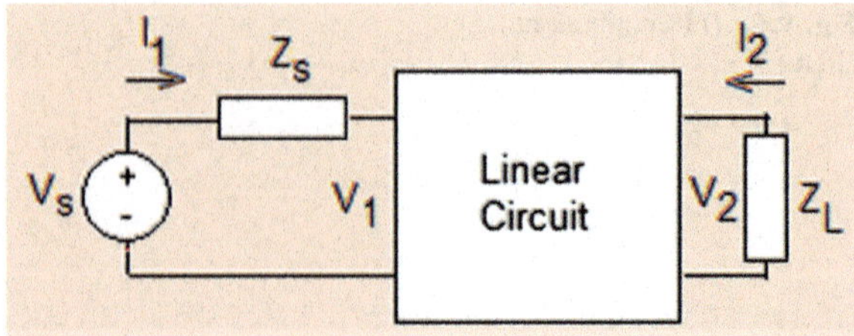

Fig. 9.8 The circuit for Problem 9.7

Problem 9.7 Thevenin equivalent of the circuit at the output terminal (without Z_L) is as follows (Fig. 9.8):

$$Z_{\text{Th}} = z_{22} - \frac{z_{12} \times z_{21}}{z_{11} + Z_s}$$

$$V_{\text{Th}} = \frac{z_{21}}{z_{11} + Z_s} V_s$$

For

$$[Z] = \begin{bmatrix} 10 & 20 \\ 30 & 200 \end{bmatrix} \Omega, \quad Z_s = R_s = 50\,\Omega, \quad V_s = 760\,\text{V(rms)}.$$

Find the load impedance for maximum power transfer and power delivered to the load.

Solution

Since the circuit is purely resistive, maximum power transfer condition is fulfilled if

$$Z_{\text{th}} = Z_L = R_L$$

$$Z_{\text{Th}} = z_{22} - \frac{z_{12} \times z_{21}}{z_{11} + Z_s} = 200 - \frac{20 \times 30}{10 + 50} = 200 - \frac{600}{60} = 190\,\Omega$$

$$P_{\max} = \left(\frac{V_{\text{th}}}{2R_{\text{th}}}\right)^2 \times R_{\text{th}} = \frac{V_{\text{th}}^2}{4R_{\text{th}}} = \frac{(760)^2}{4 \times 190} = 760\,\text{W}.$$

Problem 9.8 Use SPICE, and obtain z-parameters for the circuits in Fig. 9.9. (z1.cir), (z2.cir).

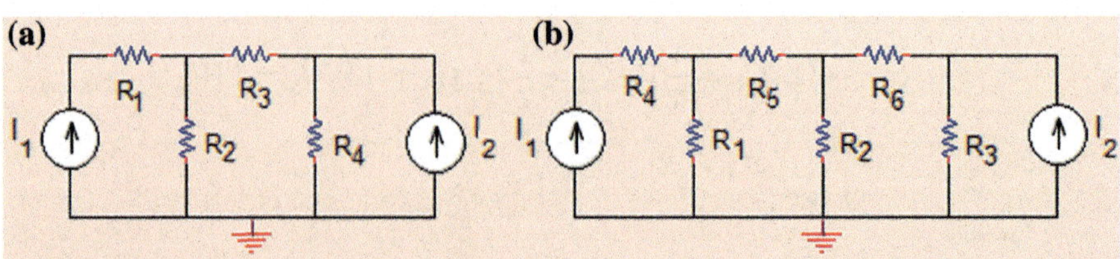

Fig. 9.9 The circuits for Problem 9.8

$$R_1 = 4\,\Omega,\ \ R_2 = 8\,\Omega,\ \ R_3 = 10\,\Omega,\ \ R_4 = 2\,\Omega,\ \ R_5 = 6\,\Omega,\ \ R_6 = 12\,\Omega.$$

Solution

The z-parameters z_{11}, z_{21} are obtained by applying a current source I_1 at the input and measuring input voltage V_1 and V_2, and then calculating

$$z_{11} = \frac{V_1}{I_1} \text{ and } z_{21} = \frac{V_2}{I_1},$$

while z_{22}, z_{12} are obtained by applying a current source I_2 at the output and measuring output voltages V_2 and V_1, and then calculating

$$z_{22} = \frac{V_2}{I_2} \text{ and } z_{12} = \frac{V_1}{I_2}.$$

Applying SPICE to the given circuits yields the following results:

(a) $[z] = \begin{bmatrix} 8.8 & 0.8 \\ 0.8 & 1.8 \end{bmatrix}$

(b) $[z] = \begin{bmatrix} 4.9916 & 0.6723 \\ 0.6723 & 6.2185 \end{bmatrix}.$

Following are SPICE netlists for z-parameters testing (z1.cir and z2.cir).

```
z parameters testing (z1.cir)
*first run (op analysis) for i1=1A,then run for i2=1A
*results are (z11, z21=z12) and then z22
i1 0 1 1
*i2 0 3 1
R1 1 2 4
R2 2 0 8
R3 2 3 10
R4 3 0 2
```

```
z parameters testing (z2.cir)
*first run (op analysis) for i1=1A,then run for i2=1A
*results are (z11, z21=z12) and then z22
*i1 0 1 1
i2 0 4 1
R1 2 0 4
R2 3 0 8
R3 4 0 10
R4 1 2 2
R5 2 3 6
R6 3 4 12
```

Problem 9.9 Synthesize two-port LC circuit such that

$$y_{22} = \frac{s(s^2+3s)}{(s^2+2)(s^2+4)}.$$

Solution

$$y_{22}(s) = \frac{s^3+3s}{s^4+6s^2+8}$$
$$Z_A = \frac{1}{y_{22}} = \frac{s^4+6s^2+8}{s^3+3s} = s + \frac{3s^2+8}{s^3+3s} = sL_1 + Z_B$$
$$L_1 = 1\,\mathrm{H}$$

(Fig. 9.10)

$$Y_B = \frac{1}{Z_B} = \frac{s^3+3s}{3s^2+8} = \frac{1}{3}s + \frac{\frac{1}{3}s}{3s^2+8} = sC_2 + Y_C$$
$$C_2 = 1\,\mathrm{F}$$

(Fig. 9.11)

$$Z_C = \frac{1}{Y_C} = \frac{3s^2+8}{\frac{1}{3}s} = 9s + \frac{24}{s}$$
$$L_3 = 9\,\mathrm{H}$$
$$C_4 = \frac{1}{24}\,\mathrm{F}.$$

Synthesized circuit is shown in Fig. 9.12.

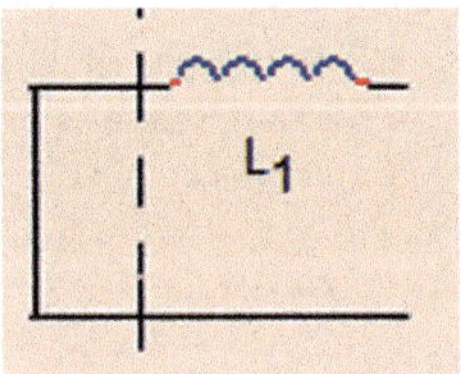

Fig. 9.10 First step of two-port LC circuit synthesis

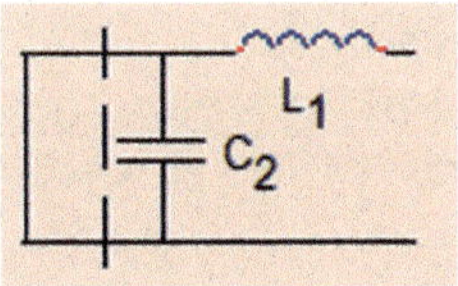

Fig. 9.11 Second step of two-port LC circuit synthesis

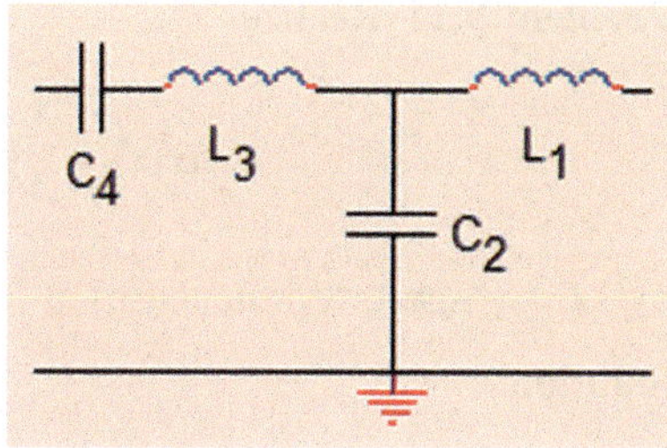

Fig. 9.12 Synthesized circuit

Problem 9.10 Design a second-order LC low-pass Butterworth filter circuit terminated with 1 Ω resistor. Determine the input impedance of resistively terminated circuit in terms of its voltage transfer function.

Solution

Normalized filter VTF is

$$H(s) = \frac{1}{s^2 + \sqrt{2}s + 1}.$$

Model circuit is shown in Fig. 9.13 (Second-order Butterworth LPF).

$$H(s) = \frac{z_{12}}{1 + z_{11}} = -\frac{y_{21}}{1 + y_{22}} = \frac{\frac{N_e}{D_o}}{1 + \frac{N_e}{D_o}} = \frac{\frac{1}{\sqrt{2}s}}{1 + \frac{s^2+1}{\sqrt{2}s}}.$$

Since all transmission zeros of $H(s)$ are at $s = \infty$, Cauer's first form can be used to realize y_{22},

$$y_{22} = \frac{s^2+1}{\sqrt{2}s} = \frac{1}{\sqrt{2}}s + \frac{1}{\sqrt{2}s} = Cs + \frac{1}{Ls} \rightarrow C = \frac{1}{\sqrt{2}}\,\mathrm{F}, L = \sqrt{2}\,\mathrm{H}$$

$$Z(s) = sL + \frac{R}{1 + sRC} = \frac{s^2L + s\frac{L}{RC} + \frac{1}{C}}{s + \frac{1}{RC}} = \frac{\sqrt{2}s^2 + 2s + \sqrt{2}}{s + \sqrt{2}} = \frac{\sqrt{2}(s^2 + \sqrt{2}s + 1)}{s + \sqrt{2}}$$

$$Z(s) = \frac{\sqrt{2}}{(s + \sqrt{2})} \cdot \frac{1}{H(s)}.$$

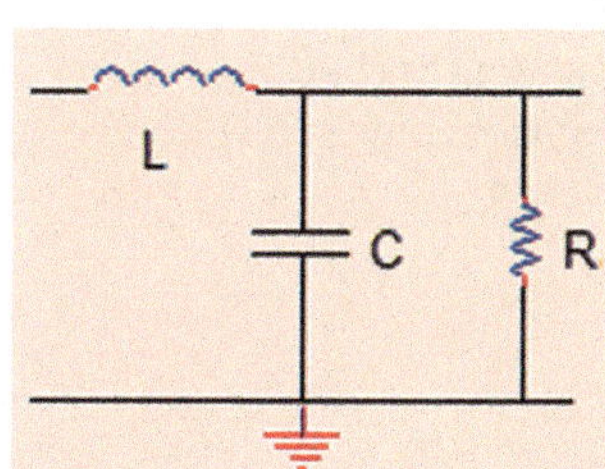

Fig. 9.13 Low-pass Butterworth filter circuit

Problem 9.11 Realize

$$H(s) = \frac{V_0(s)}{V_i(s)} = \frac{10^6}{s^2 + 10^6 s + 10^9},$$

by LC two-port, terminated by a 50 Ω load.

Solution

Let LC two-port terminated by a 1 Ω load (resistance scaling):

$$H(s) = \frac{M_1(s)}{M_2(s) + N_2(s)},$$

where M and N are even and odd parts of numerator and denominator polynomials. Since all transmission zeros of $H(s)$ are at $s = \infty$, Cauer's first form can be used to realize y_{22}

$$y_{22} = \frac{M_2}{N_2} = \frac{s^2 + 10^9}{10^6 s} = 10^{-6}s + \frac{10^3}{s} = sC + \frac{1}{sL}$$
$$C = 10^{-6}\,\mathrm{F}, L = 1\,\mathrm{mH}.$$

By unscaling,

$$R_L = 50 \times 1 = 50\,\Omega,\;\; L = 50\,\mathrm{mH},\;\; C = \frac{10^{-6}}{50}F = 20\,\mathrm{nF}.$$

The circuit is shown in Fig. 9.14.

Problem 9.12 Darlington showed that any positive real function can be realized as the impedance of a lossless (LC) network terminated in a single resistor [1]:

$$Z(s) = z_{11} \cdot \frac{R + \frac{1}{y_{22}}}{R + z_{22}}.$$

Determine the input impedance of second-order LC low-pass Butterworth filter circuit terminated with R = 1 Ω resistor (Fig. 9.15).

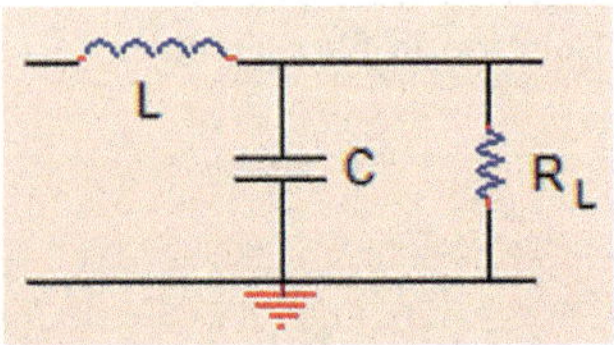

Fig. 9.14 LC two-port, terminated by a 50 Ω load resistor

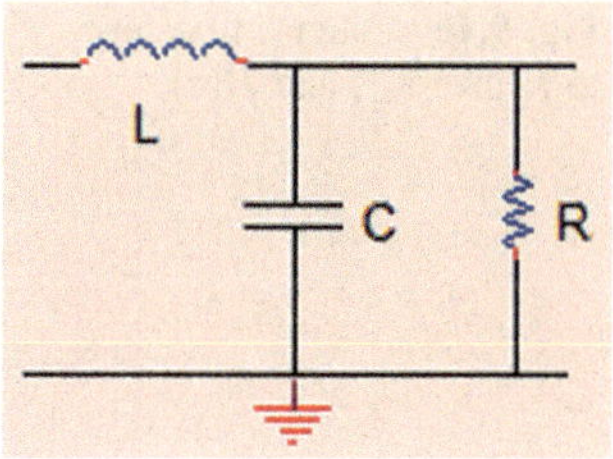

Fig. 9.15 Second-order LC low-pass filter terminated with R = 1 Ω

$$C = \frac{1}{\sqrt{2}}\,\text{F}, \quad L = \sqrt{2}\,\text{H}\cdot$$

Solution

Two-port parameters of LC circuit (without termination resistor) are

$$z_{11} = \frac{s^2LC + 1}{sC} = \sqrt{2} \cdot \frac{s^2+1}{s}, \quad y_{22} = \frac{s^2+1}{\sqrt{2}s}, \quad z_{22} = \frac{1}{sC} = \frac{\sqrt{2}}{s}.$$

Substituting these in Darlington's input impedance equation yields

$$Z(s) = z_{11} \cdot \frac{1 + \frac{1}{y_{22}}}{1 + z_{22}} = \sqrt{2} \cdot \frac{s^2+1}{s} \frac{1 + \frac{1}{\frac{s^2+1}{\sqrt{2}s}}}{1 + \frac{\sqrt{2}}{s}} = \frac{\sqrt{2}(s^2 + \sqrt{2}s + 1)}{s + \sqrt{2}}.$$

Problem 9.13 Synthesize the two-port RC VTF,

$$H(s) = \frac{k}{(s+a)(s+b)},$$

with $a = 3, b = 5\cdot$ (cauerRC1.m)

Solution

$$H(s) = \frac{z_{21}}{z_{11}} = \frac{k}{(s+a)(s+b)} \tag{9.1}$$

Select z_{11} as

$$z_{11} = \frac{(s+a)(s+b)}{(s+c)(s+d)}, \quad c < a < d < b. \tag{9.2}$$

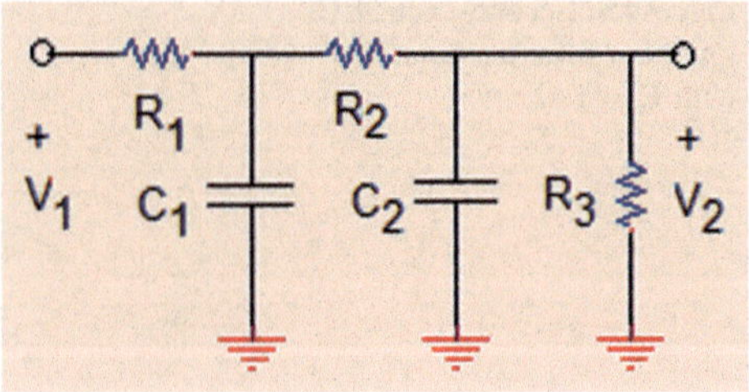

Fig. 9.16 Cauer's first form to realize z_{11}, and $H(s)$

Then in (9.1), one can identify z_{21} as

$$z_{21} = \frac{k}{(s+c)(s+d)}.$$

Using Cauer's first form (continued fraction expansion) to realize z_{11} will yield $H(s)$, and letting $c = 1, d = 4$ in Eq. (9.2):

$$z_{11}(s) = \frac{(s+3)(s+5)}{(s+1)(s+4)} = \frac{s^2+8s+15}{s^2+5s+4}.$$

The circuit in Fig. 9.16 (with the computed component values) is obtained using Cauer's first form (CauerRC1.m) to realize z_{11}, and thus $H(s)$:

$$R_1 = 1\,\Omega,\;\; C_1 = \frac{1}{3}\,\mathrm{F},\;\; R_2 = \frac{9}{4}\,\Omega,\;\; C_2 = \frac{2}{3}\,\mathrm{F},\;\; R_3 = \frac{1}{2}\,\Omega.$$

Note that the output voltage V_2 is taken across the last resistor element R_3.

Problem 9.14 Realize following voltage transfer functions (VTFs) by lossless circuits each terminated by R = 1 Ω load resistors.

(a) $H(s) = \dfrac{V_2(s)}{V_1(s)} = \dfrac{1}{3s^3+3s^2+4s+1}$,

(b) $H(s) = \dfrac{V_2(s)}{V_1(s)} = \dfrac{1}{3s^4+3s^3+5s^2+4s+1}$,

(c) $H(s) = \dfrac{V_2(s)}{V_1(s)} = \dfrac{1}{(s+3)^3}$,

(d) $H(s) = \dfrac{V_2(s)}{V_1(s)} = \dfrac{1}{s^3+\frac{1}{2}s^2+3s+1}$.

Solution

(a) $$H(s) = \frac{1}{3s^3+3s^2+4s+1} = \frac{m_1+n_1}{m_2+n_2} = \frac{m_1}{m_2+n_2} = \frac{\dfrac{m_1}{n_2}}{1+\dfrac{m_2}{n_2}} = \frac{-y_{21}}{1+y_{22}}$$

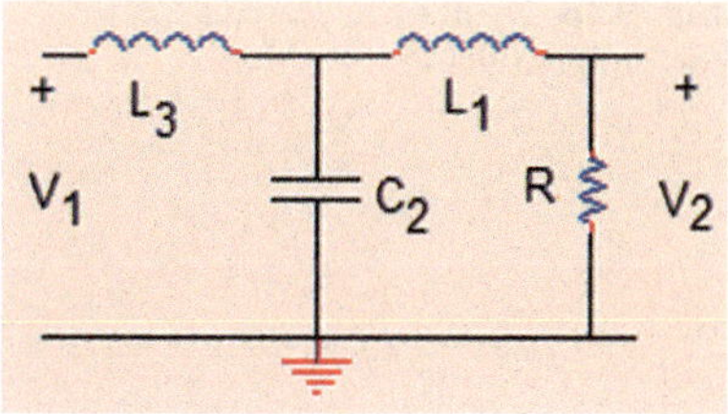

Fig. 9.17 Synthesized circuit for Problem 9.14a

$$y_{22} = \frac{m_2}{n_2} = \frac{3s^2+1}{3s^3+4s}.$$

All transmission zeros of $H(s)$ are at $s = \infty$, y_{22} can be realized using Cauer's first form.

Since denominator polynomial of y_{22} has a degree higher than the degree of numerator of y_{22},

$$Z = \frac{1}{y_{22}} = \frac{3s^3+4s}{3s^2+1} = s + \cfrac{1}{s + \cfrac{1}{3s}}.$$

Because y_{22} (and Z) "looks" from the output of "LC circuit" to the LC circuit, first element is $L_1 = 1$ H; then $C_2 = 1$ F, and $L_3 = 3$ H. Realized circuit is shown in Fig. 9.17.

$$\text{(b)} \quad H(s) = \frac{V_2(s)}{V_1(s)} = \frac{1}{3s^4+3s^3+5s^2+4s+1} = \frac{m_1}{m_2+n_2} = \frac{\frac{m_1}{n_2}}{1+\frac{m_2}{n_2}} = \frac{-y_{21}}{1+y_{22}}$$

$$y_{22} = \frac{m_2}{n_2} = \frac{3s^4+5s^2+1}{3s^3+4s}.$$

All transmission zeros of $H(s)$ are at $s = \infty$, and $y_{22}(s)$ can be realized using Cauer's first form:

$$y_{22}(s) = s + \cfrac{1}{3s + \cfrac{1}{s + \cfrac{1}{s}}}$$

$$C_1 = 1\,\text{F}, L_2 = 3\,\text{H}, C_3 = 1\,\text{F}, L_4 = 1\,\text{H}.$$

Realized circuit is shown in Fig. 9.18.

Fig. 9.18 Synthesized circuit for Problem 9.14b

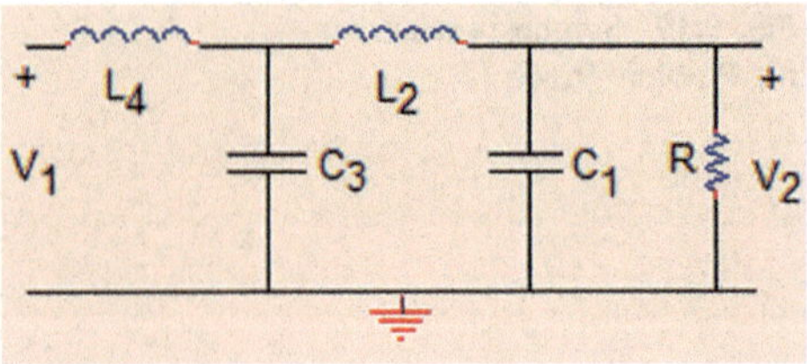

(c) $H(s)=\dfrac{1}{s^3+3s^2+3s+1}=\dfrac{m_2}{m_2+n_2}$.

Since numerator of $H(s)$ is an even function,

$$H(s)=\frac{\dfrac{m_1}{n_2}}{1+\dfrac{m_2}{n_2}}=\frac{-y_{21}}{1+y_{22}}$$

$$y_{22}=\frac{m_2}{n_2}=\frac{3s^2+1}{s^3+3s}.$$

Since the degree of denominator of y_{22} is higher than the degree of its numerator, by Cauer's first form

$$Z=\frac{1}{y_{22}}=\frac{s^3+3s}{3s^2+1}=\frac{1}{3}s+\cfrac{1}{\frac{9}{8}s+\cfrac{1}{\frac{8}{3}s}}$$

$$L_1=\frac{1}{3}\text{ H}, C_2=\frac{9}{8}\text{ F}, L_3=\frac{8}{3}\text{ H}.$$

Realized circuit has the same topology as shown in part a.

(d) $H(s)=\dfrac{V_2(s)}{V_1(s)}=\dfrac{1}{s^3+\frac{1}{2}s^2+3s+1}=\dfrac{m_1+n_2}{m_2+n_2}=\dfrac{1+0}{\left(\frac{1}{2}s^2+1\right)+(s^3+3s)}$

$$y_{22}=\frac{m_2}{n_2}=\frac{\frac{1}{2}s^2+1}{s^3+3s}.$$

Since the degree of denominator of y_{22} is higher than the degree of its numerator, by Cauer's first form:

Fig. 9.19 The circuit for Problem 9.15

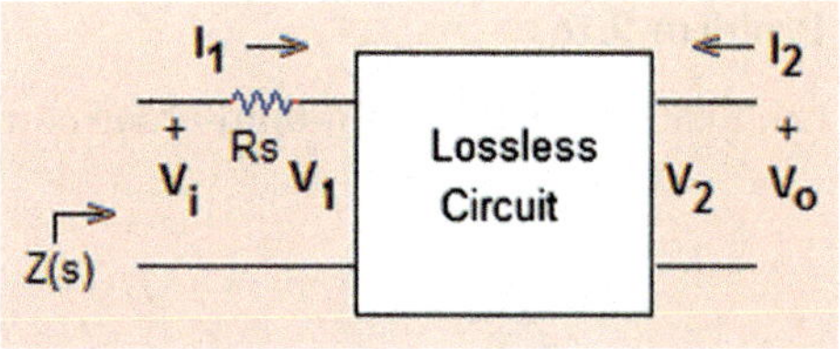

$$Y = \frac{\frac{1}{2}s^2 + 1}{s^3 + 3s} \rightarrow Z = \frac{s^3 + 3s}{\frac{1}{2}s^2 + 1} = \frac{2s^3 + 6s}{1s^2 + 2} = 2s + \frac{1}{\frac{1}{2}s + \frac{1}{s}}$$

$$L_1 = 2\,\mathrm{H}, C_2 = \frac{1}{2}\,\mathrm{F}, L_3 = 1\,\mathrm{H}.$$

Realized circuit has the same topology as shown in part a.

Problem 9.15 For a lossless two-port network terminated by a source resistor shown in Fig. 9.19

(a) Determine the voltage transfer function (between output voltage and the input voltage) in terms of source resistance and two z-parameters of lossless two-port.
(b) Determine the relationship between input impedance and VTF of the network.

Solution

(a) Dropping s-operator for convenience,

$$V_0 = V_2$$
$$V_i = R_s I_1 + V_1$$

$$H = \frac{V_0}{V_i} = \frac{V_2}{R_s I_1 + V_1} = \frac{\frac{V_2}{I_1}}{R_s + \frac{V_1}{I_1}} = \frac{z_{12}}{R_s + z_{11}}$$

(b) $Z = R_s + z_{11} = \frac{V_i}{I_1}$

$$H = \frac{V_0}{V_i} \rightarrow V_i = \frac{V_0}{H} \rightarrow Z = \frac{\frac{V_0}{H}}{I_1} = \frac{V_2}{I_1} \cdot \frac{1}{H} = z_{21} \cdot \frac{1}{H} = \frac{z_{21}}{H}.$$

Problem 9.16

(a) Realize the VTF of a third-order normalized filter

$$H(s) = \frac{k}{B_3(s)},$$

where $B_3(s)$ is a third-order Bessel polynomial, and $k = 25/2$. Use a lossless circuit, source terminated by a $R_s = 1\ \Omega$ resistor (cfe.m).

(b) Is this a Bessel type low-pass filter?

Solution

(a) $B_3(s) = s^3 + 6s^2 + 15s + 15.$

Thus,

$$H(s) = \frac{12.5}{s^3 + 6s^2 + 15s + 15} = \frac{m_1}{m_2 + n_2} = \frac{\dfrac{m_1}{n_2}}{\dfrac{m_2}{n_2} + 1} = \frac{z_{12}}{1 + z_{11}}.$$

All transmission zeros of $H(s)$ are at $s = \infty$, Cauer's first form can be used to realize z_{11}.

$$z_{11} = \frac{m_2}{n_2} = \frac{6s^2 + 15}{s^3 + 15s} = \frac{1}{6}s + \frac{1}{\dfrac{12}{25}s + \dfrac{1}{\dfrac{5}{6}s}}, \quad L_1 = \frac{1}{6}\ \text{H}, \quad C_2 = \frac{12}{25}\ \text{F}, \quad L_3 = \frac{5}{6}\ \text{H}.$$

Realized circuit is shown in Fig. 9.20.

(b) This is not a perfect Bessel LPF, since $B(0) = 15 \neq k$, where $B(0) = 15$, $k = 12.5$.

For a third-order Bessel LPF, VTF would be

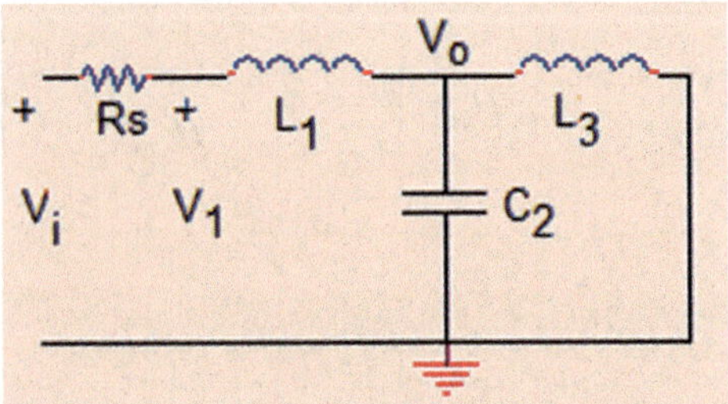

Fig. 9.20 The realized circuit for Problem 9.16

$$H_3(s) = \frac{15}{B_3(s)}.$$

Problem 9.17

(a) Why h-parameters are called hybrid parameters? Draw the h-parameter equivalent network of a two-port network.
(b) What are the z–z and h–h relationships for reciprocal networks?

Solution

(a) The z- and y-parameters of the part networks sometimes do not exist. For example, ideal transformer does not have z-parameters. This third set of parameters is called hybrid, since they are a hybrid combination of ratios:

$$\boldsymbol{h}_{11} = \left.\frac{\boldsymbol{V}_1}{\boldsymbol{I}_1}\right|_{V_2=0}, \boldsymbol{h}_{12} = \left.\frac{\boldsymbol{V}_1}{\boldsymbol{V}_2}\right|_{I_1=0}, \boldsymbol{h}_{21} = \left.\frac{\boldsymbol{I}_2}{\boldsymbol{I}_1}\right|_{V_2=0}, \boldsymbol{h}_{22} = \left.\frac{\boldsymbol{I}_2}{\boldsymbol{V}_2}\right|_{I_1=0},$$

where

h_{11} is short-circuit input impedance
h_{12} is open-circuit reverse voltage gain
h_{21} is short-circuit forward current gain
h_{22} is open-circuit output admittance.

(b) For reciprocal networks, $h_{12} = -h_{12}$ and $z_{12} = z_{21}$.

Problem 9.18 Determine the h-parameters for the circuit of Fig. 9.3 from its z-parameters.

$$(R_1 = 50\,\Omega,\ R_2 = 60\,\Omega,\ R_3 = 70\,\Omega).$$

Solution

$$[z] = \begin{bmatrix} 120 & 70 \\ 70 & 130 \end{bmatrix} \Omega$$

$$h_{11} = \frac{\Delta_z}{z_{22}},\ h_{12} = \frac{z_{12}}{z_{22}},\ h_{21} = -\frac{z_{21}}{z_{22}},\ h_{22} = \frac{1}{z_{22}}$$

$$\Delta_z = z_{11}z_{22} - z_{12}z_{21} = (120)(130) - (70)^2 = 15600 - 4900 = 10700$$

$$h_{11} = \frac{10700}{130} = 82.31\,\Omega,\ h_{12} = \frac{70}{130} = 0.538$$

$$h_{21} = -\frac{70}{130} = -0.538,\ h_{21} = \frac{1}{130} = 0.0077\,\text{S}.$$

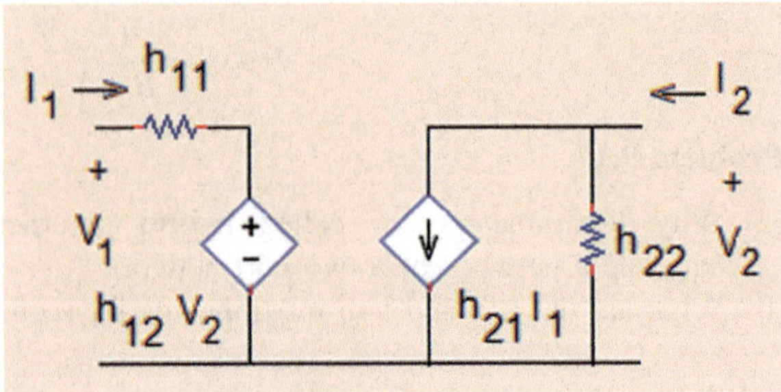

Fig. 9.21 The h-parameter equivalent network of a two-port network

Problem 9.19 A common emitter bipolar junction transistor configuration is represented by the equivalent circuit of Fig. 9.21. Manufacturer datasheet specifies

$$h_{ie} = h_{11} = 1\,\mathrm{k\Omega},\ \ h_{re} = h_{12} = 10^{-4},\ \ h_{fe} = h_{21} = 100,\ \ h_{oe} = h_{22} = 10\,\mu\mathrm{S}.$$

A signal source with $1\,\mathrm{k\Omega}$ output impedance (in series with $h_{ie} = h_{11}$) supplies $1\,\mathrm{mV}$ sinusoidal signal to the input of the transistor (base). A $10\,\mathrm{k\Omega}$ load resistor is connected to the collector of the transistor (in parallel to $h_{oe} = h_{22}$).

(a) Determine I_1, I_2, V_1, V_2.
(b) Determine voltage gain, (V_2/V_1).
(c) Determine the current gain, (I_2/I_1).
(d) Determine the input impedance of the circuit, (V_1/I_1).

Solution

(a) For the input circuit,

$$V_s = (R_s + h_{ie})I_1 + h_{re}V_2$$

$$10^{-3} = (1000 + 1000)I_1 + 10^{-4}\,V_2 \rightarrow 10^{-3} = 2000I_1 + 10^{-4}V_2.$$

For the output circuit,

$$I_2 = -\frac{V_2}{R_L} = h_{fe}I_1 + h_{oe}V_2$$
$$I_2 = -10^{-4}V_2 = 100I_1 + 10 \times 10^{-6}V_2 \rightarrow 0 = 100I_1 + \left(10^{-4} + 10^{-5}\right)V_2$$

or in matrix equation,

$$\begin{bmatrix} 10^{-3} \\ 0 \end{bmatrix} = \begin{bmatrix} 2000 & 10^{-4} \\ 100 & 0.00011 \end{bmatrix}\begin{bmatrix} I_1 \\ V_2 \end{bmatrix}.$$

Solution of this equation yields

$$I_1 = 0.52\,\mu\text{A}, V_2 = -0.476\,\text{V} = -476\,\text{mV}.$$

Other two h-parameters are

$$I_2 = -\frac{V_2}{R_L} = -10^{-4} \times (-0.476) = 47.6 \times 10^{-6}\,\text{A} = 47.6\,\mu\text{A}$$

$$V_1 = I_1 h_{11} + h_{12} V_2 = 0.52 \times 10^{-6} \times 10^3 + 10^{-5}(-0.476) = 0.515\,\text{mV}$$

(b) $A_V = \dfrac{V_2}{V_1} = -\dfrac{476}{0.515} = -923.8\,(\text{V/V})$

(c) $A_I = \dfrac{I_2}{I_1} = \dfrac{47.6}{0.52} = 91.54\,(\text{A/A})$

(d) $R_i = \dfrac{V_1}{I_1} = \dfrac{0.515\,\text{mV}}{0.52\,\mu\text{A}} = 990\,\Omega.$

Problem 9.20 The h-parameters of a bipolar junction transistor (BJT) in common emitter (CE) configuration are simplified as $h_{\text{oe}} = h_{\text{re}} = 0$ and $h_{\text{fe}} = 100, \quad h_{\text{ie}} = 1\,\text{k}\Omega$.

(a) Design an amplifier (i.e., determine the value of load resistor) so that the voltage gain $A_v = v_0/v_i$ is (-50).
(b) Design the amplifier if BJT is connected to signal source having an output resistance of $R_s = 10\,\text{k}\Omega$, and amplifier voltage gain $A_{vs} = v_0/v_s$ is to be (-50).

Solution

(a) $h_{\text{ie}} = h_{11}, h_{\text{fe}} = h_{21}, h_{\text{oe}} = h_{\text{re}} = 0 = h_{22} = h_{12}$

$$v_0 = -h_{\text{fe}} i_b = R_L, V_i = h_{\text{ie}} i_b$$

$$A_V = \frac{v_0}{v_i} = \frac{-h_{\text{fe}} i_b R_L}{h_{\text{ie}} i_b} = -\frac{h_{\text{fe}}}{h_{\text{ie}}} R_L.$$

Substituting values of parameters

$$-50 = -\frac{100}{1000} \times R_L = -0.1 R_L \rightarrow R_L = \frac{50}{0.1} = 500\,\Omega$$

(b) $V_0 = -h_{\text{fe}} i_b R_L, v_s = (R_s + h_{\text{ie}}) i_b$

$$A_{\text{vs}} = \frac{v_0}{v_s} = \frac{h_{\text{fe}} i_b R_L}{(R_s + h_{\text{ie}}) i_b} = \frac{-h_{\text{fe}} R_L}{R_s + h_{\text{ie}}}.$$

Substituting the values,

$$-50 = \frac{-100R_L}{11000} \to 50 = \frac{R_L}{110} \to R_L = 5500\,\Omega.$$

Problem 9.21 A two-port network is tested as follows:

With the output short circuited and the input current is 1 mA, the output current is 0.1 mA and the output voltage is measured as 3 V.
With the input open circuited and a voltage of 1 V applied to the output, the output current is 0.02 mA and the input voltage is measured as 0.3 mV.

Determine the h-parameters of this network.

Solution

$$h_{11} = \left.\frac{V_1}{I_1}\right|_{V_2=0} = \frac{3\,\mathrm{V}}{1\,\mathrm{mA}} = 3\,\mathrm{k}\Omega$$

$$h_{21} = \left.\frac{I_2}{I_1}\right|_{V_2=0} = \frac{0.1\,\mathrm{mA}}{1\,\mathrm{mA}} = 0.1$$

$$h_{22} = \left.\frac{I_2}{V_2}\right|_{I_1=0} = \frac{0.02\,\mathrm{mA}}{1\,\mathrm{V}} = 0.02\,\mathrm{mS}$$

$$h_{12} = \left.\frac{V_1}{V_2}\right|_{I_1=0} = \frac{0.3\,\mathrm{mV}}{1\,\mathrm{V}} = 0.3 \times 10^{-3}.$$

Problem 9.22 Determine the y- and h-parameters of a two-port, if $[z]$ has the form of

$$(a)\begin{bmatrix} 1280 & 10 \\ 10 & 0 \end{bmatrix}, (b)\begin{bmatrix} 10 & 10 \\ 10 & 10 \end{bmatrix}, (c)\begin{bmatrix} 10 & 0 \\ 20 & 0 \end{bmatrix}, (d)\begin{bmatrix} 10 & 0 \\ 5 & 1 \end{bmatrix}, (e)\begin{bmatrix} 1027.2 & 0.51 \\ 1.3 & 12.53 \end{bmatrix}$$

(z_to_y_h.m).

Solution

$$y_{11} = \frac{z_{22}}{\Delta_z}, y_{12} = \frac{-z_{12}}{\Delta_z}, y_{22} = \frac{z_{11}}{\Delta_z}, y_{21} = \frac{-z_{21}}{\Delta_z}$$

$$h_{11} = \frac{\Delta_z}{z_{22}}, h_{12} = \frac{z_{12}}{z_{22}}, h_{21} = \frac{-z_{21}}{z_{22}}, h_{22} = \frac{1}{z_{22}}.$$

(a) $[y] = \begin{bmatrix} 0 & 0.1 \\ 0.1 & -12.8 \end{bmatrix}$, $[h]$ does not exist ($z_{22} = 0$)

(b) [y] does not exist $\left(\Delta_z = 0, [h] = \begin{bmatrix} 0 & 1 \\ -1 & 0.1 \end{bmatrix}\right)$,

(c) [y] does not exist ($\Delta_z = 0$), [h] does not exist, $z_{22} = 0$·

(d) $[y] = \begin{bmatrix} 0.1 & 0 \\ -0.5 & 1 \end{bmatrix}, \; h = \begin{bmatrix} 10 & 0 \\ -5 & 1 \end{bmatrix}$

(e) $[y] = \begin{bmatrix} 0.000974 & -0.0000396 \\ -0.000101 & 0.0798126 \end{bmatrix}, \; [h] = \begin{bmatrix} 1027.147 & 0.040702 \\ -0.10375 & 0.079808 \end{bmatrix}.$

MATLAB script (z_to_y_h.m) is given below.

```
z=[1027.2  0.51; 1.3  12.53]
format long
%format short
%format rat
d=det(z)
y11=z(2,2)/d; y22=z(1,1)/d; y12=-z(1,2)/d; y21=-z(2,1)/d;
y=[y11 y12;y21 y22]
h11=d/z(2,2); h22=1/z(2,2); h21=-z(2,1)/z(2,2); h12=z(1,2)/z(2,2);
h=[h11 h12; h21 h22]
```

Reference

1. Darlington S (1939) Synthesis of reactance 4-poles which produce prescribed insertion loss characteristics. J Math Phys 18:257–353

Chapter 10
Operational Amplifiers

10.1 Basic Op-amp Circuits

Problem 10.1.1

(a) Determine the $[z]$, $[y]$, and $[h]$ parameters of the inverting op-amp circuit.
(b) In an inverting op-amp configuration of Fig. 10.1, input resistor value is 1 kΩ, while the feedback resistor value is 10 kΩ. If $v_i(t) = 10\sin(5t)\,\text{mV}$, determine the output voltage.

Solution

(a) For an inverting op-amp circuit, $\boldsymbol{V}_1 = \boldsymbol{V}_i,\ \boldsymbol{V}_2 = \boldsymbol{V}_o,$

$$z_{11} = \frac{\boldsymbol{V}_1}{\boldsymbol{I}_1} = R_i, \quad z_{22} = \frac{\boldsymbol{V}_2}{\boldsymbol{I}_2} = 0,$$
$$z_{12} = \frac{\boldsymbol{V}_1}{\boldsymbol{I}_2} = 0, \quad z_{21} = \frac{\boldsymbol{V}_2}{\boldsymbol{I}_1} = -R_f,$$

$$[Z] = \begin{bmatrix} R_i & 0 \\ -R_f & 0 \end{bmatrix}$$

[y] does not exist for an invertin op-amp circuit, since $[z]^{-1}$ does not exist (determinant = 0). Since, $\boldsymbol{Z}_{22} = 0$, [h] does not exist for an inverting op-amp circuit.

(b) $v_o(t) = -\frac{R_f}{R_i} \cdot v_i = -10 \cdot 10\sin 5t = -100\sin 5t \quad (\text{mV})$

Problem 10.1.2 For the ideal inverting op-amp configuration, calculate output voltage, the current supplied by the voltage source to input resistor of 1 kΩ, and current magnitude through the feedback resistor with a value of 10 kΩ, if the input signal is 1 mV dc voltage.

A.Ü. Keskin, *Electrical Circuits in Biomedical Engineering*,
DOI 10.1007/978-3-319-55101-2_10

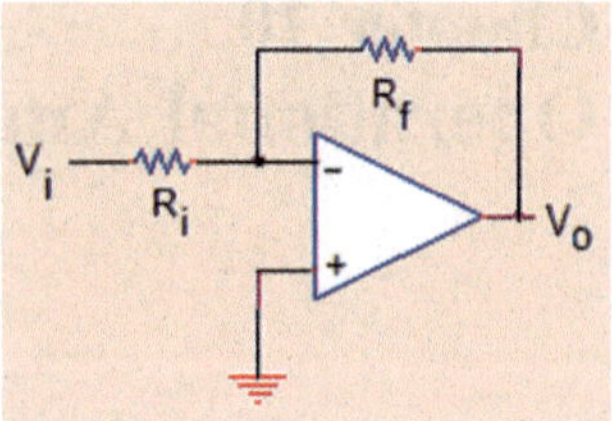

Fig. 10.1 An inverting op-amp configuration

Solution

$$V_o = -\frac{R_f}{R_i} \cdot V_i = -\frac{10}{1} \cdot 1\,\text{mV} = -10\,\text{mV}$$

Current through $1\,\text{k}\Omega$ resistor is

$$i = \frac{V_i}{R_i} = \frac{1\,\text{mV}}{1\,\text{k}\Omega} = 1\,\mu\text{A}$$

This is the same current flowing through the feedback resistor.

Problem 10.1.3 In the circuit shown in Fig. 10.2, determine the value of unknown resistor, R_x, in terms of V_i, V_O, and R. Calculate R_x if $V_i = 2.5\,\text{V}$, $V_O = 10\,\text{V}$, $R = 1\,\text{k}\Omega$. What is the current flowing through input resistor?

Solution
By inverting op-amp property;

$$V_O = -\frac{R}{R_x} V_i \quad \rightarrow \quad \frac{|V_O|}{|V_i|} = \frac{R}{R_x} \quad \rightarrow \quad R_x = \frac{|V_i|}{|V_O|} \times R$$

For given values of parameters,

$$R_x = \frac{2.5}{10} \times 1 = 0.25\,\text{k}\Omega = 250\,\Omega$$

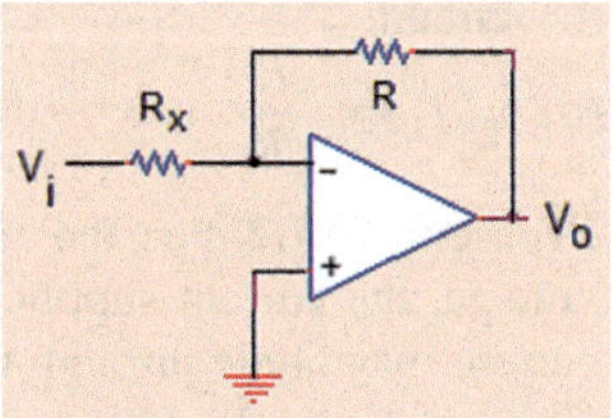

Fig. 10.2 Inverting amplifier

The current flowing through R_x (and R) is

$$i = \frac{V_0}{R} = \frac{10\,(\mathrm{V})}{1\,(\mathrm{k}\Omega)} = 10\,\mathrm{mA}$$

Problem 10.1.4 Two ideal and identical Negative Temperature Coefficient (NTC) temperature sensors (thermistors) are placed on the input and feedback path of an inverting op-amp configuration, see, Fig. 10.3. One of the sensors monitors ambient temperature, while the other one on the feedback path monitors a test chamber temperature. Determine the output voltage level relative to both temperature values, if $V_i = -1$ V.

Solution
For a negative temperature coefficient resistive sensor, the resistance, and temperature parameters are inversely proportional to each other. On the other hand, for the inverting op-amp,

$$V_0 = -V_i \frac{R_f}{R_i}$$

R_f is the resistance of NTC sensor on the feedback path, R_i is the resistance of NTC sensor on the input path. Therefore,

$$\begin{array}{lll} V_0 = 1\,\mathrm{V} & \text{if} \quad T_{\text{chamber}} = T_{\text{ambient}}, & R_f = R_i \\ V_0 > 1\,\mathrm{V} & \text{if} \quad T_{\text{chamber}} < T_{\text{ambient}}, & R_f > R_i \\ V_0 < 1\,\mathrm{V} & \text{if} \quad T_{\text{chamber}} > T_{\text{ambient}}, & R_f < R_i \end{array}$$

Problem 10.1.5

(a) Find out the expression $V_0 = f(V_1, V_2, V_3)$ for the circuit shown in Fig. 10.4.
(b) What happens if feedback resistor value R_f is greater than R?
(c) If $v_1 = \sin 2t$, $v_2 = -\sin 3t$, $v_3 = \sin 4t$, $V_0 =?$ Plot the signals. (Plot_time_function_1.m)

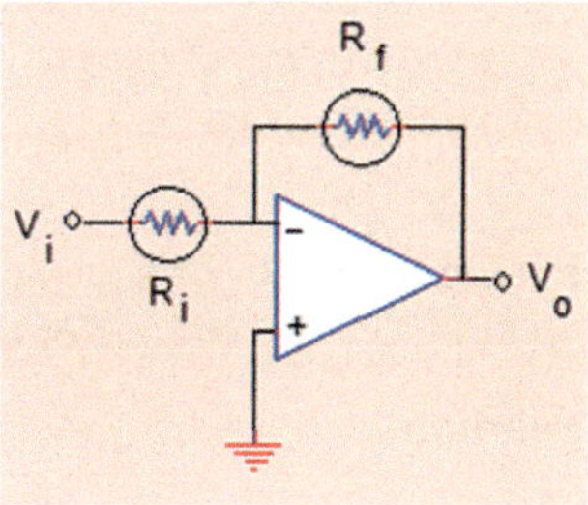

Fig. 10.3 Op-amp with two NTC thermistors

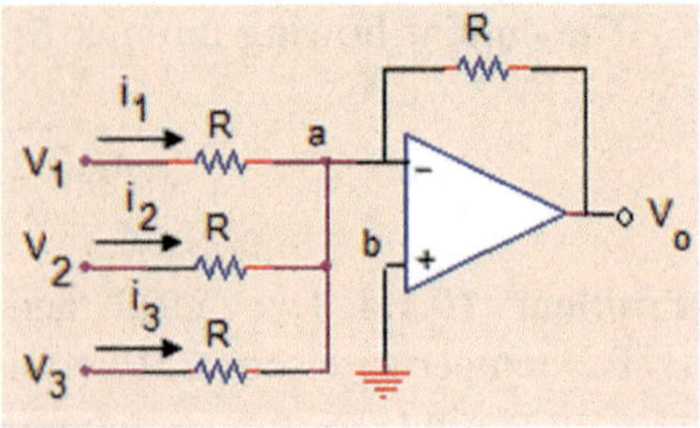

Fig. 10.4 Inverting summing amplifier circuit

Solution

(a) By KCL at node a

$$i_1 + i_2 + i_3 - i = \frac{1}{R}\left\{ \sum_{i=1}^{3} (V_i - V_a) - (V_a - V_o) \right\} = 0,$$

$$V_a = V_b = 0\,\text{V}, \quad \rightarrow \quad V_1 + V_2 + V_3 + V_o = 0 \quad \rightarrow \quad V_o = -\sum_{i=1}^{3} V_i$$

(b) This is a summer circuit with inverted output. If the feedback resistor = R_f > R there is amplification,

$$V_o = -\frac{R_f}{R}(V_1 + V_2 + V_3)$$

(c) *If* $v_1 = \sin 2t,\ v_2 = -\sin 3t,\ v_3 = \sin 4t,\ R_f = R$

$$v_o = -[\sin 2t - \sin 3t + \sin 4t] = -\sin 2t + \sin 3t - \sin 4t$$

The signals are plotted in Fig. 10.5.
A MATLAB script (Plot_time_function_1.m) is given below.

```
t=0:0.01:10;
v0=-sin(2*t)+sin(3*t)-sin(4*t);
plot(t,sin(2*t),'r',t,-sin(3*t),'g',t,sin(4*t),'k');hold on;
plot(t,v0,'linewidth',2);xlabel('t [s]');ylabel('Vo(t) [V]');
title('Vo(t)=-sin(2*t)+sin(3*t)-sin(4*t)');
legend('sin(2*t)','-sin(3*t)','sin(4*t)','v0(t)'); grid on;
```

Problem 10.1.6 Design an inverting summer circuit to provide an output $V_0 = -(4V_1 + V_2/3)$ and one of the input resistors has a value of R_b = 10 kΩ (Fig. 10.6).

Solution

$$\frac{R_a}{R_b} = 4, \quad \frac{R_a}{R_c} = \frac{1}{3}$$

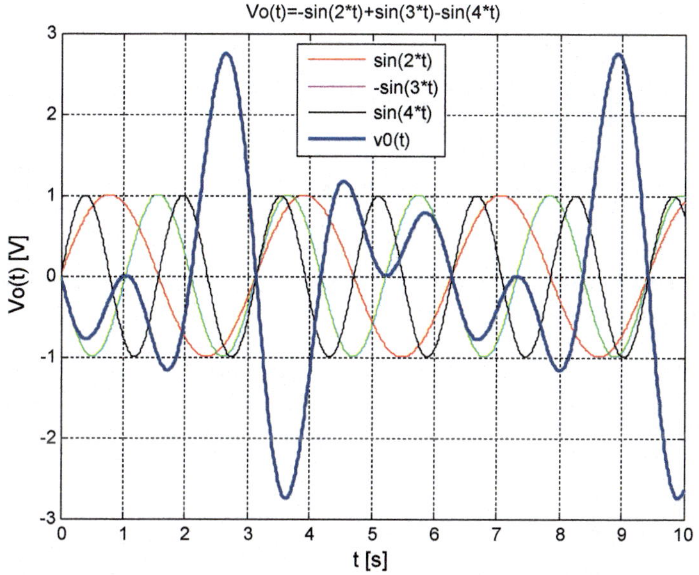

Fig. 10.5 The signals at the terminals of inverting summing amplifier circuit

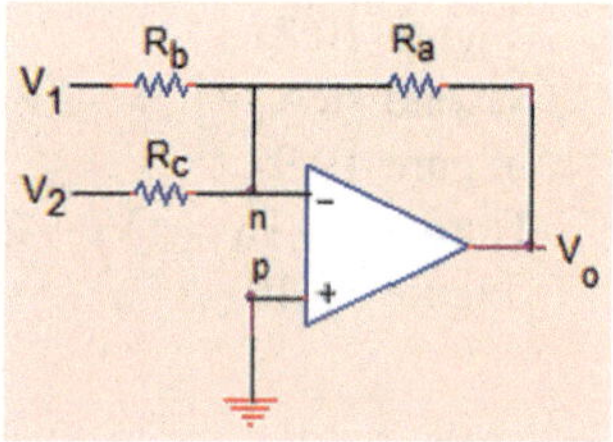

Fig. 10.6 Inverting summer circuit

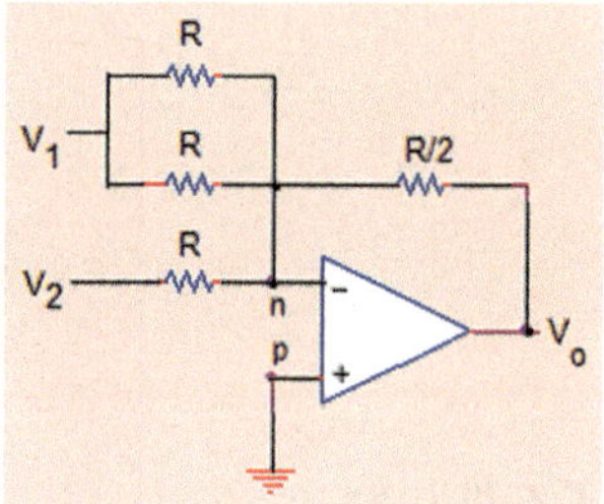

Fig. 10.7 The circuit for Problem 10.1.7

$$\text{If } R_b = 10\,\text{k}\Omega, \quad \rightarrow \quad R_a = 40\,\text{k}\Omega, \quad R_c = 3R_a = 120\,\text{k}\Omega$$

Problem 10.1.7 In the circuit of Fig. 10.7, determine output voltage if $V_1 = 3\,\text{V},\ V_2 = -3\,\text{V}$.

Solution

$$V_0 = -\left(\frac{\frac{R}{2}}{\frac{R}{2}}\right) \cdot V_1 - \left(\frac{\frac{R}{2}}{R}\right) \cdot V_2$$

$$= -V_1 - \frac{1}{2}V_2 = -\left(V_1 + \frac{1}{2}V_2\right) = -\left(3 + \frac{1}{2}(-3)\right) = -1.5\,\text{V}$$

Problem 10.1.8 Use two ideal op-amps and seven resistors at maximum, to implement the equation,

$$V_0 = V_1 + 2V_2 - 3V_3 - 4V_4$$

Solution

$$V_{01} = -\left(\frac{R_2}{R_1}\right) \cdot V_1 = -\frac{R}{R} \cdot V_1 = -V_1, \quad V_{02} = -\left(\frac{2R}{R}\right) \cdot V_2 = -2V_2$$

$$V_{03} = -\left(\frac{3R}{R}\right) \cdot V_3 = -3V_3, \quad V_{04} = -\left(\frac{4R}{R}\right) \cdot V_4 = -4V_4$$

First step, $V_0' = -V_1 - 2V_2$
(Figure 10.8).
Second step, $V_0'' = -3V_3 - 4V_4$
(Figure 10.9)
Third step, $V_0 = (V_1 + 2V_2) - 3V_3 - 4V_4$
(Figure 10.10)

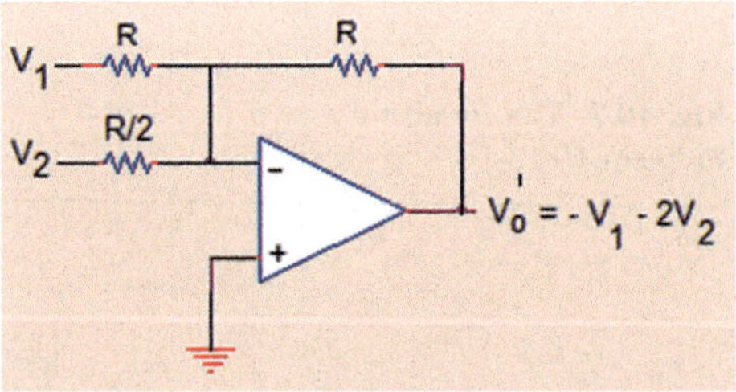

Fig. 10.8 First step

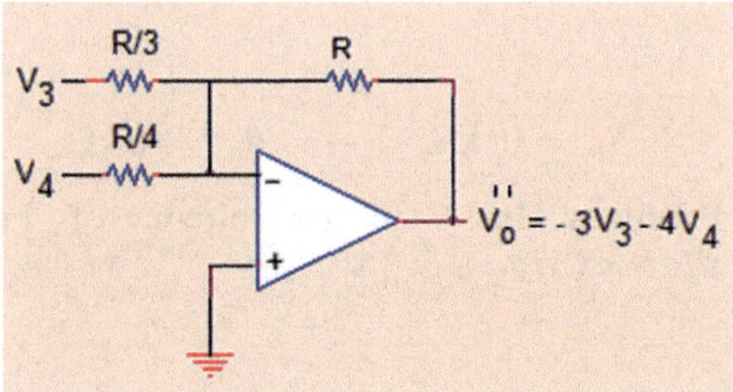

Fig. 10.9 Second step

Fig. 10.10 Third step

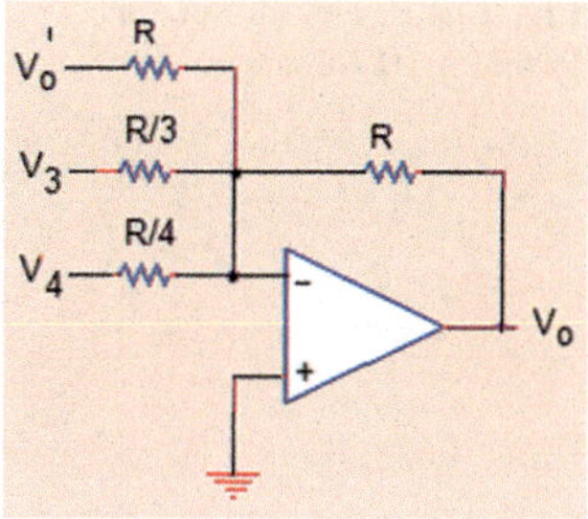

Problem 10.1.9 Two signals, $V_{dc} = 2\,\text{V}$, and $V_s = 5\sin\omega t$ are available. A signal $V_o = -(5 + 5\sin\omega t)$ is to be obtained using an op-amp inverting summer circuit. Realize this circuit.

Solution
(Figure 10.11)

$$V_{o1} = -\frac{R_1}{R_3}\cdot V_{dc} = -5$$
$$-\frac{R_1}{R_3}\cdot 2 = -5 \quad \rightarrow \quad \frac{R_1}{R_3} = 2.5(\Omega/\Omega)$$
$$V_{o2} = -\frac{R_1}{R_2}\cdot V_s = -\frac{R_1}{R_2}\cdot 5\sin\omega t = -5.\frac{R_1}{R_2}\sin\omega t$$
$$R_1 = R_2, \quad \rightarrow \quad V_{o2} = -5\sin\omega t,$$
$$V_o = V_{o1} + V_{o2} = -\left(\frac{R_1}{R_3}\right)V_{dc} - 5\sin\omega t$$

Therefore one may chose $R_1 = R_2 = 2.5\,\text{k}\Omega, \quad R_3 = 1\,\text{k}\Omega$

Problem 10.1.10 Calculate the output voltage of the circuit shown in Fig. 10.12.

Solution
Applying Kirchhoff's current Law at (n);

$$\frac{V_n - V_o}{10k} = \frac{8 - V_n}{5k}$$
$$V_n - V_o = 16 - 2V_n \rightarrow -V_o = 16 - 2V_n - V_n = 16 - 3V_n$$
$$V_o = 3V_n - 16$$

Fig. 10.11 Op-amp inverting summer circuit

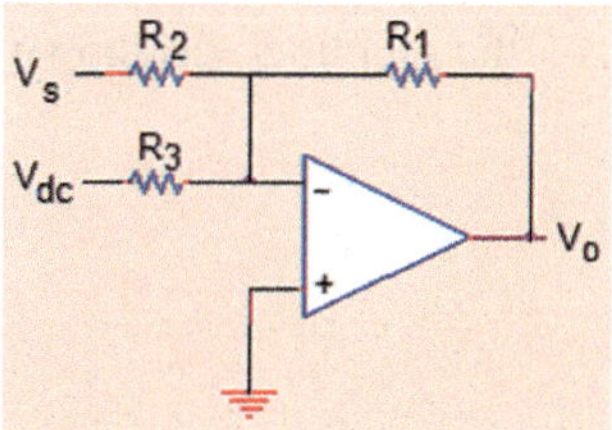

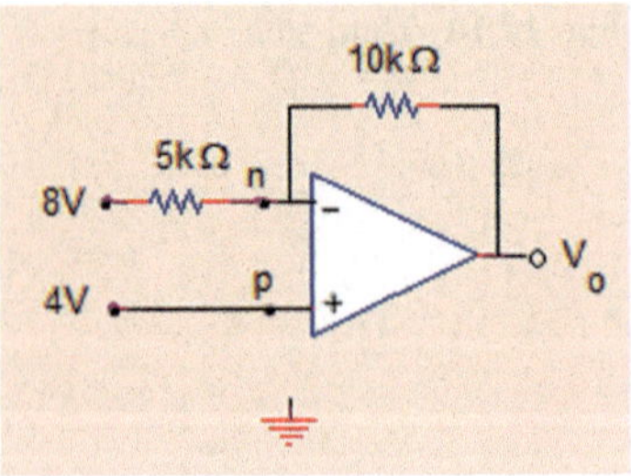

Fig. 10.12 The circuit for Problem 10.1.10

Since $V_p = V_n = 4\,\text{V}, \quad V_o = 3 \cdot 4 - 16 = 12 - 16 = 4\,\text{V}$

Problem 10.1.11

(a) What is the input resistance of an inverting op-amp circuit with a 1 kΩ resistor connected to its inverting terminal and feedback resistor value is 100 kΩ ?
(b) If input voltage source V_s has an internal resistance of R_s = 99 kΩ, what is the signal strength at the input of inverting op-amp circuit?
(c) In a practical application, the required gain of an inverting op-amp configuration is G = 200. It is required that the input resistance be 1 MΩ, or less. The resistors in stock are all less than 1 MΩ. Design an inverting amplifier to satisfy these conditions.

Solution

(a) $R_i = \frac{V_i}{i_i} = \frac{V_i}{\frac{V_i}{R_1}} = R_1$

Therefore, $R_i = R_1 = 1\,\text{k}\Omega$.

(b) Using voltage division rule,

$$V_i = V_s \cdot \frac{R_1}{R_s + R_1} = \frac{1\,\text{k}\Omega}{99\,\text{k}\Omega + 1\,\text{k}\Omega} \cdot V_s = 0.01 V_s$$

There is a considerable loss of signal strength.

(c) For the conventional op-amp inverting circuit,

$$G = -200 = -\frac{R_f}{R_i} = \frac{R_f}{1\,\text{M}\Omega} \quad \rightarrow \quad R_f = 200\,\text{M}\Omega$$

However, this is not a practical value.
An alternative realizable solution is to use a T circuit on the feedback path of the amplifier, as shown in Fig. 10.13.

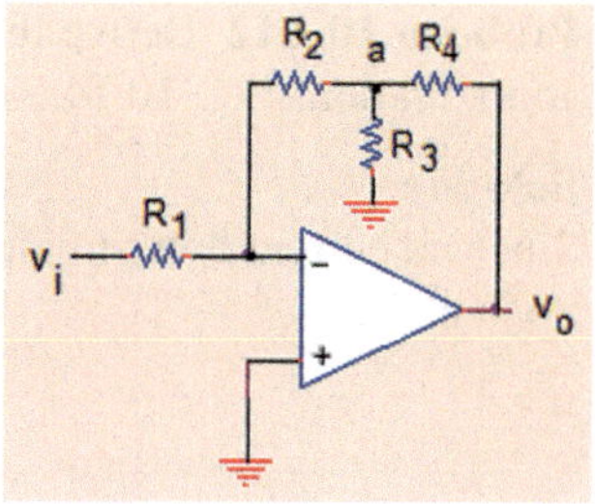

Fig. 10.13 An alternative solution is to use a *T* circuit on the feedback path of opamp

Here,

$$V_{-} = V_{+} = 0$$
$$i_i = \frac{V_i - V_{-}}{R_1} = \frac{V_i}{R_1} = i_{R2}, \quad V_a = V_{-} - i_{R2} \cdot R_2 = -\frac{V_i}{R_1} \cdot R_2 = -\frac{R_2}{R_1} \cdot V_i$$
$$i_{R3} = \frac{0 - V_a}{R_3} = -\frac{V_a}{R_3} = -\left(-\frac{R_2}{R_1} \cdot V_i\right) \cdot \frac{1}{R_3} = \frac{R_2}{R_1 R_3} \cdot V_i$$

Apply KCL at node a:

$$i_{R2} + i_{R3} - i_{R4} = 0 \quad \rightarrow \quad i_{R4} = \frac{V_i}{R_1} + \frac{R_2}{R_1 R_3} \cdot V_i$$

The output voltage is

$$V_o = V_a - i_{R4} \cdot R_4 = -\frac{R_2}{R_1} \cdot V_i - \left(\frac{V_i}{R_1} + \frac{R_2}{R_1 R_3} \cdot V_i\right) \cdot R_4$$
$$= V_i\left(-\frac{R_2}{R_1} - \frac{R_4}{R_1} - \frac{R_2 R_4}{R_1 R_3}\right)$$
$$= -V_i\left[\frac{R_2}{R_1} + \frac{R_4}{R_1}\left(1 + \frac{R_2}{R_3}\right)\right] = -V_i \cdot \frac{R_2}{R_1}\left(1 + \frac{R_4}{R_1} + \frac{R_4}{R_3}\right)$$
$$\frac{V_o}{V_i} = -\frac{R_2}{R_1}\left(1 + \frac{R_4}{R_1} + \frac{R_4}{R_3}\right)$$

Since $R_1 = 1\,\text{M}\Omega$, then $R_2 = 1\,\text{M}\Omega$,

$$-200 = \left(1 + \frac{R_4}{R_2} + \frac{R_4}{R_3}\right) \quad \rightarrow \quad 199 = \frac{R_4}{R_2} + \frac{R_4}{R_3}$$

Select $R_2 = R_4 = 1\,\text{M}\Omega$,

$$199 = 1 + \frac{R_4}{R_3} \quad \rightarrow \quad 198 = \frac{R_4}{R_3} = \frac{1\,\text{M}\Omega}{R_3} \quad \rightarrow \quad R_3 = \frac{1\,\text{M}\Omega}{198} = 5.05\,\text{k}\Omega.$$

Problem 10.1.12 Derive the voltage transfer function for the noninverting op-amp configuration, Fig. 10.14.

Solution
Since no current flows to input terminals of op-amp, by employing nodal analysis at node a,

$$-\frac{V_a}{R_1} = \frac{V_a - V_o}{R_2}$$

But at node b, $V_b = V_i$, $V_a = V_b = V_i$, therefore,

$$-\frac{V_i}{R_1} = \frac{V_i - V_o}{R_2}$$

dividing both sides by V_i and arranging one obtains the resulting equation,

$$\frac{V_o}{V_i} = 1 + \frac{R_2}{R_1}$$

Problem 10.1.13 Determine the voltages (V_0) of the circuits shown in Fig. 10.15.

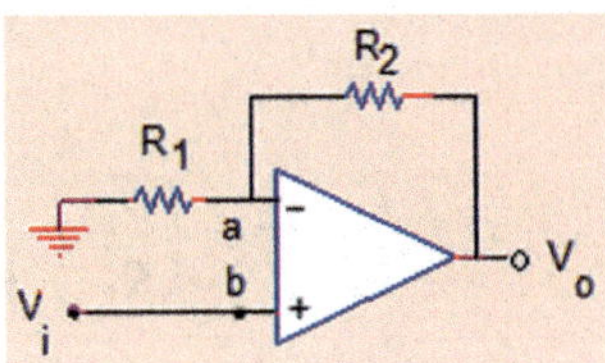

Fig. 10.14 The noninverting op-amp configuration

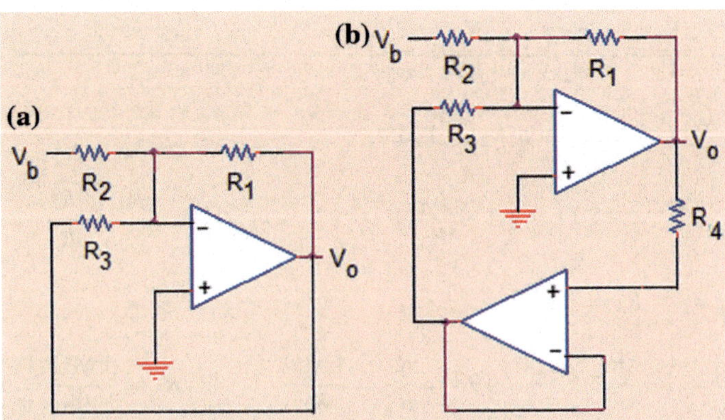

Fig. 10.15 The circuits for Problem 10.1.13a, b

Solution

(a) Inspection of circuit shown in Fig. 10.15a (left) reveals that $(R_1 \parallel R_3)$, Thus,

$$V_0 = -\left(\frac{R_1 \parallel R_3}{R_2}\right) V_b = -\frac{R_1 R_3}{R_2(R_1 + R_3)} V_b$$

(b) In the circuit shown in Fig. 10.15b (right), there is no current flow through resistor R_4. The second op-amp is a buffer circuit, thus, the voltage at the output of first op-amp has the same value as found in part (a).

Problem 10.1.14

(a) A signal $V_1 = AX - B$ is given where A, B are constant terms (voltages) and X is the measurand. Design a noninverting summer to obtain the output signal of the linear form $V_0 = AX$, without using negative reference voltages.
(b) Repeat part (a) using a differencing circuit.

Solution

(a) The output voltage of a possible circuit shown in Fig. 10.16 is,

$$V_0 = \left(1 + \frac{R_{f2}}{R_{f1}}\right)\left(\frac{1}{R_1 + R_2}\right)(V_1 R_2 + V_2 R_1)$$

If $R_1 = R_2 = R_{f2} = R_{f1} = R$,

$$V_0 = (1+1)\left(\frac{1}{2R}\right)(V_1 R + V_2 R) = 2 \times \frac{1}{2R} \times R(V_1 + V_2) = (V_1 + V_2)$$
$$V_2 = B, \quad V_1 = AX - B, \quad V_0 = (AX - B) + B = AX$$

(b) Using a difference circuit requires a negative reference: $V_0 = (AX - B) - (-B) = AX$(Fig. 10.17).

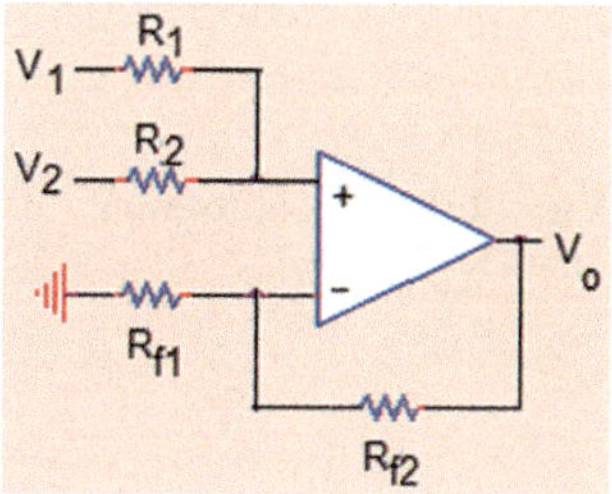

Fig. 10.16 The circuit for Problem 10.1.14a

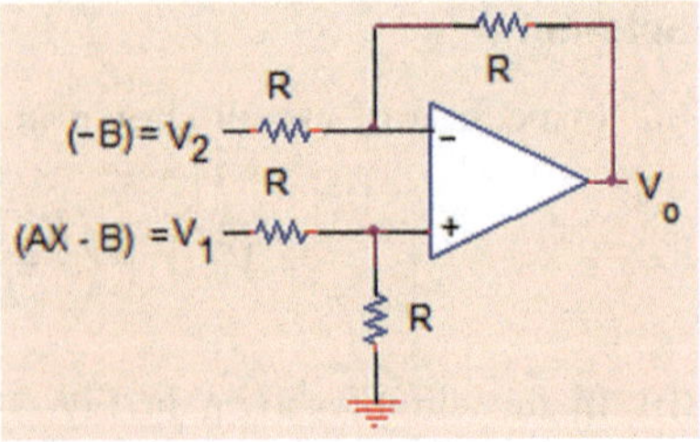

Fig. 10.17 The circuit for Problem 10.1.14b

Problem 10.1.15

(a) In the circuit shown in Fig. 10.18, determine the current through the load resistor R_L.
(b) Calculate the value of R, if the current through R_L is to be 1 mA, U = 2.5 V, R_L = 1 kΩ.
(c) Find the voltage at the output of op-amp.

Solution

(a) Due to op-amp property, $V_+ = V_-$, the current through R is $i_R = U/R$ which is also the current through R_L. The circuit acts as a constant current source for the floating R_L.
(b) Using given values,

$$R = \frac{2.5\,\mathrm{V}}{1\,(\mathrm{mA})} = 2.5\,\mathrm{k}\Omega$$

This resistance value can be realized in practice by parallel connecting four 10 kΩ resistors.
(c) The voltage at the output of op-amp is

$$V_0 = i(R_L + R) = 10^{-3}\left(10^3 + 2.5 \times 10^3\right) = 3.5\,\mathrm{V}$$

Check: Noninverting amplifier gain;

$$V_0 = U\left(1 + \frac{R_L}{R}\right) = 2.5\left(1 + \frac{1}{2.5}\right) = 1.4 \times 2.5 = 3.5\,\mathrm{V}$$

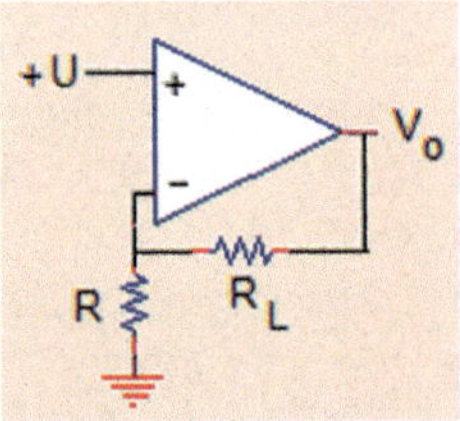

Fig. 10.18 A non-inverting amplifier circuit

Problem 10.1.16

(a) Determine the voltage at the output of op-amp circuit shown in Fig. 10.19.
(b) Determine the output voltage, if $V_1 = -V_2$

Solution

(a)

$$\text{Let } V_2 = 0; \quad \rightarrow \quad V_0 = V_1\left(1 + \frac{R_2}{R_1}\right),$$

$$\text{Let } V_1 = 0; \quad \rightarrow \quad V_0 = -\frac{R_2}{R_1}V_2$$

Summing above (output) equations (superposition !),

$$V_0 = V_1\left(1 + \frac{R_2}{R_1}\right) - \frac{R_2}{R_1}V_2 = V_1 + \frac{R_2}{R_1}V_1 - \frac{R_2}{R_1}V_2 = V_1 + \frac{R_2}{R_1}(V_1 - V_2)$$

(b) $V_1 = -V_2$;

$$V_0 = -V_2 + \frac{R_2}{R_1}(-2V_2) = -V_2\left(1 + 2\frac{R_2}{R_1}\right)$$

Problem 10.1.17 For an op-amp-based differential amplifier circuit shown in Fig. 10.20,

(a) $V_o = f(R_1, R_2, R_3, R_4) = ?$
(b) If $R_1 = R_2 = R_3 = R_4$, $V_o = ?$
(c) If $A_d = \dfrac{R_4}{R_1} = \dfrac{R_3}{R_2}$, $V_o = ?$

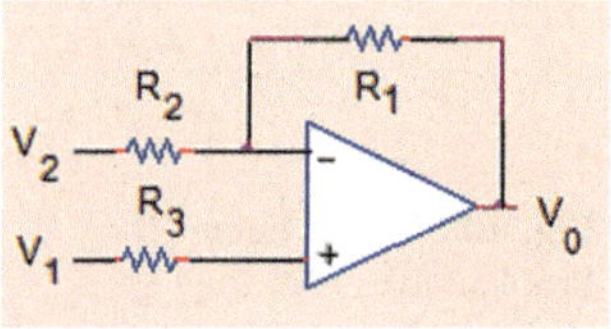

Fig. 10.19 The circuit for Problem 10.1.16

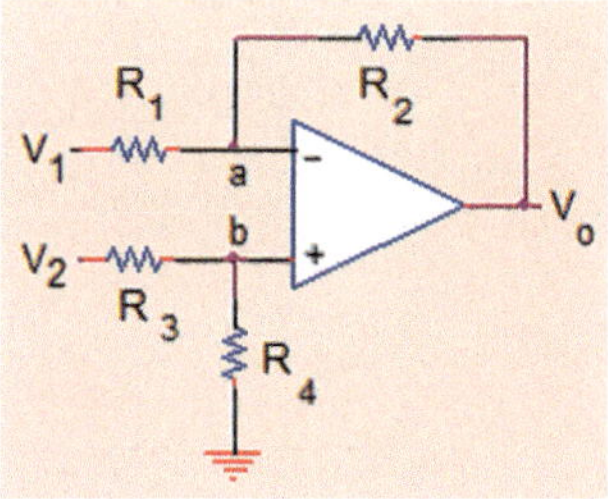

Fig. 10.20 An op-amp-based differential amplifier circuit

Solution

(a) $V_b = V_2 \cdot \dfrac{R_3}{R_3 + R_2} = V_a$

$$\frac{V_1 - V_a}{R_1} = \frac{V_a - V_o}{R_4} \quad \rightarrow \quad \frac{\left(V_1 - \dfrac{V_2 \cdot R_3}{R_2 + R_3}\right)}{R_1} = \frac{\dfrac{V_2 \cdot R_3}{R_2 + R_3} - V_o}{R_4}$$

$$\frac{V_1(R_2 + R_3) - V_2R_3}{R_1(R_2 + R_3)} = \frac{V_2R_3 - V_o(R_2 + R_3)}{R_4(R_2 + R_3)}$$

$$R_4 \cdot V_1(R_2 + R_3) - V_2R_3R_4 = V_2R_3R_1 - V_o(R_2 + R_3)R_1$$

$$V_1R_4(R_2 + R_3) - V_2R_3R_4 - V_2R_3R_1 = -V_oR_1(R_2 + R_3)$$

$$V_1\frac{R_4(R_2 + R_3)}{R_1(R_2 + R_3)} - V_2\frac{(R_3R_4 + R_1R_3)}{R_1(R_2 + R_3)} = -V_o$$

$$V_2\frac{R_3(R_1 + R_4)}{R_1(R_2 + R_3)} - V_1\frac{R4}{R_1} = V_o$$

(b) If $R_1 = R_2 = R_3 = R_4 = R, \quad V_2 - V_1 = V_o$

(c) If $\frac{R_4}{R_1} = \frac{R_3}{R_2} = A_d, \quad R_3 = A_d \cdot R_2, \quad R_4 = A_d \cdot R_1$

$$V_o = V_2\left[\frac{A_d \cdot R_2(R_1 + A_d \cdot R_1)}{R_1(R_2 + A_dR_2)}\right] - V_1 \cdot \frac{A_d \cdot R_1}{R_1} = V_2 \cdot \left[\frac{A_d \cdot R_1R_2(1 + A_d)}{R_1R_2(1 + A_d)}\right] - V_1 \cdot A_d$$

$$V_o = A_d \cdot V_2 - A_d \cdot V_1 = A_d(V_2 - V_1)$$

Problem 10.1.18 Determine the current through R_2 in the circuit shown in Fig. 10.21.

$$R_3 = R_4 = 2.2\,\text{k}\Omega, \quad R_1 = 10R_2 = 10\,\text{k}\Omega, \quad U = 1\,\text{V}$$

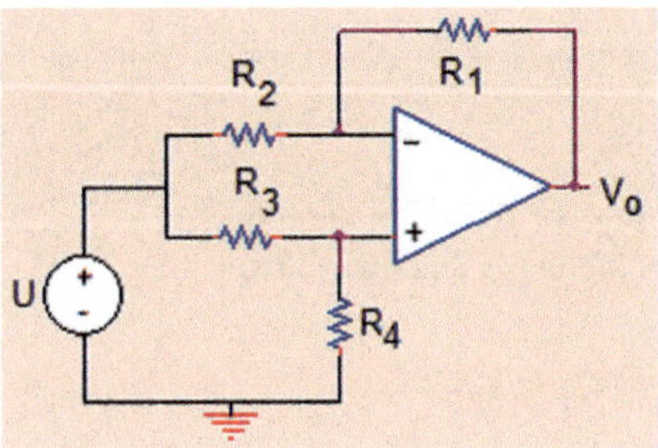

Fig. 10.21 The circuit for Problem 10.1.18

Solution
For ideal op-amp;

$$V_+ = \frac{R_4}{R_3 + R_4} \cdot U = \frac{1}{2} \times 1 = 0.5\,\text{V}, \quad V_+ = V_- = 0.5\,\text{V}$$

$$i = \frac{U - V_-}{R_2} = \frac{U - V_-}{\left(\frac{R_1}{10}\right)} = \frac{1 - 0.5}{1} = 0.5\,\text{mA}$$

Problem 10.1.19 Find the voltage transfer ratio in the circuit shown in Fig. 10.22.

Solution

$$V_+ = \frac{R_4}{R_4 + R_3} \cdot V_i = V_-$$

$$V_0 = \left(1 + \frac{R_2}{R_1}\right) \cdot V_+ = \left(1 + \frac{R_2}{R_1}\right) \cdot \left(\frac{R_4}{R_4 + R_3}\right) \cdot V_i$$

$$\frac{V_0}{V_i} = \left(1 + \frac{R_2}{R_1}\right) \cdot \left(\frac{R_4}{R_4 + R_3}\right)$$

Problem 10.1.20 In a differential amplifier shown in Fig. 10.23, determine the output signal if $R_2 = R_4 = 10\ \text{k}\Omega$, $R_1 = R_3 = 1\ \text{k}\Omega$,

$$v_1 = 10\sin(2\pi 50t) - \sin(2\pi 250t)\text{V}, \quad v_2 = 10\sin(2\pi 50t) + \sin(2\pi 250t)\text{V}$$

Fig. 10.22 The circuit for Problem 10.1.19

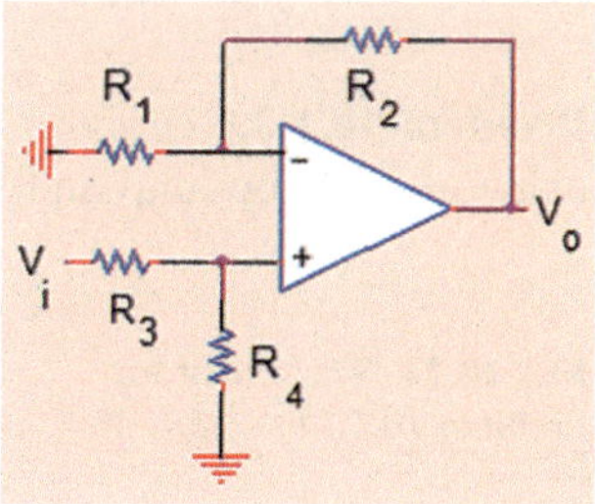

Fig. 10.23 The circuit for Problem 10.1.20

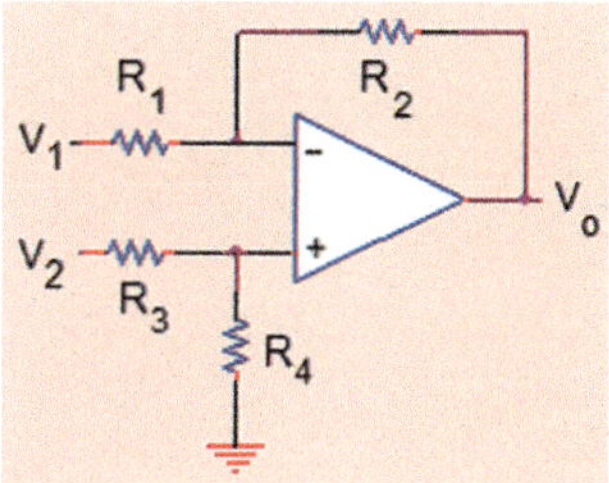

Solution

$$v_o = \frac{R_2}{R_1}(v_2 - v_1) = \frac{10}{1}(v_2 - v_1)\text{V}$$
$$= 10[10\sin(2\pi 50t) + \sin(2\pi 500t) - 10\sin(2\pi 50t) + \sin(2\pi 500t)]$$
$$v_o = 10[2\sin(2\pi 500t)] = 20\sin(2\pi 500t)\text{ V}$$

Problem 10.1.21 Find the output voltage of differential amplifier circuit shown in Fig. 10.24,

if $V_1 = 10\sin(2\pi \times 60t) - 0,1\sin(2\pi \times 1000t)$, and $V_2 = 10\sin(2\pi \times 60t) + 0,1\sin(2\pi \times 1000t)$.

Solution

$$V_0(t) = \left(\frac{20R}{R}\right) \cdot (V_2 - V_1) = 20(V_2 - V_1)$$
$$= 20[0.1\sin(2\pi \times 1000t) - [-0.1\sin(2\pi \times 1000t)]]$$
$$= 20[0.2\sin(2\pi \times 1000t)] = 4\sin(2\pi \times 1000t)V$$

Problem 10.1.22 Design (i.e., determine the resistor values for) a differential amplifier with $V_o = 10(V_2 - V_1)$, using a 10 kΩ feedback resistor.

Solution

Figure 10.25 is the circuit topology for op-amp-based differential amplifier.

Since the voltage gain is given by

$$\frac{R_2}{R_1} = \frac{R_4}{R_3} = 10, \quad \rightarrow \quad R_3 = R_1 = 1\,\text{k}\Omega, \quad \rightarrow \quad R_2 = R_4 = 10\,\text{k}\Omega$$

Problem 10.1.23 Design a noninverting amplifier with voltage gain $G = 15$ V/V using a single op-amp and maximum four resistors consisting of 15 and 1 kΩ.

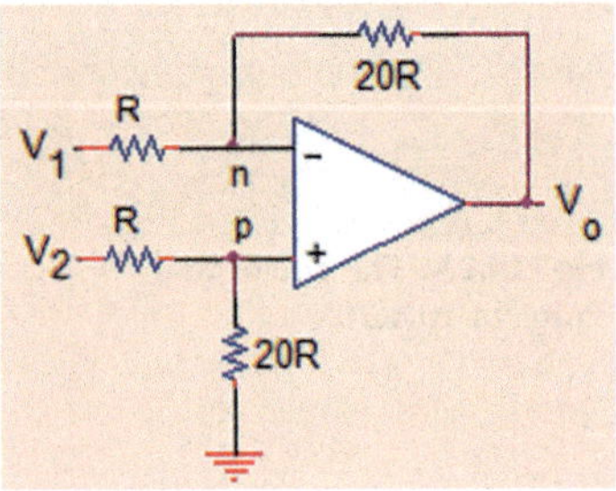

Fig. 10.24 The circuit for Problem 10.1.21

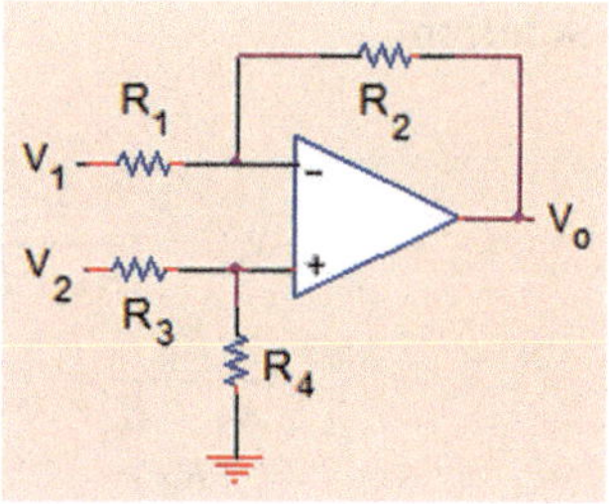

Fig. 10.25 Differential amplifier

Solution

$$G = 1 + \frac{R_f}{R_i} \quad \rightarrow \quad 14 = \frac{R_f}{R_i}$$

Conventional noninverting single op-amp topology (with a feedback resistor and an input resistor) can not be realized for this gain factor and using only resistor values given above. A possible solution is the circuit of Fig. 10.26, with $R_2 = 15\,\text{k}\Omega, \quad R_1 = 1\,\text{k}\Omega$.

Problem 10.1.24 Determine the voltage at the output of op-amp circuit shown in Fig. 10.27.

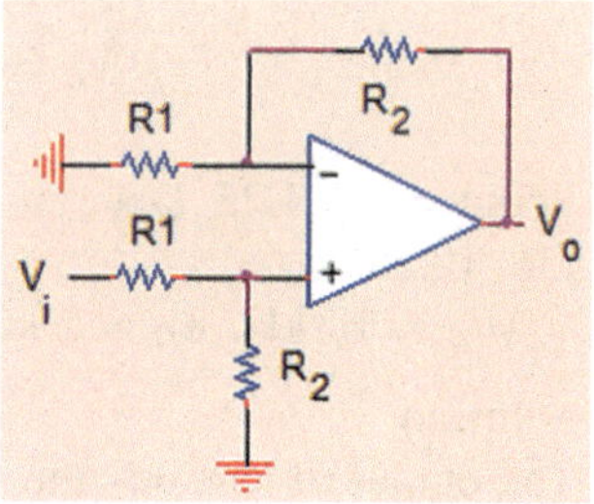

Fig. 10.26 The circuit for Problem 10.1.23

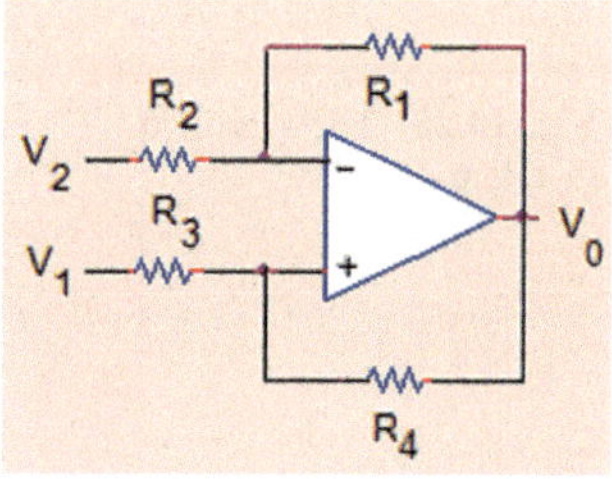

Fig. 10.27 The circuit for Problem 10.1.24

Solution

$$V_+ = \frac{R_4}{R_3+R_4}(V_1 - V_0) = V_- = \frac{R_2}{R_1+R_2}(V_2 - V_0)$$
$$\frac{R_4}{R_3+R_4}(V_1 - V_0) = \frac{R_2}{R_1+R_2}(V_2 - V_0)$$
$$\frac{R_4}{R_3+R_4}V_1 - \frac{R_4}{R_3+R_4}V_0 = \frac{R_2}{R_1+R_2}V_2 - \frac{R_2}{R_1+R_2}V_0$$

Collecting like terms of voltage together (V_0),

$$\frac{R_2}{R_1+R_2}V_0 - \frac{R_4}{R_3+R_4}V_0 = \frac{R_2}{R_1+R_2}V_2 - \frac{R_4}{R_3+R_4}V_1$$
$$V_0\left(\frac{R_2}{R_1+R_2} - \frac{R_4}{R_3+R_4}\right) = \frac{R_2}{R_1+R_2}V_2 - \frac{R_4}{R_3+R_4}V_1$$
$$V_0 = \frac{\frac{R_2}{R_1+R_2}V_2 - \frac{R_4}{R_3+R_4}V_1}{\frac{R_2}{R_1+R_2} - \frac{R_4}{R_3+R_4}} = \frac{R_2(R_3+R_4)V_2 - R_4(R_1+R_2)V_1}{R_2(R_3+R_4) - R_4(R_1+R_2)}$$

Denominator: $R_2R_3 + R_2R_4 - R_1R_4 - R_2R_4 = R_2R_3 - R_1R_4$
Finally,

$$V_0 = \frac{V_2R_2(R_3+R_4) - V_1R_4(R_1+R_2)}{R_2R_3 - R_1R_4}$$

Problem 10.1.25 Determine the output signal for the op-amp circuit shown in Fig. 10.28.
(R_{f1} = 10 kΩ, R_{f2} = 5 kΩ, R = 1 kΩ).

Solution
The output of first op-amp (left) is (inverting summer),

$$V_{o1} = -\frac{R_{f1}}{R}(V_1 + V_2)$$

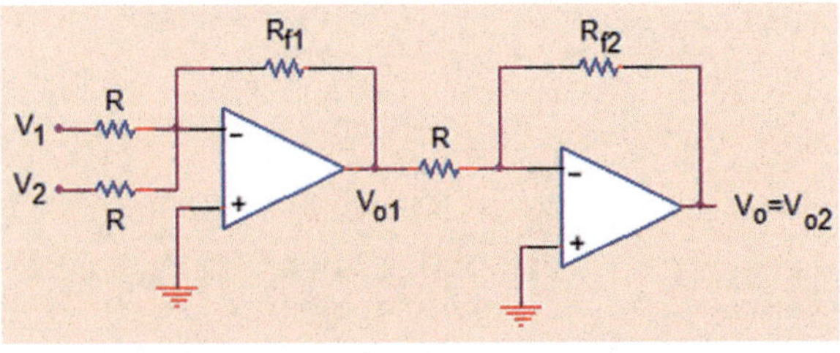

Fig. 10.28 The circuit for Problem 10.1.25

Then, the output voltage of the circuit is

$$V_o = V_{o2} = -\frac{R_{f2}}{R} \cdot V_{o1} = -\frac{R_{f2}}{R} \cdot \left[-\frac{R_f}{R}(V_1 + V_2) \right] = \frac{R_{f1} \cdot R_{f2}}{R^2}(V_1 + V_2)$$

$$V_o = \frac{10 \cdot 5}{1}(V_1 + V_2) = V_o = 50(V_1 + V_2)\text{V}$$

The circuit behaves as a noninverting summer with some gain.

Problem 10.1.26 Design a circuit using 2 inverting op-amps and equal valued resistors to separate a signal V_1 from the additive parasitic signal V_2.

Solution

The circuit of Fig. 10.29 is a solution.

The signal is $V_1 = s$, and the parasitic signal is $V_2 = s + n$, where n denotes noise.

At the output of first op-amp, $V_{01} = -V_2$.

At the output of second op-amp, $V_0 = -(-V_2 + V_1 + V_2) = V_1$

The signal is separated from parasitic V_2 and then inverted.

Problem 10.1.27 Design an instrumentation amplifier of Fig. 10.30, with voltage gain ×210.

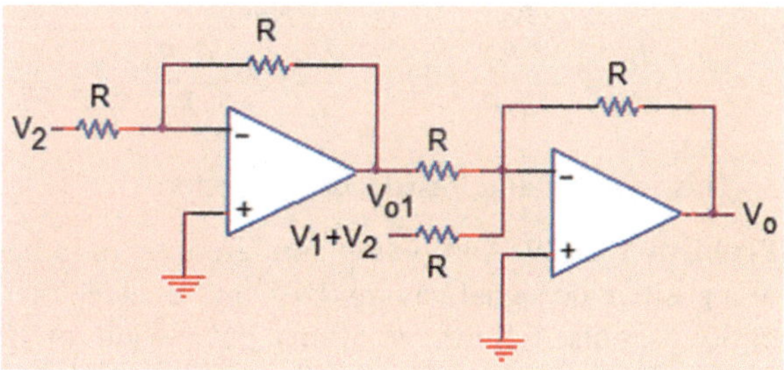

Fig. 10.29 The circuit for Problem 10.1.26

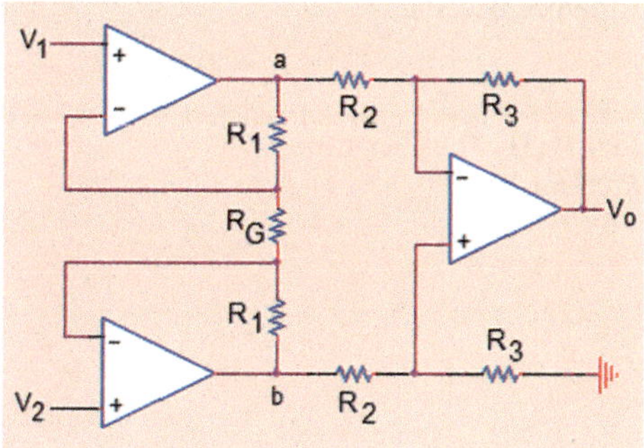

Fig. 10.30 Instrumentation amplifier

Solution

$$\frac{V_o}{V_d} = 210 = \left(1 + \frac{2R_1}{R_G}\right) \cdot \frac{R_3}{R_2}$$

Let R_1 = R_3 = 10k, R_6 = R_2 = 1k,

$$\frac{V_o}{V_d} = \left(1 + \frac{2 \times 10}{1}\right)\left(\frac{10}{1}\right) = 21 \times 10 = 210$$

Problem 10.1.28 Calculate the ratio, V_s/I_s for circuit shown in Fig. 10.31, in terms of three resistor values. What function does this circuit perform?

Solution
Noninverting amplifier output is,

$$V_o = V_s\left(1 + \frac{R_2}{R_1}\right)$$

On the other hand, current supplied by the source voltage is

$$I_s = \frac{V_s - V_o}{R_3} = \frac{V_s - V_s\left(1 + \frac{R_2}{R_1}\right)}{R_3} = \frac{V_s - V_s - V_s\frac{R_2}{R_1}}{R_3} = \frac{-V_s \cdot R_2}{R_1 R_3}$$

$$\frac{I_s}{V_s} = -\frac{R_2}{R_1 R_3} \quad \rightarrow \quad \frac{V_s}{I_s} = -\frac{R_1 R_3}{R_2}$$

This is a negative resistance converter.

Problem 10.1.29 An op-amp configuration is to be used for (−10x) amplification of a positive dc signal having 0.4 V pulse added on 0.1 V base line, pulse rise time of 30 ms, pulse fall time of 50 ms, pulse width of 100 ms, pulse period of 500 ms.

Use well-known 741 type op-amp and design this amplifier. Display the resulting signals on the same SPICE graph. Present the SPICE net list. (opamp_01.cir).

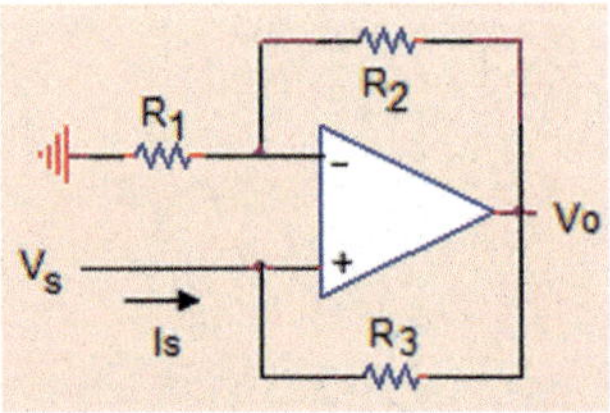

Fig. 10.31 The circuit for Problem 10.1.28

Solution
Use an inverting op-amp configuration. Since the amplification factor is (−10), one may select the feedback resistor 10 kΩ, input resistor as 1 kΩ. Figure 10.32 displays input (red) and output (blue) signals. The SPICE netlist is attached.

```
inverting operational amplifier with 741
 .lib national.lib; keep cir and lib files in same folder
VCC 10 0 DC 15 V
VEE 11 0 DC -15 V
vi 1 0  PULSE(0.1  0.5  0S 0.03S 0.05S 0.1S 0.5S)
*       PULSE(v1   v2   TD TR   TF   PW   PER)
*v1 signal base, v2 signal pulsed value, TD time delay
*TR rise time, TF Fall time, PW Pulse width, PER period
Ri 1 2 1 k
Rf 2 3 10 k
XOA1  0  2 10 11 3  LM741
*XOA1 NI I +  -  Vo  LM741
*.tran 1 m 1.5 0  v3
```

Problem 10.1.30 An op-amp configuration is to be used for 11 times the amplification of a signal of 4 V pulse added on 0.1 V base line, pulse rise time of 30 ms, pulse fall time of 50 ms, pulse width of 100 ms, pulse period of 500 ms

Use well-known 741 type op-amp and design this amplifier. Display the resulting signals on the same SPICE graph. Present the SPICE net list (opamp_01.cir).

Solution
Use a noninverting op-amp configuration. Since the amplification factor is 11, one may select the feedback resistor 10 kΩ, and the input resistor as 1 kΩ. Figure 10.33 displays input (red) and output (blue) signals. The SPICE netlist is attached.

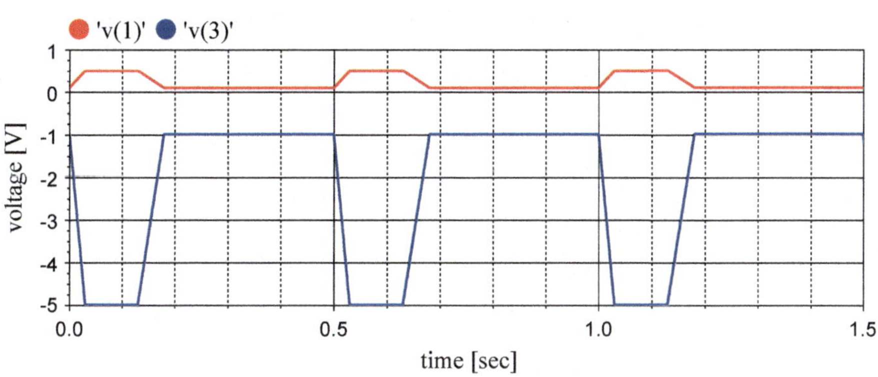

Fig. 10.32 The input (*red*) and output (*blue*) signals for the circuit of Problem 10.1.29

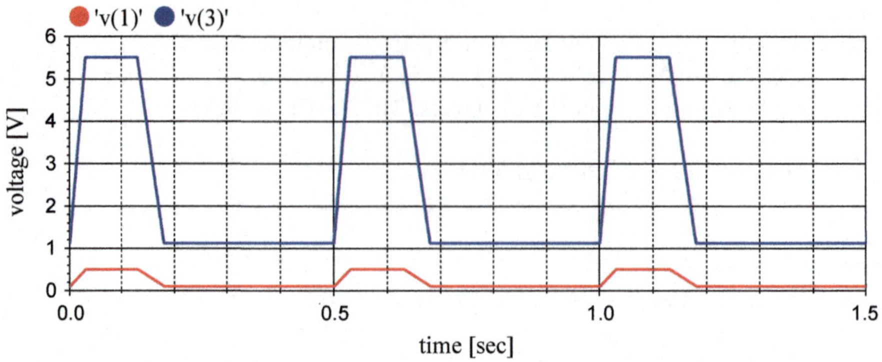

Fig. 10.33 The input (*red*) and output (*blue*) signals for the circuit of Problem 10.1.30

```
noninverting operational amplifier with 741
 .lib national.lib; keep cir and lib files in same folder
VCC 10 0 DC 15 V
VEE 11 0 DC -15 V
vi 1 0   PULSE(0.1  0.5  0S 0.03S 0.05S 0.1S 0.5S)
*        PULSE(v1  v2  TD TR   TF   PW  PER)
*v1 signal base, v2 signal pulsed value, TD time delay
*TR rise time, TF Fall time, PW Pulse width, PER period
Ri 2 0 1 k
Rf 2 3 10 k
XOA1  1  2 10 11 3  LM741
*XOA1 NI I + - Vo LM741
*.tran 1 m 1.5 0 v3
```

Problem 10.1.31 An op-amp is to be used for amplification of a 10 Hz sinusoidal signal of 1 V peak value to a peak value of 10 V and 180° phase difference between the input. Use well-known 741 type op-amp and design this amplifier circuit. Display the resulting signals on the same SPICE graph and present the SPICE net list (test 741.2.cir).

Solution

Use an inverting op-amp configuration. Since the amplification factor is (−10), one may select the feedback resistor 10 kΩ, input resistor as 1 kΩ . Figure 10.34 displays input (red) and output (blue) signals. The SPICE simulation example employs 741 op-amp (library model).

SPICE netlist (test 741.2.cir) is given below (keep .cir and .lib files in the same folder).

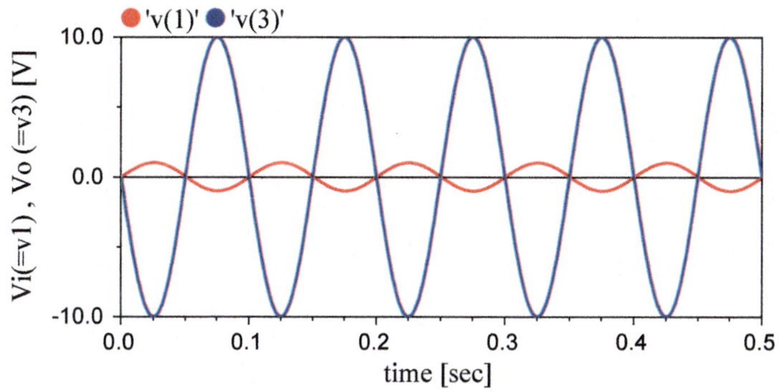

Fig. 10.34 Inverting op-amp signals. Input (*red*), output (*blue*)

```
inverting ac amplifier
.lib national.lib
VCC 10 0 DC 15 V
VEE 11 0 DC -15 V
*vi 1 0 SIN(bias  Vm  f )
vi 1 0 SIN(0    1    10)
Ri 1 2 1 k
Rf 2 3 10 k
XOA1 0  2 10 11 3 LM741
*XOA1 NI I +  -  Vo LM741
*tran .1 m 0.5 0 1 m uic
```

Problem 10.1.32 An op-amp configuration is to be used for amplification factor of (−10). Use a 741 type op-amp and design this amplifier circuit. Display input–output curve, specifying the active and saturation regions of operation and present the SPICE net list.

Solution

Use an inverting op-amp configuration. Since the amplification factor is −10, one may select the feedback resistor 10 kΩ, input resistor as 1 kΩ . Figure 10.35 displays input output curve. It is noted here that when the absolute value of input signal magnitude exceeds 1.1 V, output signal saturates. Active region of operation is the section of the piecewise linear curve with negative slope. It is also noted that output saturation level is not the "power supply value," but about 1 V less than the absolute value of symmetrical power supply voltage. The SPICE netlist is attached.

```
Analysis: DC Transfer Curves
inverting operational amplifier with 741
.lib national.lib
VCC 10 0 DC 12 V
```

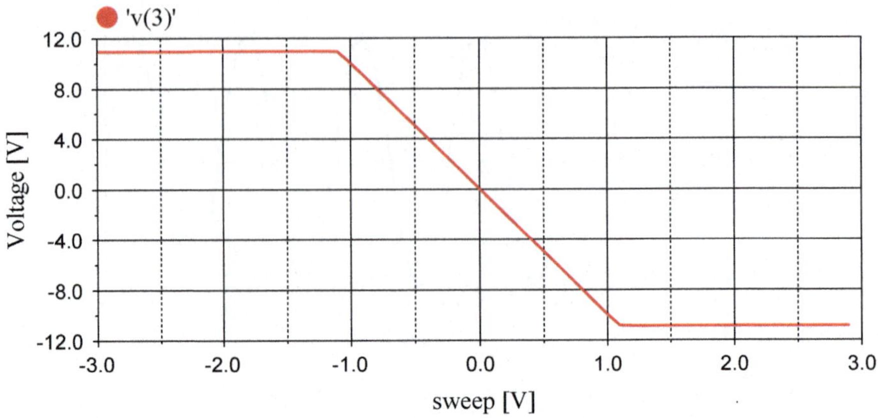

Fig. 10.35 The input output curve for the circuit in Problem 10.1.32

```
VEE 11 0 DC -12 V
vi 1 0 dc 1v
Ri 1 2 1 k
Rf 2 3 10 k
XOA1 0  2 10 11 3  LM741
*XOA1 NI I +  -  Vo LM741
*.dc vi  -3  3  .1
```

Problem 10.1.33 There's an error voltage present at the output of an op-amp due to unbalances in its internal transistors and resistors. The net error is modeled as an offset voltage, Voff, in series with op-amp's input terminals. The input offset voltage ranges from microvolts to millivolts, it can be either polarity. (The cause of input offset voltage is due to the inherent mismatch of the input transistors and components during fabrication of the silicon die, and stresses placed on the die during the packaging process (minor contribution). These effects collectively produce a mismatch of the bias currents that flow through the input circuit, and primarily the input devices, resulting in a voltage differential at the input terminals of the op-amp. It has been reduced with modern manufacturing processes through increased matching and improved package materials and assembly [1]).

Use SPICE and build a noninverting op-amp circuit with gain of (11). Apply 0 V to its input. What do you measure at the output? What is the input offset voltage for this op-amp?

Solution

Since the amplification factor is 11, one may select the feedback resistor 10 kΩ, input resistor as 1 kΩ . Figure 10.36 displays the output signal. It is noted here that when the input signal is 0 V, output signal is measured as 12 mV. Therefore the input offset voltage is calculated as 12/11 = 1.091 mV.

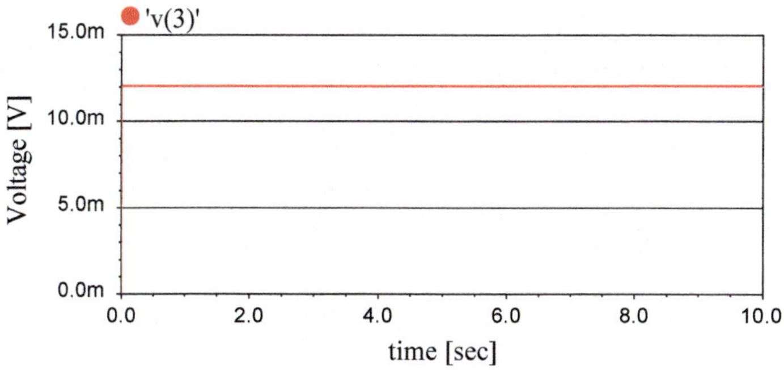

Fig. 10.36 The output signal for the circuit in Problem 10.1.33

```
Analysis: Transient
OPAMP OFFSET VOLTAGE 741
.lib national.lib
VCC 10 0 DC 12 V
VEE 11 0 DC -12 V
vi 1 0 dc 0mv
Ri 2 0 1 k
Rf 2 3 10 k
XOA1 1  2 10 11 3  LM741
*XOA1 NI I +  -  Vo LM741
*.tran 1 10 0 uic
```

Problem 10.1.34 By means of SPICE analysis, perform open-loop input offset voltage measurements of OP 27 operational amplifier, and plot its input–output graph.

Solution

Input offset voltage (Vio) measurements are performed via SPICE analysis OP27 operational amplifier, and its input–output graph is plotted in Fig. 10.37.

LTSPICE netlist for Vio_test1.cir:

```
opamp input offset voltage
vp 10 0  15
vn 11 0 -15
vi  1 0  2
R1 3 0 1G
XOA1  0 1 10 11 3 OP27
* Node assignments NI I +  -  Vo
*sub-circuit/models not listed here (see .cir files).
```

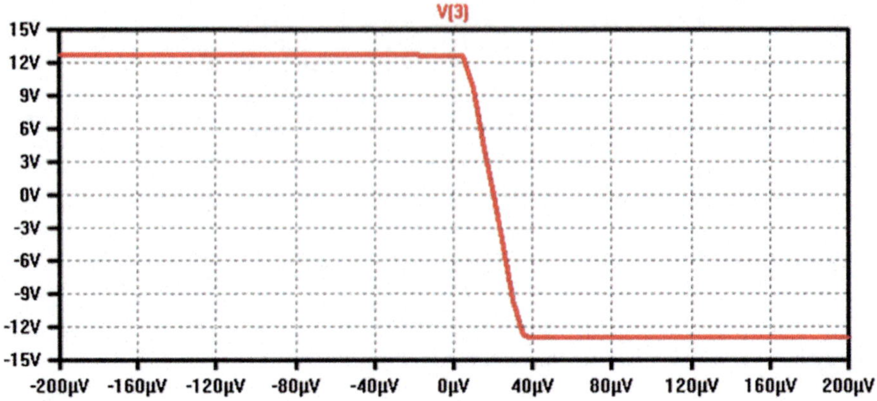

Fig. 10.37 The input–output graph

```
.step Vi -0.2 m 0.2 m 0.005 m
.op
.backanno
.end
```

Measured Vio = 20 µV.

The OP 27 operational amplifier is a higher version of OP07 [2]. Offsets down to 25 µV (SPICE Measured Vio = 20 µV) and maximum drift of 0.6 µV/°C. Low noise, en = 3.5 nV/√Hz, at 10 Hz, a low 1/f noise corner frequency of 2.7 Hz, gain = 1.8 million, gain bandwidth product = 8 MHz, slew rate = 2.8 V/µs.

Problem 10.1.35 Use a SPICE analysis software and plot the variation of input bias currents of three different op-amps as a function of input common mode voltage (sweep) between ±12 V: LT1001 [3], OP27 and LM741. (input_bias_current1 LT.cir).

Solution

The change of input bias current (Ib) as a function of input common mode voltage between ±12 V for three different op-amps (LT1001, OP27 and LM741) are displayed in Figs. 10.38, 10.39 and 10.40. Test circuit is a buffer op-amp configuration, and input current is plotted against input voltage sweep. (For the netlist, refer to .cir file).

Problem 10.1.36 Perform SPICE analysis to plot the variation of input bias currents of three different op-amps as a function of temperature at constant input voltage of 12 V: LT1001, OP27, and LM741. (input_bias_current2 LT.cir).

Solution

The change of input bias current (Ib) as a function of temperature between 0 and 100 °C for three different op-amps (LT1001, OP27 and LM741) are displayed in

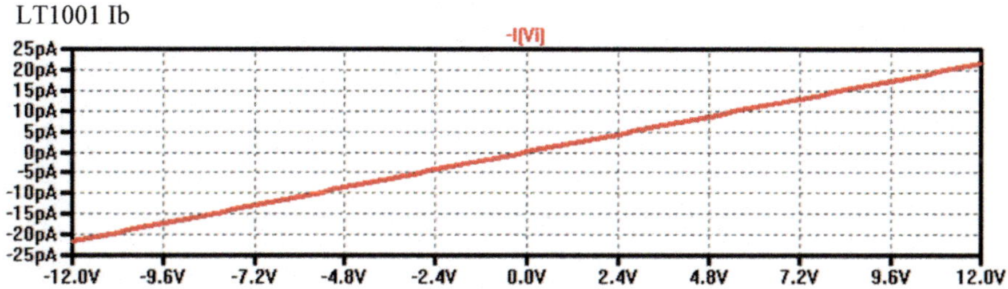

Fig. 10.38 The input bias current as a function of input CM voltage for LT1001

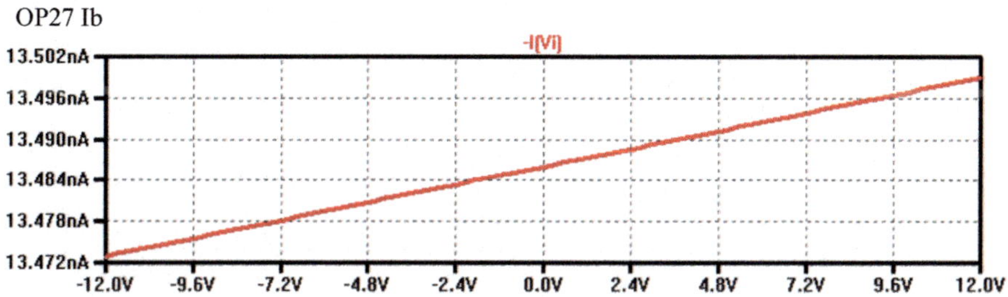

Fig. 10.39 The input bias current as a function of input CM voltage for OP27

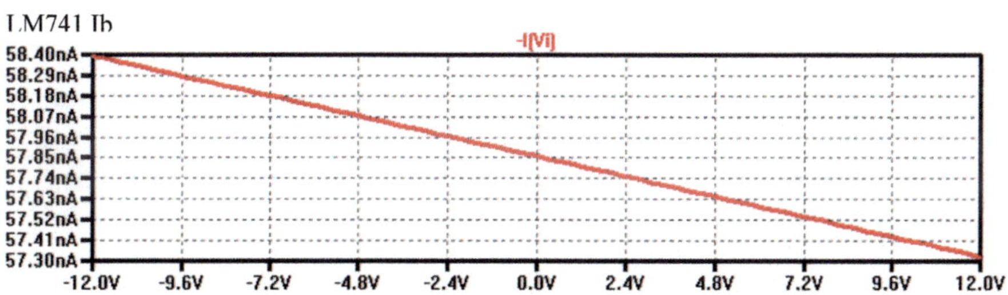

Fig. 10.40 The input bias current as a function of input CM voltage for 741

Figs. 10.41, 10.42 and 10.43. Test circuit is a buffer op-amp configuration, and input current is plotted against temperature sweep. (For the netlist, refer to .cir file).

Problem 10.1.37 A NTC Thermistor is shunted by a fixed resistor *Rp*. Determine the equation of op-amp output voltage as a function of thermistor temperature, in the circuit of Fig. 10.44.

Solution

The equation of op-amp output voltage as a function of thermistor temperature can be derived as follows: Thermistor *R*-*T* exponential relationship,

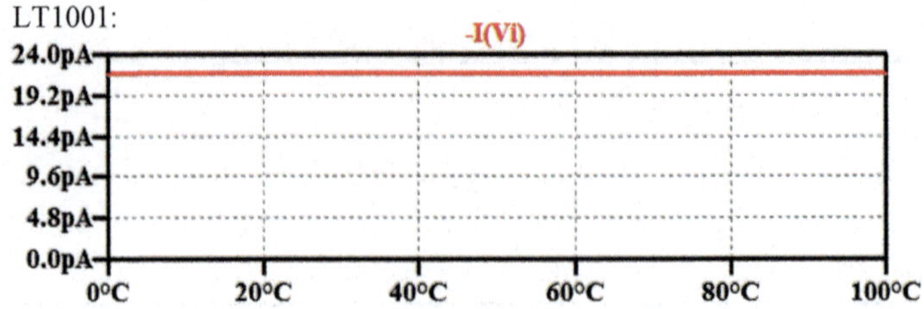

Fig. 10.41 Input bias current of LT1001 as a function of temperature

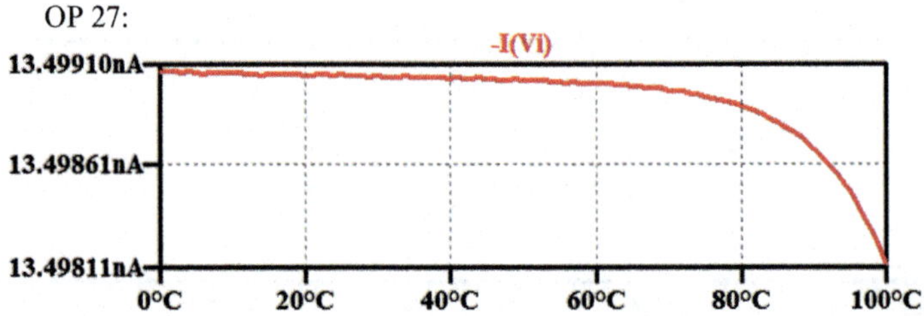

Fig. 10.42 Input bias current of OP27 as a function of temperature

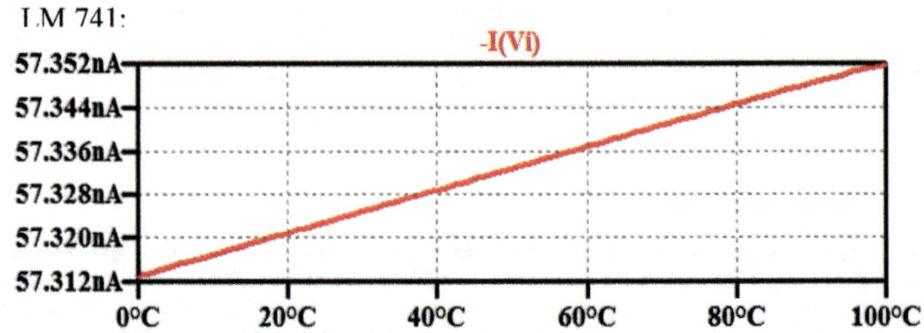

Fig. 10.43 Input bias current of 741 as a function of temperature

Fig. 10.44 The circuit for Problem 10.1.37

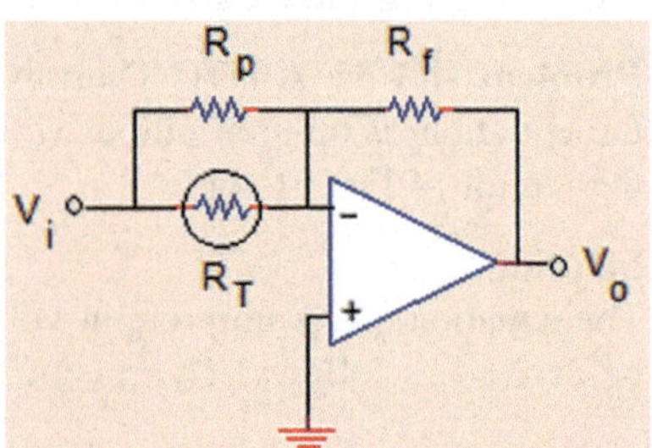

$$R_T = R_0 e^{\beta\left(\frac{1}{T}-\frac{1}{T_0}\right)} \quad (10.1)$$

$$R_i = R_p \parallel R_T = \frac{R_p R_T}{R_p + R_T} \quad (10.2)$$

Putting (10.2) into inverting op-amp gain equation gives

$$V_O(T) = -\frac{R_f}{R_i} V_i = -\frac{R_f(R_p + R_T)}{R_p R_T} V_i \quad (10.3)$$

On the other hand, it is known that the parallel linearization resistance is

$$R_p = R_{TM}\frac{\beta - 2T_M}{\beta + 2T_M} \quad (10.4)$$

where R_{TM} is the resistance of thermistor at mid-point of the working temperature, T_M. Substituting (10.4) into (10.3) yields

$$V_O(T) = -V_i \frac{R_f\left(R_{TM}\frac{\beta - 2T_M}{\beta + 2T_M} + R_T\right)(\beta + 2T_M)}{R_T R_{TM}(\beta - 2T_M)} \quad (10.5)$$

$$= -V_i \frac{R_f R_{T_i}(\beta - 2T_M) + R_f R_T(\beta + 2T_M)}{R_T R_{T_i}(\beta - 2T_M)} \quad (10.6)$$

$$V_O(T) = -V_i R_f\left[\frac{1}{R_T} + \frac{\beta + 2T_M}{R_{T_i}(\beta - 2T_M)}\right] \quad (10.7)$$

Problem 10.1.38

(a) Determine the voltage at the output of op-amp as a function of temperature in the circuit shown in Fig. 10.45. $R_T = R(T)$ denotes a NTC thermistor and Rs is the linearization resistor.

(b) Use MATLAB to plot the voltage at the output of op-amp as a function of temperature for $Vi = -2.5$ V, $R_f = 1.5$ kΩ, NTC thermistor is EPCOS

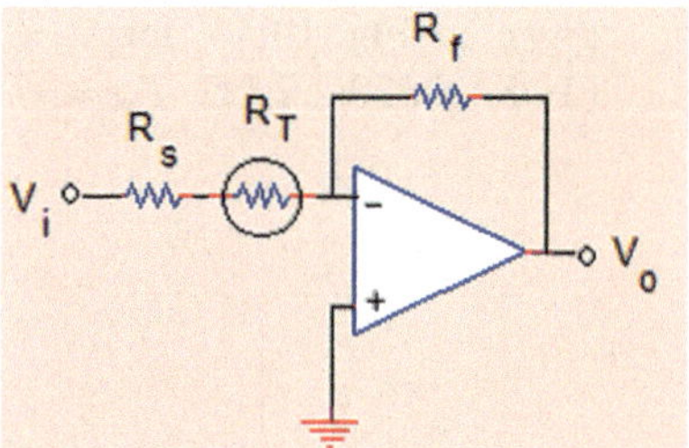

Fig. 10.45 The circuit for Problem 10.1.38

(TDK) C620/2.2 kΩ [4], T_m = 37 °C, T_L = 27 °C, T_H = 47 °C. (Use SPICE data to compute thermistor beta parameter, first). (thermlin5_series.m).

(c) V(27 °C) = ?, V(37 °C) = ?, V(47 °C) = ? Determine the sensitivity of the circuit, in mV/°C.
Repeat the same computations for R_f = 10 kΩ. What finally limits the maximum value of the linear range?

Solution

$$R_i = R_s + R_T, \tag{10.8}$$

$$R_T = R_0 e^{\beta\left(\frac{1}{T}-\frac{1}{T_0}\right)} \tag{10.9}$$

$$V_0(T) = -\frac{R_f}{R_i}V_i = -\frac{R_f}{R_s + R_T}V_i \tag{10.10}$$

It is known that series linearization resistance is

$$R_s = R_{T_M}\frac{\beta - 2T_M}{\beta + 2T_M} \tag{10.11}$$

where R_{TM} is the resistance of thermistor at mid-point of the working temperature. Substituting known quantities into (10.10) yields

$$V_0(T) = -V_i\frac{R_f(\beta + 2T_M)}{2T_M(R_T - R_{T_M}) + R_T(\beta + 2T_M)} \tag{10.12}$$

This can also be written as

$$V_0(T) = -V_i\frac{R_f(\beta + 2T_M)}{2T_M(R_T - R_{T_M}) + R_T(R_T + R_{T_M})} \tag{10.13}$$

at mid-point temperature, $T = T_M,\ R_T = R_{T_M}$;

$$V_0(T_M) = -V_i\frac{R_f(\beta + 2T_M)}{2\beta R_{T_M}} \tag{10.14}$$

(b) A plot of the voltage at the output of op-amp as a function of temperature is given in Fig. 10.46, for V_i = −2.5 V, R_f = 1.5 kΩ, NTC thermistor EPCOS (TDK) C620/2.2 kΩ, T_m = 37 °C, T_L = 27 °C, T_H = 47 °C.

A MATLAB script is also presented below.

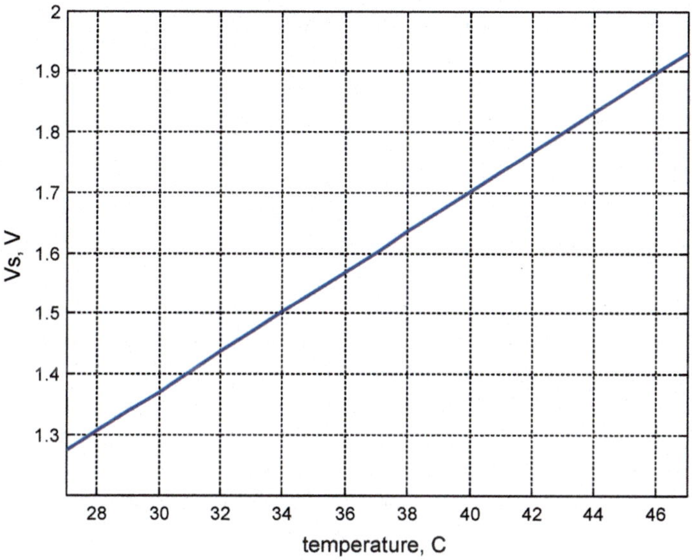

Fig. 10.46 Plot of voltage vs temperature for the Problem 10.1.38b

A MATLAB script is also presented below.

```
%Thermistor linearization using inverting opamp and series resistor
clc;clear all;
Vi=-2.5;%reference voltage
Rf=1.5e3;
%R-T measurement using SPICE data
T0=25; R0=2146;
TM=37; RTM=1386;
B=log(RTM/R0)/(1./(273+TM)-(1./(273+25)))%beta computation
Rs=RTM*(B-2*(273+TM))/(B+2*(273+TM))%Rp=Rs
TLow=27; %C
THigh=47;%C
T=(TLow:THigh)';
RT=R0*exp(B*(1./(273+T)-1./(T0+273)));
%[T RT]
%plot(T,RT);grid on;
Vs=-Vi*Rf*(B+2*(273+TM))./(RTM*(B-2*(273+TM))+RT.*(B+2*(273+TM)));
[T Vs]
plot(T,Vs,'linewidth',1.5);grid on;xlabel('temperature, C');ylabel('Vs, V');
xlim([27 47]);
S=(max(Vs)-min(Vs))/(THigh-TLow)*1000 %Sensitivity in mV/C
```

(c) Using the MATLAB script, for $R_f = 1.5$ kΩ

V(27 °C) = 1.2733 V, V(37 °C) = 1.602 V, V(47 °C) = 1.9310 V. Temperature Sensitivity is computed as 32.885 mV/°C.

If $R_f = 10$ kΩ, V(27 °C) = 8.4886 V, V(37 °C) = 10.6802 V, V(47 °C) = 12.8732 V, Temperature Sensitivity = 219.2326 mV/°C. Maximum value of the linear range is limited by the power supply voltage of the op-amp.

Problem 10.1.39

(a) Determine the linearized voltage at the output of op-amp as a function of temperature in the circuit shown in Fig. 10.47. $R_T = R(T)$ denotes a NTC thermistor and *Rs* is the linearization resistor.
(b) Plot the voltage transfer function $V_s = Vo/Vi$ as a function of temperature (without employing beta parameter of the thermistor) by exporting thermistor *R-T* data from SPICE into MATLAB, for $R_f = 1.5$ kΩ. NTC thermistor is EPCOS (TDK) C620/2200 Ω. Determine and plot the first-order linear approximation voltage transfer function *Vss* to the VTF V_s, in an overlaid plot.
(c) Plot the voltage and temperature error functions and state the maximum error magnitude, as well as the number of alternances occurring on this graph. How does the decreasing of temperature measurement interval influence the temperature measurement error? What can be the superiority of the circuit over simple "series resistor and NTC thermistor" linearization? Does it influence voltage and temperature measurement errors? (thermlin5_LT.cir), (thermlin7.m).

Solution

$$R_i = R_s + R_T, \tag{10.15}$$

$R_T = R(T)$ denotes a NTC thermistor and *Rs* is the linearization resistor value. On the other hand, it is known that the linearization resistance for a single NTC thermistor element is

$$R_s = \frac{R_M(R_H + R_L) - 2R_H R_L}{R_H + R_L - 2R_M} \tag{10.16}$$

where R_S = series resistor value, R_L = lower "zero error point" resistance, R_M = mid-point resistance, R_H = higher "zero error point" resistance. Then,

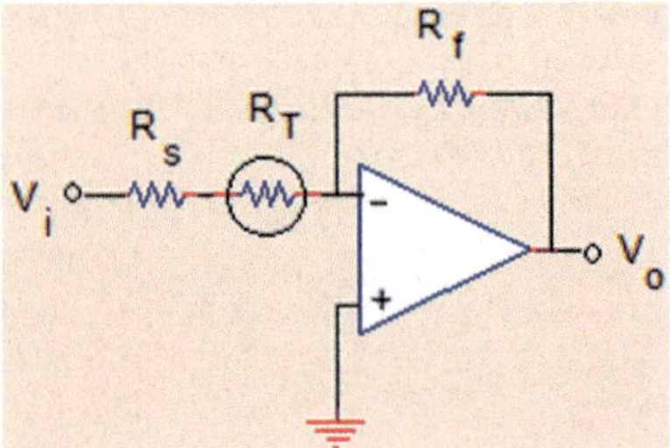

Fig. 10.47 The circuit for Problem 10.1.39

$$\frac{V_0(T)}{V_i} = -\frac{R_f}{R_i} = -\frac{R_f}{R_s + R_T} \tag{10.17}$$

At this stage, this equation is not processed further and left in its relatively implicit form.

(b) First, an *R-T* characteristic curve for the given type of NTC thermistor element (EPCOS–TDK- C620/2.2 k) is obtained (in plotted as well as in text data form) in LTSPICE, see, Fig. 10.48. Then, data (in text file) is exported into MATLAB (LTSPICE File/export/select trace to export; then go back to "all files," select relevant .txt file—open to check its contents first—then close, drag-and-drop by Mouse from LTSPICE/all files area into MATLAB current file area, then initiate file/import data procedure, after "data1" file appears in the MATLAB workspace, it is copied into current file area and renamed as TRdata1.mat for future applications). The *R-T* curve is replotted in MATLAB, as shown in Fig. 10.49. (Remark to trace name).

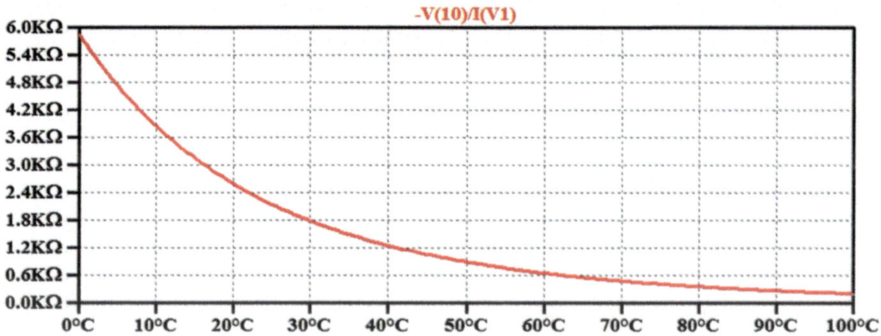

Fig. 10.48 The R-T characteristic curve for an NTC thermistor (SPICE plot)

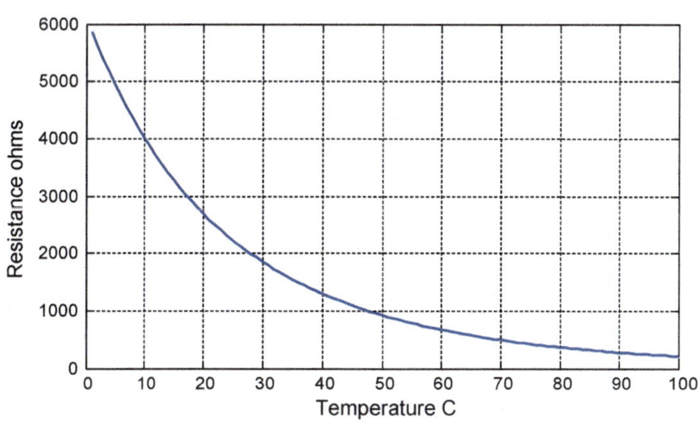

Fig. 10.49 The R-T curve, replotted in MATLAB

(c) If the slope $p1$ and offset $p2$ of the first-order function are computed via curve fitting, first-order linear approximation as a function of temperature T has the form $y = p1.T + p2$. The voltage transfer function $V_s = V_O/V_i$ (as a function of temperature) is plotted (without any attempt to use the beta parameter of given thermistor, $R_f = 1.5$ kΩ) along with the approximating first-order polynomial function ($Vss = -0.01508\text{T} - 0.1739$) as shown in Fig. 10.50. V_s is a linearized voltage transfer function using a single series resistor and Vss is the best linear approximation to this function in least square sense.

Then, first-order linearization error in voltage transfer function is

$$errV = y - V_S$$

Absolute measurement error as a function of temperature is displayed in Fig. 10.51. The maximum linearization error magnitude is less than 60 mV/V within the 100 °C temperature span, and the number of alternances occurring on this graph is four.

In most cases, one is interested in temperature measurement error rather than the voltage measurement error. If we rewrite measured temperature parameter t as

$$t = \frac{Vs - p2}{p1}$$

Then the temperature measurement error, becomes

$$errT = t - T$$

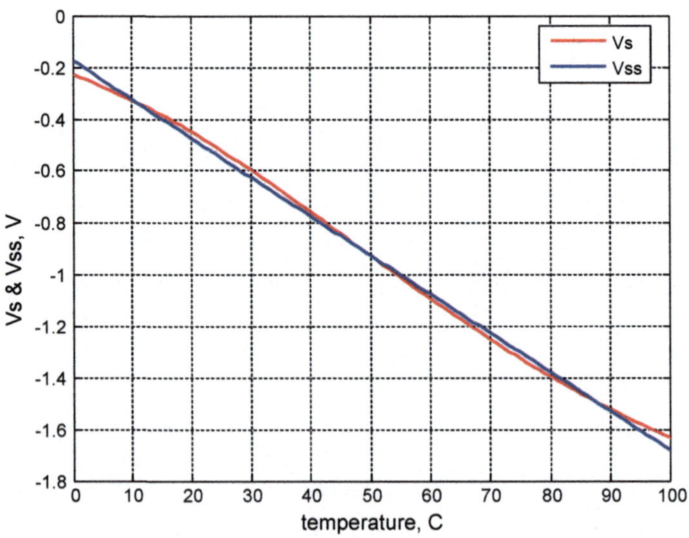

Fig. 10.50 Graph of the voltage transfer function along with the approximating first-order polynomial as functions of temperature

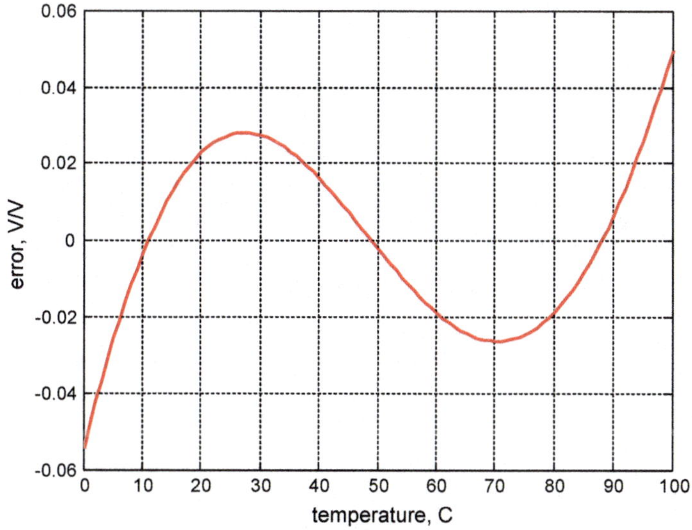

Fig. 10.51 Plot of the absolute measurement error as a function of temperature

It is noted here that larger temperature spans cause relatively large errors. Working in smaller temperature ranges considerably reduces the temperature measurement error. This is demonstrated in Fig. 10.52. It should also be pointed out that each interval change results in a different value of resistance in series with NTC thermistor element. For example, thermistor linearization within 0–25 °C temperature interval requires a series implementation of a resistor with a value of 2665.3 Ω, in contrast to R_s = 722.2 Ω for a temperature range of 0–100 °C, and Rs = 1647.4 Ω for 0–50 °C.

This circuit is advantageous as compared to bare series resistor and NTC thermistor linearization implementation, because inverting op-amp configuration provides the strength of slope adjustment possibility (which is a significant matter). This is illustrated in Fig. 10.51. As it would be expected, feedback resistor change is reflected on transfer function slope, directly proportional to the value of feedback resistor.

However, by running circuit simulation studies at different feedback resistance values at a constant temperature interval, it is noted that feedback resistance variation does not influence the temperature error, besides the voltage gain and the voltage error, all are functions of temperature (Figs. 10.52 and 10.53).

A MATLAB script (thermlin7.m) for thermistor linearization using inverting op-amp:

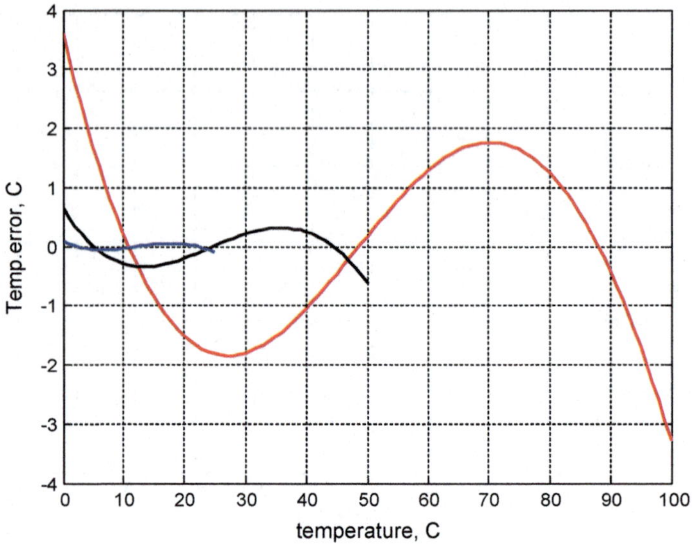

Fig. 10.52 The temperature error

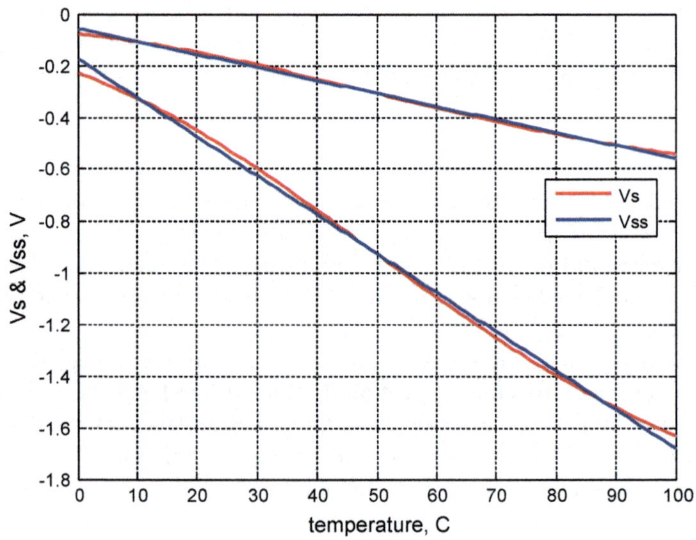

Fig. 10.53 The voltage transfer function at different feedback resistance values at a constant temperature interval

```
%select p1, p2, TX, Rf set first.
clc;clear all;
%NTC  resistances , (SPICE values imported in file named
%TRdata1.mat)
%NTC sample: EPCOS(TDK) C620/2.2k
load 'TRdata1.mat';
Rf=1.5e3;
TX=100; % 0<=TX<=100C
T=0:TX; T=T';
RT=data1(T+1); RL=min(RT);RH=max(RT);RM=median(RT);
RS=(RM*(RH+RL)-2*RH*RL)/(RH+RL-2*RM)
Vs=-Rf./(RS+RT); %Vss=curve fit T vs Vs (linear)
fo = fit(T, Vs,'poly1')
% Linear model Poly1: fo(x) = p1*x + p2, Coefficients (95% conf.
%bounds):
p1 =-0.01508  %Rf=1.5k,100C
p2 =-0.1739
Vss=p1*T+p2; plot(T,Vs,'r-',T,Vss,'linewidth',2);grid on;
legend('Vs','Vss','Location','best');
xlabel('temperature,  C');ylabel('Vs & Vss,  V');hold on;
```

Problem 10.1.40 In the circuit shown in Fig. 10.54, determine the voltage transfer ratio.

Solution
KCL at (a):

$$\frac{V_a - V_o}{R_f} + \frac{V_a}{R_2} + \frac{V_a - V_s}{R_1} = 0 \quad \rightarrow \quad \frac{V_a}{R_f} - \frac{V_o}{R_f} + \frac{V_a}{R_2} + \frac{V_a}{R_1} - \frac{V_s}{R_1} = 0$$

$$V_a\left(\frac{1}{R_2} + \frac{1}{R_f} + \frac{1}{R_1}\right) - \frac{V_s}{R_1} = \frac{V_o}{R_f} \tag{10.18}$$

KCL at (b) and noting that $V_a = V_b$ due to op-amp,

$$\frac{V_b}{R_4} + \frac{V_b - V_s}{R_3} = 0 \quad \rightarrow \quad \frac{V_a}{R_4} + \frac{V_a}{R_3} - \frac{V_s}{R_3} = 0$$

$$V_a\left(\frac{1}{R_3} + \frac{1}{R_4}\right) = \frac{V_s}{R_3}$$

Fig. 10.54 The circuit for Problem 10.1.40

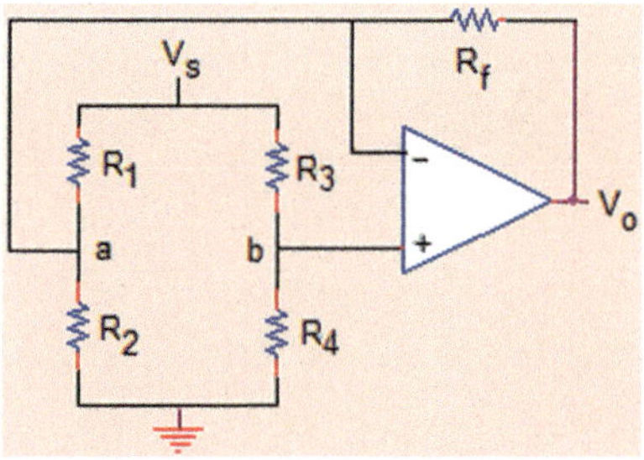

$$V_a = \frac{V_s}{R_3\left(\frac{1}{R_3} + \frac{1}{R_4}\right)} = \frac{V_s \cdot R_3 R_4}{R_3(R_3 + R_4)} = \frac{R_4}{R_3 + R_4} \cdot V_s \tag{10.19}$$

(10.19) → (10.18):

$$\left(\frac{R_4 V_s}{R_3 + R_4}\right)\left(\frac{1}{R_2} + \frac{1}{R_f} + \frac{1}{R_1}\right) - \frac{V_s}{R_1} = \frac{V_o}{R_f}$$

$$\frac{V_o}{R_f} = V_s\left[\left(\frac{R_4}{R_3 + R_4}\right)\left(\frac{1}{R_2} + \frac{1}{R_f} + \frac{1}{R_1}\right) - \frac{1}{R_1}\right]$$

$$\frac{V_o}{V_s} = A = R_f\left[\left(\frac{R_4}{R_3 + R_4}\right)\left(\frac{1}{R_2} + \frac{1}{R_f} + \frac{1}{R_1}\right) - \frac{1}{R_1}\right] \quad \text{V/V}$$

Problem 10.1.41 Is the circuit shown in Fig. 10.55, always suitable for a biomedical temperature or pressure sensing application?

Solution

Diffamp used in this circuit has low-input impedance. It will load the Wheatstone bridge.

$$R_\alpha = R_1 + R_2 \,\|\, (R_i + R_f)$$

$$i_\alpha = \frac{V_s}{R_\alpha}$$

$$V_a = i_\alpha \cdot \frac{R_2 \cdot (R_i + R_f)}{R_2 + R_i + R_f} = V_s \cdot \frac{R_2 \cdot (R_i + R_f)}{R_1 R + R_2 \cdot (R_i + R_f)} \tag{10.20}$$

Note that for high-input impedancevoltage differencing circuit,

$$V_a = \frac{R_2}{R_1 + R_2} \cdot V_s \tag{10.21}$$

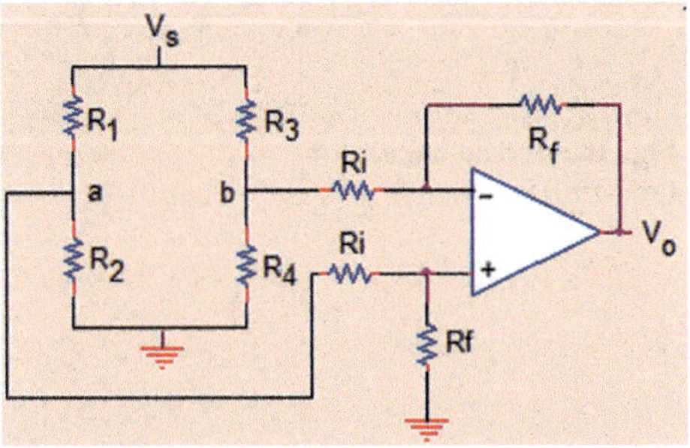

Fig. 10.55 The circuit for Problem 10.1.41

Comparing Eqs. (10.20) and (10.21) shows the loading effect of R_i, R_f. It is recommended that a high–input impedance circuitry (preferably an instrumentation amplifier) be used for sensor measurements.

Problem 10.1.42 A circuit for linearizing a single-element varying bridge is shown in Fig. 10.56. Op-amp output can be used as a low-impedance output point for the bridge measurement. The output is linear, even for large values of sensor resistance variance. (This bridge is usually followed by a second amplifier due to the small output signal). Determine the output voltage.

Solution

By applying voltage division rule at the positive terminal of the op-amp, and op-amp property,

$$V_+ = \frac{U}{2} = V_-$$

Current flowing through the feedback resistor,

$$i = \frac{U - \frac{U}{2}}{R} = \frac{\frac{U}{2} - V_0}{R + \Delta R} \quad \rightarrow \quad \frac{U}{2R} = \frac{U - 2V_0}{2(R + \Delta R)}$$

$$U(R + \Delta R) = R(U - 2V_0) = UR - 2V_0R \quad \rightarrow \quad U(R + \Delta R) - UR = -2V_0R$$

$$V_0 = -\frac{U}{2R}\Delta R$$

Problem 10.1.43 Determine the voltage at the output of circuit shown in Fig. 10.57.

Solution

$$V_+ = \frac{R}{R + R + \Delta R} = \frac{R}{2R + \Delta R} = V_- \tag{10.22}$$

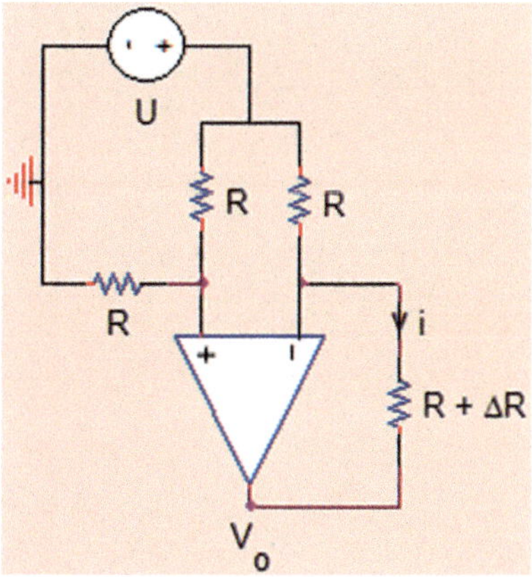

Fig. 10.56 Circuit for linearizing a bridge

$$i = \frac{U - V_-}{R} = \frac{V_- - V_0}{R + \Delta R} \tag{10.23}$$

$$\frac{U}{R} - \frac{V_-}{R} - \frac{V_-}{R + \Delta R} = \frac{-V_0}{R + \Delta R} \tag{10.24}$$

$$\frac{-V_0}{R + \Delta R} = \frac{U}{R} - V_-\left(\frac{1}{R} + \frac{1}{R + \Delta R}\right) \tag{10.25}$$

Substitute (10.22) into (10.25)

$$\frac{-V_0}{R + \Delta R} = \frac{U}{R} - \left(\frac{UR}{2R + \Delta R}\right)\left(\frac{2R + \Delta R}{R(R + \Delta R)}\right) = U\left(\frac{1}{R} - \frac{1}{R + \Delta R}\right) = \frac{U\Delta R}{R(R + \Delta R)}$$

$$V_0 = -\frac{U \cdot \Delta R}{R}$$

This is a circuit for linearizing a two-element varying voltage-driven bridge. It is similar to the circuit for linearizing a single element varying bridge, but has twice the sensitivity. Additional gain (op-amp amplifier stage) may be necessary, because it can provide an output signal with relatively small amplitude.

Problem 10.1.44 In Fig. 10.58, input of an instrumentation amplifier with the gain $G = 22$ is connected to (a, b) terminals of a bridge circuit.

(a) $R_1 = 12\,\text{k}\Omega$, $R_2 = 20\,\text{k}\Omega$, $R_3 = 2\,\text{k}\Omega$, $R_4 = 10\,\text{k}\Omega$. $V_o/V_i = ?$
(b) Determine the value of R_G if AD620 type Instrumentation amplifier IC is used.

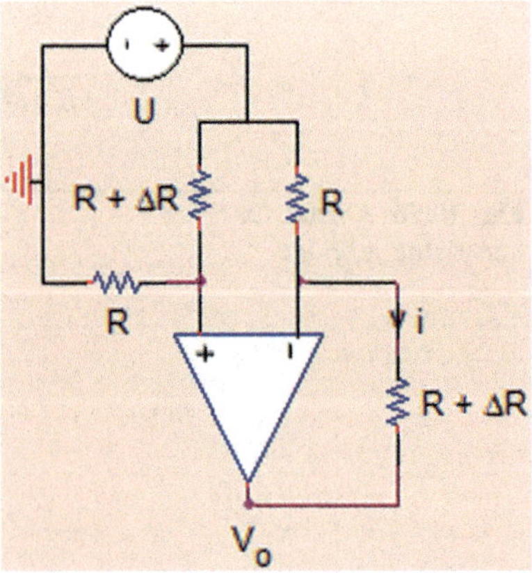

Fig. 10.57 The circuit for Problem 10.1.43

Solution

$$V_a = V_i \cdot \frac{R_4}{R_3 + R_4} = V_i \cdot \frac{10}{2+10} = 0.833V_i, \quad V_b = V_i \cdot \frac{R_2}{R_1 + R_2} = V_i \cdot \frac{20}{12+20} = 0.625V_i$$

$$V_o = G(V_a - V_b) = 22 \times (0.833 - 0.625)V_i = 4.583 \cdot V_i$$

$$\frac{V_o}{V_i} = 4.583 \text{ (V/V)}$$

(b) Inspection of the AD 620 data sheet given by its manufacturer [5] (Analog Devices Inc., Norwood, MA, U.S.A), shows that AD620 is a monolithic instrumentation amplifier based on a modification of the classic three op-amp approach. Absolute value trimming allows the user to program gain with a single resistor. The gain is given by

$$G = \frac{R_1 + R_2}{R_G} + 1$$

where the internal gain resistors, R_1 and R_2, are trimmed to 24.7 kΩ, then, the gain equation becomes

$$G = \frac{49.4\,\text{k}\Omega}{R_G} + 1$$

The gain adjusting resistor value is calculated by substituting given gain value into this equation,

$$R_G = \frac{49.4\,\text{k}\Omega}{G-1} = \frac{49.4\,\text{k}\Omega}{22-1} = 2.35\,\text{k}\Omega$$

Problem 10.1.45 Design a Wheatstone bridge to measure the voltage over a Pt 100 type temperature sensor placed in one of its grounded arms using a commercially available instrumentation amplifier (IA) with overall gain of 10 V/V. Run a

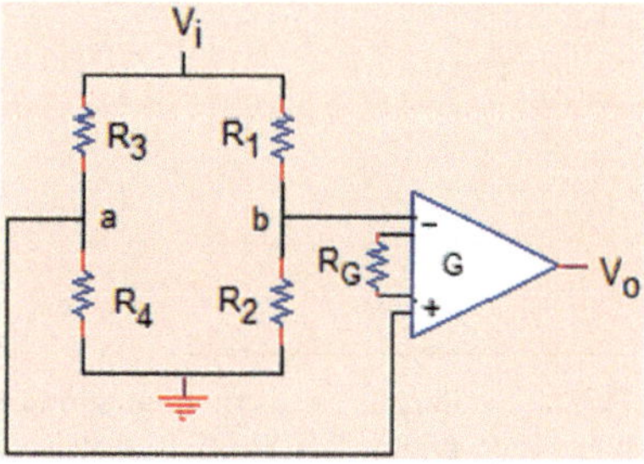

Fig. 10.58 The circuit for Problem 10.1.44

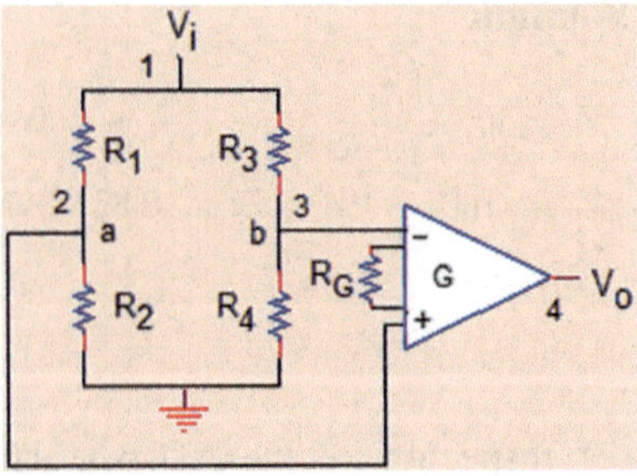

Fig. 10.59 Wheatstone bridge and instrumentation amplifier

simulation using SPICE, and plot the voltage at the output of IA against resistance of the sensor (wheatstone2.cir, ad620.cir).

Solution

Let R_4 represent the sensor, and select the values of the other 3 resistors equal to 100 Ω. As the IA, AD620 (Analog Devices Inc.) can be used in this circuit. SPICE Node assignments are shown in Fig. 10.59. The gain adjusting resistor value for the IA is calculated by substituting given gain value ($G = 10$) into the following equation,

$$R_G = \frac{49.4\,\text{k}\Omega}{G-1} = \frac{49.4\,\text{k}\Omega}{10-1} = 5.488\,\text{k}\Omega$$

While the resistor R_4 of the Wheatstone bridge is linearly varied between 0 and 100 Ω, the output of instrumentation amplifier changes as shown in Fig. 10.60. Due to supply voltage limiting, output voltage of the instrumentation amplifier saturates for $R_4 < 69\ \Omega$.

(1 ms corresponds 1Ω on the horizontal axis, e.g., 82 ms corresponds a resistance value of $R_4 = 82\ \Omega$[1]).

SPICE net list:

```
wheatstone bridge with variable resistor
.lib ad620.cir
VCC 1 0 DC 12 V
VEE 5 0 DC -12 V
R1 1 2 100
R2 2 0 100
```

[1]Voltage-controlled resistor simulation can also be performed by using "step parameter" statement (e.g., .step param R 100 60 1 = sweep R from 100 to 60 by 1 Ω increments, see, for example, LTSPICE). This requires no further effort to make axis manipulations. Introducing R parameter as the variable component name is sufficient for a straightforward analysis.

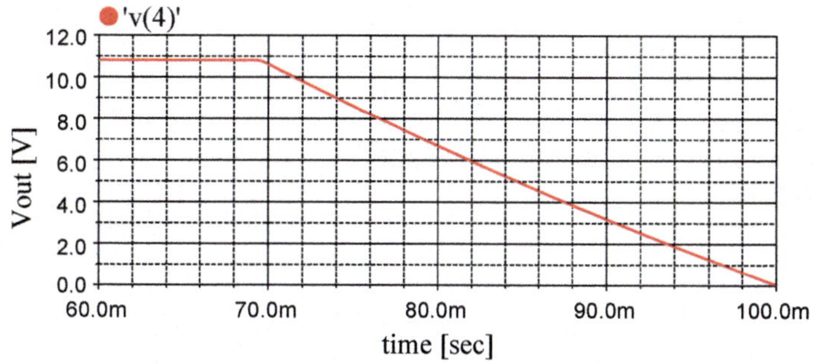

Fig. 10.60 Voltage at the output of IA against resistance of the sensor. Note that, horizontal axis (in time) corresponds to the resistance of the sensor, e.g., 82 ms corresponds a resistance value of $R_4 = 82\ \Omega$

```
R3 1 3 100
*decreasing R4 causes increasing voltage at the output v4
R5 4 0 10Meg
R6 7 8 5.48888 k
*R6 is adjusted for gain=10 ;Formula: Rg=49.4/(G-1), in kiloohm
*XOA1 NI  I  +  -  Vo  0  Rg1  Rg2  ad620
XOA1  2  3  1  5  4  0  7   8  ad620
*VOLTAGE-CONTROLLED RESISTOR
* Control Voltage
Vcont    10    0    PWL(0mS 0 V  100mS 1.0 V)
R10  10    0    1MEG
* VCR subcircuit call
XVC1    3 0    10 0    VCR
.Subckt VCR  1 2  4 5
*Resistor terminals: 1,2    Control terminals: 4,5
eRes  1  3 Poly(2) (6,0) (4,5) 0 0 0 0 1
Vsense 3  2  DC 0 V
Fcopy  0  6  Vsense 1
RRes  6  0  100
.ends
*v(4) is plotted
*.tran 1 ms 100 ms 0 1 ms
```

Problem 10.1.46

(a) Determine the voltage at the output of noninverting amplifier of Fig. 10.61.
(b) Comment on the form of relationship between the output voltage and the change of sensor resistance.

(c) Calculate the output voltage at 100 °C, if R_3 is a Pt 100 type sensor (use first-order approximation), R_6 = 10 kΩ, R_5 = 1 kΩ, U = 5 V.

Solution

The voltage at (a) is at virtual ground due to op-amp. Then, the current through R_1 and R_3 becomes

$$I = \frac{U}{R_1} = -\frac{U}{R_3} \tag{10.26}$$

The current through R_3 is the same as above, but its direction is negative with respect to virtual ground . Since no current flows to the noninverting terminal of op-amp through R_5, the voltage difference V_{ab} is;

$$V_{ab} = [U - (-U)] \times \left(\frac{R_3}{R_1 + R_3} - \frac{R_2}{R_2 + R_4}\right) \tag{10.27}$$

Let R_3 be a sensor. If its value changes

$$R_3' = R_3 + \Delta R_3 \tag{10.28}$$

$$IR' = I(R_3 + \Delta R_3) = [-U + I(-\Delta R_3)] = -U - \Delta U \tag{10.29}$$

From Eq. (10.27);

$$\frac{U - (U - \Delta U)}{R_2 + R_4} R_2 = \frac{2U + \Delta U}{R_2 + R_4} R_2 = \frac{2U}{R_2 + R_4} R_2 + \frac{\Delta U}{R_2 + R_4} R_2 \tag{10.30}$$

But when bridge is balanced,

$$\frac{2U}{R_2 + R_4} R_2 = 0 \tag{10.31}$$

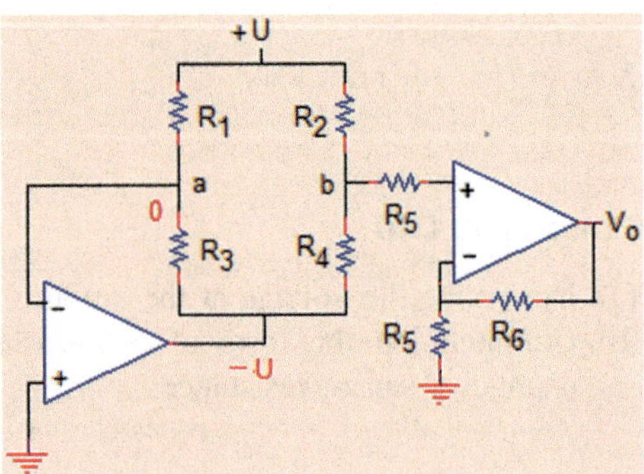

Fig. 10.61 The circuit for Problem 10.1.46

Then,

$$V_b = \frac{\Delta U}{R_2 + R_4} R_2 = \frac{\Delta U}{1 + \frac{R_4}{R_2}} \tag{10.32}$$

Since in a balanced bridge

$$\frac{R_4}{R_2} = \frac{R_3}{R_1} \tag{10.33}$$

Equation (10.32) can be rewritten as

$$V_b = \frac{\Delta U}{1 + \frac{R_3}{R_1}} \tag{10.34}$$

If $R_1 = R_2 = R_3 = R_4$,

$$\Delta U = I \times \Delta R_3 \tag{10.35}$$

Using Eq. (10.26)

$$V_b = \frac{U}{2} \times \frac{\Delta R_3}{R_3} = \frac{U}{2} \times \frac{\Delta R}{R} \tag{10.36}$$

The output voltage is $V_{ab} \times A = (0 - V_b) \times A$

$$V_o = -V_b \left(1 + \frac{R_6}{R_5}\right) \tag{10.37}$$

$$V_o = -\left(\frac{U}{2} \times \frac{\Delta R}{R}\right) \times \left(1 + \frac{R_6}{R_5}\right) \tag{10.38}$$

(b) The output is a linear function of ΔR,

$$V_0 = -k \times \Delta R \tag{10.39}$$

where;

$$k = \frac{U}{2R}\left(1 + \frac{R_6}{R_5}\right) \tag{10.40}$$

(c) If R_3 is a Pt 100 type sensor a first-order approximation, with $\alpha = 3.9 \times 10^{-3}$, gives

$$R_3 = R_0(1 + \alpha T) = R_0 + \alpha R_0 T = R_0 + \Delta R \tag{10.41}$$

and $R_3 = R_0$ (in balanced bridge case), then,

$$V_0 = -\frac{U}{2} \cdot A \cdot \frac{\Delta R}{R} = -\frac{UA\alpha R_0 T}{2R_0} = -\frac{UA\alpha T}{2} \tag{10.42}$$

Here,

$$A = 1 + \frac{R_6}{R_5}$$

$$R_6 = 10\,\text{k}\Omega, \quad R_5 = 1\,\text{k}\Omega, \quad U = 5\,\text{V}, \quad \alpha = 3.9 \times 10^{-3}$$

$$V_0 = -\frac{5 \times (1 + 101) \times 3.9 \times 10^{-3}}{2} \times T = -107.25 \times 10^{-3} \times T$$

At $T = 100°C$,

$$V_0 = -0.10725 \times 100 = -10.725V$$

Problem 10.1.47

(a) Determine the voltage at the output of the circuit shown in Fig. 10.62.
(b) Comment on the form of relationship between the output voltage and the change of sensor resistance [6].
(c) Calculate the output voltage at 100 °C, if R_3 is a Pt 100 type sensor (use first-order approximation), R_5 = 10 kΩ, $R_1 = R_4$ = 1 kΩ, U = 15 V, $\alpha = 3.9 \times 10^{-3}$

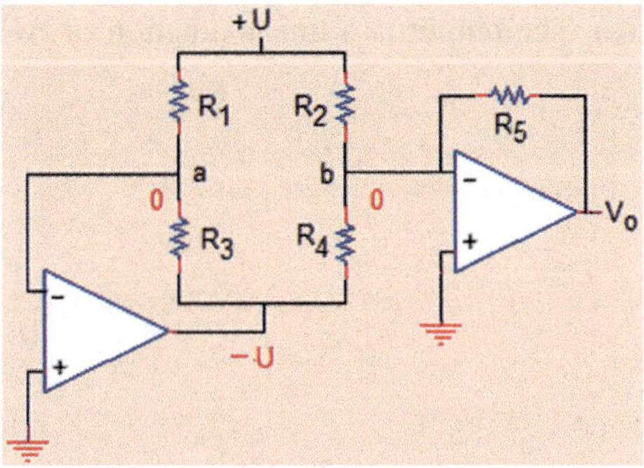

Fig. 10.62 The circuit for Problem 10.1.47

Solution
The voltage at (a) is at virtual ground due to op-amp. Then, the current through R_1 and R_3 becomes

$$I = \frac{U}{R_1} \tag{10.43}$$

The current through R_3 is the same as above, but its direction is negative with respect to virtual ground. Due to inverting behavior of the first (left) op-amp circuit, the voltage at node c becomes

$$V_c = -\frac{R_3}{R_1} \cdot U \tag{10.44}$$

Second op-amp to the right side of the bridge is wired as an inverting summer circuit,

$$V_o = -\frac{R_5}{R_2} \cdot U - \frac{R_5}{R_4} \cdot V_c \tag{10.45}$$

Substituting Eq. (10.44) into Eq. (10.45),

$$V_o = -\frac{R_5}{R_2} \cdot U + \frac{R_5}{R_4} \cdot \left(-\frac{R_3}{R_1} \cdot U\right) = -\frac{R_5}{R_2} \cdot U - \frac{R_5}{R_4} \cdot \frac{R_3}{R_1} \cdot U = -R_5 U\left(\frac{1}{R_2} - \frac{R_3}{R_1 R_4}\right)$$
$$V_o = -\frac{R_5 U}{R_4}\left(\frac{R_4}{R_2} - \frac{R_3}{R_1}\right) \tag{10.46}$$

In balanced case,

$$\frac{R_4}{R_2} = \frac{R_3}{R_1}$$

Which causes the output voltage to be zero volt. Let R_3 be a sensor and a variation take place in its value so that its new resistance becomes

$$R_3' = R_3 + \Delta R_3 \tag{10.47}$$

Then,

$$V_o = -\frac{R_5 U}{R_4}\left(\frac{R_4}{R_2} - \frac{R_3 + \Delta R_3}{R_1}\right) = -\frac{R_5 U}{R_4}\left(\frac{R_4}{R_2} - \frac{R_3}{R_1} - \frac{\Delta R_3}{R_1}\right) \tag{10.48}$$

First two terms within the last parenthesis cancel out (balance condition),

$$V_o = \frac{R_5 U}{R_4}\left(\frac{\Delta R_3}{R_1}\right) = k\,\Delta R_3, \quad \text{where} \quad k = \frac{R_5 U}{R_1 R_4} \tag{10.49}$$

(b) This equation indicates that the output voltage is directly proportional to the change in transducer resistance.
(c) If R_3 is a Pt 100 type sensor a first-order approximation, with $\alpha = 3.9 \times 10^{-3}$, gives

$$R_3 = R_0(1 + \alpha T) = R_0 + \alpha R_0 T = R_0 + \Delta R_3 \tag{10.50}$$

then,

$$V_o = k\,\Delta R_3 = k\alpha R_0 T = \left(\frac{R_5 U}{R_1 R_4}\right)\alpha R_0 T \tag{10.51}$$

Here, $R_5 = 10\,\text{k}\Omega, \quad R_1 = R_4 = 1\,\text{k}\Omega, \quad U = 15\,\text{V}, \quad \alpha = 3.9 \times 10^{-3}$

$$V_0 = \left(\frac{10k \times 15\,\text{V}}{1k \times 1k}\right) \times 3.9 \times 10^{-3} \times 0.1k \times T = 58.5 \times 10^{-3} \times T$$

At 100 °C,

$$V_0 = 0.0585 \times 100 = 5.85\,\text{V}$$

Problem 10.1.48

(a) In the circuit shown in Fig. 10.63, R_4 is a transducer with a linear dependency on the measure and x (temperature, pressure, strain…), i.e., $R_4 = R_N(1 \mp \alpha x)$, $R_1 = R_3$ and balanced bridge condition is obtained when $R_2 = R_4 = R_N$.
Find the voltage at the output of the instrumentation amplifier.
(b) State the condition under which the output voltage is a linear function of measurand [7].
(c) Calculate V_0 under the condition found in part (b), if $g = 1$, $G = 9$ V/V, $V_R = 1.2$ V, $T = 100$ °C for a RTD Pt 100 type temperature sensor.

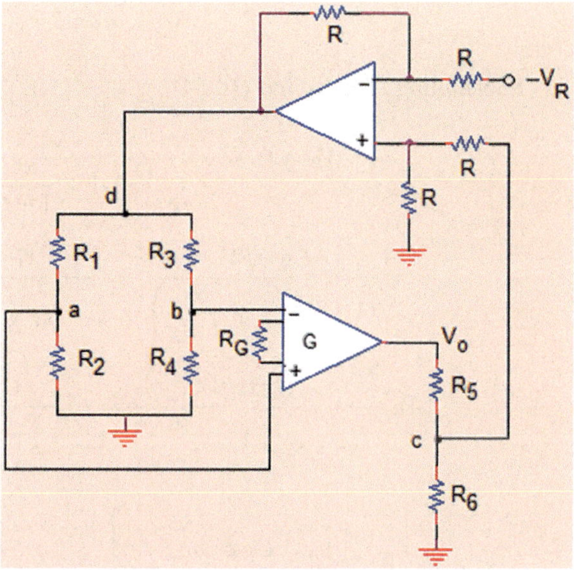

Fig. 10.63 The circuit for Problem 10.1.48

Solution

(a) Assume that only Wheatstone bridge with an external supply voltage at V_d exists. Then

$$V_{ab} = \frac{g}{(1+g)^2} \times \frac{\alpha x}{1 + \frac{\alpha x}{1+g}} \tag{10.52}$$

here,

$$g = \frac{R_1}{R_2} = \frac{R_3}{R_N} \tag{10.53}$$

Now, include the interface circuit. The voltage at the output of instrumentation amplifier is

$$V_0 = G \times V_{ab} = G \times \frac{g}{(1+g)^2} \times \frac{\alpha x}{1 + \frac{\alpha x}{1+g}} \times V_d \tag{10.54}$$

This output is reduced by a factor of β, by voltage divider

$$V_c = \beta V_0 \tag{10.55}$$

It is then fed to the noninverting input of unity gain differential amplifier circuit whose other input is supplied by the reference voltage $(-V_R)$. The output of differencing circuit is

$$V_d = \beta V_0 + V_R \tag{10.56}$$

Substituting (10.56) into (10.54),

$$V_0 = G \times \frac{g}{(1+g)^2} \times \frac{\alpha x}{1 + \frac{\alpha x}{1+g}} \times \beta V_0 + V_R \tag{10.57}$$

$$= \frac{Gg\alpha x\beta V_0}{(1+g)^2\left(1+\frac{\alpha x}{1+g}\right)} + \frac{Gg\alpha x V_R}{(1+g)^2\left(1+\frac{\alpha x}{1+g}\right)} \tag{10.58}$$

$$V_0 - \frac{Gg\alpha x\beta V_0}{(1+g)^2\left(1+\frac{\alpha x}{1+g}\right)} = \frac{Gg\alpha x V_R}{(1+g)^2\left(1+\frac{\alpha x}{1+g}\right)} \tag{10.59}$$

$$V_0\left[1 - \frac{Gg\alpha x\beta}{(1+g)^2\left(1+\frac{\alpha x}{1+g}\right)}\right] = \frac{Gg\alpha x V_R}{(1+g)^2\left(1+\frac{\alpha x}{1+g}\right)} \tag{10.60}$$

$$V_0 = \frac{\dfrac{Gg\alpha x V_R (1+g)^2\left(1+\frac{\alpha x}{1+g}\right)}{(1+g)^2\left(1+\frac{\alpha x}{1+g}\right) - Gg\alpha x\beta}}{(1+g)^2\left(1+\frac{\alpha x}{1+g}\right)} = \frac{Gg\alpha x V_R}{(1+g)^2\left(1+\frac{\alpha x}{1+g}\right) - Gg\alpha x\beta} \tag{10.61}$$

(b) Factorize denominator,

$$V_0 = \frac{Gg\alpha x V_R}{(1+g)^2\left[1 + \frac{\alpha x}{1+g} - \frac{Gg\alpha x\beta}{(1+g)^2}\right]} \tag{10.62}$$

$$V_0 = \frac{Gg\alpha x V_R}{(1+g)^2\left[1 + \frac{\alpha x}{1+g}(1+g-Gg\beta)\right]} \tag{10.63}$$

If

$$\beta G = \frac{1+g}{g} \tag{10.64}$$

The term in large brackets in denominator becomes unity;

$$V_0 = \frac{Gg\alpha x V_R}{(1+g)^2\left[1 + \frac{\alpha x}{1+g}(1 + g - \frac{1+g}{g}g)\right]} \tag{10.65}$$

$$V_0 = \frac{Gg\alpha x V_R}{(1+g)^2} = kx \tag{10.66}$$

This result indicates that the voltage obtained at the output of instrumentation amplifier is linearly proportional to measurand value x under the condition that the Eq. (10.64) is satisfied. The proportionality constant is

$$k = \frac{Gg\alpha V_R}{(1+g)^2} \tag{10.67}$$

(c) If the sensor is a RTD Pt 100, $\alpha = 3.9083 \times 10^{-3}$, $G = 9\,\mathrm{V/V}$, $g = 1$, $V_R = 1.2\,\mathrm{V}$, the output voltage at $x = T = 100$ °C is calculated by substituting these values into Eq. 10.66,

$$V_0 = \frac{9 \times 1 \times 3.9083 \times 10^{-3} \times 1.2}{(1+1)^2} \times 100 = 1.055\,\mathrm{V}$$

Problem 10.1.49 An 8 bit DAC covers a voltage range of 0–10 V. Calculate how much voltage each bit is worth.

Solution

$$\frac{100}{255} = 0.0392\,\mathrm{V} = 39.2\,\mathrm{mV}$$

Problem 10.1.50 Design a 5-bit Digital-to-Analog converter (DAC) using R-2R circuit at the input of an op-amp.

Solution
The circuit of Fig. 10.64 is the solution.

$$V_o - iR_f$$

if $R_f = 2R$

$$D = \frac{b_1}{2^1} + \frac{b_2}{2^2} + \frac{b_3}{2^3} + \frac{b_4}{2^4} + \frac{b_5}{2^5}$$

$$I_o = \frac{V}{64R}, \quad V_o = -U \cdot D$$

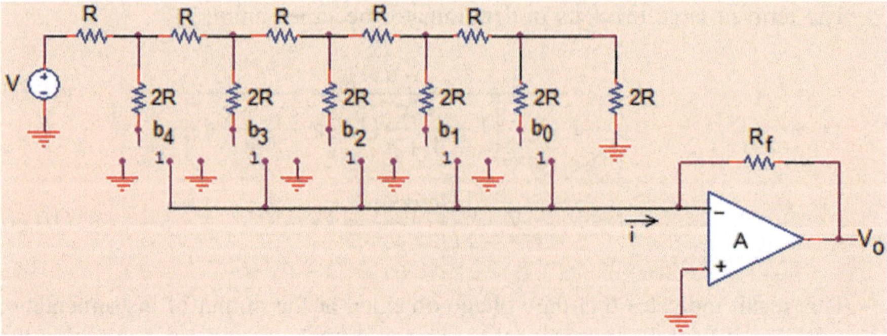

Fig. 10.64 The circuit for Problem 10.1.50

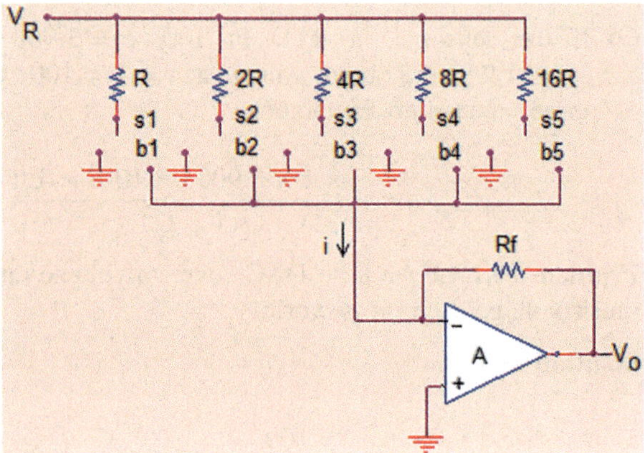

Fig. 10.65 The circuit for Problem 10.1.51

Problem 10.1.51 A 5-bit digital-to-analog using binary weighted resistor configuration is shown in Fig. 10.65.

(a) What is the expression for the output voltage in term of switch positions and the reference voltage (V_R)?

(b) What is the value of output voltage for a digital input of $B_5 = 00101$, $B_{31} = 11111$? (Let $V_R = 5$ V, $b_1\ b_2\ b_3\ b_4\ b_5 = B$, $b_1 = \text{MSB}$, Most significant bit).

(c) What is the maximum resistor ratio at the input of op-amp?

Solution

$$i = \frac{V_R}{R} \cdot b_1 + \frac{V_R}{2R} \cdot b_2 + \frac{V_R}{4R} \cdot b_3 + \frac{V_R}{8R} \cdot b_4 + \frac{V_R}{16R} \cdot b_5$$
$$i = \frac{2 \cdot V_R}{R} \left(\frac{b_1}{2} + \frac{b_2}{2^2} + \frac{b_3}{2^3} + \frac{b_4}{2^4} + \frac{b_5}{2^5} \right) = \frac{2 \cdot V_R}{R} \cdot B$$

Let $R_f = R/2$,

$$V_o = -i \cdot R_f = \left(2 \cdot \frac{V_R}{R} \right) \cdot B \cdot \frac{R}{2} = -V_R \cdot B$$
$$B_0 = \frac{0}{2} + \frac{0}{2^2} + \frac{0}{2^3} + \frac{0}{2^4} + \frac{0}{2^5} = 0$$
$$B_1 = \frac{0}{2} + \frac{0}{2^2} + \frac{0}{2^3} + \frac{0}{2^4} + \frac{1}{2^5} = \frac{1}{2^5} = \frac{1}{32}\ \mathrm{V}$$
$$B_2 = \frac{0}{2} + \frac{0}{2^2} + \frac{0}{2^3} + \frac{1}{2^4} + \frac{0}{2^5} = \frac{2}{32}\ \mathrm{V}$$
$$B_3 = \frac{0}{2} + \frac{0}{2^2} + \frac{0}{2^3} + \frac{1}{2^4} + \frac{1}{2^5} = \frac{3}{32}\ \mathrm{V}$$
$$B_4 = \frac{0}{2} + \frac{0}{2^2} + \frac{1}{2^3} + \frac{0}{2^4} + \frac{0}{2^5} = \frac{4}{32}\ \mathrm{V}$$
$$B_5 = \frac{1}{2} + \frac{1}{2^2} + \frac{1}{2^3} + \frac{1}{2^4} + \frac{0}{2^5} = \frac{30}{32}\ \mathrm{V}$$

$\ldots$

$$B_{30} = \frac{1}{2} + \frac{1}{2^2} + \frac{1}{2^3} + \frac{1}{2^4} + \frac{0}{2^5} = \frac{30}{32}\ \mathrm{V}$$
$$B_{31} = \frac{1}{2} + \frac{1}{2^2} + \frac{1}{2^3} + \frac{1}{2^4} + \frac{1}{2^5} = \frac{31}{32}\ \mathrm{V}$$

$$\textit{if}\ (V_R = 5\,\mathrm{V}), \quad B_5 = \frac{5}{32}$$

$$V_o = -5 \cdot \frac{5}{32} = -\frac{25}{32}\ \mathrm{V} = 0.78125\ \mathrm{V}$$

if $B_{31} = \dfrac{31}{32} \quad V_o = -5 \cdot \dfrac{31}{32} = 4.84375\ \mathrm{V}$

(c) Resistor spread = Max. resistor ratio = $2^{N-1} = 2^{5-1} = 2^4 = 16$

$$\left(\text{note that, Max. resistor ratio} = \frac{16R}{R} = 16 \right),$$

i.e., if $R = 1\,\mathrm{k\Omega}, \quad b_1 \rightarrow 1\,\mathrm{k\Omega}, \quad b_5 \rightarrow 16\,\mathrm{k\Omega}$

Problem 10.1.52 General multi-input single-output (MISO) op-amp circuit design: One of the basic analog signal processing tool is so-called multi-input single-output MISO general summing op-amp having the characteristic of the form

$$V_o = \sum_{i=1}^{m} b_i V_{bi} - \sum_{j=1}^{n} a_i V_{aj}$$

Here, a, b are the constant (gains). The number of inputs is $(n + m)$. Note to the presence of one extra element on the inverting input of the op-amp, which may be included or not needed for a particular design.

Design steps [8] (see Fig. 10.66),

(1) Set $G_f = 1$ S, $G_{ai} = a_i.S$, $i = 1,2, \ldots n$, $G_{bj} = b_j S$, $j = 1, 2 \ldots m$.
(2) Let $\delta = (1 + a_1 + a_2 + \cdots a_n) - (b_1 + b_2 + \cdots b_m)$

(a) If $\delta > 0$, $G_g = \delta$, $\Delta G = 0$
(b) If $\delta \leq 0$, $G_g = 1S$, $\Delta G = |\delta| + G_g$

(3) Multiply all conductance values by some constant, k. This will scale computed elements values to realize particular ones.

Design a MISO op-amp circuit to realize the relationship

$$V_o = 4V_{b1} + V_{b2} - 5V_{a1} - V_{a2}$$

(Here, indices a are for inverting, indices b are for noninverting terminal connections).

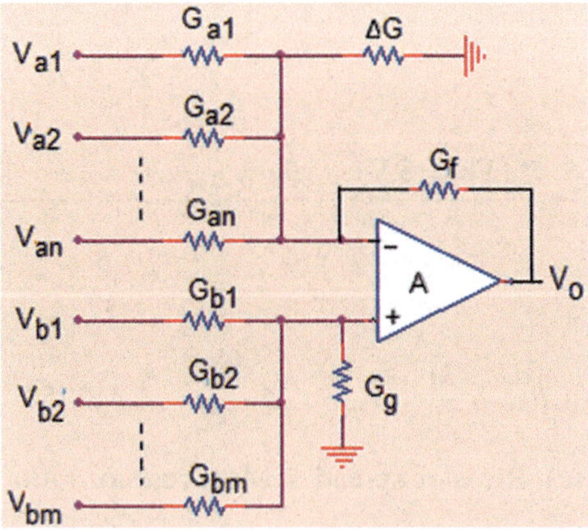

Fig. 10.66 General multi-input single-output (MISO) op-amp circuit

Solution

Let $G_f = 1S$, $G_{a1} = 5S$, $G_{a2} = 1S$, $G_{b1} = 4S$, $G_{b2} = 1S$, then,

$$\delta = (1 + a_1 + a_2) - (b_1 + b_2) = (1 + 5 + 1) - (4 + 1) = 2 > 0,$$

Set $\Delta G = 0,\ G_g = \delta = 2S$

Let $k = 10^{-3} = 0.001$, this scaling yields

$$R_f = \frac{1}{G_f} = 1\,\text{k}\Omega, \quad R_{a1} = \frac{1}{G_{a1}} = \frac{10^3}{5S} = 0.2\,\text{k}\Omega$$

$$R_{a2} = 1\,\text{k}\Omega, \quad R_{b1} = \frac{1}{4S \times 10^{-3}} = 0.25\,\text{k}\Omega, \quad R_{b2} = 1\,\text{k}\Omega$$

$$R_g = \frac{1}{2S \times 10^{-3}} = 0.5\,\text{k}\Omega$$

Figure 10.67 displays the circuit topology.

10.2 Filters and Other Op-amp Circuits

Problem 10.2.1

(a) What is apassive filter?
(b) What is an active filter?
(c) What are the types of filters according to their frequency selectivities?
(d) What are the major limitations of passive filters?
(e) What are the advantages of active filters over passive ones?
(f) What are the disadvantages of active filters over passive ones?

Solutions

(a) A filter is called a passive filter if it consists of only *R*, *L*, and *C*.
(b) A filter is called an active filter if it consists of transistors, op-amps and other types of active elements (such as operational trans-conductance amplifiers, current conveyors, etc.) in addition to *R*, *L*, and *C* elements.

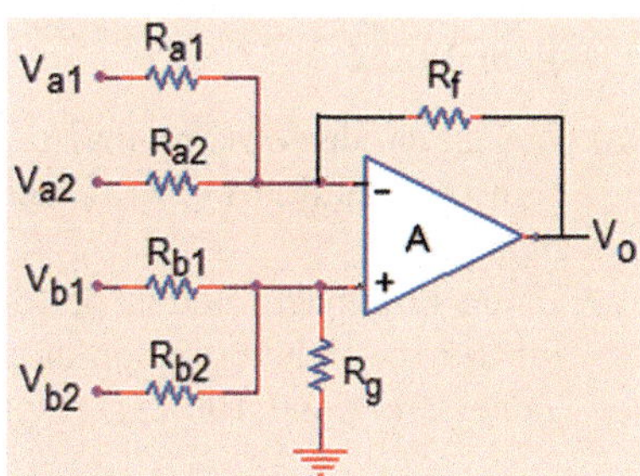

Fig. 10.67 MISO op-amp circuit

(c) There are basically four types of filters according to their frequency selectivities, no matter if they are passive or active:

1. A low-pass filter (LPF) passes low frequencies and stops high frequencies,
2. A high-pass filter (HPF) passes high frequencies and rejects low frequencies,
3. A band-pass filter (BPF) passes frequencies between two (corner) frequencies and stops or attenuates frequencies outside this band,
4. A bandstop (notch) filter (BSF) blocks or attenuates frequencies between two (corner) frequencies and passes frequencies outside this band,
(Yet there exists a fifth type of filter, called an all-pass filter (APF) which passes all frequencies but shifts the phase on a defined manner).

(d) Major limitations of the passive filters are,

1. They have a maximum gain of 1,
2. Passive filters may require inductors that costs and need larger volumes in a printed circuit board.
3. Their frequency performance at low frequencies is relatively poor.

(e) Advantages of active filters over passive ones are,

1. Active filters occupy less space and they are less expensive, (they do not require inductors). Therefore, their integrated circuit realizations are common in practice.
2. Active filters can provide gain (greater than unity).
3. Active filters can be buffered to isolate each stage of the filter from source and load impedance effects, allowing independent stage design and cascading flexibility.

(f) Major disadvantages of active filters over passive ones;

1. Active filters are generally of limited use at high frequencies, while passive filters may operate at very high frequencies.
2. Active filters may be less stable and less reliable with respect to passive ones.
3. Active filters may demand more power, especially a point of concern in battery operated instruments. (This problem is circumvented by using MOSFET structures.)

Problem 10.2.2

(a) Design an idealop-amp macromodel with a single-pole at 10 Hz and open loop gain of 1 million (V/V). Determine its gain-bandwidth product (= unity gain frequency).
(b) Using this macromodel, perform SPICE analysis to plot the voltage transfer function of an inverting op-amp configuration, if the input and feedback resistor values are 1 and 100 kΩ, respectively.

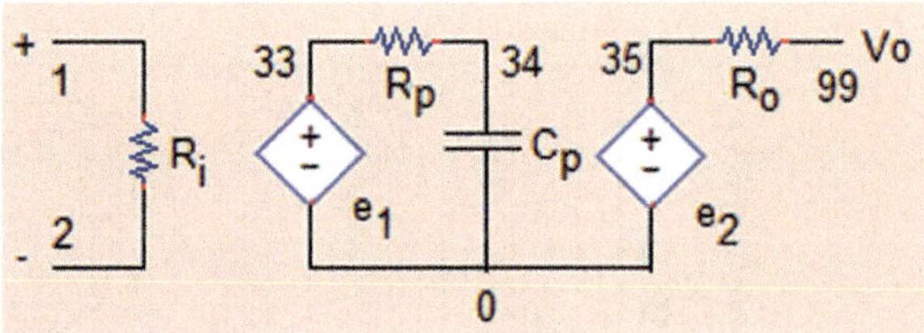

Fig. 10.68 Op-amp macromodel

Solution

(a) Figure 10.68 is an op-amp macromodel.
Open-loop gain = 1×10^6 V/V

$$f_p = \frac{1}{2\pi R_p C_p} = 10\,\text{Hz} = \frac{f_0}{A_0} = \frac{f_0}{10^6}$$

Unity gain frequency, $f_0 = 10 \times 10^6 = 10^7\,\text{Hz} = 10\,\text{MHz}$
Let; $R_p = 100\,\text{k}\Omega$,

$$C_p = \frac{1}{2\pi R_p f_p} = \frac{0.15915}{10^5 \times 10} = 0.15915\,\mu\text{F}$$

(b) A semi-logarithmic plot of the voltage transfer function of an inverting op-amp configuration (using op-amp ideal single pole macromodel) when the input and feedback resistor values are 1 and 100 kΩ, respectively, is shown in Fig. 10.69. SPICE netlist is:

```
Vi 10   0   ac 1
R1   10     1   1 k
R2   1      3   100 k
XOP  0  1  3   OAM
Rload    3  0      1meg
* Single pole opamp macromodel
*connections:   (NI  I  output)
.Subckt OAM  1  2  99
Ri 1 2 1G; input resistance
*polefrq = 1/(Rp.Cp), rad/sec
*unity gain frequency = (dcgain).(polefrq)
e1 33 0 1 2 1meg; dc gain
Rp 33 34 100 k
Cp 34 0 159.15n
e2 35 0 34 0 1
Ro 35 99 1; output resistance
.ends
*.AC  dec 10 1 10MEG ;v3
```

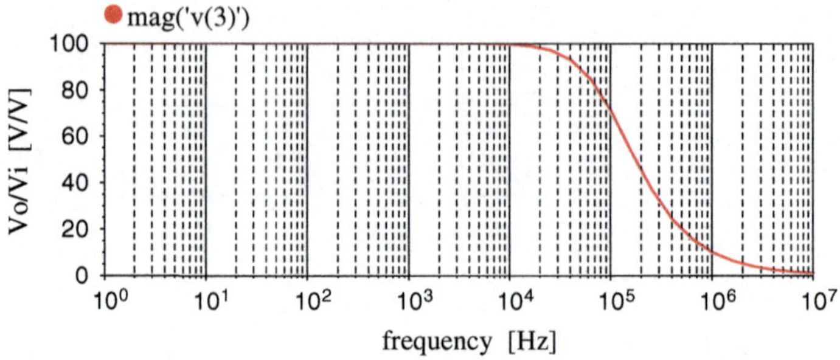

Fig. 10.69 Semi-logarithmic magnitude vs frequency plot

Fig. 10.70 A first order low pass filter using operational amplifier

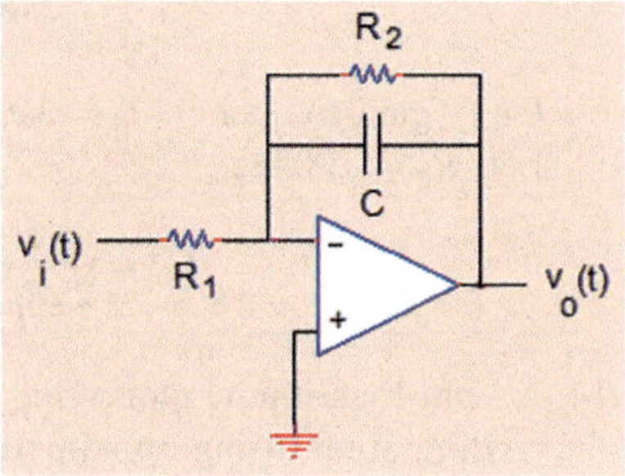

Problem 10.2.3

(a) Determine the voltage transfer (gain) function for the circuit shown in Fig. 10.70, and describe its frequency performance.

(b) Design a first-order filter having a dc gain of 60 dB, cutoff frequency of 100 Hz and input resistance of 1 kΩ. Determine its unity gain frequency, and the phase shift at that frequency.

Solution

(a) $H(s) = -\frac{R_2 \parallel (1/sC)}{R_1} = \frac{-\frac{R_2}{1+sR_2C}}{R_1} = -\frac{R_2}{R_1}\frac{1}{1+sR_2C}$

This is a single time constant circuit with dc gain $K = -R_2/R_1$, and cutoff frequency of $\omega_0 = 1/CR_2$, therefore performing as a first-order low-pass filter.

(b) $R_1 = 1\,\text{k}\Omega$,

$$|K| = 60\,\text{dB} = \frac{R_2}{R_1} = 1000 \quad \rightarrow \quad R_2 = |K|R_1 = 1000 \times 1\,\text{k}\Omega = 1\,\text{M}\Omega$$

$$\omega_0 = \frac{1}{R_2C}$$

$$C = \frac{1}{\omega_0 R_2} = \frac{1}{628 \times 10^6} = 1592.4\,\mu\text{F}$$

The gain falls off at a rate of −20 dB/decade, thus it reaches 0 dB in three decades at $f = 1000$, $f_0 = 100\,\text{kHz}$, where the phase shift will be almost +90°

Problem 10.2.4 A single op-amp-based filter is to be designed as a band-pass filter for ECG signals between 0.1 and 100 Hz and provide inverted gain of 20 dB.

(a) Draw the circuit,
(b) Determine its Voltage Transfer Function,
(c) Determine its component values.

Solution

(a) The single op-amp-based circuit of Fig. 10.71. can be used as a band-pass filter for ECG signals and provide inverted gain.

(b) $Z_2 = R_2 \parallel \dfrac{1}{sC_2} = \dfrac{R_2}{1 + sC_2R_2}, \quad Z_1 = R_1 + \dfrac{1}{sC_1} = \dfrac{1 + sC_1R_1}{sC_1}$

$$\frac{V_o(s)}{V_i(s)} = -\frac{Z_2}{Z_1} = \frac{-sC_1R_2}{(1 + sC_2R_2)(1 + sC_1R_1)} = -\frac{R_2}{R_1} \cdot \frac{1}{(1 + sC_2R_2)\left(1 + \dfrac{1}{sC_1R_1}\right)}$$

(c) $20\,\text{dB} = 10\,\text{V/V} = \left| -\dfrac{R_2}{R_1} \right|$

Since two cutoff frequencies are two decades apart, midband gain can be calculated in this manner.
let $R_2 = 10\,\text{k}\Omega$, then $R_1 = 1\,\text{k}\Omega$

$$\omega_H = \frac{1}{C_2R_2} = 2\pi \times 100 \quad \rightarrow \quad C_2 = \frac{1}{2\pi \times 100 \times 10 \times 10^3} = 159\,\text{nF}$$

$$\omega_L = \frac{1}{C_1R_1} = 2\pi \times 0.1 \quad \rightarrow \quad C_1 = \frac{1}{2\pi \times 0.1 \times 10^3} = 1592\,\mu\text{F}$$

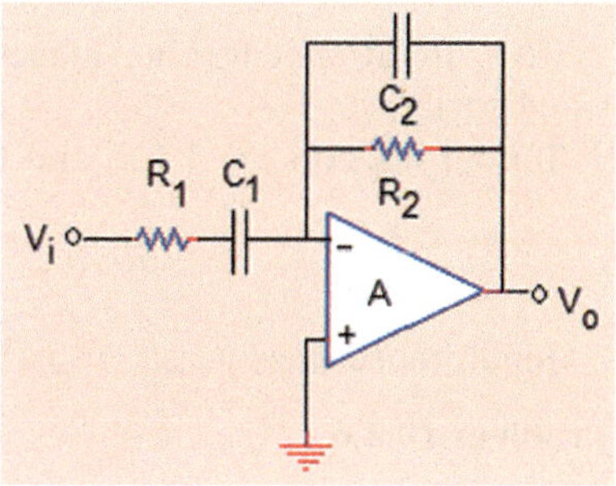

Fig. 10.71 Band-pass filter

Fig. 10.72 The circuit for Problem 10.2.5

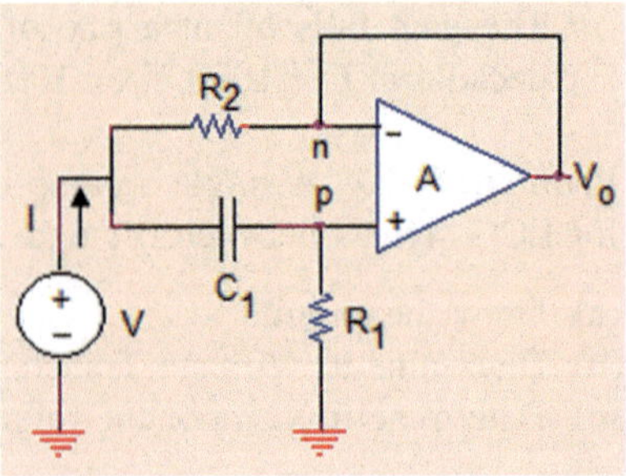

Problem 10.2.5 Calculate the ratio $Z = V/I$ in the circuit of Fig. 10.72. What does the circuit simulate?

If both resistors are 1 kΩ, and $C = 1$ μF, what is the value of simulated quantity?

Solution

$$V_p = \frac{R_1}{R_1 + \frac{1}{sC}} \cdot V = V_n$$

$$I = \frac{V - V_n}{R_2} = \frac{V - V\left(\frac{R_1}{R_1 + 1sC}\right)}{R_2} = \frac{V\left(1 - R_1R_1 + \frac{1}{sC}\right)}{R_2} = \frac{V\left(1 - \frac{sR_1C}{sR_1C + 1}\right)}{R_2}$$

$$I = \frac{V(sR_1 + 1 - sR_1C)}{R_2(sR_1C + 1)} = V \cdot \frac{1}{sR_1R_2C + R_2}$$

$$\frac{I}{V} = \frac{1}{sR_1R_2C + R_2}$$

$$\frac{V}{I} = R_2 + sR_1R_2C = Z$$

$$L = R_1R_2C$$

The circuit simulates an inductor in series with a resistor (Lossy inductance simulator).

If both resistors are 1 kΩ, and $C = 1$ μF,

$$L = R_1R_2C = 10^3 \times 10^3 \times 10^{-6} = 1H$$

Input impedanceconsists of a 1H inductor in series with 1 kΩ resistor.

Problem 10.2.6

(a) In circuit shown in Fig. 10.73 (switched capacitor inverting integrator), switches are driven by a nonoverlapping two-phase clock. Let the clock frequency $f_c = 1/T_c$ be much higher than the maximum frequency content of the

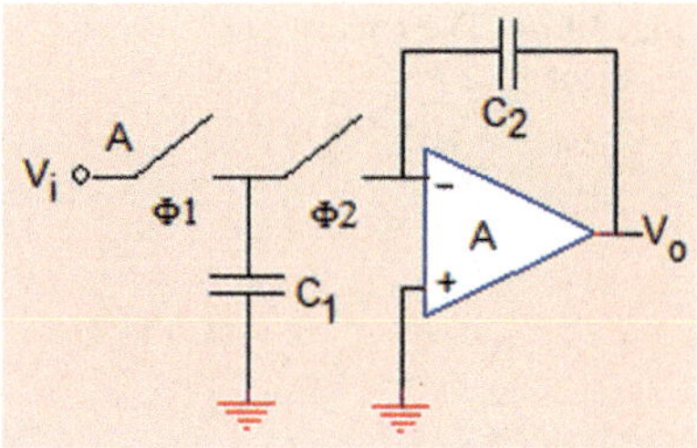

Fig. 10.73 A switched capacitor inverting integrator

signal $v_i(t)$. Determine the time constant of the circuit, in terms of the clock frequency and capacitor values.

(b) Determine the time constant if C_2 = 40 pF, C_1 = 4 pF, f_c = 100 kHz.

Solution

During φ_1, (on) C_1 is connected to V_i and it charges, $q_{c1} = C_1 V_i$

During φ_2, C_1 is connected virtual ground (negative terminal of op-amp) and it transfers its electrical charge (q_{c1}) to C_2.

The average current flow from input (A) to virtual ground is:

$$i_{av} = \frac{q_{c1}}{T_c} = \frac{C_1 \times V_i}{T_c} \quad \rightarrow \quad \frac{i_{av}}{V_i} = \frac{C_1}{T_c}$$
$$\frac{V_i}{i_{av}} = \frac{T_c}{C_1} = R$$

In other words, an equivalent resistance appears between input node and the virtual ground. Time constant of the circuit is:

$$\tau = RC_2 = T_c \times \frac{C_2}{C_1} = \frac{C_2}{f_c \times C_1}$$

(b) The time constant of the integration is

$$\tau = \frac{40}{10^5 \times 4} = 10^{-4}\,\text{s} = 0.1\,\text{ms} = 100\,\mu\text{s}$$

(c) Input resistance is

$$R = \frac{T_c}{C_1} = \frac{1}{10^5 \times 4 \times 10^{-12}} = 2.5\,\text{M}\Omega$$

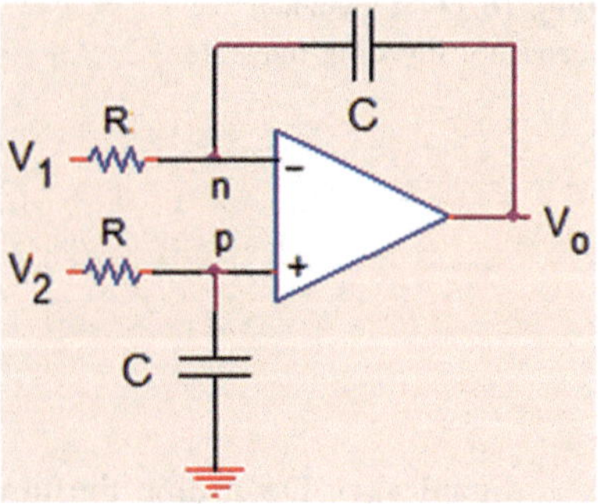

Fig. 10.74 The circuit for Problem 10.2.7

Problem 10.2.7 Determine the voltage at the output of op-amp in the circuit shown in Fig. 10.74.

Solution

$$V_O = -\frac{1}{C}\int_{-\infty}^{t} I_1 \mathrm{d}\tau + V_+ \tag{10.68}$$

$$V_+ = \frac{1}{C}\int_{-\infty}^{t} I_2 \mathrm{d}\tau \tag{10.69}$$

Substitute Eq. (10.69) into Eq. (10.68) yields

$$V_O = \frac{1}{C}\int_{-\infty}^{t} (I_2 - I_1)\mathrm{d}\tau = \frac{1}{C}\int_{-\infty}^{t} \left(\frac{V_2}{R} - \frac{V_1}{R}\right)\mathrm{d}\tau = \frac{1}{RC}\int_{-\infty}^{t} (V_2 - V_1)\mathrm{d}\tau$$

Problem 10.2.8 Implantable electronic circuits are designed to meet rigorous technical specifications especially regarding size and power consumption requirements. For example, realization of large time constants in small chip area is required for integrated filters. One solution is to employ a capacitance multiplier [9–15].

(a) Assuming ideal op-amp performance, determine the input impedance for the circuit given in Fig. 10.75.
(b) Determine the equivalent capacitance at the input of the circuit. Is this a pure capacitive reactance? Comment.
(c) Find the numerical component values at the input of the circuit if $R_1 = 10\ \mathrm{k\Omega}$, $R_2 = 100\ \Omega$, $C = 100\ \mathrm{pF}$. What is the percent error in making some approximation?

Fig. 10.75 Capacitance multiplier

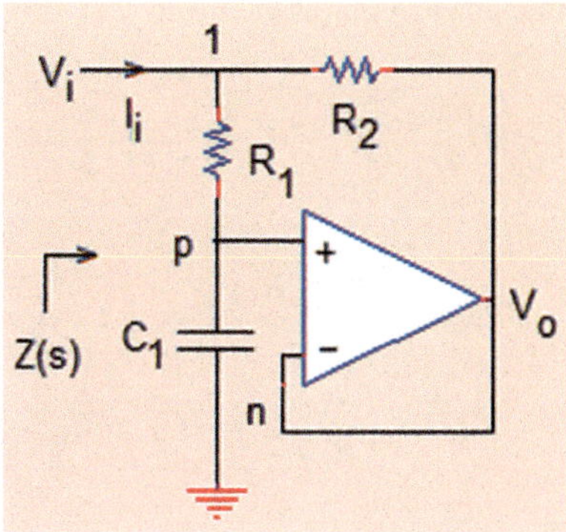

Solution

(a) By KCL at node 1,

$$I_i = \frac{V_i - V_o}{R_1} + \frac{V_i - V_o}{R_2} = (V_i - V_o)\left(\frac{1}{R_1} + \frac{1}{R_2}\right)$$

Using voltage division rule at node p,

$$V_o = V_i\left(\frac{\frac{1}{sC}}{R_1 + \frac{1}{sC}}\right) = V_i\left(\frac{1}{sCR_1 + 1}\right)$$

Substituting this equation into the first one,

$$I_i = \left[V_i - V_i\left(\frac{1}{sCR_1 + 1}\right)\right]\left(\frac{1}{R_1} + \frac{1}{R_2}\right) = V_i\left(1 - \frac{1}{sCR_1 + 1}\right)\left(\frac{1}{R_1} + \frac{1}{R_2}\right)$$

$$\frac{I_i}{V_i} = \left(1 - \frac{1}{sCR_1 + 1}\right)\left(\frac{1}{R_1} + \frac{1}{R_2}\right) = \left(\frac{sCR_1}{sCR_1 + 1}\right)\left(\frac{1}{R_1} + \frac{1}{R_2}\right)$$

Inverting this equation gives the input impedance,

$$\frac{V_i}{I_i} = Z_i = \left(\frac{sCR_1 + 1}{sCR_1}\right)\left(\frac{R_1R_2}{R_1 + R_2}\right)$$

$$= \left(1 + \frac{1}{sCR_1}\right)\left(\frac{R_1R_2}{R_1 + R_2}\right) = \frac{R_1R_2}{R_1 + R_2} + \frac{1}{sCR_1}\left(\frac{R_1R_2}{R_1 + R_2}\right)$$

$$Z_i = \frac{R_1R_2}{R_1 + R_2} + \frac{R_2}{sC(R_1 + R_2)}$$

(b) This input impedance consists of series connected resistance and capacitance, as in Fig. 10.76, it is not a pure capacitive reactance (in other words, the circuit is a lossy capacitance simulator),

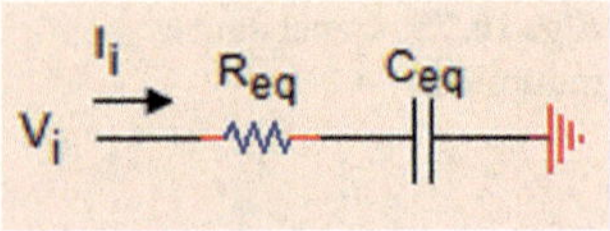

Fig. 10.76 Equivalent circuit

$$C_{eq} = \frac{(R_1 + R_2)}{R_2}C = \left(1 + \frac{R_1}{R_2}\right)C \cong \frac{R_1}{R_2}C, \quad R_{eq} = \frac{R_1 R_2}{R_1 + R_2} \cong R_2, \quad R_1 \gg R_2$$

$$C_{eq} = \left(\frac{R_1}{R_2}\right)C = kC, \quad R_{eq} = R_2$$

The circuit can be used to simulate a much larger grounded capacitor by employing much smaller one. In fact, the circuit simulates a series connected (R_{eq}),(C_{eq}) circuit, the resistor R_2 is about the same size as the resistor in the simulated circuit (R_{eq}), but the capacitor C_{eq} is k (=R_1/R_2) times larger than C. The voltages across the simulated and actual capacitors are the same, although their currents are not the same. The op-amp keeps the input voltage at the same voltage as the voltage across the capacitor, C. This means the voltage across R_2 has the same voltage as simulated resistance, R_{eq}. Therefore, R_2 and R_{eq} carry the same amount of current.

(c) Using given component values, R_1 = 10 kΩ, R_2 = 100 Ω, C = 100 pF,

$$C_{eq} = \frac{(R_1 + R_2)}{R_2}C = \left(1 + \frac{10}{0.1}\right) \times 100 = 101 \times 100 = 10{,}100\,\text{pF} = 10.1\,\text{nF}$$

$$R_{eq} = \frac{R_1 R_2}{R_1 + R_2} = \frac{10 \times 0.1}{10 + 0.1} = \frac{1}{10.1} = 0.099\,\text{k}\Omega = 99\Omega$$

Approximate result is,

$$C_{eq} = \left(\frac{R_1}{R_2}\right)C = \left(\frac{10}{0.1}\right) \times 100 = 10\,\text{nF}, \quad R_{eq} = R_2 = 100\Omega$$

The error of approximation for simulated capacitance value is

$$\text{error} = \frac{|10 - 10.1|}{10} = 1\%$$

Problem 10.2.9 A disturbance often seen in biomedical practice due to power line interference is the generation of common-mode voltage, V_{CM}, on a patient. It is reduced by means of an instrumentation amplifier. This effect can be further reduced by means of a feedback circuit, as shown in Fig. 10.77.

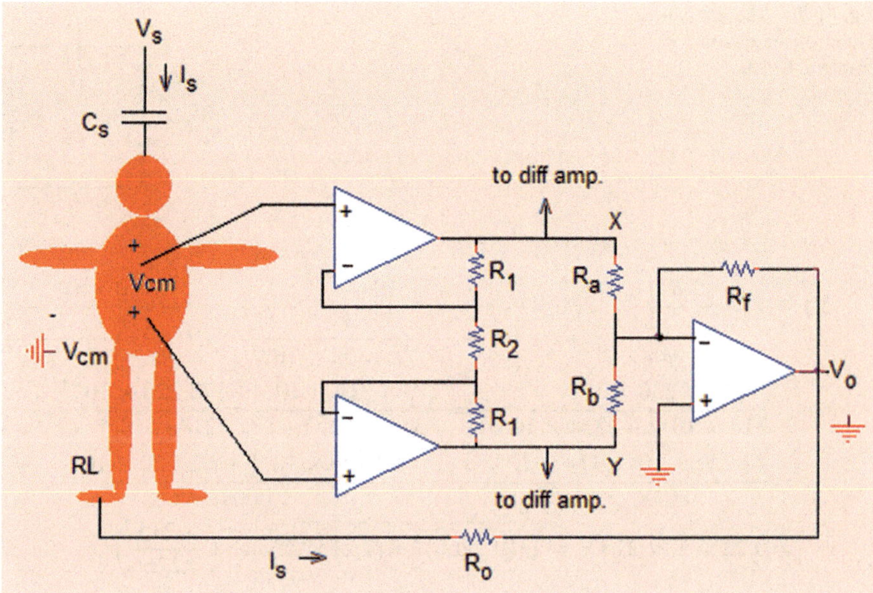

Fig. 10.77 Common-mode voltage reduction

(a) Assume that the patient is coupled by a 10 pF stray capacitance to power line and a differential amplifier is used without buffering and input of each port having resistance of $R = 10\ \text{M}\Omega$. Find the voltage values at the input of op-amp. Draw a conclusion.

(b) If the body is connected to a driven-right-leg feedback system and insulated from ground, but coupling capacitance has the same value as given previously, determine the common mode voltage equation.

(c) Assuming a European power-line system (230 V, 50 Hz), let $R_a = 40\ \text{k}\Omega = 2R_b$, $R_f = 5\ \text{M}\Omega$ compute the V_{CM}. Compare this result with the one found in part (a).

(d) Using MATLAB, plot the curve V_{CM} as a function of R_f for $1\ \text{M}\Omega \leq R_f \leq 10\ \text{M}\Omega$ and repeat the same computations for the public electrical network in the USA (120 V, 60 Hz) and Japan (100 V, 60 Hz). Plot three curves on the same graph. (Vcm1.m)

Solution

(a) The equivalent circuit can be drawn as in Fig. 10.78 (for one terminal of differential amplifier, the other one is the same).
By the voltage division rule,

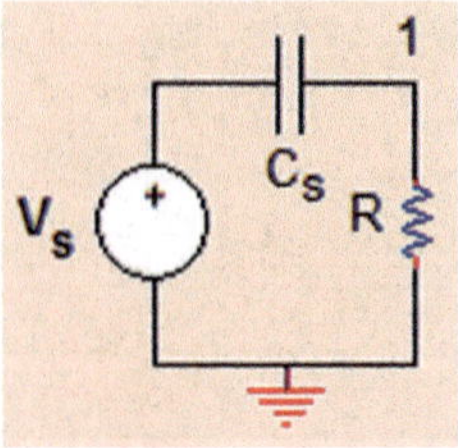

Fig. 10.78 The equivalent circuit of the circuit for Problem 10.2.9

$$\begin{aligned}
\boldsymbol{V}_1 &= \boldsymbol{V}_s \cdot \frac{R}{R + \frac{1}{j\omega C}} = 230 \times \frac{10^7}{10^7 + \frac{1}{j \times 2\pi \times 50 \times 10^{-11}}} = \frac{230 \times 10^7}{10^7 - j\frac{1}{314 \times 10^{-11}}} \\
&= \frac{230 \times 10^7}{314 \times 10^{-4} - j314 \times 10^{-11}} = \frac{230 \times 314 \times 10^{-4}}{314 \times 10^{-4} - j} = \frac{72{,}220 \times 10^{-4}}{314 \times 10^{-4} - j} \\
&= \frac{72{,}220 \times 10^{-4}(314 \times 10^{-4} + j)}{98596 \times 10^{-8} + 1} = \frac{72{,}220 \times 10^{-8} + j72{,}220 \times 10^{-4}}{1.00098596} \\
&= 0.2266 + j7.215\,\text{V} = \sqrt{(0.2266)^2 + (7.215)^2}\ \angle \text{atan}\left(\frac{7.215}{0.2266}\right)
\end{aligned}$$

$$\boldsymbol{V}_1 = 7.219\ \angle 88.2^\circ\ \text{V}$$

A common mode voltage value of 7219 mV exists on the patient which is much larger than the biopotential value, on the order of 1 mV.

(b) Applying KVL to the "patient's body"

$$V_s - I_s\frac{1}{j\omega C} - V_{CM} = 0 \tag{10.70}$$

The common mode voltage appears on nodes X, Y. By KCL on the negative mode of the inverting amplifier,

$$\frac{V_{CM}}{R_a} + \frac{V_{CM}}{R_b} + \frac{V_O}{R_f} = 0 \tag{10.71}$$

Voltage drop over the resistor R_O,

$$(V_{CM} - V_O) = I_s R_O \tag{10.72}$$

Substituting (10.72) into (10.70) for I_s,

$$V_s - \left(\frac{V_{CM} - V_O}{R_O}\right)\frac{1}{j\omega C} = 0 \tag{10.73}$$

Solving (10.73) for V_0,

$$V_s - V_{CM} = \frac{V_{CM} - V_0}{j\omega R_0 C_s} \tag{10.74}$$

$$-j\omega R_0 C_s (V_s - V_{CM}) + V_{CM} = V_0 \tag{10.75}$$

Putting this equation into (10.71) and solving for V_{CM}

$$\begin{aligned}
&\frac{V_{CM}}{R_a} + \frac{V_{CM}}{R_b} + \frac{V_{CM} - (V_s - V_{CM}) j\omega R_0 C_s}{R_f} = 0 \\
&\frac{V_{CM}}{R_a} + \frac{V_{CM}}{R_b} + \frac{V_{CM}}{R_f} - \frac{(V_s - V_{CM}) j\omega R_0 C_s}{R_f} = \frac{V_{CM}}{R_a} \\
&+ \frac{V_{CM}}{R_b} + \frac{V_{CM}}{R_f} - \frac{V_s j\omega R_0 C_s}{R_f} + \frac{V_{CM} j\omega R_0 C_s}{R_f} = 0 \\
&V_{CM}\left(\frac{1}{R_a} + \frac{1}{R_b} + \frac{V_{CM}}{R_f} + \frac{1}{R_f} + \frac{j\omega R_0 C_s}{R_f}\right) = \frac{V_s j\omega R_0 C_s}{R_f} \\
&V_{CM} = \frac{j V_s \omega R_0 C_s}{\frac{R_f}{R_a} + \frac{R_f}{R_b} + 1 + j\omega R_0 C_s}
\end{aligned} \tag{10.76}$$

This equation shows that common mode voltage is a function of frequency and the resistor values.

(c) For European public power network, using given values in (10.76),

$$\begin{aligned}
&V_{CM} = \frac{230 \times j \times 314 \times 5 \times 10^6 \times 10^{-11}}{\frac{5 \times 10^6}{40 \times 10^3} + \frac{5 \times 10^6}{20 \times 10^3} + 1 + j \times 314 \times 5 \times 10^6 \times 10^{-11}} \\
&|V_{CM}| = 9.6\,\text{mV},\ \theta = 90^\circ
\end{aligned}$$

It is seen that common mode voltage is reduced from 7219 to 9.6 mV by using the feedback circuit.

(d) The variation of the induced common mode voltage on a patient as a function of feedback resistance of an ECG front-end circuit, in European, the U.S. and Japanese public electrical networks are displayed in Fig. 10.79 $1\ \text{M}\Omega \leq R_f < 10\ \text{M}\Omega$.

A MATLAB script is given below. (Vcm1.m).

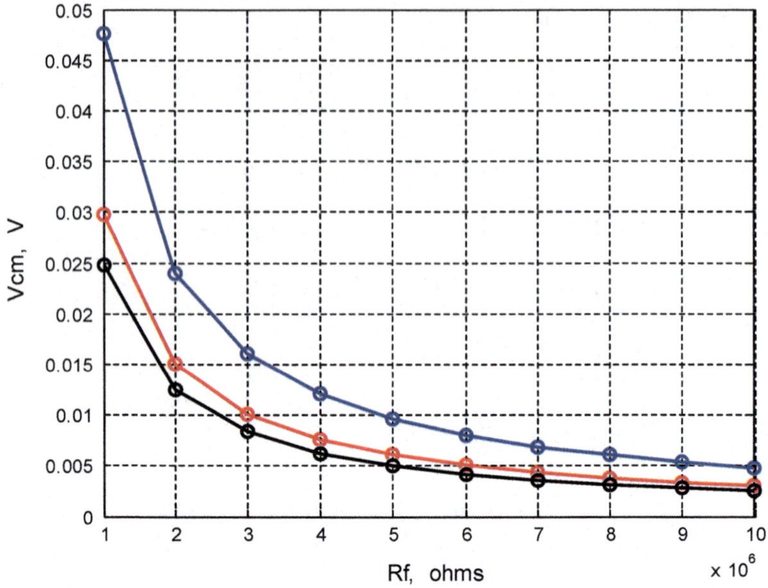

Fig. 10.79 The graph of V_{CM} *vs* R_f. Legend: *Blue line* European, *Red line* USA, *Black line* Japanese (60 Hz) network

```
%Common mode voltage computation, Vcm
%Vs(230V,120V,100V) and line frequency(50-60Hz) are changed
manually
Vs=230; %rms voltage
Rf=5e6;Ra=40e3;Rb=20e3; Ro=5e6;w=2*pi*60;
Cs=1e-11; %stray capacitance,F
d2=1i*w*Ro*Cs;
Rfmin=1e6;Rfmax=1e7;dRf=1e6;
Rf=(Rfmin:dRf:Rfmax)';
d1=Rf*(1/Ra+1/Rb); d=d1+d2+1; n=1i*Vs*w*Ro*Cs;
Vcm=n./d;MagVcm=abs(Vcm)theta1=angle(Vcm);%phase angle in radians
%theta=theta1*90/(pi/2)%phase angle in degrees
[Rf/1e6 MagVcm]%list
plot(Rf,MagVcm,'bo-','linewidth',2);grid on;
xlabel('Rf,ohms');ylabel('Vcm,V')
hold on; %fig. is repeated for other nework parameters
%Vs(230V,120V,100V) and frequency(50-60Hz) are changed manually.
```

Problem 10.2.10 Two RC circuit sections, (with capacitors grounded) are separated by an ideal op-amp buffer. Input signal $v_i(t)$ is applied to first R_1C_1 section, producing the signal $v_i(t)$ at node 1, it is buffered by the op-amp to the input of the second R_2C_2 section (node 2), the output of which is $v_o(t)$.

(a) Derive the differential equation relating the circuit output $v_o(t)$ to its input, $v_i(t)$.
(b) Determine its characteristic equation and find its roots.
(c) Can this circuit produce underdamped natural response?

Solution

(a) Due to the buffering action of op-amp, node 2 follows the signal at node 1 (Fig. 10.80).
Since input to op-amp (node 1) is connected to its noninverting terminal, there is no current flowing into this terminal. The inverting input of op-amp is shorted to its output terminal. Therefore, output terminal of op-amp follows its input signal at noninverting terminal.
KCL at nodes 1 and 2:

$$C_1\frac{\mathrm{d}v_1(t)}{\mathrm{d}t}=\frac{v_i(t)-v_1(t)}{R_1} \tag{10.77}$$

$$C_2\frac{\mathrm{d}v_0(t)}{\mathrm{d}t}=\frac{v_1(t)-v_0(t)}{R_2} \tag{10.78}$$

From (10.77),

$$v_1(t)=C_2R_2\frac{\mathrm{d}v_0(t)}{\mathrm{d}t}+v_0(t) \tag{10.79}$$

Substitute (10.79) into (10.78),

$$R_1C_1\frac{\mathrm{d}}{\mathrm{d}t}\left[C_2R_2\frac{\mathrm{d}v_0(t)}{\mathrm{d}t}+v_0(t)\right]=v_i(t)-\left[C_2R_2\frac{\mathrm{d}v_0(t)}{\mathrm{d}t}+v_0(t)\right]$$
$$R_1R_2C_1C_2\frac{\mathrm{d}^2}{\mathrm{d}t^2}v_0(t)+R_1C_1\frac{\mathrm{d}}{\mathrm{d}t}v_0(t)=v_i(t)-C_2R_2\frac{\mathrm{d}v_0(t)}{\mathrm{d}t}-v_0(t)$$
$$R_1R_2C_1C_2\frac{\mathrm{d}^2}{\mathrm{d}t^2}V_0(t)+(R_1C_1+R_2C_2)\frac{\mathrm{d}V_0(t)}{\mathrm{d}t}+V_0(t)=V_i(t)$$

(b) The characteristic equation is obtained from differential equation as;

$$R_1R_2C_1C_2s^2+(R_1R_2+C_1C_2)s+1=0$$

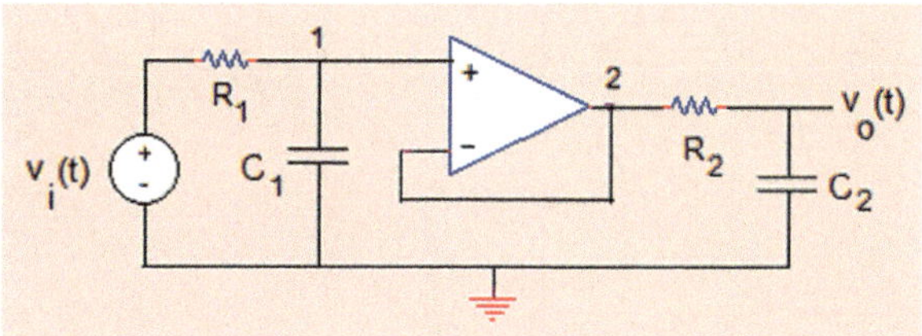

Fig. 10.80 The circuit for Problem 10.2.10

This equation has roots at

$$s_1 = -\frac{1}{R_1 C_1}, \quad s_2 = -\frac{1}{R_2 C_2}$$

(c) Since the roots of characteristic polynomial are always real, it does not produce any underdamped response.

Problem 10.2.11 An noninverting operational amplifier has voltage gain of $G = 100$, and output resistance, $R = 200\ \Omega$. A pure capacitive load of 1000 μF is connected to its output terminal. Determine the voltage at the output of this amplifier, if voltage applied to its input terminal is $v_i(t) = 0.1\ u(t)$, neglecting all internal capacitances.

Solution

$$\tau = RC = 200 \times 10^{-3} = 0.2\ \text{s}.$$
$$v_0(t) = v_0(\infty) + [v_0(0) - v_0(\infty)] \cdot \mathrm{e}^{-t/\tau}$$
$$v_0(\infty) = G \cdot v_i(\infty) = (100)(0.1) = 10\,\text{V}, \quad v_0(0^+) = 0\,\text{V} = v_0(0^-),$$
$$v_0(t) = v_0(\infty) \cdot (1 - \mathrm{e}^{-t/\tau}) = 10\,(1 - \mathrm{e}^{-5t}) \cdot u(t)$$

Problem 10.2.12 For the op-amp circuit in Fig. 10.81, find v_0 for $t > 0$. $C = 1\ \mu F$, $R = 1\ \text{k}\Omega$.

Solution
For $t < 0$, by voltage division,

$$v_1(0^-) = 10 \cdot \frac{R}{2R} = 5\,\text{V}$$

For $t > 0$,

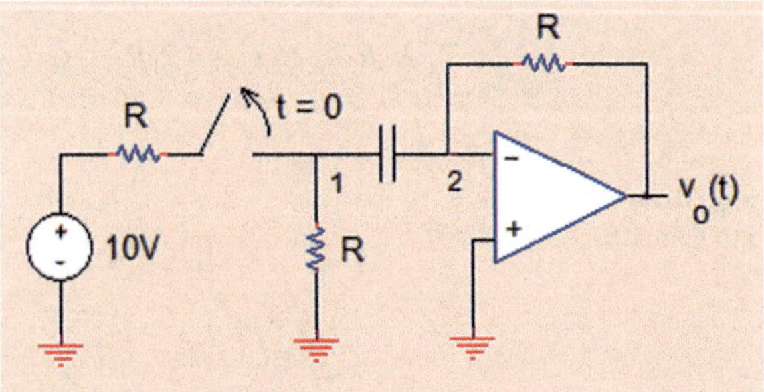

Fig. 10.81 The circuit for Problem 10.2.12

$$v_1(0^+) = v_1(0^-) = V_0 = 5\,\text{V}$$

Since inverting input of op-amp is at virtual ground potential, $v_2 = 0,\ v_1 = v_c$. Apply KCL at node 1:

$$-C\frac{\mathrm{d}v_1}{\mathrm{d}t} + \frac{0 - v_1}{R} = 0 \quad \rightarrow \quad C\frac{\mathrm{d}v_1}{\mathrm{d}t} + \frac{v_1}{R} = 0 \quad \rightarrow \quad \frac{\mathrm{d}v_1}{\mathrm{d}t} + \frac{v_1}{RC} = 0$$

The solution of this differential equation is

$$v_1(t) = V_0 \cdot \mathrm{e}^{-t/\tau}, \quad \tau = RC = 10^3 \times 10^{-6} = 10^{-3}\,\text{s}.$$

KCL at node 2 :

$$C\frac{\mathrm{d}v_1}{\mathrm{d}t}\left(\frac{0 - v_0}{R}\right) = 0 \quad \rightarrow \quad v_0(t) = -RC\left(\frac{V_0}{R.C}.e^{-\frac{t}{R.C}}\right) = V_0 \cdot \mathrm{e}^{-\frac{t}{R.C}},$$

Substituting numerical values yields,

$$v_0(t) = 5\mathrm{e}^{-1000t}\,\text{V}, \quad t > 0$$

Problem 10.2.13 Determine $v_0(t), \quad t > 0$, in the circuit of Fig. 10.82. $R = 1\ \text{k}\Omega$, $C = 1\ \mu\text{F}$, $U = 10\ \text{V}$.

Solution

$$v_0(t) = v_1(t) - v_C(t), \tag{10.80}$$

$$v_C(t) = v_C(\infty) + [v_C(0) - v_C(\infty)] \cdot \mathrm{e}^{-t/\tau}, \quad t > 0 \tag{10.81}$$

$$\tau = RC = 10^3 \times 10^{-6} = 10^{-3}\,\text{s}.$$

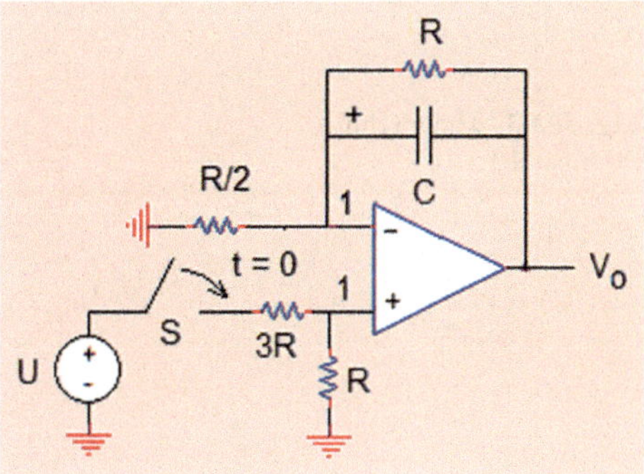

Fig. 10.82 The circuit for Problem 10.2.13

$v_1(0^-) = 0\,\mathrm{V}$, The voltage at node 1+ is found by voltage division,

$$v_1(0^+) = \frac{R}{4R}U = \frac{10}{4} = 2.5\,\mathrm{V} = v_1(\infty) = v_1(t), \quad t > 0$$

The op-amp is a noninverting amplifier, therefore,

$$v_O(\infty) = v_1\left(1 + \frac{R}{\frac{R}{2}}\right) = v_1(1+2) = 3v_1 = 3 \times 2.5 = 7.5\,\mathrm{V}$$

$$v_C(\infty) = v_1(\infty) - v_O(\infty) = 2.5 - 7.5 = -5\,\mathrm{V}$$

From (10.81):

$$v_c(t) = -5 + [0-(-5)] \cdot \mathrm{e}^{-1000t} = 5 \cdot (\mathrm{e}^{-1000t} - 1)V, \quad t > 0$$

From (10.80):

$$v_O(t) = v_1(t) - v_C(t) = 2.5 - 5 \cdot \mathrm{e}^{-1000t} + 5 = 7.5 - 5 \cdot \mathrm{e}^{-1000t} V, \quad t > 0$$

Problem 10.2.14 If the input to the circuit in Fig. 10.83 is $v_i(t) = A\cos(\omega t)\,V$, the output is known to be $v_o(t) = B\cos(\omega t + \varphi)\,V$.

(a) If $A = 10$, Find values for R, C and resulting value of B for $\varphi = -60°$, at $f = 100$ Hz.
(b) Determine the values of components such that $B = A/3$. Find the phase angle ($f = 100$ Hz). (RC3.cir)

Solution
The voltage transfer function (gain) can be written as

$$H(j\omega) = \frac{1}{1 + j\left(\frac{\omega}{\omega_0}\right)}$$

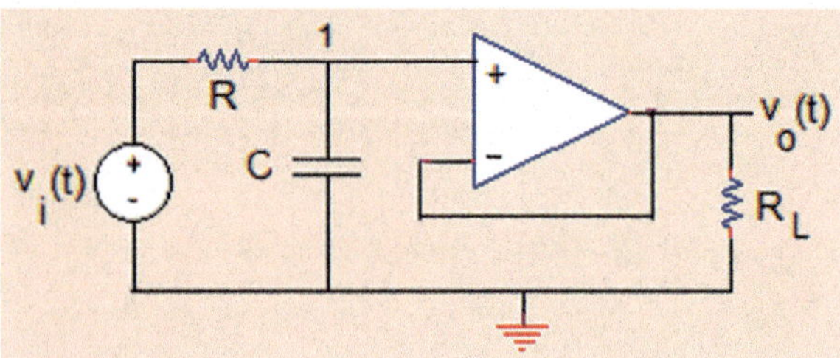

Fig. 10.83 The circuit for Problem 10.2.14

Here, $\omega_0 = 1/RC$. The magnitude response is

$$|H(j\omega)| = \frac{1}{\sqrt{1+\left(\frac{\omega}{\omega_0}\right)^2}}$$

The phase response is

$$\varphi(\omega) = -\tan^{-1}\left(\frac{\omega}{\omega_0}\right)$$

The role of op-amp here is to act ass a buffer to prevent loading of the RC circuit.

(a) $\omega = 2\pi f = 2\pi \cdot 100 = 628\,\text{rad}\cdot\text{s}^{-1}$

$$60^\circ = \tan^{-1}\left(\frac{628}{\frac{1}{RC}}\right) = \tan^{-1}(628RC)$$

$$\tan 60^\circ = \sqrt{3} = 628RC \quad \rightarrow \quad RC = \frac{\sqrt{3}}{628} = 0.00276$$

Let $C = 1\ \mu\text{F}$,

$$R = \frac{0.00276}{10^{-6}} = 2760\,\Omega$$

$$\frac{B}{A} = |H(j\omega)| = \frac{1}{\sqrt{1+(628\cdot RC)}} \quad \rightarrow \quad B = \frac{10}{\sqrt{1+(628)(2760)10^{-6}}} = 5\,\text{V}$$

(b) $$H(j\omega) = \frac{1}{\sqrt{3}} = \frac{1}{\sqrt{1+\left(\frac{\omega}{\omega_0}\right)^2}} \quad \rightarrow \quad 3 = \sqrt{1+(\frac{\omega}{\omega_0})^2} \quad \rightarrow \quad \frac{\omega}{\omega_0} = \frac{628}{\frac{1}{RC}} = \sqrt{8}$$

$$RC = \frac{\sqrt{8}}{628}$$

Let $C = 1\ \mu\text{F}$, then,

$$R = \frac{\sqrt{8}}{628 \times 10^{-6}} = 4504\,\Omega$$

Therefore, when $B = A/3,\ R = 4504\,\Omega,\ C = 1\,\mu\text{F}$, the phase angle is,

$$\varphi = \tan^{-1}\left(\frac{\omega}{\omega_0}\right) = \tan^{-1}(628\,RC) = \tan^{-1}(628 \times 4504 \times 10^{-6}) = -70.53^\circ$$

The phase response is plotted in Fig. 10.84.

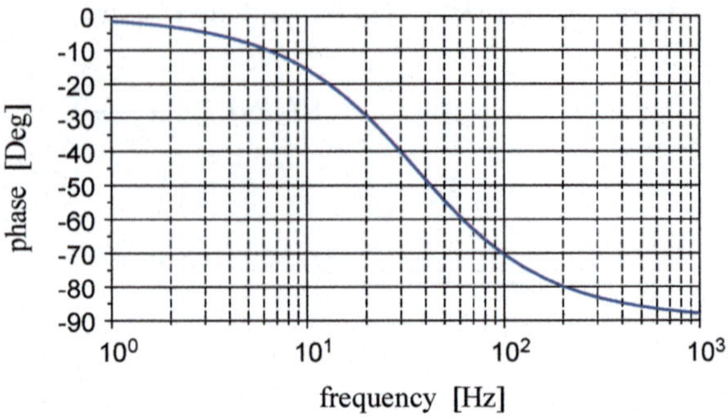

Fig. 10.84 For $B = A/3$, $R = 4504\ \Omega$, $C = 1\ \mu\text{F}$, $f = 100$ Hz, phase angle is (−70.53°)

Problem 10.2.15 In the circuit of Fig. 10.85, determine output voltage for > 0, if $R_3 = 2R_1 = 2R_2 = 1\ \text{k}\Omega$, $C = 1000\ \mu\text{F}$.

Solution

For $t < 0$, $\quad v_2(0^-) = v_0(0^-) = 10\,\text{V}$

For $t > 0$, $\quad v_2(0^+) = v_0(0^+) = 10\,\text{V}$

KCL at node 1, $V_0(s) = V_C(s)$

$$\frac{V_0 - V_1}{R_3} - \frac{V_1}{R_1} - \frac{V_1 - V_0}{R_2} = 0 \quad \rightarrow \quad \frac{V_0 - V_1}{R} - \frac{2V_1}{R} - \frac{2(V_1 - V_0)}{R} = 0$$

$$3V_0 - 5V_1 = 0 \quad \rightarrow \quad V_1 = \frac{3}{5}V_0 \tag{10.82}$$

KCL at node 2, $V_0(s) = V_C(s)$,

$$\frac{V_1 - V_0}{R_2} - \frac{V_0}{\frac{1}{sC}} = 0 \quad \rightarrow \quad \frac{2(V_1 - V_0)}{R} - sCV_0$$

$$2V_1 - 2V_0 - sRCV_0 = 0 \quad \rightarrow \quad 2\left(\frac{3}{5}V_0\right) - 2V_0 - sRCV_0 = \frac{4}{5}V_0 + sRCV_0 = 0$$

$$s + \frac{4}{5RC} = 0 \quad \rightarrow \quad s = -\frac{4}{5RC}$$

$$v_0(t) = v_0 e^{st}u(t) = V_0 e^{-\frac{4}{5RC}t}u(t).$$

Substituting $R = 1\,\text{k}\Omega$, $\quad C = \mu\text{F}$, $\quad V_0 = 10V$, $\quad v_0(t) = 10e^{-0.8t}u(t)$.

Fig. 10.85 The circuit for Problem 10.2.15

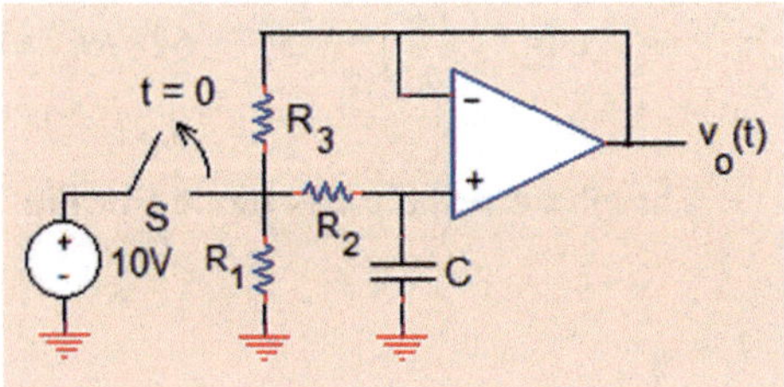

Problem 10.2.16 Design analog computer circuits to solve homogenous parts of the following differential equations with the same initial conditions,

(a) $\frac{d^2x}{dt^2} + \frac{1}{3}\frac{dx}{dt} + \frac{1}{4}x = -f$, (b) $\frac{d^2x}{dt^2} + \frac{5}{6}\frac{dx}{dt} + \frac{1}{6}x = -f$,

$x(0) = 0.4$, $dx(0)/dt = 0$
where f is the forcing function. (analog-computer.cir).

Solution

(a) Let

$$\ddot{x} \rightarrow \frac{d^2x}{dt^2}, \quad \dot{x} \rightarrow \frac{dx}{dt}$$

where the prime refers to the derivative. Select the highest order derivative term,

$$\ddot{x} = -\frac{1}{3}\dot{x} - \frac{1}{4}x - f = -\left(\frac{1}{3}\dot{x} + \frac{1}{4}x + f\right)$$

Integrate the get $\dot{x}$,

$$\dot{x} = -\int_0^t \left(\frac{1}{3}\dot{x} + \frac{1}{4}x + f\right) dt$$

Let $RC = 1$, ($R \rightarrow M\Omega$, $C \rightarrow \mu F$) integrate the $\dot{x}$ bearing in mind that integration using an inverting op-amp configuration introduces a minus sign.
The resulting analog computer circuit is displayed in Fig. 10.86a All resistors except R_2 and R_3 are equal in value (=R).
Since the characteristic equation is

$$s^2 + \frac{1}{3}s + \frac{1}{4} = 0$$

Its roots are $s_{1,2} = -0.1667 \pm j0.4714$, indicating that the system shows an underdamped behavior, as shown in Fig. 10.86b, with the output equation of $x(t) = e^{-\frac{t}{6}}\left[\frac{1}{5\sqrt{2}}\sin\left(\frac{\sqrt{2}}{3}t\right) + \frac{2}{5}\cos\left(\frac{\sqrt{2}}{3}t\right)\right]$.

(b) Repeating similar procedure as in part (a) results in the analog computer circuit as shown in Fig. 10.19a, only differences being $R_2 = 6R$, $R_3 = (6/5)R$, replacing $R_2 = 4R$, $R_3 = 3R$, respectively, all else remaining the same.
This system has the following characteristic equation;

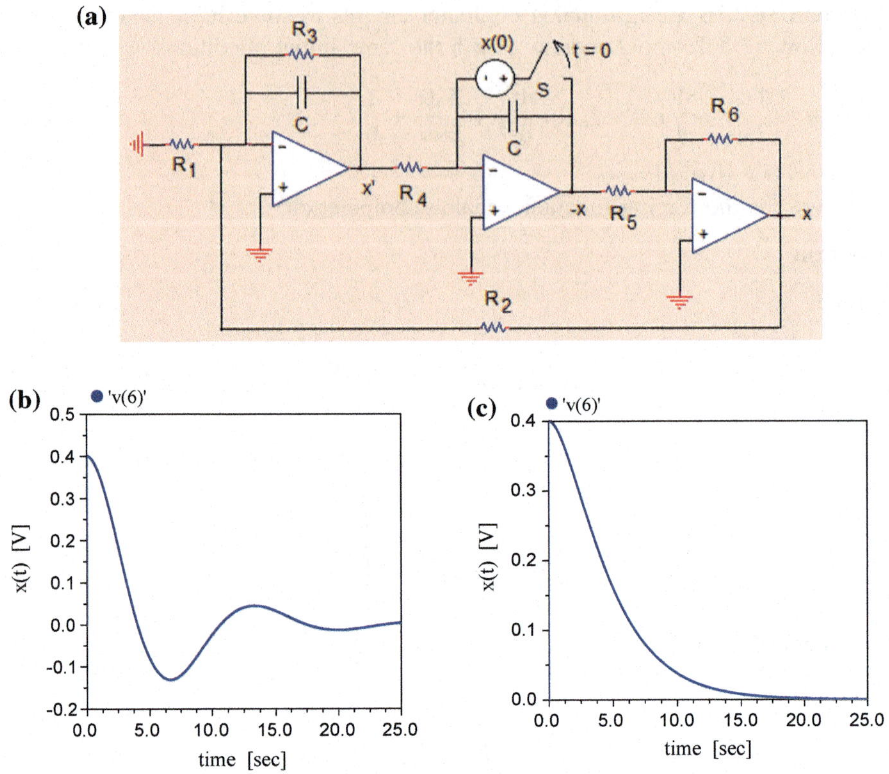

Fig. 10.86 **a** Analog computer circuit for the solution of homogenous part of a second-order differential equation with initial condition (switch opens at time zero). All resistors except R_2 and R_3 are equal in value (=R). **b** Analog computer output $x(t)$ for $x(0)$ = 0.4 V, (part a). **c** Analog computer output $x(t)$ for $x(0)$ = 0.4 V (part b)

$$s^2 + \frac{5}{6}s + \frac{1}{6} = 0$$

with roots of $s_1 = -1/2$ and $s_2 = -1/3$ both real and negative. Therefore, the system described by such a differential equation has an overdamped nature, as shown in Fig. 10.86c. with $x\,(t) = \frac{1}{5}\left(6\mathrm{e}^{-t/3} - 4\mathrm{e}^{-t/2}\right)$.

Note that for homogenous solution of given differential equations, f = 0. (This is realized by connecting resister R_1 to ground). If a complete solution is desired for a differential equation with a forcing function, then the ground connection is removed from resister R_1 and forcing function injected at this point. It should also be noted here that, the selection of op-amps is important in analog computer applications, as they have different input offset voltages and bias currents. (In the examples described here in, an ideal single pole op-amp model is employed).

SPICE netlist (analog-computer.cir) is given below.

```
vi 7 0 sin(0 1e-6 1 0 0)
R1 1 7 1MEG
R2 1 6 6MEG
R3 1 2 1.2MEG
R4 2 3 1MEG
R5 4 5 1MEG
R6 5 6 1MEG
C1 1 2 1U
C2 3 4 1U IC=0.4
XOA1 0 1 2 OA1
XOA2 0 3 4 OA1
XOA3 0 5 6 OA1
*OPAMP MACRO MODEL
.SUBCKT OA1    1  2  6
*XOA           NI  I  Vo OA1
RIN  1 2 100MEG
EGAIN  3 0  1 2  100000
RP1  3  4  10 K
CP1  4  0  1.6UF
EBUFF  5 0  4 0  1
ROUT  5  6  1
.ENDS
*.tran .1 m 25 0 10 m uic
```

Problem 10.2.17

(a) What is the expression for the general form of the magnitude of Butterworth low-pass filter (LPF) with pass-band gain K and cutoff at ω_c?

(b) Determine the first-order Butterworth LPF magnitude response and its transfer function if pass-band gain is 20 dB and $\omega_c = 100\,\text{Hz}$.

(c) Determine analytically the required number of poles (=degree of Butterworth filter) if $|H_{\min}/H_{\max}|$ is given, where $H_{\min}$, $H_{\max}$ are stop band gain and pass-band gain, respectively.

(d) A LPF with $f_c = 100\,\text{Hz}$ and pass-band gain of +40 dB is to be designed with minimum gain of +20 dB at 400 Hz. Determine the minimum order of LP Butterworth filter.

Solution

(a) All Butterworth low-pass filters produce magnitude (gain) response of the form

$$|H_n(j\omega)| = \frac{|K|}{\sqrt{1+\left(\frac{\omega}{\omega_c}\right)^{2n}}}$$

where n is the order of filter (=number of poles), ω_c is the cutoff frequency, and K is pass-band (maximum) gain.

(b)
$$|H_1(j\omega)| = \frac{10}{\sqrt{1+\left(\frac{\omega}{314}\right)^{2}}}$$

The general form of first-order Butterworth LPF function with given values is

$$H_1(s) = \frac{K}{D\left(\frac{s}{\omega_c}\right)} = \frac{10}{1+\frac{s}{314}} = \frac{3140}{s+314}$$

(c) Since $|K| = H_{\max}, \quad |H(j\omega_{\min})| \leq H_{\min}$

$$|H(j\omega_{\min})| = \frac{H_{\max}}{\sqrt{1+\left(\frac{\omega_{\min}}{\omega_c}\right)^{2n}}} \leq H_{\min}$$

Solution of this equation for n yields

$$n \geq \frac{1}{2}\frac{\ln\left[\left(\frac{H_{\max}}{H_{\min}}\right)^2 - 1\right]}{\ln\left(\frac{\omega_{\min}}{\omega_c}\right)}$$

(d) $\frac{\omega_{\min}}{\omega_c} = \frac{400}{100} = 4, \quad \left|\frac{H_{\max}}{H_{\min}}\right| = \frac{40\text{ dB}}{20\text{ dB}} = 20\text{ dB} = 10\text{ V/V}$

$n \geq \frac{1}{2}\frac{\ln\left[(10)^2 - 1\right]}{\ln(4)} = \frac{1}{2} \times \frac{4.595}{1.386} = 1.66 \quad n = 2$ is the minimum order Butterworth LPF.

Problem 10.2.18 Determine the transfer function of a third-order Butterworth low-pass filter with a cutoff frequency of 10 Hz.

Solution

$$\omega_c = 2\pi f = 6.28 \times 10 = 62.8\,\text{rad/s}$$

Normalized Butterworth filter transfer function is for $\omega_c = 1\,\text{rad/s}$,

$$H_3(s) = \frac{1}{D(s)} = \frac{1}{(s+1)(s^2+s+1)}$$

Frequency scaling is performed by replacing each s by (s/ω_c),

$$H(s) = \frac{\omega_c^3}{(s+\omega_c)(s^2+\omega_c s+\omega_c^2)} = \frac{(62.8)^3}{(s+62.8)\left(s^2+62.8s+(62.8)^2\right)}$$

$$H(s) = \frac{247{,}673.152}{(s+62.8)(s^2+62.8s+3943.84)}$$

Problem 10.2.19 Design a second-order Buterworth type low-pass filter (LPF) with cuttoff frequency of 10 Hz,

(a) using RLC series filter topology.
(b) using the Sallen–Key filter structure.

Solution

$$\omega_0 = \omega_c = 2\pi.10 = 62.8\ \text{rad/s}$$

Second-order Butterworth filter normalized voltage transfer function is (see, Appendix, Butterworth polynomials in factored form),

$$H_2(s) = \frac{1}{D_2(s)} = \frac{1}{s^2+\sqrt{2}s+1} = \frac{k\omega_0^2}{s^2+\left(\frac{\omega_0}{Q}\right)s+\omega_0^2}$$

(a) RLC series circuit has a LP transfer function

$$H(s) = \frac{\frac{1}{LC}}{s^2+\left(\frac{R}{L}\right)s+\frac{1}{LC}} = \frac{k\omega_0^2}{s^2+\left(\frac{\omega_0}{Q}\right)s+\omega_0^2}$$

with

$$\omega_0^2 = \frac{1}{LC} \tag{10.83}$$

$$k = 1, \quad Q = \frac{\sqrt{2}}{2} = 0.707 \quad \text{(Butterworth second order filter}$$

$$Q = \frac{1}{R}\sqrt{\frac{L}{C}} = \frac{\sqrt{2}}{2} = 0.707 \tag{10.84}$$

Let $C = 1000\ \mu\text{F}$, From (10.83),

$$L = \frac{1}{\omega_0^2 C} = \frac{1}{(62.8)^2 10^{-4}} = \frac{2.54 \times 10^{-4}}{10^{-4}} = 2.54\ H$$

From (10.84),

$$R = \frac{\sqrt{\frac{L}{C}}}{Q} = \frac{\sqrt{\frac{2.54}{10^{-4}}}}{0.707} = 22.54\ \Omega$$

Figure 10.87 displays the filter structure with $R = 22.54\ \Omega$, $L = 2.54\ H$, $C = 100\ \mu\text{F}$, so that $f_c = 10\ \text{Hz}$.

(b) Sallen–Key LPF structure design parameters are,

$$\omega_0 = \frac{1}{RC}, \quad Q = \frac{1}{3 - A}, \quad k = A, \quad R_f = (A - 1)R$$

Since $\omega_0 = 62.8\ \text{rad/s}$, let $C = 10\ \mu\text{F}$,

$$R = \frac{1}{\omega_0 C} = \frac{1}{(62.8)(10^{-5})} = 1592.4\ \Omega$$

For a second-order Buterworth filter, $Q = \sqrt{2}/2 = 0.707$

$$A = 3 - \frac{1}{Q} = 3 - \frac{1}{0.707} = 1.586 = k$$

$R_f = (A - 1)R = (1.586 - 1) \times 1592.4 = 932.5\ \Omega$. Figure 10.88 displays the filter with

$$C = 10\ \mu\text{F}, \quad R = 1592.4\ \Omega, \quad R_f = 932.5\ \Omega$$

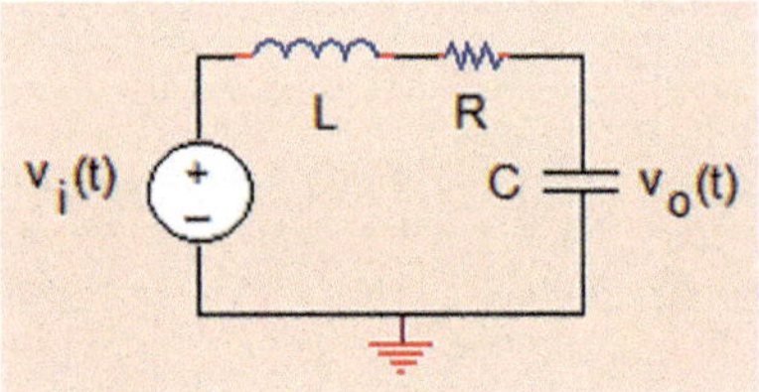

Fig. 10.87 RLC low-pass filter structure

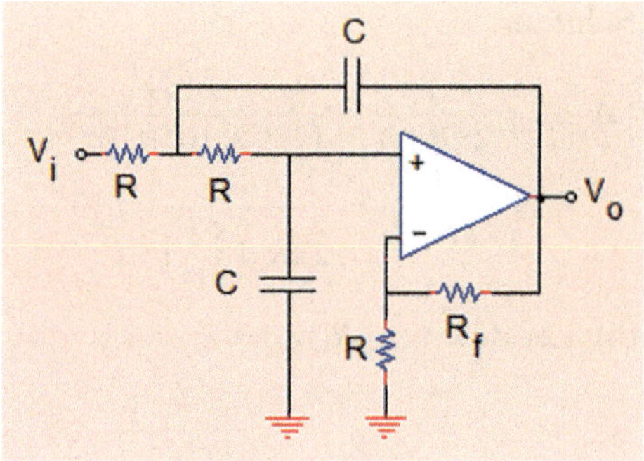

Fig. 10.88 Sallen–Key low-pass filter structure

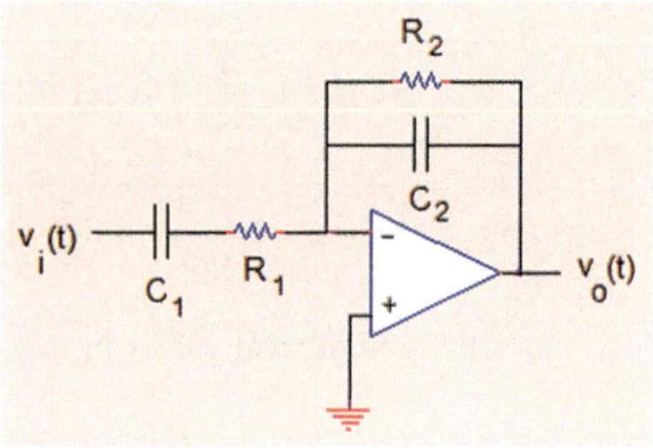

Fig. 10.89 Band-pass filter

Problem 10.2.20 The circuit shown in Fig. 10.89 is a simple band-pass-filter which is part of a galvanic skin response measurement and recording circuit. Lower cutoff frequency is required to remove base line drift while higher cutoff frequency is for eliminating high frequency noise.

(a) What are the values of lower and upper cutoff frequencies?
(b) Determine voltage transfer function of circuit.
(c) Plot log–log magnitude transfer function (frequency in Hz).
(d) Determine the grounded resistor value of a preceding noninverting op-amp circuit, if is overall gain of the filter and amplifier cascade is to be 9 V/V. Use magnitude plot obtained in part (c) and let the feedback resistor value of the noninverting op-amp circuit be $R_f = 100$ kΩ.
$R_1 = R_2 = 3.3$ MΩ, $C_1 = 10C_2 = 100$ nF. (plot_transfer_function5.m)

Solution

(a) $f_H = \dfrac{1}{2\pi R_1 C_1} = \dfrac{0.159}{(3.3\times 10^6)(10^{-7})} = \dfrac{0.159\times 10}{3.3} = 0.48\,\text{Hz}$

$$f_L = \frac{1}{2\pi R_2 C_2} = \frac{1}{2\pi R_2\left(\dfrac{C_1}{10}\right)} = 10 f_H = 4.8\ \text{Hz}$$

(b) Let $Z_f = Z_c \parallel R,\ Z_s = Z_c + R,$

$$H(s) = -\frac{Z_f}{Z_s} = -\frac{\dfrac{R}{sRC_2+1}}{\dfrac{sRC_1+1}{sC_1}} = -\frac{sRC_1}{(sRC_2+1)(sRC_1+1)}$$

$$H(s) = \frac{-sRC_1}{RC_2\left(s+\dfrac{1}{RC_2}\right)RC_1(s+1RC_1)} = -\frac{s}{RC_2\left(s+\dfrac{1}{RC_2}\right)\left(s+\dfrac{1}{RC_1}\right)}$$

$$H(s) = \frac{-s}{RC_2\left[s^2+s\left(\dfrac{1}{RC_1}+\dfrac{1}{RC_2}\right)+\dfrac{1}{R^2C_1C_2}\right]}$$

Substituting numerical values in this equation yields

$$H(s) = \frac{-30.3s}{s^2+33.33s+91.83}$$

(c) Fig. 10.90 shows the log–log magnitude response as a function of frequency (in Hz)

MATLAB script (plot_transfer_function5.m) is given below.

```
%Galvanic skin response BPF
R=3.3e6; C1=100e-9; C2=10e-9;
a=1/(R*C1); b=1/(R*C2); A=a+b; w0=1/(R^2*C1*C2);
num=[-b  0]; den=[ 1  A  w0]; H=tf(num,den)
L=400; w=1;
for k=1:L;
s=i*w*k; N=-s*b; D=s.^2+A*s+w0; H1(k)=abs(N/D);
end
k=1:L; f=k*w*0.159;
loglog(f,H1);grid on;xlabel('frequency [Hz]');
ylabel('Gain [V/V)');
```

(d) Let overall gain be

$$A = A_{\text{filter}}A_{\text{nia}} = (0.9)(A_{\text{nia}}) = 9 \quad\rightarrow\quad A_{\text{nia}} = 10 = 1+\frac{R_f}{R_g} = 1+\frac{100}{R_g}$$

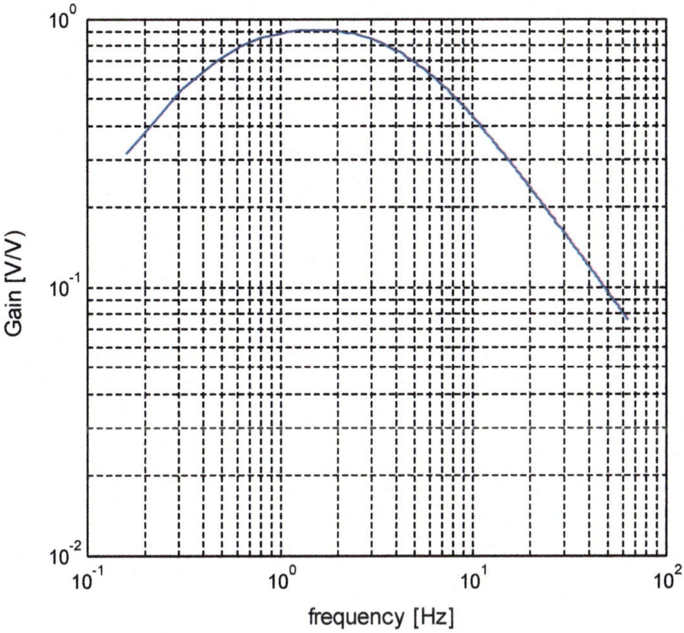

Fig. 10.90 The log–log magnitude response as a function of frequency (Hz)

Solving this equation for R_g yields $R_g = 11.11\,\text{k}\Omega$.
Note that filter mid-band gain is obtained from the magnitude plot. The voltage gain function can be written as

$$\frac{V_o(s)}{V_i(s)} = -\frac{R_2}{R_1} \cdot \frac{1}{(1 + sC_2R_2)\left(1 + \dfrac{1}{sC_1R_1}\right)}$$

where R_2, R_1 are the feedback and input resistances, C_2, C_1 are the feedback and input capacitance values, respectively. From this last equation it can be concluded that the mid-band gain is expected to be $K = -R_2/R_1$.
Since $R_2 = R_1$, $K = 1$. However, in many biomedical applications, two corner frequencies are usually close to each other. On the other hand, the gain equation is valid in (flat) midrange if the two corner frequencies are distant apart from each other at least by two decades.
Therefore, computational mid-band gain value ($K = 1$) is not to be used here, because two corner frequencies are only one decade apart from each other.

Problem 10.2.21 A sensor output signal with spectral power concentrated between 0 and 100 Hz is to be sampled by a frequency of 500 Hz. An anti-alias filter is required with pass-band gain of +10 dB and stop band gain of −20 dB in aliasing range. Construct and realize this filter in Butterworth form, use Sallen–Key structure for quadratic sections. (LPF3.cir).

Solution

The degree of required filter is

$$n \geq \frac{1}{2}\frac{\ln\left[\left(\frac{H_{max}}{H_{min}}\right)^2 - 1\right]}{ln\left(\frac{\omega_{min}}{\omega_{max}}\right)}$$

$$f_c = 100\,\text{Hz}, \quad f_s = 500\,\text{Hz}, \quad f_{min} = f_s - f_c = 400\,\text{Hz}$$

$$\left|\frac{H_{max}}{H_{min}}\right| = 30\,\text{dB} = 20\log x \quad \rightarrow \quad x = 10^{3/2} = \left|\frac{H_{max}}{H_{min}}\right|$$

$$n \geq \frac{1}{2} \times \frac{\ln\left[\left(10^{3/2}\right)^2 - 1\right]}{ln\left(\frac{2\pi \times 400}{2\pi \times 100}\right)} = \frac{1}{2} \times \frac{6.907}{1.386} = 2.49$$

Hence a third-order Butterworth response meets given stop-band requirements.

The third-order Butterworth polynomial (see, table of normalized Butterworth polynomials in Appendix),

$$D_3(s) = (s+1)\left(s^2 + s + 1\right)$$

using frequency scaling, $K = 10\,\text{dB} = 10^{1/2} = 3.16$

$$H(s) = \frac{K}{D_3\left(\frac{s}{\omega_c}\right)} = \frac{3.16}{\left(\frac{s}{\omega_c} + 1\right)\left[\left(\frac{s}{\omega_c}\right)^2 + \frac{s}{\omega_c} + 1\right]}$$

By partitioning this equation, one obtains

$$\frac{H_2(s)}{3.16} = \frac{1}{\left(\frac{s}{\omega_c} + 1\right)} \cdot \frac{1}{\left(\frac{s}{\omega_c}\right)^2 + \frac{s}{\omega_c} + 1}$$

This means that, a first-order LPF is cascaded with a second-order LPF.

(i) Second-order stage design[2] using Sallen–Key topology (Fig. 10.88): Applying equal component values method [16], the transfer function has the form

[2]Note that this second-order section should not to be confused with the second-order Butterworth type filter, because the latter has linear (s) coefficient term of $\sqrt{2}$ in denominator polynomial.

$$H_2(s) = \frac{K}{\left(\frac{s}{\omega_c}\right)^2 + 2\zeta\frac{s}{\omega_c} + 1} = \frac{K}{\left(\frac{s}{\omega_c}\right)^2 + \frac{s}{\omega_c} + 1}$$

Stage parameters are $\omega_c = 2\pi f_c = 2\pi \times 100 = 628\,\text{rad/s}$.
Zeta parameter $= \zeta = 1/2$

$$\frac{R_{f2}}{R} = 2(1-\zeta) = 2\left(1 - \frac{1}{2}\right) = 1 \quad \rightarrow \quad R_{f2} = R$$

Let $R = 10\,\text{k}\Omega$, then,

$$C = \frac{1}{\omega_c R} = \frac{1}{628 \times 10^4} = 160\,\text{nF}$$

$$K_2 = 1 + \frac{R_{f2}}{R} = 1 + 1 = 2 \quad (\text{stage gain})$$

Stage parameters in the alternative form:

$$\omega_0 = \frac{1}{RC}, \quad Q = \frac{1}{3 - K_2}$$

Since $Q = 1$ here, $K_2 = 2$

$$K_2 = 2 = 1 + \frac{R_{f2}}{R} \quad \rightarrow \quad \frac{R_{f2}}{R} = 1 \quad \rightarrow \quad R_{f2} = R$$

(ii) First-order stage design (using noninverting op-amp configuration following an RC LPF);

$$H_1(s) = \frac{K_1}{\frac{s}{\omega_c} + 1}, \quad \omega_c = \frac{1}{RC} = 628\,\text{rad/s}$$

Let $R = 10\,\text{k}\Omega$, then $C = 160\,\text{nF}$
Since $K = K_2 \cdot K_1 = 2K_1 = 3.16 \quad \rightarrow \quad K_1 = 1.58$

$$1.58 = 1 + \frac{R_{f1}}{R} \quad \rightarrow \quad \frac{R_{f1}}{R} = 0.58 \quad \rightarrow \quad R_{f1} = 0.58R = 5.8\,\text{k}\Omega$$

Finally, third-order Butterworth LP filter circuit and its SPICE generated magnitude vs frequency response are shown in Figs. 10.91 and 10.92.

SPICE netlist (LPF3.cir) is also given below.

SPICE Netlist (LPF3.cir): Butterworth third- order LPF

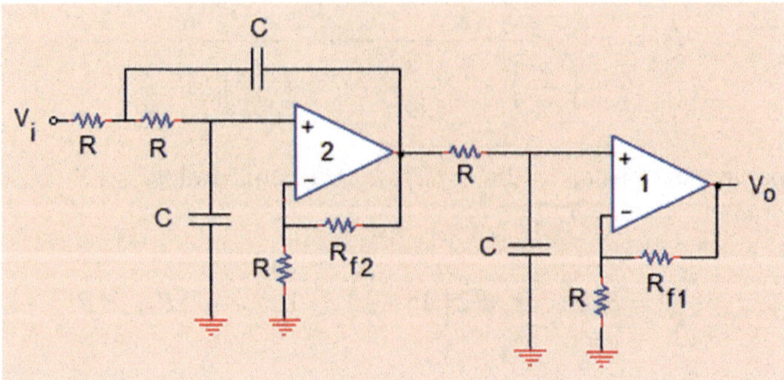

Fig. 10.91 Third-order Butterworth LP filter

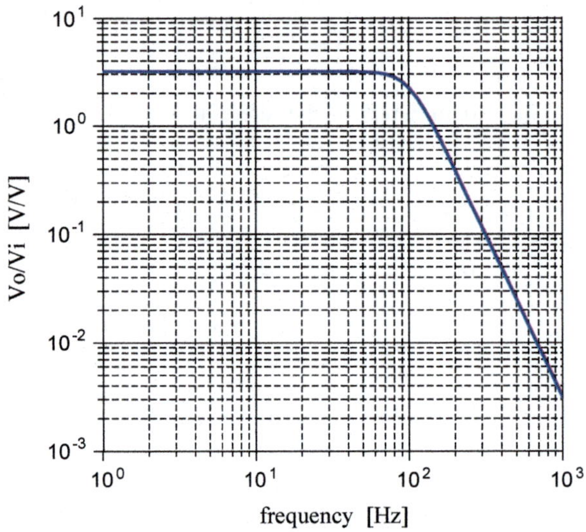

Fig. 10.92 Third-order Butterworth LP filter magnitude versus frequency plot

```
vi 1 0 1 AC
R1 1 2 10 k
R2 2 3 10 k
R3 4 0 10 k
RF1 4 5 10 K
R5 5 6 10 k
R6 7 0 10 k
RF2 7 8 5.8 K
C1 2 5 160 N
C2 3 0 160 N
C3 6 0 160 N
XOA1 3  4  5  OA1
XOA2 6  7  8  OA1
```

```
.SUBCKT OA1       1  2  6
*XOA             NI  I  Vo OA1
RIN 1 2 1000MEG
EGAIN 3 0 1 2 100000
RP1 3 4 10 K
CP1 4 0 1.6UF
EBUFF 5 0 4 0 1
ROUT 5 6 1
.ENDS
*.ac dec 20 1 1000
```

Problem 10.2.22

(a) Design an active band-pass filter (BPF) for an ECG monitoring unit front-end application with input resistance of 0.1 MΩ, to pass frequencies between 0.1–100 Hz by combining a LPF and HPF (both with unity gain and single time constant) and an inverting op-amp circuit with some gain, so that overall gain is 10 at the center frequency.

(b) Determine the pass-band gain of cascaded LPF-HPF-inverter topology of Fig. 10.93, and express the gain function in standard form in terms of critical frequencies and pass-band gain.

Solution

(a) Fig. 10.93. displays the circuit.
Upper corner frequency is set by the LPF while lower corner frequency is set by the HPF.

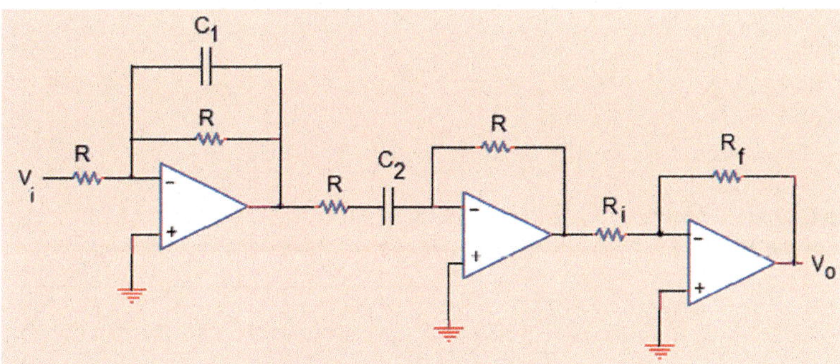

Fig. 10.93 The BPF using cascaded LPF-HPF-Inverter sections

Pass-band gain is determined by the inverting amplifier gain.
BPF voltage transfer function is

$$H(s) = \frac{V_O(s)}{V_i(s)} = \left(\frac{-1}{1+sRC_1}\right)\left(\frac{-sRC_2}{1+sRC_2}\right)\left(-\frac{R_f}{R_i}\right)$$

Lower corner frequency is

$$f_1 = \frac{1}{2\pi RC_2} \quad \rightarrow \quad C_2 = \frac{1}{2\pi Rf_1}$$

Upper corner frequency is

$$f_2 = \frac{1}{2\pi RC_1} \quad \rightarrow \quad C_1 = \frac{1}{2\pi Rf_2}$$

Let $R = 100\,\text{k}\Omega$ (this value will satisfy input resistance requirement), then,

$$C_2 = \frac{1}{2\pi \times 10^5 \times 10^{-1}} = 15.9\ \mu\text{F}$$
$$C_1 = \frac{1}{2\pi \times 10^5 \times 10^2} = 15.9\,\text{nF}$$

Select $R_i = R = 100\,\text{k}\Omega$ for the input resistance of inverting amplifier circuit. Using SPICE, filter mid-band gain of 10 (= +20 dB) at f = 3 Hz is realized when $R_f = 1\,\text{M}\Omega$.
Figure 10.94 displays the voltage transfer function of the filter in dB units.
SPICE netlist (BPF.cir) is given below.

```
; BPF (AC analysis)
vi 1 0 1 AC
R1 1 2 100 k
R2 2 3 100 k
R3 3 4 100 k
```

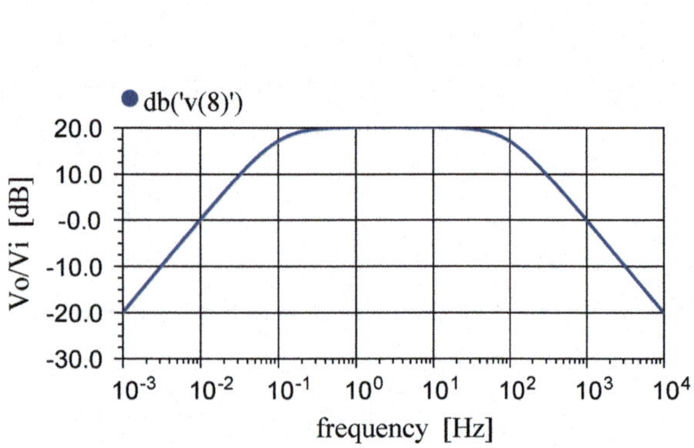

Fig. 10.94 Voltage transfer function of the BPF filter

```
R4 5 6 100 K
Ri 6 7 100 k
Rf 7 8 1000 k
C1 2 3 15.9 N
C2 4 5 15.9u
XOA1 0 2 3 OA1
XOA2 0 5 6 OA1
XOA3 0 7 8 OA1
*OPAMP MACRO MODEL
.SUBCKT OA1 1  2  6
RIN    1    2    1000MEG
EGAIN    3 0    1 2    100000
RP1    3    4    10 K
CP1    4    0    1.6UF
EBUFF    5 0    4 0    1
ROUT    5    6    1
.ENDS
```

(b) With $\omega_L > \omega_H$,

$$\omega_0 = (\omega_L \omega_H)^{\frac{1}{2}}, \quad BW = \omega_L - \omega_H, \quad Q = \frac{\omega_0}{BW}$$

$$\boldsymbol{H}(\omega) = -\frac{R_f}{R_i} \frac{j\omega\omega_H}{(1 + \dfrac{j\omega}{\omega_H})(1 + \dfrac{j\omega}{\omega_L})}$$

$$\text{here,} \quad \omega_L = \frac{1}{RC_1}, \quad \omega_H = \frac{1}{RC_2}$$

$$\boldsymbol{H}(\omega_0) = -\frac{R_f}{R_i} \frac{j\omega_0\omega_L}{(\omega_H + j\omega_0)(\omega_L + j\omega_0)}$$

$$|\boldsymbol{H}(\omega_0)| = \frac{R_f}{R_i} \frac{\omega_L}{\omega_L + \omega_H} = k$$

The gain function can be rewritten as

$$\boldsymbol{H}(\omega) = -\frac{k(\omega_L + \omega_H)}{\omega_L} \cdot \frac{1}{1 + j\omega + C_1 R} \cdot \frac{j\omega C_2 R}{1 + j\omega C_2 R}$$

where k is the pass-band gain. In standard form,

$$\boldsymbol{H}(\omega) = k\left(\frac{\omega_L + \omega_H}{\omega_L}\right) \frac{\dfrac{j\omega}{\omega_H}}{(1 + j\dfrac{\omega}{\omega_H})(1 + j\dfrac{\omega}{\omega_L})}$$

Since
$\frac{\omega_L + \omega_H}{\omega_L} = 1 + \omega_H \omega_L$, if $\omega_L \gg \omega_H$, then

$$k = |\boldsymbol{H}(\omega_0)| = \frac{R_f}{R_i}$$

Note that the value of R_f found in part (a) is modified here,

$$\frac{R_f}{R_i} = \frac{k(\omega_L + \omega_H)}{\omega_L} = \frac{k(f_L + f_H)}{f_L}$$
$$R_f = \frac{k(f_L + f_H)}{f_L} \cdot R_i = \frac{10(100 + 0.1)}{100} \cdot 100 = 1001\,\text{k}\Omega = 1.001\,\text{M}\Omega$$

Problem 10.2.23

(a) Determine the gain function of the circuit shown in Fig. 10.95.
(b) Evaluate the value of gain at $\omega = 30$ rad/s if $R = R_1 = R_2 = R_3 = R_4 = 1\,\text{k}\Omega$ and $C_2 = 1000\ C_1, \quad C = C_1 = 1\,\mu\text{F}$.
(c) Determine the gain function of (b) in s-domain.
(d) Determine the phase shift at the center frequency.
(e) A front-end of an diagnostic electrocardiography unit includes a BPF with corner frequencies at $f_1 = 0.159$ Hz and $f_2 = 159$ Hz, and mid-band gain of 20 dB. Adapt the circuit of Fig. 10.95 with minimal modifications to these requirements. (complex 3.m), (BPF2.cir).

Solution

(a) For HPF,

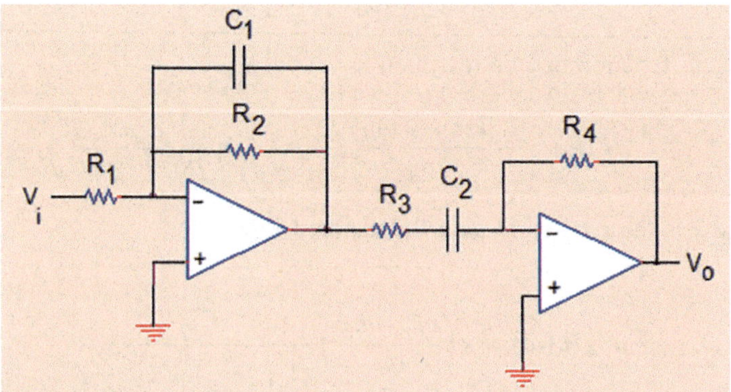

Fig. 10.95 Cascaded LPF and HPF (BPF)

$$H_H(\omega) - k_\omega \frac{j\omega}{1+j\dfrac{\omega}{P_H}}, \quad k_H = R_4C_2 = 100RC$$

$$p_H = \frac{1}{R_3C_2} = \frac{1}{1000RC}$$

For LPF,

$$H_L(\omega) = -k_L \frac{1}{1+j\dfrac{\omega}{p_L}}, \quad k_L = \frac{R_2}{R_1}, \quad p_L = \frac{1}{R_2C_1} = \frac{1}{RC}$$

$$H(\omega) = 1000RC \cdot \frac{j\omega}{(1+j1000RC\omega)(1+jRC\omega)}$$

(b) $RC = 10^3 \times 10^{-6} = 10^{-3}\,\mathrm{s}$

$$H(\omega) = \frac{j\omega}{(1+j\omega)(1+j10^{-3}\omega)}, \quad H(30) = 0.999 + j0.0033$$

$$|H(30)| = 0.999$$

(c) $s = j\omega$

$$H(s) = \frac{s}{(1+s)(1+10^{-3}s)} = \frac{1000s}{(s+1)(s+1000)}$$

(d) Bode phase plot shows that at the center frequency (= geometric mean of corner frequencies), $f_0 = \sqrt{1000} = 31.62$ rad/s phase shift is 0° (Fig. 10.96).

(e) $p_H = \omega_H = 2\pi f_H = 2\pi \times 0.159 = 1\,\mathrm{rad/s}\ (f_1 = f_H)$

$p_L = \omega_L = 2\pi f_L = 2\pi \times 159 = 1000\,\mathrm{rad/s}\ (f_2 = f_L)$

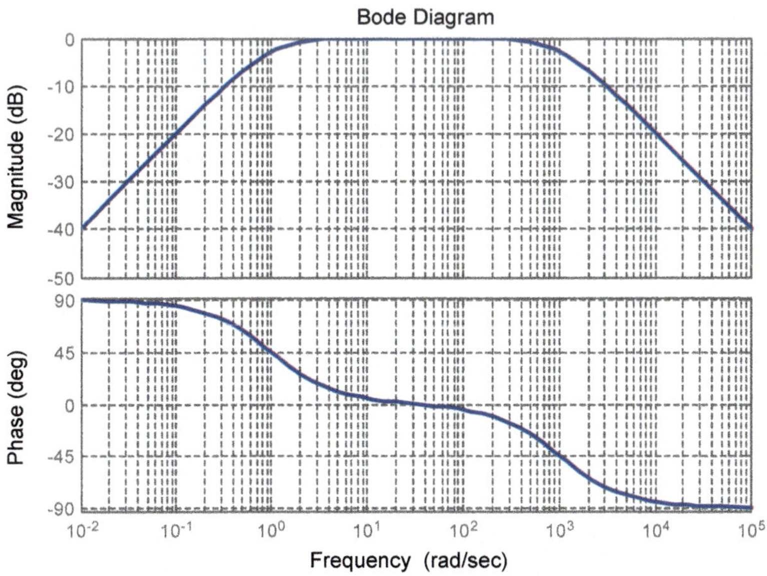

Fig. 10.96 Bode gain and phase plots

The gain of the previous circuit is to be increased from 1 to 10 (= 20 db) without changing corner frequencies. This is possible by increasing HPF gain by 10,

$$k_H = R_4C_2 = 1000RC \rightarrow k'_H = R_4C'_2 = R\,(10{,}000C)$$

and compensating the increase of the value of C_2 on p_H,

$$p'_H = \frac{1}{R'_3C'_2} = \frac{1}{(0.1R)(10{,}000C)} = \frac{1}{1000RC} = p_H$$

Summarizing; $C'_2 = 10C_2$ and $R'_3 = R_3/10$ will introduce a mid-band gain of 20 dB while keeping corner frequencies the same as before (see, Fig. 10.97). MATLAB script (complex3.m) is given below.

```
w=30; h=(1i*w)./((1+1i*w).*(1+1i*w*0.001))
H=abs(h)
a=[1 1]; b=[1 1000]; c=conv(a,b); G=tf([1000 0],c)
bode(G); grid on;
```

SPICE netlist (BPF2.cir) is as follows;

```
; BPF2
vi 1 0 1 AC
R1 1 2 1 k
R2 2 3 1 k
R3 3 4 .1 k
R4 5 6 1 k
C1 2 3 1u
```

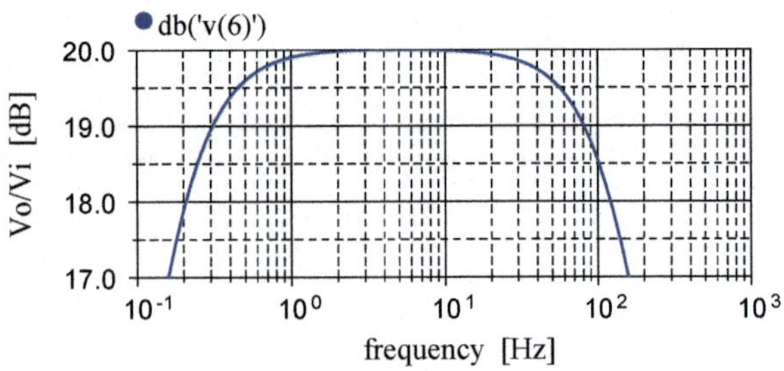

Fig. 10.97 Checking corner (−3 dB) frequencies and the new gain of BPF with SPICE

```
C2 4 5 10000u
XOA1 0 2 3 OA1
XOA2 0 5 6 OA1
XOA3 0 7 8 OA1
.SUBCKT OA1 1 2 6
RIN 1 2 1000MEG
EGAIN 3 0 1 2 1000000
RP1 3 4 10 K
CP1 4 0 1.6UF
EBUFF 5 0 4 0 1
ROUT 5 6 1
.ENDS
```

Problem 10.2.24 Design a BPF using Sallen–Key topology to filter out β waves from an EEG device. (BPFSK.cir).

Solution

β waves are in a range of 16–31 Hz. Transfer function of BPF is

$$H(s) = \frac{k\frac{\omega_0}{Q}s}{s^2 + \left(\frac{\omega_0}{Q}\right)s + \omega_0^2}$$

or

$$H(\omega) = \frac{k}{1 + jQ\left(\frac{\omega}{\omega_0} - \frac{\omega_0}{\omega}\right)}$$

Design equations for *S-K* topology are

$$\omega_0 = 1/RC, \quad Q = 1/3 - A \quad \text{and} \quad k = AQ$$

$$\omega_0 = 2\pi\sqrt{31 \times 16} = 44.54\pi \text{ rad/s}$$

$$BW = \omega_L - \omega_H = 2\pi(31 - 16) = 2\pi x 15 = 30\pi \text{ rad/s}$$

$$Q = \frac{\omega_0}{B\omega} = \frac{44.54\pi}{30\pi} = 1.485$$

$$A = 3 - \frac{1}{Q} = 3 - \frac{1}{1.485} = 2.326$$

$$\omega_0 = \frac{1}{RC}$$

Let $C = 1\ \mu\text{F}$, then

$$R = \frac{1}{\omega_0 C} = \frac{1}{(44.54\pi)10^{-6}} = 7.15\,\mathrm{k\Omega}$$

The mid-band gain is

$$k = AQ = (2.326)(1.485) = 3.454$$
$$R_f = (A - 1)R = (2.326 - 1)x7.15 = 9.48\,\mathrm{k\Omega}$$
$$R_{30} = 2R = 2 \times 7.15 = 14.3\,\mathrm{k\Omega}$$

The filter is shown in Fig. 10.98. Its gain-frequency performance is displayed in Fig. 10.99.

SPICE Netlist (BPFSK.cir) is given below.

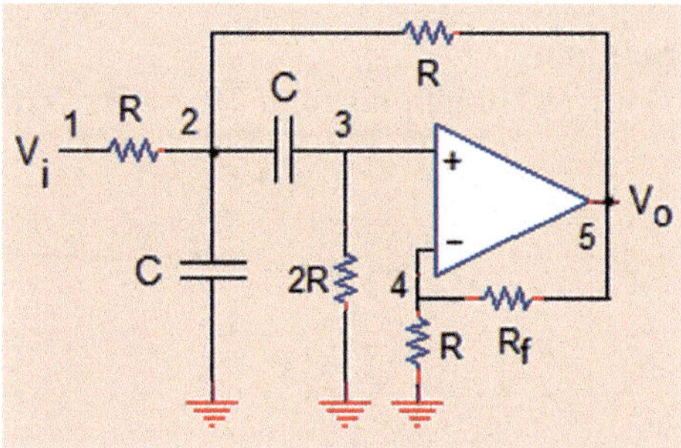

Fig. 10.98 Sallen–Key BPF topology

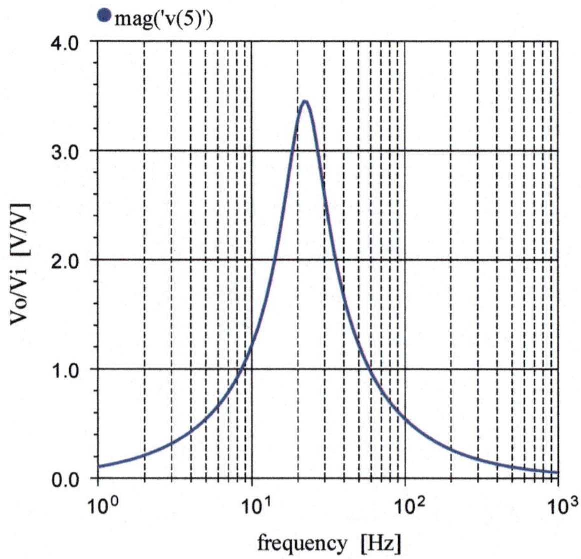

Fig. 10.99 Sallen–Key BPF gain-frequency performance

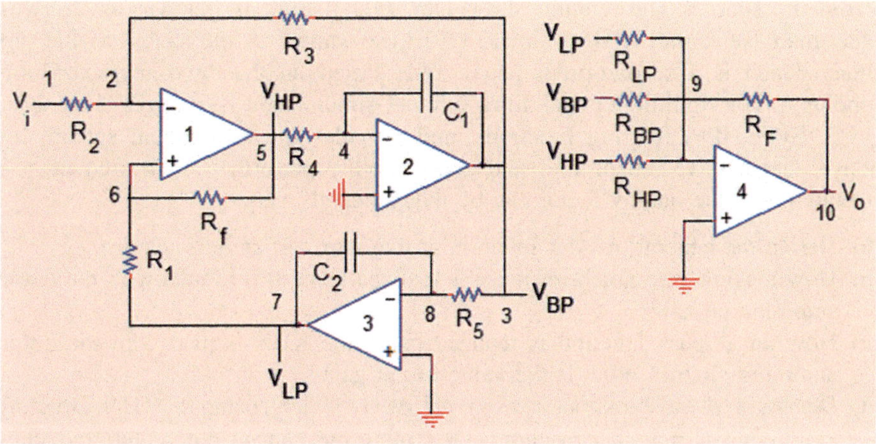

Fig. 10.100 KHN biquad multifunctional filter

```
; BPF Sallen-Key
vi 1 0 1 AC
R1 1 2 7.15 k
R2 2 5 7.15 k
R3 3 0 14.3 k
R4 4 0 7.15 k
Rf 4 5 9.48 k
C1 2 0 1u
C2 2 3 1u
XOA1 3 4 5 OA1
*OPAMP MACRO MODEL
.SUBCKT OA1      1  2  6
RIN  1  2  1000MEG
EGAIN  3 0  1 2  1000000
RP1  3  4  10 K
CP1  4  0  1.6UF
EBUFF  5 0  4 0  1
ROUT  5  6  1
.ENDS
```

Problem 10.2.25 The circuit shown in Fig. 10.100 is known as Kerwin–Huelsman–Newcomb (KHN) biquad [17] (also known as the state variable type filter) which is a second-order active filter based on the "two-integrator-loop" topology. This multifunctional filter circuit[3] simultaneously realizes three basic (LPF, BPF, HPF) filtering functions, and provides low component spread, low active, and passive sensitivities and good stability performance. Its frequency of operation and the quality factor can be independently varied.

(a) Derive the general transfer function at the output of summer circuit.
(b) How a symmetric notch response is realized using this biquad with the output summing circuit?
(c) How an all-pass function is realized using the KHN biquad with the output summing circuit? What is the value of flat gain?
(d) Design a β-wave extractor from an EEG device, using a KHN structure, employing $C = 1\ \mu\text{F}$ capacitors. Determine the gain at the center frequency. Plot band-pass, high-pass, low-pass, and band-stop filter outputs, simultaneously (KHN1.cir).

Solution
For a general second-order HPF function,

$$\frac{V_{HP}}{V_i} = \frac{Ks^2}{s^2 + \left(\frac{\omega_0}{Q}\right)s + \omega_0^2} \tag{10.85}$$

Here, K = high frequency gain. This can be rewritten as

$$V_{HP} = KV_i - \frac{1}{Q} \cdot \frac{\omega_0}{s} V_{HP} - \frac{\omega_0^2}{s^2} V_{HP} \tag{10.86}$$

Let $C_1 = C_2 = C, \quad R_4 = R_5 = R$
For a Miller integrator, $CR = 1/\omega_0$
The signal at the output of the first integrator is a BPF,

$$H_{BP}(s) = \frac{\frac{-\omega_0}{s} \times V_{HP}}{V_i} = -\frac{K\omega_0 s}{s^2 + \left(\frac{\omega_0}{Q}\right)s + \omega_0^2} = \frac{V_{BP}}{V_i} \tag{10.87}$$

[3]This type of filters appealed much interest in technical literature, with designs including other active components besides opamps [18–24]. They are also commercially available in 14-pin plastic DIP and SOIC-16 surface-mount packages, specified for the −25 to +85 °C temperature range [25].

Center frequency gain is

$$H\left(s\right)=\frac{-K\omega_0 s}{s^2+\left(\frac{\omega_0}{Q}\right)s+\omega_0^2}$$

$$H(j\omega_0)=\frac{-K\omega_0(j\omega_0)}{-\omega_0^2+\frac{\omega_0}{Q}j\omega+\omega_0^2}=\frac{K\omega_0^2}{\left(\frac{\omega_0^2}{Q}\right)}=KQ$$

$$|H_{BP}(\omega_0)|=KQ \tag{10.88}$$

The output of the second integrator realizes a LPF,

$$H_{LP}(s)=\frac{\left(\frac{\omega_0}{s}\right)^2}{V_i}V_{HP}=\frac{K\omega_0^2}{s^2+\left(\frac{\omega_0}{Q}\right)s+\omega_0^2}=\frac{V_{BP}}{V_i} \tag{10.89}$$

Using superposition, the output of the summer (opamp-1) ($=V_{HP}$) is expressed in terms of

$$V_{BP}=-\frac{\omega_0}{s}V_{HP},\quad V_{LP}=\left(\frac{\omega_0}{s}\right)^2V_{HP}, \tag{10.90}$$

it is shown that

$$R_f=R_1 \tag{10.91}$$

$$\frac{R_3}{R_2}=2Q-1 \tag{10.92}$$

$$K=2-\frac{1}{Q} \tag{10.93}$$

The summer (opamp-4) output,

$$V_0=-\left(\frac{R_f}{R_{HP}}V_{HP}+\frac{R_F}{R_{BP}}V_{BP}+\frac{R_F}{R_{LP}}V_{LP}\right) \tag{10.94}$$

Substituting each filter value of V_{HP}, V_{BP}, v_{LP} in (10.94) yields

$$\frac{V_0}{V_i}=-K\frac{\left(\frac{R_F}{R_{HP}}\right)s^2-s\left(\frac{R_F}{R_{BP}}\right)\omega_0+\left(\frac{R_F}{R_{LP}}\right)\omega_0^2}{s^2+\left(\frac{\omega_0}{Q}\right)s+\omega_0^2} \tag{10.95}$$

(b) A symmetric notch response can be obtained if R_{BP} is eliminated (opened), and $R_{HP} = R_{LP}$,

$$\frac{V_0}{V_i} = -\left(K\frac{R_F}{R_{LP}}\right)\frac{s^2+\omega_0^2}{s^2+\left(\frac{\omega_0}{Q}\right)s+\omega_0^2} \tag{10.96}$$

(c) An all-pass filter (APF) function is defined as

$$H(s) = A\frac{s^2-\frac{\omega_0}{Q}s+\omega_0^2}{s^2+\frac{\omega_0}{Q}s+\omega_0^2}$$

where A is the flat gain, and circuit provides a phase shift of $-180°$ at ω_0 ($0°$ at dc to $-360°$ at very high frequencies). Therefore, an APF is realized if $R_{LP} = R_{HP} = R_{BP}/Q$

The value of flat gain becomes

$$A = K\frac{R_F}{R_{LP}}$$

(d) For β-waves, $16\,\text{Hz} \le f \le 31\,\text{Hz}$,

$$BW = 31 - 16 = 15\,\text{Hz},$$
$$f_0 = \sqrt{f_H \times f_L} = \sqrt{16 \times 31} = 22.3\,\text{Hz}$$
$$BW = \frac{\omega_0}{Q} \quad \rightarrow \quad Q = \frac{\omega_0}{BW} = \frac{22.3}{15} = 1.487$$
$$\omega_0 = \frac{1}{RC} \quad \rightarrow \quad R = \frac{1}{\omega_0 C} = \frac{1}{2\pi \times 22.3 \times 10^{-6}} = 7.13\,\text{k}\Omega$$

Let $R_f = R_1 = 1\,\text{k}\Omega$,

$$\frac{R_3}{R_2} = 2Q - 1 = 2 \times 1.487 - 1 = 1.974$$

Select $R_2 = 1\,\text{k}\Omega$,

$$R_3 = 1.974 \times 1 = 1.974\,\text{k}\Omega$$

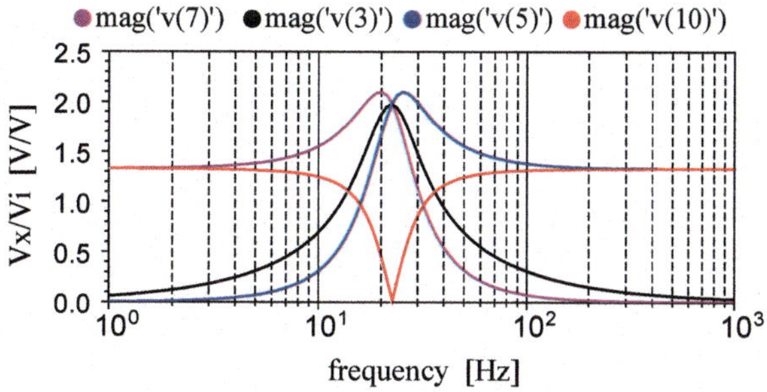

Fig. 10.101 Semi-log plot for simultaneous outputs of KHN biquad

Summarizing;

$$R = 7.13\,\text{k}\Omega,\ C = 1\,\text{nF},\ R_f = R_1 = R_2 = 1\,\text{k}\Omega,\ R_3 = 1.974\,\text{k}\Omega$$

The gain at center frequency is calculated using (10.88),$|H(j\omega_0)| = A = KQ$ Since

$$K = 2 - \frac{1}{Q} = 2 - \frac{1}{1.487} = 1.328$$
$$|H(j\omega_0)| = A = KQ = 1.328 \times 1.487 = 1.974$$

Figure 10.101 displays band-pass, high-pass, low-pass, and band-stop filter outputs.

SPICE netlist (KHN1.cir) is given below.

```
;KHN1 AC Analysis
vi 1 0 1 AC
R1 6 7 1 k
R2 1 2 1 k
R3 2 3 1.974 k
R4 5 4 7.13 k
R5 8 3 7.13 k
Rf 5 6 1 k
C1 4 3 1u
C2 8 7 1u
RLP 9 7 1 k
RHP 9 5 1 k
RBP 9 3 1G
RFF 9 10 1 k
XOA1 6 2 5 OA1
XOA2 0 4 3 OA1
XOA3 0 8 7 OA1
```

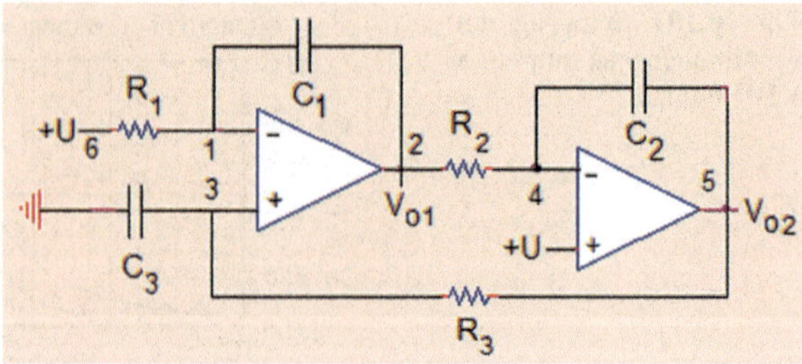

Fig. 10.102 Sinusoidal quadrature oscillator

```
XOA4 0 9 10 OA1
.SUBCKT OA1 1 2 6
RIN  1  2  1000MEG
EGAIN  3 0  1 2  1000000
RP1  3  4  10 K
CP1  4  0  1.6UF
EBUFF  5 0  4 0  1
ROUT  5  6  1
.ENDS
```

Problem 10.2.26

(a) What is Barkhausen criterion?
(b) How many "poles" are required to design an oscillator using single time constant (passive) circuits?
(c) Why LC oscillators using op-amps are not common in practice?
(d) What determines the frequency stability of an oscillator? What is the maximum phase shift of a cascade of buffered RC sections?
(e) For the quadrature oscillator[4] circuit shown in Fig. 10.102, determine the loop gain (βA) when $R = R_1 = R_2 = R_3$ and $C = C_1 = C_2 = C_3$. What is the oscillation frequency of this circuit?
(f) Design and simulate a sinusoidal quadrature oscillator which will produce 10 Hz and 1 V (peak-to-peak) oscillations (QO1.cir).

Solution

(a) The performance of a feedback system is described by

$$\frac{V_O(s)}{V_i(s)} = \frac{A(s)}{1 + \beta A(s)}$$

[4]Applications of quadrature oscillators are abundant in biomedical and telecommunication industries. They are also designed for different mode of operations, see references [26–28].

This equation shows that a feedback system becomes unstable if denominator of its transfer function vanishes, since A(s)/0 is an undefined state/
The condition to design an oscillator is called the Barkhausen criterion,

$$\beta A(j\omega) = 1\angle - 180^\circ,$$

where the phase shift is introduced by the components. Amplifiers should not contribute phase shift at the oscillation frequency.

(b) A single time constant passive circuit contributes maximum of 90° phase shift, but 180° is required for oscillation. Therefore, at least two poles must be used.
(c) In spite of the fact that an LC circuit has two poles, they are designed for high frequency applications usually above the unity gain frequency of operational amplifiers.
(d) For buffered RC sections, the phase shift multiplies by the number of sections, therefore
$\varphi_{\max} = -n \times 90^\circ$, where n is the number of sections.
The phase changing rate as a function of frequency, $\mathrm{d}\varphi/\mathrm{d}t$, determines the frequency stability.
(e) In this circuit, due to integrator action,

$$V_Y = -V_Z \frac{1}{sR_2C_2} \tag{10.97}$$

By voltage division,

$$V_X = \frac{1}{1 + sR_3C_3} V_Y = \frac{1}{1 + sR_3C_3} \left(-\frac{V_Z}{sR_2C_2} \right) \tag{10.98}$$

By voltage division,

$$V_X = V_Z \frac{R_1}{R_1 + \dfrac{1}{sC_1}} = V_Z \frac{sR_1C_1}{1 + sC_1R_1} \tag{10.99}$$

From (10.98) and (10.99),

$$\frac{-V_Z}{sR_2C_2(1 + sR_3C_3)} = \frac{V_Z sR_1C_1}{1 + sR_1C_1} \tag{10.100}$$

$$A\beta = -1 = \frac{1 + sR_1C_1}{sR_1C_1.sR_2C_2(1 + sR_3C_3)} \tag{10.101}$$

When equal valued resistors and capacitors are used, loop gain is

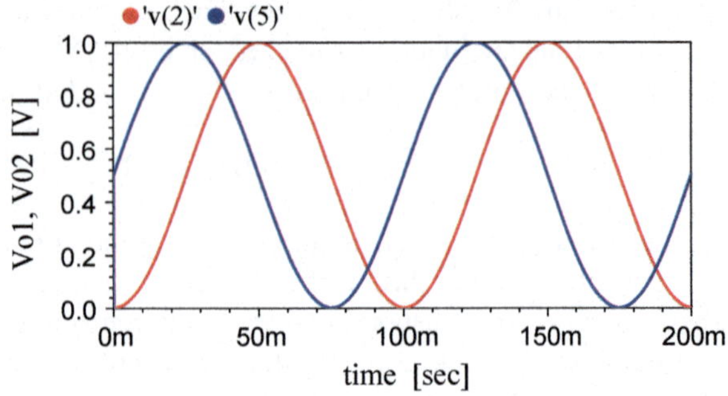

Fig. 10.103 Quadrature sinusoidal oscillations

$$A\beta = -1 = \frac{1}{(sRC)^2} \tag{10.102}$$

$$\frac{1}{(j\omega)^2 R^2 C^2} = -1 \quad \rightarrow \quad \omega = \frac{1}{RC} \tag{10.103}$$

(f) For equal valued resistors and capacitors, $R = R_1 = R_2 = R_3$, $C = C_1 = C_2 = C_3$,

$$f_o = 10 = \frac{1}{2\pi RC}$$

Let $R = 1590\,\Omega$, then

$$C = \frac{1}{2\pi \times 1590 \times 10} = 10\,\mu\text{F}$$

and for $U = 0.5$ V bias voltage, sinusoidal quadrature oscillator generates 10 Hz and 1 V peak-to-peak oscillations, as shown in Fig. 10.103.
SPICE netlist (QO1.cir) is given below.

```
R1 1 6 1.59 k
R2 2 4 1.59 k
R3 3 5 1.59 k
C1 1 2 .01 m
C2 4 5 .01 m
C3 3 0 .01 m
XOA1 3  1  2  OA1
XOA2 6  4  5  OA1
v1 6 0  0.5  dc
.SUBCKT OA1  1   2   6
```

```
RIN  1  2  1000MEG
EGAIN  3 0  1 2  100000
RP1  3  4  10 K
CP1  4  0  1.6UF
EBUFF  5 0  4 0  1
ROUT  5  6  1
.ENDS
```

References

1. Palmer R (2001) DC parameters: input offset voltage (V_{IO}). Texas instruments application report, SLOA059
2. Analog Devices Inc. Norwood, MA, USA. http://www.analog.com/
3. Linear Technology, USA. http://www.linear.com/product/LT1001. Last accessed on 27 Dec 2016
4. TDK-EPCOS https://en.tdk.eu/sensors. Last accessed on 27 Dec 2016
5. http://www.analog.com/media/en/technical-documentation/data-sheets/AD620.pdf. Last accessed on 27 Dec 2016
6. Rehman M, Ahmed MT, Arif M (1990) Critical study and applications of a self-balancing bridge. In: IEE proceedings, (137 Part A 1):25
7. Madhu Mohan N, Geetha T, Sankaran P, Jagadeesh Kumar V (2008) Linearization of the output of a wheatstone bridge for single active sensor. In: Proceedings of 16th IMEKO TC4 Symposium 09/2008
8. Lin PM (1985) Simple design procedure for a general summer. Electron Eng 57:37–38
9. Stotts LJ (1989) Introduction to implantable biomedical IC design. IEEE Circ Dev Mag:12–18
10. Kafe F, Psychalinos C (2014) Realization of companding filters with large time-constants for biomedical applications. Analog Integr Circ Signal Process: 217–231
11. Li Y, Wong AKY, Zhang YT (2010) Fully-integrated transimpedance amplifier for photoplethysmographic signal processing with two-stage Miller capacitance multiplier. Electron Lett 46(11):745–746
12. Bustos SS, Silva H, Sanches E (2000) A 60-dB dynamic-range CMOS sixth-order 2.4-Hz low-pass filter for medical applications. IEEE Trans Circ Syst Analog Dig Sig Process 47:1391–1398
13. Martinez JS, Alejandro Vázquez A (1998) Impedance scalers for IC active filters. In: IEEE international symposium on circuits and systems: 51–154
14. Kulej T (2009) Regulated capacitance multiplier in CMOS technology. In: International conference on mixed design of integrated circuits and systems, pp 316–319
15. Darweesh HY, Farag FA, Khalaf YA (2007) New active capacitance multiplier for low cutoff frequency filter design. In: Proceedings of 19th International Conference on Microelectronics (ICM'07), pp 381–384
16. Karki J (2002) Analysis of the Sallen-Key architecture. Texas Instruments Inc., Application Report SLOAO24B: p 5
17. Kerwin W, Huelsman L, Newcomb R (1967) State variable synthesis for insensitive integrated circuit transfer functions. IEEE J Solid-State Circ SC-2:87–92
18. Soliman AM (2008) History and progress of the Kerwin–Huelsman–Newcomb filter generation and Op Amp realizations. J Circ Syst Comput 17(4):637
19. Soliman AM (2011) Generation of Kerwin-Huelsman-Newcomb biquad filter circuits using nodal admittance matrix expansion. Int J Circuit Theory Appl 39(7):697–717

20. Senani R, Singh VK (1995) KHN-equivalent biquad using current conveyors. Electron Lett 31:626–628
21. Salama KN, Soliman AM (2000) Voltage mode Kerwin-Huelsman-Newcomb circuit using CDBAs. Frequenz 54:90–93
22. Toker A, Ozoguz S, Acar C (1999) Current-mode KHN-equivalent biquad using CDBAs. Electron Lett 35:1682–1683
23. Keskin AU, Cam U (2007) Insensitive high-output impedance minimum configuration SITO-type current-mode biquad using dual-output current conveyors and grounded passive components AEU-Int. J Electron Commun 61:341–344
24. Keskin AU, Biolek D, Hancioglu E, Biolkova V (2006) Current-mode KHN filter employing current differencing transconductance amplifiers. AEU-Int J Electron Commun 60(6):443–446
25. Universal Active Filter Data Sheet (2010) SBFS002B UAF 42 Texas Instruments Inc
26. Mancini R (2000) Design of op amp sine wave oscillators. Analog Appl J Texas Instrum Inc.:33–37
27. Keskin AU, Biolek D (2006) Current mode quadrature oscillator using current differencing transconductance amplifiers (CDTA). IEE Proc Circ Devices Syst 153(3):214–218
28. Keskin AU, Aydın C, Hancıoğlu E, Acar C, (2006) Quadrature oscillators using current differencing buffered amplifiers. Frequent (J RF-Eng Telecommun) 60(3–4):57–59

Appendix

A.1. Decimal Prefixes of SI Units

International System of units (SI) prefixes used to form decimal multiples and submultiples of SI units are given below:

Factor	Name	Symbol
10^{15}	Peta	P
10^{12}	Tera	T
10^{9}	Giga	G
10^{6}	Mega	M
10^{3}	Kilo	k
10^{-3}	Milli	m
10^{-6}	Micro	μ
10^{-9}	Nano	n
10^{-12}	Pico	p
10^{-15}	Femto	f

A.2. Standard Resistance Values (Preferred Values)

"E" series specify the preferred resistance values for various tolerances. The number following the "E" specifies the number of logarithmic steps per decade. E48, E96 series values are needed for higher accuracy and close tolerance requirements. Derivation is based on $\boldsymbol{M} = 10^{\frac{\boldsymbol{m}-1}{\boldsymbol{E}}}$, where M is the nominal resistance value at m position, E is a coefficient related to tolerance.
Example The third multiplier in E24 series is $M = 10^{\frac{3-1}{24}} = 10^{\frac{1}{12}} = 1.21 \rightarrow 1.2$

A.Ü. Keskin, *Electrical Circuits in Biomedical Engineering*,
DOI 10.1007/978-3-319-55101-2

E12 series multipliers (10%)

1.0	1.2	1.5	1.8	2.2	2.7	3.3	3.9	4.7	5.6	6.8	8.2

E24 series multipliers (5%)

1.0	1.1	1.2	1.3	1.5	1.6	1.8	2.0	2.2	2.4	2.7	3.0
3.3	3.6	3.9	4.3	4.7	5.1	5.6	6.2	6.8	7.5	8.2	9.1

E48 series multipliers (2%)

1.00	1.05	1.10	1.15	1.21	1.27	1.33	1.40
1.47	1.54	1.62	1.69	1.78	1.87	1.96	2.05
2.15	2.26	2.37	2.49	2.61	2.74	2.87	3.01
3.16	3.32	3.48	3.65	3.83	4.02	4.22	4.42
4.64	4.87	5.11	5.36	5.62	5.90	6.19	6.49
6.81	7.15	7.50	7.87	8.25	8.66	9.09	9.53

E96 series multipliers (1%)

1.00	1.02	1.05	1.07	1.10	1.13	1.15	1.18
1.21	1.24	1.27	1.30	1.33	1.37	1.40	1.43
1.47	1.50	1.54	1.58	1.62	1.65	1.69	1.74
1.78	1.82	1.87	1.91	1.96	2.00	2.05	2.10
2.15	2.21	2.26	2.32	2.37	2.43	2.49	2.55
2.61	2.67	2.74	2.80	2.87	2.94	3.01	3.09
3.16	3.24	3.32	3.40	3.48	3.57	3.65	3.74
3.83	3.92	4.02	4.12	4.22	4.32	4.42	4.53
4.64	4.75	4.87	4.99	5.11	5.23	5.36	5.49
5.62	5.76	5.90	6.04	6.19	6.34	6.49	6.65
6.81	6.98	7.15	7.32	7.50	7.68	7.87	8.06
8.25	8.45	8.66	8.87	9.09	9.31	9.53	9.76

A.3. Mathematical Formulas and Tables

Exponential Identities

$$i^2 = -1$$

$$\mathrm{e}^{iA} = \cos A + i \sin A \text{ (Euler's formula)}$$

$$\cos A = \frac{\mathrm{e}^{iA} + \mathrm{e}^{iA}}{2}, \quad \sin A = \frac{\mathrm{e}^{iA} - \mathrm{e}^{iA}}{2i}$$

$$\log_{10} x = \frac{\ln x}{\ln 10}, \quad x^y = \mathrm{e}^{y \cdot \ln x}$$

$$\sin hx = \frac{\mathrm{e}^x - \mathrm{e}^{-x}}{2}, \quad \cos hx = \frac{\mathrm{e}^x + \mathrm{e}^{-x}}{2}, \quad \tan hx = \frac{\sin hx}{\cos hx}$$

Trigonometric Identities

$\sin(A+B) = \sin A \cos B + \cos A \sin B, \quad \sin(A-B) = \sin A \cos B - \cos A \sin B$
$\cos(A+B) = \cos A \cos B - \sin A \sin B, \quad \cos(A-B) = \cos A \cos B + \sin A \sin B$

$$\tan(A+B) = \frac{\tan A + \tan B}{1 - \tan A \tan B}$$

$$\tan(A-B) = \frac{\tan A - \tan B}{1 + \tan A \tan B}$$

$$\sin A \cos B = \frac{1}{2}(\sin(A+B) + \sin(A-B))$$

$$\cos A \cos B = \frac{1}{2}(\cos(A+B) + \cos(A-B))$$

$$\cos A \sin B = \frac{1}{2}(\sin(A+B) - \sin(A-B))$$

$$\sin A \sin B = \frac{1}{2}(\cos(A+B) - \cos(A-B))$$

$$\sin A + \sin B = 2\sin\frac{A+B}{2}\cos\frac{A-B}{2}$$

$$\cos A + \cos B = 2\cos\frac{A+B}{2}\cos\frac{A-B}{2}$$

$$\sin A - \sin B = 2\cos\frac{A+B}{2}\sin\frac{A-B}{2}$$

$$\cos A - \cos B = -2\sin\frac{A+B}{2}\sin\frac{A-B}{2}$$

Following table (Table A.1) lists values of some angles.

Table A.1 Some angles and values

Angle degrees	Angle radians	$\sin\theta$	$\cos\theta$	$\tan\theta$
0	0	0	1	0
30	$\frac{\pi}{6}$	$\frac{1}{2}$	$\frac{\sqrt{3}}{2}$	$\frac{1}{\sqrt{3}}$
45	$\frac{\pi}{4}$	$\frac{\sqrt{2}}{2}$	$\frac{\sqrt{2}}{2}$	1
60	$\frac{\pi}{3}$	$\frac{\sqrt{3}}{2}$	$\frac{1}{2}$	$\sqrt{3}$
90	$\frac{\pi}{2}$	1	0	Undefined
180	π	0	-1	0

$$\sin^2 A + \cos^2 A = 1$$

$$\cos^2 A = \frac{1 + \cos 2A}{2}$$

$$\sin^2 A = \frac{1 - \cos 2A}{2}$$

$$\sin\left(\frac{A}{2}\right) = \mp\sqrt{\frac{1 - \cos A}{2}}$$

$$\cos\left(\frac{A}{2}\right) = \mp\sqrt{\frac{1 + \cos A}{2}}$$

$$\sin 2A = 2\sin A\,\cos A$$

$$\cos 2A = \cos^2 A - \sin^2 A = 2\cos^2 A - 1 = 1 - \sin^2 A$$

Some Power Series Expansions

$$e^x = 1 + x + \frac{x^2}{2!} + \frac{x^3}{3!} + \frac{x^4}{4!} + \cdots$$

$$\cos x = 1 - \frac{x^2}{2!} + \frac{x^4}{4!} - \frac{x^6}{6!} + \cdots$$

$$\sin x = x - \frac{x^3}{3!} + \frac{x^5}{5!} - \frac{x^7}{7!} + \cdots$$

$$\ln x = 2\left[\frac{x-1}{x+1} + \frac{1}{3}\left(\frac{x-1}{x+1}\right)^3 + \frac{1}{5}\left(\frac{x-1}{x+1}\right)^5 + \cdots\right]$$

Table of Standard Derivatives

$f(x) = x^n, \quad f'(x) = nx^{n-1}$

$f(x) = e^x = f'(x)$

$f(x) = a^x, \quad f'(x) = a^x \ln a \quad (a > 0)$

$f(x) = \sin x, \quad f'(x) = \cos(x)$

$f(x) = \cos x, \quad f'(x) = -\sin(x)$

$$f(x) = \sin^{-1} x, \quad f'(x) = \frac{1}{\sqrt{1 - x^2}} (-1 < x < 1)$$

$$f(x) = \cos^{-1} x, \quad f'(x) = -\frac{1}{\sqrt{1 - x^2}} (-1 < x < 1)$$

$$f(x) = \tan^{-1} x, \quad f'(x) = \frac{1}{1 + x^2}$$

L'Hopital's Rule

If $\lim_{x \to A} f(x) = \lim_{x \to A} h(x) = 0$,

$$\lim_{x \to A} \frac{f(x)}{h(x)} = \lim_{x \to A} \frac{f'(x)}{h'(x)}$$

prime indicating differentiation operation. Also,

If $\lim_{x \to \infty} f(x) = \lim_{x \to \infty} h(x) = \infty$,

$$\lim_{x \to \infty} \frac{f(x)}{h(x)} = \lim_{x \to \infty} \frac{f'(x)}{h'(x)}$$

Table of Standard Integrals

$$\int x^n \, dx = \frac{x^{n+1}}{n+1} + c \quad n \neq -1$$

$$\int \frac{1}{x} \, dx = \ln x + c$$

$$\int e^x \, dx = e^x + c$$

$$\int a^x \, dx = \frac{a^x}{\ln a} + c \quad (a > 0)$$

$$\int \sin x \, dx = -\cos x + c$$

$$\int \cos x\,dx = \sin x + c$$

$$\int \tan x\,dx = \ln|\sec x| + c$$

$$\int \sin^2 x\,dx = \frac{x}{2} - \sin\frac{2x}{4} + c$$

$$\int \cos^2 x\,dx = \frac{x}{2} + \sin\frac{2x}{4} + c$$

$$\int \frac{1}{a^2+x^2}dx = \frac{1}{a}\tan^{-1}\frac{x}{a} + c \quad (a > 0)$$

$$\int \frac{1}{\sqrt{a^2-x^2}}dx = -\sin^{-1}\frac{x}{a} + c \quad (-a<x<a)$$

$$\int \frac{1}{\sqrt{a^2+x^2}}dx = \sin h^{-1}\frac{x}{a} + c \quad (a > 0)$$

Some Laplace Transform Pairs

$$f(t) = \delta(t), \quad F(s) = 1$$

$$f(t) = u(t-a), \quad F(s) = \frac{e^{-as}}{s}$$

$$f(t) = u(t), \quad F(s) = \frac{1}{s}$$

$$f(t) = t^n, \quad F(s) = \frac{n!}{s^{n+1}}$$

$$f(t) = e^{at}, \quad F(s) = \frac{1}{s-a}$$

$$f(t) = t^n \cdot e^{-at}, \quad F(s) = \frac{n!}{(s+a)^{n+1}}$$

$$f(t) = \sin \omega t, \quad F(s) = \frac{\omega}{s^2+\omega^2}$$

$$f(t) = \cos \omega t, \quad F(s) = \frac{s}{s^2+\omega^2}$$

$$f(t) = \mathrm{e}^{-at}\sin\omega t, \quad F(s) = \frac{\omega^2}{(s+a)^2+\omega^2}$$

$$f(t) = \mathrm{e}^{-at}\cos\omega t, \quad F(s) = \frac{s+a}{(s+a)^2+\omega^2}$$

$$f(t) = t\,\sin\omega t, \quad F(s) = \frac{2\omega s}{(s^2+\omega^2)^2}$$

Laplace Transforms, Some Properties

$f(t)$ $\quad F(s) = \int_0^\infty \mathrm{e}^{st} f(t)\mathrm{d}t$ (definition)

$af(t)+bg(t)$ $\quad aF(s)+bG(s)$ (Linearity)

$\mathrm{e}^{kt}f(t)$ $\quad F(s-k)$ (shift in s)

$f'(t)$ $\quad sF(s)-f(0)$ (first derivative)

$f''(t)$ $\quad s^2F(s)-sf(0)-f'(0)$ (second derivative)

$\int_0^t f(u)\mathrm{d}u$ $\quad \frac{1}{s}F(s)$ (integral)

$H(t-a)f(t-a)$ $\quad \mathrm{e}^{as}F(s)$ (shift in t)

$f(t+T)=f(t)$ $\quad \frac{1}{1-\mathrm{e}^{-sT}}\int_0^T \mathrm{e}^{-st}f(t)\mathrm{d}t$ (Periodic Function)

$\lim_{t\to 0} f(t)$ $\quad \lim_{s\to\infty}\{sF(s)\}$ (initial value)

$\lim_{t\to\infty} f(t)$ $\quad \lim_{s\to 0}\{sF(s)\}$ (final value)

$tf(t)$ $\quad -\frac{\mathrm{d}F(s)}{\mathrm{d}s}$ (Frequency differentiation)

$\frac{f(t)}{t}$ $\quad \int_0^\infty F(\sigma)\mathrm{d}\sigma$ (Frequency integration)

$\int_0^t f(t-\tau)g(\tau)\mathrm{d}\tau$ $\quad F(s)G(s)$ (Convolution)

Cramer's Rule for Solving Equations of the Form [A] . [X] = [Y]
If [A] is a symmetric matrix having a nonzero determinant, and the vector

$$[X] = [x_1\ x_2\ x_3 \ldots x_n]^T$$

is the column vector of unknowns, then the system has a unique solution, whose individual values for the unknowns are

$$x_j = \frac{\det(A_j)}{\det(A)} \quad j = 1, 2, \ldots, n$$

A_j is the matrix formed by replacing the *j*th column of [A] by the column vector [Y].

Example

$$\begin{bmatrix} a & b \\ c & d \end{bmatrix}\begin{bmatrix} x_1 \\ x_2 \end{bmatrix} = \begin{bmatrix} e \\ f \end{bmatrix}, \quad x_1 = \frac{\det(A_1)}{\det(A)} = \frac{ed - bf}{ad - bc}, \quad x_2 = \frac{\det(A_2)}{\det(A)} = \frac{af - ec}{ad - bc}$$

Solving Linear System of Simultaneous Equations of the Form [A] . [X] = [Y] in MATLAB
Column vector of unknowns, $[Y]$ is a column vector;

```
%A=(5*5) example
A=[1 2 -1 3 1;0 2 -2 1 2;3 1 -2  1 -1;1 1 0 -1 1;1 0 2 3 -2]
Y=[1;-1;0;2;1];
X=A\Y;
Y=Y'
X=X'
---------------------------------------------------------------------------
The print out of the resulting solution for 5x5 linear equations:
(Y and X vectors are transposed for space saving reason)
A =
        1         2        -1         3         1
        0         2        -2         1         2
        3         1        -2         1        -1
        1         1         0        -1         1
        1         0         2         3        -2

Y =     1        -1         0         2         1

X =    2.3333    -2.5000    1.1667    0.6667    2.8333
```

Partial Fraction Expansion in MATLAB

Consider a transfer function, which is represented by a ratio of two polynomials in s-domain.

$$H(s) = \frac{b(s)}{a(s)} = \frac{b_1 s^m + b_2 s^{m-1} + b_3 s^{m-2} + \cdots + b_{m+1}}{a_1 s^n + a_2 s^{n-1} + a_3 s^{n-2} + \cdots + a_{n+1}}$$

Such a function can be represented by two vectors, one of them specifying the coefficients of the numerator polynomial, and the other vector specifying the coefficients of the denominator polynomial. For example, assuming that both of these polynomials are fourth-order polynomials, then numerator polynomial coefficients vector is

b = $[b_1 b_2 b_3 b_4 b_5]$

and denominator polynomial coefficients vector is

a = $[a_1 a_2 a_3 a_4 a_5]$

These vectors specify the coefficients of the polynomials in descending powers of s, and the orders of these polynomials can be different. Partial fraction expansion of this rational function $H(s)$ is

$$H(s) = \frac{b(s)}{a(s)} = \frac{R_1}{s - p_1} + \frac{R_2}{s - p_2} + \frac{R_3}{s - p_3} + \frac{R_4}{s - p_4} + k(s)$$

R_1, R_2, R_3 and R_4 are the residues, and p_1, p_2, p_3 and p_4 are the poles. The term k (s) is a polynomial in s. MATLAB representation of these vectors are

$$\mathrm{R} = [\mathrm{R_1\ R_2\ R_3\ R_4}], \quad \mathrm{p} = [\mathrm{p_1\ p_2\ p_3\ p_4}], \quad \mathrm{k} = [\mathrm{C_2\ C_1\ C_0}]$$

R and **p** are column vectors, while **k** is a row vector.

The command below finds the partial fraction expansion of the ratio of two polynomials.

```
[r, p, k] = residue(b, a)
```

This command calculates the poles and residues from $H(s)$. On the other hand, the command

```
[b2, a2] = residue(r, p, k)
```

calculates the coefficients of polynomials if the poles and residues are given, and the result is normalized for the leading coefficient in the denominator.

Example

$$H(s) = \frac{b(s)}{a(s)} = \frac{s^4 + 10s^3 + 40s^2 + 75s + 50}{s^4 + 10s^3 + 35s^2 + 50s + 24}$$

```
b = [1 10 40 75 50]
a = [1 10 35 50 24]
```

its partial fraction expanded form is computed as

```
[r, p, k] = residue(b, a)
r = [-1 -2 2 1]
p = [-4 -3 -2 -1]
k = 1
```

This means,

$$H(s) = \frac{-1}{s+4} + \frac{-2}{s+3} + \frac{2}{s+2} + \frac{1}{s+1} + 1$$

The partial fraction expansion for multiple poles:
Example

$$H(s) = \frac{b(s)}{a(s)} = \frac{3}{s^3 + s^2}$$

```
b = [3]
a = [1 1 0 0]
```

its partial fraction expanded form is computed as

```
[r, p, k] = residue(b, a)
r = [3 -3 3]
p = [-1 0 0]
k = []
```

This means,

$$H(s) = \frac{3}{s+1} - \frac{3}{s} + \frac{3}{s^2}$$

Note that if a transfer function has multiple poles, then small changes in the data or round-off errors can cause large variations in the resulting poles and residues.

Rules to find Thevenin's Equivalent Circuit

Thevenin's theorem helps to reduce any one-port linear electrical network to a single voltage source and a single impedance (Fig. A.1).

(a) When the circuit contains resistors and independent sources:

(1) Find open-circuit voltage $V_{oc} = V_{Th}$
(2) Find Thevenin's resistance R_{Th} by deactivating all independent sources (open circuit the current sources and short circuit the voltage sources).

(b) When the circuit contains resistors, dependent sources and independent sources

(1) Find $V_{oc} = V_{Th}$
(2) Short circuit a–b (output terminals) and determine the current through a–b ($I_{sc} = I_{ab}$)

$$R_{Th} = \frac{V_{oc}}{I_{sc}}$$

(c) When the circuit has resistors and dependent sources (no independent sources)

(1) Find $V_{oc} = V_{Th}$
(2) Connect a 1 A current source flowing from terminal b to terminal a;
(3) $R_{Th} = \frac{V_{oc}}{I} = \frac{V_{oc}}{1\,\mathrm{A}}$

The equivalent circuit consists of only R_{Th} (there is neither a current nor a voltage source).

Voltage Dividers

A voltage V_i is applied to two series connected impedances, Z_1, Z_2. Let Z_2 has a connection to reference (ground) and Z_1 has a connection to the ungrounded terminal of the voltage source, V_i. The output voltage V_o is obtained at the junction of Z_1, Z_2 (Laplace operator s is omitted) (Table A.2).

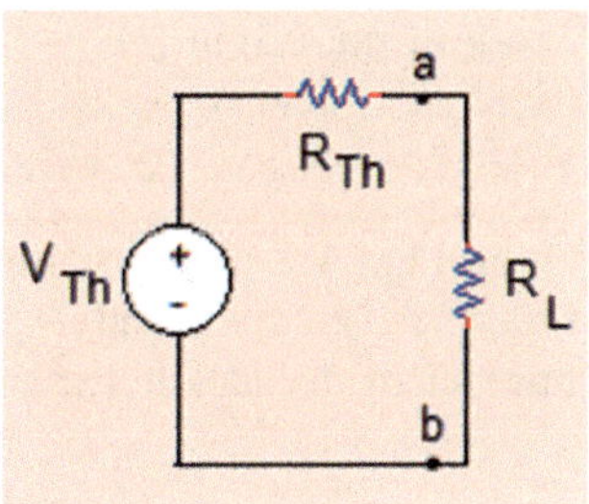

Fig. A.1 Linear electrical network to a single voltage source and a single impedance

Table A.2 Voltage dividers

Type of voltage divider	Voltage transfer function
Resistive	$\frac{V_o}{V_i} = \frac{R_2}{R_2 + R_1}$
Inductive	$\frac{V_o}{V_i} = \frac{L_2}{L_2 + L_1}$
Capacitive	$\frac{V_o}{V_i} = \frac{C_1}{C_2 + C_1}$

Magnitude and Phase of Transfer Functions

$$H(s) = \frac{N(s)}{D(s)} = \frac{a_n s^2 + a_{n-1}s^{n-1} + \cdots + a_o}{b_m s^m + b_{m-1}s^{m-1} + \cdots + b_o} = K\frac{(s - z_1)(s - z_2)\cdots(s - z_n)}{(s - p_1)(s - p_2)\cdots(s - p_m)}$$

where, $a_n \neq 0$, $b_m \neq 0$, all a_i, b_i, are real.

The magnitude of $H(j\omega)$ in decibels is defined as

$$20\log_{10}|H(j\omega)| = 20\log_{10}|K| + \sum_{i=1}^{n} 20\log_{10}|j\omega - z_i| - \sum_{i=1}^{m} 20\log_{10}|j\omega - p_i|$$

The phase in degrees (radians) is defined as

$$\varphi = \tan^{-1}\frac{\mathrm{Im}[H(j\omega)]}{\mathrm{Re}[H(j\omega)]} = \sum_{i=1}^{n}\tan^{-1}\frac{\mathrm{Im}(j\omega - z_i)}{\mathrm{Re}(j\omega - z_i)} - \sum_{i=1}^{n}\tan^{-1}\frac{\mathrm{Im}(j\omega - p_i)}{\mathrm{Re}(j\omega - p_i)}$$

Bode Plots

Exact manual calculation of magnitude and phase is a laborious process. Approximate sketches of these functions can be easily performed using so called Bode plots, noting that numerator and denominator of a transfer function (in factored form) are made up of the following terms:

(i) Constant term (K)
(ii) A root of the origin (s)
(iii) A real root ($s + p$)
(iv) Complex conjugete ($s^2 + as + b$)

Following Fig. A.2 displays drawing rules for magnitude and phase graphs for constant term, s, $1/s$, $s + z$ and $1/(s + p)$.

The number of decades between two frequencies is given as

$$\Delta f_{10} = \log_{10}\left(\frac{f_2}{f_1}\right), \quad f_2 > f_1$$

The number of octaves between two frequencies is

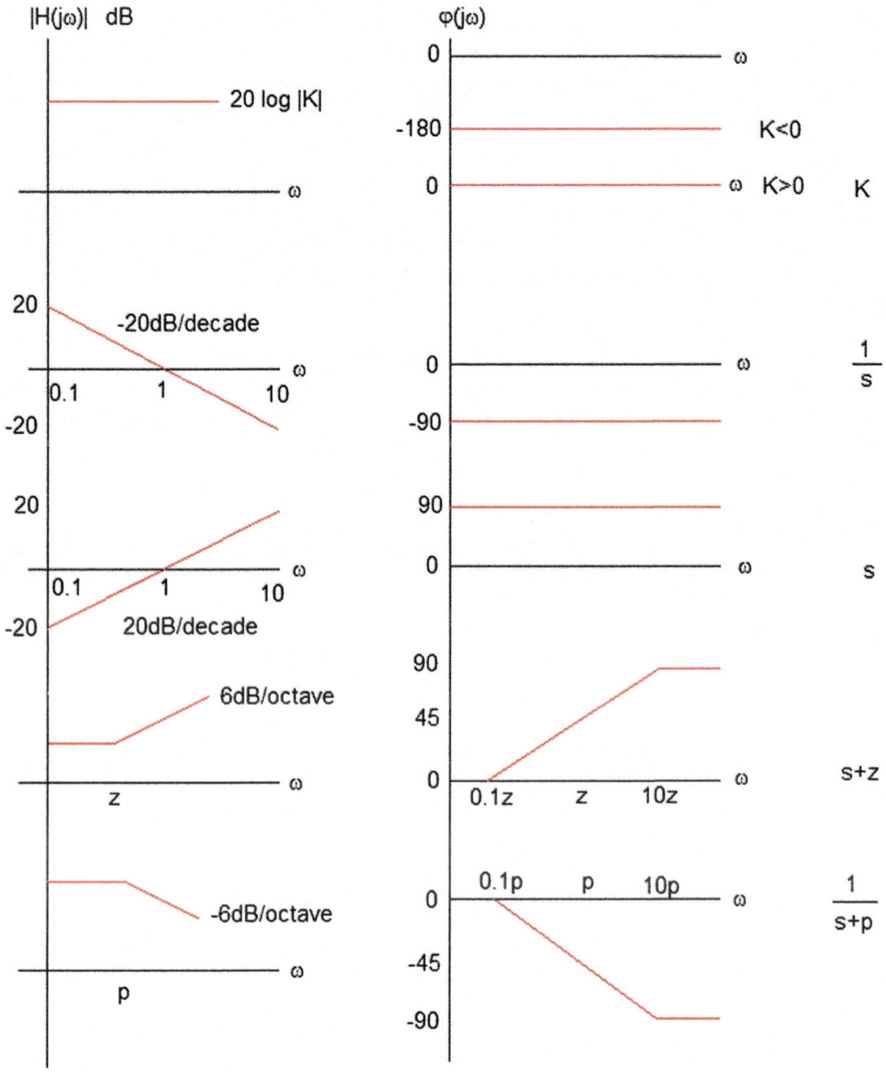

Fig.A.2 Rules for magnitude and phase graphs for constant term

$$\Delta f_2 = \log_2\left(\frac{f_2}{f_1}\right), \quad f_2 > f_1$$

Duality

Dual circuits are the ones which are described by the same characteristic equations with dual quantities interchanged. A dual of a relationship can be written by interchanging voltage and current in an expression. The dual expression produced is of the same form as the original equation.

Some duals: Open circuit–Short circuit, Switch turns on–switch turns off, Current–Voltage, Parallel connected elements–Serial connected elements, Voltage Generator–Current Generator, Node voltage–Mesh current, Branch–Branch, Resistance–conductance, Impedance–Admittance, Inductance–Capacitance, Reactance–Susceptance, Kirchhoff's Voltage Law (KVL)–Kirchhoff's Current Law (KCL), Thévenin's Theorem–Norton's Theorem, Faraday's Law–Ampére's Law, Permittivity–permeability, Piezoelectricity–Piezomagnetism, Permanent magnet–Electret.

A dual circuit is not the same thing as an equivalent circuit. For example, the dual of a star (Y) network of inductors is a delta (Δ) network of capacitors, which is not the same thing as a star-delta (Y-Δ) transformation; the transformation results in an equivalent circuit.

The dual of a mutual inductance cannot be formed directly, since there is no corresponding capacitive element.

In case the circuit configuration is not parallel or series or it contains dependent sources, following steps can be used to construct graphically the dual of a planar circuit:

(a) At the center of each mesh, place a node for the dual circuit.
(b) Reference node of the dual circuit is placed outside of the given circuit.
(c) Draw lines between nodes and reference line in such a way that each line crosses an element of the given circuit. Then, this element is replaced by its dual.
(d) Assign polarities of sources. A voltage source producing clockwise mesh current has its dual current source pointing from ground to non-reference node.

A Table A.3 for dual circuit elements and relationships is shown.

Table A.3 Dual circuit elements and relationships

Ohm's law	$v(t) = i(t)\,.\,R$	$i(t) = v(t)\,.\,G$
Capacitor–Inductor differential expression	$i_C(t) = C\frac{\mathrm{d}}{\mathrm{d}t}v_C(t)$	$v_L(t) = L\frac{\mathrm{d}}{\mathrm{d}t}i_L(t)$
Capacitor–Inductor integral expression	$v_C(t) = V_0 + \frac{1}{C}\int\limits_0^t i_C(\tau)\mathrm{d}\tau$	$i_L(t) = I_0 + \frac{1}{L}\int\limits_0^t v_L(\tau)\mathrm{d}\tau$
VCCS–CCVS	$i_m = xv_n$	$v_m = xi_n$
VCVS–CCCS	$v_m = xv_n$	$i_m = xi_n$
Resistor conductor	$R = x\Omega$	$G = x\mathrm{S}$
Capacitor inductor	$C = x\mathrm{F}$	$L = x\mathrm{H}$
Voltage–Current source	$v = x\mathrm{V}$	$i = x\mathrm{A}$
Voltage–Current division	$v_{R_a}(t) = \frac{R_a}{R_a + R_b}\cdot v$	$i_{G_a}(t) = \frac{G_a}{G_a + G_b}\cdot i$

SPICE Models For Dependent (Controlled) Sources

Voltage controlled voltage source VCVS: `Ename N1 N2 +C1 -C2 Value`

Example

```
E2 3 4 6 0 12
* load control voltage
vc 6 0
```

Current controlled voltage source CCVS: `Hname N1 N2 Vcontrol Value`

Example

```
H4 1 3 Vm 12
Vm 4 0 dc 0
```

Voltage controlled current source VCCS: `Gname N1 N2 +C1 -C2 Value`

Example

```
G1 3 5 4 6 12
* load control voltage
vs 4 6
```

Current controlled current source CCCS: `Fname N1 N2 Vcontrol Value`

Example

```
F1 0 3 Vm 5
Vm 4 0 dc 0
```

N1 and N2 are the positive and negative terminals of the dependent source, respectively.

+C1 and −C2 are the positive and negative terminals of the controlling voltage source, respectively.

Vcontrol is the zero value voltage source used to measure the controlling current (the positive current flows into the positive terminal of the controlling voltage source).

Operational Amplifier

Operational amplifier (op-amp) is a versatile active element that behaves like a voltage-controlled voltage source. It is used to perform many mathematical operations, filtering and signal processing.

Key Assumption: The op-amp operates in the linear range (away from saturation) (Fig. A.3).

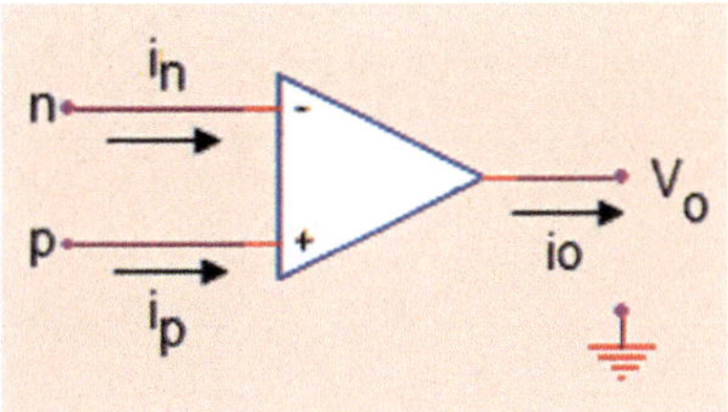

Fig.A.3 The op-amp operates in the linear range

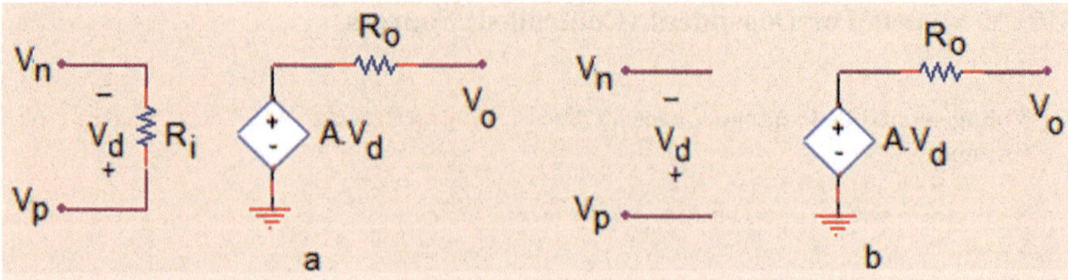

Fig.A.4 Opamp model

Ideal op-amp:

$$V_o = A \cdot V_d = A(V_p - V_n)$$
$$i_n = i_p = 0, \quad V_p = V_n$$
$$R_o = 0\,\Omega, \quad R_i = \infty\,\Omega$$

A = Open loop voltage gain (sometimes expessed in dB, $x\,\text{dB} = 20\log_{10} x$)
R_o = Output resistance, R_i = Input resistance (Fig. A.4).

The inverting input terminal in this particular configuration is at zero volts which is referred to as virtual ground. Note that this pin is not actually grounded. The input terminals are not shorted together.

Instrumentation Amplifier

The instrumentation amplifier is an essential circuit in biomedical electronics. For example, a two terminal sensor produces a signal but neither of its terminals may be connected to the same ground level as with the measuring network. These terminals may be DC biased at relatively large voltages or added to the noise. The differential amplifier acts seletively on measuring the difference between the input terminals.

Addition of 2 buffers between the sensor output and the differential amplifier prevents the loading of both sensor and the measuring electronics. The circuit configuration shown here provides gain, as well (Fig. A.5).

$$\frac{V_o}{V_d} = \left(1 + \frac{2R_1}{R_G}\right) \cdot \frac{R_3}{R_2}, \quad V_d = V_1 - V_2$$

Butterworth Polynomials in Factored Form

Order	Denominator, $D(s)$
1	$s+1$
2	$s^2+\sqrt{2}s+1$
3	$(s^2+1)(s+1)$
4	$(s^2+0.765s+1)(s^2+.848s+1)$
5	$(s+1)(s^2+0.618s+1)(s^2+1.618s+1)$
6	$(s^2+0.518s+1)(s^2+\sqrt{2}s+1)(s^2+1.932s+1)$

Fig.A.5 Instrumentation amplifier

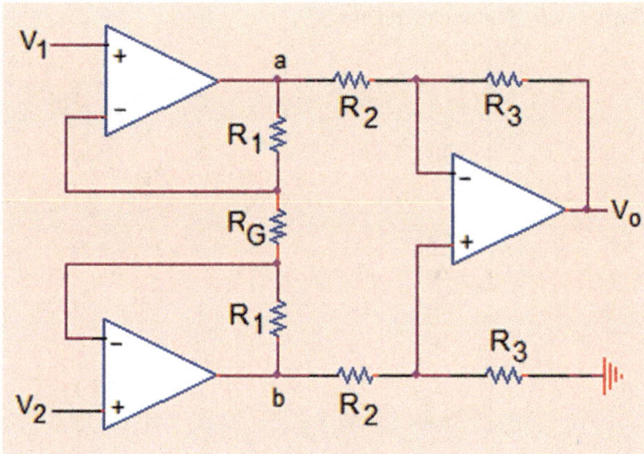

Second-Order (Biquad) Filter Transfer Functions (Table A.4)
A biquadratic filter transfer function is defined as

$$H(s) = \frac{a_2 s^2 + a_1 s + a_0}{s^2 + \left(\frac{\omega_0}{Q}\right) s + \omega_0^2}$$

The poles are

$$p_1, p_2 = -\frac{\omega_0}{2Q} \pm j\omega_0 \sqrt{1 - \frac{1}{4Q^2}}$$

Conversion of Two-Port Parameters
(Table A.5).

Table A.4 Second order (biquad) filter transfer functions

Filter type	Transfer function	Gain
Low-pass	$\frac{a_0}{s^2 + \frac{\omega_0}{Q} s + \omega_0^2}$	$\frac{a_0}{\omega_0^2}$
High-pass	$\frac{a_2 s^2}{s^2 + \frac{\omega_0}{Q} s + \omega_0^2}$	a_2
Band-pass	$\frac{a_1 s}{s^2 + \frac{\omega_0}{Q} s + \omega_0^2}$	Center frequency $= a_1 . \frac{\omega_0}{Q}$
Band-stop	$a_2 \frac{s^2 + \omega_0^2}{s^2 + \frac{\omega_0}{Q} s + \omega_0^2}$	dc (= high freq.) gain a_2
All-pass	$a_2 \frac{s^2 - \frac{\omega_0}{Q} s + \omega_0^2}{s^2 + \frac{\omega_0}{Q} s + \omega_0^2}$	Flat gain a_2

Table A.5 Parameter relations

	z	y	h
z	$\begin{matrix} z_{11} & z_{12} \\ z_{21} & z_{22} \end{matrix}$	$\begin{matrix} \frac{y_{22}}{\Delta_y} & -\frac{y_{12}}{\Delta_y} \\ -\frac{y_{21}}{\Delta_y} & \frac{y_{11}}{\Delta_y} \end{matrix}$	$\begin{matrix} \frac{\Delta_h}{h_{22}} & \frac{h_{12}}{h_{22}} \\ -\frac{h_{21}}{h_{22}} & \frac{1}{h_{22}} \end{matrix}$
y	$\begin{matrix} \frac{z_{22}}{\Delta_z} & -\frac{z_{12}}{\Delta_z} \\ -\frac{z_{21}}{\Delta_z} & \frac{z_{11}}{\Delta_z} \end{matrix}$	$\begin{matrix} y_{11} & y_{12} \\ y_{21} & y_{22} \end{matrix}$	$\begin{matrix} \frac{1}{h_{11}} & -\frac{h_{12}}{h_{11}} \\ \frac{h_{21}}{h_{11}} & \frac{\Delta_h}{h_{11}} \end{matrix}$
h	$\begin{matrix} \frac{\Delta_z}{z_{22}} & \frac{z_{12}}{z_{22}} \\ -\frac{z_{21}}{z_{22}} & \frac{1}{z_{22}} \end{matrix}$	$\begin{matrix} \frac{1}{y_{11}} & -\frac{y_{12}}{y_{11}} \\ \frac{y_{21}}{y_{11}} & \frac{\Delta_y}{y_{11}} \end{matrix}$	$\begin{matrix} h_{11} & h_{12} \\ h_{21} & h_{22} \end{matrix}$

$\Delta_z = z_{11}z_{22} - z_{12}z_{21}$, $\Delta_y = y_{11}y_{22} - y_{12}y_{21}$, $\Delta_h = h_{11}h_{22} - h_{12}h_{21}$

Note that there are other parameter sets to characterize two-port networks other than the three parameter types presented here. However, they are not used in this book.

Historical Profiles

...in the belief that remembrance adds more human values of respect, appreciation, and progress.

Alessandro Giuseppe Volta, 1745–1827

Volta was an Italian physicist and chemist. He invented the first electrical battery, the Voltaic pile, in 1799. Volta's invention led to the development of the field of electrochemistry. The SI unit of electric potential is named in his honor. In 1778 he managed to isolate methane. Volta studied capacitance, and it was for this work that the unit of electrical potential has been named the volt.

Andre-Marie Ampère, 1775–1836

Ampère, a French physicist, is the man who created the science of electrodynamics. The unit of electrical current—the Ampere (A)—is named in his honor.

A.Ü. Keskin, *Electrical Circuits in Biomedical Engineering*,
DOI 10.1007/978-3-319-55101-2

Léon Charles Thévenin, 1857–1926
Thévenin was a French electrical engineer. He developed his famous theorem, known as Thévenin's theorem.

Joseph Henry, 1797–1878
Henry was an American scientist who discovered electromagnetic induction independently of and at about the same time as Michael Faraday. Henry discovered the self-inductance, Unit of inductance is named in his honor.

Michael Faraday, 1791–1867 Faraday invented the electric motor in 1821. He discovered the induction of electric current by magnetism in 1831, and the ability of magnetic fields to change the polarization of light in 1845. The unit of electrical capacitance—the farad (F) —is named in his honor. Faraday discovered the carbon compound benzene, and in 1823, he was the first scientist to liquefy a gas. He was also first to introduce terms such as "electrode," "cathode" and "ion".

Georg Simon Ohm, 1789–1854

A German physicist and mathematician. Ohm found that there is a direct proportionality between the potential difference applied across a conductor and the resultant current. This relationship is known as Ohm's law. The work of Ohm marked the beginning of circuit theory; the unit of resistance is named in his honor.

Gustav Robert Kirchhoff, 1824–1887

Kirchhoff was a physicist. He contributed to the fundamental understanding of electrical circuits, spectroscopy, and the emission of black-body radiation by heated objects. Kirchhoff formulated his circuit laws (KCL, KVL) in 1845. He worked at University of Heidelberg in 1854, collaborating in a spectroscopic work with Robert Bunsen, where they discovered cesium and rubidium in 1861.

Ernst Werner von Siemens, 1816–1892

Siemens's name has been honored as the SI unit of electrical conductance. He invented electrically charged sea mines, worked on perfecting technologies that had already been established, and invented a telegraph that used a needle to point to the right letter, instead of using Morse code. Siemens is the father of the trolleybus, which was first introduced in 1882.

Bernard D.H. Tellegen, 1900–1990

B. Tellegen was a Dutch electrical engineer. He is the inventor of the pentode tube in 1926 and the gyrator in 1948. Tellegen held 41 US patents and he is also known for a theorem in circuit theory. He received MSEE degree from Delft University in 1923.

Hendrik Wade Bode, 1905–1982

Bode was a Dutch-American engineer. Bode received his Ph.D. from Columbia University in 1935. He is a pioneer of modern control theory and electronic telecommunications. He made important contributions to the design, guidance and control of anti-aircraft systems missiles and anti-ballistic missiles. He also made important contributions for the analysis of stability of linear systems. He is known for the graph that honored on his name, the Bode plot. He held 25 US patents.

Alan Lloyd Hodgkin, 1914–1998

AL Hodgkin was an English physiologist and biophysicist, 1963 Nobel Prize winner in Medicine. He worked in Cambridge University, also held additional administrative posts in the University of Leicester, from 1971 to 1984, and Trinity College, Cambridge, from 1978 to 1985. He worked on experimental measurements and developed an action potential theory representing one of the earliest applications of voltage clamping technique.

Andrew Fielding Huxley, 1917–2012

A.F. Huxley was a 1963 Nobel Prize-winning physiologist and biophysicist for his studies on the action potentials with A.L. Hodgkin. He developed interference microscopy to study muscle fibers. He discovered in 1954 the mechanism of muscle contraction, so called the sliding filament theory.

William T. Bovie, 1882–1958

W.T. Bowie was an American biophysicist. He invented the electrosurgical generator. He completed a Ph.D. in plant physiology from Harvard University. The first use of the electrosurgical device in an operating room was on October 1, 1926, by Harvey W. Cushing at Peter Bent Brigham Hospital in Boston, Massachusetts.

Willem Einthoven, 1860–1927

Einthoven was a Dutch doctor and physiologist. He was the inventor of the first electrocardiogram in 1903. Einthoven received a medical degree from the University of Utrecht. He became a professor at the University of Leiden in 1886. ECG equipment he first used had a string galvanometer and moving roll of photo-sensitive paper, weighting about 270 kg. He received the Nobel Prize in Medicine in 1924 for his contributions in this field.

Nikolai Sergeyevich Korotkov, 1874–1920

A Russian inventor of auscultatory technique for blood pressure measurement which is considered a "gold standard" for blood pressure measurement. The name "Korotkoff sounds" are given in his honor for pulse-synchronous circulatory acoustic signals observed through the stethoscope in auscultation of blood pressure using a sphygmomanometer.

Hans Berger, 1873–1941

Berger was a German neurologist. Berger received his medical degree from Jena in 1897. He became Rector of Jena University in 1927. He recorded the electrical brain waves, the electroencephalogram (EEG) in 1924. Berger also described the different waves or rhythms which were present in the normal and abnormal brain.

Ian Donald, 1910–1987

Ian Donald was a Scottish and educated in Edinburgh, graduated from the Diocesan College in Cape Town. He then studied medicine and was awarded MB BS at London University in 1937. In 1954, he introduced echo-sounding (the term from sonar) and searched its possible medical applications.

Godfrey Newbold Hounsfield, 1919–2004
He was an English electrical engineer who received the 1979 Nobel Prize with Allan McLeod Cormack for developing the X-ray computed tomography. Hounsfield built a prototype head scanner and tested it on himself. In 1971, the first head scanner was operated in a London hospital. His name is given to Hounsfield scale, a quantitative measure in evaluating CT scans.

Paul C. Lauterbur, 1929–2007
Lauterbur was an American chemist. In 2003, the Nobel Prize in Physiology or Medicine was awarded to Paul Lauterbur and Sir Peter Mansfield for their research related to MRI. He received a B.S. in chemistry from the Case Institute of Technology he obtained his Ph.D. in 1962 from the University of Pittsburgh.

Raymond V. Damadian, 1936–
Damadian was credited as the originator of the whole-body magnetic resonance imaging. He earned his degree in mathematics from the University of Wisconsin–Madison in 1956, and an M.D. degree from the Albert Einstein College of Medicine in New York City in 1960. In 1974, he received the first patent in the field of MRI.

William Bennett Kouwenhoven, 1886–1975
Kouwenhoven invented the first cardiac defibrillator. He received his BSEE from Brooklyn Polytechnic in New York, and his Ph.D. in electrical engineering from the Karlsruhe Technische Hochschule in Germany in 1913. He joined the faculty of the Johns Hopkins University School of Engineering in 1914. He tested his device on a dog. In 1947, Professor Claude Beck used his device on a 14-year-old boy at Case Western Reserve University.

Paul M. Zoll, 1911–1999
Zoll was a practicing physician and recognized as a Pioneer in Cardiac Pacing. In 1952, Paul M. Zoll described cardiac resuscitation via electrodes on the bare chest with 2-millisecond duration pulses of 100–150 V across the chest, at 60 stimuli per minute. This became the basis for future clinical pacing developments. In 1956, he published a transcutaneous approach to terminate ventricular fibrillation with a shock voltage up to 750 V, and later described similar termination of ventricular tachycardia.

Karl William Edmark, 1924–1994
A cardiovascular surgeon, inventor. Edmark developed a defibrillator that utilized direct current (DC), which provided lower-energy and more effective shocks. Edmark's invention, known as the Edmark Pulse Defibrillator, was first used to save the life of a 12-year-old girl in 1961.

James Francis Pantridge, 1916–2004
Professor Pantridge and Dr. John Geddes of the Royal Victoria Hospital in Belfast produced the first portable defibrillator in 1965. A mains (AC) powered defibrillator was powered by an inverter, which converted a 12 V car battery to 230 V. The unit weighed 70 kg. By 1968 he had designed an instrument weighing only 3 kg, incorporating a mini capacitor manufactured for NASA.

Bernard Lown, 1921–
A Lithuanian-American, 1985 Nobel Peace Prize laureate, Professor of Cardiology Emeritus at the Harvard School of Public Health, Boston, developed the direct current defibrillator for cardiac resuscitation, and introduced a new use for lidocaine to control heartbeat disturbances. In 1961 Lown, Baruch Berkowitz, and coworkers proved that a specific current waveform (Lown waveform) reversed ventricular fibrillation, without injuring heart.

John G. Webster
J.G. Webster is a pioneer in biomedical engineering. (1953 BSEE, Cornell University, 1965 MSEE, University of Rochester, 1967 Ph.D., Elec. Eng. University of Rochester). He first proposed the idea of electrical impedance tomography in 1978, and published many books on biomedical engineering. Prof. Webster was a professor emeritus in the College of Engineering at the University of Wisconsin–Madison (2015).

Selected Bibliography

Following is a list of selected books for further reading.

Circuit Analysis

Alexander CK, Sadiku MNO (2013) Fundamentals of electric circuits, 5 edn. McGraw Hill

Balabanian N, Bickart TA, Seshu S (1969) Electrical network theory. Wiley

Basso CP (2016) Linear circuit transfer functions: an introduction to fast analytical techniques. Wiley-IEEE Press

Bobrow LS (1987) Elementary linear circuit analysis, Holt Rinehart and Winston

Boylestad R (2013) Introductory circuit analysis, 12th edn. Pearson New International Edition

Davis AM (1998) Linear circuit analysis. PWS Publishing Company

Dorf RC, Svoboda JA (2013) Introduction to electric circuits, 9th edn. International Student Version, Wiley

Floyd T (2013) Principles of electric circuits, 9th edn. Pearson New International Edition

Hayt WH, Kemmerly JE (2001) Engineering circuit analysis, 6th edn. McGraw-Hill Book Company Inc.

Irwin JD, Nelms RM, Patnaik A (2015) Engineering circuit analysis, 11th edn. International Student Version, Wiley.

Johnson DE, Hilburn JL, Johnson JR, Scott PD (1995) Basic electric circuit analysis, 5th edn. Prentice-Hall, Inc.

Nahvi M, Edminister JA (2014) Schaum's outline of electric circuits, 6th edn. McGraw-Hill Education

Nilsson J, Riedel S (2014) Electric circuits with mastering engineering, 10th edn. Pearson Global Edition

O'Malley J (1992) Schaum's outline of basic circuit analysis, 2nd edn. McGraw-Hill

Reddy HC (2002) The circuits and filters handbook, 2nd edn. CRC Press

A.Ü. Keskin, *Electrical Circuits in Biomedical Engineering*,
DOI 10.1007/978-3-319-55101-2

Spence R (2008) Introductory circuits. Wiley
Thomas RE, Rosa AJ, Toussaint GJ (2015) The analysis and design of linear circuits. Wiley
Vlach J (2014) Linear circuit theory: matrices in computer applications. Apple Academic Press
Wing O (2009) Classical circuit theory. Springer

Circuit Synthesis and Design

Anderson BDO, Vongpanitlerd S (1973) Network analysis and synthesis: a modern systems approach. Prentice-Hall
Baher H (1984) Synthesis of electrical networks. Wiley, New York
Bakshi UA, Bakshi AV (2009) Fundamentals of network analysis and synthesis. Technical Publications Pune
Daryanani G (1976) Principles of active network synthesis and design. Wiley, New York
Glisson TG (2011) Introduction to circuit analysis and design. Springer
Guillemin EA (1977) Synthesis of passive networks: theory and methods appropriate to the realization and approximation problems (Reprint). Huntington, N.Y., R. E. Krieger Pub. Co.
Kuo F (1966) Network analysis and synthesis, 2nd edn. Wiley
Lamm HY-F (1979) Analog and digital filters: design and realization. Prentice Hall, Inc.
Schaumann R, Valkenburg MEV (2001) Design of analog filters. Oxford University Press.
Temes GG, Lapatra JW (1977) Circuit synthesis and design. McGraw-Hill
Valkenburg MEV (1960) Introduction to modern network synthesis. Wiley
Weinberg L (1962) Network analysis and synthesis
Yarman BS (2010) Gewertz design of ultra-wideband power transfer networks. Wiley, Chichester, UK

Electronics

Heumann K (2012) Basic principles of power electronics. Springer Science & Business Media
Kandaswamy A, Pittet A (2009) Analog electronics. Prentice Hall India, Learning Pvt. Ltd.
Khanchandani S (2007) Power electronics. Tata McGraw-Hill Education
Liu Y (2012) Power electronic packaging: design, assembly process, reliability and modeling. Springer Science & Business Media
Peyton A, Walsh V (1993) Analog electronics with op-amps: a source book of practical circuits. Cambridge University Press
Sedra AS, Smith KC (2004) Microelectronic circuits, 5th edn. Oxford University Press

Biomedical Engineering

Aston R (1991) Principles of biomedical instrumentation and measurement. Merrill Publishing Company (Macmillan)

Barsoukov E, Macdonald JR (2005) Impedance spectroscopy, theory, experiment and applications. Wiley Interscience

Bruce EN (2001) Biomedical signal processing and signal modelling. Wiley

Carr JJ, Brown JM (2001) Introduction to biomedical equipment technology, 4th edn. Prentice Hall

David Y, Maltzahn WW, Neuman MR, Bronzino JD (2003) Clinical engineering. CRC Press

Enderle J, Blanchard S, Bronzino J (2005) Introduction to biomedical engineering, 2nd edn. Elsevier Academic Press

Saltzman WM (2015) Biomedical engineering: bridging medicine and technology. Cambridge University Press

Semmlow JL (2011) Signals and systems for bioengineers. 2nd edn. A MATLAB-Based Introduction (Biomedical Engineering) Academic Press.

Street LJ (2011) Introduction to biomedical engineering technology, 2nd edn. CRC Press

Weiss TF (1996) Cellular biophysics, electrical properties, vol. 2. The MIT Press

Webster JG (ed) (1998) Medical instrumentation: application and design, 3rd edn. Wiley, New York

Webster JG (ed) (2004) Bioinstrumentation. Wiley

SPICE, MATLAB and Others

Banzhaf W (1989) Computer-aided circuit analysis using SPICE. Prentice Hall

Butt R (2009) Introduction to numerical analysis using MATLAB. Jones & Bartlett Learning

Hahn BD (2002) Essential MATLAB for scientists and engineers, 2nd edn. Butterworth-Heinemann

Rashid MH, Rashid HM (2006) SPICE for power electronics and electric power, 2nd edn. Taylor and Francis

Sedra AS, Roberts GW, Smith KC (1992) SPICE for microelectronic circuits. Saunders College Pub.

Smythe WR (1989) Static and dynamic electricity, 3rd edn. Taylor and Francis

Thorpe TW (1992) Computerized circuit analysis with SPICE: a complete guide to SPICE, with applications. Wiley

Yang X-S (2006) An introduction to computational engineering with MATLAB. Cambridge Int. Science Publishing

Index

A.Ü. Keskin, *Electrical Circuits in Biomedical Engineering*,
DOI 10.1007/978-3-319-55101-2

Printed by Printforce, the Netherlands